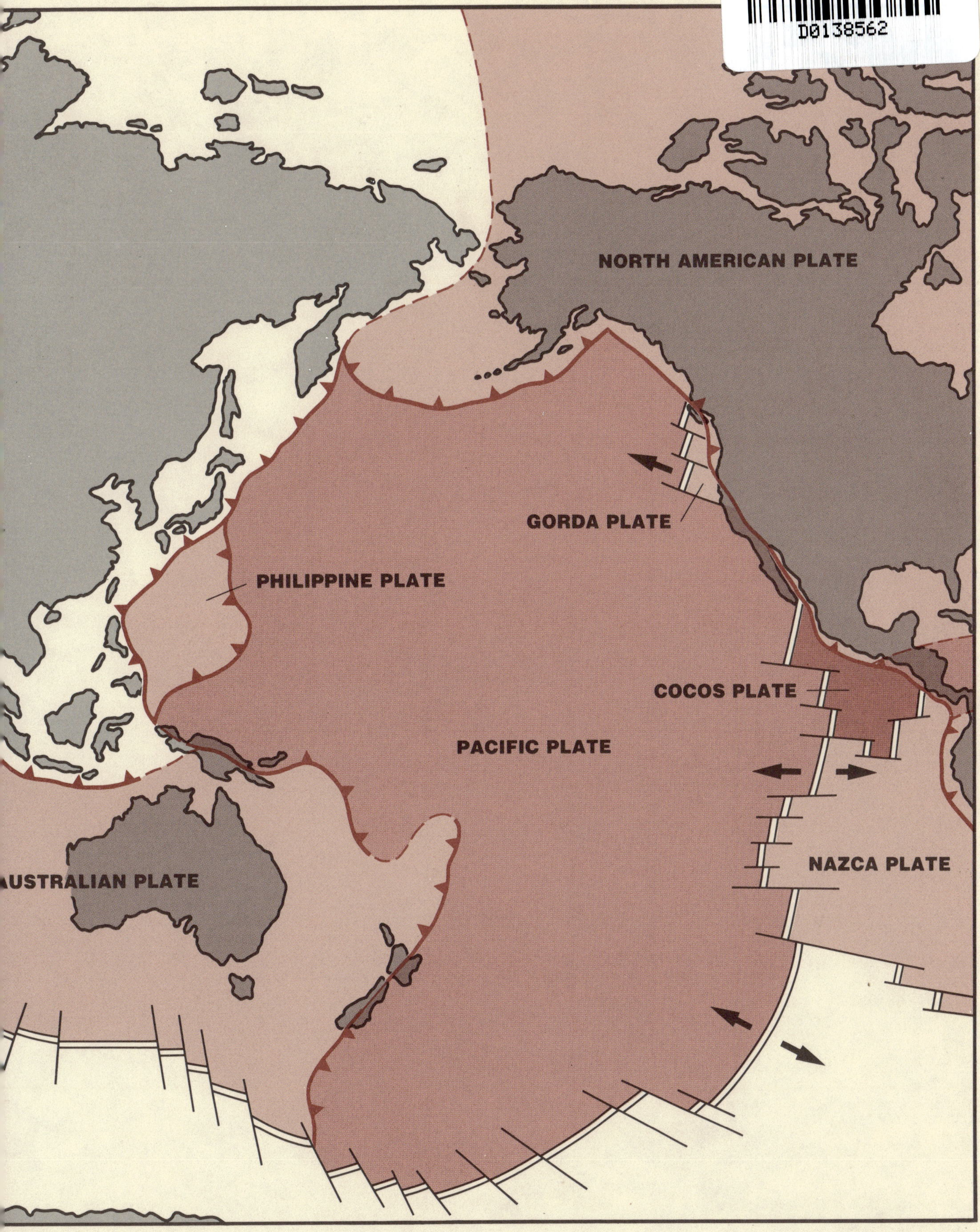
NORTH AMERICAN PLATE
GORDA PLATE
PHILIPPINE PLATE
COCOS PLATE
PACIFIC PLATE
NAZCA PLATE
AUSTRALIAN PLATE

[redacted]

Physical Geology

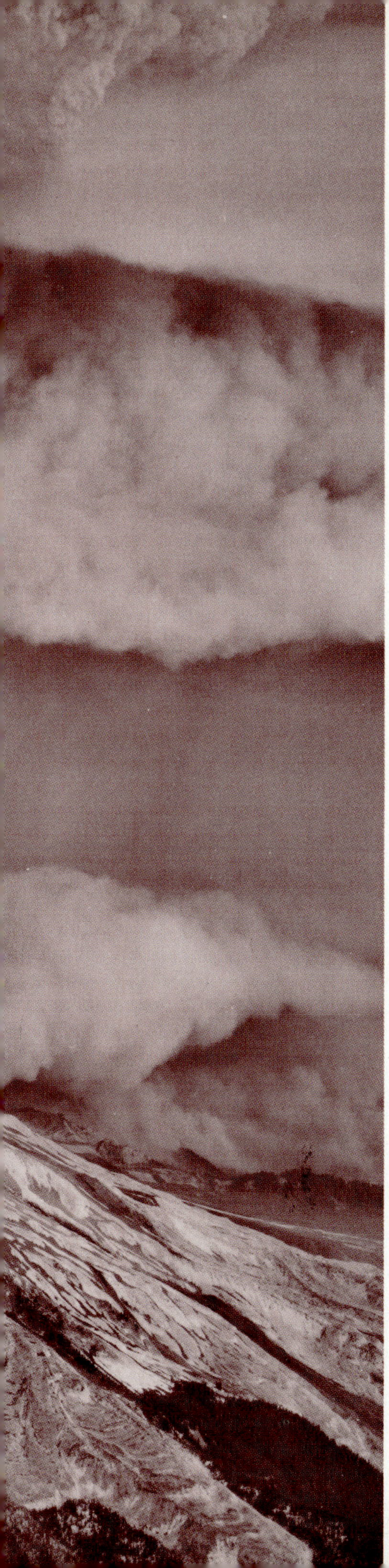

Physical Geology

EDGAR W. SPENCER
Washington and Lee University

ADDISON-WESLEY
PUBLISHING COMPANY

Reading, Massachusetts · Menlo Park, California
London · Amsterdam · Don Mills, Ontario · Sydney

Sponsoring Editor: Nancy J. Kralowetz
Developmental Editors: Katharine Gregg and Caroline L. Eastman
Production Editor: Margaret Pinette
Text Designer: Vanessa Piñeiro
Art Development Editor: Arthur Ciccone
Illustrators: Marilee Bailey and Oxford Illustrators
Cover Designer: Richard Hannus, Hannus Design Associates
Cover Painting: Alan Magee
Art Coordinator: Dick Morton
Production Manager: Herbert Nolan

The text of this book was set in Palatino by Monotype Composition Company.

Library of Congress Cataloging in Publication Data

Spencer, Edgar Winston.
Physical geology.

Includes index.
1. Physical geology. I. Title.
QE28.2.S646 1983 551 82-18415
ISBN 0-201-06423-5

ISBN 0-201-06423-5
ABCDEFGHIJ-HA-89876543

PREFACE

After nearly a quarter of a century of teaching physical geology, I hope to approach the subject by conveying some of its fascination to students. I believe that beginning students in geology, whether they continue in the subject or not, need to understand what science is and how scientists approach problems, as well as the basic content of geology and how geology relates to other disciplines and to our society.

This book attempts to satisfy these objectives. The first step toward understanding the earth is learning to identify and describe what we see at the earth's surface. But as a science, geology involves much more. As in other sciences, the student must learn to observe, formulate questions about what he or she sees, and design hypotheses that can be tested. Through this process our understanding of the earth grows. We can see this method at work as we study the evolution of ideas about such questions as the origin of basalt or continental drift. I have emphasized this methodology in a number of places in the text. For example, I discuss the theory of plate tectonics at length both because it provides insight into the evolution of geologic thinking and because it effectively synthesizes a large part of what we know about the structure of the earth and the dynamics of the earth's interior.

The great variety of forms we see at the surface of the earth attract our curiosity and invite inquiry; and we find that they are the products of processes that are continuing to shape the earth today as they have for millions of years. It soon becomes apparent that these same processes constitute an important component of our environment, and that we often come into conflict with them when we alter natural systems. For these reasons I have placed special emphasis on water supply and hydrologic systems. I am convinced that we must understand these processes if we are to make intelligent decisions about many of the environmental problems facing our society.

I have organized the subject matter of physical geology into five parts. Part I introduces the student to the science of geology, provides an overview of the field, and describes minerals. Most of Part II is devoted to a discussion of the origin and interpretation of rocks. Part III treats internal processes, rock structure, and plate tectonics. Part IV covers surficial processes and their environmental consequences. A single chapter reviews natural resources in Part V. Because many teachers prefer to treat surficial processes before internal processes, Parts III and IV are designed so they can be interchanged. Students who have studied Parts I and II should have no difficulty going into either Part III or Part IV.

I have placed emphasis on the natural processes that govern and shape the earth. Many dramatic case histories are used to illustrate the operation of these processes and to give the student a feeling for the geology of many parts of North America. The text presentation is largely descriptive, but I have included a few boxes that go into geophysical processes in greater detail. These boxes should challenge and stimulate those students who are interested in the more theoretical aspects of geology. At the end of every chapter I have included selected readings and study questions to stimulate students to think and pursue topics further.

I have made every effort to define geologic terms fully. Each chapter contains a list of key terms. There is also a glossary at the end of the book so students may recheck difficult terms that have been defined in an earlier chapter. I have included identification charts for rocks and minerals as appendices.

I am especially indebted to Elizabeth Spencer, my wife, for preparing the art for the line drawings, for reading and suggesting ways to improve the manuscript, and for her encouragement through the long period of tedious work involved in completing this book. A special note of thanks is also due Susan White for typing, editing, securing permissions, and other work with the manuscript; and to Peggy Riethmiller and Betty Brewbaker for helping with these tasks. My colleagues and students at Washington and Lee have been a continuing source of help and stimulation throughout this project. Reviews of the manuscript by Richard Brown, Walter Brownfield, R. V. Fodor, Jerry D. Horne, Ernst H. Kastning, George Kosovic, Jere H. Lipps, Douglas McDowell, Gene C. Ulmer, and Kenneth J. Van Dellen have been most helpful. I am grateful for the many constructive suggestions they offered. I appreciate the help of Nancy Spencer and Patrick Hinley, who printed many of the photographs. I wish to acknowledge with thanks the work of Caroline Eastman who edited the manuscript and prepared the chapter summaries. Thanks are also extended to the editorial staff, Nancy Kralowetz, Stephen Quigley, and Katharine Gregg; to the production staff, Earle Pitts, Art Ciccone, Margaret Pinette, Richard Morton, and Vanessa Piñeiro; and to the artist Marilee Bailey, for their efforts to make this a worthwhile book.

Lexington, Virginia E. W. S.
December 1982

CONTENTS

Physical Geology

INTRODUCTION: THE PLANET EARTH

Part 1

Geology is like other sciences in that it involves the use of observations, experiments, and hypotheses to obtain a better understanding of natural phenomena. Geology differs from other sciences in that understanding what we observe depends to a great degree on discovering how things change through time. The earth is a dynamic body that is continually evolving. Evidence of this change is found everywhere around us. The shape of the land surface is changing; volcanoes and earthquakes are evidence of internal changes, as is the changing magnetic field; even the continents and ocean basins are shifting position.

Part I is an introduction to the study of the earth. Included in it are descriptions of the nature and scope of geology, the basic materials of which the earth is composed, and the large-scale features of the earth. With this background, the reader will be prepared to turn next either to the evolution of the earth and its materials or to the processes that are shaping the surface of the earth.

GEOLOGY: THE STUDY OF THE EARTH

Chapter 1

Events like the dramatic eruption of Mount St. Helens in May 1980 (Fig. 1.1), the return of rock samples from the moon, the exploration of the oceanic ridges, and the energy crisis have focused attention on geology in recent years. This young science is devoted to the study of the earth and its origin, and the processes that have shaped it and govern its evolution.

Central to the study of the earth is a knowledge of the materials that form it and the physical and chemical processes that occur on and within the planet. In addition to this traditional coverage, several themes are developed throughout the book. The first of these is the nature of science and the methods used in geology, which illustrate the way scientists approach problems. The second concerns the relationship between the earth, the moon, and other planets. Space exploration has given us a better understanding of the place of the earth in the solar system, and of the origin and early history of the earth. The third is the large-scale structure of the earth, called **tectonics.** Discoveries made since the early 1950s concerning the sea floor and the earth's interior have led to the development of a theory called **plate tectonics.** According to this theory, the outer shell of the earth can be subdivided into a mosaic of thin rigid plates. The interaction of these plates is responsible for such diverse phenomena as earthquakes, volcanic activity, and the formation of mountains. The conclusions derived from this theory have had such far-reaching implications for our understanding of geology and the history of the earth that the development of the theory is sometimes referred

U.S.G.S

Figure 1.1
Mount St. Helens in southern Washington during the eruption of May 1980. Volcanoes provide dramatic proof that the processes that produce molten rock in the earth's interior are still active.

to as the "revolution of the earth sciences." The fourth theme of this text involves the relationship among the development of natural resources, the understanding of natural processes, and the quality of our environment. We see this most dramatically in the shortages of energy resources and in the devastation that results from natural hazards.

This opening chapter sets forth a few of the most widely held views of what science is and how it works. The science of geology is presented, and we survey briefly the four main themes of the book.

1.1 THE NATURE OF SCIENTIFIC INQUIRY

Many colleges and universities require students to take a course in science in the hope that students will not only become familiar with a small part of the content of scientific knowledge but will also learn something of how scientists approach and solve problems using the scientific method. The achievements of science and technology in the last twenty-five years would have been unimaginable only a few decades ago. We need think of relatively few of these accomplishments, such as space exploration or computer development, to appreciate how significant they have been. Growth and application of scientific discoveries have transformed our lives and have confronted society with problems we have yet to resolve. As changes escalate, every thinking person has an increasing need to understand the basic methods and genuine accomplishments of science. We will examine many scientific discoveries and methods of geology in detail; but first let us consider the general question of what science is and what its methods are.

The word **science** is derived from a Latin word meaning knowledge. In our usage it refers to specialized knowledge—the understanding of natural phenomena—and it implies a method of study that includes hypotheses, observation, and experimentation. This definition places emphasis on science as a process by which our understanding grows toward the truth, we hope, rather than as an accumulation of the end products of truth itself. In his book *The Sciences—Their Origins and Methods* (see the bibliography for this chapter), R. Harré adds one further aspect of science, the knowledge of causes: "Anyone can accumulate facts, scientists accumulate explanations"(*1*).

Although the term *scientific method* is often used, it is difficult to define in any widely acceptable way. Scientists usually go about their work without giving much thought to the existence of any pattern of approach or method that should guide their steps, and many will deny that there *is* a method. Yet most scientists do agree that science involves an interplay of observation, hypothesis, and experiment, and that certain qualities characterize good scientific work. These characteristics include precise use of words and symbols, care and precision in making observations and conducting experiments, determination of the scientist to examine and interpret the results without bias and without undue influence of preconceived notions, and finally a willingness of the scientist to face criticism, to consider alternative interpretations, and when necessary to discard an idea that has been refuted. The history of science contains many examples of prominent scientists who failed in one or more of these qualities, especially the latter ones, but the progress of science is clearly related to these qualities in scientists as a group. The test of good observational and experimental work is that the results can be duplicated by others working independently; that is, the reproducibility of a result.

The interaction of observation, experiment, and hypotheses follows no prescribed pattern in science. Perhaps it is more common for an observation to lead to formulation of a hypothesis, which in turn suggests an experiment or other possible observations that may confirm or disprove the hypothesis, but other sequences are certainly not uncommon. The role of the **hypothesis** is clearly central. A good hypothesis not only explains or correlates known facts; it should also suggest new experiments or observations from which still more facts can be established. Hypotheses may be of great value to science if they suggest new connections between facts, connections that lead to new experiments. Even if these experiments ultimately show the hypothesis is not valid, it has served to move up gradually toward a better understanding of natural phenomena and their causal mechanisms. The test of a good hypothesis is that it explains observations and experimental results; the test of the best hypothesis is that it can be used to predict accurately the results of experiments and observations that were not made at the time the hypothesis was formulated. Many would insist that the best hypothesis is the simplest one that will explain observed phenomena.

The Use of Multiple Working Hypotheses

Research is often most fruitful when not one but **multiple working hypotheses** are pursued at the same time. Writing in the late nineteenth century,

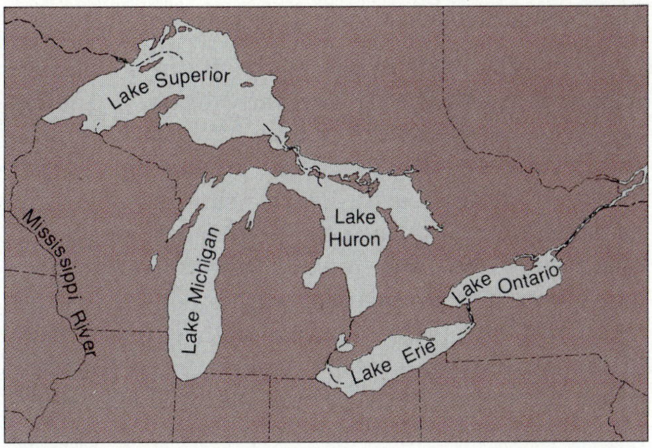

Figure 1.2
The Great Lakes formed mainly as a result of a glaciation that affected most of Canada and the northern United States in the last two million years. But other factors also help determine their exact location and shape. Geologists often find that natural features have complex causes.

the American geologist T.C. Chamberlin stressed the advantage of considering several possible explanations (2). He noted that the differences among hypotheses tend to suggest observations or experiments to distinguish between them, and that often a single hypothesis oversimplifies nature. "An adequate explanation often involves the coordination of several agencies (factors), which enter into the combined result in varying proportions." As an example, Chamberlin discussed the question of the origin of the Great Lake basins (Fig. 1.2). It could not, he said, be attributed solely either to a glacier excavating river valleys or to downward bowing of the earth's crust in that region. "All of these must be taken together, and possibly they must be supplemented by other agencies. The problem, therefore, is the determination not only of the participation, but of the measure of the extent of each of these agencies in the production of the complex result."

1.2 GEOLOGY AS A SCIENCE

Birth of Modern Geology

The study of the earth emerged as a distinct scientific discipline sometime in the eighteenth century. Many scholars, including such famous men as the artist Leonardo da Vinci (1452–1519), the philosopher Nicolas Steno (1631–1687), and the mathematician Robert Hooke (1635–1703), made observations and developed ideas about various geological phenomena, but growth of geology as a subject in university curricula and the appearance of scholars who devoted their careers to the study of the earth came later in the 1700s. These scholars were concerned with great controversies, such as the origin of fossils and the question of whether crystalline rocks like granite and basalt are formed by sedimentation or by volcanic activity. These questions may seem trivial today, but although our present ideas about such questions are so firmly established, we must not forget that the discoveries of this early period are fundamental to our understanding of modern geology.

James Hutton (1726–1797) is considered by some historians of science to be the founder of modern geology. He was clearly one of the most important and influential of the earlier geologists. Hutton, born in Edinburgh, Scotland, became interested in chemistry while a boy and went on to study medicine in Paris and Leyden. Around 1750, his interest in chemistry drew him back to London where his experiments led to discoveries and a business interest that made him financially independent. As a result of his association in Edinburgh with several other brilliant scientists and of his wide travels through England, the Scottish Highlands, and the Continent, Hutton became interested in geology. In the remaining thirty years of his life he studied rocks, minerals, landforms, and the operation of natural processes. A keen observer and an imaginative and creative interpreter of all these diverse phenomena, he correctly recognized that granite and basalt are formed by heat underground; that rivers form the valleys in which they flow; that the consolidated rocks composed of gravel, sand, and lime fragments were originally deposited under water; that those layers of sediment have been tilted and strongly deformed in places; that the surface of the earth is constantly but slowly changing; and much more. While plausible to us, Hutton's conclusions were not inevitable. At about the same time the extraordinarily enthusiastic, convincing, and productive leader of many geologists on the continent, Abraham Werner (1749–1817), concluded that granite and basalt had a sedimentary origin and that vertical strata were upright because they had been deposited in that position.

Why was Hutton so successful where others failed? The answer seems to lie in two important aspects of Hutton's approach. First, his interest in geology grew from the observations he made while traveling, and he generally reasoned from his observations in the formulation of ideas or hypotheses. Many others started with a grand idea or scheme regarding the origin of things and inferred other

Figure 1.3
The dark gray rock band in this picture is composed of basalt, a rock that has solidified from a molten state. This outcrop is located in Colorado. Like other igneous rocks, this body formed as a result of the cooling and consolidation of molten material injected into the surrounding rock. Note the trees for scale.

hypotheses from this. Hutton collected extensive field evidence that demonstrated that basalt (Fig. 1.3) and granite invade or intrude other rocks; in contrast, Werner reasoned that the world had been covered by an ocean from which all the rocks we see were deposited.

Secondly, Hutton's basic assumption was that the past history of the earth can best be explained in terms of what is happening now or what has recently happened. This idea had been expressed much earlier by a French physician, Georges Fushsel (1722–1773) who stated it as follows: ". . . the manner in which nature at the present time is still acting and producing things must be assumed as the rule in our explanation." This is known as the **principle of uniformitarianism** and is popularly expressed as "the present is the key to the past."

Of course, many others played important roles in the development of geology, and we will consider the contributions of some of them; but Hutton, by his use of rigorous observations and supported in his interpretations by his knowledge of chemistry and physical priciples, placed geology on a firm foundation.

The Earth Sciences

Geology as it exists today only faintly resembles the science that emerged in the early 1800s. In the last two decades, the earth sciences have grown rapidly and become highly specialized. In part this growth has been stimulated by heavily subsidized government programs in space exploration and oceanography and by the push toward both development of new energy resources and protection of the environment. Dramatic increases in the number of scientists, in the sophistication of the equipment at their disposal, and in the effective application of mathematics, physics, and chemistry to the problems of geology have all had great impact on the study of the earth.

Those who imagine that geology is primarily concerned with the description and classification of rocks, minerals, and fossils will be surprised to see the diversity of interests expressed in an outline of the field of earth science specialization (Table 1.1). The word *geology* literally means the study of the earth, but it has become customary to associate the field of geology with certain particular aspects of the earth—its materials, landforms, structure, and history. Geologists do describe and classify the materials, landforms, and internal structures of the earth; but the goals of geology are broader, directed more toward understanding natural processes and the functioning of natural systems and toward unraveling the evolution of the earth through its long history than toward merely cataloging its contents.

The variety of subject matter and methods of study in the earth sciences is reflected in the subdivisions that have developed in the field (Table 1.1). These subdisciplines are intimately related, but each either requires a specialized educational background or is concerned with topics, problems, or methods that set it apart from the others. Most geologists are trained in several of these subdisciplines, but very few, if any, are able to keep up with all of the new discoveries and developments.

1.3 REVOLUTION IN SCIENCE: THE THEORY OF PLATE TECTONICS

The words *theory* and *hypothesis* are sometimes used as synonyms, but more often the term **theory** represents a more mature stage in our understanding of a process. It is often used to distinguish conceptual schemes that demonstrate the relationship of a number of hypotheses to one another or to some common

Table 1.1. The Earth Sciences

Atmospheric Sciences	**Oceanography**	**Geophysics**	**Geochemistry**
Meteorology *Climatology*	*Marine Geology* *Biological Oceanography* *Physical Oceanography* *Chemical Oceanography*	*Geodesy* *Gravity* *Geomagnetism* *Seismology*	*(abundance and distribution of elements)*

Geology	
Earth Materials and Their Origin	
Mineralogy	description, identification, and occurrence of minerals
Petrology	
Igneous	description, origin, and occurrence of rock melts and rocks that form from them
Metamorphic	study of the causes, processes, and products of metamorphism
Sedimentology	study of the processes of sedimentation, sediments, and sedimentary rocks
Landforms and Processes	
Volcanology	study of processes and products of volcanic activity
Geomorphology	study of the form of the land and processes that shape it
Hydrology	study of running water, including underground water
Glaciology	glacial forms, cause of glaciation
Configuration of earth's Crust	
Structural Geology	mechanics of rock deformation, geometry of rock bodies, causes of structure
Tectonics	origin of the large scale features of the crust
Historical Geology	
Paleontology	study of ancient life forms
Stratigraphy	study of the stratified rocks and the record of the history they contain
Applied Geology	
Petroleum Geology	exploration and development of oil and gas
Hydrology	flood control and water supply
Engineering Geology	application of geology to engineering earthen works, e.g., dams
Coal Geology	exploration and development of coal
Economic Geology	exploration and development of ore (metal) deposits and economically important rocks and mineral deposits

explanation. It can be viewed in much the same way as hypothesis—as a proposition to explain some set of observations. Thus, the theory of evolution, the theory of relativity, and the atomic theory of matter are conceptual schemes that incorporate, tie together, and provide central explanations for many hypotheses. When a theory is invented that makes it possible to unify many hypotheses and the observational and experimental evidence on which they are based, it may stimulate a very rapid growth of new ideas and new hypotheses. At such times, old hypotheses may be reevaluated and discarded to be replaced by new ideas. When major changes in the accepted interpretations of current knowledge take place in a short time, the period may be referred to as a scientific revolution.

Such a revolution has taken place in geology since the early 1960s. This revolution is best represented by the theory known as plate tectonics—a theory that describes the dynamics of the outer part of the earth. **Plate tectonic theory** is the proposition that the outer hardened shell of the earth can be subdivided into a number of large and relatively thin plate-like units, and that the structure, form, and most processes in this outer shell can be explained in terms of the movements of these plates relative to one another. This theory has passed the test of providing a relatively simple explanation of a large body of observational data; it has tied together a number of important earlier hypotheses and theories, and it has been used successfully to predict the results of new studies. We will later trace some of the most important discoveries that ultimately led to this exciting new theory, and the theory itself will be a focal point throughout the first half of this book.

The extent to which plate tectonic theory applies to other planets in our solar system has been one focus of planetary studies of the last decade.

1.4 GEOLOGY AND THE STUDY OF THE PLANETS AND THEIR MOONS

Even the earliest rockets carried equipment that collected information about the earth's atmosphere, giving us a much better understanding of the atmosphere's temperature structure, the geomagnetic field, and the abundance and type of electrically charged particles in the higher altitudes of the atmosphere. Measurements made by instuments carried in these rockets produced some significant surprises—such as the existence of radiation belts high in the atmosphere—but what really transformed our views of the planetary system and even of the earth itself has been the development of satellite and spacecraft photography and other imagery, the results obtained by unpiloted landing craft, and especially the successful moon landing, which made possible the collection and return of samples and informed observations by scientists.

This mass of new data about the moon and planets has revolutionized our understanding of these bodies in much the same way that large amounts of new data about the ocean basins changed our thinking about the movements—the dynamics—of the earth's crust. Although no single new theory comparable to plate tectonic theory has yet been put forward with regard to planets, it seems obvious that a scientific revolution is in progress in planetary science. Because the methods used in geology and our knowledge of the earth are readily applied to the study of other planets, geologists have played an important role in these new developments.

The Pioneer, Viking, Mariner, and Voyager flights have opened a whole new field of study of planets and their satellites. It is now possible to examine pictures (Fig. 1.4) and describe the surface features of these bodies. Using our knowledge of the earth and processes here, we can devise working hypotheses to explain the origin of these features and the processes that have caused them. Similarly, we can make inferences about the interior of these planets based on their density, size, their magnetic fields, and other physical characteristics.

NASA

Figure 1.4
Oblique photograph of the surface of Mars. Analysis of photographs such as this have provided important insights to the origin and early history of the solar system. Using photographs like these, we can now compare the structure, history, and processes that shape the planets.

Among the most exciting and eagerly sought discoveries from the moon are those that provide insight to the early history of the solar system and indirectly to the origin and evolution of the earth. One may reasonably ask why we should have to look to the moon or other planets to study the early history of the earth. It would appear that the cost and ease of such exploration would be far more reasonable here on the earth than on the moon. Unfortunately, we have been unable to find rocks on the earth that might have formed in the earliest periods. The earth's rocky outer crust is constantly changing. The earth's atmosphere and its abundant water continually modify the materials and shape of the land surface. Water and carbon dioxide especially react with the minerals in rocks at the earth's surface, causing them to decompose. The products of decomposition and mechanical disintegration at the surface are then washed down slopes into streams and carried out into the oceans where they are deposited as sediment. By these and other processes we will consider later, the materials at the earth's surface are slowly recycled.

In contrast, the moon has no atmosphere. Therefore, its surface is changing exceedingly slowly. In a sense, the surface of the moon is almost static—frozen in the form it had several billion years ago. Thus, we can study the conditions that existed on its surface at a time before the origin of the oldest rocks we have found on earth. If the moon was a satellite of the earth at that time, conditions that caused what we see there are our best source of information about the primitive earth.

The study of the other planets is still a young science. Many new discoveries lie ahead and perhaps many of our present notions about these planets and the early history of our own planet will have to give way to new and better conceptions.

1.5 GEOLOGY, RESOURCES, AND THE ENVIRONMENT

The science of geology is most directly related to the needs and goals of our society in the fields of natural resource development and environmental control. The basic role of natural resources in our economy and the way the supply of new materials affects our daily lives have been dramatically illustrated by the effects of gasoline shortages during the 1970s. Our dependency on an abundant and cheap supply of fuel is now, probably for the first time, apparent to most of the people of the western world. The extent to which this fuel supply has shaped the economy and society of the United States is still probably greater than most of us realize. It is certainly a basic ingredient in the growth of our use of automobiles, and these have prompted us to abandon railway systems in favor of road systems, to abandon centralized communities in favor of suburban growth, to abandon downtown shopping centers in favor of strip development and suburban shopping. Cheap fuel has made long-distance commuting feasible. In short, the availability of cheap fuel over a long period of time coupled with development of the automobile has influenced where we live, where we work, where we shop, the types of houses and buildings constructed, and many other aspects of our lives.

The effects of interruptions of fuel supply are felt more quickly and more widely than shortages of most other materials, mainly because we use such vast quantities of fuel that storage of supplies adequate for long periods of time is not feasible. Nor can we quickly substitute one fuel for another—cars can, of course, be built to operate on fuels other than gasoline, but a long period of time and great expense both for the manufacturer and the consumer are involved in making such a change. It is much easier to stockpile adequate supplies or to find substitutes for many other raw materials, and the cost and dislocations involved in changing from one raw material to another are not so great as they are for fuel. Nevertheless, our society consumes vast quantities of a long list of raw materials, and we have allowed and even encouraged wasteful use of the goods made from these materials. Only in recent years have we begun to think of recycling materials to any significant extent. As our demand for all types of raw materials has increased, the best-known and most readily available sources of these materials have been rapidly depleted. Some natural resources are present in many places, so that it is not difficult to open a new quarry or mine to replace the supply from an older one that is exhausted or too expensive to operate; however, many other resources occur only under special and sometimes rare geological conditions.

Geologists play a key role in exploration for and development of deposits of rocks, minerals, and fuels that are used to produce the needed raw materials (Fig. 1.5). Perhaps the most widely used method of exploration involves looking for places where the special geological conditions that caused one deposit to be formed are duplicated. Other methods may entail looking for more direct evidence, such as trace quantities of the needed element in water or soil samples, or for some abnormal (anomalous) magnetic, seismic, or gravitational indication. The application

Kennecott Corporation

Figure 1.5
The open pit mine in Bingham Canyon, Utah. Mines like this one supply the natural resources used by our society. This mine is a major producer of copper, but many other resources, including iron, coal, and limestone, are mined in this way. Such mines are used when large quantities of rock must be removed in the mining operation and when the ore is not too deep. For scale, note the train on one of the terraces at far right.

of some of these methods to natural resource exploration will be pointed out as these topics are developed.

The expression *environmental control* is used with some misgivings to characterize a second important area in which geology is applied to problems that confront our society. The need to understand our environment has become abundantly clear, but the best solution to the problems is not always to control the environment, but to learn to live with it and adjust to it. With increasing frequency the solution to environmental problems seems to be for us to learn to control ourselves rather than the environment. Many of the problems with which we are concerned in geology have become familiar to us as the tragic events that result in large numbers of lost lives and property damage. The loss of an estimated quarter of a million lives in an earthquake in China in 1976 is an extreme example, but major and disastrous earthquakes, volcanic eruptions, and floods are frequent occurrences; and the loss of valuable land due to stream, wave, current, and wind erosion occurs continually. Losses due to instability of steep slopes are also widespread and economically significant. For example, mudslides destroy homes in California almost every year. Gradually, we are becoming aware of the great cost involved when ground water supplies become polluted, regardless of whether that pollution is caused by movement of salt water inland as a result of excessive pumping of wells or as a result of leaks from nuclear reactors or buried chemical wastes.

Still other problems, some with highly significant consequences, may arise so slowly that we are only barely aware of them. Changes in the balance between water and ice on the earth will in the long run be critically important to coastal areas where changes in the sea level will be felt. The ice–water balance shifts in response to climatic changes. It may also be that we, through burning of fossil fuels, are affecting climatic trends. In all of these areas of environmental concern it is crucial that we learn enough about the earth and the operation of natural systems to evaluate

U.S.G.S.

Figure 1.6
Damage in Anchorage, Alaska, caused by an earthquake in 1964. Geologists can help identify potential natural hazards through their understanding of the processes that cause them.

hazards that exist naturally as well as those that may come into existence as a result of our own activities. Armed with this knowledge we can move toward reducing the economic and human losses (Fig. 1.6) caused by those natural conditions over which we have little control, and we can reduce the cost and bring greater efficiency in managing those natural processes over which we can exercise control.

SUMMARY

Geology, the study of the earth's materials and processes, proceeds as other sciences do by a combination of hypothesis, observation, and experimentation, often called the scientific method. In the case of geology this has led to the formulation of the theory of plate tectonics, which holds that the outer shell of the earth is composed of thin rigid sheets that move across the surface, causing earthquakes, volcanic activity, mountain formation, and other geologic effects. In the last two decades space flights have made it possible to study the surfaces and processes on other planets of the solar system and to compare the processes there with what we observe on the earth. These observations supplement our knowledge of the solar system in times before formation of the earliest known earth rocks.

Two important applications of geology today are in assessing the effects that human civilization and the physical environment have on each other and in locating within the earth the natural resources on which modern civilization depends.

KEY TERMS

hypotheses
multiple working hypotheses
plate tectonics
plate tectonics theory
principle of uniformitarianism
science
tectonics
theory

STUDY QUESTIONS

1. Look up the definition of the word *science* in a dictionary. In what ways does this differ from the definition developed here?
2. Based on your understanding of other fields, compare the methods of science, as described here, with the methods used in history or the social sciences.
3. Compile a list of practical problems you think geology may help resolve.
4. What distinctions are drawn between observation, hypotheses, and theories?

REFERENCES

1. R. Harré, *The sciences: Their origins and methods*. London: Blackie and Sons, 1967, p. 2.
2. T. C. Chamberlain, "The method of multiple working hypotheses," *Science* (1st series), 15 (1890).

SUGGESTED READINGS

Bates, R. L., and J. A. Jackson, *Glossary of Geology*, 2nd ed. Falls Church, Va.: American Geological Institute, 1980.

Brown, G. B., *Science—Its Method and Its Philosophy*. London: Allen & Unwin, 1950.

Conant, J. B., *On Understanding Science: An Historical Approach*. New Haven: Yale University Press, 1951.

Davies, J. T., *The Scientific Approach*, 2nd ed. New York: Academic Press, 1973.

Geike, Sir Andrew, *The Founders of Geology*, 2nd ed. New York: Dover Publication, 1962.

Griggs, G. B., and J. A. Gilchrist, *The Earth and Land Use Planning*. North Scituate, Mass.: Duxbury Press, 1977.

Lapedes, D. N., ed., *McGraw-Hill Encyclopedia of Geological Sciences*. New York: McGraw-Hill, 1978.

Vitaliano, D. B., *Legends of the Earth*. Bloomington: Indiana University Press, 1973.

Whewell, William, *History of Scientific Ideas*, 3rd ed. London: J. W. Parker & Sons, 1858.

THE MATERIALS OF THE EARTH

Chapter 2

The image many people have of a geologist at work is that of someone battering away vigorously on a rock outcrop with a hammer, carefully inspecting one of the freshly broken pieces with a magnifying glass, and then moving on to examine the next outcrop. This is one method, and perhaps the simplest and most common method used to study the materials that make up the earth's crust. But today, it represents only the first stage in the process. The geologist's samples are likely to be carried back to a modern laboratory where more sophisticated tests, employing high-powered microscopes to take pictures like the one shown in Fig. 2.1, x-ray examinations, and chemical analysis all take place out of public view. Fortunately, many materials can be identified without complex equipment, and a surprising amount of information can be obtained by the student who can make observations, reason, and use imagination in asking the right questions. As in other sciences, advanced equipment and technology make it possible to pose new and different questions and provide another dimension to our understanding.

Most of what we know about the earth has come from studying, analyzing, and interpreting rocks, minerals, fossils, and other materials. Consequently, one of the first skills the student of geology needs to acquire is the ability to identify and distinguish the different types of earth materials. The variety of these materials is far greater than a casual observer might imagine, but finding out how they originated and what has happened to them is often limited only by our ability to formulate intelligent and probing questions.

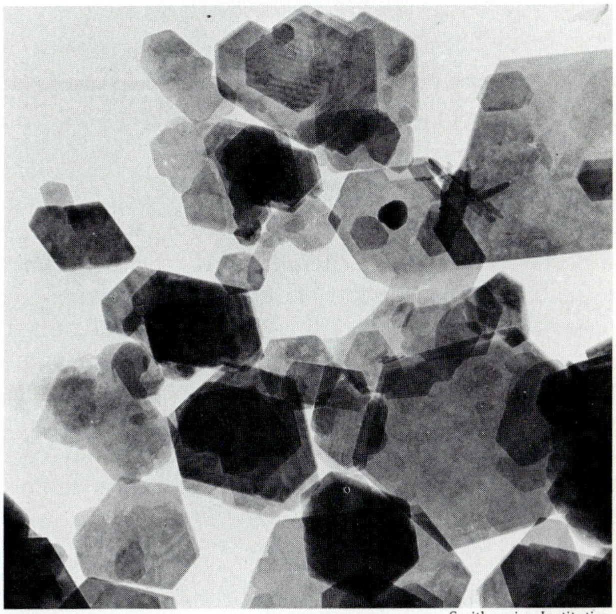

Smithsonian Institution

Figure 2.1
Electron micrograph of clay, one of the common materials found on the earth. The individual pieces of clay are seen enlarged nearly 80,000 times.

Understanding the composition of materials is fundamental to almost every aspect of geology. Economic geologists, for example, are most directly concerned with which minerals contain valuable elements (gold, silver, iron, tin, and zinc, for example) and what natural conditions cause concentrations of those particular minerals. We have already seen that critically important questions involving the origin of the planets hinge on their compositions. Comparison of the compositions of the rocks on the moon with those on the earth provides important clues to the origin of both. Later we will see that the lavas that come from volcanoes located along oceanic ridges are distinctly different in composition from those in subduction zones, and that these differences can be explained in plate tectonic theory. Even differences in the composition of rocks can explain why one layer of rock is more prominently in the landscape than others. Thus, the topics covered in this chapter will be important in many later discussions.

In order to understand the various types of materials that compose the earth, we will review some of the elementary concepts of the atomic structure of matter and chemical compounds in this chapter. The methods used to identify minerals are introduced, and a few of the most important rock-forming minerals are discussed.

2.1 THE MATERIALS EXPOSED AT THE EARTH'S SURFACE

Johann Lehmann, a teacher of mineralogy and mining at Berlin in the eighteenth century, devised one of the first generalized schemes used to classify and explain the distribution of the different types of materials found at the earth's surface. He recognized three major groups of materials: surficial layers composed of loose, uncemented soil fragmental materials such as sand carried by and settled out of water, the air, or ice, called sediment; cemented and layered rocks in which fossils could be found; and crystalline rocks containing no fossils (Fig. 2.2). He went on to interpret these as rocks formed after Noah's Flood, during the Flood, and at the beginning of the earth. No geologist would accept Lehmann's interpretations today, but most materials do fall into one of his three categories.

The earth's surface is almost everywhere covered by a relatively thin layer of loose matter called soil (Fig. 2.3). Most soil is composed of a mixture of clay, organic matter, and sand or silt, but it varies in composition and texture from place to place. The relationship of the soil to the underlying material (usually rock) is sometimes exposed in road cuts and other types of excavations. In such places the gradual transition between the soil and underlying rock is visible. Partially decomposed fragments of the rock often appear in the lower part of the soil, indicating that the soil formed by the breakdown of the rock.

In other places a sharp boundary occurs between the loose, or what geologists call unconsolidated, material above and dissimilar material below. In these cases the unconsolidated material is usually a **sediment,** material that was brought in and deposited on the underlying rock. Deposition of this type can occur where fine material settles out of streams, lakes, ocean currents, or the wind. Most of the sea floor is covered by such unconsolidated sediment.

Where sediment accumulations become thick, the lower layers are compacted, and they may become cemented to produce solid **sedimentary rocks** (Fig. 2.4). Thus sand may become sandstone and loose shell fragments may become limestone. Such rocks constituted Lehmann's second category of materials. He thought that the third class of materials, the crystalline rocks, always occurred under consolidated sediments. Geologists agree that crystalline rocks do underlie all sedimentary rocks deep in the earth, but

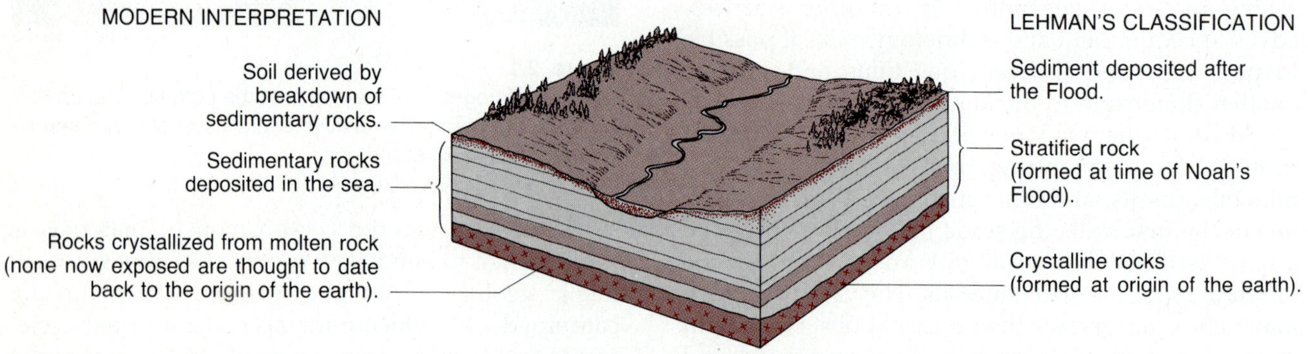
MODERN INTERPRETATION
Soil derived by breakdown of sedimentary rocks.
Sedimentary rocks deposited in the sea.
Rocks crystallized from molten rock (none now exposed are thought to date back to the origin of the earth).
LEHMAN'S CLASSIFICATION
Sediment deposited after the Flood.
Stratified rock (formed at time of Noah's Flood).
Crystalline rocks (formed at origin of the earth).

Figure 2.2
This schematic diagram depicts the three major divisions of rock units recognized by Lehmann (described at right) and modern interpretations of these divisions (at left).

Figure 2.3
Sediments. (a) Unconsolidated soil with plants growing in the upper layer. (b) This stream, flowing out of the New Zealand Alps, is heavily loaded with sediment derived from the erosion of the high terrain. (c) A close-up of stratified (layered) sediment composed of sand and gravel sizes. The inclined layers were formed by water currents.

(a) U.S.G.S.

(c)

(b) V. C. Browne

Figure 2.4
Like most sedimentary rocks, these in Arizona were deposited in layers, or strata.

we now know that crystalline rocks also occur as masses that penetrate and cut across fossil-bearing sedimentary layers.

From the perspective of modern geology, we continue to recognize soil and sedimentary rocks in much the sense that Lehmann described them. What Lehmann observed to be crystalline rocks we recognize today as either igneous or metamorphic, according to their method of formation. **Igneous rocks** formed from the cooling of a rock melt, which is known as **magma** if it occurs below the surface or as **lava** if it reaches the surface (Fig. 2.5). **Metamorphic rocks** like those of Fig. 2.6 form when either sedimentary or igneous rock is subjected to temperature or pressure, or both, high enough to alter the original rock. All three types of rocks are aggregates of minerals, naturally occurring inorganic solids. To understand the nature of minerals and the rocks that contain them, we shall begin by looking at the fundamental particles of all matter—atoms and molecules.

(a) U.S.G.S.

(b) U.S.G.S.

(c)

Figure 2.5
(a) Molten rock is sprayed into the air during the 1960 eruption in Hawaii. These fountains of lava are about 60 meters high. This lava is nearly 1000°C. (b) Solidified lava that flowed across the surface of the ground and into a fissure in Hawaii. Lava wrapped around the trees indicates the height of the flow. (c) An igneous rock, showing flow structure. The magma rose on the right and spread toward the left. This is rock that cooled and solidified below the surface.

(a) (b)

Figure 2.6
(a) Many metamorphic rocks retain the layered appearance of the sediments from which they were derived, but the minerals present and the texture are different. (b) This metamorphic rock was heated to a temperature so high that some of the rock began to melt. It is transitional between metamorphic and igneous rocks.

2.2 THE BUILDING BLOCKS OF EARTH MATERIALS

Atoms, Elements, and Ions

The **atom** is the basic building block of matter. The Greek philosophers Democritus (460–356 B.C.) and Lucretius (99–55 B.C.) thought of atoms as indivisible particles, but modern physicists know they are composed of electrons, protons, neutrons, and other smaller particles. Atoms are infinitesimal; enlarged 100 million times, an atom would still be only about the size of a pea, and even at this size the electrons, protons, and neutrons would not be visible with a high-powered microscope. Physicists have devised highly successful models of atomic structure, and they have measured the size, mass, and electrostatic charge on most of the component parts.

The three largest component parts of atoms are electrons, protons, and neutrons. Electrons have one negative electrostatic charge, commonly written -1; protons have one positive charge, $+1$; and neutrons are neutral. A neutron has mass approximately equal to that of a proton, about 1800 times as massive as an electron. Consequently, most of the mass of atoms is contained in the neutrons and protons.

According to modern models, a typical atom contains a **nucleus** composed of protons and neutrons and surrounded by orbiting electrons. Atoms differ in the number of particles they contain. The number of protons, called the **atomic number,** is used as a means of identifying the different **elements**—substances that cannot be broken down to simpler substances by ordinary chemical methods. A neutral atom contains as many orbiting electrons as there are protons in the nucleus. Each element, of which more than 100 are known, possesses certain unique physical and chemical properties. Each differs in the number of electrons and protons it contains.

A simple model of atomic structure that satisfactorily explains many observations is the one devised by the Danish physicist Niels Bohr in 1913. Bohr postulated a model for the internal structure of the atom that resembles a planetary system with electrons moving in nearly circular orbits around a nucleus composed of neutrons and protons. According to the Bohr model, electrons are restricted to only a few orbital pathways around the nucleus. Selection of these orbits is carried out according to well-defined rules. Electrons occupy elliptical orbits only at certain specific distances from the nucleus, and the number of electrons in each of these orbits is strictly limited. No more than two electrons may occupy the innermost orbit, the **first shell.** When atoms have more than two electrons, the additional electrons are located in other shells. Figure 2.7 shows the electron configuration of some common elements. The second shell has two subshells, with an inner subshell that may

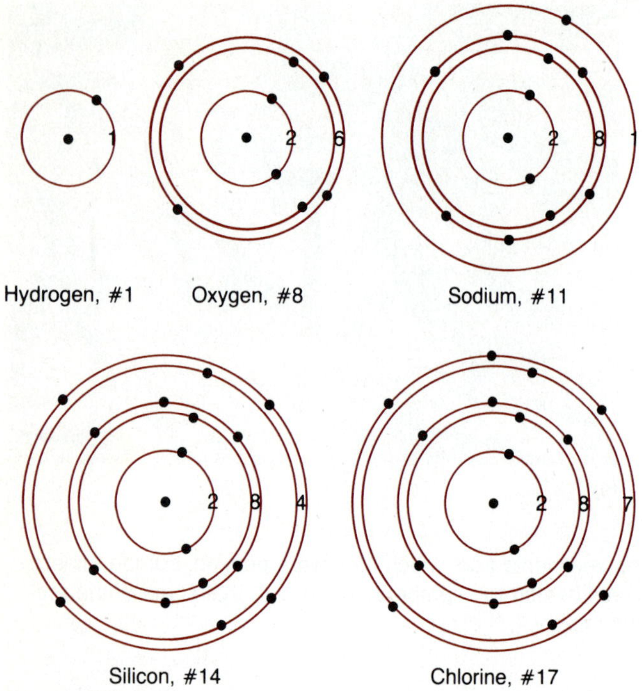

Figure 2.7
These schematic drawings depict the orbits in which electrons occur in some of the common elements, according to a simple model of atomic structure. The number of electrons in the outer shell determines whether the element is likely to become a positive or negative ion.

contain up to two electrons and an outer subshell that may contain up to six electrons, giving a total of eight electrons in the second shell. The third shell may have up to three subshells, which are occupied by a maximum of two, six, and ten electrons, respectively. The fourth shell may have up to four subshells, occupied by a maximum of two, six, ten, and fourteen electrons respectively.

Certain elements, known as the **noble** or **inert gases,** are so remarkably stable that, until recently, no compounds of these elements were known. They are helium, atomic number (Z) 2 (symbolized $_{2}$He); neon, $Z=10$ ($_{10}$Ne); argon, $Z=18$ ($_{18}$Ar); krypton, $Z=36$ ($_{36}$Kr); xenon, $Z=54$ ($_{54}$Xe); and radon, $Z=86$ ($_{86}$Ra). Because these gases are inert, the grouping of the electrons in these elements must be stable. With the exception of helium, which has two electrons, each of the inert gases has an electron configuration that contains eight electrons in the outermost orbit. Elements that have this configuration do not lose or gain electrons from their outer orbits. Except for hydrogen, other elements tend either to lose or gain electrons until eight electrons occupy the outer shell.

The presence of eight electrons in the outer orbit of an atom makes it particularly stable, and atoms have a tendency to attain this stable form. If an atom loses an electron, it has a net positive charge; if it loses two electrons, it will have an excess of two positive charges. Atoms which have net electrostatic charges are called **ions.** They may be either positive or negative, depending on whether they gain or lose electrons. Which atoms are most likely to become positive ions and which negative ions are easily predicted. If a given atom has eight electrons in the next to the outermost shell and one in the outermost shell, it is more likely to lose that one electron than it is to gain seven additional ones in order to achieve the stability of eight in the outer shell. Likewise, an atom with two electrons in its outer shell is more likely to lose these than it is to gain six. On the other hand, an atom with six or seven electrons in its outer orbit is more likely to gain one or two, forming a negative ion, than it is to lose six or seven.

Minerals as Chemical Compounds

The crystalline solids of which rocks are composed, **minerals,** are formed by the combination of elements. Most rocks and minerals are composed of the relatively small number of elements that are listed in Table 2.1. This estimate is made on the basis of analyses of rocks exposed at the surface, plus geologists' best judgements about the make-up of the deep interior. With the exception of some of the "native" metals, such as gold, silver, platinum, and copper, few elements occur in pure, uncombined form. Once combined, minerals have physical and chemical properties different from those of their component elements. Their physical properties provide one of the most useful ways of identifying minerals, and we will consider how to use these later in this chapter. But first we will review some of the

Table 2.1. Composition of the Earth

Element	Weight Percent in the Earth
Iron	35
Oxygen	28
Magnesium	17
Silicon	13
Nickel	2.7
Sulfur	2.7
Calcium	0.6
Aluminum	0.4

Figure 2.8
Crystals of the mineral halite, sodium chloride, exhibit the smooth, flat surfaces that are characteristic features of crystals.

different ways that compounds form. The bonds holding elements together in the compounds of most minerals are of three types—ionic, covalent, and metallic.

Ionic compounds. Ions are the charged particles formed when an atom gains or loses one or more electrons. The net charge of an ion is the difference between the number of protons in the nucleus and the number of electrons in its orbits. Ions obey the laws of static electricity: they are attracted to ions with an opposite charge and repelled by those of a like charge. As the negative and positive ions come together, they pack around one another in a close-fitting structure that grows until it appears as a solid with a regular internal arrangement of atoms. This orderly internal arrangement may be expressed in a regular geometric external form called a **crystal** (Fig. 2.8).

The shape of a crystal depends mainly on the relative sizes of the ions of which it is composed. We may visualize ions or atoms as spheres that have a radius determined partly by the number of electron shells. The electrostatic forces acting on the ions tend to pull them as close together as possible. Thus, the number of positive and negative ions that can fit around one another depends on the relative sizes of the two, and the way they pack together determines the shape of the crystal.

The ratio of the radius of a positive ion to that of the negative ion with which it combines is a convenient way to express the relative sizes of ions. This **radius ratio** is significant, too, because it serves as a guide to possible configurations a given negative-positive ion combination may have. The number of negative ions that are in contact with a positive ion in any given packing arrangement is called the **coordination number.** Ions of any radius can be packed together if they are packed in coordination of one or two, but three negative ions can fit around one positive ion of the corners of an equilateral triangle only if the radius ratio falls in a certain range (between 0.155 and 0.225). Coordination of four to form a tetrahedron, of six to form an octahedron, of eight to form a cube, and of twelve to form a cube are possible only with certain limits of radius ratios (Fig. 2.9). Thus, knowledge of the ionic radius of any ion combination should enable us to predict both the probable coordination and configuration of a crystal containing the two.

One of the most common ionic compounds is sodium chloride, the mineral halite, or what we ordinarily call salt. Sodium, atomic number 11, has

Coordination number	Radius ratio + ion/− ion	Configuration of atoms
1	0–∝	Side by side
2	0–∝	On two sides
3	0.15–0.22	Corners of an equilateral triangle
4	0.22–0.41	Corners of a tetrahedron
6	0.41–0.73	Corners of an octahedron
8	0.73–1	Corners of a cube
12	1	Center of cube edges

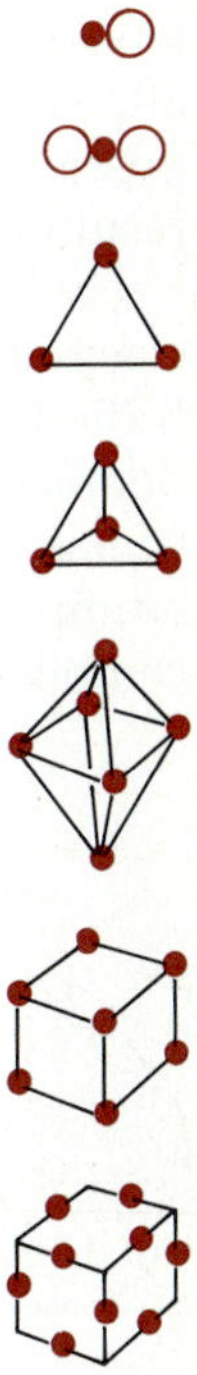

Figure 2.9
The packing arrangements usually found for the most common coordination numbers.

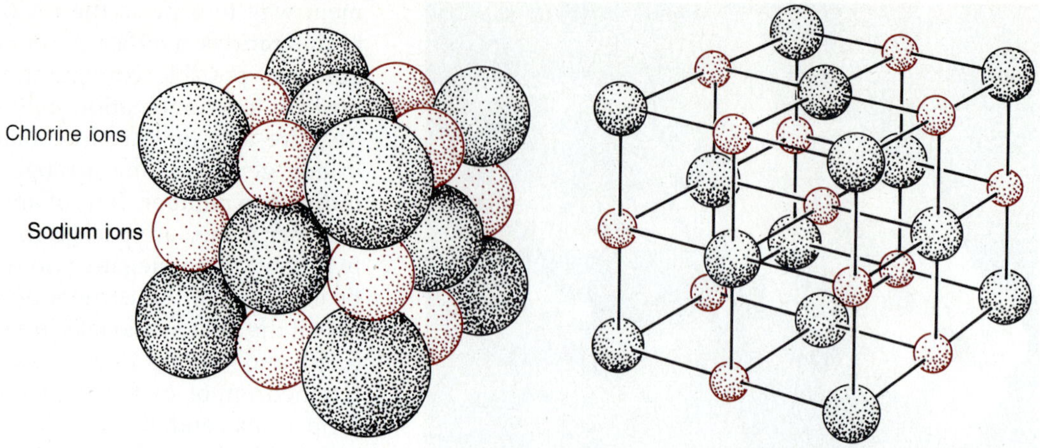

Figure 2.10
Packing of sodium and chlorine ions in the ionic compound sodium chloride (NaCl). The structure is enlarged at right.

two electrons in the first shell, eight in the second, and one in the third. Therefore, it has a tendency to lose the single electron in the outer shell and become a sodium ion with a positive charge (Na^+). Chlorine, atomic number 17, has two electrons in its first shell, eight in its second, and seven it its third shell. It tends to gain one electron in the third shell to become a chlorine ion with a negative charge (Cl^-). The Na^+ ion and the Cl^- ion, having different charges, are attracted toward one another and pack as closely as possible in the configuration shown in Fig. 2.10. The resulting crystal is cubic in shape.

Most metals, like sodium, tend to lose their outer electrons, whereas nonmetals tend to gain electrons. If the metal involved loses only one electron but the nonmetal picks up two, the ionic compound may contain two metal ions to each nonmetal ion (for example, Na_2S), but many ionic compounds are more complicated structures in which electrons from metals are given up to groups of nonmetals (as shown in Fig. 2.11) rather than to single atoms. For example, the ion CO_3^{2-}, carbonate, occurs commonly in minerals. When three oxygens, each of which has six electrons in its outer shell, combine with one carbon that has four electrons in its outer shell, the combination still needs two electrons in order to provide the stability of eight electrons in the outer shell of each atom. If those two are picked up, a net charge of -2 exists on the combined carbon and oxygen. This often attracts calcium, Ca^{2+}, to become a stable compound $CaCO_3$, the mineral calcite.

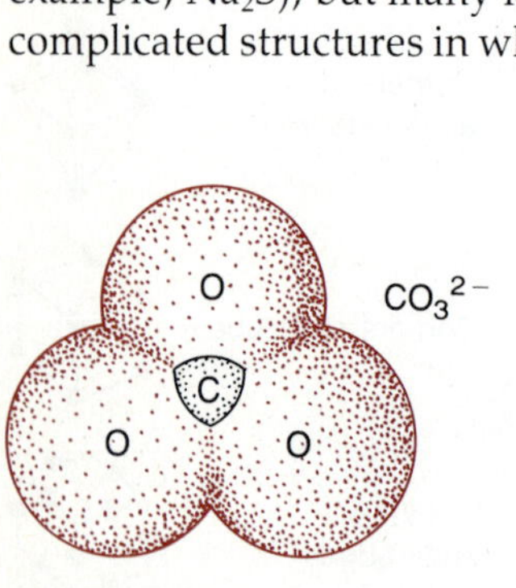

Figure 2.11
Three oxygens may fit around and share electrons with one carbon atom. This combination has two extra negative charges (electrons), CO_3^{2-}. In nature it is most often found combined with calcium (Ca^{2+}) or magnesium (Mg^{2+}) ions.

Covalent compounds. A second type of bond is formed when elements share electrons rather than completely transfer them. The element does not become ionized; the electrons remain with their respective atoms, but the electrons in the outer orbit are shared by the adjacent atoms. It is thus possible for each of the atoms to achieve the stability of having eight electrons in its outer orbit by sharing those of one or more other atoms. Diamond, in which each crystal is essentially a single molecule of carbon atoms, is a good example of a mineral with covalent bonding. Methane, a compound of carbon and hydrogen, also exhibits covalent bonding. A free hydrogen atom has one electron in its orbit. Carbon has two electrons in its first orbit and four in its outer orbit. A stable compound is formed when a carbon atom is able to share the electrons of four hydrogen atoms. Thus, methane is expressed as CH_4. Another common compound held together by covalent bonds is water, H_2O. Hydrogen has one electron and oxygen ($_8O$) needs two electrons in its outer orbit to have eight.

Thus one oxygen atom and two hydrogen atoms can share electrons in such a way that there are eight electrons in the outer orbit of the oxygen atom and two in each hydrogen's orbit. This configuration produces the water molecule.

In some covalent compounds the molecules behave somewhat like small magnets; these are called **polar covalent** compounds. When the distribution of electrons in the molecule is uneven, one part becomes slightly positive and the other slightly negative. Water is one such molecule. In water, the two hydrogens are slightly positive and the oxygen slightly negative (Fig. 2.12). The **polar nature** of the water molecule explains why it is such an effective solvent. When a grain of salt or another ionic compound is placed in water, the H_2O molecules, acting as small magnets, become oriented with their positive ends toward negative ions in the solid and their negative ends toward positive ions. The effect of the attractions between water molecules and ions is to loosen the bonds between atoms in the solid and allow the ions to move into solution. The amount of such compounds that will go into a solution is limited, however, because the number of free water molecules is reduced as more ions are freed or dissolved in the solution.

Covalent bonds, polar and nonpolar, are important in many complex minerals. Silicon and oxygen are held together by covalent bonds in silicon–oxygen tetrahedra, described later, and aluminum, sulfur, and carbon also form covalent groups.

Metallic compounds. Metallic bonding, which is characteristic of pure metals, is different from ionic and covalent bonding in that bonding electrons are shared throughout the entire metal crystal. The atoms, all of the same size, pack tightly together so that each has twelve others around it. Twelve spheres of identical size can always be packed around a thirteenth sphere of the same radius coordination number (12). The atoms of the metals lose their outer electrons easily, and the electrons can move through the crystal lattice. For this reason, metals can conduct electricity. The character of metallic bonds also makes it possible for the atoms in a crystal of pure metals to be rearranged with little loss of strength; as a result, metals are ductile in contrast with ionic crystals (like salt), which tend to be brittle.

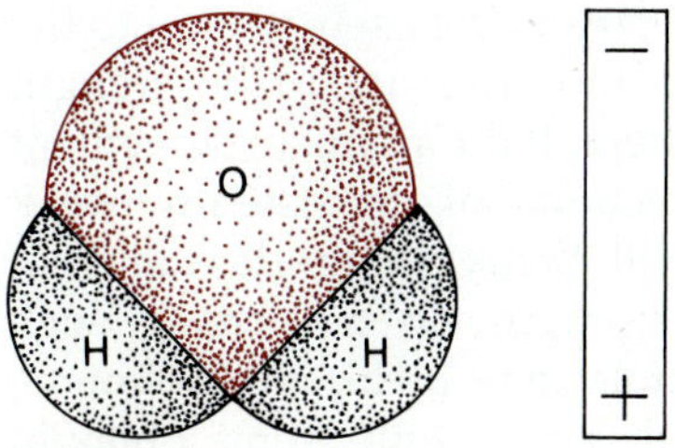

Figure 2.12
Because the hydrogen atoms are close together, the water molecule acts somewhat like a small bar magnet.

Dating the Materials of the Earth

Isotopes. Elements are defined by the number of protons in the nucleus; all atoms of an element have the same number of protons in the nucleus. But two atoms of the same element may differ in the number of neutrons they contain and consequently in atomic mass. Atoms with the same atomic number but different atomic masses are **isotopes.** Two isotopes of an element have chemical and physical properties that are almost identical, but they may differ in one important way. Some isotopes are unstable and break down spontaneously by the emission of particles. Such isotopes are said to be **radioactive.**

Through the emission of these particles, radioactive isotopes disintegrate and form either different elements or occasionally a different isotope of the same element. The rate of disintegration is constant for each isotope, regardless of its physical state or its chemical bonding. The length of time required for half the mass of the isotope to break down is called the isotope's **half-life.** The new daughter isotope or element may be radioactive also and break down again, but eventually a stable isotope or element is formed. Because the half-life is a constant, and because a stable element is formed as an end product, it is possible to use radioactive isotopes as geologic clocks.

Radiometric age determination. Radioactive isotopes commonly occur in many igneous and metamorphic rocks. Because the rate of disintegration of these isotopes is so precisely known, the ratio of the amount of the daughter to the amount of the parent isotope in these rocks can be used to determine the length of time the process has been going on.

This method is thought to be reliable if three conditions are met. First, the rock must be essentially a closed system. That is, neither the parent nor daughter elements have been introduced or removed from the rock since the time at which the parent originated. For igneous rocks, that is the time it crystallized from magma; for metamorphic rocks, it is the time the metamorphic mineral crystallized. Secondly, the amount of the parent and daughter elements present must be sufficient for accurate processing. Finally, the analytical procedure must be accurate.

Of the many radioactive isotopes only a few are widely used in geology. Among the most important of these are isotopes of uranium, potassium, and rubidium. The uranium isotope of mass 238, written uranium-238, decays to form a stable substance, lead-206, only after passing through a number of other radioactive isotopes. The half-life of uranium-238 is 4.56×10^9 years. One million grams of uranium-238 produces 1/7600 gram of lead-206 per year.

Let us suppose that one of the minerals that crystallizes from a magma contains uranium-238. As soon as it has crystallized, the radioactive nuclei of uranium-238 become frozen in the rock. A certain percentage of the atoms disintegrate every minute. After 4,560 million years, half of them will have disintegrated. As the uranium-238 disintegrates, helium and lead-206, both of which are stable and undergo no further change, form as end products of decay. Thus the ratio of helium and lead-206 to the amount of uranium-238 bears a definite relationship to the age of the mineral in which it is found and, therefore, to the time of its formation.

Another uranium isotope, uranium-235, is also radioactive, decaying eventually to the stable daughter product, lead-207. Since U-235 and U-238 occur together, the age for a sample can be determined by two methods, allowing a good check. The age equations are:

$$T_{238} = 6.50 \times 10^9 \, ln\left(1 + \frac{\text{Pb-206}}{\text{U-238}}\right)$$

$$T_{235} = 1.03 \times 10^9 \, ln\left(1 + \frac{\text{Pb-207}}{\text{U-235}}\right)$$

where T = absolute age, Pb-207/U-235 is the ratio of the number of atoms of these two isotopes, and *ln* stands for *natural logarithm*.

The element potassium is present in several of the most abundant minerals on earth, notably the micas, clay, and feldspar. One isotope of potassium, potassium-40, is radioactive, leading ultimately to the argon isotope of mass 40. The time it takes potassium to decay to the inert gas argon is now one of the most widely used standards for radiometric age determination. The formula for age is:

$$T = 1.88 \times 10^9 \, ln\left(1 + 9.10 \frac{\text{Ar-40}}{\text{K-40}}\right)$$

A major problem with this method is that argon may be lost if the rock has been reheated. Some argon is lost from clay minerals at 50 °C, from biotite mica at 150 °C, and from muscovite mica at 200 °C. Thus heating after the minerals crystallize may cause the age to be underestimated. We will examine the use of two other radioisotopes, rubidium and carbon-14, in later chapters.

2.3 MINERAL IDENTIFICATION

Physical Properties

Minerals may be identified by their chemical composition, by their atomic structure, or by their physical properties. Some physical properties can be determined easily and quickly and those described below are commonly used in making identification of hand specimens.

Specific gravity of a mineral is the ratio of its weight to the weight of an equal volume of water. Graphite is more than twice as dense as water; that is, one cubic centimeter of graphite weighs 2.3 times as much as a cubic centimeter of water. The specific gravity is, therefore, 2.3.

Some minerals tend to split along smooth planes when they break. This property is called **cleavage.** If atomic bonds are strong between atoms in some directions but relatively weak in other directions, the mineral will tend to break in the direction of the weaker bonds. In some minerals, differences in bond strength are great; in others, there is little difference. Cleavage planes are always parallel to a crystal face or to a possible crystal face, and there may be one or as many as six directions of cleavage. Some minerals break to form smooth cleavage surfaces, as shown in Fig. 2.13, which are said to be perfect; others that do not break so easily have imperfect cleavage. Still others break along cleavages only with difficulty. Three directions of cleavage at right angles to one another are evident in the cleavage fragments of halite (salt). Another familiar mineral group, the micas, have one perfect cleavage.

The way a mineral breaks is closely related to the internal configuration of the atoms and to the strength of the bonds between them. If the atomic arrangement results in certain directions of consistently weak bonds, the mineral will break along cleavages. If differences in bond strength do not create such weaknesses, the minerals may break into splinters; others fracture smoothly and evenly; some break in rough, irregular surfaces. A few, such as quartz, break with a smooth, curved fracture known as a

Smithsonian Institution

Figure 2.13
The cleavage fragment or clear (optical quality) calcite at right shows three perfect cleavages. A single near-perfect cleavage is seen on the feldspar fragment at left.

Smithsonian Institution

Figure 2.14
Most minerals and some rocks break with a characteristic fracture. Quartz breaks with a type of fracture known as a conchoidal fracture, which is also typical of glass, such as the natural glass, obsidian, shown here.

conchoidal (shell-like) **fracture,** which is illustrated in Fig. 2.14.

The appearance of a mineral's surface in reflected light is called its **luster.** Minerals fall into two major groups, metallic (such as copper or pyrite) and nonmetallic (such as quartz or calcite) on the basis of luster.

Color may be used to identify most metallic minerals, but color in nonmetallic minerals may be due to very small amounts of impurities. It is not uncommon for a single nonmetallic mineral such as quartz or calcite to occur in as many as five or six different colors.

Streak, the color of the powder of a mineral, is usually more consistant than the color of the mineral. Although the color of many nonmetallic minerals is highly varied, most of them have a white or colorless streak.

Hardness refers to the ability of one mineral to scratch another. A scale of ten minerals, called **Moh's hardness scale,** shown in Table 2.2, is used as a basis for hardness comparisons. The scale is nearly linear (that is, a mineral of hardness 5 is almost five times as hard as a mineral of hardness 1) for the first nine minerals; however, diamond is much harder than corundum.

These simple physical properties, cleavage, hardness, streak, and color, can be observed easily in the field. In combination, they often serve to identify minerals. The laboratory technique of x-ray analysis

Table 2.2. Moh's Hardness Scale. The minerals in this chart are arranged in order of increasing hardness. A mineral can be scratched by those with higher numbers, and it can be used to scratch those with lower numbers.

Hardness	Representative Mineral
1	Talc
2	Gypsum
	a fingernail
3	Calcite
	a copper penny
4	Fluorite
5	Apatite
	a steel knife
	plate glass
6	Feldspar
	a file
7	Quartz
8	Topaz
9	Corundum
10	Diamond

offers more detailed information about the internal structure of materials.

X-Ray Analysis

X-ray analysis of minerals is based upon a discovery made in 1912 by German physicist Max von Laue, who was studying the separation of beams of light by diffraction gratings, highly polished surfaces with very precisely spaced etched lines. As the wavelength of light to be dispersed decreased, von Laue required finer and finer gratings. When the gratings required became too fine to be manufactured by any technique known at that time, he turned to natural crystals. Crystals were thought to contain closely spaced layers of atoms that would act as a three-dimensional grating. Von Laue hypothesized that the waves would be diffracted within a crystal in such a way that separate beams of radiant energy would emerge at specific angles. His experiments consisted of placing a crystal in the path of a beam of x-rays, which are very short wavelength, high energy radiation. The x-rays passed through the crystal, and their paths were recorded on a photographic plate behind the crystal. When developed, the plate showed a dark spot in the center, where the x-rays had traveled through the crystal undeflected, and a large number of small dark spots arranged around the center in a regular geometrical pattern, located where x-rays reflected from various layers were reinforced (Fig. 2.15). Thus, von Laue proved that crystals do possess an orderly internal arrangement of atoms.

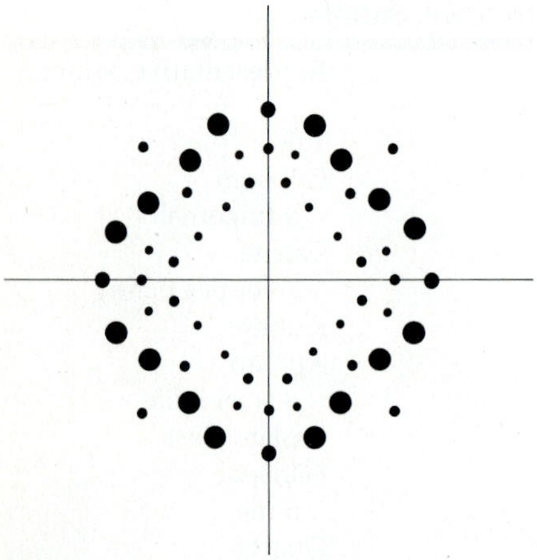

Figure 2.15
Pattern formed by x-rays that passed through a piece of halite onto a photographic plate.

In the same year, 1912, Bragg and Bragg (father and son) began using x-ray signatures to identify minerals. The use of x-ray techniques developed rapidly and has proved to be a powerful tool for analysis of the internal structure of crystals. The spacing between layers of atoms in a mineral's structure can be determined by x-ray analysis, and this spacing is an excellent guide to the identification of minerals, even of those materials in which the crystals are too small for microscopic examination.

A newer technique for x-ray analysis uses the x-rays actually emitted by a substance to identify it. With an **electron probe** a beam of electrons is focused on a point on the surface of a sample. Electron bombardment causes the sample to emit x-rays. The distribution of x-ray wavelengths emitted, the **x-ray spectrum,** is characteristic of the elements present; the intensity of various wavelengths is a measure of the concentration of the elements. Because the electron probe makes it possible to determine chemical composition within very small areas (as small as one-thousandth of a millimeter, or 5 microns), the probe can be used to study the arrangement of atoms within individual crystals and even to study the surface between two crystals.

2.4 IMPORTANT MINERALS AND MINERAL GROUPS

Mineral Groups

More than two thousand different minerals have been identified; so, it is desirable to use some system of classification. The most convenient and useful system is one that groups minerals according to chemical composition. The "native" elements are minerals that are all composed of a single element, and they are put in a class together. All other minerals are compounds of several elements or groups of elements; they are put in classes defined by the negative ion or ion group. Thus, the sulfur compounds make up a group called sulfides, and the compounds with the complex negative ion of carbon and oxygen (CO_3^{2-}) are known as carbonates. A list of the groups and some common examples of each follow:

Native elements: gold, silver, copper, platinum, arsenic, sulfur, diamonds.

Sulfides: sulfur compounds, such as galena, (PbS), sphalerite (ZnS), pyrite (FeS_2).

Oxides: compounds of oxygen, such as ice

(H_2O), hematite (Fe_2O_3), corundum ($A1_2O_3$), magnetite (Fe_3O_4).

Halides: compounds of chlorine or fluorine, such as halite (NaC1), fluorite (CaF_2).

Carbonates: compounds of carbonate (CO_3), such as calcite ($CaCO_3$), dolomite ($CaMg[CO_3]_2$).

Sulfates: compounds of sulfate (SO_4), such as gypsum ($CaSO_4 \cdot 2H_2O$).

Phosphates: compounds of phosphate (PO_4), such as apatite ($Ca_5[PO_4]_3[F]$).

Silicates: compounds of silicon and oxygen, such as quartz (SiO_2), feldspar ($KAlSi_3O_8$), kaolinite ($A1_2[Si_2O_5][OH]_4$).

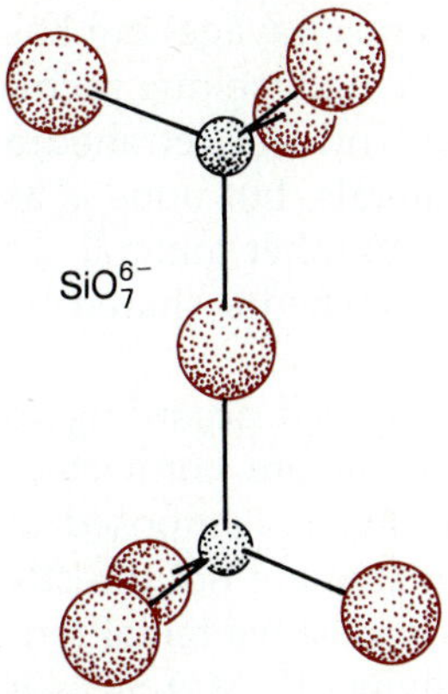

Figure 2.17
Two silicon–oxygen tetrahedra linked.

The Silicates

Most of the material near the earth's surface is made up of only a few elements. Of these, oxygen, and silicon are far more abundant than any other element. These, together with aluminum, iron, calcium, sodium, potassium, and magnesium, account for about 99 percent of the earth outside the central core. Their abundances alone would lead us to expect compounds of silicon and oxygen, the **silicates,** to be important constituents of rocks, and they are indeed by far the most important group of rock-forming minerals. For this reason we will study their structure and some common minerals of this group in greater detail.

Structure of the silicates. The chief structural component, sometimes called the building block of the silicates, is the silicon–oxygen tetrahedron (Fig. 2.16). Silicon, atomic number 14, has four electrons in its outer orbit, and oxygen, atomic number 8, has six electrons in its outer orbit. These two elements commonly come together and share electrons. The size of silicon is such that four oxygen atoms can cluster around it to form a tetrahedrom. Oxygen is bound to silicon by a single electron-pair bond, leaving each of the four oxygens with seven electrons in its outer shell free to bond to another atom or to accept one more electron. Thus, the silicon–oxygen tetrahedron has a net negative charge of minus four ($SiO_4{}^{-4}$).

In ionic compounds the charges on ions and some ionic groupings, such as the silicon oxygen tetrahedra, are equalized or neutralized by opposite charges of other ions in the compound; thus, the compound has no net electrostatic charge. In all silicates, this is accomplished in whole or in part by the sharing of one or more of the oxygen atoms by adjacent tetrahedra. Thus, although a single tetrahedron has a net charge of -4, two adjacent tetrahedra sharing an oxygen atom as shown in Fig. 2.17 have a total net charge of -6 (-3 each for each tetrahedron). If each of the oxygens of a tetrahedron is shared by another tetrahedron, the net charge is reduced to zero, and the resultant ratio of silicon atoms to oxygen atoms is 1:2. This structure is found in one of the minerals, quartz, SiO_2, a common constitutent of many rocks.

Because the bonds are strong in every direction through quartz, the mineral is hard. It does not break

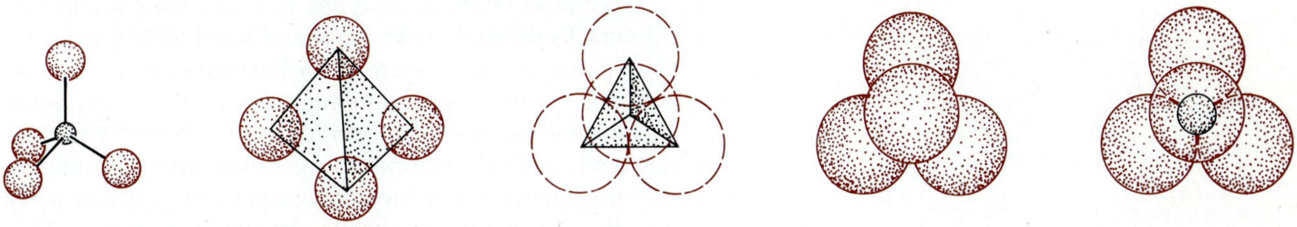

Figure 2.16
Various ways of representing the silicon–oxygen tetrahedral grouping. The silicon atom is located in the center, as shown in the enlarged view at left. The four oxygens occupy the corners of a tetrahedron.

in any preferred direction (it has no cleavage) but has a conchoidal fracture, and it has a high melting point, 1470 °C. Three dimensional networks of tetrahedra occur in a few other silicate minerals, but none is as simple as that of quartz, and in no other mineral are all of the oxygen atoms in the tetrahedra shared by other tetrahedra.

The silicate minerals are classified according to how the silicon–oxygen tetrahedra are connected. The simplest group includes minerals composed of independent tetrahedral groups that are not directly linked together but are bound to positive metal ions such as magnesium or iron. Olivine, $(FeMg)_2SiO_4$, is such a silicate. The next more complex silicate is composed of double tetrahedra that share one oxygen (as in Fig. 2.17) and are bound by other positive ions. Silicate ring structures are produced by the sharing of oxygens by three, four, or six tetrahedra in such a way as to form a ring (Fig. 2.18). An example is the mineral beryl. Chains linking tetrahedra into either single chains (Si:O ratio 1:3), or double chains consisting of cross-linked chains with Si:O ratios of 4:11 (Fig. 2.19) are the basic structures of the minerals pyroxene and amphibole, respectively. Sheet structures produced by the sharing of each of three oxygens with adjacent tetrahedra form an extensive sheet, such as that found in the mica minerals. Three-dimensional networks like that of quartz constitute the last group of silicates.

Metal ions fit into the silicate structures to satisfy the net electrical charges on these tetrahedral groupings and to bind the sheets, rings, and chains together. Water molecules (H_2O), hydroxyl groups (a combination of one oxygen with one hydrogen, OH), and additional oxygen atoms also are found with some of the silicate tetrahedral groupings. Aluminum is especially important. It is abundant, and because it is similar in size to silicon it can replace silicon in the SiO_4 tetrahedra if the charge balance can be maintained when Al^{+3} is substituted for Si^{+4}.

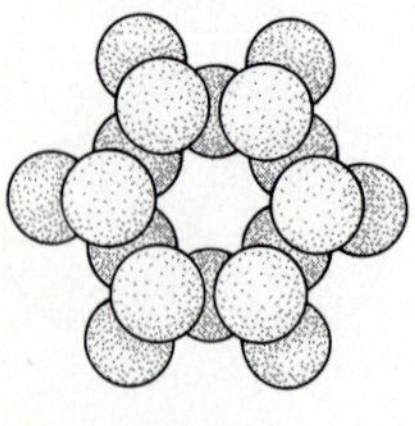

Figure 2.18
A ring structure formed by silicon–oxygen tetrahedra.

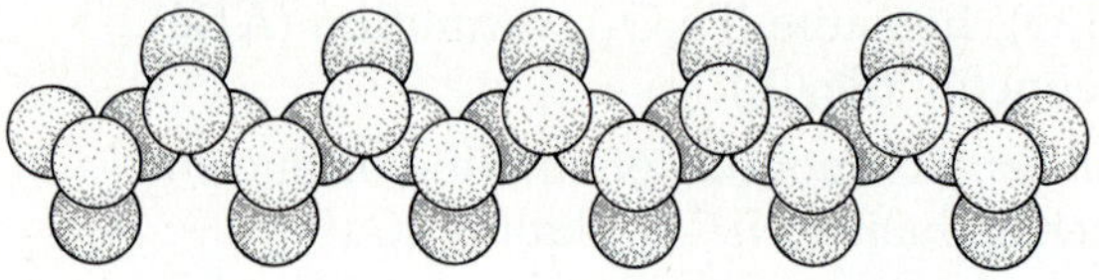

(a)

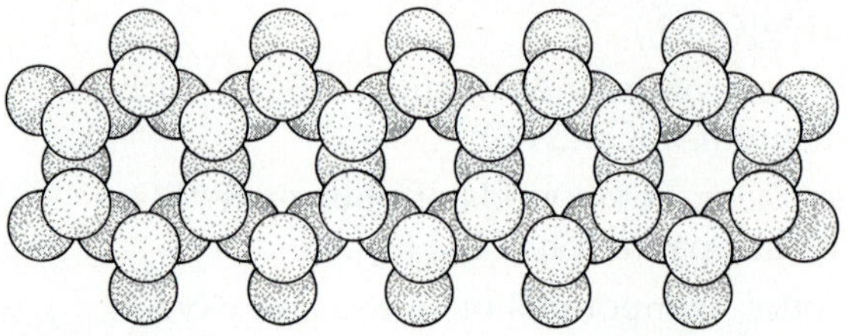

(b)

Figure 2.19
(a) A chain formed by sharing of one oxygen atom by two tetrahedra producing a single chain structure is typical of the pyroxene group. (b) A double chain structure of silicon–oxygen tetrahedra of the type found in amphiboles.

Quartz. As the second most abundant mineral in the earth's crust, quartz (SiO_2) is a common constituent of igneous, metamorphic, and sedimentary rocks. As these rocks break down, pieces of quartz or silica in solution are freed. Because quartz is nearly insoluble and very hard and durable, it remains as an important component of sediment. Sometimes it is the main mineral in sediment, as is frequently the case with the sand on beaches and in sand dunes; more commonly it is mixed with clay and other products of the decay and disintegration of rocks.

Quartz probably occurs in a greater variety of forms than any other mineral. In addition to its fragmental forms, the silica that goes into solution as rock slowly breaks down is often deposited chemically or by organisms. Some animals build their hard skeletal parts out of silica, and occasionally the carbonate shells of invertebrate animals are found replaced by silica, as is the woody matter of petrified wood. Many of the sedimentary rocks have their fragmental parts cemented together by quartz. It may occur as veins in all types of rocks, and sometimes cracks through rocks are found lined with quartz and quartz crystals where silica has been chemically deposited from water moving through the crack. Some springs are sites of deposition of silica from hot waters, and nodules composed of silica occur in marine sediments where the quartz apparently formed while the sediments were being changed into rock.

The larger crystals of quartz, such as those shown in Fig. 2.20, will be familiar to most readers. Aggre-

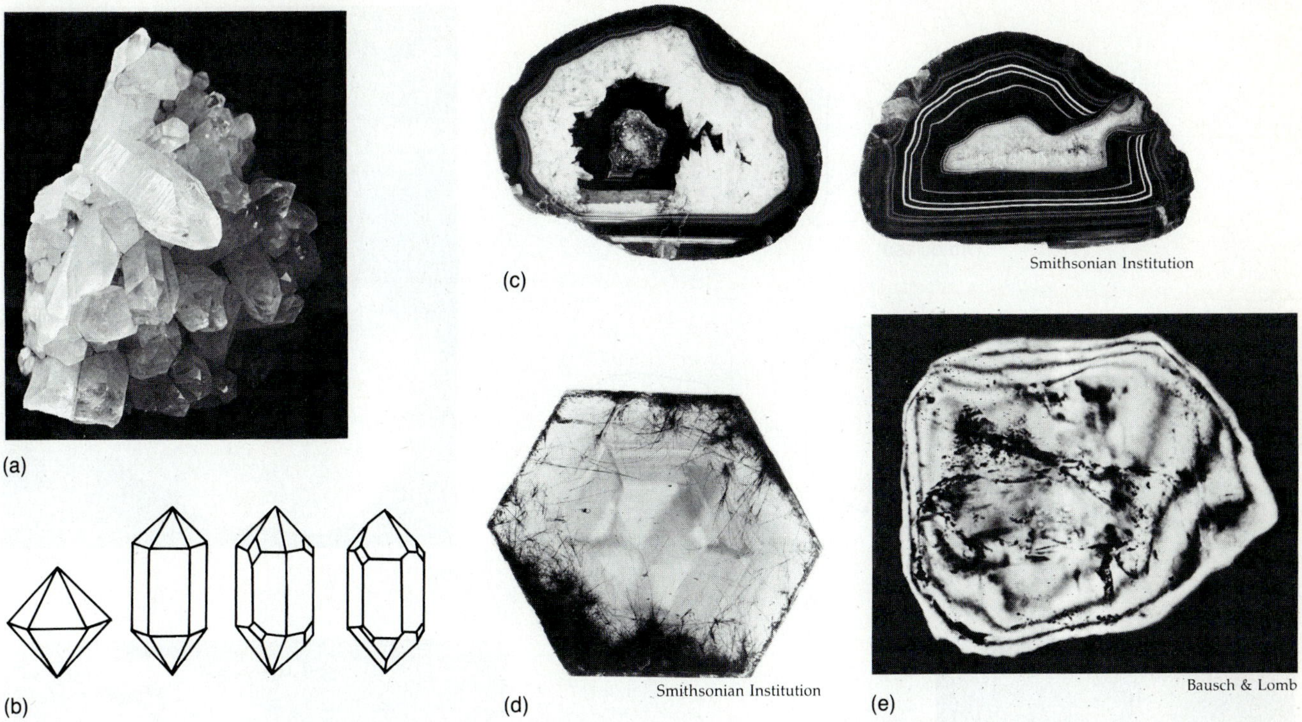

Figure 2.20
(a) A cluster of quartz crystals. (b) Sketches of quartz crystals showing common crystal shapes. (c) Quartz deposited in a cavity. The banded parts are exceedingly small crystals. Larger crystals line the cavity. (d) Section of a quartz crystal containing hairlike crystals of the mineral rutile. (e) A grain of quartz sand, about 2 mm across.

gates of crystals, called cryptocrystalline, so small they cannot even be seen with a microscope, make up the forms of quartz known as flint (chert). In other forms some water is present with the silica and oxygen producing a cryptocrystalline form of quartz known as opal.

Quartz can generally be identified by its physical properties. It has a hardness of 7; its specific gravity is 2.6. It has no cleavage because the bonds in its three-dimensional network of tetrahedra are strong in all directions through the structure. Its fracture is conchoidal. Color is not an accurate criterion for identification because minute quantities of impurities give quartz a wide range of shades of color.

Feldspar group. The group of minerals known as the feldspars are the most abundant minerals in the earth's crust. They are especially prominent as components of many igneous and metamorphic rocks frequently appearing as large crystals in these rocks (Fig. 2.21). Unlike quartz, however, the feldspars slowly decompose to form clay when they come in contact with water and carbon dioxide at the ground surface. For this reason they may be completely absent even from sediments derived from rocks like granite that contain abundant feldspar. Other sediments that were buried rapidly may contain large quantities of the feldspars. Like quartz the feldspars are silicates; but unlike quartz, they may differ in composition. All feldspars are aluminum silicates, bound together in a three-dimensional network of silicon–oxygen and aluminum–oxygen tetrahedra. Positive ions of potassium (K), sodium (Na), and/or clacium (Ca) fit into the spaces created by the tetrahedra. Those feldspars that contain potassium or sodium are called the alkali feldspars, and those containing sodium and/or calcium are called plagioclase feldspars (Fig. 2.22). The potassium feldspar (orthoclase) has a composition $KAlSi_3O_8$. The plagioclase feldspars contain varying amounts of Na and Ca. One end member of this series is albite ($NaAlSi_3O_8$) but sodium and calcium ions are about the same size and can substitute for one another in the crystal structure. However, because the ion of calcium has two positive charges

Figure 2.21
(a) Feldspar crystals. (b) Crystals of potassium feldspar. (c) Photomicrograph of feldspar in an igneous rock (× 10). (d) Plagioclase feldspar showing fine straight lines called striations (approximately natural size). (e) Outcrop of metamorphic rock intruded by a mass composed mainly of white feldspar.

while the sodium ion has a single charge, this substitution must be accompanied by a substitution of aluminum (Al^{+3}) for silicon (Si^{+4}) in order for the crystal structure to maintain a neutral electrostatic charge.

The feldspars are nonmetallic and have colors ranging from white (usually albite), pink (orthoclase), to pearl gray (Ca–Na plagioclase). The hardness is 6, and two cleavages are present at right angles. Plagioclase exhibits twinning, a systematic pattern of reversed atomic arrangement, which appears as a set of fine straight lines (**striations**) on some cleavage faces (Fig. 2.21*d*). Orthoclase usually has white stringlike intergrowths in it, but does not show twinning striations.

Olivine group. Olivine is an especially important mineral because it is a component, and possibly the main one, of the outer half of the earth. Olivine is also found in stony meteorites thought to be derived from the breakup of planetlike bodies, and it is found in many rocks, especially those igneous and volcanic

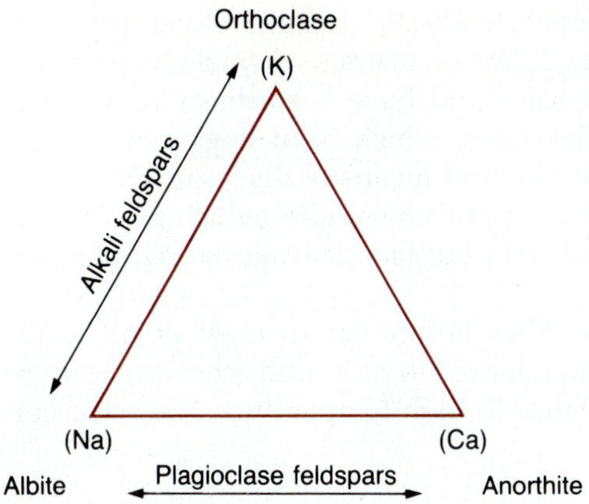

Figure 2.22
The compositional variations in the feldspar group can be represented on a triangular diagram like this one. The plagioclase feldspars show a continuous variation in the relative amounts of positive ions Na and Ca present from 100 percent Ca to 100 percent Na. The alkali feldspars do not exhibit a continuous variation, but both albite and orthoclase belong to this group.

rocks that contain large quantities of iron and magnesium (gabbro, dunite, basalt). Only rarely are quartz and olivine found in the same rock. Olivine does occur, though not commonly, in metamorphic rocks, but because it breaks down rapidly in water it is rarely found in sediments or sedimentary rocks.

The olivine minerals range in composition from Mg_2SiO_4 (forsterite) to Fe_2SiO_4 (fayalite). Single SiO_4^{4-} tetrahedra are linked by the iron or magnesium ions. Because they are nearly the same size, magnesium and iron can substitute for each other in the structure, so the general formula is $(Mg,Fe)_2SiO_4$. The expression $(Mg,Fe)_2$ indicates that there are a total of two ions in any combination. Small amounts of other ions, especially calcium, may also be present.

Olivine (Fig. 2.23) is nonmetallic; it has a characteristic greenish color, which in combination with its glassy appearance and occurrence as masses of grains (granular texture) makes it easy to identify. The hardness is 6.5.

Pyroxene group. The pyroxenes are more frequently found in igneous and metamorphic rocks than in sediments or sedimentary rocks, for, like olivine, they tend to weather rapidly, decomposing to clay minerals. They are most abundant in the igneous rocks that crystallize from magmas rich in magnesium and iron; so, they are present in gabbro, basalt, peridotite, and ultramafic rocks, described later in this chapter. One of the pyroxenes, diopside, is often formed when rocks rich in carbonate minerals are thermally metamorphosed.

The pyroxenes are silicate minerals that contain varying amounts of magnesium, iron, and calcium. Lithium, sodium, and aluminum may also be present. These metal ions join chains of silicon–oxygen tetrahedra that share oxygen atoms on two corners. The general formula for the pyroxenes is $(Ca,Mg,Fe)Si_3O_9$.

Pyroxene's hardness is 5.5, and one of the most common pyroxenes, augite, is black to greenish in color. The two cleavages of pyroxenes define prisms that are nearly square in section (Fig. 2.24).

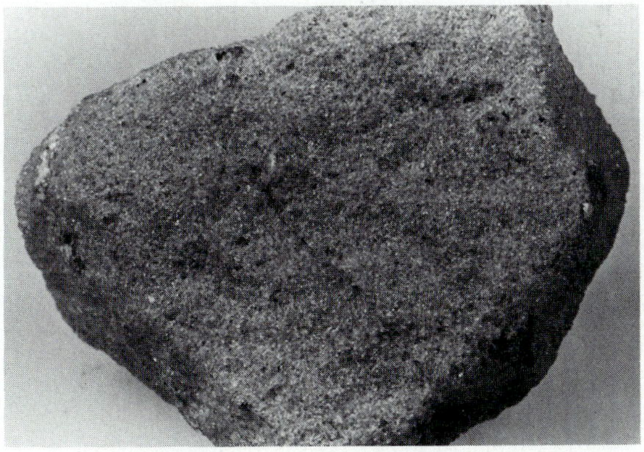

(a)

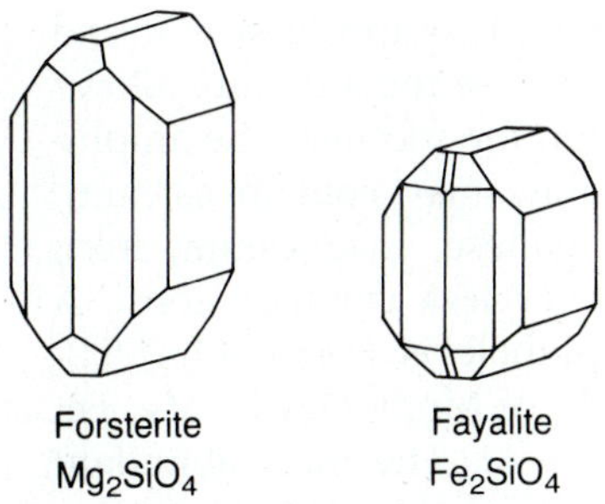

(b)

Figure 2.23
(a) Granular form in which olivine often occurs (approximately natural size). (b) Form exhibited by individual crystals.

(a)

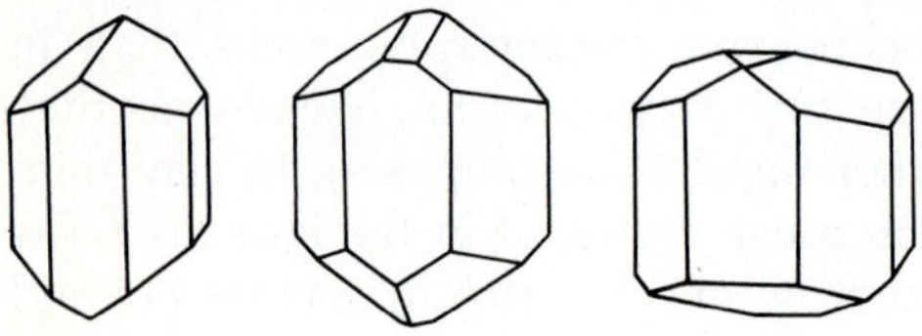
(b)

(c)

Figure 2.24
(a) The mineral augite is one of the most common minerals of the pyroxene group. (b) Common crystal forms of augite. (c) Typical cleavage of pyroxene.

Amphibole group. The amphiboles are common constituents of igneous and metamorphic rocks. They occur especially in those igneous rocks that are intermediate or rich in silica content and in metamorphosed sedimentary rocks.

A wide range of chemical composition is found among the amphiboles, because the structure allows easy substitution of one ion for another. The amphiboles are silicates, and they may contain calcium, sodium, potassium, manganese, magnesium, iron, aluminum, and some other less common ions. A general formula for amphibole is $(x)(y)(z)O_{22}(OH)_2$ where x may be Ca, Na, K, or Mn; y may be Mg, Fe, or Al; and z may be Si or Al. The most abundant amphibole, hornblende, is $NaCa_2Mg_4Al_3Si_6O_{22}(OH)_2$.

The main feature of amphibole structure is a double-linked chain in which silicon–oxygen and silicon–aluminum tetrahedra are bound by sharing oxygens. The double chains are joined by the various positive ions named above, and OH ions are present.

Hornblende tends to occur as long prismatic crystals (Fig. 2.25*b*) or masses of small crystals that are black in color and have a hardness of 5.5. The two good cleavages, which form angles of 56° and 124°, provide the best means of distinguishing hornblende from the pyroxene augite, which is similar in hardness and color but has cleavages at right angles.

Mica group. Mica is familiar to most of us as the sheetlike material used in fuses and other applications where resistance to high temperature or electric cur-

(a)

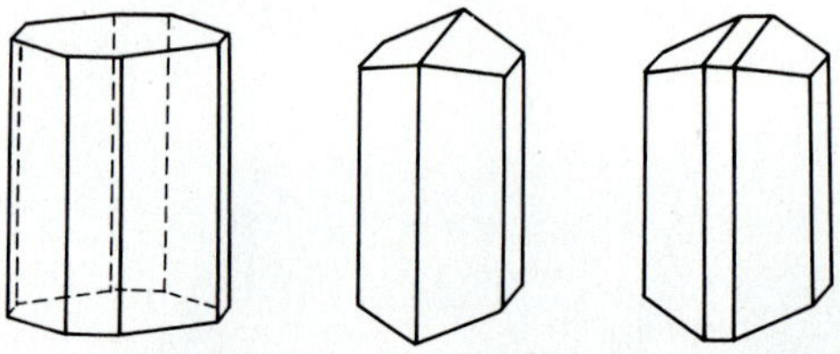
(b)

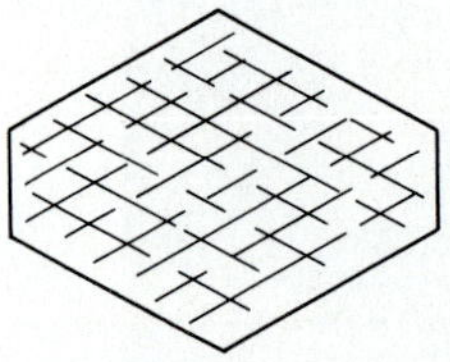
(c)

Figure 2.25
(a) Hornblende, a common member of the amphibole group. (b) The crystal forms of hornblende. (c) The cleavage shown by hornblende. The double chain structure of amphibole is illustrated in Fig. 2.19.

rent is required. The micas are common constituents of igneous, metamorphic, and sedimentary rocks. The group includes black mica, biotite, which is found in many different rocks, and white mica, muscovite, which is rare in sediments but crystallizes from some melts at 700 °C and may form in a solid state at lower temperatures. A third mica, glauconite or greensand, is found in marine sediments and occurs as rounded, fine-grained aggregates.

The essential feature of mica structure is the presence of large sheets of silicon–oxygen tetrahedra formed by the sharing of oxygens. These sheets are joined together by a layer of positive ions as is shown in Fig. 2.26*c*. In addition to silicon and oxygen, micas may contain a great variety of other elements, so a number of chemical compositions are possible. The general formula for mica is: $x_2y_{4-6}z_8O_{20}(OH,F)_4$ where x is usually K, Na, or Ca; y is Al, Mg, or Fe; and z is Si or Al.

Mica is easily identified by its sheetlike structure, which is due to a single perfect cleavage. The hardness is from 2.5 to 3, and the color varies.

The Carbonates

The essential structural unit of minerals in the carbonate group is CO_3^{-2} ion. Although it occurs compounded with manganese, iron, barium, strontium, and other positive ions, by far the most abundant carbonates are those of calcium, $CaCO_3$ (calcite), and a combination of calcium and magnesium, $CaMg(CO_3)_2$, a mineral known as dolomite. Because these are such abundant minerals among the rocks found at the earth's surface, special attention is focused on them here.

Calcite. Calcite and quartz probably occur in a greater variety of forms than any other minerals. In the case of calcite, this is attributable in part to the fact that calcium carbonate is soluble and moves around as a constituent of ground water and even surface water. Calcite, $CaCO_3$, is most commonly found in sediment and sedimentary rocks, but it may also be found in some igneous and metamorphic rocks. Many sea organisms use calcium carbonate derived from sea water to build shells and other hard parts that eventually become constituents of sediment; for example, the chalk cliffs of Dover are composed of billions of minute shells. Calcium carbonate may also be chemically precipitated from sea water under favorable conditions. Limestones are made up of compacted and consolidated deposits of calcium carbonate, and when these have been subjected to heat and pressure, marble is formed. Calcite may occur in veins, collect in pockets, or form coatings on plants where a fine mist of water is produced or around springs, to list but a few of the ways it is formed through chemical precipitation. Most cave formations are also created by chemical precipitation of calcite from ground water.

The structure of calcite resembles that of sodium chloride described earlier, except that the crystal is rhombohedral (Fig. 2.27) instead of cubic in shape. Calcite has a hardness of 3, is nonmetallic, varies greatly in color, and is often nearly clear. It has

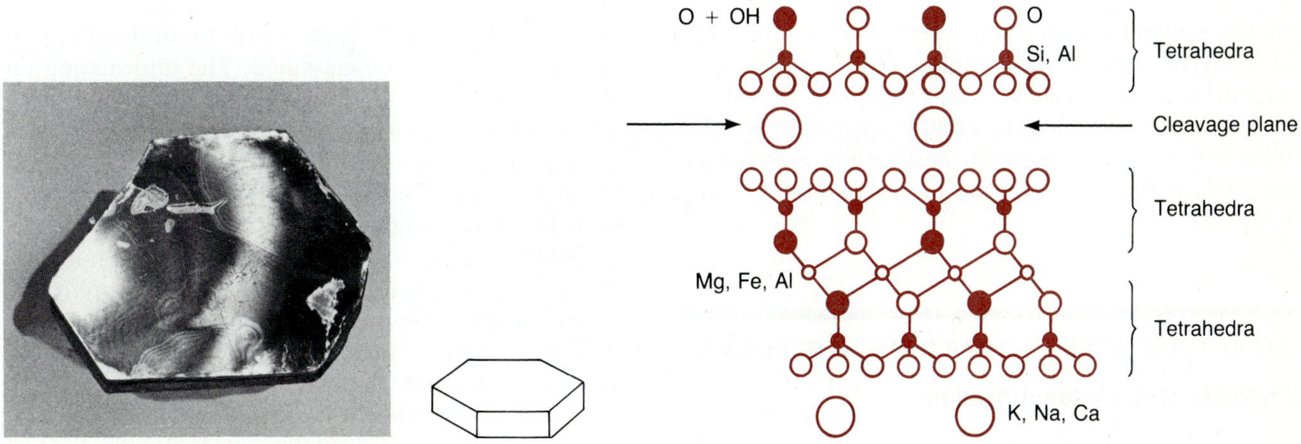

(a) (b) (c)

Figure 2.26
(a) Photograph of mica. (b) The most common crystal form of micas. (c) A sectional sketch illustrating the way the SiO_4 tetrahedra are bound together in sheets that are joined by K, Na, or Ca ions.

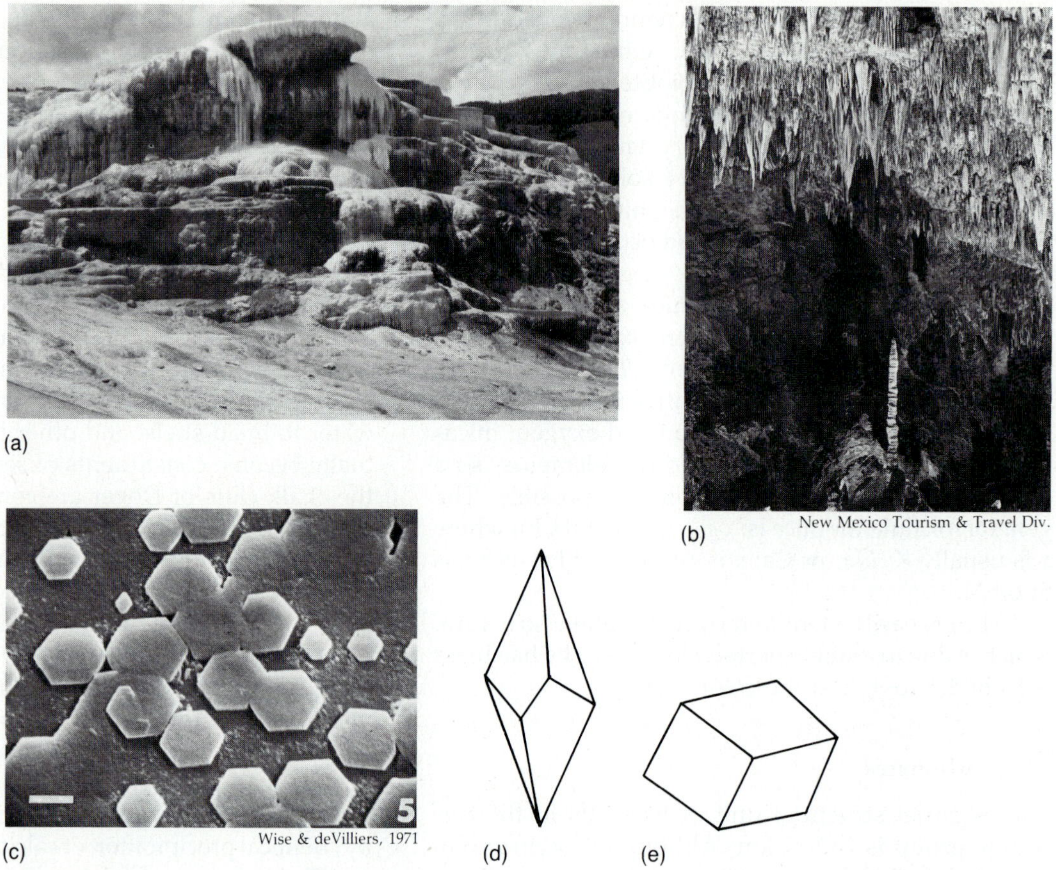

Figure 2.27
(a) Calcite deposits nearly 10 meters high formed from hot water in Yellowstone Park. (b) The unusual formations seen on the ceiling and floor of most caves are composed of calcite. These formations are in Carlsbad Caverns in southern New Mexico. (c) Electromicrograph of calcite in a molluscan shell. The bar is 2μ long (0.002 mm). (d) Common crystal shape for calcite. (e) Cleavage rhomb of calcite, shown in a sketch here, is also seen in Fig. 2.13. Note that the corners are not square.

rhombohedral cleavage—three perfect cleavages, none of which intersect at right angles. Treated with dilute hydrochloric acid, calcite fizzes as carbon dioxide is released. Its solubility is greatly increased by the presence of carbon dioxide, and unlike most other minerals it tends to become more soluble at lower temperatures.

2.5 AGGREGATES OF MINERALS: THE ROCKS

Methods of Rock Identification

Much of the record of the earth's past and the processes it has undergone is in the **rocks** of the earth. For this reason it will be helpful in reading the following chapters to be familiar with a few of the most abundant rock types and to understand in general how rocks are classified. The information on minerals in the preceeding section and some fairly simple observations can go a long way toward identifying rocks. Two characteristics, composition and texture, are used in making identifications and as a basis for classification. Composition is usually determined by making a rough appraisal of what minerals are most abundant and their approximate proportions in the sample. Even this rough estimate is usually sufficient; if more detailed information about composition is needed, that can be obtained by actually counting the number of grains on the surface of a small slab of the rock. Texture refers to the geometrical aspects of the mineral components of a rock and their size, shape, and arrangement. Sometimes the texture of a rock is so fine—the sizes of mineral crystals in a rock are so small—that identification is impossible. Then a section of the rock must be analyzed by use of a petrographic microscope, by x-ray analysis, or one of the other techniques.

Microscopic study of rocks. Beyond field observations, the petrographic or polarizing microscope is perhaps the chief method of rock identification. Used to examine thin sections of rock, to study the texture of the rock, and to determine the composition of the minerals it contains, a petrographic microscope is equipped with a polarizing device above and below a revolving stage. The polarizing devices are filters that transmit only light waves vibrating in a particular direction. When two such devices are crossed, no light is transmitted.

The petrographic microscope can be used like any other microscope to obtain an enlarged image of a part of a rock, to study texture, shape of crystals, grain boundaries, or internal structure (Fig. 2.28). The sections of rock are ground so thin, 0.01 to 0.1 mm, that light will pass through the nonmetallic minerals, and it is possible to measure the index of refraction (capacity of the crystal to bend light rays that pass through it) and other optical properties of the mineral. Not only does the value of the index of refraction differ from one mineral to another, but some minerals have different indices of refraction when light passes through them in different directions. Minerals with more than one index of refraction produce interference colors when viewed in polarized light. The colors change intensity as the stage is rotated, and in certain positions, the light is extinguished entirely. These and other optical properties of minerals can be measured and used as a means of identification.

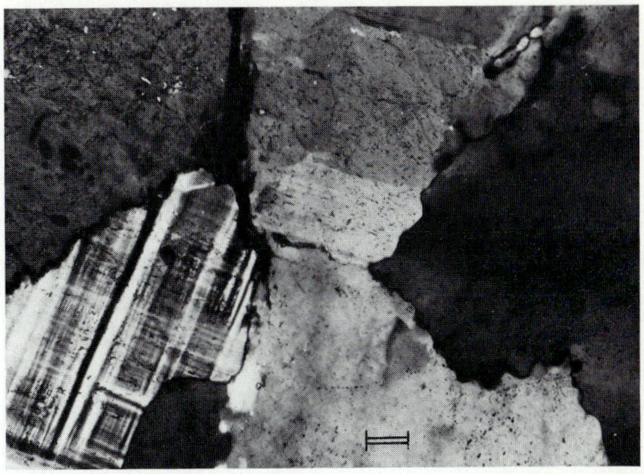

Figure 2.28
Photograph of a thin section of an igneous rock taken with a petrographic microscope. The crystals at right are quartz; the mineral with the grid pattern is feldspar. The bar is about 1 mm long. The grains form an interlocking mosaic, which is quite clear at this scale.

Simplified Classification of Major Rock Types

Igneous rocks. Most igneous rocks can be classified according to texture and composition, as illustrated in Table 2.3. The important mineral constituents are

Table 2.3. Simplified Igneous Rock Classification

Composition (Common Minerals Present)		**Rock Names**	
Major Constituents	*Minor Constituents*	*Grained (Coarse Texture)*	*Glassy (Fine Texture)*
Quartz K Feldspar	Amphibole Mica	Granite	Rhyolite (fine-grained) Pumic (glassy froth) Obsidian (natural glass)
Quartz K Feldspar Plagioclase	Amphibole Biotite	Diorite	Andesite (fine-grained)
Plagioclase Feldspar Pyroxene	Olivine	Gabbro	Basalt (fine-grained) Scoria (contains gas pockets)
Pyroxene Olivine	Plagioclase	Peridotite	none known

Table 2.4. Classification of Sediment According to the Size of the Fragments It Contains

Boulder	anything above 256 mm in diameter
Cobble	64 mm–256 mm
Pebble	4 mm–64 mm
Granule	2 mm–4 mm
Sand	$\frac{1}{16}$ mm–2 mm
Silt	$\frac{1}{256}$ mm–$\frac{1}{16}$ mm
Clay	$\frac{1}{256}$ mm or less

listed at the left and the major rock types are placed in one of two textural groups. Thus, a rock that is coarse-grained and composed mainly of orthoclase feldspar, quartz, some hornblende, and biotite is called granite. A coarse-grained rock composed mainly of plagioclase feldspar, pyroxene, and olivine is called gabbro. A rock with the same composition as gabbro but having such fine grains that they cannot be distinguished by visual inspection is called basalt.

Sedimentary rocks. A great many sedimentary rocks are composed of small fragments of minerals formed by the breakdown of other rocks (Tables 2.4 and 2.5). The sizes of these fragments may range from boulders many meters across to clay, defined as particles less than $\frac{1}{256}$ mm in diameter. The term **clastic** is applied to these fragmental materials regardless of size. The coarse clastic rocks often have compound names: one that indicates the composition of the rock, and a second that indicates the size of the particles in the rock: for example, a quartz sandstone or a calcareous sandstone. Sedimentary rocks formed by chemical or biochemical deposition may have textures that resemble the interlocking mosaic of crystals common in igneous and metamorphic rocks. These crystalline sedimentary rocks are usually known by a name that indicates their composition or mode of origin. For example, a travertine is a limestone deposited in a cave.

Metamorphic rocks. Metamorphic rocks are formed by alteration of other rock types through the effects of heat and pressure. Sometimes the alteration takes place without the addition or removal of material, and in such instances the bulk chemical composition remains unchanged, but the mineral composition may be substantially altered. If the original rock contained nothing but silicon dioxide (quartz) or calcium carbonate (calcite or limestone), the resulting metamorphic rock will be a quartz or calcite rock, but if a number of different elements are present before metamorphism, new minerals may be formed as elements from the original minerals combine in new ways to form different minerals. If new minerals that develop during metamorphism are long or platelike, the crystals frequently grow in one particular direction. Hornblende or micas especially may show this strong directional quality. The preferred orientation is usually at right angles to the direction of pressures acting on the rock during the formation of new crystals, during recrystallization. This alignment of crystals, called **foliation,** is characteristic of many, but not all, metamorphic rocks. The names of some of the most common metamorphic rocks and of the sedimentary rocks from which they may be derived are listed in Table 2.6.

Table 2.5. Simplified Classification of Common Sedimentary Rocks

Fragmental (clastic) rocks		
	conglomerate	consolidated gravel
	sandstone	contains sand size fragments, often quartz
	arkose	sandstone-like but with high feldspar content
	grawacke	gray clastic containing mixtures of quartz feldspar and volcanic minerals (olivine, pyroxene, and the like)
	shale	consolidated clay
	limestone	fragments of shells
Crystalline		
	limestone	calcium carbonate deposited by precipitation at springs (tufa), in caves (travertine), as small spheres (oolitic).
	rock salt	

Table 2.6. Sediments, the Sedimentary Rock Formed from Them, and the Metamorphic Equivalent

Sediment	Sedimentary rock	Metamorphic rock
sand	sandstone	quartzite
limy mud	limestone	marble
marl	sandy limestone	siliceous marble
calcareous sand	calcareous sandstone	quartzite
clay, mud	shale	slate phyllite schist gneiss
volcanic tuffs	tuff	slate phyllite schist gneiss
graywacke	graywacke	schist gneiss

2.6 THE ROCK CYCLE

Permanent as they may seem, the rocks and other materials that make up the outer shell of the earth are constantly being recycled. The complex system of natural processes through which they pass are known as the **rock cycle** (Fig. 2.29). The most obvious parts of the cycle take place at or near the ground surface where we can see that rocks of all types, when exposed to the atmosphere, decompose and disintegrate. A number of chemical and mechanical processes, collectively known as weathering, accomplish this. We see the products of the weathering processes picked up and carried by sheet wash that follows a heavy rain, by streams, and by glaciers; blown into the air in places; and washed away by wave and current action along coasts; only to settle out as sediments somewhere else. Some of these sediments form layers on lake beds, dunes, beaches, deltas, and other sites, but vast quantities find their way into the oceans where layer after layer of sediment is deposited on continental margins. The sediments are eventually compacted as they are buried, usually becoming cemented to form sedimentary rock.

Viewed at another point in time or with less information, the story might end here—an incomplete cycle—but metamorphic rocks that contain fossils and the types of primary features seen in sediments have been found. Clearly these metamorphic rocks were once sediments now transformed by heat and pressure into a new, very different rock. Furthermore, it is also possible to examine exposed sections of metamorphic rock showing increasing indications of high temperature, until the rock begins to show signs of having flowed and intruded other rock as an igneous rock. These types of observations demonstrate that sediment can be altered to a metamorphic rock, and that metamorphism grades at high temperatures into igneous activity. Melting of rock produces both volcanic activity and masses of magma that eventually cool to form rock inside the earth. Some of these igneous rocks may be raised by movements of the crust until they are exposed by weathering and erosion and become a source of new sediment, thus completing the cycle.

The evidence proves that recycling does take place, but this does not mean that all materials in the crust are recycled. Indeed, we would not find really ancient rocks if the recycling were rapid and complete. Large areas of the continents are underlain by rocks more than a billion years old, and some are more than three times that age. Apparently, much continental material is unlikely to be recycled. Obviously, recycling can be slow, and we are fortunate that it is, otherwise we would have little hope of reconstructing the history of the earth.

SUMMARY

The material at the earth's surface is of several types. The loose, unconsolidated material containing clay, sand, silt, and organic matter is soil. The solid rocky material is

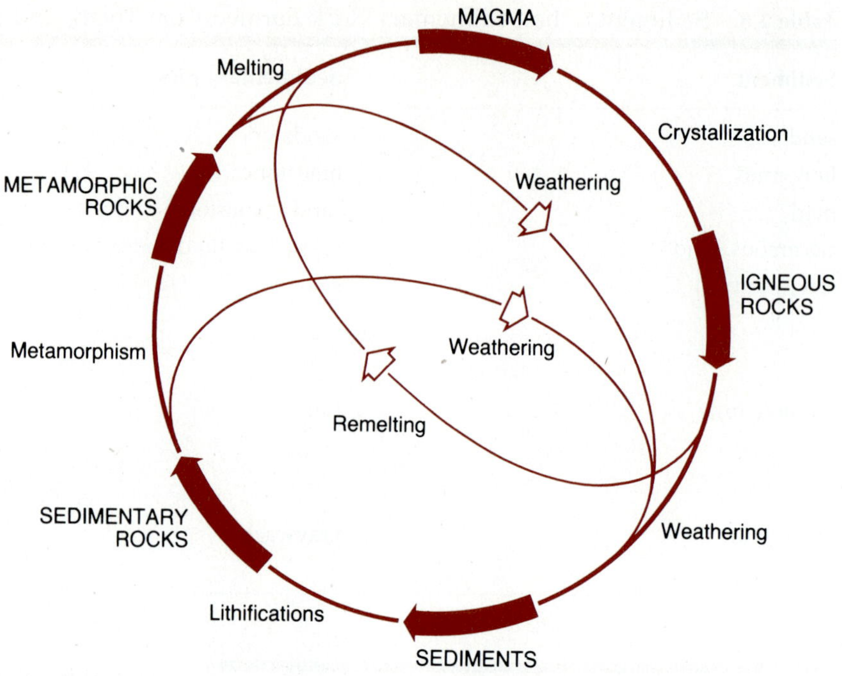

Figure 2.29
The rock cycle is schematically depicted here in terms of the major types of materials and the processes by which one type of material is transformed to another.

classifed as sedimentary, igneous, or metamorphic, according to how it formed. All rocks are solid mixtures of chemical compounds called minerals that occur naturally on earth. The physical and chemical properties of minerals, like those of any compound, depend on the elements of which they are composed and on the type of bonding—ionic, covalent, polar covalent, or metallic—between atoms.

Some elements in minerals exist as radioactive isotopes. The amounts of these isotopes and of the daughter elements formed by radioactive decay can be used to determine how long ago the mineral in which they occur crystallized.

Identification of minerals is based upon physical properties such as specific density, cleavage, fracture, hardness, luster, and streak. X-ray analysis yields detailed information on the composition and on the arrangement of atoms in crystals.

All minerals that are not pure elements are grouped according to their negative ions. Of these the most common are the silicates (including quartz, the feldspars, olivine, and pyroxene) and the carbonates (especially calcite and dolomite).

Minerals are found in a wide variety of combinations in rocks. The composition of a rock—what minerals it contains and in what proportions—and its texture serve to classify rocks within the larger framework defined by their origin.

Igneous rocks, those that solidified from a melt, range from coarse-grained (such as granite and gabbro) to fine-grained (rhyolite and basalt). Sedimentary rocks formed by compacting and cementing of particles have compound names based upon the size of the particles and their composition. Metamorphic rocks, rocks recrystallized from igneous or sedimentary ones under altered conditions of temperature and pressure, often show distinctive textures produced by the metamorphism.

Rocks are constantly being recycled on the earth. Igneous rock formed from rising magma is gradually weathered into small particles. Transported and redeposited by wind, water, or ice, these particles may eventually be converted to sedimentary rocks. If these rocks are carried down into the earth they may be metamorphosed or even remelted, ultimately to form new igneous rocks.

KEY TERMS

atom
atomic number
beds
clastic
cleavage
color
conchoidal fracture
coordination number
crystal
electron probe
elements
first shell
foliation
half-life
hardness
igneous rock
ion
isotopes
lava
luster
magma
metamorphic rock
minerals
Moh's hardness scale
noble or inert gases
nucleus
polar covalent
polar nature
radioactive
radius ratio
rock
rock cycle
sediment
sedimentary rock
silicates
soil
specific gravity
streak
striations
x-ray spectrum

STUDY QUESTIONS

1. If oxygen and silicon are the most abundant elements in the crust, why aren't they the most common elements in the earth?
2. What rock types occur in the area where you live?
3. Describe the types of surficial deposits you have seen in the area where you live.
4. Prepare a sketch of a model of argon, atomic number 18.
5. Predict the type of cleavage salt will have from the model of its atomic structure, Fig. 2.10.
6. Why is water such a powerful solvent?
7. How can you distinguish pyroxene from amphibole?
8. Make a list of the various ways quartz, calcite, and feldspar may occur.
9. How do geologists identify minerals that are too small to be seen?
10. Compare and contrast the textures of igneous, metamorphic, and sedimentary rocks.

SUGGESTED READINGS

Ernst, W. G., *Earth Materials.* Englewood Cliffs, N.J.: Prentice-Hall, 1969.

Holden, A., and P. Singer, *Crystals and Crystal Growing.* Garden City, N.Y.: Doubleday & Co., 1969.

Landers, I., and P. F. Kerr, *Mineral Recognition.* New York: John Wiley & Sons, 1967.

Mason, B., and L. G. Berry, *Elements of Mineralogy.* San Francisco: W. H. Freeman, 1968.

Pearl, R. M., *Rocks and Minerals.* New York: Barnes and Noble, 1963.

Pough, F. H., *A Field Guide to Rocks and Minerals.* Cambridge, Mass.: Houghton Mifflin, 1960.

Simpson, Brian, *Rocks and Minerals.* Oxford, England: Pergamon Press, 1966.

Singh, Raman J., and Jonathan Bushee, "The rock cycle," *Journal of Geological Education* 25:5, (1977), 146–147.

Tindall, J. R., and R. Thornhill, *The Collector's Guide to Rocks and Minerals.* New York: Van Nostrand Reinhold Co., 1975.

EARTH AS A PLANET

Chapter 3

3.1 EARTH: THE UNIQUE PLANET

The spherical shape of the earth was recognized as early as 230 B.C. by Eratosthenes, and the position of the earth in the solar system was established by Copernicus in the sixteenth century. But probably only in recent times, when we have been able to see the earth from space in views like that of Fig. 3.1, have we been able clearly to visualize the earth as an isolated planet and to appreciate its most distinctive features.

NASA

Figure 3.1
This view of the earth greeted the Apollo 8 astronauts as they came from behind the moon during a lunar orbit.

Viewed from distant space, the earth might be described as one of the small inner planets of the solar system located 150 million kilometers from the sun. We have known the dry statistics of the earth's size and motion for years—or in some cases, for centuries: earth has a diameter of 12,700 km and a mean density of 5.5 gm/cm^3, and it revolves around the sun in a nearly circular orbit that is close to the equatorial plane of the sun. The earth completes its revolution about the sun in a period of $365\frac{1}{4}$ days. The earth also rotates once in 23 hours and 56 minutes, about an axis that is inclined $76\frac{1}{2}°$ to the plane in which it revolves around the sun. Earth is the closest planet to the sun to possess a satellite. The moon is about one-quarter the diameter of the earth and has a mass about $\frac{1}{80}$ that of the earth. It travels in a nearly circular orbit that is inclined 5° to the plane of the earth's orbit and is about 400,000 km in radius. One of the remarkable features of the moon is that its period of rotation about its axis (27.32 days) is the same as its period of rotation about the earth; so, the same side always faces the earth. The moon has no atmosphere and lacks the varicolored surface features of the earth. It does have dark patchy areas called seas, some of which are nearly circular. These areas are separated by irregular, lighter-colored areas and ray-like patterns.

Seen from the moon, many subtler features of the planet earth are revealed. Its surface is partly obscured by a thin, slowly moving cloud cover characterized by large eddylike swirls that transfer heat from the equatorial regions toward the poles. These swirls gradually dissipate in an atmosphere composed primarily of nitrogen and oxygen with small amounts of carbon dioxide, water vapor, and trace amounts of other gases. The polar regions are covered by highly reflective solids, ices of water, but at lower latitudes the surface is characterized by various colored materials. Large desert areas have yellow to brown hues; others with vegetative cover have green tints. But most of the surface is covered by the large, irregularly shaped blue areas of the oceans. From space the vastness of the oceans that cover nearly three-quarters of the earth's surface becomes apparent, and the relief features, especially the great mountain chains along the west side of North and South America, and the system that crosses southern Europe into southeast Asia, stand in stark contrast to the cratered surface of the moon, Mars, and other inner planets. Finally, the existence of richly diverse and abundant plant and animal life confirm the uniqueness of earth in the solar system.

3.2 DIMENSIONS OF THE EARTH

The Pythagorean brothers, a monastic brotherhood founded in one of the Greek colonies in southern Italy about 500 B.C., taught that the earth is a sphere. The fact that other celestial bodies could be seen to be spherical suggested to them that, for reasons of symmetry, the earth should also be spherical. They argued that an earth of any other shape would fall in on itself. The concept did not gain wide acceptance,

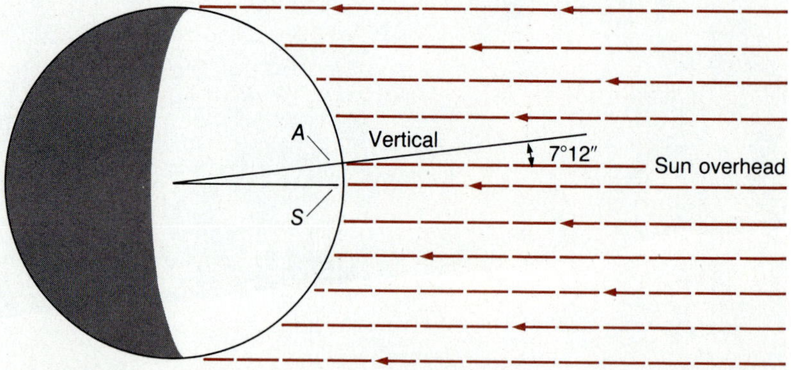

Figure 3.2
Eratosthenes measured the diameter of the earth by observing the angle a vertical pole at Alexandria (*A* on diagram) made with the incident sunlight. Alexandria was a known distance north of Syene (*S* on diagram) where a shadow is cast. The angle between *A* and *S* at the center of the Earth is 7°12″, the same as the angle between the vertical and the incident sunlight.

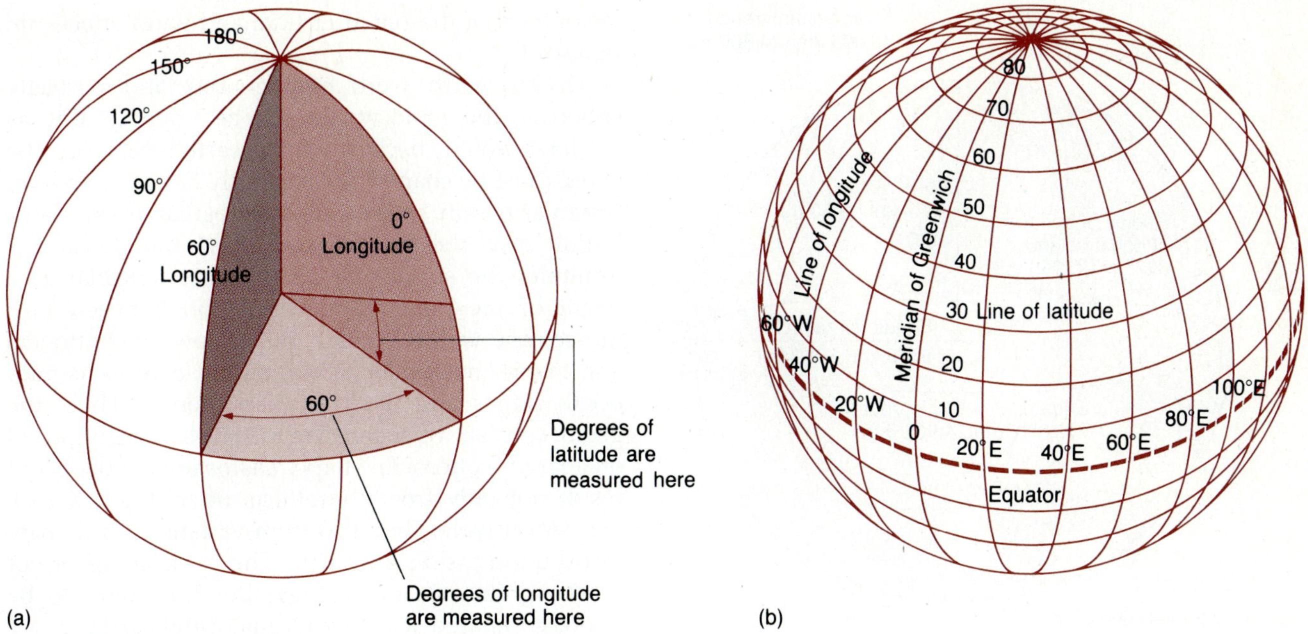

Figure 3.3
(a) A cutaway view of the earth showing how latitude and longitude are measured. (b) An external view of the globe showing lines of latitude and longitude.

however, until Plato (428–348 B.C.) argued that the sphere was the most perfect shape for a body, and hence, the earth, at the center of the universe, must be a sphere. Aristotle (ca. 340 B.C.), however, presented a much more convincing argument: during an eclipse of the moon, when the earth comes into a position directly between the sun and the moon, the shadow cast by the earth on the moon is circular. He also pointed out that anyone who travels even a short distance north or south will find a changed star pattern. These observations, plus the familiar experience of seeing a ship disappear over the horizon, laid a firm qualitative basis early in the history of science for the idea of a spherical earth.

A remarkable accurate estimate of the size of the earth was made by Eratosthenes about 230 B.C. He observed that at noon in midsummer, a vertical pole at Syene (Aswan), Egypt, does not cast a shadow. (Some historians say he observed that the bottom of a well was illuminated; the principle is the same.) At the same time a vertical pole at Alexandria, 800 kilometers north, does cast a shadow. If the pole at Syene is in a truly vertical position, the shadow cast by the pole at Alexandria indicates that the sun there is displaced by 7°12″ from the vertical (Fig. 3.2). It was either known or assumed that Alexandria and Syene are on the same longitude (Alexandria is due north of Syene). If the distance from Alexandria to Syene could be accurately measured, the circumference of the earth, and in turn its diameter, could be calculated. Recognizing this, Eratosthenes determined that the earth's diameter is 12,560 kilometers, only 112 kilometers less than modern estimates and a figure so close that both historians and astronomers have speculated that Eratosthenes may have been very lucky.

To a first approximation, the earth is a sphere. We might expect, therefore, that it has a uniform radius and that if we measure arcs along any great circle produced where a plane that passes through the center of the earth intersects the surface (Fig. 3.3), we would find all one degree arcs to be the same length. The first attempts actually to make this measurement were made in Newton's time. In 1669, J. Picard measured the length of one degree of latitude at Paris. This work was extended by Cossini to the south into Spain, and much to the surprise of those involved, the length of one degree of latitude turned out to be 111,017 meters in the north and 111,284 meters in the south. Two explanations appeared possible: (1) that the earth is not spherical, but is flattened toward the north; or (2) that there is some anomaly in the shape of the earth in the region of France. To settle this question, expeditions were sent

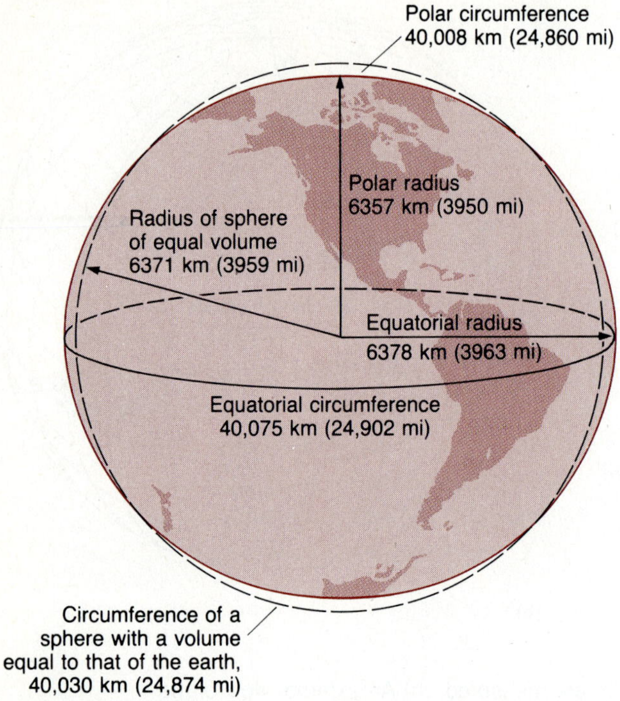

Figure 3.4
The major dimensions of the earth. Note especially the difference in the polar and equatorial radii.

to Peru to measure the length of a degree of latitude near the equator and to Lapland to find the length of a degree at high latitudes. The results showed that the earth is not spherical, but is flattened toward the poles; its figure comes close to being an oblate spheroid—a slightly flattened sphere. Because of this flattening, the radius of the earth at the equator is, as shown in Fig. 3.4, about 21 kilometers greater than that through the poles. This flattening is caused by the rotation of a not-so-solid earth, which bulges around the equator as it rotates.

Detection of Irregularities in Earth's Shape

To a hiker, the earth's surface is a maze of irregularities, but these irregularities must be smoothed out in making an approximation of the shape of the earth as a whole. To accomplish this, geologists use a smoothed figure of the earth called the geoid. The **geoid** is the earth viewed as a hypothetical ellipsoid with the surface represented as sea level, or the level at which the sea would stand if all the oceans were connected by a system of canals. **Sea level** is the mean level of the sea after tidal and wind effects are removed.

If the earth were homogenous and perfectly spherical, the geoid would also be a sphere. But, as we have noted, because of the earth's rotation, the geoid must be shaped like a slightly flattened sphere. Irregularities in shape due to irregular mass distribution near the surface also affect the geoid. For example, the surface of the ocean is irregular as a result of the gravitational attraction between rock masses and water. Water is pulled toward continents and toward such major topographic features as mid-ocean ridges and the Hawaiian Islands. Thus, the geoid consists of many irregularities superimposed on a larger spheroidal shape. Distortions of the geoid result not only from the effects of continental rock masses on water, but also from variations in density distribution inside the earth. These variations are of great significance in geology, but they tend to be minor compared with the radius of the earth.

Satellite orbits provide important clues to the shape of the earth. If the earth were a uniform sphere, satellite orbits would follow circular or elliptical paths around it. But because the attraction between earth and satellite varies from place to place as a result of uneven distribution of mass in the earth, satellites are drawn into slightly irregular paths. The actual paths of satellites can be observed and plotted with high precision. These paths are clearly affected by the equatorial bulge due to flattening of the earth, but the amount of flattening is slightly less than had been predicted earlier. Satellite data also indicate the presence of large regional bulges of the geoid with amplitudes of up to 80 meters. Because of them, some writers have referred to the earth as being pear-shaped. However, it is important to remember that the bulges are minor when compared with the overall flattening.

Earth's Gravity Field

One of the best clues we have about features on and below the earth's surface comes from studying the earth's gravitational field. The law that describes gravitational attraction was formulated by Sir Isaac Newton (1642–1727) to explain the origin of the forces that cause planets to travel in elliptical paths around the sun. Newton saw the answer to the origin of these forces in the attraction of objects to the earth. He generalized this observation to conclude that all objects attract one another with a gravitational force.

MEASURING THE FORCE OF GRAVITY ON THE EARTH

Instruments designed to measure the force of gravity on the earth are among the most sensitive yet developed. Meters for measuring earth tides are read to one part in a billion, and instruments used to measure the earth's gravitational field are accurate to one part in 100 million. One of the earliest instruments used for this purpose was the pendulum. An ideal pendulum completes its full swing in a time (period) T, given by the following equation:

$$T = 2\pi\sqrt{L/g}$$
$$g = (2\pi)^2 L/T^2$$

In this equation, L is the length of the pendulum and g is the acceleration caused by gravity, and π is a constant equal to 3.1417. (Note that the dimensions of g, cm/sec^2, are those of accelerations, or distance/time^2.) Thus, the gravitational attraction at a given location can be determined by measuring the length of the pendulum and the period of the swing. Other methods of measuring gravity involve timing the rate of fall of a body in a near vacuum and the use of supersensitive spring balances called gravimeters.

Measurements of gravity are expressed as **gals,** units named in honor of Galileo. One gal is an acceleration of 1 cm/sec^2; 1 **milligal** is 1/1000 gal. The force of gravitational attraction between the earth and objects on its surface is the product of the mass of the object and the acceleration due to gravitational attraction of earth, g. The value of g varies slightly from place to place but it is about 980 cm/sec^2 (32 feet/sec^2) near sea level. Accuracy of 0.1 milligal is considered necessary in most modern studies.

Thus, he reasoned, the central force on the planets is the gravitational attraction to the sun. In his famous book, *Principia,* he set forth the mathematical explanations of the forces of gravitational attraction between bodies, and applied these to the motions of the planets, comets, the moon, and the tides.

Newton established the magnitude of the force as well as its direction. And he was able to show that the force (F) upon each planet is inversely proportional to the square of the distance (R) from the center of the planet to the Sun, and that the magnitude of the force is proportional to the masses (M_1 and M_2) of the objects. His **law of universal gravitational attraction** can be stated as follows:

$$F = \frac{GM_1M_2}{R^2}$$

The value of the **universal gravitational constant,** G, was established by Cavendish (1731–1810) as being 6.754×10^{-8}. Although Newton's theory of gravitation grew from studies of planetary motions, the principle has been applied with equal success to the mutual attraction of all matter.

Determining the Mass and Density of the Earth

Newton's law of universal gavitational attraction provides the basis for calculating the mass and density of the earth. According to this law, the force of attraction exerted by the earth on an object at the surface can be calculated if we know the mass of the object (a measure of a quantity of matter), the universal gravitational constant (G), and the distance from the object to the center of mass of the earth. For an object on the surface, this distance is the radius of the earth at that point. The gravitational force of attraction is what we normally call the weight of an object. According to laws Newton developed to describe the relationship between force and motion, this weight or force of gravity of the object must be equal to the product of the object's mass, M_1 and the acceleration due to gravity (g):

$$\text{force of attraction} = M_1 g.$$

Therefore, we can set up an equation as follows:

$$\text{force of attraction} = \frac{G M_1 M_2}{R^2}$$
$$= M_1 g = \text{weight of the object.}$$

M_1 appears on both sides of the equation and thus cancels out. So, we can simply rewrite the equation to give the mass of the earth as follows:

$$M_2 = \frac{gR^2}{G}.$$

The radius of the earth is determined independently, as described in the following section, and the value of g can be determined by use of a pendulum. Using R = 6,380 kilometers and g = 980 cm/sec^2, then the mass of the earth is 5.98×10^{27} grams.

The volume of the earth can also be calculated ($V = 4/3\pi R^3$), and from this we can find the average density of the earth (density = mass/volume), which is about 5.5 gm/cm^3.

3.3 A MODEL OF THE EARTH'S INTERIOR

Most of what we know about the interior of the earth is based on interpretation of indirectly observed data. Even the rocks and magmas that originate below the surface come from depths that are relatively shallow compared to the radius of the earth. The principal tools that we use to study the deeper parts of the earth are analysis of the gravity and magnetic fields and especially interpretation of shock waves generated by earthquakes. How these methods are used to develop the present model of the interior is discussed in Chapter 9.

The results of this study are depicted in the model of the earth shown in Fig. 3.5. The interior is comprised of a number of concentric zones of material that differ from one another in physical properties and in composition. The outermost and by far the thinnest of these divisions, the **crust,** is composed of the rocks we can examine at the surface and in wells. A great variety of rocks is exposed on the continents, most of them sedimentary. These form a veneer covering igneous and metamorphic rocks, most of which are granitic in composition. In the oceans, generally a much thinner veneer of unconsolidated sediment covers a solid crust composed mainly of basalt lava flows and intrusions.

The crust is thicker under continents than it is in the oceans, but everywhere it covers the underlying **mantle,** the region of the earth between the crust and core. The rocks of the mantle, like those of the crust, are composed mainly of minerals belonging to the silicate group. But unlike crustal rocks, those of the mantle are mainly composed of minerals such as olivine, garnet, and pyroxene, which are denser than most crustal rocks.

At depths between 100 and 200 kilometers, the rock in the mantle appears to become plastic. That is, it becomes weaker and capable of flow and may be partially molten. This zone is known as the **asthenosphere,** or low-velocity zone. The rock above this zone all the way to the surface, including the crust, is strong and brittle and is known as the **lithosphere.**

The mantle extends almost halfway to the center of the earth. At that depth, several remarkable changes take place, and the deeper part of the Earth, the **core,** is strikingly different. The outer part of the core appears to be hot molten iron, possibly with nickel or other minor constituents. The inner part of this core is probably solid metal, and it is almost certainly in the core that the earth's magnetic field originates. The high density of the material in the core accounts for the high average density of the earth.

The interior has an important bearing on what happens at the surface of the earth, and it is with the surface that we are most directly concerned. In later chapters we will consider how processes in the interior are connected with the creation of surface features such as those described in the following section.

3.4 THE MAJOR SURFACE FEATURES OF THE EARTH

Continents and ocean basins are the most prominent physical features of the earth's surface. The differences between these two would be even more pronounced if the ocean basins were emptied. The great difference in elevation would then be most impressive; Fig. 3.6 shows the percentage of the earth's surface at different elevations. Twenty-three percent of the surface lies between four and five kilometers below sea level, and twenty-one percent lies between sea level and one kilometer in elevation. Thus, it is

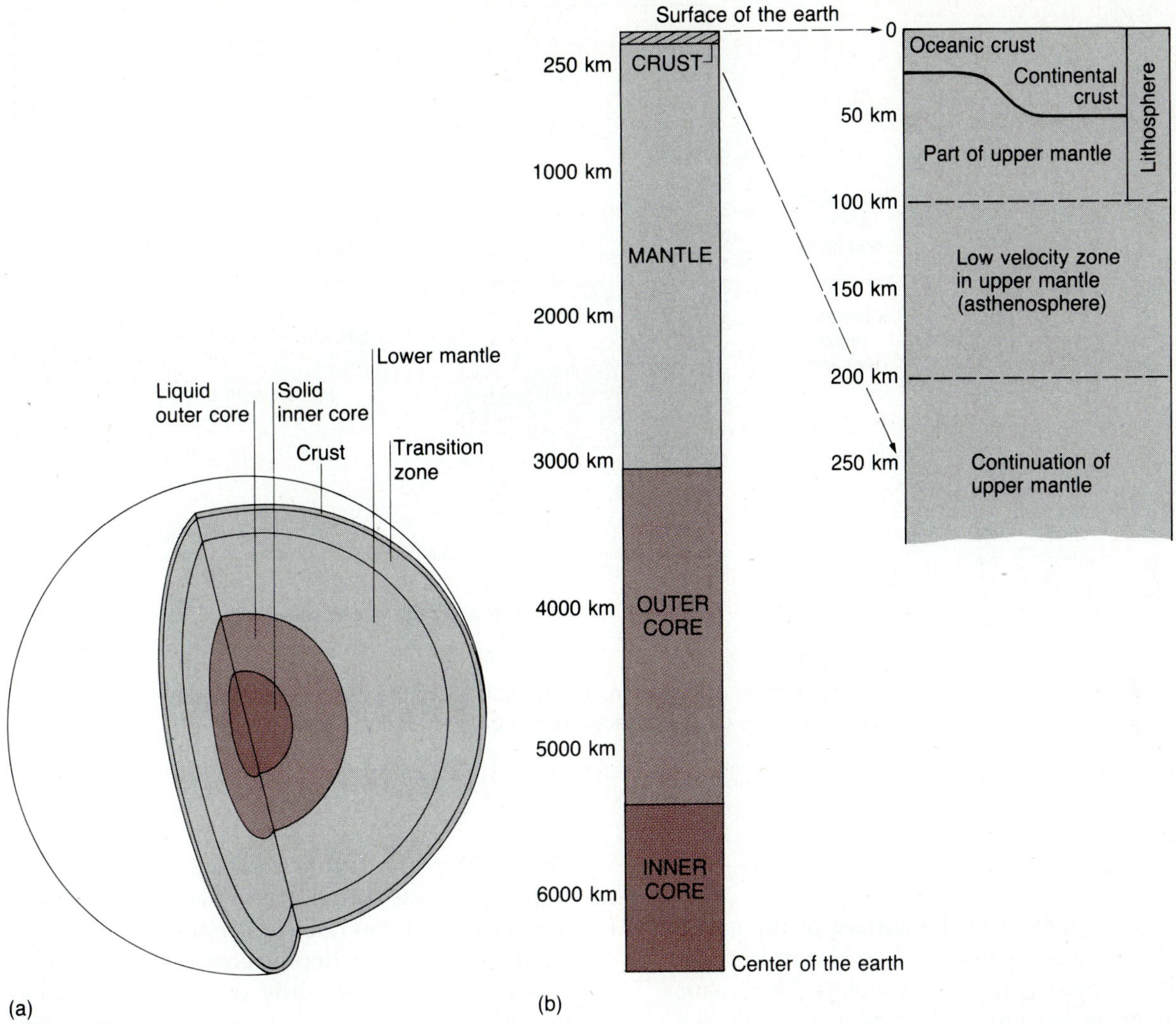

Figure 3.5
(a) A cutaway view of the earth's interior. (b) The main divisions and approximate depths to the boundaries between the divisions.

clear that most surface elevations fall in one of two major elevation ranges. The average elevation of the oceans is about 3.7 kilometers below sea level.

The distribution of surface elevations is but one of the important differences between continents and oceans. The types of rocks that predominate, their ages, and the structure of the rock masses are other important distinctions. In general, the continents are composed of rocks that are less dense and richer in silicon and aluminum (sialic) than the rocks in the sea floor. Continental rocks are characterized as being granitic, although many rock types are present. Rocks beneath the sea floor are mainly basaltic and formed by the extrusion and intrusion of basaltic lavas. The history recorded in the rocks of the continents is long, covering events starting more than three billion years ago. In contrast, the floor of the oceans appears to be much younger, most of it having formed in the last hundred million years.

In terms of structure, the complexly deformed, intruded, and partially metamorphosed rocks of the mountain systems are characteristic features of continents. In contrast, the mountains on the sea floor are composed primarily of only slightly deformed, solidified basaltic lava flows. The following sections provide a brief introduction to the features of the ocean basins and the continents. We will examine them in greater detail and explore some of the ideas geologists have formulated to explain what we know in Part III.

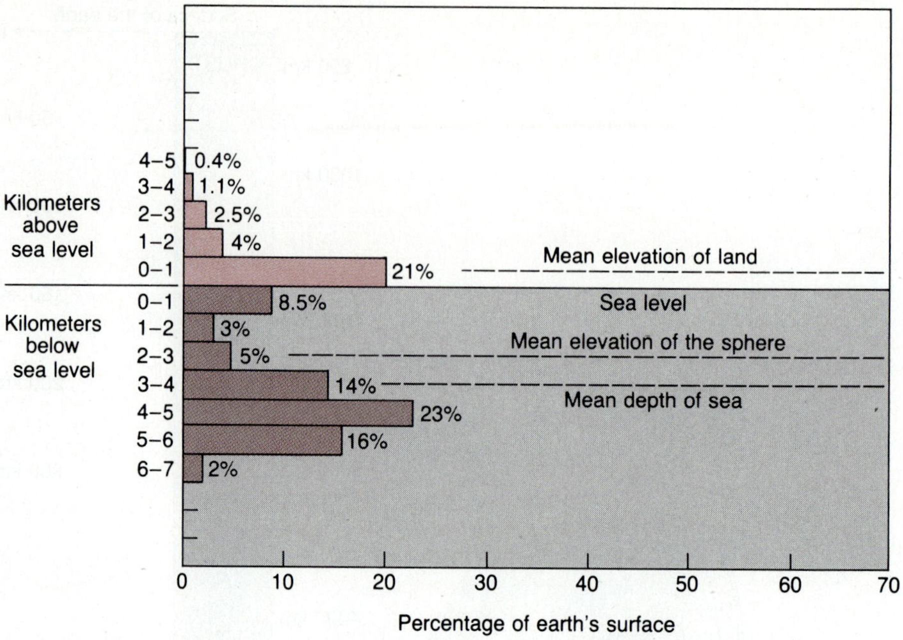

Figure 3.6
The distribution of elevations on the surface of the earth. The bar graph at left shows the percentage of the surface area in each elevation range.

The Ocean Basins

Slightly over 70 percent of the surface of the earth, or 361 million square kilometers, is covered by sea water to an average depth of 3,730 meters. The oceans contain about 86 percent of the water on earth; the remainder is held in sediments and sedimentary rocks, in the atmosphere, in lakes and rivers, and in ice on land. This vast quantity of water is distributed in an irregular pattern that has slowly but continually changed through the history of the earth. Today, nearly 67 percent of the earth's land area is located in the northern hemisphere. This leaves only 33 percent of the land in the southern hemisphere, and, of that, almost none occurs between latitudes 40° and 65° south. The Pacific Ocean basin is so large that it is possible to divide the earth into hemispheres, one of which is almost entirely covered by water, while the other is largely land (Fig. 3.7). Not only is the water distributed unevenly, but the various ocean basins are far from uniform in shape, area, and depth. The Arctic is almost completely landlocked; the Atlantic is a long, relatively narrow basin; and the Pacific and Indian Oceans are more nearly oval.

Only within the last few decades have we had more than a vague notion of the deeper portions of the ocean basins. Even now, observations are widely scattered. But precision depth-recording instruments have revolutionized the study of submarine topography, giving us new insights to the nature of the deep sea floor.

The deep-sea floor. Parts of each ocean basin are composed of vast featureless plains. Because of their great depth, usually 5,000 to 6,000 meters, they are called **abyssal plains.** As the profile in Fig. 3.8 shows, they are flat—much flatter, in fact, than most of the Great Plains of North America—and irregular in shape, ranging in width from a few to several hundred kilometers. At the edges of abyssal planes, occasional isolated hills, many composed of volcanic rocks, project above the otherwise extremely flat surface.

Although all oceans contain abyssal plains, the plains do not occupy central parts of any ocean. In the North Atlantic, the plains are located on both sides of the Mid-Atlantic Ridge (Fig. 3.9). A similar, almost symmetrical arrangement is found in the South Atlantic. Deep basins also occur on either side of a

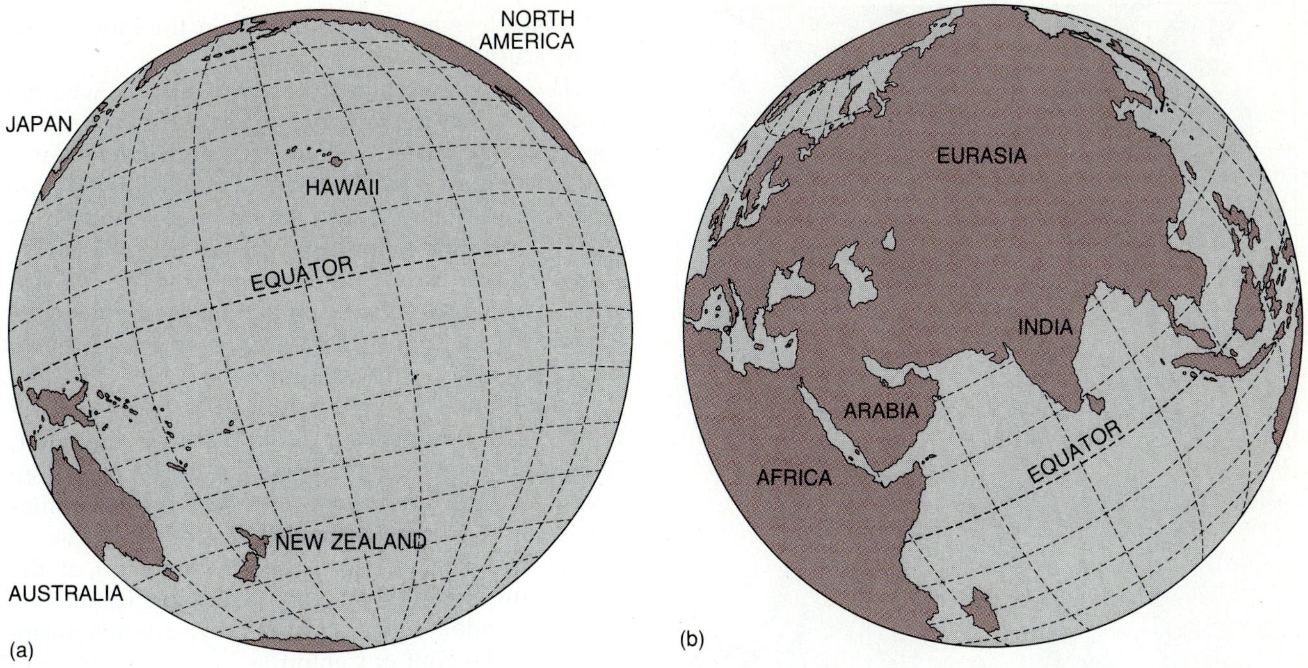

Figure 3.7
These two views of the earth illustrate that viewed from one angle nearly an entire hemisphere is water-covered. From another the vast land areas of Eurasia and Africa cover most of a hemisphere.

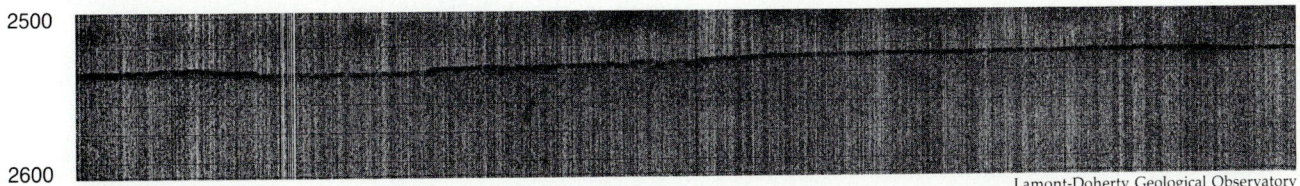

Figure 3.8
Echo-sounding profiles of the abyssal plains off the western coast of France. The nearly horizontal line across the profile is a reflection of the shape of the sea floor. Depths are given in fathoms. This profile depicts the topography along a line nearly 20 kilometers long.

central ridge in the Arctic Ocean. In contrast, the abyssal plains of the Pacific occupy a large part of the North Pacific where their flat surfaces are broken by volcanic submarine mountains or **seamounts,** the Hawaiian Island Ridge, and a number of long cliffs called escarpments. These escarpments are **faults,** regions on the sea floor across which one part of the crust has been displaced relative to the other. Some of these faults can be traced for hundreds and even thousands of kilometers across the deep ocean basin. The occurrence of near-surface earthquakes along them indicates that they are still active. The location of the volcanic islands and submarine ridges in the Pacific suggests that they too may be related to faults in the sea floor.

Oceanic ridges and rises. The most surprising result of the program of sea floor mapping that got under way in the 1950s was the discovery of a long connected submarine ridge system on the sea floor. The portion of it called the Mid-Atlantic ridge lies almost exactly along the center of the Atlantic Ocean. This is connected to ridges in the Indian and Pacific Oceans (Fig. 3.10) and occasionally projects above sea level to form islands. We now know this mid-ocean ridge to be one of the major features of the earth's crust,

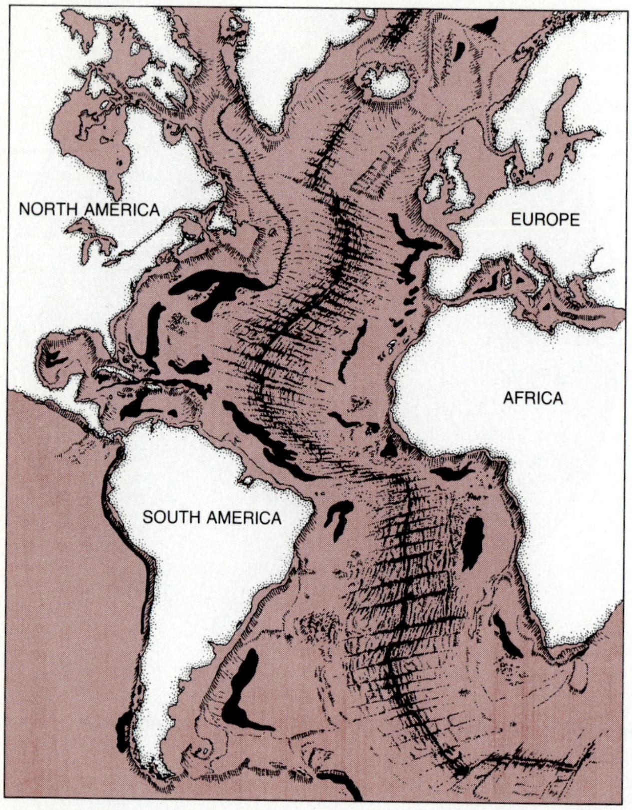

Figure 3.9
Physiographic features of the Atlantic Ocean basin. The abyssal plains, crest of the Mid-Atlantic ridge, and fault and fracture zones across the ridge are shown in black.

one that plays an important role in the movements of the crust.

The ridge system is not very distinctive where it passes through the Greenland Sea across Iceland in the relatively shallow ocean of the northern Atlantic, but it emerges as a prominent and distinct feature farther south. Through the Atlantic the ridge occupies a central position following a pattern that reflects the shape of the African and South American coasts. From the South Atlantic, the ridge system can be traced into the Indian Ocean where one long, linear ridge extends northward into the Gulf of Aden. Another branch, also characterized by rugged topography, splits off toward the southeast, but the **relief,** the difference between highest and lowest elevation, decreases along this branch of the system. It continues across the southern Pacific as a broader arch known as the East Pacific rise. This rise branches at the equator into two sections; the Galapagos Islands are located on one branch, and the other extends northward into the Gulf of California.

A long depression or trough, often called the **median valley,** runs along the crest of the ridge system throughout most of its extent, but is best known along the Mid-Atlantic Ridge. It shows as a notch or cleft on most profiles (Fig. 3.11). Often, this trough is about 30 to 45 km wide and as much as 2,000 meters lower than the peaks on either side of it. The trough appears to be a dropped section of the crust, bounded on either side by steeply inclined faults. Geologists think this valley marks a zone where the oceanic crust is being extended—literally pulled apart. This explains why many shallow earthquakes occur in this trough.

The continuity of the ridge is broken by yet another type of fault, which forms a series of narrow, steep-sided troughs across the ridge. Earthquakes

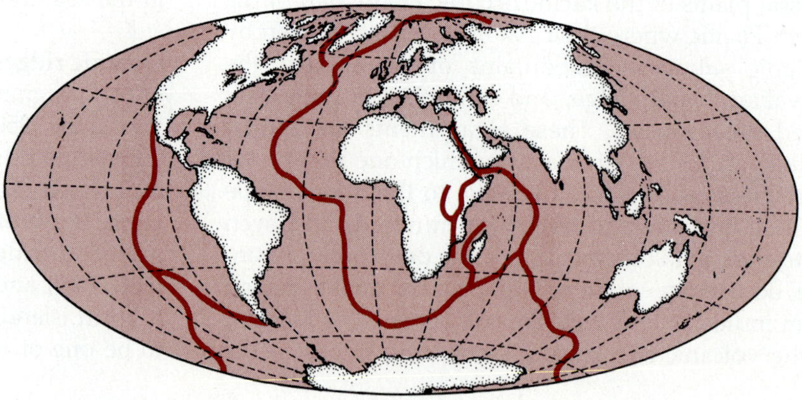

Figure 3.10
A system of submarine ridges, some as high as the Rocky Mountains, can be traced through the oceans as shown here. The true extent of this ridge system was not recognized until the 1950s. A more recent map of the ridge system appears on the inside of the front cover.

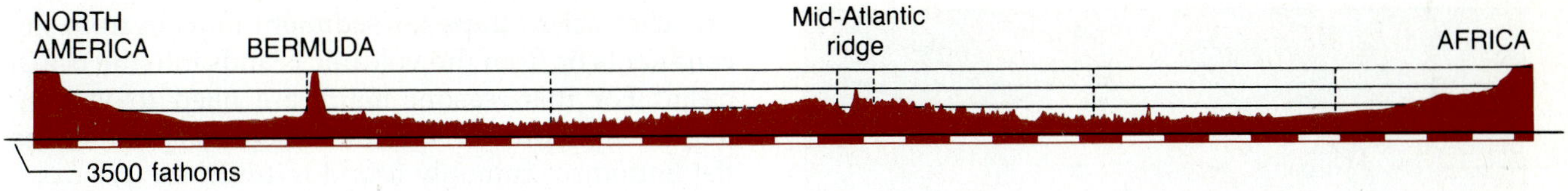

Figure 3.11
Profile across the Atlantic Ocean from Cape Henry, Virginia, through Bermuda, and across the Mid-Atlantic ridge to North Africa. Vertical exaggeration is 40:1.

occur along both the ridge crest and these transverse faults, undoubtedly caused by movements of crust at the faults. Volcanic activity is present along the ridge crest but usually not on the transverse faults.

Volcanic islands located along the oceanic ridges include the Azores in the northern Atlantic and Sao Paulo, Ascension, Tristan da Cunha, and Gough in the South Atlantic. Some of them, like the Azores, have been sites of volcanic eruptions within recorded history—some have erupted in recent years. The lavas are predominantly of basaltic composition, as are the igneous and volcanic rocks dredged and recovered by drilling on the ridges. Although basalt predominates, several other rock types have been found, including dunite (100 percent olivine), gabbro, and serpentine, a rock formed when gabbro or dunite is altered by hot water. Significantly, all of these are rich in magnesium and iron, and all are rock types that could originate in the mantle. Granitic rocks and thick, consolidated sedimentary rocks, like the sequences found on continents, are almost completely absent in the ocean basins. But a sedimentary cover is usually present. The sediment is usually thin over the ridge crest where bare volcanic rock is exposed on the sea floor, but thicker sediment (about 0.5 km) occurs on the flanks of the ridges and on the deep abyssal plains.

According to plate tectonic theory, the volcanic and igneous rock forming along the oceanic ridge system creates new sea floor. The oceanic crust spreads away from the ridges. This slowly moving crust ultimately reaches places where the crust sinks back into the interior of the Earth. Most of these places are characterized by arcuate chains of volcanoes.

Volcanic arcs and deep-sea trenches. Scientists have been intrigued by the arc shape common to many of the volcanic island chains of the Pacific ever since the peculiar distribution and orientation of these islands were first recognized. In some, such as the Aleutians (Fig. 3.12), the volcanic islands lie along perfectly circular arcs. In others, such as the islands northeast of New Guinea (Fig. 3.13), the islands are more scattered. Lavas erupted from the volcanoes in these arcs are andesites (intermediate in composition between granite and basalt). Because this composition is so different from the basalts erupted in the ocean basins and along the oceanic ridges, we are sure the two lavas originate in different ways.

Deep-sea trenches are elongate depressions in the sea floor. Figure 3.14 shows these trenches as well as andesite volcanoes that are associated with most island arcs. The deepest of these trenches, located in the Mariana Island group, is over 9,000 meters deep. This one is probably deeper than others because it is located far from major land areas.

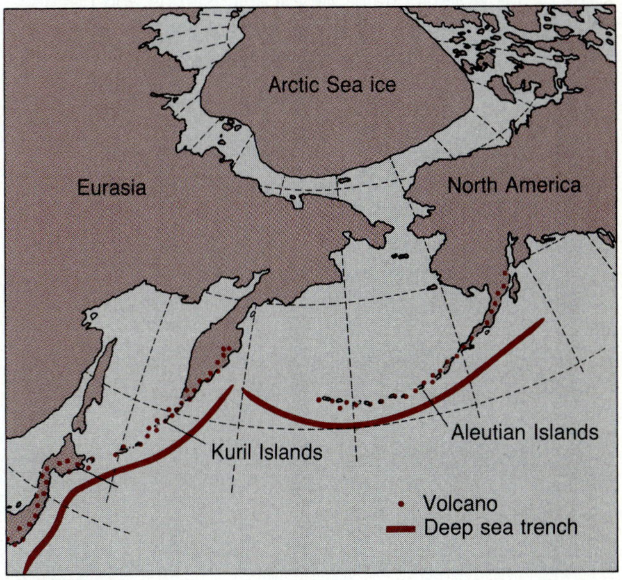

Figure 3.12
The volcanic islands of the Aleutians and the Kuril Islands form nearly perfectly arcuate chains. Note that the deep-sea trench is located on the oceanward side of the volcanoes.

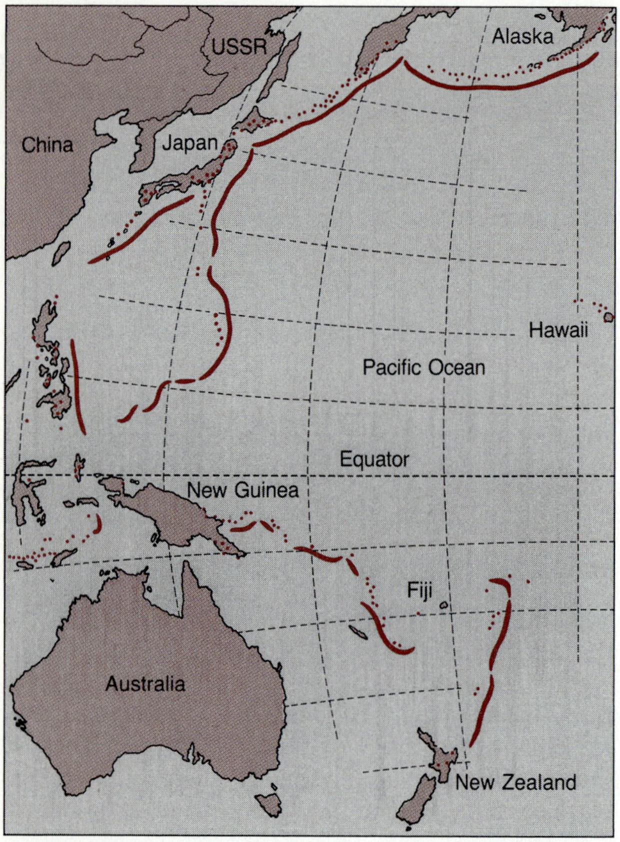

Figure 3.13
Most of the world's chains of volcanic islands and deep-sea trenches are located in the western Pacific. These form an almost continuous belt from Alaska to New Zealand.

Trenches act as traps for sediment moving from the continents or from the volcanic islands into the ocean basin. For this reason, many are filled to varying degrees by sediment, and these deposits produce the flat bottoms commonly found in trenches.

Island arcs are the most earthquake prone parts of the earth's crust. Shallow earthquakes are concentrated in the sea floor on the oceanward side of and under the volcanic islands. Earthquakes occur at greater depth below the volcanic arcs, and the deepest earthquakes occur behind the volcanic arcs. The deepest known earthquakes occur in these zones at depths of about 700 kilometers—far below the base of the crust. These seismic zones provide evidence that deep-sea trenches and island arcs are caused by processes acting in the upper mantle.

Continental Margins: The Transition from Ocean Basins to Continents

The continental margins constitute the zone of transition from the topographically high, low density, **sialic rocks** (those rich in silicon and aluminum) of the continental plates to the much lower, higher density, **mafic rocks** (those rich in magnesium and iron) of the deep ocean basins. These marginal zones between continents and ocean basins are of several distinctly different types.

A long and almost continuous belt of young mountains and active faults lies near and is roughly parallel to the continental margin of North America all the way from Alaska to Central America. A chain of active volcanoes is found on the narrow strip of land that links North and South America. Farther south another young mountain system, the Andes, lies along the continental margin the length of South

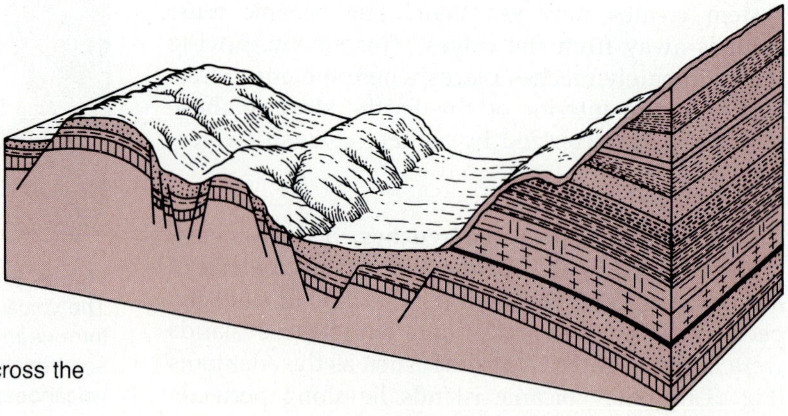

Figure 3.14
Schematic diagram of the subsurface geology across the Mid-America trench off Guatemala.

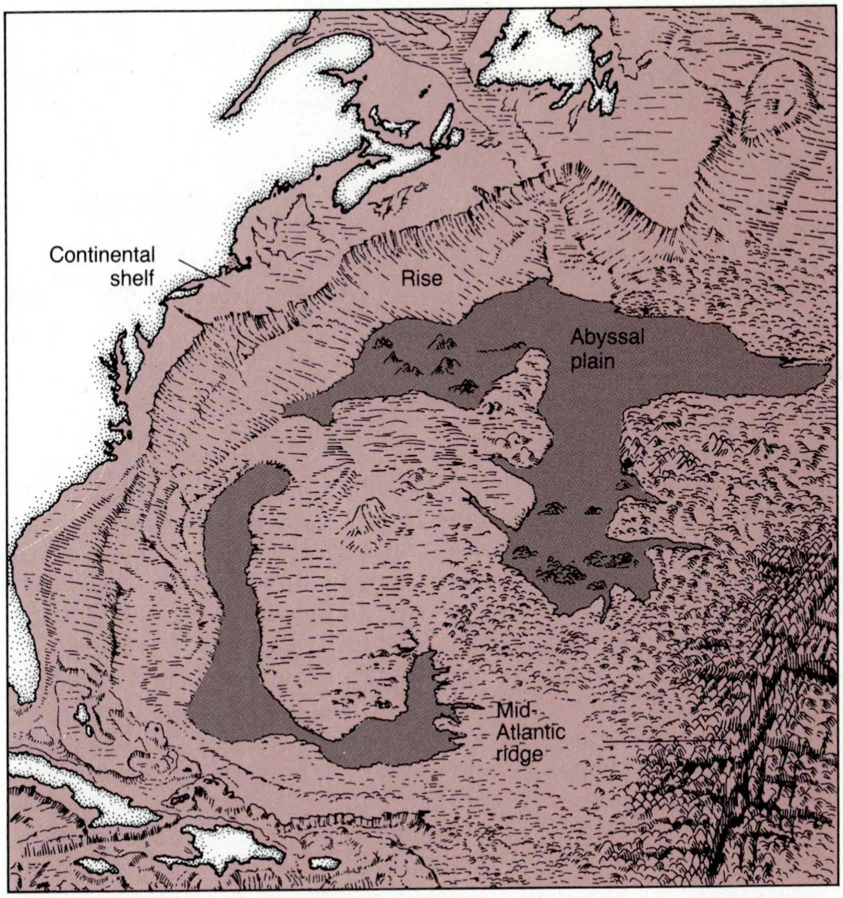

Figure 3.15
Physiographic diagram of the North Atlantic ocean basin.

America. Deep trenches in the sea floor occur close to the coast off Central America and along the Chile–Peru coasts (Fig. 3.14). All along this chain of mountains and volcanoes, we find evidence of modern crustal activity—earthquakes, volcanic eruptions, recent uplift of the mountains, and displacements along active faults. The pattern and type of movement varies from place to place, but this is clearly one of the more active zones in the lithosphere.

The western margin of the Pacific is also a belt of great crustal instability. A long chain of interconnected volcanic islands can be traced all the way from the Aleutians to New Zealand (Fig. 3.13). These islands, aligned in arc-shaped chains, are the sites of most of the major deep focus earthquakes.

The margins of the Atlantic Ocean are quiet by comparison with those of the Pacific. Few earthquakes occur near the Atlantic margins; no active volcanoes are aligned along the continental coasts; the mountain systems near the margins of the Atlantic are much older than those in the Pacific and none is aligned close to the coastlines. The transition from land to deep sea along these **Atlantic-type margins** is usually gradual (Figs. 3.15 and 3.16). The continent is bordered by a shelf-like extension covered by less than 200 meters of water. At the edge of this **continental shelf,** the sea floor becomes steeper, forming what is called the **continental slope.** This slope turns into the very gentle **continental rise,** before passing into the broad, almost flat, and nearly featureless plains of the deep sea.

Continents

Although the continents differ greatly in size, shape, and geography, they are similar in several fundamentally important ways. The crust of continents is much thicker (40 to 50 km) than the crust of the oceans (10 km). Each continent contains one or more belts of strongly deformed rocks that are now or once were a mountain system. And each continent contains at least one large area of ancient igneous and metamorphic rocks that has been stable for hundreds of millions of years.

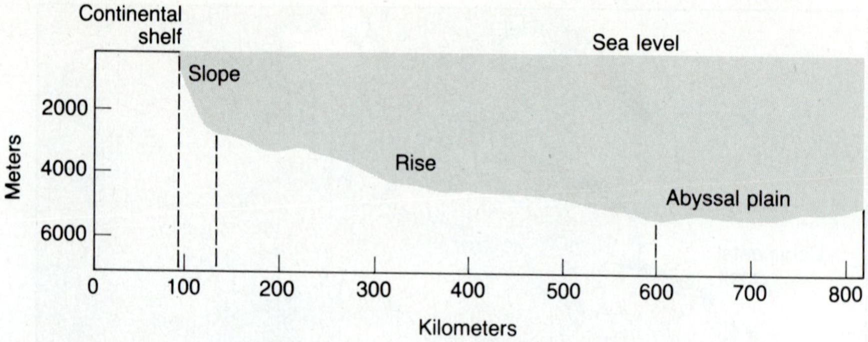

Figure 3.16
Profile across the Atlantic continental margin of North America.

Continental shields. The large stable areas of ancient rocks exposed or buried by relatively thin sedimentary cover on every continent in the world (Fig. 3.17) are known as **cratons** or continental **shields.** These areas are stable; few earthquakes occur within them; volcanic centers there are rare; and they have not been strongly uplifted or deformed into mountain belts for hundreds of millions of years. Shallow earthquakes are occasionally recorded, and steep (making a large angle with the plane of the horizon) faults are sometimes found. But active faults inclined at small angles to the horizon along which crustal rocks move laterally, and **folds,** bends in rock layers, are not present. Although the evidence indicates that these shields have been stable for more than half a billion years, the areas they encompass were not stable earlier in the earth's history. Every shield contains evidence of the existence of ancient mountain systems in which the rocks were strongly deformed and metamorphosed. Large-scale volcanic activity also occurred in every shield during the early history of the earth.

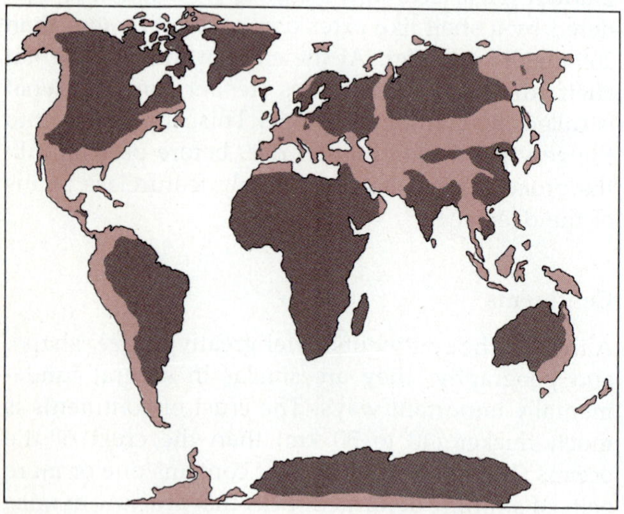

Figure 3.17
The continental shield areas and adjacent platforms are colored on this map. Ancient igneous and metamorphic rocks underlie most of these areas.

Most of the rocks in the shields and all of the oldest rocks on earth have been metamorphosed. Large portions are granitic in composition, and although some of these are now known to be highly altered sedimentary rocks, others are of igneous origin. The ancient mountain belts have been reduced by erosion to surfaces of low relief; consequently, the terrain of the shields is typically flat, although gently rolling hills and isolated features etched out by many years of erosion are found. The margins of the shields have subsided in many places, making it possible for sediment to be deposited over them. These sediments often increase in thickness away from the shield, especially in the relatively young mountain belts, such as the Appalachians and the Rocky Mountains.

A vast shield is located in central Canada. Portions of this shield extend into the north-central part of the United States, where it is covered by sedimentary rocks. Thus, an essentially stable platform underlies the Great Plains and central midwestern states of the United States.

Mountain belts. Mountains hold a special fascination for most geologists, and it is easy to see why when we view the dramatic effects of the processes acting to build and shape the mountains. The downslope movement of materials, the cascading streams, the grinding of glaciers are all evident and so effective that we must wonder how it is that mountains can exist at all in view of the long spans of time involved in their creation and history. Mountains are being formed today, and probably about as rapidly as they have been at any time in the past. We can also see mountains in various stages of decay in different parts of the earth. The younger mountains like the Alps, Himalayas, and Rockies, which have been uplifted in the last few million years, show great differences in elevation, while older chains like the Appalachians, where mountain building ceased 200 to 300 million years ago, have been worn to much lower relief and gentler slopes. Evidence of still older mountains is preserved in the shield areas of the world where the high relief of the mountains has been completely removed by erosion, leaving only the internal structure of the mountains.

The internal structure of mountains reveals a great deal about how they formed. If we travel across the Great Plains of North America, we can see the sedimentary strata exposed as almost horizontal layers along streams and in roadcuts. If it were possible to obtain a cross-sectional view of the plains at depth, we would see a sedimentary pile of rocks of many ages, more or less flat-lying and looking much like the layers of a cake (Fig. 3.18). Some of these layers can be traced for hundreds of kilometers laterally. At the front ranges of the Rockies, however, these same layers are first tilted and then turned up at a steep angle. The underlying rocks are successively exposed until the oldest igneouus and metamorphic complexes are exposed in the highest parts of the mountains. In other parts of the mountain system, the layers of sedimentary rock are bent and folded into belts that resemble the folds formed in a table cloth pushed across the table. Generally, these same strata are broken by faults inclined at low angles (Fig. 3.19) along which the layers have slipped laterally, often as much as a kilometer and sometimes as much as tens of kilometers. Thus, the internal structure of most mountains systems is composed of features created by the uplift from a great depth of large volumes of rock and by the strong deformation, usually by compression, of the overlying sedimentary cover. Having once seen the contrast between the flat-lying layers of the plains and these same layers crushed, folded, and uplifted in the mountains, it is easy to understand why the question of how mountains originate holds the interest of so many people.

Little remains of some of the mountains that originated during the early parts of the earth's history, but the younger mountain systems are well exposed, and the processes that caused them, collectively known as **orogeny,** are still acting. Two major systems of young mountains are prominent features of the continents at the present time. One of these systems includes the mountains of western North and South America, the Cordilleran Mountains, and the Andes. The Cordilleran Mountain System starts in Alaska and includes the broad belts of mountains in both Canada and the United States, including the region from the Rocky Mountains at its eastern edge to the California Coast Ranges. The second major young mountain system extends from southeastern Asia, through the Himalayas, across Turkey and Greece,

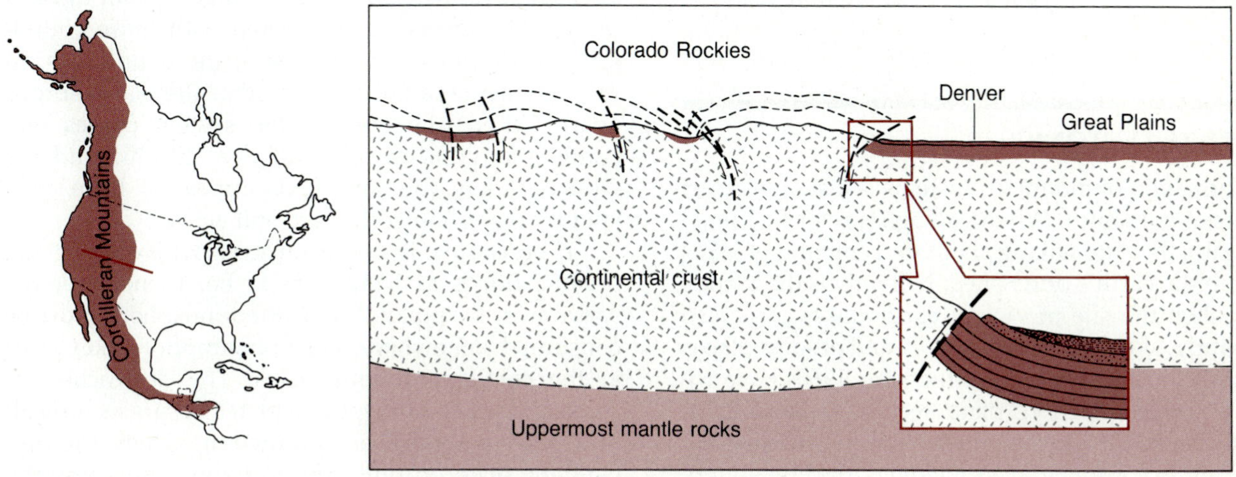

Figure 3.18
Cross section across the Great Plains and Colorado Rockies. The high mountains in Colorado are eroded from huge blocks of crust that have been uplifted along high-angle faults. The relative movement on these faults is indicated by arrows. Compare this with the lateral movement indicated in Fig. 3.19.

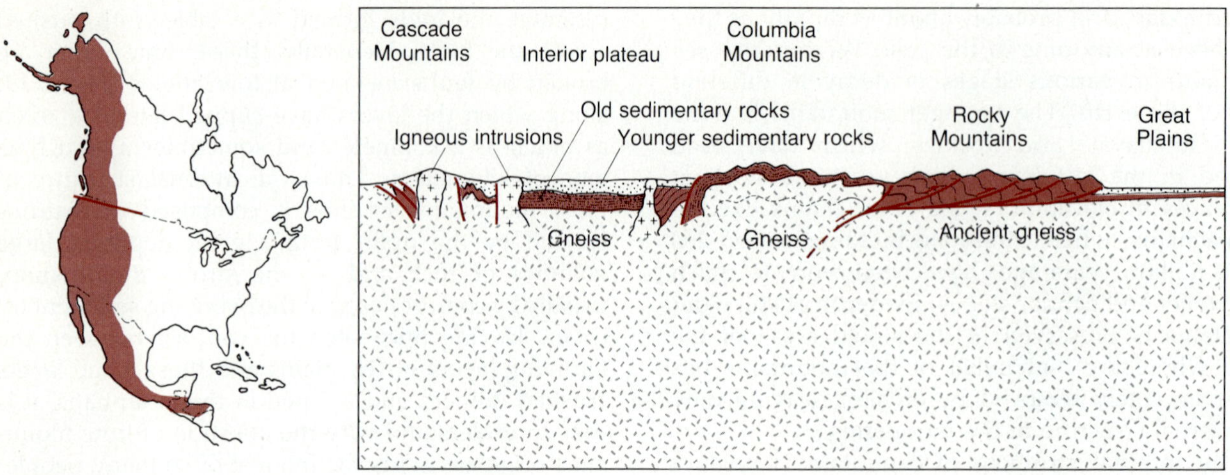

Figure 3.19
Generalized cross section across the Canadian Cordillera. The earth's crust is intensely deformed and contorted in mountain belts like this one.

into the Alpine system of southern Europe, and finally into the mountains of Spain and North Africa. Both of these mountain systems are long—each extends nearly half the circumference of the earth—but their geographic settings are strikingly different. The Cordilleran–Andean system is situated along the western margin of the North and South American continental shields, but the Alpine–Himalayan system is located between shields—between the Indian and Asian shield in the east, and between the African and European shields in the west. We will find later that quite different ideas have been advanced to explain the origin of these two types of mountain systems.

3.5 PLATE TECTONICS

The formation of mountain systems is just one aspect of crustal dynamics. When geologists refer to crustal dynamics they are talking about movement that takes place in the earth's outer shell. Not only is the earth's outer shell the site of violent earthquakes, it is also constantly, slowly moving laterally and in some places vertically as well. Only since the early 1960s have we had a unified theory, plate tectonics, capable of explaining all the movements for which evidence can be found. According to plate tectonics, the lithosphere is divided into a mosaic-like pattern of relatively thin plates as shown on the inside front cover. Although ocean basins and continents are the most important subdivisions of the crust, plate boundaries do not always coincide with the boundaries between continents and ocean basins. Plate boundaries are of three distinctly different types according to the type of movement along them: divergent boundaries, convergent boundaries, and fault boundaries.

The plates on either side of **divergent boundaries** are moving away from one another. The junction between the plates is thus a zone in which the crust is spreading. For this reason shallow earthquakes abound here; faults caused by the extension are present; and volcanic activity attests to the rising magma beneath the ridge crest. The divergent boundaries coincide with the oceanic ridge system in most places. The intrusions associated with these boundaries create new oceanic crust at the boundaries. In this way new sea floor is formed as the plates spread apart. Where the ocean ridge system passes into continental crust in the Gulf of California and the Gulf of Aden, the continents appear to have been split by major faults that are still active.

If spreading is going on everywhere along the oceanic ridges as it appears to be, then either the earth must be expanding or the lithosphere must be consumed somewhere. This consumption takes place at the **convergent boundaries.** The destruction or consumption of lithospheric plates occurs as a result of a process known as **subduction**, where the lithospheric plate on one side of a zone sinks into the mantle (Fig. 3.20). Subduction takes place in the island arcs and along the western edge of parts of both South and North America. The subducting plate causes earthquakes deep in the mantle, and also is a cause of volcanic activity in the island arcs along so many of these zones.

Oceanic crust can move down subduction zones with relative ease, but continents, because they are largely composed of less dense rock, cannot readily be carried down into the denser mantle. Thus if a subducting plate carries a continent toward a subduction zone, the continent can collide with an island arc. If the subduction takes place along a continental margin, two continents may collide. One result of such collisions is the formation of mountain belts. The Alpine–Himalayan belt is the result of collision among Eurasia, Africa, and India. Most geologists now think that the Appalachians formed as a result of collisions that took place long ago, before the Atlantic Ocean opened. Parts of the Cordilleran system also appear to be the result of collisions of large islands with western North America.

Faults constitute the third type of plate boundary. These fault boundaries are invariably steep zones that cut deep into or through the lithosphere and along which plates move laterally past one another. Similar faults cut across the oceanic ridges in places and some of these bound similar plates. To Americans, the best-known fault boundary is one that runs the length of California, the San Andreas fault. Across the fault plane, the Pacific plate and the North American plate move past each other at a rate of about one centimeter per year. We will consider the consequences of this motion in Chapter 13.

One of the many appealing attributes of plate tectonic theory is how well it explains the rock cycle. Through plate theory we can relate the causes of igneous activity, the reasons why metamorphic rocks are generated, the conditions that produce uplift and mountain building, and the tectonic environments in which various types of sediment are deposited.

According to plate theory, magma is generated where rising hot mantle materials come near the surface at oceanic ridge crests. This magma is added to the crust as spreading of the sea floor takes place. Marine processes cover the sea floor with sediment as it spreads. So, as the sea floor approaches a subduction zone, the composition of the lithosphere is different from that at the ridge crest. Within the subduction zone the lithosphere descends, carrying its sediment cover with it. As it moves down, heat and pressure increase, and new magma is generated that rises into the overlying crust, part eventually finding its way to the surface via volcanoes, which add their own products to the sediment in their vicinity. Thus, we have an effective framework within which we can examine sedimentary environments, the generation of magma, and metamorphism, three important components of the rock cycle.

Plate theory also provides a unified framework within which we can explain the cause and distribution of volcanic activity, earthquakes, the structure of the lithosphere, and the origin of mountains. In the next part of this book we will turn to the origin of plates and the rocks that compose the earth. Plate theory will be used to explain many aspects of the origin of our lithosphere. Then in Part III we will examine the structure of the earth and the origin of plate tectonics in more detail. Throughout the text we shall see that the theory can be used to explain many features of the earth. It is by far the most powerful and fruitful theory ever devised in the earth

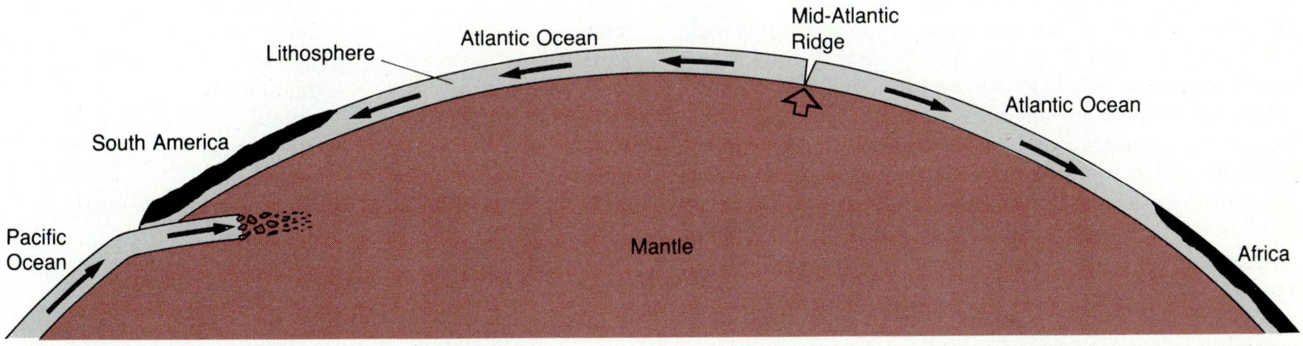

Figure 3.20
Cross section of the lithosphere showing the mid-Atlantic ridge spreading in the center and the subduction zone along the west coast of South America. The continents (black) move with the lithospheric plates of which they are a part.

sciences. Even if it ultimately fails as the best explanation for some phenomena for which it has been advanced, it will have served us well. At the present time, it clearly has far more proponents than any alternative under consideration. But if the best traditions of the scientific approach prevail, we will continue to evaluate plate theory, to question its application, to test it, and to seek better alternatives or refinements of the theory.

SUMMARY

The earth is a generally spherical body, but it is slightly flattened at the poles, bulges at the equator, and is marked by irregular surface depressions and elevations.

Like all massive bodies, the earth attracts other bodies with a gravitational force that is proportional to the mass of both bodies and inversely proportional to the square of the distance between them. This law of gravitational attraction enables us to calculate the diameter and density of the earth and to detect many subsurface irregularities in its structure.

The earth is divided into concentric zones, the outermost zone being the crust, which is thinnest under the oceans and thicker under the continents. The region from the crust to the earth's core is called the mantle. The crust and outer most mantle form a rigid outer shell, the lithosphere, underlain by a plastic zone where rock is capable of flow, the asthenosphere. About halfway to the center, rocky material gives way to a core of much denser material believed to be largely molten iron.

Less than one-third of the earth's surface is land area. The remainder is ocean, of average depth 3.8 km. The ocean floor is generally composed of relatively young, dense, basaltic rock. The continents are composed mostly of much older, less dense, sialic rock. The two areas differ also in degree of deformation, the continental bedrock being much more strongly deformed than that undersea.

The ocean floor shows important topographic features. The most striking are the mid-ocean ridges of basaltic rock that stretch the length of the Atlantic and Arctic oceans and through the South Pacific and Indian oceans. These ridges are flanked by deep basins and show long cliffs formed by faulting. Low, flat-lying areas called abyssal plains lie farther from the ridge. The depth of sediment on these features increases from little or none over the ridge crest to perhaps half a kilometer on the abyssal plains where sediment has accumulated for many millions of years.

Within the oceans, volcanic island chains formed of an andesitic lava are very different from the oceanic ridges. These arc-shaped groups of islands are associated with nearby deep-sea trenches and with considerable earthquake activity.

Along the coast of the Americas, the continental margin is marked by a young and active mountain system, the Cordillera, where faulting, earthquake activity, and volcanism are all common. The Atlantic margin is quiet by contrast. The mountain system there, the Appalachian Mountains, is older, and the ocean floor drops off more slowly, forming first a continental shelf, then a steeper continental slope, which gives way to the continental rise and abyssal plains.

The continents themselves all contain at least one broad stable area of underlying igneous or metamorphic rocks, which is granitic. These continental shields are quiet, showing very little earthquake activity. At the shield's edges, this rock becomes folded and uplifted to form the bordering mountains.

Mountains continue to be formed today, especially in two belts: the Cordillera of the western Americas, and the long mountain system that includes the mountains of southern Europe, Turkey, and the Himalayas.

These features of the earth's surface are the product of plate tectonics. Rising magma at the ocean ridges marks a zone of diverging plates where the crust is spreading. In convergence zones where plates are moving together, one must be carried down into the mantle. Subduction of this type is taking place today in island arcs and along the western coast of the Americas. Along fault boundaries two plates move past each other without creation or loss of crust. These three plate boundary types are characterized by different patterns of rock composition, earthquake, and volcanic activity.

KEY TERMS

abyssal plain
asthenosphere
Atlantic-type margin
continental rise
continental shelf
continental slope
convergent boundary
core
craton
crust
deep-sea trench
divergent boundary
faults
folds
gal
geoid
law of universal gravitational attraction
lithosphere
mafic rock
mantle
median valley
milligal
orogeny
relief
sea level
seamounts
shield
sialic rock
subduction
universal gravitational constant

STUDY QUESTIONS

1. In what ways do continents differ from ocean basins?
2. Compare the eastern and western sides of the Pacific.

3. How do the continental margins around the Pacific differ from those of the Atlantic?
4. Draw a distinction between the universal gravitational constant, G, and the acceleration due to gravity on earth, g.
5. How does the earth's crust differ from its lithosphere?
6. How do the mountains in the ocean basins differ from those on the continents?
7. Why do we need to know about the earth's interior in order to understand the features at its surface?
8. Why doesn't an object that weighs one kilogram on earth have the same weight on any other planet?
9. If the earth were removed from the gravitational attraction of the sun and moon, what processes on earth would be most affected?
10. Why doesn't a degree of latitude have the same length everywhere on earth?

SUGGESTED READINGS

Emery, K. O., "The continental shelves," *Scientific American* offprint 882 (1969).

Fisher, R. L., and Roger Revelle, "The trenches of the Pacific," *Scientific American* offprint 814 (1955).

Heezen, B. C., Marie Tharp, and W. M. Ewing, "The North Atlantic—text to accompany the physiographic diagram of the North Atlantic, part one of the floors of the oceans," *Geological Society of America Special Paper* 65, 1959.

King-Hele, Desmond, "The shape of the earth," *Scientific American* offprint 875 (1967).

Stetson, Henry C., "The continental shelf," *Scientific American* offprint 808 (1955).

Takeuchi, H., S. Uyeda, and H. Kanamori, *Debate About the Earth*, rev. ed. San Francisco: Freeman, Cooper, and Co., 1970.

THE ORIGIN OF THE EARTH AND ITS MATERIALS

Part 2

Our best evidence suggests that the earth originated about 4.5 billion years ago from a cloud of gases and small bodies surrounding the sun. For evidence of the earliest parts of earth history we are forced to look to the moon and other planets, because the surface of the earth has been so drastically changed by the effects of large quantities of water, and by plant and animal life. During its early history the earth was bombarded by extraterrestrial material that shattered its thin crust and gave rise to volcanic activity. As our atmosphere evolved from volcanic gases, rains fell, the oceans formed, and a sedimentary veneer spread over the earth's surface. Most of the processes by which this sedimentary cover was created continue to the present time, as do the processes that generate molten material and alter the crust. Because these processes are still going on, we have a unique opportunity to observe them in action.

The processes we see acting today at the earth's surface and the processes we know about indirectly in the interior appear to be the same processes that have shaped the earth throughout geological history. Thus, the knowledge we gain about the earth today is a key to unraveling its history. A large part of this knowledge comes from study of the materials, the rocks, that make up the earth's crust.

THE ORIGIN OF THE PLANETS

Chapter 4

The age of space exploration truly began when the spacecraft Pioneer I first escaped the hold of the earth's gravity field in October 1958. In the years since then, trips to the moon and other planets came in quick succession. Until these first historic voyages into space, study of our moon and the other planets could be only at long distance by astronomers with telescopes. The arrival on earth of the extraordinary, detailed pictures, first of the moon and later of the other planets and their satellites, opened up the way for the use of photo interpretation techniques long used by geologists to study the earth.

It quickly became apparent that our knowledge of the surficial processes acting on the earth could be applied successfully to other planets. Some of the more impressive surficial features of other planets can be compared with similar forms on the earth, as we shall see in Part IV of this book. An even more compelling objective for space exploration grasped the imagination of the scientific community: elsewhere in the solar system we may find evidence that will help us better understand the origin and evolution of the earth and of the solar system as a whole.

Evidence of this type has been found, especially on the moon, where the absence of an atmosphere and water has left the surface essentially unchanged except for the effect of meteorite impacts almost since the time it and the earth were formed. The surface of the moon is today much as it was several billion years ago. Study of samples returned from the moon enables us to go far beyond the earlier speculations about its character and its history. This knowledge, in combination with what we are learning about the sun, other planets, their satellites, and even meteorites, is helping us narrow the critical gaps in our understanding of the origin of the earth.

Most astronomers currently favor a theory for the origin of the universe that pictures all of the matter of the universe as being concentrated in a relatively small space 8 to 25 billion years ago. Extremely high pressures and temperatures within that initial cloud of matter made it impossible for elements to exist; matter consisted instead of a soupy cloud of neutrons. From this state, the matter of the universe began to expand, and with this expansion and subsequent reduction in temperature, elements began to form, often as diffuse clouds of matter called **nebulae.** Vast, turbulent cells existed in the initial masses of gases, but these cells of matter began to form separate large clouds that eventually became the systems of stars called galaxies (Fig. 4.1). The Milky Way, the galaxy in which the solar system is located, is one of these. Presumably smaller cells of turbulence, destined to become stars, existed within each of these galaxies. Our sun is one such star.

The sun and planets formed when the matter in the nebula began to be drawn together by gravitational attraction. Gravitational contraction of the matter made it hot near its center. The planets and their satellites formed in this hot rotating cloud of matter. Away from the center, temperatures were cool enough for the gases to condense into liquids, and finally these cooled and crystallized into solids. These solids accumulated to form the planets.

Mount Wilson

Figure 4.1
Our own galaxy, the Milky Way, probably resembles the galaxy Andromeda pictured here.

4.1 THE SOLAR SYSTEM

The Sun

Stars differ from planets, their satellites, and other smaller bodies in the universe because their high temperatures and extreme size create such great pressures in their interiors that the material of which they are composed cannot exist in a solid or even liquid state. Atoms break down at these temperatures and pressures, forming clouds of incandescent gases composed of electrons and the atomic nuclei. The sun at the center of our solar system is a medium-sized star. Like many other stars, the sun has an extremely hot, dense center grading to a less dense outer region, the sun's atmosphere.

The sun's atmosphere. The vast envelope of gases that comprise the sun's atmosphere is divided into three regions: the **photosphere,** the **chromosphere,**

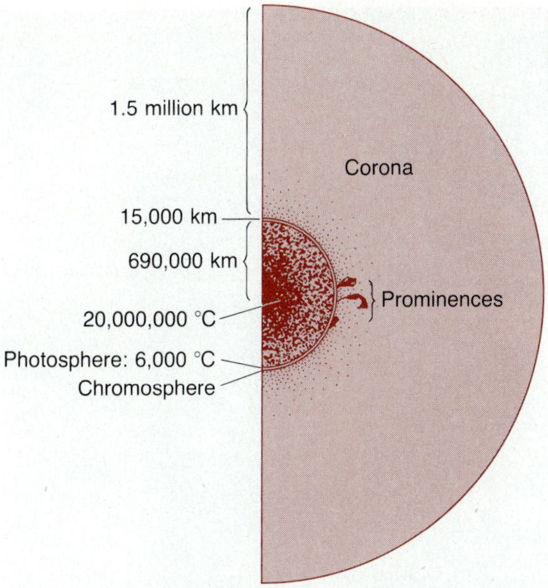

Figure 4.2
The solar atmosphere consists of three main divisions: the corona, the chromosphere, and the photosphere.

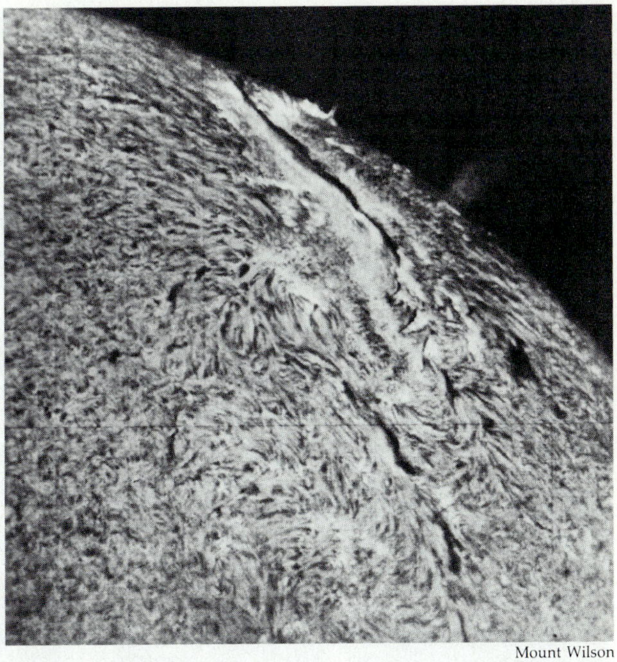

Mount Wilson

Figure 4.4
Convection cells caused by the high temperatures and boiling motion of gases near the surface of the sun.

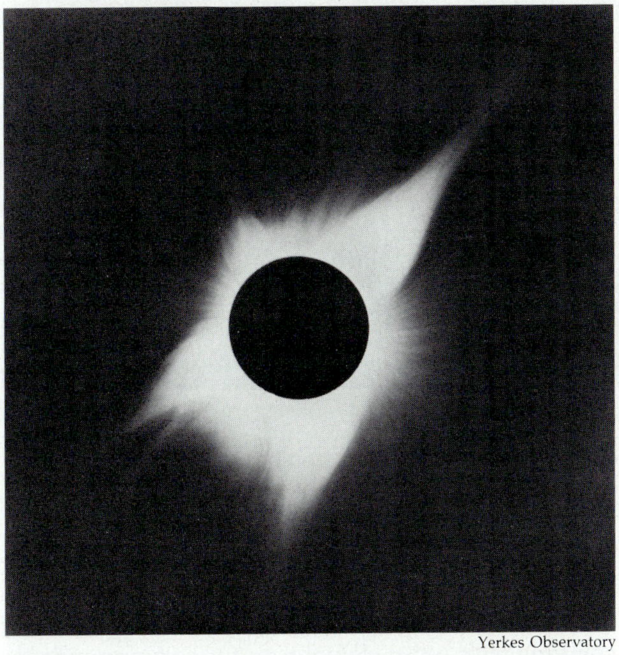

Yerkes Observatory

Figure 4.3
The sun's corona. This thin gaseous layer of the sun is usually visible only when the moon covers the photosphere during total eclipses. Parts of the corona extend over a million kilometers above the photosphere. Even in its denser parts, the corona is only about one million-millionth of the density of the earth's atmosphere at sea level.

and the **corona** (Figs. 4.2 and 4.3). We normally see the photosphere, the relatively thin (15,000 km) layer where most solar radiation originates. Temperatures in the photosphere are estimated to be about 6000 °C, but these are low compared to temperatures in the sun's interior, thousands of times higher. The great temperature difference between the surface and the interior of the sun causes convection, a slow boiling motion in the gases, and accounts for the granular appearance of the sun's surface (Fig. 4.4). The gases outside the photosphere (Fig. 4.3) are thin and are visible on earth only during eclipses when the moon passes between the earth and the sun, blocking out the intense light from the photosphere. This outer part of the solar atmosphere is composed primarily of hydrogen gas heated to extremely high temperatures (5,000 to 35,000 °C in the chromosphere and up to 1,000,000 °C in the corona) by the intense radiation from the photosphere.

During solar eclipses, huge, cloud-like, arch-shaped protrusions, called **prominences** or solar flares, can be seen extending far out from the surface of the sun, at times penetrating into the corona (Fig. 4.5). These flares are similar to gigantic atomic explosions.

Composition of the sun. Hydrogen is by far the most abundant element in the sun. It is about ten times

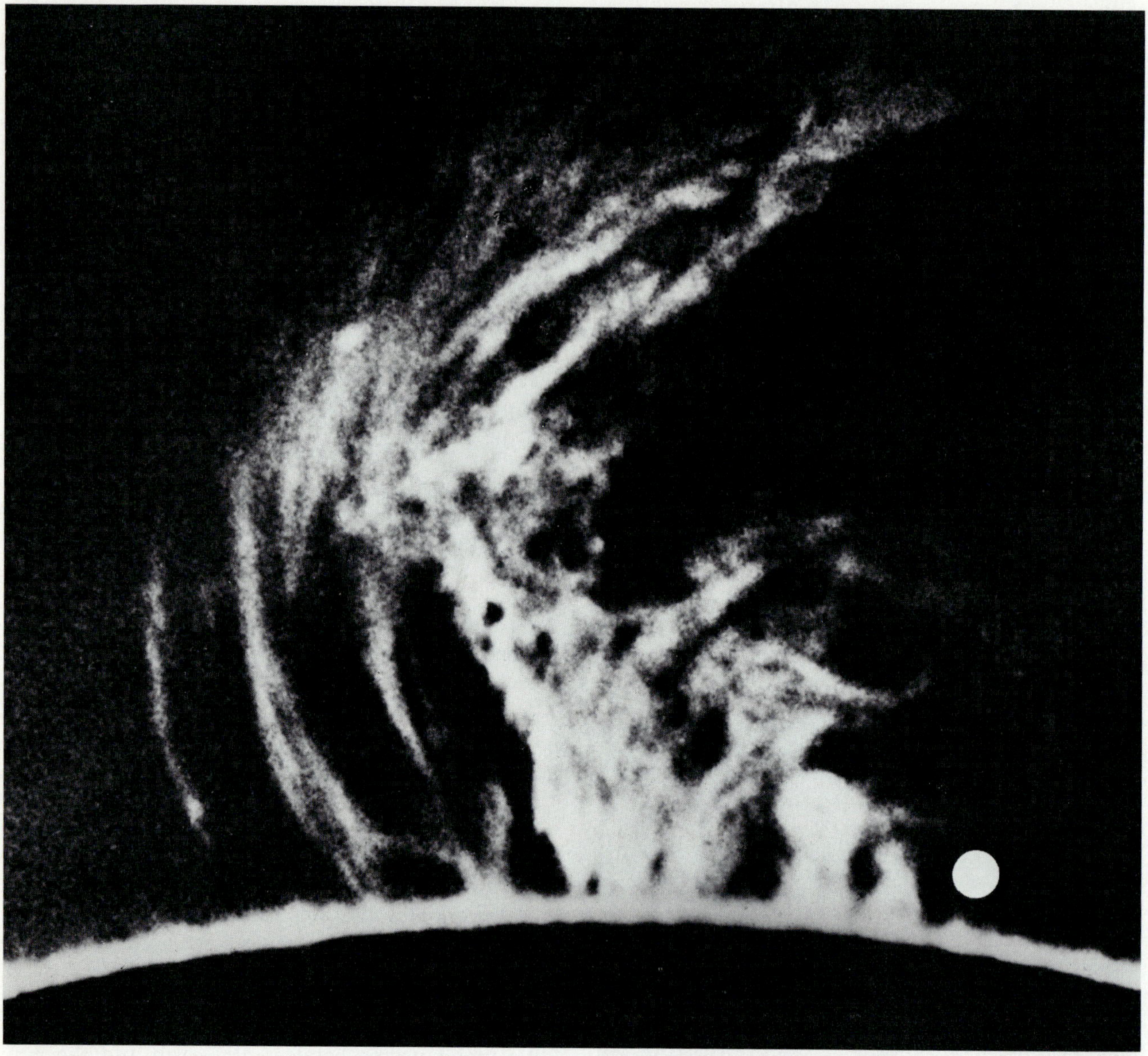

Mount Wilson

Figure 4.5
Solar flare. This large, active prominence is over 200,000 kilometers high. Explosions of this kind occur above the photosphere in the lower part of the corona. The small white circle at right represents the size of the earth.

more plentiful than the next most abundant element, helium, which was discovered in the corona even before it was known on earth. These two elements combined constitute well over 99 percent of the material in the sun's atmopsphere. Table 4.1 lists the lesser components as well. Neither hydrogen nor helium is especially abundant on the earth or on our close neighbors Mercury, Venus, and Mars. In part this is because they become gases at very low temperatures (−269 °C for helium) and are so extremely light that they are not held to the earth by its gravity field. In addition, helium is inert and does not combine with other elements to form compounds. If, as seems likely, the planets and sun all originated from a single mass of material, we must consider why these inner planets are so different from the Sun in the relative proportions of these gases and other heavier elements.

Table 4.1 Comparison of the composition of the sun's atmosphere with an estimate of the abundance of these elements on earth. The elements that make up the sun's atmosphere are listed below in order of decreasing quantity. All of these elements are found on earth, but neither hydrogen nor helium is abundant here.

Element	Weight on sun (%)	Weight on earth (%)
Hydrogen	74.0	—
Helium	23.0	—
Oxygen	0.8	30.0
Iron	trace	35.0
Nitrogen	trace	—
Magnesium	trace	13.0
Silicon	trace	15.0
Carbon	trace	—
Sulfur	trace	2.0
Aluminum	trace	1.0
Sodium	trace	.6
Calcium	trace	1.0
Nickel	trace	2.4

The Planets

One of the main objectives of geologists and astronomers alike is to find evidence that will help us understand how the solar system originated and evolved to its present condition. Much of the information we need to succeed in this goal can be obtained through studies of the planets.

The planets can be divided into two groups based on their locations, compositions, and physical characteristics. Those planets closest to the sun—Mercury, Venus, Earth, and Mars—are much smaller both in mass and radius than are the outer planets, Jupiter, Saturn, Uranus, and Neptune (Fig. 4.6 and Table 4.2). The inner group, known as the terrestrial planets, is also much more dense, and presumably all of them are composed of materials quite similar to the earth's; the much larger, less dense, outer planets are thought to contain large quantities of hydrogen, helium, ammonia, and methane, some of which may be ices. The planetary atmospheres of the two groups also differ. The inner planets have atmospheres composed of such gases as nitrogen, oxygen, carbon dioxide, and water vapor. The main constituents in the atmospheres of the outer planets are ammonia and hydrocarbons such as methane, compounds that are rare in the atmospheres of inner planets.

In a following section we will examine some of the proposed explanations for the origin of the planets. Any acceptable theory for their origin must provide a satisfactory explanation for, or at least be compatible with, a large body of observation data, especially that concerning the location and motion of the planets. Here are some significant constraints we must consider in evaluating models of planetary evolution.

The information in Table 4.2 shows, for example, that the mass of the solar system is concentrated in the sun, which contains about 99.8 percent of the total mass. The planets contain the remaining 0.2 percent. We know too that the planets move in orbits about the sun; all except Venus revolve in the same direction that the sun rotates. The orbits of the planets are nearly circular—the orbits of Pluto and Mercury have the greatest departures from a circle—and they all lie in approximately the same plane, close to the plane in which the sun rotates. Only Pluto has an orbit more than 8° out of the plane of the other orbits (Pluto's orbit is inclined 17°). As they revolve around the sun, the planets also rotate about axes, and most rotate in the same direction that the sun does. The distance of the planets from the sun is systematic—each planet is about twice as far from the Sun as the next closer planet.

The spacing of the planets is also regular rather than random. We have seen that the planets form two groups: an inner group that includes the small dense planets, Mercury, Venus, Earth, and Mars; and an outer group that includes the giant less-dense planets, Jupiter, Saturn, Uranus, and Neptune, and the smaller planet, Pluto.

The arrangement and movement patterns of the planetary satellites is similar to that of the planets around the sun. Most of the satellites lie in the plane of planetary rotation. Most rotate in the same direction about their axes, and revolve around the planets in the same direction as they rotate about their axes.

The oldest rocks found on the earth are in the range of 3 to 4 billion years old.

What emerges from this information on the solar system is a picture of a disc-shaped arrangement of sun, planets, and satellites, in which rotation and revolution are both predominantly in one direction. The speed of movement and the distribution of matter is such that the **angular momentum** (mass × velocity × distance to the center of rotation, squared) of the solar system is concentrated in the planets. One familiar effect of angular momentum is seen when a ball tied to a string is twirled around your finger. As

Figure 4.6
Exploration of the solar system during the last two decades has revealed an impressive variety of features on the major planets and their satellites. Methods of study initially developed to analyze the origin of landforms found on earth are being applied to understand the features we can now see in detail on other planets. The surface of our closest neighbor in space, Venus (a), is obscured by clouds, and the most distant planets, Uranus, Neptune, and Pluto (not shown), have yet to be photographed at close range. But pictures taken by Viking, Mariner, and Voyager I and II show us details of the other planets. Saturn (b) and Jupiter (c) are giant planets compared with the others (see Table 4.2). Their surfaces are obscured by clouds, but the surface features of most of their satellites, such as Jupiter's giant moons Ganymede (d), and Callisto (e), both of which are nearly 0.4 the diameter of earth, and Saturn's moon Enceladus (f) can be seen. All three of these show cratered surfaces like that of the earth's satellite, the moon. Craters can also be seen on Mercury (g) and Mars (h). Even some of the smallest satellites, such as Mars' satellite Phobus (i), only about 22 kilometers across, are cratered. The cause of this cratering throughout the solar system is discussed in later sections of this chapter. Mars has long been of special interest because of the unusual surface markings seen by means of telescope. An ice cap seen in (h) resembles the earth's own icy polar covering.

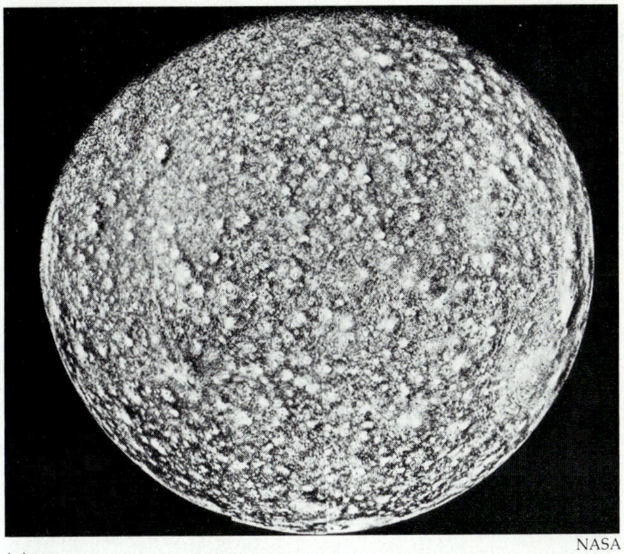

NASA

(e)

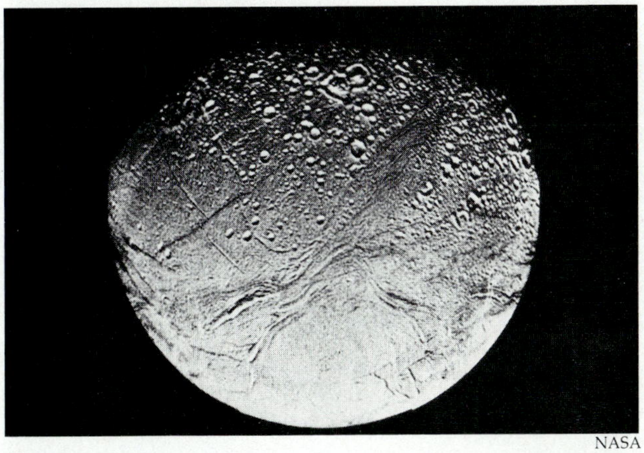

NASA

(f)

NASA

(g)

NASA

(h)

NASA

(i)

Table 4.2 Characteristics of the Planets.

Inner Planets

Relative sizes of sun and planets

	Sun	Mercury	Venus	Earth
Distance from sun (relative to earth)	0	0.4	0.7	1
Size (relative to earth)	109	0.38	0.95	1
Mass (relative to earth)	332,000	0.55	0.8	1
Density (relative to water)	1.4	5.5	5.2	5.5
Number of satellites		0	0	1
Rotation (earth days)	25.4	58.6	243 (reverse)	1
Revolution (earth years)	—	.24	.61	1
Composition of atmosphere*	Hydrogen, helium	No atmosphere; Mercury does not have sufficiently strong gravity to hold gases	Carbon dioxide, nitrogen, helium, hydrogen, argon, clouds of H_2O, O_2, sulfuric acid	Nitrogen, oxygen, carbon dioxide, argon
Surface features		Heavily cratered; scarps, ridges, lava flows	Craters, scarps, volcanoes, high mountains	Rare craters, fault valleys, continents, ocean basins, lava flows, glaciers, streams, deserts
Temperature (approximate)	6000 °C	430 °C (day side); −170 °C (night side)	480 °C	

* These are subject to change as analysis of satellite data continues.

the string becomes shorter, the angular momentum remains constant and the ball revolves faster. The sun rotates so slowly that nearly 98 percent of the angular momentum is held in the planets. Jupiter alone accounts for nearly two thirds of the angular momentum in our solar system.

Any model proposed for the evolution of the solar system must be evaluated within these constraints and one other based on findings here on earth—that the oldest earth rocks ever found are three to four billion years old. In addition to these planetary constraints, data from solid fragments—the meteorites that fall to earth from the sky—also put important limitations on any theory.

Meteorites

Meteorites are bodies of stone and metal that fall through the atmosphere and strike the earth. As they pass through the atmosphere at speeds of 10 to 15 km/sec, friction within the atmosphere causes them to glow or burn, producing bright fiery trails. Burning starts at approximately 150 km above the earth's surface, and small particles, many of them no larger than peas, are completely burned. The larger meteorites have surfaces sculptured by this burning (Fig. 4.7), but they are still cold internally when they hit the ground, presumably because they have extremely

Outer Planets

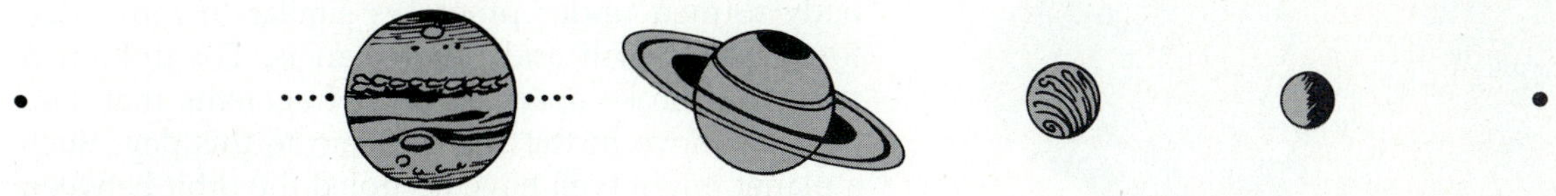

	Mars	Jupiter	Saturn	Uranus	Neptune	Pluto
	1.5	5.2	9.5	19.2	30	39.4
	0.53	11.23	9.41	4.06	3.88	0.2
	0.11	318	95	15	17	0.002
	3.9	1.4	0.7	1.3	1.6	—
	2	16*	16*	5	2	1
	1	0.4	0.4	0.45 (reverse)	0.6	6.4
	1.88	11.86	29.46	84	164.8	284.4
	Carbon dioxide, argon	Hydrocarbons, ethane, acetylene	Largely nitrogen	Methane	Methane	Unknown; probably methane and water
	Craters, volcanoes, stream valleys, fault scarps, lava flows	Obscured by clouds	Obscured by clouds	Unknown	Unknown	Unknown
	−50°C (average)	−130 °C	−185 °C	−200 °C	−215 °C	−230 °C

low temperature in space and because the burning is of such short duration.

Most meteorites can be classed as either metallic or stony. The metallic meteorites, called **siderites,** are composed of about 90 percent iron and 10 percent nickel. The stony meteorites, called **chondrites,** are composed almost entirely of aggregates of the minerals olivine and pyroxene. Still other meteorites are mixtures of iron, nickel, olivine, and pyroxene. The stony meteorites are thought to be more abundant than the nickel–iron bodies because, among the meteorites actually seen falling, most are chondrites. But, because they resemble rocks, they are less likely to be identified as meteorites unless they are seen to fall. The largest known nickel–iron meteorite, measuring nearly four meters in diameter and weighing about fifty thousand kilograms, fell in southwest Africa. Most stony meteorites are small by comparison, the largest of them weighing about a thousand kilograms.

When a large meteorite hits the surface of the earth, the impact forms a crater; sometimes the meteorite explodes, excavating a much larger crater. Huge craters are found on the moon and on other planets and their satellites. The famous meteor crater of Arizona, over a kilometer in diameter and 200 meters deep, is an example of an explosion crater. The meteorite that formed this crater struck relatively

Figure 4.7
This meteorite has been cut to show the metallic crystals inside. The uncut edges show the smooth sculpturing produced by burning when the meteorite passed through the atmosphere. This meteorite is about 12 cm long.

horizontal sedimentary rocks, shattering them and folding the layers around the edges of the crater into an upright position. A similar pattern is produced by near-surface underground explosions. Some of the sand grains in the rock in the bottom of the meteor crater have been fused under extreme pressure, forming an unusual variety of silica, and fragments of the meteorite are found scattered around the crater.

Meteorites are of special interest for a number of reasons. They provide one of our best clues to the relative abundance of elements in the solar system. We believe that they are fragments of the materials from which the planets formed, and that they all formed at the same time—at the time of origin of the solar system. Evidence we will examine later indicates that the density of the earth's core, described in Chapter 3, is approximately that of a nickel–iron alloy, and the outer mantle of the earth is probably composed largely of olivine and pyroxene. It appears that the earth is composed of basically the same materials as meteorites, and the density of the other inner planets suggests that they may also be made of these materials.

According to radiometric dating methods, the age of the crystallization of the minerals in all of the meteorites so far tested is 4.5 to 4.7 billion years. These data agree with the 4.6-billion-year age of the moon, suggesting that the meteorites and the planets are products of the same cosmic event.

Numerous ideas have been advanced to explain the origin of meteorites. Some scientists think they are fragments of a disrupted planet. This planetary body formed under processes similar to those that formed the earth and other planets. For unknown reasons it broke apart, forming fragments that continue to move in the solar system to this day. Such a planet might well have occupied the orbit between the inner and outer planets, where a large number of small planet-like masses called asteroids are found. Other scientists think meteorites are left over, remnants of the materials from which the inner planets formed.

Armed with this background of knowledge about the sun, the planets, and meteorites, we can now examine some of the ideas that have been advanced to explain how the planets formed.

4.2 THE FORMATION OF THE PLANETS

Most theories of the formation of the solar system fall into one of two groups—evolutionary or catastrophic. According to the **evolutionary theories,** the planets and sun developed from a single mass of materials undergoing a progressive evolution. The **catastrophic theories** suggest that the formation of the planets resulted from some chance incident. The differences between these two theories have an important implication: according to evolutionary theories, planets should be found around other stars; but, if a catastrophic theory is correct, planets might be unique to our solar system.

Early Evolutionary Theories

The first seeds of modern views about the origin of the solar system are found in the writing of the French philosopher René Descartes (1596–1650), who envisioned the solar system developing from a vast cloud of gas within which eddy-like vortices swirled. He proposed that the sun condensed at the center of one of these large gaseous vortices, and that the planets condensed at smaller vortices radially distributed around the sun.

The German philosopher Immanuel Kant (1755) and the French mathematician the Marquis de Laplace (1796) independently developed rather similar evolutionary models for the solar system. Both conceived the system as initially consisting of a large, flattened rotating cloud of hot gases and particles surrounding the sun—a nebula similar to the one in Fig. 4.8. They pointed out that gravitational attraction would cause contraction in the denser portions of the nebula and that the contracting mass would rotate even more rapidly, just as a spinning ice skater does when arms and legs are brought closer to the body. This rapid rotation would tend to cause the contracting mass to flatten eventually into a disc-shaped body. Kant envisioned the formation of a number of secondary concentrations of condensing matter within the nebula, which would become planets, but Laplace concluded that rings of matter would be left around the outer edge of the contracting nebula as its rate of rotation increased. Later the matter in these rings would condense and coalesce into the planets and their satellites.

Early nebular hypotheses failed to explain why the angular momentum of the solar system is today concentrated in the planets and not in the sun. The fact that the sun does not rotate fast enough to contain most of the system's angular momentum is incompatible with the idea that the nebula rotated faster as it condensed. This failure was a major factor prompting the exploration of catastrophe hypotheses, the best known of which was the **dynamic encounter hypothesis.**

Palomar Observatory

Figure 4.8
Many scientists think that when the planets were forming the solar system may have resembled a nebula similar to this spiral nebula in Virgo, seen edge on. The bright mass in the center is similar to the sun; the planets formed in the ring of dark, cooler material.

Catastrophic theory—The dynamic encounter. The idea that planets formed as a result of the near collision of the sun and a comet was proposed in 1749 by the French philosopher Buffon. We know now that comets are of too low density and mass to produce the effect Buffon described, but a more massive body might. In the early 1900s, an American geologist, T. C. Chamberlin, and an astronomer, F. R. Moulton, elaborated the idea in what is known as the dynamic encounter theory. According to this theory, gravitational attraction of an approaching star caused tides or long filaments of hot gases to be pulled out from the sun and given a rotational motion as the star swung past the sun.

According to this theory, the gaseous matter in the filaments condensed into solid particles and gradually came together to form planets after the star had passed. Chamberlin and Moulton proposed that two filaments similar in size and shape developed and that in each of these filaments planets of similar size formed at about the same distances from the sun. They concluded that the size of the planets should be a function of the amount of matter distributed in the filament at different distances from the sun. Thus, their theory explains the occurrence of pairs of planets such as Mercury and Mars, Venus and Earth, Jupiter and Saturn, and Uranus and Neptune, of approximately the same size and in adjacent orbits around the sun. The rotation of the filaments of gas would also account for the concentration of angular momentum in the planets rather than in the sun itself. Since matter did not gradually concentrate in the sun according to the hypothesis, it does not require a rapidly rotating sun.

Chamberlin and Moulton's theory had some major difficulties that no subsequent modifications have managed to solve. One significant unanswered question was why the material, once separated from the sun, did not expand and dissipate in space rather than contracting and condensing to form planets. Also, the near approach of stars was viewed as a chance occurrence with extraordinarily low probability. (We now know that some stars are actually pairs of stars relatively close together in space.) Analysis of the forces needed to draw out filaments of gases indicate that the sun and the approaching star would almost have to touch each other. In order for the star to separate and keep moving away from the sun, it would have to approach the sun at a velocity greater than is known among any existing stars. For these reasons, the theory steadily lost favor, and revised versions of the evolutionary theory gained acceptance.

The modern nebular hypothesis. Most scientists now believe that the sun and planets formed from a nebula, and that the present configuration of planets is a natural outgrowth of the evolution of that nebula. Most of the original mass of that cloud of matter is now concentrated in the sun. Just how this concentration occurred is not clear, but similar central concentrations of matter surrounded by disc-shaped clouds of gases and dust are familiar sights to astronomers who study star formation in other parts of our universe (Fig. 4.9).

If our sun and planets did originate from a single nebula, it seems likely that the materials in that cloud would have started out being uniform in composition, although the matter was concentrated near to the center. If this is correct, we must explain what events in the evolution of the nebula ultimately brought about the marked contrast between the inner terrestrial ("earth-like") planets, largely composed of silicates and iron, and the outer giant planets, composed mainly of ices of water, ammonia, and hydrocarbon compounds such as methane.

If the solar nebula was originally of uniform composition, the present planets must be composed of only a small fraction of their original, much larger mass of material. Assuming that the nebula originally had the same composition as the sun, we can calculate the amount of each element needed by each planet to re-establish the proportions of the elements found in the sun. For example, the inner planets, which are highly deficient in hydrogen, helium, methane, and ammonia, would require large amounts of these to balance the huge proportions of silicates and heavy metal they now contain. To account for the quantity of these latter materials present in the inner planets, an initial cloud 500 times the present mass of these planets would be necessary. In contrast, Jupiter could be formed from a cloud only five times its present mass. Thus, if the nebular hypothesis is to fit the solar system as we now know it, it must include some process by which mass—especially composed of light elements—can be lost from the solar system or drawn back into the sun.

Astrophysical evidence indicates that nebulae are initially cool and very diffuse clouds of matter with only a few atoms per cubic centimeter. However, they begin to contract as the matter in them collapses under gravitational forces. The density of matter increases tremendously during the process, so nebulae are exceedingly hot when stars form; after that, they cool very slowly. Temperature, pressure, and the character of the materials in the nebula determine how the nebula evolves. If the solar nebula were hot

Palomar Observatory

(a)

Figure 4.9
The nebulae seen here are viewed (a) from above the disc shaped spiral nebulae in *Cares Venatici* and (b) at an oblique angle to the Great Nebula in *Andromeda*.

Palomar Observatory

(b)

enough in the initial stages, it was entirely gaseous. When temperatures in the solar nebula dropped to 1660 °C, iron could condense, forming small droplets of grains. Condensation from gases generally takes place on some particle that serves as a condensation nucleus. If no nuclei for condensation are present, temperatures may drop well past the normal condensation point without any droplets forming. Under these supercooled conditions, once growth does begin, it is rapid, and the structure of many iron-rich meteorites suggests that, in fact, they did crystallize rapidly. When temperatures dropped to 1400 °C, silicates could also condense probably as coatings on the iron. We might expect heavy high-temperature condensates, such as iron and silicates, to be concentrated in the inner part of the nebula where temperatures were high and where the density of matter was greatest. It seems probable that since meteorites contain mainly materials that condense at temperatures above 600 °C, they originated in a position relatively close to that of the inner planets. Lower temperature condensates formed initially at greater distances from the sun.

Once the process of condensation began, the solar system consisted of a star, the sun, surrounded by clouds of droplets formed by condensation of gases. Cooling of the gases speeded up as a cloud of hot condensates formed, blocking out sunlight in the more distant reaches of the nebula. Far away from the sun, temperatures were low enough (−70 °C) to allow crystallization of ices not only of water but also of methane and ammonia. It was from these ices that the outer planets are thought to have formed. Presumably, the inner planets are not composed of these ices because most of these constituents were either driven off by solar radiation before the planets formed or were later transformed to gases and lost.

After condensates of silicates, metals, and ices had formed, growth presumably changed from processes of condensation to one of lumping together of condensed particles, a process known as **accretion** (Fig. 4.10). At this stage, the small grains of ices and silicates moved within great cells of rotating turbulent gases, rather like wind-driven snow. The whole mass of gases and particles moved rapidly, but because they moved in the same direction, the relative velocity between particles was not great. This is important because growth by accretion requires the sticking together of solid particles that would be fragmented by any collisions at high velocity.

Accretion of grains of hot silicates and metals might well have started in the region of the inner

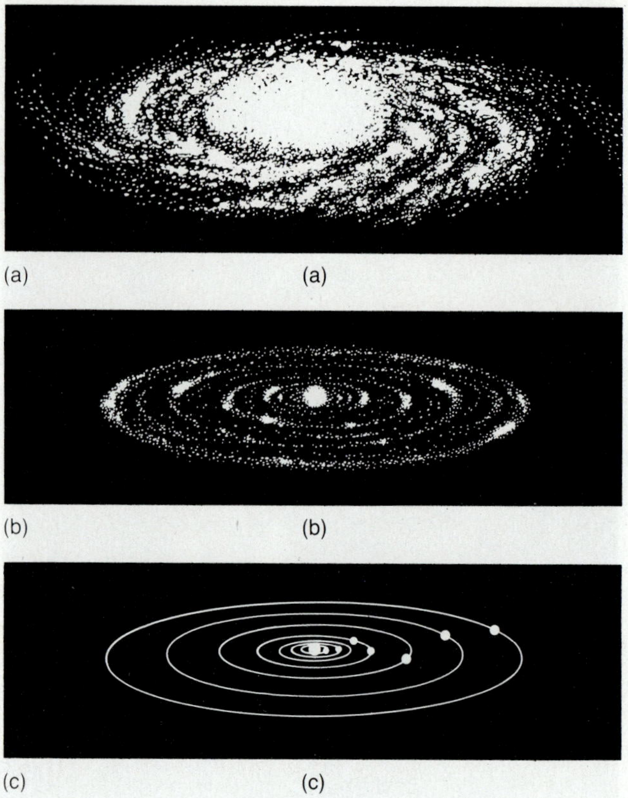

Figure 4.10
Stages in the formation of planets. (a) Condensing matter distributed in a flattened disc around the sun. (b) Bodies grow by the accretion of material. A cloud of droplets and ices remains at this stage. (c) The nebula begins to clear as the larger planets sweep up other smaller bodies.

planets while the particles were still hot enough to weld together on contact. At a greater distance, where the temperatures were lower, the ices could stick together as wet snowflakes do to form rafts. In either case the particles slowly grew larger.

Rates of growth by accretion have been calculated on the basis of collision rates. Current estimates indicate a growth rate of .001 cm/year, which would produce a body ten meters in diameter in one million years. As these larger bodies, called **planetesimals,** formed, they moved toward and became concentrated in a plane (Fig. 4.10). Eventually, most material from the nebula was concentrated either in the sun or in more or less solid planetesimals. At this stage, the nebula began to clear. Once sunlight penetrated the cloud, the remaining light elements, like hydrogen and helium, were driven off the inner planets by the pressure of the solar radiation, and ices on the innermost bodies evaporated. Thus, the inner planets lost their lighter and more volatile constituents, which were driven farther out in the solar system.

Another critical phase in the evolution of the system was reached when the planetesimals—now freed of the turbulent clouds that spawned them—grew large enough that their gravitational fields could effectively attract material. Growth at this stage was rapid because planets above a critical size would sweep up the smaller bodies moving in nearby orbits. Some of these bodies presumably grew more rapidly than others. These larger bodies soon dominated the orbital zone in which they moved. Initially, many smaller planets fell into the dominant ones, and the resulting impact created the large craters we now see on the moon, Mercury, Venus, and Mars (Figs. 4.6 and 4.11; *see also* Fig. 1.4). This growth process was probably completed in the first billion years of the solar system's history. Thus this theory predicts that most of the Martian and lunar craters will prove to be about the same age, and indeed lunar studies already indicate that this is true on the moon. Presumably, the earth's surface once bore similar markings as its most distinctive features.

The impact of two planetesimals could either lead to the fusion of the two into a single body—as appears to be the usual condition—or, if the momentum of the impact was great enough, one or both of the bodies might be broken up. It is especially interesting here to note that the belt of asteroids located between the inner and outer planets is in an orbit predicted by astronomers for a planet. It has been suggested that the planetesimals in this orbit were accreted by the gravitational influence of the giant planet Jupiter, and that they collided at velocities too great for fusion. Jupiter may also have had another important effect—that of disrupting the orbits and thus ejecting from the solar system those planetesimals that came close but not close enough to fall into it.

The systems of satellites we now find around the planets have been explained both as a result of the capture of smaller planets by a larger one moving in a nearby orbit and as a result of the splitting off of a part of a larger planet to form a satellite. We will examine the evidence for these ideas with special interest in the relationship of the earth and its satellite, the moon.

NASA

Figure 4.11
The existence of impact craters on the surface of the moon has been known for decades, but the presence of similar craters on the surfaces of Mars, Mercury, Venus, and many of the satellites of the outer planets is a recent discovery. This vertical view of Mercury resembles the surface of the moon, as shown in Fig. 4.12. (*See also* Fig. 1.4.)

4.3 ORIGIN OF THE EARTH

A variety of geologic evidence, which we shall review in the later chapters, suggests that the earth has a nickel–iron core covered by a silicate mantle and a highly differentiated outer lithosphere about 100 kilometers thick. If the earth did grow by accretion of cool planetesimals, as the modern nebular hypothesis would have it, it must have passed through a later stage when it was sufficiently molten to allow the heavy nickel–iron compounds to separate from the lighter silicates. Such a phase is needed to account for the fact that the earth is made up of concentric zones that are stratified according to density (see Fig. 3.5).

An alternative theory accounts for the stratification of the earth's interior, postulating that the earth formed at the same time as condensation of the materials in the nebula. According to this hypothesis, initial accretion of solid materials occurred while nebular temperatures near the earth were about 1200 °C and nickel–iron was condensing. As the earth grew, it became hotter as a result of the impact of material falling into it. The nickel–iron condensate of which it consisted formed a molten surface layer. As the nebula cooled further, olivine and pyroxene started to condense and fall into the growing earth, but liquids composed of metals did not mix with silicate melts. Because the metals are heavier, a sharp boundary formed between core and mantle. During this high-temperature stage at the earth's surface, materials that are easily made volatile were given off to become part of a primitive atmosphere; and the lightest constituents were lost to space. Only when the temperatures on earth began to drop did many of these constituents become part of the outer most layers of the earth.

According to both of these theories, the earth was molten at an early stage in its history. Thus, all rocks on the earth are in a sense derived ultimately from rock melts and from the materials that crystallized near the surface during this earliest and most obscure stage in the earth's history. To be sure, these primary materials have been altered, recycled, and greatly affected as conditions on earth changed, and especially as the atmosphere, water, and living organisms formed and evolved.

The accretion of the earth was completed long before the oldest known earth rocks were formed. We have been unable to find rocks older than about 3.9 billion years; and in every case these oldest rocks have been recrystallized, apparently by high temperatures. This fact suggests to some geologists that the earth underwent a major thermal event that altered all earlier rocks with the effect that the radioactive minerals used to determine ages were recrystallized, making it impossible to find radioactively any older dates. If this is correct, we will never be able to decipher, by conventional geological techniques, the interval of time before this strong heating occurred. But fortunately, rocks older than 3.9 billion years have been discovered on the moon, and their evidence can be used to infer conditions on earth.

The Moon: A Guide to Early Earth History

Study of the moon rocks brought back to earth by Apollo astronauts and of the landforms seen on the moon have enabled geologists to outline the major events in the history of the moon. Four major stages in lunar history have been identified:

1. Formation and early melting.
2. Formation of a lunar crust.
3. Formation of the lunar seas, the maria.
4. Formation of new impact craters, minor flows, and some volcanoes.

Initial formation and early melting. How the moon originated is still uncertain, despite the tremendous efforts directed to finding an answer to this question. Several possibilities are being actively investigated. According to one, the moon formed from materials derived from the earth before the outer part of earth became as rigid as it is now. Some envision the moon breaking away from a great tidal bulge on earth. A second alternative is that the moon and the earth formed about the same time, but the materials of the moon were never part of the earth itself. A third possibility is that the moon formed separately from the earth, essentially as a separate planet that was later captured by the earth.

Although it is still too early to say which theory is correct, a number of geological constraints can now be placed on these theories. One is related to the age of the moon, which is thought to be approximated by the age of the oldest lunar samples we have collected. These samples have been dated radiometrically at 4.6 billion years. This figure is close to the age of most meteorites, and it is the estimated age of the earth and the solar system. It would be a remarkable coincidence if the moon originated simultaneously with the rest of the solar system, but outside it. Secondly, a considerable amount of evidence indicates that the moon formed in a relatively short period of time. Current estimates of the length of time involved range from 200 million years down to a few hundred years. Thirdly, the moon almost certainly either formed as a hot body or became hot enough to melt rock at an early stage in its history. The evidence for this conclusion is that the oldest rocks found on the moon have low gas content, suggesting that the gases were lost from the lunar crust as a result of high temperatures. A fourth constraint on the theory of origin of the moon is found in the observation that the moon rocks bear a strong resemblance to iron- and magnesium-rich igneous rocks found on earth. This supports the view that either the earth and moon originated close to one another or they were a single body that split at an early stage.

Formation of the lunar crust. The second stage of the moon's history is also obscure—not because lunar rocks of this age are rare but because they are exposed in the most rugged and least-explored portion of the moon—the highlands. The highlands are composed of the primitive lunar crust, characterized by rough, mountainous, cratered terrain. It was in this severely battered crust (dated at 4 to 4.5 billion years) that the large lunar seas, called the **maria** (singular form, **mare**), were formed. This old crust is composed of

Yerkes Observatory

Figure 4.12
The complex cratered terrain of the lunar highlands is emphasized by oblique lighting on the moon's surface. The oldest rocks on the moon are in the highland regions.

Mount Wilson

Figure 4.13
This view of the moon shows Copernicus crater at the bottom and the vast nearly flat surface of a lunar sea, Imbrium. Highland areas lie around the edge of the sea.

rocks like some found on earth known as basalt, gabbros, and anorthosite (a rock composed of plagioclase feldspar). The primitive lunar crust was subjected to intense bombardment by falling material, creating the closely spaced, sometimes overlapping craters shown in Fig. 4.12. Although much of the old crust was broken up by these impacts, some relatively smooth areas remain. This second stage was obviously one of accretion as many large bodies fell into the moon, creating huge impact basins destined to become the sites of the maria.

Formation of the lunar seas. The formation of the maria—the vast seas of iron-rich basaltic lava flows similar to those that compose the Hawaiian Islands—was one of the most dramatic events in lunar history. These lavas filled large impact craters and covered parts of the earlier topography (Figs. 4.13 and 4.14). They provide a valuable benchmark by which to distinguish older landforms and rock structures from those that formed after the mare basalts solidified.

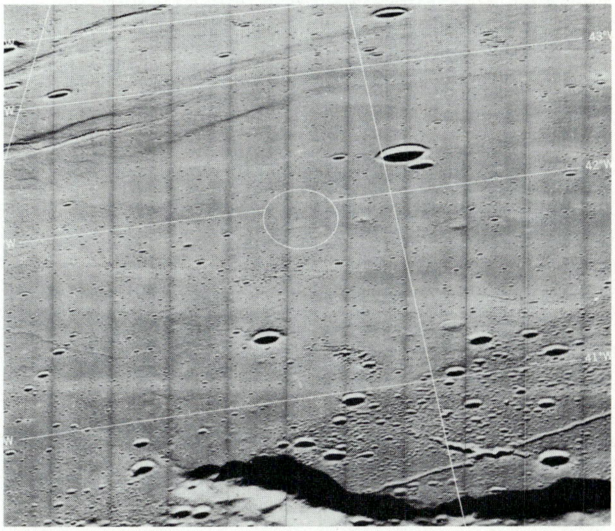

NASA

Figure 4.14
This low oblique view across a lunar sea shows irregularities caused by young craters and ridges, which may mark the edges of lava flows or breaks in the surface.

These lavas are so vast and continuous that they must mark a major event in which large volumes of the outer part of the moon were partially melted. This melting may have been caused by meteorite impacts, which removed enough of the outer crust to reduce pressure and thus melt underlying rock. The temperature of similar lavas in Hawaii is about 1200 °C. Because these flows were hot and fluid, they spread evenly over immense areas, filling the previously formed impact craters. Dating of lava flows from the Sea of Tranquility indicate that these events took place about 3.2 to 3.7 billion years ago.

Post-mare history. Most of the major features of the moon landscape had formed by the time the last of the mare lava flows filled in the remaining low areas. However, some meteorites continued to fall into the moon, producing the youngest and many of the most perfectly shaped impact craters; craters with radiating ridges, called rays, such as the Copernicus crater (Fig. 4.15), are among these. Some volcanism accompanied these last major impacts, and faults formed in and around the mare. Although the infall of large fragments became rare, many small meteorites have continued to fall onto the moon just as they have on earth, but the resulting effects on the two landscapes are dramatically different. Few scars or traces of meteoroids remain on earth because most of the impacts occurred many millions of years ago and have long since been altered beyond recognition by the weathering processes that decompose and break down rocks when they come in contact with the atmosphere. In addition, impact features are modified by streams and glaciers; and, of course, many meteorites have fallen into the oceans. In contrast, on the moon, impacting meteorites crushed surface rock into a soil-like veneer of dust and fragments. This covering has remained almost unchanged since its creation because the Moon has no atmosphere, streams, glaciers, or oceans that act upon it.

Following what most geologists believe was a similar early history, the moon and the earth evolved in strikingly different ways. In the next chapter we will consider how these differences came about.

NASA

Figure 4.15
Photomosaic of lunar crater Copernicus and associated ray pattern and secondary craters. Copernicus is about 100 km in diameter.

SUMMARY

Much of the evidence used to deduce the origin of the earth and the rest of the solar system comes from study of the sun, meteorites, and the other planets. The sun is a large, extremely hot cloud of incandescent gas composed of dissociated electrons and atomic nuclei. Its atmosphere is of three parts: the photosphere, in which most light is generated; the outer chromosphere; and the corona. The sun is composed almost entirely of the two lightest elements, hydrogen and helium, all other elements combined making up less than one percent.

According to composition and physical characteristics as well as location, the planets fall into two classes. The small dense terrestrial planets—Mercury, Venus, Earth, and Mars—travel in the orbits nearer the sun. The giant outer planets—Jupiter, Saturn, and Neptune—have very different compositions and atmospheres.

The sun and planets form a disc-shaped array in which most motion of revolution and rotation is in the same direction and angular momentum is concentrated in the planets.

All the meteorites that fall to earth are of about the same age, 4.3 to 4.7 billion years, our best estimate of the earth's and the moon's age. Metallic meteorites (siderites) are believed to have the same composition as the earth's core, and the stony ones (chondrites) are composed of the same minerals that predominate in the Earth's mantle. All these data suggest similar processes formed both the earth and the meteorites. According to one widely held theory of origin of the solar system, a gaseous nebula gradually evolved into the sun and planets. As the nebula cooled, first metals and then lighter silicates condensed in particles.

These particles grew by accretion into planetesimals, some of which eventually became large enough to attact matter by gravity, thus sweeping up and accreting smaller bodies. In its early history, the earth was a hot body composed mainly of nickel–iron condensate from the nebula. As the nebula cooled, silicates accumulated as an outer layer.

According to evidence from moon rocks, the moon is of about the same age as meteorites and the earth and is composed of material much like the iron- and magnesium-rich igneous rock on earth. The moon's surface shows scars of many of the impacts of falling planetesimals that accreted to form it. Some of the impact craters later filled with lava flows of similar composition to those found in Hawaii. The flat areas formed by these have left new craters and a soil-like veneer of dust and fragments that remains virtually unchanged today.

KEY TERMS

accretion
angular momentum
catastrophic theories
chondrite
chromosphere
corona
dynamic encounter hypothesis
evolutionary theories
mare (plural, maria)
meteorite
nebula
photosphere
planetesimal
prominence
siderite

STUDY QUESTIONS

1. What observations about the planetary positions and their patterns of movement around the sun suggest that they grew within the solar system rather than outside the system?
2. If the earth formed from the materials in the sun, why aren't the relative proportions of the elements in the two bodies closer in value?
3. Why should the ages of the meteorites be our best guides to the age of the planets?
4. Why are the outer planets so much less dense than the inner planets?
5. Why should the rocks on the moon contain valuable information not found on earth about the early history of the solar system?
6. How have geologists succeeded in establishing the relative ages of events on the moon?
7. What geological processes are shaping the surface of the moon at present?
8. If the earth and the moon originated in the same way, why are they so different at the present time?

SUGGESTED READINGS

Bok, B. J., "The Milky Way galaxy." *Scientific American, 244:* 3 (1981), 92–120.

Gurin, Joel, "In the beginning." *Science 80, 1:*5 (1980), 44–51.

Hartmann, W. K., "Moons of the outer solar system become real, although weird places." *Smithsonian, 10:* 10 (1980), 36–46.

King, E. A., *Space Geology: An Introduction.* New York: John Wiley & Sons, 1976.

Mutch, T. A.; R. E. Arvidson; J. W. Head; K. L. Jones; and R. S. Saunders, *The Geology of Mars.* Princeton, N.J.: Princeton University Press, 1976.

Wood, J. A., *The Solar System.* Englewood Cliffs, N.J.: Prentice Hall, 1979.

York, D., *Planet Earth.* New York: McGraw-Hill Book Company, 1975.

EVOLUTION OF THE EARTH AND FORMATION OF ITS SEDIMENTARY VENEER

Chapter 5

The earth's history has been one of constant change. Many of the changes take place so slowly as to be imperceptible, but the long-term record has been dramatic. The rocks of the earth preserve this record, and they tell many fascinating stories about the earth's continuing development.

Today most of the surface of the earth is covered by highly varied sediment and sedimentary rock (Fig. 5.1) but in the earth's earliest days, as the new crust was solidifying, the first sediments probably consisted mainly of fragments created as planetesimals pounded the hardening surface. Volcanic activity must also have spread a blanket of ash over vast areas. As the earth evolved, water condensed, rains fell, and streams and the first oceans began to form. These oceans became not only one of the most dramatic features of the earth's surface, but also repositories of the sediments that have accumulated throughout the planet's history, now constituting a rock record that geologists have gradually learned to read and interpret. This record contains most of the evidence we have concerning the origin of life and its gradual evolution; but the history of life on earth is but one of the stories in the rock record.

Geologists have also learned to reconstruct ancient environments by comparing the texture and composition of present-day sediments with those laid down in ages past. By making comparisons and correlations we can deduce what the earth was like in the past from our understanding of the environments in which sediments form today. We can, for example, identify old stream channels by recognizing

Union Pacific

Figure 5.1
Layers of sediment deposited in a lake millions of years ago are now exposed in the strange eroded forms seen here in Bryce Canyon, Utah. These are among the varied sediment types found on earth but not on the moon. Note the trees for scale.

the types of sediments left in those channels. Similarly, ancient deltas, seashores, swamps, and deep-sea floor can be recognized. Even mountains long since eroded away leave their imprint in the sediments that settled in the bodies of water around them.

Our primary objective in this chapter will be to understand the physical and chemical processes that lead to the formation of the various sedimentary rocks. In this context we will consider the major changes that have influenced the development of the earth's sedimentary veneer and some of the environments where sediments are being deposited today. Finally, we will consider how the sedimentary rock record has been interpreted to enable geologists to construct a geologic time scale.

5.1 EARLY STAGES OF EARTH'S HISTORY

Before the Oldest Rocks Formed

According to radiometric dating techniques, the oldest rocks so far found on earth are nearly 4 billion years old; yet the meteorites were formed 4.5 to 4.7 billion years ago, and the oldest rocks found on the moon also fall in the 4.5 to 4.7 billion year range. These ages strongly suggest that, in the region of the inner planets, conditions favorable for crystallization of the materials existed about 4.5 billion years ago. By that time the earth was probably a distinct body with at least a solid outer region or lithosphere. Because none of the rocks formed during these early phases of earth's history have been preserved near its surface (at least we have never found any) we are unable to make direct inferences about the conditions on earth during that time, approximately a quarter of the entire history of the planet. This early unrecorded era is often referred to as "pregeological."

Fortunately, we have a record of this time interval preserved on the moon. That record should serve as a reasonable indication of some of the events which took place on earth during its early history, especially if the earth and moon formed as close in space as they are now. Based on our knowledge of the moon, we postulate that the initial crust on earth was composed almost entirely of solidified igneous rock. It seems likely that this crust was forming while the planet was growing by the accretion of falling planetesimals. The impact of those bodies broke through the crust, shattered it, and forced fragments down into the molten interior. The high-velocity impacts must have created a layer of crushed rock and dust, such as that still present over much of the surface of the moon.

One of the early concerns about exploration of the moon centered on the possibility that its surface might be covered by a thick layer of fine dust in such an unconsolidated state that a spacecraft would sink into it and be buried on landing. Of course, the soil did not turn out to be that soft, although it is unconsolidated and, in some places, several meters thick. It did prove to be formed in the way we anticipated—by the crushing impact of meteorites, which, over a period of millions of years, pulverized previously formed rocks.

The lunar soils formed by impact are generally mixtures of small bits of iron meteorites, fragments of basalt and gabbro, and some fragments that cooled so quickly that they lack any crystal structure. A solid of this kind is a **glass,** and its texture is said to be glassy. The dust contains fragments of **microbreccia** (a rock consisting of fine angular fragments) mechanically mixed by the impact of meteorites. The glasses appear to have been formed as a result of local melting with splashing, evaporation, and condensation following impact. They occur as small spheres, as stringlike pieces of glass in intensely fractured mineral fragments, as frothy rope-like glass spattered on mineral fragments, and as glass linings on depressions of the kind seen in Fig. 5.2. Some glass forms show evidence of having suffered later impacts that shattered the original spheres and other forms. Many of the moon rocks brought back by the astronauts formed as a result of high-velocity impact. From photographs like that of Fig. 5.3, we now know that the surfaces of Mars and Mercury are also covered in places by fragmental rocks, presumably formed in the same way.

If we assume that the earth's surface was once bombarded as heavily by meteorites as the moon has been, then the crust of the earth must have been badly broken up and many of the rocks fragmented just as those on the moon are today. Clouds of hot gases rose from the thin crust and especially from craters and cracks formed by impacts. These breaks through the crust allowed the molten rock below the surface to flow out as hot lava. On the moon, gases produced in this way drifted off into space because the moon's force of gravity is too weak to hold them. Some gases escaped from the earth, too, but others must have accumulated to form a primitive atmosphere. They may have had a composition similar to the gases that escape today from volcanoes like those

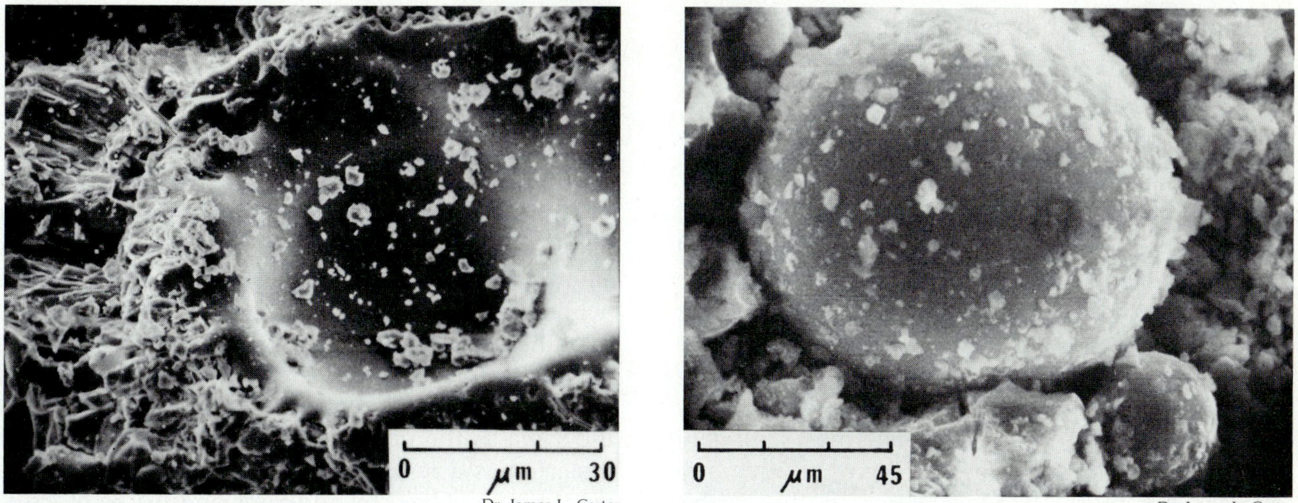

Figure 5.2
Scanning electron micrographs of soil samples collected on the moon. (a) A glass-lined hole with splashed rim and glass fragments. (b) A sphere with adhering dust fragments in an angular microbreccia. One micron (μ) is one millionth of a meter.

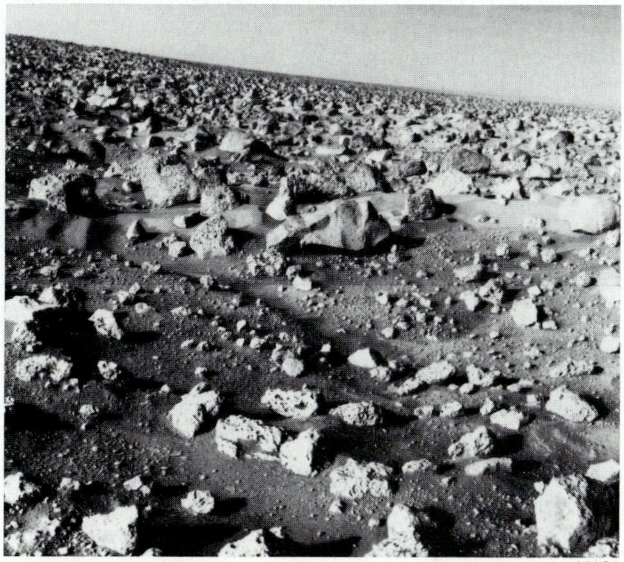
NASA

Figure 5.3
Shattering impacts and volcanic explosions are believed to have formed these rock fragments on the surface of Mars. Large rocks in the foreground are 10 to 20 centimeters across.

Table 5.1 Hawaiian Volcanic Gases Include the Constituents Shown Below.

		Percent of total volume
Water vapor	H_2O	67.8
Carbon dioxide	CO_2	12.7
Nitrogen	N_2	7.65
Sulfur dioxide	SO_2	7.03
Sulfur	S_2	1.04
Hydrogen	H_2	0.75
Carbon monoxide	CO	0.67
Chlorine	Cl_2	0.41
Argon	A	0.20

in Hawaii (Table 5.1). When the temperature of this early atmosphere cooled enough for water to condense at the then-prevailing atmospheric pressure, rains fell on the hot crust, eventually accumulating as large bodies of water. Even at this early stage, the earth's evolution was markedly different from that on the moon, where all gases were lost and no rains fell. The role of the atmosphere in the evolution of the earth's surface is clearly so important that we will now briefly consider some theories of its origin.

Origin and Evolution of the Atmosphere

Two contrasting hypotheses have been advanced to explain the origin of the earth's atmosphere. According to one, the present atmosphere is a residue of gases that enveloped the earth when it formed. The second, and more widely held, hypothesis postulates that the atmosphere is the result of **degassing** of the earth's interior. Both hypotheses recognize that the atmosphere has changed as a result of chemical reactions with the solid earth and, as organic materials evolved, of biochemical processes.

The Primitive Atmosphere

If the earth formed in a gaseous state as is proposed by the **dynamic encounter** theories, the primitive atmosphere should have been similar to that of the sun. Although the elements found in the sun are known on the earth, the relative abundance of the gases in the earth's present atmosphere is notably different from that on the sun. The most abundant elements on the sun are hydrogen, helium, and oxygen; the most abundant gases in the earth's atmosphere are nitrogen, oxygen, argon, and carbon dioxide (Table 5.2). Hydrogen and the inert gases (helium, xenon, and krypton) are all far less abundant in the earth's atmosphere than they are on the sun.

The amounts of hydrogen and helium in the sun greatly exceed all other elements in abundance. Thus, a large quantity of these two elements was presumably present near the earth when it formed. Because both hydrogen and helium are too light to be held to the earth by its gravitational field, they would have been lost from the atmosphere. Nevertheless, as the earth cooled compounds, such as water, containing hydrogen did form.

Table 5.2. Composition of the Earth's Atmosphere.

		Percent by volume
Constituents that are nearly constant in their relative abundances	nitrogen	78.1
	oxygen	20.9
	argon	.9
	carbon dioxide	.03
Constituents that vary in abundance	water vapor	
	ozone	
	sulfur dioxide	
	nitrogen dioxide	
	carbon monoxide	
Constituents that occur in trace amounts	neon	
	helium	
	krypton	
	xenon	
	hydrogen	
	methane	
	nitrous oxide	
	radon	

Hydrogen-containing compounds, ammonia (NH_3), and hydrocarbons such as methane (CH_4) are also presumed to have been present in the primitive earth's atmosphere, because these are still abundant as ices in the atmospheres of several of the largest planets of the solar system. Both hydrocarbons and ammonia are stable in the presence of hydrogen and helium. Thus advocates of the "residual" atmosphere hypothesis postulate that it contained hydrocarbons, ammonia, nitrogen, water vapor, hydrogen, and helium. It is clear that our present atmosphere is not the same as the postulated primitive atmosphere. If it started as a residual atmosphere, the atmosphere must have evolved by additions and losses of gases over time.

Opponents of the residual atmosphere hypothesis point to the low concentration of inert gases in the earth's present atmosphere compared to the concentrations on the sun. If the earth formed by the accretion of planetesimals, as described in Chapter 4, it probably never had very large concentrations of hydrogen and helium, since these would be lost even more readily from small planets than from large ones. Even heavier inert gases would be lost from planetesimals, and this is the most probable explanation for the relative deficiency of krypton and xenon on the earth. Both of these inert gases are thousands of times less abundant on earth than are some of the nongaseous elements that are close to them in weight.

According to the more widely accepted degassing hypothesis, our atmosphere is primarily composed of gases emitted from volcanoes. Measurements around modern volcanoes, such as the one at Mount Aso (Fig. 5.4), indicate the main gases produced are water vapor, carbon dioxide, nitrogen, sulfur dioxide, carbon monoxide, hydrochloric acid, and some argon. The first three of these are major constituents of the earth's atmosphere, but we must look elsewhere to explain the origin of one of the most abundant and important gases, oxygen.

The evolution of the atmosphere. At the present time the earth's atmosphere is composed of the following constituents (percent by volume): nitrogen 78 percent, oxygen 20.9 percent, argon 0.9 percent, carbon dioxide 0.03 percent. No other constituent makes up as much as 0.01 percent. The nitrogen in the atmosphere is easily explained by either the residual atmosphere or degassing theory: it is present in meteorites and in volcanic gases, and it can be

Japan National Tourist Organization

Figure 5.4
Volcanic gases are thought by many geologists to be the main source of our atmosphere. Volcanoes continue to add their contents to the atmosphere. Gases can be seen here emanating from the crater of Mount Aso in Japan.

derived from the decomposition of ammonia. But neither theory accounts for the large amount of oxygen. Oxygen does not appear to be a residue from a primitive atmosphere, nor is it a volcanic gas.

The first free oxygen in the atmosphere was probably formed when water molecules were split by the action of solar radiation in the upper layers of the atmosphere, a process called **photochemical dissociation.** This process is taking place today at a rate which, if continued throughout geologic time, would provide several times the amount of oxygen presently found in the earth's atmosphere. Of course, much of the oxygen produced in the past has been incorporated in oxygen compounds in the minerals of the earth's crust. Oxidation of iron-bearing minerals is the most significant chemical reaction, removing oxygen from the atmosphere and incorporating it in minerals.

Photosynthesis by green plants is a second source of oxygen content in the atmosphere. This process may be responsible for maintaining the present oxygen content of the atmosphere, but it almost certainly did not account for oxygen in the primitive atmosphere, because the plants producing oxygen by photosynthesis were not present in abundance until much later in the earth's history—well after the atmosphere became oxygen-rich. Oxygen-producing plants became significant contributors to the earth's atmosphere about 2 to 3.5 billion years ago. By 2 billion years ago, the quantity of oxygen in the atmosphere was sufficient to cause oxidation of iron compounds. Before that time, the atmosphere contained so little oxygen it was a reducing atmosphere (one which lacks sufficient oxygen to cause oxidation).

Through geologic time many interchanges have taken place between the atmosphere and the earth's crust, mainly through what are called **weathering processes.** The atmosphere and rocky crust of the earth react, and these reactions tend to remove certain constituents from the atmosphere as they combine to form compounds. For example, when the amount of carbon dioxide in the atmosphere reached a certain critical level, a complex series of reactions began to take place in which carbon dioxide and water reacted with the minerals of igneous rocks. Some of the liberated constituents of the igneous rocks, especially calcium, recombined with carbonate ions in water to form limestone. The net effect of reaction was to remove CO_2 from the atmosphere and tie it up in the lithosphere. The absence of large quantities of limestone in the oldest known rocks on earth suggests that the atmosphere did not originally contain very much carbon dioxide. Another mechanism for extracting carbon dioxide from the atmosphere is through its incorporation in plants and some animals. The remains of these same organisms may eventually be transformed into coal, petroleum, or disseminated carbon in rocks.

The present atmosphere is clearly the product of an evolutionary process, which is continuing. Most geologists think it began with the emanation of volcanic gases from the earth's interior, and was dramatically changed by the addition of oxygen formed by photochemical dissociation of water vapor, by photosynthesis, and by interaction with the rocks of the earth's surface and with the waters of the oceans.

Origin of the Oceans

The origin of water in general and of the oceans specifically is closely related to the origin of the earth's atmosphere. Liquid water probably was present on earth as soon as the temperature of the surface cooled down to the critical temperature for water. At that time rains began to fall, and the face of the earth undoubtedly bore both the scars of impact craters

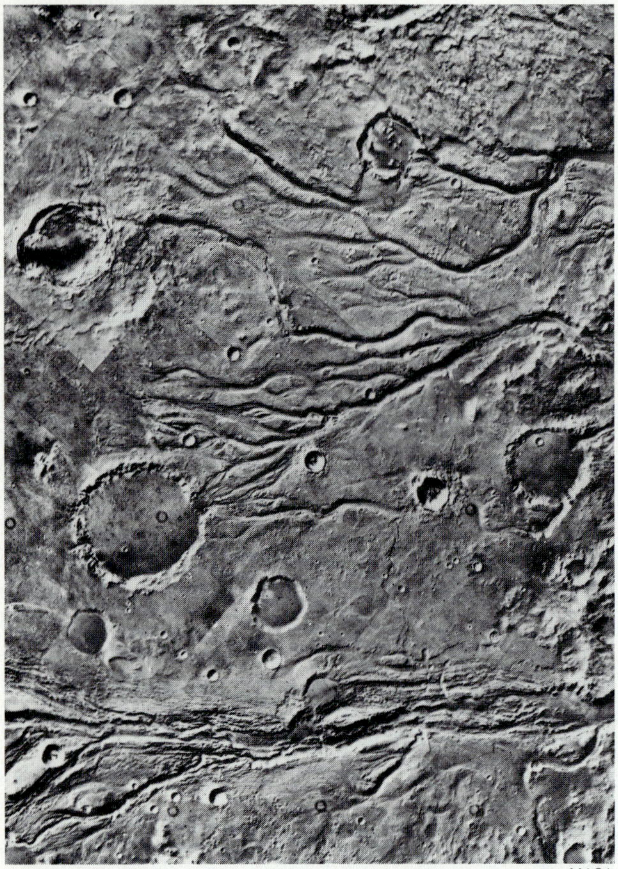
NASA

Figure 5.5
This photograph of the surface of Mars shows a type of topography that might well have existed on the earth early in its history, when the surface was marked by both impact craters and stream channels. Some impact craters are thus older than the channels; others are younger. Why water no longer flows in these channels on Mars is uncertain; possibly the water is now frozen and remains on the planet as ice.

and stream channels, a combination now seen on Mars (Fig. 5.5).

If the earth initially possessed a primitive atmosphere from which the ocean waters were derived, the oceans formed as the earth cooled may have been as extensive as those we know today. But if the more-favored idea that the atmosphere evolved from the degassing of the earth's interior is correct, the oceans accumulated gradually as water came out of volcanoes in the form of steam. The rate at which water was produced through degassing is uncertain, and very little evidence is to be found in the geological record for these early times. Large quantities of water were probably generated from the interior of the earth at first, while the crust was thin and volcanic activity widespread. Indeed, large quantities of steam continue to be brought to the surface of the earth through volcanoes today. Steam comprises about 98 percent of the total gas content of volcanic emanations.

Of these volcanoes, many are located near the margins of the oceans or within the ocean basins. Part of the steam that comes from them may be water that is being recycled, but even if no more than 0.8 percent of the water brought out of volcanoes today is new water derived from magma, this would, over geologic time, account for the amount of water now found in the oceans.

Unfortunately, we have no firm evidence to clarify the question of what those early oceans were like during the pregeological history of earth. Some geologists suggest that continent-like masses of low-density minerals formed during crystallization of the crust, producing relief on the earth's surface. If this assumption is correct, water undoubtedly accumulated in the low areas first, leaving some dry land as we have today. Others think that the crust was more nearly uniform in composition and thickness, and that the newly formed oceans covered almost the entire surface of the earth. Regardless of whether the ocean was universal or restricted in size, it was within this watery envelope that the earth's most distinctive feature, living organisms, originated and evolved.

Origin of Life

The origin of life on earth is certainly one of the most significant events in the earth's history. At present, we have no proof that life exists anywhere else in the universe, though the probability that it does seems great. Living organisms have significantly modified the effects of natural processses on earth, and this is nowhere more evident than in the sedimentary rock record.

A plausible early history of life on earth has been outlined, but no record exists to prove the details. This theory suggests that organic molecules evolved through a chemical reaction three to four billion years ago, and that the first extremely simple plants came into existence after that. During the next billion years, blue-green algae lived in the oceans. Although they produced oxygen by photosynthesis, the amount of oxygen present was not sufficient to oxidize sediments. About two billion years ago, more advanced organisms, the protozoans, became abundant, and at approximately this same time, the first significant amounts of limestone ($CaCO_3$) appeared on earth—

evidence of considerable amounts of calcium and carbonate ions in the oceans. Calcium ions are derived from weathering of igneous rocks, and carbonate is one of the ions formed when carbon dioxide dissolves in water. This carbon dioxide in the oceans presumably resulted from decay of organic matter. Thus, it seems probable that the buildup of organic material on earth started two billion years ago. The more advanced multicellular animals evolved slowly. They eventually became abundant, and a rich record of their remains, in the form of fossils, is preserved in rocks deposited during the last 600 million years of earth history.

Most fossils are preserved in sedimentary rocks deposited in the oceans. Organisms have played an extremely important role in the formation of sediments. Most limestones are formed by organic processes; all carbonaceous sediment such as coal is organic in origin, and bacteria have been important in carrying out the chemical reactions that formed certain sedimentary iron ores. The importance of organisms will become clear as we study the character of the earth's sedimentary veneer. Later we will see they are also critically important in analyzing the earth's geological history.

5.2 EARTH'S SEDIMENTARY VENEER

Today sedimentary rocks cover about three-quarters of the land surface to an average depth of 2300 meters (Fig. 5.6), and they are spread almost continuously across the ocean floors. Sediments contain most of the world's fuel and water resources. It is largely from these rocks that the history of the earth has been deciphered.

These history-telling sediments can be highly varied in both composition and manner of deposition. They may be derived from any of the following components: sands and gravels found in streams; sands, silts, or cobbles on beaches; muds on lake bottoms; decaying plants and mud in swamps; evaporite deposits in desert lakes; sands, dust, and ashes transported by the wind; debris from meltwater of glaciers; and calcite deposited from water dripping in a cave. If conditions prove favorable for their preservation, these sediments may be compacted and cemented to form sedimentary rocks. Those sediments deposited on land are likely to be eroded and ultimately transported toward the sea, where chances of burial and long-term preservation are much greater.

Sources of Sediment

From the beginning, the earth's oceans have been repositories for sediments produced by the breakdown of rocks and minerals exposed on the continents. The running water in streams moves materials toward the oceans in numerous ways. Some fraction of the stream's load is carried as dissolved materials in solution, but a much greater amount of fine fragmental materials is carried in suspension, while larger rocks roll and bounce slowly along the channel of the stream.

Petroleum Information Corp.

Figure 5.6
Stratified sedimentary rocks such as these cover most of the surface of the earth.

The early conditions that produced sediments on earth were strikingly different from those at work today. The impact of meteorites and volcanic activity produced large amounts of fine fragmental debris, and the land areas had no plant cover. Thus, these fine materials must have been easily eroded and subject to chemical alteration. From early times, rocks and minerals exposed to the atmosphere have been weakened and decomposed by chemical reactions with water and other atmospheric constituents. The precise reactions that take place are determined by the composition of both the mineral and the atmosphere. Collectively, these processes are called **chemical weathering.**

Many of the common minerals of igneous rock will react with water, especially if the water contains carbon dioxide. These reactions normally occur slowly and often leave a residue of insoluble clay minerals. Other products of the reaction remain dissolved in water and may be carried off to be deposited elsewhere. Silicon is such a product, and chemical weathering is one way silicon dioxide is liberated from the crustal rocks, allowing it to become concentrated as the mineral quartz.

Weathering of rocks also involves dissolving of soluble constituents. Minerals break down into their component ions when they dissolve—a process that is generally aided by the presence of acids or bases in the water. Because solution takes place at the surface of a solid, a greater exposed surface area means that solution can take place faster. Thus it is easier to dissolve a fine material than a coarse one, and solution is accelerated by mechanical breakdown of the material being dissolved. For most of the common rock-forming minerals, which are nearly insoluble, decomposition resulting from chemical reactions and mechanical disintegration are far more important than solution in breaking rock into sizes that can be moved by streams.

Types of Sedimentary Materials

The materials of which most sediments are composed can be classed as ions, organic materials, or fragmental materials. Ions are produced by the process of solution, and they often result from chemical reactions, which may leave them free to move in water. Organic matter, including waste products of organisms, solid parts such as shells, and plant remains, all may become part of the sediment. Some sediments, such as coal, are made almost entirely of just such remains of plants and animals.

Most fragmental materials are pieces of minerals or rock produced by the disintegration of preexisting rock. Volcanic ash and meteoritic particles fall in this category as well. Unlike most other sediments, they represent new additions of material to the crust. All such fragmental materials are referred to as **clastic.** Several examples are illustrated in Fig. 5.7.

The finest clastic materials are called **colloids.** This term refers to a particular size of particles—those in the range of 10^{-5} to 10^{-7} centimeters (0.00001 to 0.0000001 cm) in diameter. Colloidal particles that make up mud and clay are so small that they cannot be distinguished individually, even with a high-powered optical microscope. If clay is stirred in water until the particles move in suspension through the solution, the heavier and larger aggregates will settle out, but colloidal-sized particles will remain in suspension indefinitely and will move more or less at random through the solution.

Processes of Sedimentation

Settling of particles. Many sediments are transported suspended in either water or air and eventually settle out of the fluid medium. These solid particles may originate in many different ways. Sand and dust are blown from the ground surface by the wind; grains and fragments freed from rocks by weathering move in streams and ocean currents until they settle out. Extraterrestrial material, meteoritic dust, falls to earth through the atmosphere and sinks through ocean waters; volcanic ash and debris settle through air and often through water as well; clay and other colloids form aggregates in salt water and settle out; shells and other remains of animals or plants living in water settle when the organisms die; and fecal pellets, mainly invertebrate excreta, also settle through water.

Although much of the sediment carried by streams into the sea is deposited as layers in the shallow water of the continental shelves, some sediment also finds its way into the deep sea. Many of the particles that settle on the deep sea floor originate in the upper layers of the sea—for example, the remains of organisms that live near the surface where light penetrates the water. Both ash and clay colloids may be carried long distances and into deep water by slow but steady oceanic surface currents. Other sediments are transported into the deep sea as clouds of suspended sediment resulting from submarine landslides.

Chemical precipitations. Water is by far the most common solvent on earth. Ocean and lake water typically

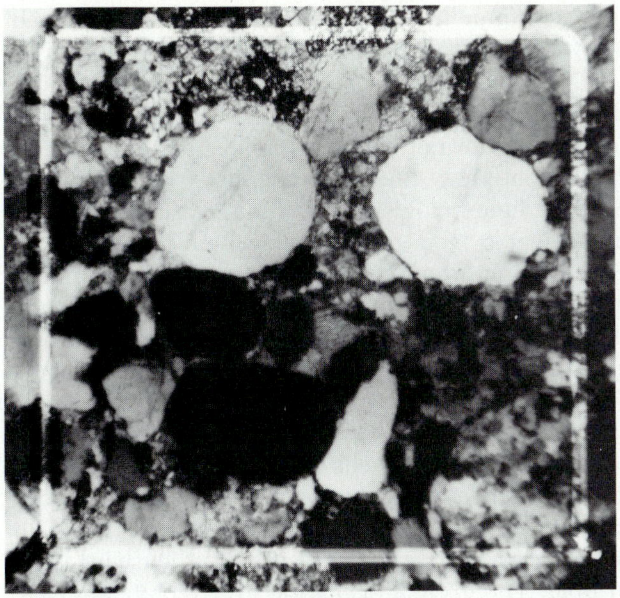

(a)

Bausch & Lomb

(b)

Smithsonian Institution

(c)

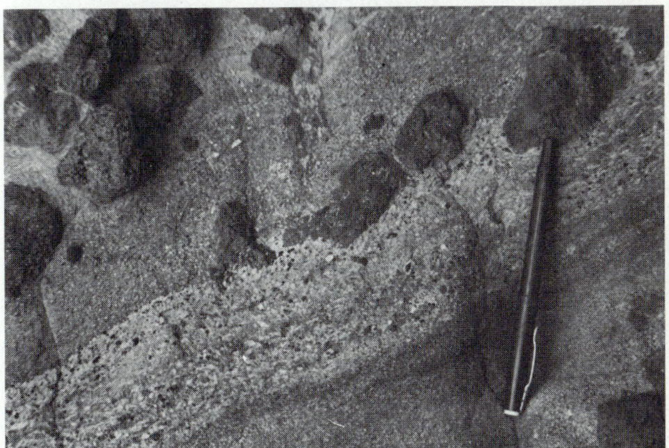

(d)

Figure 5.7
Examples of various types of clastic materials. (a) Micrograph of tightly cemented sandstone with grains 2 to 4 mm in diameter. (b) Loose angular quartz sand fragments 2 to 4 mm in diameter. (c) Fossil fragments 1 to 3 cm across. (d) Fragments blown out of a volcanic crater in southern Colorado.

contain dissolved gases, solids (ions), and liquids. The process of precipitation involves mainly the ions. Most natural solutions are not single-component systems; they contain many different ions in solution, so the chemical reactions involved may be complex.

Precipitation may be brought about in several ways. A solution is said to be saturated with respect to a particular compound when no more of that compound can be dissolved in a given amount of solvent. The amount that can be dissolved generally decreases as the temperature of the solution decreases. A solution may, for example, become saturated as it cools. As a result the ions combine to form solids. Deposits around hot springs are often formed in this way. Even more commonly, evaporation of the solution causes saturation and brings about precipitation. The deposits of calcium carbonate found along some streams and near waterfalls or cascades (Fig. 5.8*a*) are a result of precipitation from evaporating spray. The unusual egg-shaped balls of calcium carbonate, called **oolites** (Fig. 5.8*b*), are also formed by precipitation induced by evaporation. Modern examples of oolite formation are found in the Bahama Island region where calcium carbonate precipitates on nuclei of sand grains or fecal pellets. The waters become saturated with calcium carbonate because the rate of evaporation is high. The rounded shapes result from continual rolling of the oolites in the agitated seawater.

Biochemical deposition. The slight changes in acidity found in seawater near certain plants and animals prove that they affect the chemistry of the water. Some sedimentologists believe that most precipitation in seawater is a result of biochemical processes; certainly important changes do result where animals in the water take in oxygen and give off carbon dioxide and where plants take in carbon dioxide and give off oxygen as a result of photosynthesis. Some sediments are composed mainly of shells or skeletons of animals. Calcium carbonate is the most common constituent of the shells of almost all invertebrate animals, but microscopic plants, the **diatoms** and a group of single-celled animals, radiolarians (Fig. 5.9), take silica out of seawater and build their hard parts of opal. The excreta of marine invertebrate animals, fecal pellets, are usually composed of calcium carbonate and impurities held together by a mucus material.

Formation of Rock from Sediment

The sediments on the floor of a lake or in the upper layers of the sediment on the sea floor are generally

New Mexico Tourist Bureau

(a)

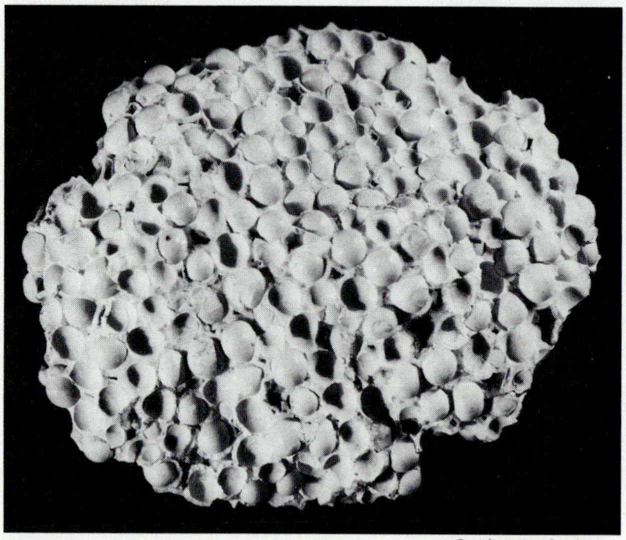

Smithsonian Institution

(b)

Figure 5.8
(a) Calcium carbonate precipitated from spray along a stream. The porous rock is called tufa. (b) Oolitic limestone from Carlsbad, Bohemia. Oolites are about 4 mm in diameter.

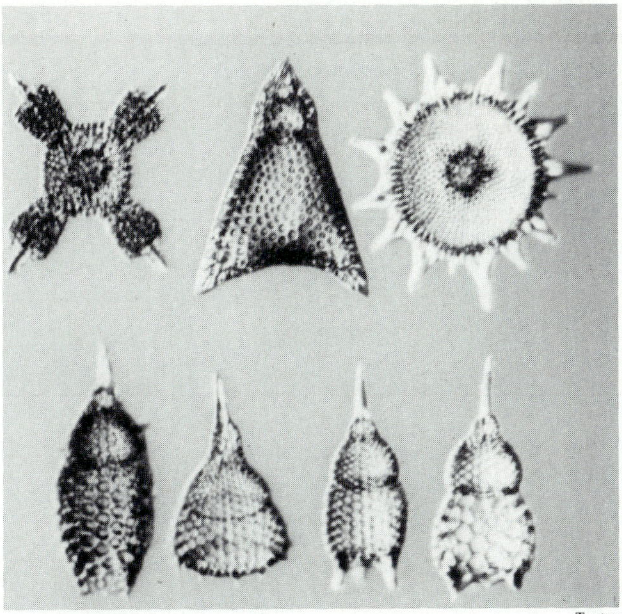

Turtox

Figure 5.9
The siliceous hard parts of radiolarians are shown greatly magnified here. A single radiolarian might be less than 1 mm long.

composed of loose sandlike deposits or of soft muds and clays. Yet, most of the ancient sediments are preserved as hard rocks, much denser than their modern sediment equivalents. Most sediments originate as unconsolidated materials like the familiar sand beaches, the less familiar layers of organic oozes found in the deep sea, or the layers of clay formed where clay colloids come together to form small lumps that sink in salt water. All of these materials may be solidified into rock by the action of a group of processes collectively called **lithification.** This process usually involves some degree of compaction and cementation of the unconsolidated sediment.

Compaction is usually the first stage in lithification, especially where sediments have accumulated over a long period of time, as in the sea. The weight of the overlying sediment compresses the deeper sediment, forcing water out, eliminating open spaces, and pressing soft materials together. Colloids are highly compressible—the volume of clay can be reduced by 40 percent or more—but sands and other clastic sediments are only slightly compressible.

When mud is present it may act as a cementing material, but more commonly cement is formed when solutions passing through a sediment carry ions. These ions precipitate in pore spaces, cementing the particles of sediment. The most common cementing materials are calcite, quartz, iron carbonate (siderite), and iron oxide (hematite). These may be derived from the sediment itself, or they may be introduced from outside by the water flowing through. Some cements are important economically. For example, the iron oxide that cements some marine sandstones in the Appalachians is rich enough to make the sandstone an iron ore. It is the principle ore for the Alabama iron industry.

Classification of Sedimentary Rocks

Sedimentary rocks are usually classified according to their texture and chemical or mineralogical composition, but a few are classed according to their mode of origin. The two major textural divisions, fragmental materials and chemical or biochemical precipitates, are generally easy to recognize. However, some very fine fragmental sediments, clay-sized particles for example, are so small that individual fragments are not easily seen. They can, therefore, be confused with chemically precipitated rocks.

Among the fragmental rocks, more precise identification is based on the size and shape of the fragments and on their composition. The sizes used to separate the fragmental materials are listed in Table 5.3. Most fragmental sediments consists of particles with a relatively narrow range of sizes, but it is not unusual to find sediments that contain particles that fall in two or more of the size ranges listed in the table. How consistent the particle sizes are is described in terms of the degree of **sorting;** a well-sorted sand is one in which the range of sizes is not great, whereas a poorly sorted sand contains a broad range of sediment sizes, most of which fall within the sand size range. Some fragmental sedimentary rocks are further subdivided according to the composition of the fragments. The following adjectives may be used

Table 5.3. Wentworth Scale of Fragment Sizes for Sediments (after C.K. Wentworth, "A scale of grade and class terms for clastic sediments," *J. Geol.* 30 (1922), 381).

Name of the fragment	Size (mm)
Boulder	
	256
Cobble	
	64
Pebble	
	4
Granule	
	2
Sand	
	1/16
Silt	
	1/256
Clay	

Table 5.4. The Most Abundant Sediments and Sedimentary Rock Types.

	Sediment	Sedimentary rock
Mechanically deposited fragmental materials	coarse rounded fragments	conglomerate
	angular fragments	breccia
	sand-sized fragments (general term)	sandstone
	with abundant feldspar	arkose
	with dark minerals	graywacke
	silt-sized fragments	siltstone
	clay-sized fragments	shale
Chemically deposited sediments	silica deposits	chert (flint)
	calcium carbonate	tufa (deposited near streams or springs)
		travertine
		oolitic limestone (open ocean)
		bedded limestone
	calcium magnesium carbonate	dolomite
	iron-bearing sediment	sedimentary hematite
	salts	salt
Organically formed sediments	radiolarian ooze	radiolarian chert
	diatom ooze	diatomaceous earth
	globigerina ooze	limestone (chalk)
	shell fragments	coquina (loose aggregation)
		limestone
	plant matter	coal (peat, anthracite)

to indicate specific composition: **siliceous,** containing silica; **carbonaceous,** containing carbon; **ferruginous,** containing iron; **calcareous,** containing calcium carbonate; and **argillaceous,** containing clay.

Many of the fragmental rocks (Table 5.4) are actually composed of the remains of plants or animals, especially shell fragments; some sandstones are entirely made up of sand-sized fragments of the shells of marine invertebrates. Such rocks may be called calcareous sandstones, but most sedimentary rocks that are formed largely of calcium carbonate are called limestones, or they are given special names (such as chalk, a limestone consisting of the shells of microscopic-sized animals).

Chemical precipitates tend to be much more compact and uniform in texture than fragmental sediments are. Many precipitates have a fine banding or laminated structure, and examination with a microscope may reveal a mosaic of intergrown crystals resembling igneous rock textures. These and other characteristics of sedimentary rock texture and composition enable us to deduce the environment in which the rock formed.

Primary Features of Sedimentary Rocks

Many features seen in sediments form at the time sedimentation takes place. Most nearly universal of these is stratification, the formation of sediment in layers or what we commonly call beds. This layering is apparent because it is reflected in the texture, the composition, or even the color of the rock. Bedding surfaces mark the position of the interface between the rock and the medium (such as water or air) in which sedimentation was taking place. The processes of sedimentation and the environment of deposition may be clearly indicated by the character of the features formed at this interface.

Ripple marks (Fig. 5.10*a*) are an excellent example of this. Long rows of ridges and depressions are usually visible in the sand on beaches. These ripples are due to the movement of sand by currents and

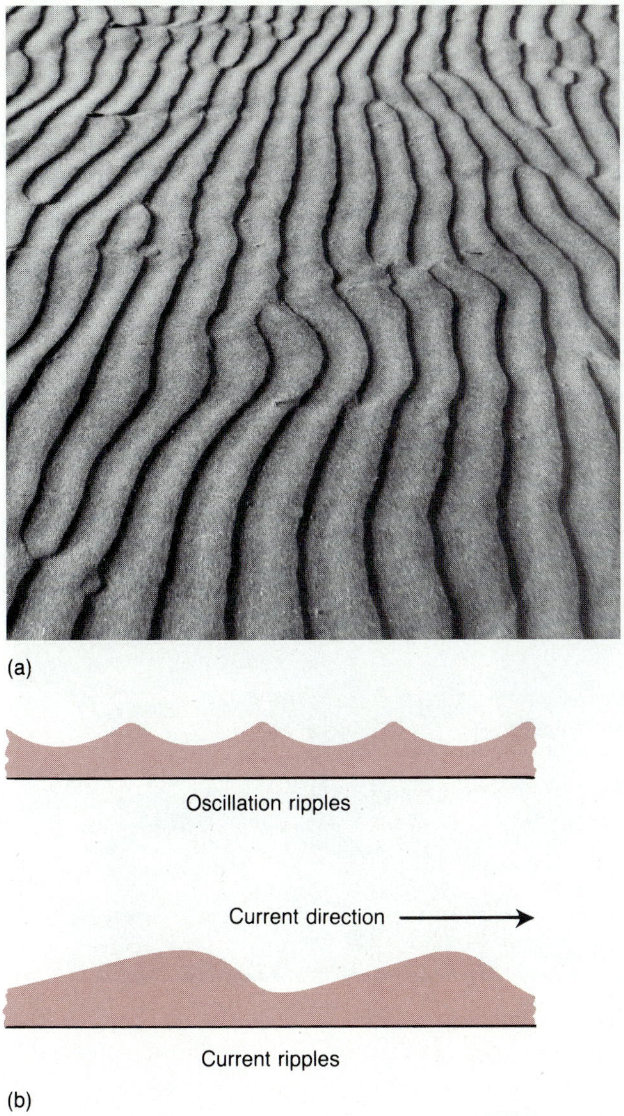

Figure 5.10
(a) These ripples formed in shallow water on a sand beach. The crests are about 10 cm apart. (b) Their assymetric profile is caused by current action. Those found in deep water where currents are absent are often symmetrical.

may be found in streams as well as on beaches. Ripples fomed by current action are asymmetrical in cross section, as seen in Fig. 5.10*b*, and this asymmetry indicates the direction of movement of the current.

A second type of ripple, called an **oscillation ripple mark,** is formed in deeper water as a result of the back and forth movement of water under waves. The ripples formed under these conditions are symmetrical and have pointed crests.

Cross-bedding is the name given to layers in which a second set of laminations or beds can be seen inclined within the layers (Fig. 5.11). This feature, like ripples, forms where currents are moving fragmental materials. Depressions scoured out in the bottom of a stream are frequently later filled in with more sediment (Fig. 5.11*a*). Cross-bedding often forms in these depressions. The process of filling such scoured-out depressions is much like that which occurs where a delta is built on the edge of a steep slope. At the edge of the slope sediment slips down the slope forming a steeply inclined layer called a **foreset bed.** Fine sediment is carried in suspension farther out and deposited on the bottom ahead of the inclined layer as a **bottom-set bed.** The edge of the slope gradually builds farther out. As more sediment is brought out to the edge of the slope, the top of the inclined beds is continually swept clean. Note how sharp the upper edge of the cross-beds is in Fig. 5.11(*b*) and 5.11(*c*). This sharp upper surface and the curved surface produced where the foreset beds pass into the bottom-set beds provide evidence of the top of cross-bedded beds. This can be used to identify bed tops if these layers are later turned up on end as a result of deformation.

Mud cracks (Fig. 5.12) also provide evidence of the tops of beds in which they occur; the cracked mud forms irregularly shaped layers, which curl upward. Of course, mud cracks also indicate clearly that the sediment was deposited in an environment that was out of water at least part of the time.

Many other unusual impressions may be left on the surface of sediment. Animals leave tracks, trails, and sometimes burrows in sediment. Raindrops and hail may leave characteristic impressions when they fall on mud. Drifting plants may drag limbs across the sediment (Fig. 5.13), leaving marks. Even bubbles of gas may be held on the bottom long enough for clay to accumulate on the surface of the bubble leaving a pit and mound-shaped feature.

Some unusual features form between beds after sedimentation occurs but before the rock is solidified. Rolls, lobate ridges, or other bag-like projections called **load casts** (Fig. 5.13) are an example. Load features are produced where a clastic sediment (such as sand) fills a depression in an underlying bed, usually composed of a soft sediment such as clay. The loads are downward protrusions of sand produced by the plastic yielding of the underlying, unevenly loaded soft sediment.

Although most of the features we have considered are formed on the surface or between layers, other

(a)

(b)

(c)

U.S.G.S.

Figure 5.11
Cross-bedding. (a) The cross-beds in this ancient sandstone were formed by currents flowing from left to right. (b) These cross-beds were caused by scouring and filling in a stream channel. (c) These cross-beds in modern sediment show the typical concave shape and sharp upper boundary.

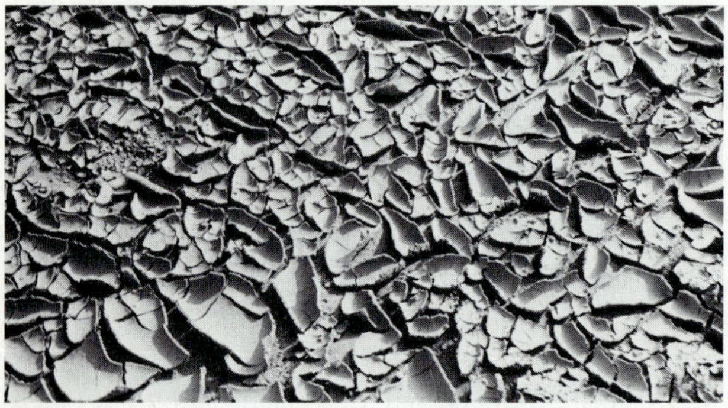

(a)

(b)

Figure 5.12
(a) Mud cracks in a dried lake floor. Most of these are only 5 to 10 cm across. (b) Sample of consolidated rock showing mud cracks formed long ago. The cracks are filled with a slightly more resistant sediment than the mud which is now hardened to a shale.

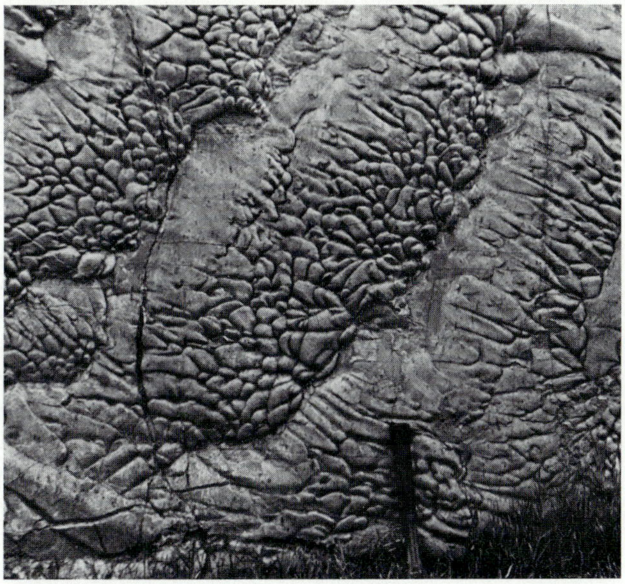

Figure 5.13
These unusual protrusions on the bottom of a sandstone layer are due to loading effects. These load fractures are superimposed on large ripple marks.

features, such as the sequence of beds, provide valuable clues to processes of sedimentation. For example, in a lake that is frozen part of the year sedimentation tends to follow a set pattern. During the summer a thick, coarse, light-colored layer forms as a result of the settling of sediment carried into the lake by streams. During the winter when the lake is frozen over this supply of sediment is cut off and a thin, fine, and usually dark-colored layer of material slowly settles. Thus over a period of years the sediment on the lake floor builds up as alternating layers of fine and coarse layers. Such seasonal deposits are called **varves** (Fig. 5.14).

5.3 SEDIMENTARY ENVIRONMENTS

As we have seen in preceding sections, sediments have compositions, textures, and structural features that reveal the type of environment in which they formed. By studying the characteristics of large bodies of sediment we can likewise determine the large-scale tectonic setting in which they occur.

For the purpose of establishing the connection between plate tectonics and sedimentary environ-

U.S.G.S.

Figure 5.14
Varved clay excavated in a lake bed deposit.

ments, we will consider three of the largest and most important sedimentary environments: the deep sea, deep-sea trenches, and an Atlantic-type continental margin as seen along the eastern coast of North America. These examples are intended to bring out characteristic features of the sedimentary environment related to the tectonic setting. It will become apparent that other factors and conditions unrelated to crustal dynamics have also left their imprint on the sedimentary record in each of these environments.

Sediments of the Deep-Sea Floor

Before the twentieth century, our picture of the deep seas was founded almost entirely on speculation. There was little information on which to base any concrete understanding. The early geomorphologists reasoned that running water, glaciation, wind, and groundwater are effective agents of erosion only above sea level. At sea level, glacial ice ceases to scour the surface and begins to float, and to them it appeared that running water also ceased to transport and erode. Samples of seawater collected throughout the oceans were always free of suspended sediment if they were taken far from shore. It seemed logical to assume that the deep-sea basins were untroubled by either currents or erosion and, therefore, were subject to very little change. Such a view was expressed by one of America's first oceanographers, Admiral M. F. Maury (*1*):

> The geological clock may, we thought, strike new periods; its hands may point to era after era; but, so long as the ocean remains in its basin—so long as its bottom is covered with blue water—so long must the deep furrows and strong contrasts in the solid crust below stand out bold, ragged, and grand. Nothing can fill up the hollows there; no agent now at work, that we know of, can descend into its depths, and level off the floors of the sea . . .

Later, when surface waters of oceans were recognized to contain vast quantities of small free-floating and swimming plants and animals, this view of the deep sea changed. Instead of bare exposed rock, the ocean floor was visualized as covered with a thick homogeneous blanket of sediment composed of remains of microscopic marine plants and animals. Rachel Carson (*2*) describes it:

> When I think of the floor of the deep sea, the single overwhelming fact that possesses my imagination is the accumulation of sediments. I see always the steady, unremitting, downward drift of materials from above, flake upon flake, layer upon layer—a drift that has continued for hundreds of millions of years, that will go on as long as there are seas and continents.

When samples of the sediments on the continental slopes were obtained in the early 1900s, they showed, much to the surprise of some geologists, that the most recent sediments were stratified. This meant that, contrary to the prevailing view, changes in sedimentation occur even in the deep sea. Until then it was thought that the processes of gradation that cause stratification on land were not effective in the deep seas. That scientists had been laboring under a misconception became even more apparent when it was discovered that layers on the bottom were composed not only of layers of soft mud containing shell fragments of microscopic marine animals, but that these muds or oozes, as they are called, were sometimes found interbedded with layers of silt and sand, sediments thought to be formed only at shallow depths on the continental shelves.

In recent years, a program of drilling by the JOIDES (Joint Oceanographic Institutions for Deep Earth Sampling) organization has provided samples of the sediment that covers the igneous rocks of the oceanic crust in most parts of the ocean basins. Many of these cores provide samples through the entire thickness of the sediment cover. Other shorter cores (10 to 30 m long), obtained by coring devices made of weighted pipe (Fig. 5.15), provide samples of the sediments formed during the last million years.

These samples are the basis for our present understanding of sedimentation in the deep sea. They give us a clearer picture of processes at work there than the early oceanographers could hope for. The variety of sediment types we know to exist on the ocean's floor would certainly surprise these first students of the seas, but some of them would be pleased to know that organic sediments are, as they suggested, a very important part of the sedimentary record.

Organic sediments in deep water. The upper layer of sea water, the zone penetrated by sunlight, is far more densely populated than most of us would imagine. This is the domain of the plankton, the drifting plants and animals shown in Fig. 5.16, few are more than 0.05 centimeters in diameter. Marine plants are the base of the food chain of marine life; of these plants two groups, **diatoms** and **flagellates,** are most important. Between 15,000 and 20,000 species of diatoms inhabit the widely varying marine

(a)

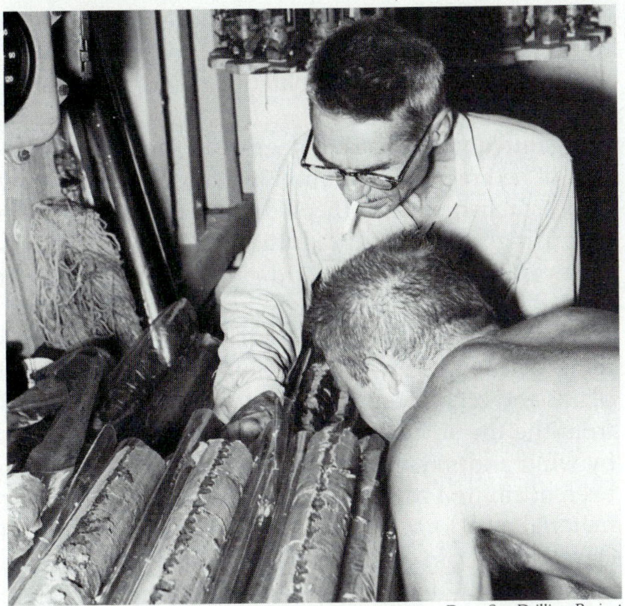

Deep Sea Drilling Project

(b)

Figure 5.15
(a) A piston coring tube over the side of the ship ready to be lowered to the sea floor. The weight is used to drive the pipe into the soft sediments on the sea floor. This is a scene on the research vessel *Vema* of Columbia University's Lamont-Doherty Geological Observatory.
(b) Scientists examine cores of deep-sea sediment obtained with the piston corer.

Turtox

Figure 5.16
Diatoms, such as those shown here, are one of the most important groups of plankton. Diatoms are plants with siliceous remains. Individual diatoms range from 1 μ to 2000 μ in diameter.

environment. They are simple plants, algae, blobs of protoplasm containing the chlorophyll necessary for photosynthesis. They reproduce by cell division, and under favorable conditions the single cell may divide several times a day. The rise of nutrient-rich water from the ocean depths and plenty of sunlight may result in a population explosion of diatoms, discoloring the water and producing what is often called the **red tide.** As many as several million diatoms may be contained in a single liter of water. Although they have many different shapes each contains a silicon dioxide framework that sinks to the sea floor when the plant dies, becoming part of the sediment. **Diatom ooze,** an extensive deposit consisting primarily of siliceous remains of diatoms, is exposed between 40 °S and the Antarctic Circle. These beautiful minute plants contain droplets of oil and thus may be the primary source of petroleum.

The flagellates are a second important group of microscopic plants; they vary in shape, some resembling a balloon with a string attached, others a pot or fancy mask; all have whiplike projections that move. Because they can move, the flagellates are sometimes classed as animals; but, like plants, they contain chlorophyll and manufacture food by photosynthesis. The flagellates may be even more abundunt than diatoms; 60 million were counted in one liter of sea water collected off Florida. Most of them, however, do not possess hard parts that are preserved in sediment. One exception, the **coccolith,** has a minute clacareous shieldlike plate.

Diatoms and flagellates are the food source of the single-celled animals of the sea, notably the **radiolarians** and the **foraminifers.** Radiolarians (look back to Fig. 5.9) have an internal skeleton with radiating projections composed of silica; their skeletal remains are important constituents of sediment in a few areas of the ocean. Foraminifers, shown in Fig. 5.17, possess small shells composed of calcite. One of the most abundant of this group, the **globigerinas,** have globular shells that resemble snail shells. These shells are the main constituent of chalk deposits such as the famous White Cliffs of Dover, where some 50,000 shells may be counted in a single spoonful of chalk.

The most extensive sedimentary deposit in the deep sea is composed of globigerina ooze, a calcareous mud formed from the shells of globigerinas. Clays are usually mixed with globigerina shells in these sediments and may constitute more than 50 percent of the deposit. These types of sediment cover nearly half of the deep-sea floor to depths of 4500 meters. Below that depth calcite is soluble due to the high pressure and low temperature. Any settling shells dissolve on their way down or soon after they reach the ocean floor.

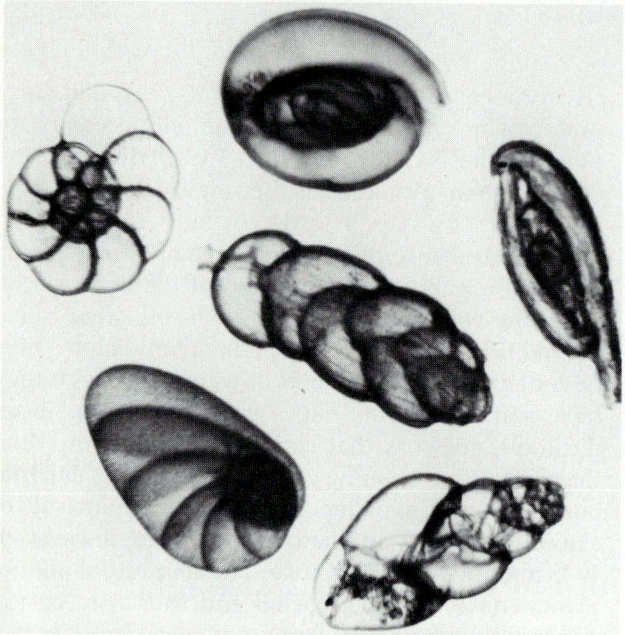

Ward's

Figure 5.17
Micrograph of a group of foraminifera. Their remains make up a large part of the sediment on the ocean floor in middle latitudes.

The relative solubility of calcareous and siliceous material is an important factor determining the dominance of silicon-rich sediments over calcium-rich ones in deep water. Calcium carbonate is more soluble than silicon dioxide in deep, cold waters; consequently, siliceous remains of diatoms and radiolarians dominate these bottom sediments in trenches and other deeps.

Diatoms are most abundant near the surface in high latitudes where water rich in dissolved nutrients rises from great depth. Radiolarians and foraminifers are most abundant toward the equator, favoring the warmer water. They are also more abundant near shore than farther out, because their food supply is greatest in shallow waters. Although the skeletons of these tiny creatures, along with diatoms, are the main constituents of deep-water sediment, they are of little importance in shallow water where their remains make up only a small fraction of sediment. Sediments there are dominated by sand and silt.

Inorganic deposits in the deep sea. The general pattern of sediment distribution on the sea floor is now well established, and the sediment types found are surprisingly diverse. In addition to sands and muds of several types there is a distinctive brownish-red clay.

Red clay, a brown to reddish clay that contains films of manganese, is widespread in the abyssal plains and other deep waters. Most red clay is derived from the finest clay fractions carried into the oceans by wind and streams from land areas, but it has also been attributed to submarine weathering of volcanic material and to accumulation of meteoric dust. Red clays cover vast areas in many parts of the ocean, especially where other types of sediment are lacking. Possibly these clays are deposited uniformly everywhere but are inconspicuous when mixed with large amounts of other sediment. Rates of accumulation of red clay are the lowest of any sediment—as low as 0.1 centimeter per 1000 years.

Sand and mud deposits accumulate much more rapidly in most areas. Several types can be recognized in deep waters. Some of these are relatively localized and can be traced to their source with ease—for example, the sands derived from the breakdown of coral reefs by wave action. Distinctive muds can also be traced toward the mouths of such major rivers as the Amazon and Yangtze Kiang. Still others are of glacial origin, apparently carried into deep water by currents that are stronger than once suspected. Micrometeorites accumulate in the sediment on the ocean floor. It is estimated that millions of tons of meteorites are added each year. But one of the surprising discoveries has been the widespread distribution of sands, silts, and clays carried into deep water by turbidity currents.

Turbidity current deposits. Early sounding methods revealed submarine canyons located near the mouths of rivers and traceable across the continental shelves. They are seaward extensions of the rivers formed during the ice ages when the water levels of the oceans were lower. But with improved sounding devices, it was later found that many of these canyons extend down the slopes and across rises, and a few can be traced into the abyssal plains. How can such features be explained? Could sea level have been low enough to account for channels 5,500 meters below sea level? No evidence supports such a hypothesis.

Breaks in transatlantic cables laid across the deep-sea floor provided the clue not only to how these canyons are extended into deep water, but to how sands and silts expected to occur only in shallow water may find their way into the deep ocean basins. In November 1929, following a small earthquake, a series of breaks occurred in cables on the sea floor off the Grand Banks of Newfoundland (Fig. 5.18). A full-scale investigation of the breaks was made in the early 1950s by the Lamont Geological Observatory. Deep-sea cores taken throughout the area near the cable breaks revealed a widespread blanketlike deposit of sand and silt about a meter thick, covering deep sea sediments such as red clay. Microfossils found in the sand suggested that the sand was originally deposited on the continental shelf. The deeper sands are also graded (Fig. 5.19); they are stratified with the coarse layers on the bottom, just as you might find if you shook a jar with mixed sediment sizes in it and let the mixture settle. Ripple marks on the top of the sand indicate strong current action as well. Bruce Heezen and Maurice Ewing, the scientists in charge of the investigation, suggested that the breaks were caused by what is known as a **turbidity current**—a mass of water made dense by a great deal of suspended material that flows with

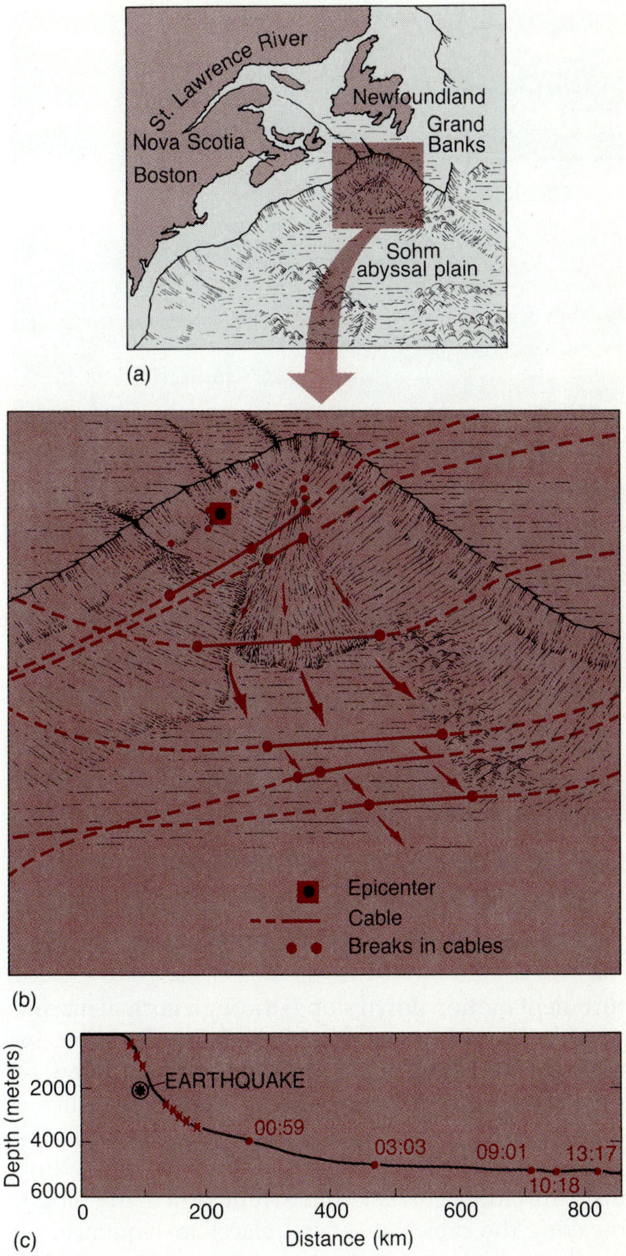

Figure 5.18
(a) The geological setting and (b) the location of the cables broken by turbidity currents in 1929. (c) This profile across the continental margin shows the time that elapsed between the initial breaks, indicated by × marks, and the breaks of other cables in deeper water. It is possible to calculate the rate of movement of the currents from these data.

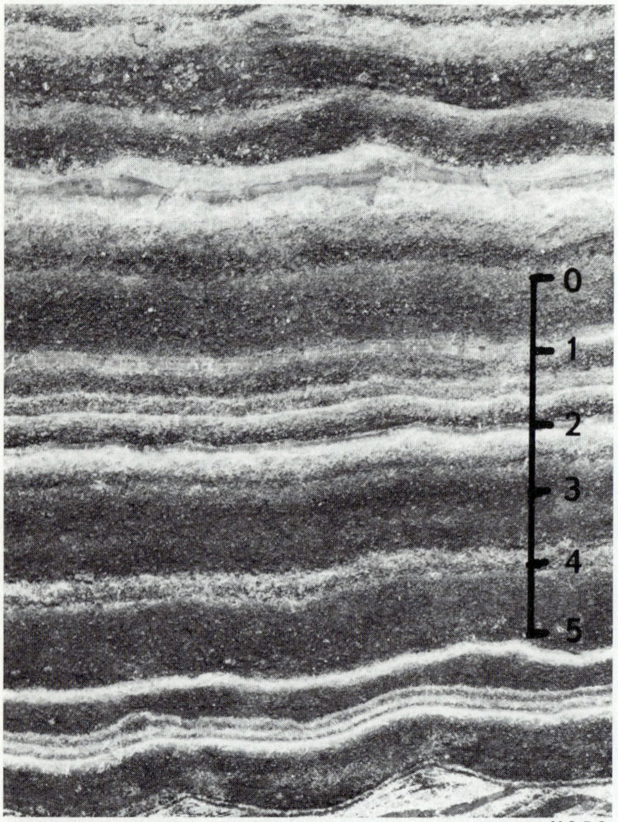

U.S.G.S.

Figure 5.19
Graded bedding.

turbulent motion down slopes through normal marine waters. Turbidity currents flow beneath water by virtue of their greater density. They sink and flow as a turbid mixture of water and sediment rather like a mudflow. A mass of shallow-water sediment broke loose from the edge of the continental shelf, setting up a turbidity current that swept down the slope, breaking the cables at several places in sequence.

Turbidity currents have proved to be a unifying explanation for what were previously thought to be unrelated problems of cable breaks, submarine canyons in the deep sea, and the presence of clastic sediment in the abyssal plains. The idea of turbidity currents is old, but its application to the erosion and deposition of the deep seas dates from these studies.

Further evidence of the importance of turbidity currents comes from a number of other sites. Cable breaks off the mouth of the Magdalena River of Colombia occurred year after year at just the same season. Invariably, the breaks took place during the highest flood stages of the river. At these times the river carried large quantities of debris out from the continent, deposited it in the waters off the mouth of the river, and usually built up a fan-shaped delta of debris in the submarine canyon of the river. During floods, it was noted that the top of these deposits disappeared just before the cable broke. In 1935, when the cable was brought up from its position 19 kilometers offshore and nearly 1,500 meters under water, it had green grass wrapped around it—dramatic proof that the river debris had been carried out to the cable and almost certainly played a major role in its failure.

Sediment transported into the ocean basins by turbidity currents may be the main factor in the formation of the abyssal plains. Deposition there occurs by the combined processes of slow accumulation of shell fragments and extraterrestrial dust and rapid deposition of mud and sand transported by turbidity currents.

Sedimentation in Deep-Sea Trenches

Soundings of deep-sea trenches that almost completely encircle the Pacific Ocean show that fan- or cone-shaped deposits are accumulating along the edges of the flat bottoms of some trenches. The trenches apparently are collecting much of the sediment that otherwise would be transported out into the deep abyssal plains of the ocean by turbidity currents. This is verified by the absence of turbidity-type deposits in the central Pacific basin. The fact that scattered nodules of manganese, estimated to form at rates in excess of 1 millimeter in 1,000 years, cover large areas of the Pacific Ocean bottom makes it clear that those areas of sea floor are not sites of turbidity current deposits.

The deep-sea trenches mark the places where oceanic sea floor is being subducted according to plate tectonic theory. Thus, the sediments deposited here are dragged down into the mantle. If the rate of subduction is rapid, only small quantities of sediment are likely to accumulate in the trench. But if subduction is slow compared to the rate at which sediment comes into the trench, the trench may be filled. Parts of the Aleutian and Peru trenches show signs of filling. A filled trench may lie off the coast of Washington and Oregon.

The clastic sediments in deep-sea trenches originate from the flanking volcanic islands. Thus, vol-

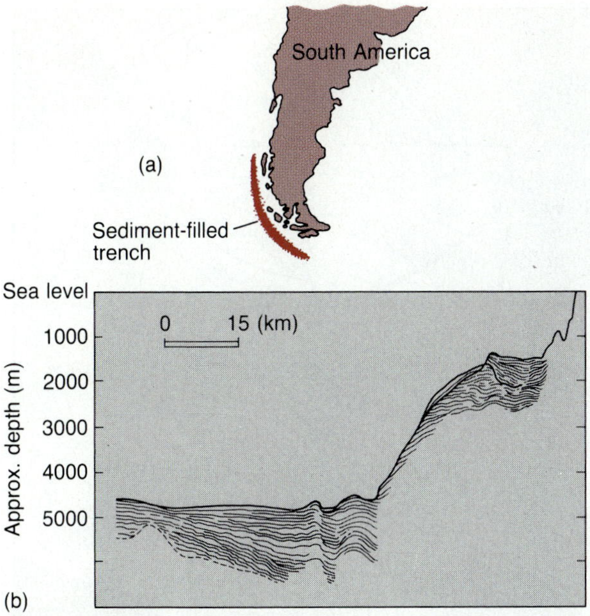

Figure 5.20
(a) Location and (b) profile of the sediment-filled southern end of the Chile trench.

canic debris, such as ash and fragments of volcanic rock, is a prominent component of the trench fill (Fig. 5.20). In addition, organic deposits composed of silica, mainly the remains of radiolarians, are characteristic of trench sediments. These siliceous remains are preserved in the deep waters of the trenches although calcareous hard parts of other organisms dissolve in the cold high-pressure water.

Sedimentation on Atlantic-type Margins

The eastern margin of North America is a continental margin that is not a site of subduction. The North American Plate has one margin along the Mid-Atlantic ridge and the other along the Pacific coast. Thus the continent rides passively on the plate, and the Atlantic margin is situated in the central part of the plate. This margin, like those of South America and Africa, formed when the larger continental mass of which they were part split and started to drift apart. At that time, the margin had much more volcanic and earthquake activity because the continental margin was so close to the oceanic ridge over which the split took place. Today these margins are far from the activity found at the oceanic ridge—they are passive trailing edges of continents that are moving away from the oceanic ridges.

The rift between North America and Africa took place nearly two hundred million years ago. Since that time, the continental margin has been quiet; volcanic and seismic activity has been very limited, and the region has been unaffected by mountain-building or major uplift. The contrast between the east and west coasts of North America in all these respects is dramatic. The Atlantic margin has been a site of sediment deposition by a number of processes, and these have modified the original rifted margin to its present form.

As we saw in Chapter 3, in most places the Atlantic margin is characterized by a broad, flat, continental shelf, which passes into a somewhat steeper continental slope and in turn into a lower sloping continental rise. All of this is covered by a sedimentary veneer, which varies greatly in thickness with the greatest accumulation located under the slope and rise. The part of the continental shelf north of Long Island has been subject to repeated glaciation during the past two million years. The land and shelf areas there have been affected by glacial erosion and deposition (Fig. 5.21). Some of these glacial deposits were laid down from glaciers far out onto the shelf. When the ice melted, the meltwater carried large quantities of finer sediment into the northern part of the ocean.

That section of the continental margin from New York to northern Florida has been the site of long continuous deposition of sediments carried by streams draining the Appalachian Mountains and emptying onto the shelf. Farther south, where few rivers enter the ocean, the sediment—mainly limestones of various types—is largely composed of organic materials and some chemical precipitates.

The thick sedimentary piles along the central and northern part of the continental margin have been formed by the combined effects of turbidity currents and deep currents flowing along the margin at depth. Shallow currents carry silts, sands, and clays from the continent onto the continental shelf where they accumulate. Those deposited near the edge of the shelf often slide. The sediment then moves into deeper water as part of a turbidity current and may then be spread (Fig. 5.22) by currents that are part of the general oceanic circulation.

In summary, the sediments show considerable variation in composition and mode of origin from place to place along the Atlantic margin. They are composed predominantly of quartz sands, limestones, and shales. Because they contain little or no

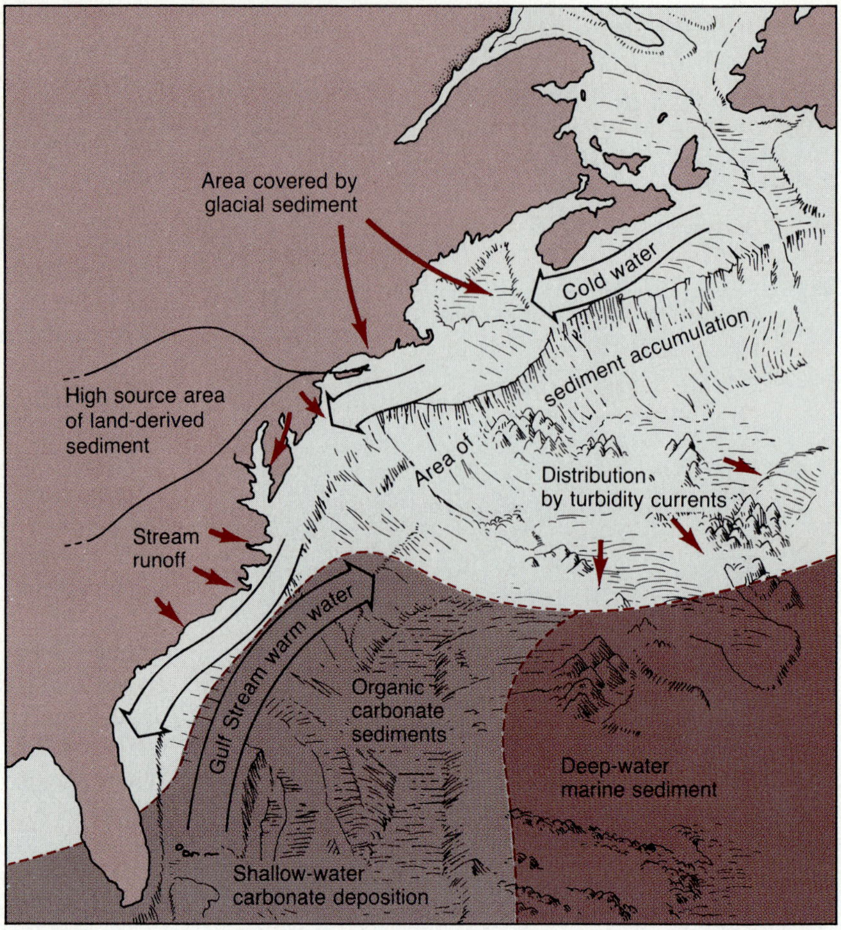

Figure 5.21
Sediment now being deposited off the east coast of North America is greatly influenced by the factors shown here. Sediments of glacial origin predominate in the north; streams in the mid-Atlantic states carry sediment from high land source areas onto the shelf. In the south most of the sediment is produced by organisms.

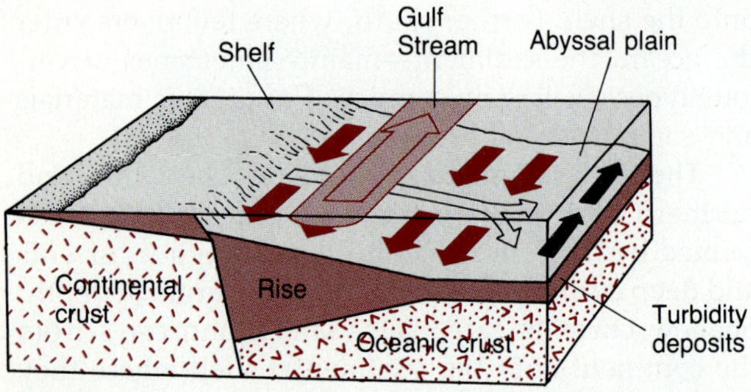

Figure 5.22
The continental rise is shaped by currents (indicated by arrows) flowing near the bottom and parallel to the slope. Turbidity currents move sediment downslope, to be spread by the bottom currents, eventually building up the wedge-shaped layers shown under the continental rise.

volcanic material, they are very unlike sediments found near volcanic island arcs. They are also distinctly different from typical deep-sea sediment.

5.4 INTERPRETING THE SEDIMENTARY ROCK RECORD

Sedimentary Facies

As shown above, sediments usually contain clues to their mode of origin. These clues consist of the composition, texture, and features formed at the time of sedimentation. This rock record of a particular depositional environment is called a **sedimentary facies.** Thus the rich organic muds, containing fossils of brackish water animals and plants, constitute a swamp facies quite distinct from the clay formed in some deep marine environments or the cross-bedded and ripple-marked sands of a beach. Most of these diagnostic features are well preserved when sediment is transformed into a sedimentary rock. Consequently we can unravel the conditions and the geological setting of sedimentation that prevailed millions of years ago by studying the rocks formed at that time.

A number of sediments we have now studied indicate rather specific environmental conditions. For example, most oolitic limestones form in shallow agitated marine waters containing a high calcium carbonate content; the presence of salt suggests high rates of evaporation, usually in isolated bodies of water; the presence of feldspar in sediment indicates that burial of the sediment was either so rapid or the environment was so dry that the feldspar did not break down to form clay. Fossil remains are valuable clues to the environment of deposition also. Many species are restricted to marine waters, others to brackish or fresh water. In the case of organisms that live attached to the sea floor, as oysters and corals are, their presence indicates not only the marine environment but water depth as well.

Some sedimentary environments cover vast areas. The strata resulting from deposition in such an environment will also have great lateral extent, but when one environment passes into another, as ordinarily happens near shorelines, the resulting strata also show this change as one facies grades into another.

The Law of Superposition

We enjoy all the advantages of having the accumulated knowledge of more than two hundred years of work done by generations of dedicated scientists who spent most of their lives probing the secrets held in the sedimentary rock record. Even though today much remains to be learned from this record, we can now look back on the first ideas that were advanced with amusement and on others with profound admiration for the careful work, the inspired insights, and the keen observations of some of the first geologists.

A number of the early philosophers recorded observations about the earth long before geology was recognized as a field of study, and many of the prinicples we now use in our interpretation of sedimentary rocks were stated long ago. One of those "laws," the **law of superposition,** was formulated by Nicholas Steno (1638–1687), who recognized that in a normal, undisturbed sequence of layered sedimentary rocks, older layers lie below younger layers. When we look at the layered or stratified rocks exposed in the Grand Canyon (Fig. 5.23) we infer that the younger layer is at the top and that successive older strata continue in sequence below with the oldest rocks at the bottom of the gorge. Steno also noticed that strata are not always "undisturbed." We

Union Pacific

Figure 5.23
The Grand Canyon of the Colorado River. These strata provide a record of the history of this region for a period of more than a billion years of earth history. Each layer in the canyon is younger than the ones below. Thus a trip into the canyon allows us to inspect rocks of increasing age as we descend.

IGS Photo

Figure 5.24
Outcrop of a sedimentary layer that has been turned upside down.

know today that they can be folded into arch-like features, broken by faults, and even completely inverted so layers are upside down (Fig. 5.24). These features are commonly encountered in many of the modern mountain belts of the world.

Fossils and the Law of Faunal Succession

Fossil-bearing sedimentary rocks, being younger than the ancient rocks formed during the earth's early history, cover much of the older rock and are better exposed for our inspection. **Fossils** may consist of the actual remains of the animal (such as a shell or bone), but more often, and especially in older rocks, the fossils are casts or impressions of the actual hard parts. Figure 5.25 shows both types. Sometimes the bone or shell was replaced by something hard and durable, such as silica. Other fossil remains consist of nothing more than casts or molds of tracks and trails.

Fossils have been exceedingly important because many of the plant and animal groups found have modern living counterparts, which in turn provide us with clues to the past. The conditions that existed in the locality where a fossil is found can be inferred from the ecology of the modern representatives of the same group. Thus, from the particular groups of fossils that they contain, we can infer whether a sedimentary rock formed in deep marine water, in shallow sunlit waters, in fresh water, or on land. Fossils provide the key to the type of environment, the age, and sometimes even the water temperature and depth of the sediment in which they are encased. Although these interpretations of fossils are now widely accepted, they were at one time the subject of heated debate.

The fifteenth century marked the start of a long debate about the origin of fossils. Some scientists believed that they grew from seeds, others interpreted them as marine organisms laid on the surface of the

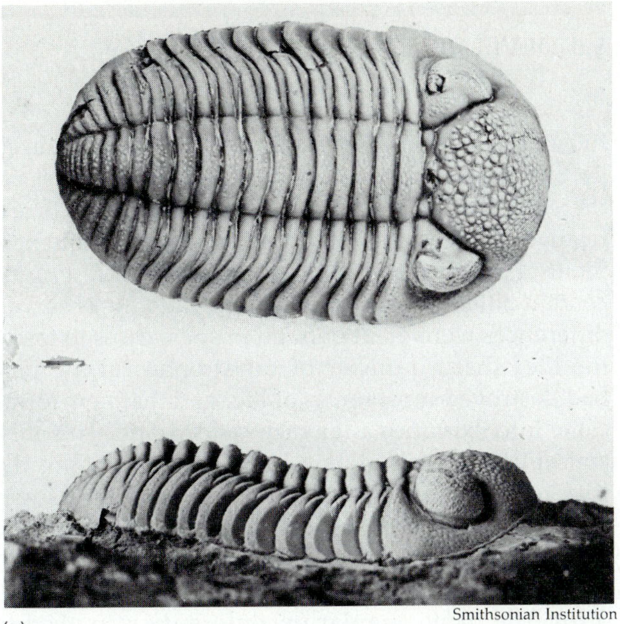

Smithsonian Institution

(a)

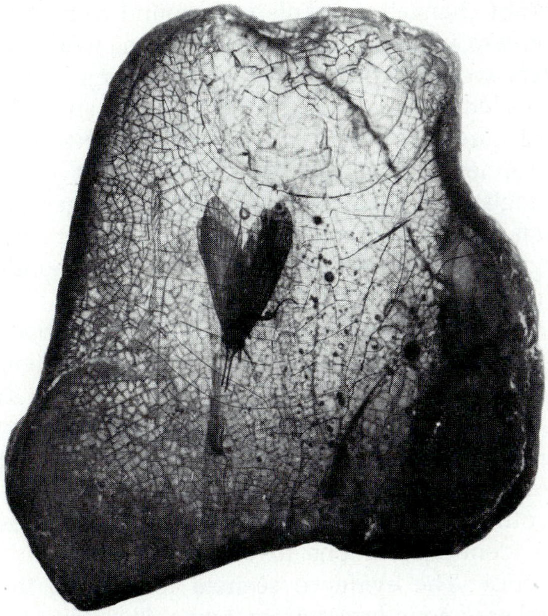

Smithsonian Institution

(b)

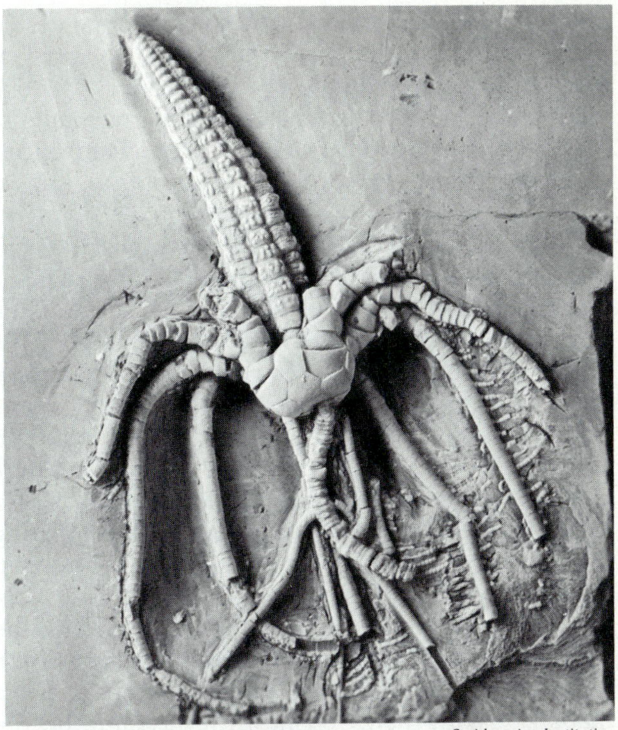

Smithsonian Institution

(c)

Smithsonian Institution

(d)

Smithsonian Institution

(e)

Figure 5.25
The remains of plants and animals may be preserved in several different ways. The trilobite (a) and crinoid (c) are casts; the insect (b) is preserved in amber (pine pitch); the plant (d) is preserved as a carbon imprint on the rock; and the gastropod (e) is the original shell.

earth during Noah's Flood, and still others suggested that the fossils were placed in rocks by Satan to confuse human beings. Leonardo da Vinci was one of the first to recognize that fossils are remnants of marine organisms deposited in the seas and incorporated in what were formerly sediments on the sea floor.

John Woodward (1665–1722), an English paleontologist, strongly opposed the then-reigning religious dogma that held that all fossils were relics of the Flood. He considered fossils to be remnants of the plants and animals that lived in the distant past, and his writings reveal that he recognized that some sedimentary strata cover large areas and that certain layers can be traced or recognized throughout large regions. This important insight provides the basis for correlating events in different areas and for comparing the sequence of events in one area with those in another.

The most significant conclusions regarding the value and use of fossils were offered by William Smith (1769–1839), a man with an insatiable curiosity and a habit of taking notes on nearly everything he observed. His life's work was engineering, and he brought an immense amount of imagination and skill to it. Employed to drain swamps, locate canals, and restore springs, he realized that an understanding of rock units was of great value in accomplishing these tasks. His success in these projects put him in such great demand that he was called to all parts of England. Often he traveled as much as 16,000 kilometers a year, and on all these trips he kept careful notes of the types of rocks he saw. Probably his most important contribution was the observation that certain rock units can be identified by the particular assemblages of fossils they contain, an observation that led him to the formulation of the **law of faunal succession.**

This law states in effect that different groups of plants and animals lived at different times, and that for this reason one will not find exactly the same assemblage of fossils in rock layers of different age. Some fossils occur in rocks of such a short age range that they can be used as guide or index fossils to identify rocks of that age. Other fossils occur in rocks of many ages. Once the fossils of a given age are recognized and differentiated from those that are younger or older, it becomes possible to identify the relative age of the rock by use of the fossil assemblage in it. Smith's observations and deductions formed the basis for developing a geologic time scale.

5.5 DEVELOPMENT OF A GEOLOGIC TIME SCALE

In any one locality only a few of the strata recognized by Smith are likely to occur at the ground surface; therefore, it took a long time to establish the correct relative positions of some units in the whole sequence. Smith had recognized that different rock groups contain different fossil assemblages, and the observed differences were great enough in some cases to foster the idea that a number of catastrophes in the past had destroyed most forms of life, and that new forms came into existance after each catastrophe. This misconception was used as a basis for subdividing the rock record. It was noted that these marked changes in the forms of life frequently occurred where the rocks of one sequence were deposited on those of an earlier sequence that had been folded or warped (Fig. 5.26). The name **angular unconformity** is applied to breaks of this type. The unconformity is an erosion surface separating deformed beds below from undeformed or less deformed beds above. Other significant changes in fossils were found to be related to breaks in a stratigraphic succession, where some layers were missing either because they were not deposited or because they were eroded away. Such breaks are called **disconformities.** The idea that catastrophes separated these divisions lost favor as geologists began to realize that the changes in life forms and the unconformities were developed over long time spans—millions of years rather than hundreds or thousands.

Geologists soon discovered that it was possible to correlate groups of strata by their position in relation to the unconformities (Fig. 5.27). The same rock units containing almost identical fossils were found immediately above an angular unconformity in many localities in continental Europe and England. Eventually, it was shown that similar units and fossils had identical relations to unconformities in North America. The evidence seemed to suggest that a number of periods of deformation of the crustal layers had occurred in the past and that these took place at about the same time in many parts of the crust.

Many geologists still consider such widespread deformations to be the ultimate basis for correlation of geologic events throughout the world. Although this point is still argued by geologists, we do know that worldwide correlation on this basis is not a simple process.

(a) IGS Photo

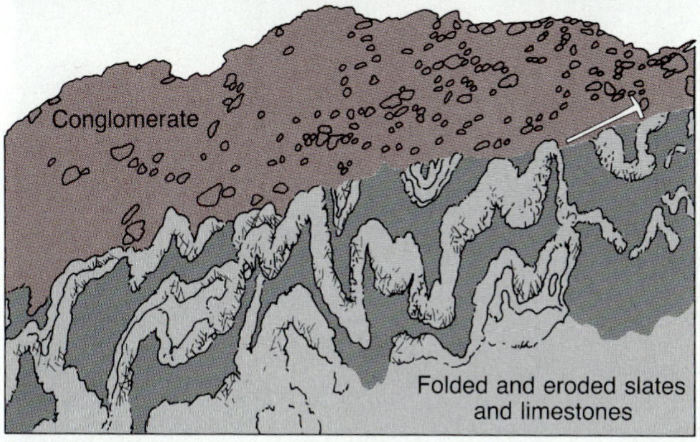

(b)

Figure 5.26
(a) An angular unconformity between conglomerates above and folded layers of shale and limestone below, in Argyll, Scotland. (b) Schematic diagram.

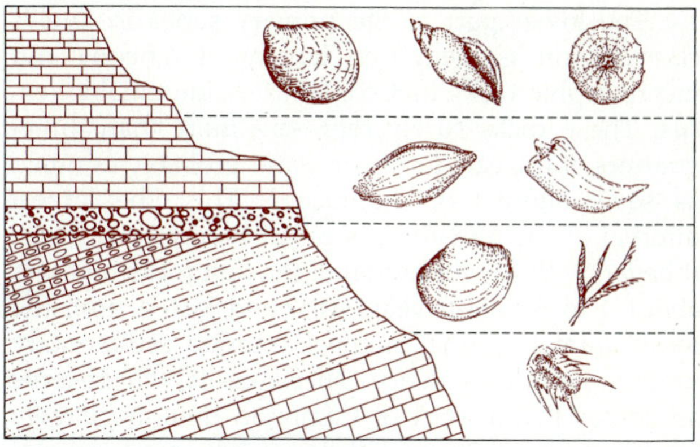

Figure 5.27
William "Strata" Smith learned to recognize various rock units by the particular fossils they contained. He also pointed out that the same fossils were repeatedly found in the strata above prominent angular unconformities like the one depicted here.

The Modern Geologic Time Scale

The first appearance of a geologic time scale similar to that used today was in 1833 in a textbook by Charles Lyell, *Principles of Geology*. Lyell's classification has a threefold subdivision, primary (or Paleozoic), secondary (or Mesozoic), and tertiary (or Cenozoic). In Britain and Europe where the scale was formulated, prominent unconformities separate the primary, secondary, and tertiary groups of Lyell's classification. **Primary rocks** were found to consist of two parts: crystalline rocks bearing no fossils, that are igneous or metamorphic; and sedimentary rocks characterized by fossils of invertebrates and fishes. **Secondary rocks** were characterized by remains of certain mollusks (cephalopods and ammonites) and reptiles, including the dinosaurs; and the **Tertiary rocks** were characterized by mollusk and mammal remains. The prominence of the unconformities and the impressive changes in fossils between these divisions led to their acceptance as the largest subdivisions of geologic time, called **eras.** They were given names that reflect the stages of development of life in each—**Paleozoic** (old life), **Mesozoic** (middle life), and **Cenozoic** (modern life).

Other unconformities and noteworthy changes in fossils are found within the rock units of each era, and many of these breaks can be recognized in different parts of the world. This led to the subdivision of the strata deposited during each era. These subdivisions, called **periods** of time, are named for localities where the rocks formed during that period were first recognized, or where an unusually complete section of rocks of that age is known. Other rocks that are classified in the same period must be correlated with the rocks in the original area where their age was first recognized.

The names of the eras and periods of the modern geologic time scale are shown on the inside back cover, along with the ages of the rocks as determined by radiometric dating and some of the important events in the history of the earth. Younger rock units generally overlie older strata and tend to cover them; so, younger units are more widespread and better exposed, and can be studied in greater detail than older units. Thus they can usually be subdivided more readily.

The lower part of the primary series in Lyell's classification is composed largely of igneous and metamorphic rocks that contain few identifiable fossils. These rocks surely represent more than three-quarters of the earth's history, and probably as much as seven-eighths of the record. This, the largest interval of earth history, is called Precambrian time. It began with the formation of the first rocks on earth about 4 to 5 billion years ago, and lasted until the fossil animals found in the Cambrian sedimentary units came into existence. These fossils, which serve as guides to the start of Cambrian time are largely trilobites (Fig. 5.28) and they are advanced invertebrate animals. We can conclude, therefore, that life had a long history of development in Precambrian time, but, so far, relatively few of the older Precambrian fossils have been found. Most of these fossils are remains of algae, worms, and similar low-level life forms. This difference between Cambrian and Precambrian fossils is due in part to the great time gap frequently found between the ages of Cambrian and Precambrian rocks where they are in contact.

Although the **Quaternary period,** in which we now live, is the shortest of all, it is highly significant because its history is known in great detail, and it is the one with which we deal most often in the study of physical geology. The Quarternary is subdivided into two parts, the Holocene (recent) and the Pleistocene.

The beginning of the **Holocene** period is set at a time about 11,000 years ago when sea level began to rise—an event that coincided with a warming of the surface layers of the ocean and rapid retreat of the continental glaciers that had occupied the continents in the Northern Hemisphere. The **Pleistocene** era is the name given to the interval of time during which large parts of the earth's surface were glaciated. The start of the Pleistocene era is still not fixed precisely in absolute dates, but is defined as the time when the ice ages began, when sea level fell, the surface of the earth cooled, and continental glaciers grew. These glaciers advanced and retreated a number of times during the Pleistocene era; it may be that Holocene time is nothing more than an interglacial interval in the Pleistocene era.

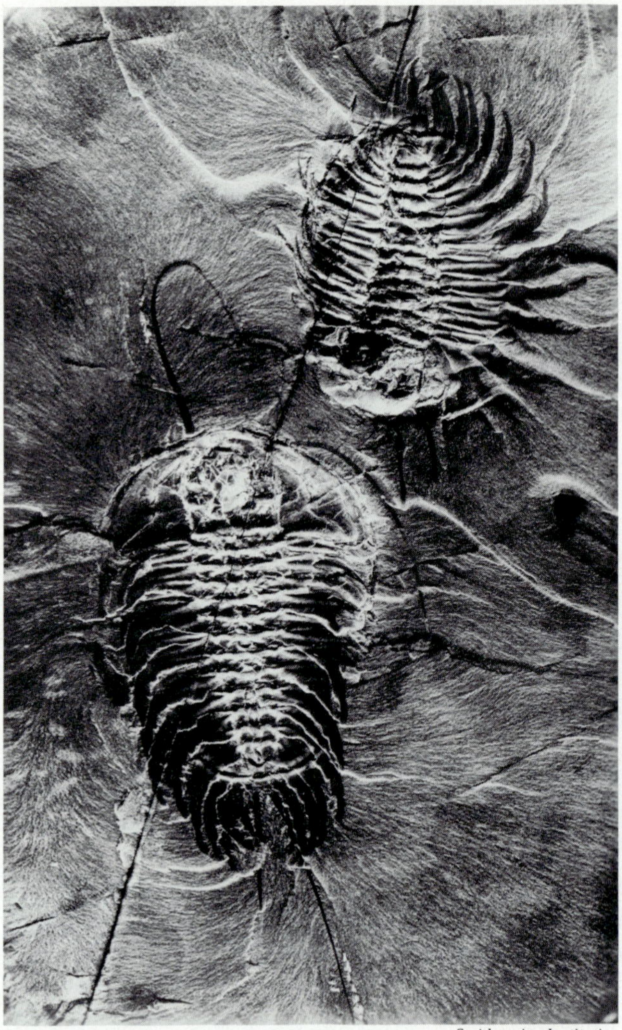

Smithsonian Institution

Figure 5.28
Trilobites, such as the ones shown here, are among the most widely used index fossils for rocks of Cambrian age. Trilobites ranged in size from a few centimeters to as much as 50 centimeters in length.

Measuring Geologic Time

The geologic time scale was developed long before the techniques of radiometric age determination were developed. The scale shown inside the back cover was based on the relative ages of events, and no errors have been found in the dating done in this way. Ideally, we would like to know how much time passed between events in addition to knowing just the order in which they occurred. Two important techniques are used to measure geologic time.

The first of these involves identification and use of processes that are unidirectional and nonreversible, like the movement of the hands of a clock. A clock would be of no use if one could not depend on the hands to move in one direction all the time. It has the added advantage of moving at constant rate. Some "geologic clocks" do not proceed at constant rates; therefore, they cannot be used to tell absolute time. Nevertheless, they are useful in fixing the time of one event in relation to the time of some second event because they process in only one direction. The primary unidirectional nonreversible processes in nature include the decay of radioactive substances, as discussed in Chapter 2. This is the best geologic clock because the rate of decay is constant so that the dates obtained are **absolute dates.** Next most important has been the evolution of life. Evolution has proceeded since the first forms of life appeared on earth, and evidence of the process is found in the fossil record. Many of the same types of animals are represented through long periods of time, and some fossils occur only in rocks of a particular sedimentary facies; but, as a group, fossils of any one time are a unique assemblage. These assemblages provide a reference system for determining the order of succession of geologic events. Later we will see that on a shorter scale the gradual accumulation of sediment, the growth of plant matter as exemplified by tree rings, and even the extent to which the land is eroded are sometimes used.

Events that recur in a regular way, such as the annual cycle of summer and winter, offer a second technique by which geologic time can be measured. Such processes are not uncommon, but geologists disagree about the existence of such processes on a worldwide scale. Locally, seasonal variations in climate influence the types of sediments deposited in lakes and the rate of their deposition. Some geologists believe evidence supports the existence of periodic changes of greater extent such as periodic changes in the world climate, bringing periods of glaciation in high latitudes, and periodic movements within the earth's crust, causing deformation and formation of mountain ranges. More must be learned about the periodicity of such processes before we can use them to interpret the geologic record.

Despite continuing arguments, we have succeeded in piecing together the broad outlines of the earth's history, and in some instances a detailed history, by analyzing the history recorded in rocks, the stratigraphic record.

The Geologic Concept of Time

A variety of evidence points to the earth's being more than four billion years old. It is hard to comprehend the implications that this vast span of time holds for the processes acting on the surface of the earth and within its interior. For example, rises in sea level too small to be detected within a lifespan could flood most of the continents if continued over a period of millions of years. Seas have invaded the continents in the past and have, in fact, covered almost all the land surface of the world. Some areas are being slowly submerged today; others are emerging, perhaps to rise eventually as mountains.

The importance of long time spans on the action of geological processes can be illustrated by changes in the land. How could a canyon like the Grand Canyon, several kilometers wide and as much as 1.6 kilometers deep, have been formed? Without consideration of the time involved, it is a perplexing problem. But for those who know how streams erode and transport rock debris, it is not difficult to visualize the deepening of the canyon over a period of hundreds of thousands of years as the rocks exposed in the walls of the canyon gradually crumble, break away, and fall into the Colorado River, which transports them to the sea. The canyon is being enlarged today in about the same way and perhaps as fast as it ever has been in the past. The face of the earth has never before looked exactly the way it does right now; it will never be the same again. It is not even the same from day to day, but the changes are so slow that we generally fail to perceive them all. In the short term, these changes seem insignificant; in the long term, they may convert areas occupied by seas into mountains, and mountains into plains. The cumulative effects of natural processes over periods of geologic time have brought about the record of change that is the history of the earth. In the next chapter we shall see what information igneous rocks, formed as magma cools, can contribute to this record.

SUMMARY

Most of the earth's surface is covered by a layer of sediment and sedimentary rock that constitutes the best existing record of the earth's history and the evolution of life.

Geologists infer earth's pregeological history from the rock record preserved on the moon. There a dusty soil of igneous rock fragments was formed by meteorite impacts and volcanic explosions. In he absence of water or atmosphere, the moon's surface preserves a record of its earliest history. On earth, a primitive atmosphere formed as the result either of volcanic activity or of volatiles that remained as the earth solidified. In either case, the original atmosphere gradually evolved to its present composition by adding oxygen from photochemical dissociation and later from photosynthesis, and by losing atmospheric carbon dioxide through weathering of rocks and incorporation in living organisms.

Weathering of surface rock by the action of the atmosphere and water breaks it down into fragments. Fracture of preexisting rock and volcanic activity produce broken, or clastic, sediments. Reactions of ions dissolved in water produce chemical precipitate sediments, and the activity of organisms caused biochemical deposition of sediments. All three types find their way to low places like the ocean floor, where they are gradually compacted and cemented to form sedimentary rock.

Classification of sedimentary rock is based on texture, composition, and occasionally on mode of deposition as well. Rock texture and structure is often indicative of the environment in which the sediment was deposited. From study of modern sites of sedimentation, geologists can identify the environment in which many ancient sedimentary rocks were laid down.

Study of the continental shelves disclosed surprising deep canyons eroded in the sea floor. The stratification of nearby sediments and their microfossils indicated that the canyons and the sediment deposits are both the result of turbidity currents set in motion by earthquakes and accompanying undersea landslides.

In the trenches formed at convergent plate boundaries, debris from adjacent volcanic islands is trapped and accumulates together with biological sediments—a mixture that characterizes such trench sediments. Deposition on tectonically quiet continental margins shows much different features; sediments there contain largely nonvolcanic clastic sediments that originated on the continent.

The sedimentary rock record is read by recognizing first that in undisturbed sequences, younger rocks overlie older ones. This law of superposition, together with the fossils found in many sedimentary rocks, allows the development of a geologic time scale of relative ages. When information from such techniques as radiometric dating is added, we can begin to develop an absolute time scale for geologic events.

KEY TERMS

absolute date
angular unconformity
argillaceous
bottom-set bed
calcareous
carbonaceous
Cenozoic era
chemical weathering
clastic
coccolith
colloid
cross-bedding
degassing
diatom
diatom ooze
disconformity
dynamic encounter
era
ferruginous
flagellate
foraminifer
forest bed
fossils
glass
globigerina
Holocene
law of faunal succession
law of superposition
lithification
load casts
Mesozoic era
microbreccia
oolites
oscillation ripple mark
Paleozoic era
period
photochemical dissociation
Pleistocene
primary rock
Quaternary period
radiolarian
red clay
red tide
secondary rock
sedimentary facies
siliceous
sorting
Tertiary period
turbidity current
varves
weathering processes

STUDY QUESTIONS

1. Briefly outline the evolution of the atmosphere according to the residual and degassing hypotheses.
2. What types of geological evidence could indicate the composition of the atmosphere at various times in the past?
3. Compare the textures of clastic sediments with those of sediments formed as a result of chemical precipitation.
4. How could you recognize an ancient turbidity current deposit now preserved on land?
5. What materials serve as cement in sedimentary rocks, and what are their sources?
6. Make a list of the different types of environments in which sediment is being deposited in the region where you live.
7. Prepare a detailed account of the events you can recognize or infer from the photograph of an unconformity in Fig. 5.26.
8. How do absolute and relative dates differ? What could cause absolute dates to be wrong?

REFERENCES

1. M. F. Maury, *The Physical Geography of the Sea.* London: Thomas Nelson and Sons, 1893, p. 314.
2. Rachel Carson, *The Sea Around Us.* New York: Oxford University Press, 1951, p. 74.

SUGGESTED READINGS

Berry, W. B. N., *Growth of a Prehistoric Time Scale.* San Francisco: W. H. Freeman 1968.

Block, Joel, "The Bible and science on creation," *Journal of Geological Education,* 24:2 (1976), *58–60.*

Eicher, D. L., *Geologic Time,* 2nd ed. Englewood Cliffs, N.J.: Prentice-Hall, 1976.

Friedman, G. M., and J. E. Sanders, *Principles of Sedimentology*. New York: Wiley, 1978.

Gurin, Joel, "In the beginning," *Science 80*, 1:5 (1980), *44–51*.

Hume, J. D., "An understanding of geologic time," *Journal of Geologic Education*, 26:4 (1978), *141–143*.

Janos, L., "Timekeepers of the solar system," *Science 80*, 1:4 (1980), *44–55*.

Ojakangas, R. W., and D. G. Darby, *The Earth, Past and Present*. New York: McGraw-Hill Book Co., 1976.

Pettijohn, F. J., *Sedimentary Rocks*, 3rd ed. New York: Harper & Row, 1975.

Selley, R. C., *Ancient Sedimentary Environment*, 2nd ed. Ithaca, N.Y.: Cornell University Press, 1978.

Stewart, R. W., "The atmosphere and the ocean," *Scientific American* offprint #881 (1969)

ROCKS BORN OF FIRE: IGNEOUS ACTIVITY

Chapter 6

The first rocks to be formed on earth were of igneous origin; they crystallized as rock melts, or magmas, cooled. Some magmas erupt on the earth's surface as lava. Others crystallize at depth and are exposed only after uplift of the crust and long periods of erosion remove the original cover. In this chapter we shall see how information obtained from the study of igneous rocks and their geological setting helps us to understand how they are generated and why they occur where they do.

The serious study of igneous rocks began as a major controversy over the origin of the rock called basalt. Basalt is the most common rock type in the sea floor, and it is abundant on continents as well. It has been studied in detail in the field and in the laboratory, and may be a rock from which many other igneous rocks are derived. Certainly basalt is one of the most important igneous rocks, and we will begin this chapter with a look at this early controversy over its origin.

6.1 THE ORIGIN OF BASALT

Basalt is a dark, fine-grained rock that contains plagioclase feldspar, pyroxene, and often olivine. For more than thirty years during the late eighteenth and early nineteenth centuries the origin of basalt was argued so vehemently that it virtually split the geologists of that time into two camps—the Vulcanists, who argued for a volcanic origin, and the Neptunists, who argued for a sedimentary origin. This argument would never have taken place at all if geologists had been able to travel around the world as freely as we do today. The center of the arguments was northern and western Europe and Great Britain, far away from the active volcanic centers of the world. Consequently, the disputants never had an opportunity directly to observe the eruption of basaltic lava.

A French geologist, Jean Guettard (1715–1786), holds the distinction of having provided arguments for both the Vulcanist and the Neptunist camps. After studying the Auvergne region of central France, Guettard concluded that the hills of this region, which contain basalt, were extinct volcanic centers. He then compared them with the territory around Vesuvius, an area about which he had read. Despite this experience in Auvergne, he later wrote a paper explaining why basalt could not be a volcanic rock. In doing so, he obviously overlooked his own field evidence.

Another French geologist, Nicholas Desmarest (1725–1815), presented the first and one of the most definitive arguments for the Vulcanist (*1*): "On the way back from the Puy de Dôme [a volcanic center in France], I followed the thin sheet of black stone and recognized in it the characters of a compact lava." He went on to point out the observational evidence:

1. There was an underlying bed of scoria—a recognized volcanic rock.
2. The basalt could be traced back to the base of a hill he believed to be an ancient volcano.
3. He could observe a streamlike flow within the thin basalt layer.

IGS Photo

Figure 6.1
The Giant's Causeway. This outcrop of basalt shows distinctive columnar shapes formed as a result of the cooling and contraction of the magma. Individual columns are about 25 cm across.

Desmarest also described prism-shaped columnar masses in the basalt identical to those found at the Giant's Causeway in Ireland (Fig. 6.1), a feature we now interpret as resulting from the cooling of lava or melts. Although these observations, reported about 1774, were known to the German professor Abraham G. Werner, Werner nevertheless pronounced that in regard to basalt, he ". . . found not a trace of volcanic action, or the smallest proof of volcanic origin" (2). Werner had already begun to develop his theory of a universal ocean that covered the world and from which most rocks had been deposited. He interpreted basalt as a compact sedimentary rock. To support his position, he pointed to layers of basalt that are bounded above and below by sedimentary layers. Werner, through his forceful personality and extraordinary speaking ability, succeeded in convincing many scientists of the day. Desmarest stayed out of the debate; he simply advised those who asked his opinion to go to Auvergne and see for themselves.

It was the Scotsman, James Hutton, who made and reported the observations that eventually convinced most geologists that basalt is of igneous origin (Fig. 6.2). He pointed out that basalt never contains fossils; that it is not stratified like sedimentary rocks; that it does not have a fragmental character; that it is always insoluble, hard, and crystalline; and, perhaps most important, that layers and masses of basalt often cut across other rocks. This last feature is more definitive than most of the others, and is referred to as a **cross-cutting** relationship. Magma is highly mobile and always moves along the lines of least resistance. In the case of rocks, this route is generally cracks in the rocks or between layers of sedimentary rock. For this reason, basalt and most other igneous rocks can be found not only on the surface of the ground as a lava flow, but also as layers injected between sedimentary rocks, a form called a **sill,** and as thin planar masses that cut across sedimentary layers, a form known as a **dike** (Fig. 6.3). Formations

IGS Photo

Figure 6.2
This famous castle at Northumberland, England, is built on a foundation of basalt that cuts across a sandstone layer. Observations at such locations convinced James Hutton that the basalt was once molten.

IGS Photo

Figure 6.3
The dark rock is a vertical dike, composed of basalt, cutting across nearly horizontal layers of sediment.

of this kind are all **intrusions,** bodies formed when molten rock intrudes into existing rocks below the surface and crystallizes there. Rock formations that result from lava that reaches the surface and cools there are, by contrast, **extrusions.**

Several additional criteria should be added to those mentioned by Hutton and Desmarest as ways of identifying igneous intrusions. Because magma is usually very hot compared to the surrounding or **country rock** which it intrudes, the magma at the margins of an igneous body is cooled rapidly, producing a very fine-grained and sometimes glassy contact zone or **chilled margin.** If the magma contains gases or watery fluids, these may invade the country rock, altering it. Even if the melt contains little water, the magma's heat alone generally causes thermal effects in the intruded rock. These changes in the composition of the rock around the magma are referred to as **contact metamorphism;** they are discussed in greater detail in Chapter 7. Because some sediments, especially limestone and shale, are easily altered by the heat and solutions found in magmas, contact effects are an excellent indication of intrusions.

6.2 IGNEOUS ROCKS

Minerals

The common igneous-rock-forming minerals include quartz, olivine, amphibole (hornblende), pyroxene (augite), the feldspars orthoclase and plagioclase, and the micas, all of which were described in Section 2.4. Many other minerals occur in igneous rocks, but the classification of the most common igneous rocks is based on the relative proportions of the minerals listed above.

Unless its magma cooled very slowly, the individual minerals in a rock are generally very small, and it may be difficult to make some distinctions in hand specimens. The following guidelines may help. Quartz resembles glass and may be clear enough to transmit light. It often appears to have a grayish cast. Olivine also resembles glass, but it is usually brownish or greenish in color and never occurs in the same rock with quartz. If mica is present, flakes can be picked off the rock. Biotite mica can be easily mistaken for amphibole or pyroxene, but neither of these will produce flakes. Amphibole and pyroxene are very difficult to distinguish. Both are ordinarily black or dark in color, and their cleavages are not often seen in small crystals. Orthoclase is commonly pink in color; plagioclase tends to be white or gray. Large plagioclase feldspar crystals may show striations.

Textures

The common igenous rock textures are described and shown in Table 6.1 and Fig. 6.4. Because igneous rocks crystallize from liquids, their textures tend to be distinctively different from those of most sedimentary rocks. Certain members of these groups can, however, be confused for two reasons. First, in order for large crystals to form, giving a rock a coarse-grained **phaneritic** texture, the magma must cool very slowly. If magma cools rapidly, coarse and easily recognized crystalline textures do not have sufficient time to form. The resulting rock is **aphanitic,** very fine-grained, with small crystals. It lacks distinctive texture and may resemble a shale, fine sandstone, or some limestones. However, igneous rocks are never composed of calcium carbonate or of clay minerals; so, if these can be identified, the rock is sedimentary. Because the minerals that make up most igneous rocks are hard, fine-grained igneous rocks are generally much harder than fine-grained sedimentary rocks such as shale or limestone. The second reason that igneous texture can be confused with sedimentary textures is that some sediments are crystalline. The two can usually be distinguished because the common crystalline sedimentary rocks have distinctive mineralogy (such as salts, gypsum, calcite, and quartz).

The best way to see and study the texture of igneous rock is with the aid of a microscope. Sections of rocks thin enough to transmit light allow excellent definition of the grain sizes and crystal boundaries. A mosaic-type intergrowth of different minerals is typical of all igneous rocks except glasses, which show no crystal structures.

Table 6.1 Textures of Igneous Rocks.

Texture	Examples	Characteristics
Natural glass	Glass	Amorphous (without orderly internal atomic structure) and noncrystalline
	Obsidian	A compact, silicon-rich glass
	Pumice	Vesicular (with bubble-like spaces) glass rich in silicon
Fine grain size	Aphanitic	Crystal grains too small to be identified without magnification
	Scoria	Dark, very fine-grained, vesicular rock
	Basalt and rhyolite	Common fine-grained extrusive and shallow intrusive rocks
Coarse grain size	Phaneritic	Grains large enough to be seen without magnification; sizes vary greatly; crystals several centimeters across are common
Mixed textures	Prophyry	Crystals (called phenocrysts) are present in a finer-grained or glassy groundmass

Natural glasses are formed where magma reaches the surface and is cooled before crystals grow. Glasses formed in this way are called obsidian. When the lava is charged with gases during eruption, bubbles are usually formed in the glass or fine-grained rock, creating scoria or pumice (see Table 6.1 and Fig. 6.4).

Classification of Igneous Rocks

Igneous rocks are classified according to composition and texture. Composition is estimated on the relative proportions of the common minerals present. Once composition is known, reference to a chart like that shown in Fig. 6.5 gives the rock type. For example, a rock composed of 70 percent plagioclase, 10 percent pyroxene, and 20 percent hornblende is a diorite.

It is important to note that these compositional varieties are gradational, so the distinction between

Figure 6.4
Textures of the most common igneous rocks: (a) coarse-grained; (b) fine-grained (aphanitic); (c) porphyry; (d) scoria; (e) pumice; (f) glass. These specimens are about 10 cm wide.

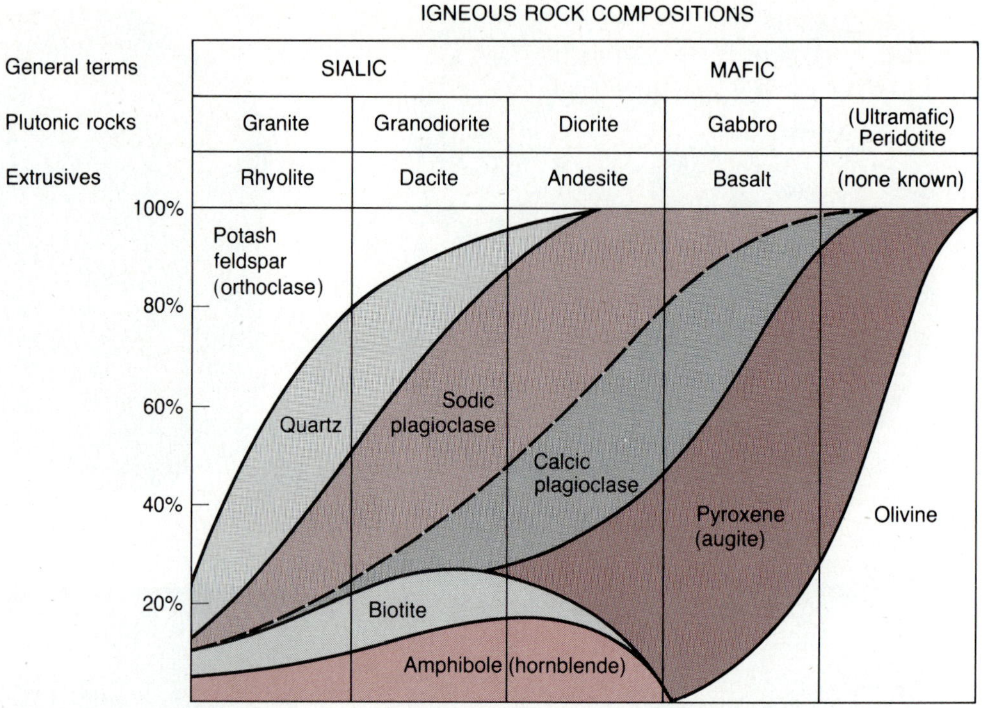

Figure 6.5
This chart depicts the relative proportions of common minerals in plutons and their eruptive equivalents (such as granite and rhyolite). Read compositions along any vertical line. Note that all rocks of a given name do not have exactly the same composition, but their composition must fall within given limits. For example, any plutonic rock with mineral compositions in the left column is called granite.

rock types that are adjacent in the chart may not always be sharp. Grain sizes also show a continuous gradation.

Despite the existence of these gradations, most igneous rocks can be easily classified. The average compositions (Table 6.2) of the most common extrusions (eruptions of magma at the earth's surface) and their intrusive (plutonic) equivalents are distinctively different. Mineral grains of some of the common plutonic rocks are depicted in Fig. 6.6.

Table 6.2 Representative Compositions of Some Common Extrusive Igneous Rocks.

Composition	Basalt (weight %)	Andesite (weight %)	Rhyolite (weight %)
SiO_2	51	57	74
Al_2O_3	13	18	13
FeO	12	5	1
CaO	9	8	1
MgO	5	4	0.3
Na_2O	3	3	3
TiO_2	3	1	0.2
K_2O	1	1	5

6.3 THE FORMATION OF IGNEOUS ROCKS

The events that happen during the change of magma into igneous rock can be studied in three quite different ways. In a few places it has been possible actually to observe the complete cooling and crystallization of a lava in its natural setting. In many other localities we can examine igneous rock bodies that were formed long ago and are exposed as a result of erosion. We can study these rocks and their geologic setting, and from these infer the conditions under which they are formed. And finally samples of rock can be melted in the laboratory where the conditions under which they solidify can be controlled. Each of these approaches has contributed to our knowledge of the origin of the igneous rocks. We will first consider the cooling of lava in the Hawaiian Islands. What we learn there will be compared with the results of experimental studies, and later we will examine the older igneous rocks exposed in the Sierra Nevada Mountains.

Crystallization of Hawaiian Basalt: A Case Study

On the island of Hawaii a number of roughly circular depressions have formed near the centers of volcanic eruptions. Some of these depressions have been filled by lava during later eruptions, forming lava lakes and creating ideal conditions for the study of the cooling and crystallization of the basalt lava that erupts in Hawaii. Volcanologists from the U.S. Geological Survey observatory station in Hawaii have undertaken long-term systematic monitoring and sampling of several of these lava lakes formed during eruptions in 1959, 1963, and 1965. In each of these cases it was possible to sample the lava soon after it reached the surface and subsequently to trace its history by drilling holes through the surface of lava lakes into the melt below the solidified crust. The core samples collected from these holes were analyzed to determine their mineralogy and chemistry. The temperature in the crust and its thickness can be directly measured.

When it first enters these lakes (Fig. 6.7), the lava is close to 1200 °C. The surface cools very rapidly where it is in contact with the atmosphere. Movies show the almost immediate color changes as the surface is chilled and turns nearly black. Some of this cooled surface rock sinks, reducing the temperature in the upper part of the lake, but a month after the eruption, the temperatures at a depth of only 10 feet are still about 1130 °C. The boundary between the rigid crust and the melt is sharp and occurs where the temperature is about 1065 °C. Samples from below the crust–melt interface can be collected by pushing a container down into the melt. Figure 6.8 shows the cooling history recorded for one lake from its eruption in 1965 until the time in 1969 when a second eruption reflooded the lake with new lava. The lines of the graph show the depth at which different temperatures were found. At first the cooling migrated downward linearly, almost as a function of the square root of the elapsed time, but later cooling became more rapid. The rather abrupt changes near the surface are related to periods of heavy rainfall.

Petrographic studies of the grain size and composition of the newly crystallized lava were made by preparing thin sections and chemical analyses at two-foot intervals along the cores. The surface of the lava

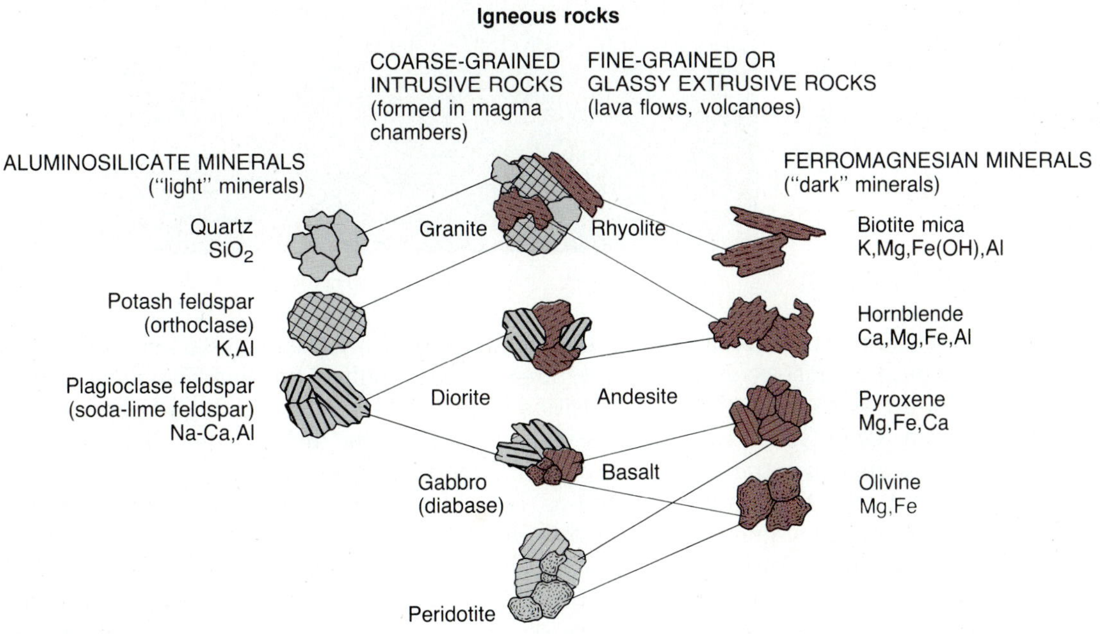

Figure 6.6
Coarse-grained igneous rocks and their main mineral constituents. Representative mineral grain shapes and growth patterns are depicted. The grains tend to form an interlocking mosaic pattern. Individual crystals are commonly several millimeters, sometimes several centimeters across.

(a)

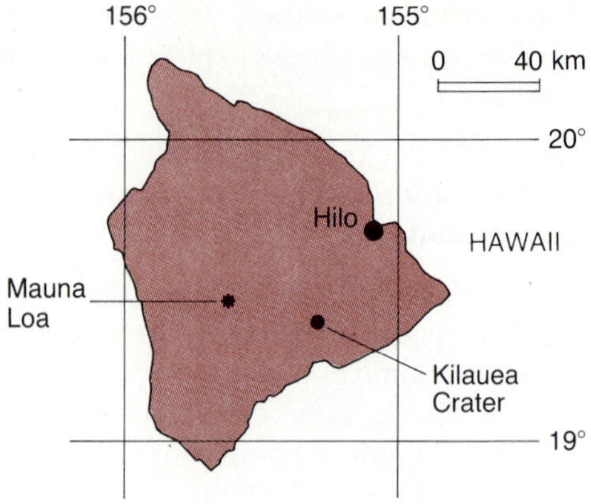

(b)

Figure 6.7
(a) Basaltic lava cascades into a lava lake several hundred meters long in Hawaii. Note the ripples formed on the surface of the lake. (b) Locator map.

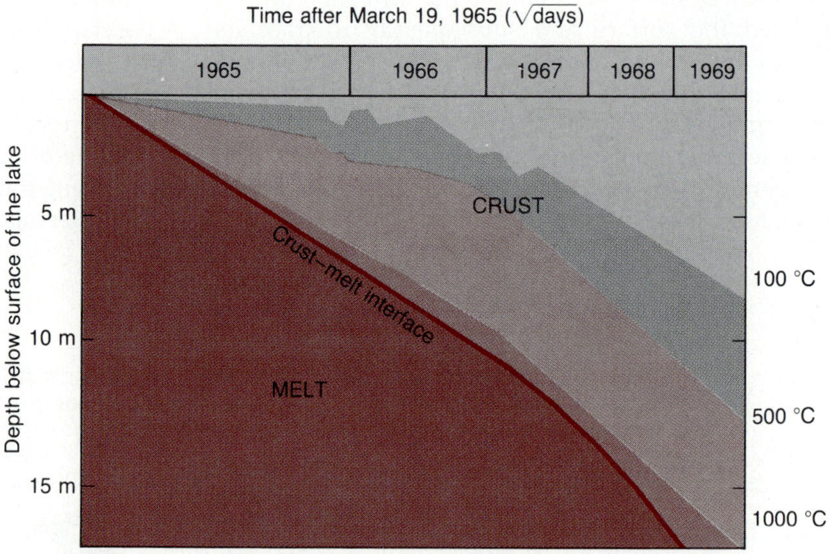

Figure 6.8
A plot of the depth of the interface between magma and solidified basalt in a Hawaiian lava lake. For example, the crust–melt interface was about 9 meters deep in January 1967. Depths to temperatures of 100 °C, 500 °C, and 1000 °C are also shown. The irregularities in the 100 °C curve are due to heavy rains that cooled the upper part of the crust.

lake is made up of glassy basalt and scoria (basalt containing holes left by gas bubbles) because cooling of the lava at this air–lava interface is so rapid that crystals do not have time to form. Grain size measurements reveal that the basalt is very fine-grained through the upper six meters, but after that, grain size increases with depth. This typical fine-grained border is a feature of many igneous intrusions and is attributed to rapid crystallization of the melt with insufficient time for crystals to grow. At greater depth the magma is liquid long enough for crystal growth to build larger grains.

Olivine is the first mineral to form as a basaltic lava cools. It grows as small perfectly shaped crystals in the melt. Not only are these crystals of olivine the first minerals to crystallize, but they also have a high density and can sink through the high-temperature, less-viscous part of the melt. If the olivine settles through the melt rapidly, none is left by the time the melt is cool and viscous enough to keep olivine crystals in place. Thus the resulting rock contains little or no olivine. For this reason crystals of olivine are found only in the upper part of the cores. Olivine decreases both in amount and in the size of the crystals at depth, and none is found in the cores taken below 10 meters. Olivine found in the top of the lava lake indicates that lava must have cooled before the olivine could sink.

Interpreting Texture

Field studies in Hawaii and elsewhere indicate that uncrystallized glassy material forms when a lava cools very quickly. This is true both of the basalt found in Hawaii and of lavas found in other areas and composed of very different materials. The sizes of the crystal grains are apparently affected by the rate of cooling, which determines the amount of time available for crystal growth. We have noted that in the basaltic lava of Hawaii, olivine is the first mineral to crystallize from basalt lava, and that it has a perfect crystal shape. This suggests that it should be possible to determine the sequence of mineral formation in a rock by studying the relationship of the mineral textures. The first-formed minerals have a good chance to grow into perfect forms because they are surrounded by liquid, but those crystallizing later have to fit into the spaces left among the previously formed minerals. It should also follow that any mineral found inside another mineral formed before the enclosing mineral. Sometimes we find a crystal that has the crystal shape of one mineral but the composition of another. Two quite different mechanisms explain these crystals. Sometimes the first-formed mineral is later replaced, molecule by molecule, as a result of a chemical reaction. Thus we find the mineral limonite in the cubic form of pyrite. In the second mechanism, the composition of the magma changes while the crystal is forming. This explanation is confirmed by the discovery that some crystals are zoned, like the plagioclase crystal in Fig. 6.9; they have a core of one composition and an outer rim of another.

If fine grain size is indicative of rapid cooling and coarse grain size slow cooling, it follows that a rock composed of minerals of two very different grain sizes may have cooled at two different rates. Rocks that are made up of relatively large and usually well-formed crystals (called **phenocrysts**) in a groundmass of glass or finer-grained minerals are known as porphyries (refer back to Fig. 6.4*c*). Many porphyries are thought to have formed in magmas that cooled slowly at first, perhaps at depth, followed by rapid cooling. This pattern might appear, for example, if the partially solidified crystal mush, containing large well-formed crystals, moves to a cooler shallow depth or is extruded onto the surface where it completes crystallization rapidly.

The cooling history of magma—both the rate at which it cools and whether it cools uniformly or experiences changes in cooling rates—is clearly a factor in determining the texture of the resulting rock. The presence or absence of nuclei of condensation and the composition of the melt, which determines at what temperature crystals will start to grow, also influence the final texture of the rock.

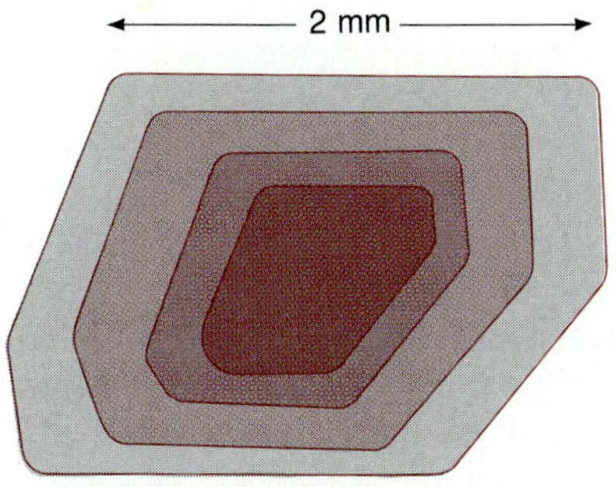

Figure 6.9
Zoned plagioclase showing a core surrounded by concentric bands of different compositions.

Experiments with Rock Melts

The objective of experimental petrology is to learn exactly what happens when a lava or magma cools and becomes tranformed into rock. It is not enough to know the temperature and composition of the melt and the texture and mineral composition of the end product. Experimentalists are trying to understand the physical and chemical processes that take place at every stage of this complex phenomena.

Magmas are multicomponent silicate melts, usually containing some solid minerals, liquids at high temperature, and gases under pressure. Before attempting to understand anything as complicated as a multicomponent silicate melt, we shall consider what happens in a simple single component system as the temperature and pressure are changed. We are all familiar with temperature-induced changes in the state of water. In normal atmospheric pressure, water passes into a gaseous state (it boils) at 100 °C and freezes to form the mineral ice at 0 °C. The temperatures of these phase changes are affected by variations in pressure. All minerals undergo similar changes of phase, but the conditions necessary for the changes differ greatly for different minerals.

The processes by which a melt, even one with a single component, becomes solid are complex. If the cooling is fast, the melt freezes before crystals can be built; but given sufficient time, the necessary atoms or ions will assemble to build an orderly crystalline material as the temperature falls. This may take place through the process of chemical precipitation. A melt

CRYSTALLIZATION OF A MIXTURE OF PLAGIOCLASE AND PYROXENE

Crystallization of the two-component melt composed of the minerals diopside (a pyroxene, $CaMgSi_2O_6$) and anorthite (a plagioclase feldspar, $CaAl_2Si_2O_8$) shows characteristics of a **simple eutectic system**. This type of system deserves special attention, because these two minerals are important constituents of basalts and gabbros. The curve (Fig. 6.10) shows the temperature at which crystals first form in a melt, the composition of which is given on the horizontal axis. At temperatures above the solid curve, the system is in a molten state. Thus, a system composed of 50 percent diopside and 50 percent anorthite is totally molten at 1400 °C, but a system of anorthite alone is solid at that temperature. If a melt of any given composition is cooled, the composition remains constant until the curve of crystallization is reached.

Assume, for example, that a rock containing 25 percent anorthite and 75 percent diopside is heated until all solid crystals are melted, and then the melt is allowed to cool slowly. As the temperature drops, the melt reaches a point at which the first solid appears at a temperature of 1310 °C, when diopside starts to crystallize. The resulting crystals are pure diopside, not a mixture of diopside and anorthite as you might assume. This temperature is unique for a magma composed of 25 percent anorthite and 75 percent diopside. As crystals of diopside continue to grow, diopside is removed from the melt so that the new composition of the melt is shifted along the crystallization curve progressively toward anorthite. The temperature necessary to crystallize this composition is slightly lower, so more diopside forms, the composition of the melt shifts further, and the temperature of crystallization is lowered correspondingly. This process continues until a temperature of 1265 °C, also unique for these two minerals, is reached; this temperature is called the eutectic point. At the eutectic point both diopside and anorthite crystallize; so, as the cooling process continues, the relative amounts of the two minerals in the melt do not change, and crystallization continues until the entire melt is a solid of the original bulk composition. The resulting rock is composed of large crystals of diopside and a groundmass composed of both diopside and anorthite, which formed last. If, on the other hand, the composition of the melt lies on the anorthite side of the eutectic, the first crystals to form are anorthite, and diopside crystallizes only after the eutectic is reached.

that contains an ionic compound is said to be saturated with respect to that compound when no more of the compound can be dissolved in the melt without causing some of the ions to combine and precipitate as a solid. Because solubility generally decreases with a decrease in temperature, precipitation may be induced by decreases in temperature. It may also occur as a result of changes in the composition of the melt or changes in the proportions of the various substances in the melt. In the following sections and in the boxed notes on crystallization of feldspars, examples of these types of reactions are described.

Crystallization of basalt: Bowen's reaction series. Of the many experiments with basalt, the pioneering work of N. L. Bowen is especially important. He demonstrated that the principles of physical chemistry can be successfully applied to the problems of rock genesis. Bowen melted samples of basalt and allowed them to crystallize under carefully controlled conditions of temperature and pressure. He stopped the cooling of samples at various stages by suddenly quenching (rapidly cooling) the sample. He could then determine what minerals had already formed and what reactions were taking place between the

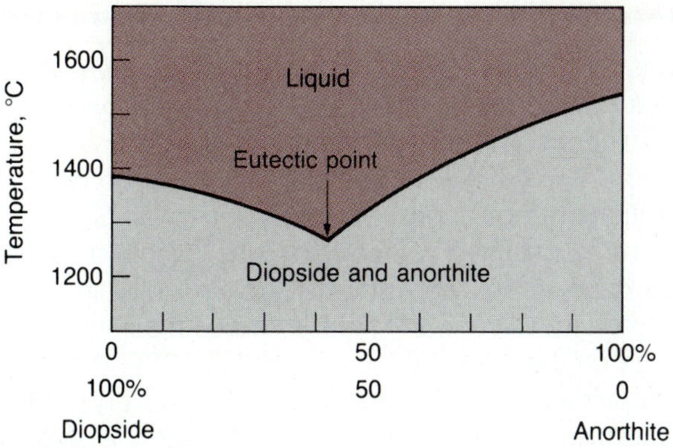

Figure 6.10
Simple eutectic system, diopside–anorthite. Pressure effects are not shown.

minerals and the melt at the temperature just before he quenched the sample. Bowen prepared a diagram that depicts his discoveries (Table 6.3), now known as Bowen's reaction series. He found that the first mineral to form in a basalt melt is olivine, followed by pyroxene and calcium plagioclase feldspar. Pyroxene does not begin to crystallize until olivine crystallization is complete, but then a reaction may occur between the olivine and the melt that can alter some or even all of the olivine to pyroxene. The pyroxene may react with the remaining melt to produce amphibole, and sodium replaces calcium in the plagioclase. Experiments with the residual fluids show that at lower temperatures, biotite and muscovite micas are formed and finally some quartz may crystallize. The remaining solutions may possibly yield still other less common minerals.

Bowen himself recognized this reaction series as a simplification of what was actually happening. The complexity becomes apparent as we realize that basalt generally contains at least three minerals: olivine, pyroxene, and plagioclase feldspar; that these minerals are made up of the elements silicon, oxygen, aluminum, iron, magnesium, calcium, and sodium; and also that some potassium and titanium are usually present. All these elements are found in basaltic magma at 1200 °C, and they are free to form many possible combinations. As soon as the first minerals crystallize, some of the elements are removed from the melt, leaving it enriched in the remaining ones. The behavior of a melt that has the elements needed to form two minerals is quite different from that of a melt that contains only the elements of a single mineral. One notable difference is the effect on the temperature at which crystallization starts. For example, a magma formed by melting the pyroxene, diopside, will crystallize at just under 1400 °C, but if molten plagioclase is added, the crystallization temperature of both is lowered. In addition, various types of reactions can take place between the previously formed crystals and the remaining melt. Two of the types of interactions involved in basaltic magma are described in the boxes on crystallization of a melt composed entirely of plagioclase feldspar and of a melt composed of plagioclase and a pyroxene (diopside). As reactions occur, the compositions of both the solid or mineral phases and liquid phases change, and this process continues until the last bit of melt

Table 6.3 Bowen's Reaction Series; Minerals Forming in a Basaltic Magma as It Cools.

Minerals		State of the magma	Temperature
		Magma is totally molten	1200 °C
olivine	calcium-rich plagioclase	First crystals form in the melt	1125 °C
↘	↘		
pyroxene		Reactions between crystals and the melt lead to the formation of minerals as shown	
↘			
amphibole			
↘			
biotite	sodium-rich plagioclase		
potash feldspar muscovite quartz	}	minerals which might form from residual fluids (note: these minerals . . . . are not usually present in basaltic rocks)	600 °C

CRYSTALLIZATION OF PLAGIOCLASE FELDSPAR: AN EXAMPLE OF A CONTINUOUS REACTION SERIES

Because the feldspars constitute about 60 percent of all silicate minerals, they are the most important rock-forming minerals. The group is divided into two distinct types of minerals: the potash feldspars, which have a single chemical composition, $KAlSi_3O_8$; and the plagioclase feldspars, whose chemical compositions range from $CaAl_2Si_2O_8$ to $NaAlSi_3O_8$. Calcium ions (Ca^{2+}) can be substituted for sodium ions (Na^+) in the plagioclase because they have about the same radius; but aluminum (Al^{3+}) also must be substituted for silicon (Si^{4+}) at the same time in order to maintain a neutral charge on the network.

Plagioclase feldspar provides a good example of a type of crystallization system known as a solid-solution series. Albite ($NaAlSi_3O_8$) and anorthite ($CaAl_2Si_2O_8$) are the end members of this series, as shown in the phase diagram in Fig. 6.11. Below the experimentally determined lower curve, the solidus, only crystals exist; and above the upper curve, everything is molten. If a melt of composition 25 percent anorthite and 75 percent albite is cooled (the dashed line in the figure), the first crystals will form when the melt is cooled to about 1,350 °C (point *a* in the figure); and these crystals will have a composition indicated by point *b* (that is, a liquid at point *a* is in equilibrium with crystals at point *b*). These first crystals contain more calcium than sodium; thus the liquid becomes enriched in sodium. Therefore, as the liquid cools and crystallizes, its composition shifts toward albite. The early formed plagioclase crystals will react with the sodium-enriched melt and sodium will be substituted for calcium in the feldspar crystal structure during this reaction. The composition of the liquid plagioclase moves along the liquidus curve. If the reactions go to completion, the final plagioclase will have the same relative amounts of calcium and sodium as the melt had initially; but usually the reaction is incomplete, and only the outer parts of the early formed crystals react. This produces a zoned crystal that has a core of the initial composition and a rim that has the later composition.

is crystallized. Solidification of melts follows this series even if they are so deficient in iron and magnesium that pyroxene and olivine can not form. In such cases, the lower part of the series applies.

Magmatic Differentiation

Careful examination of every large igneous rock mass reveals that it is composed of a number of different, though often closely related, igneous rocks. They vary not only in grain size and texture, both related to cooling rates, but usually in the mineral composition as well. The causes of some of the mineralogical

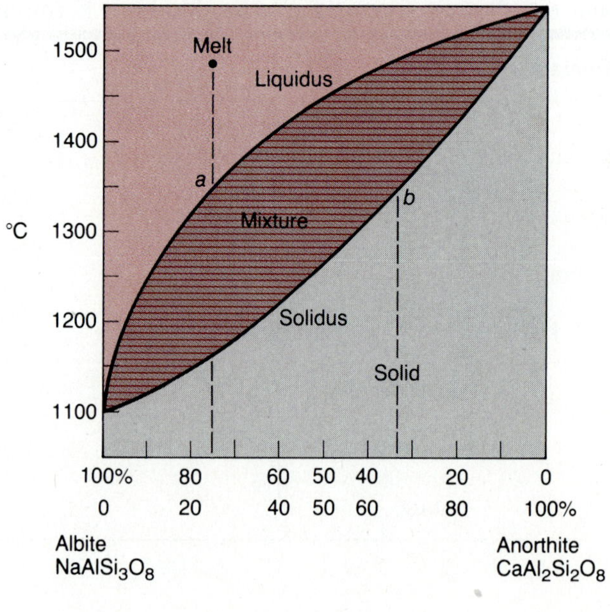

Figure 6.11
Phase diagram for the solid–solution series typical of the plagioclase feldspars. See the text for description of the reactions involved.

Table 6.4 Specific Gravity of Common Igneous Rock-Forming Minerals.

Mineral	Specific Gravity*
Zircon	4.6 –4.7
Olivine	3.2 –4.4
Pyroxene	3.2 –3.96
Amphibole (hornblende)	3.05–3.50
Mica (biotite)	2.7 –3.3
Plagioclase	2.62–2.76
K Feldspar	2.56–2.63
Quartz	2.27–2.65

* Values vary in some cases because composition is not uniform or because structure is not constant (as in quartz).

variations can be inferred from the way the different minerals are distributed in the intrusion. For example, olivine and chromite are often found as layers or in pocket-like masses near the bottom of intrusions of ultramafic (high magnesium and iron) composition. Both of these minerals crystallize at high temperatures early in the cooling of magmas in which they occur, and both have high density (see Table 6.4). It seems clear that they become segregated because, being denser than the rest of the melt, they sink, a process called **gravitational separation** (Fig. 6.12). In contrast, a few minerals, including some feldspars, are less dense than the melts in which they occur and rise toward the surface of the magma.

Veins that occur around the edges of most intrusions can be traced back to the main intrusive mass. Their compositions are often different from that of the main intrusion, and the minerals in them may be larger. Veins of quartz are commonly found in the country rock around granitic intrusions; some of these veins contain traces of native gold or silver. Other irregular veins, called pegmatites or pegmatite dikes (Fig. 6.13), contain quartz, mica, and potash feldspar accompanied by small quantities of relatively rare minerals such as beryl or tourmaline. These quartz veins and pegmatites can be traced back into the granitic intrusions. The fact that pegmatites cut across the larger bodies with which they occur proves that the intrusions were in place before the dikes were intruded. Therefore, the dikes must be younger than the intrusions. These relations, as well as their mineralogy, suggest that they are formed from the last fluids left as the magma crystallizes.

Still other less discernable processes may account for the differentiation of one magma into many rock types. It seems probable that some magma chambers become squeezed as the regions around them are deformed. If this happened when the magma is partially solidified, the fluids may be squeezed out

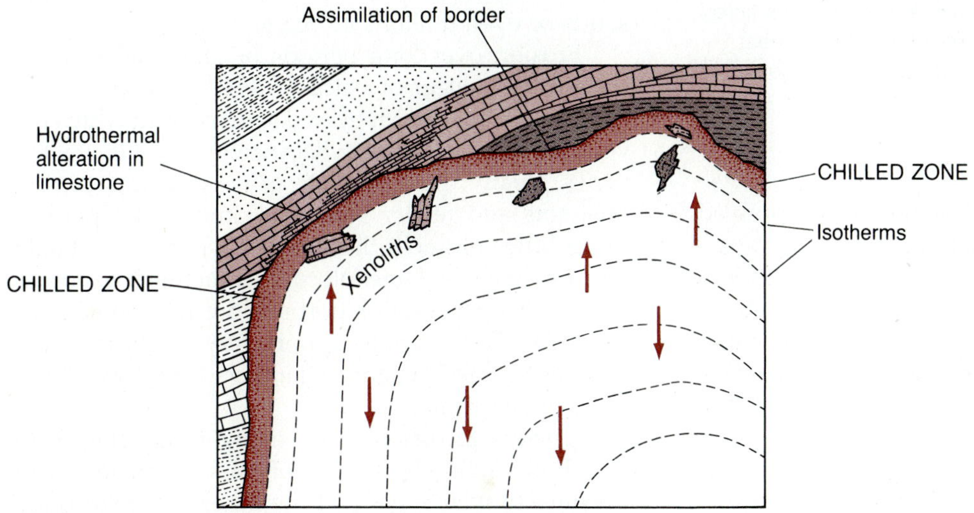

Figure 6.12
Schematic representation of several common processes by which magmas become differentiated. The isotherms are lines of equal temperature. One of the common processes is the gravitational separation of heavy minerals. Arrows indicate the sinking and rising of heavy and light minerals as they crystallize in the magma. In addition, pieces of the country rock are shown sinking into the magma. If the temperature in the magma is high enough, these may melt. The edge of the magma is chilled, and fluids in the magma may alter the surrounding country rock.

Figure 6.13
Students examine a coarse pegmatite dike that cuts across and encloses blocks of metamorphic country rock near Craggy Gardens, North Carolina, on the Blue Ridge Parkway.

as the crystal mush in the chamber is compressed. These fluids are injected into the country rock where they crystallize as an intrusion. Experiments also show that some melts, especially sulfides, split into two separate phases that will not mix. One phase floats on the other, like oil on water. The separation of such liquids eventually leads to the crystallization of different rock types.

Variation in the composition of igneous rock masses near their borders can sometimes be related to contamination of the magma as a result of melting of the country rock. Such melting must be expected if the magma has a temperature much higher than the fusion temperature of the country rock. Blocks of sedimentary or metamorphic rock from the borders may also break off and sink into the intrusion to form **xenoliths,** which are usually highly altered.

6.4 THE OCCURRENCE OF IGNEOUS ROCKS

Volcanic Rocks

Volcanoes are the surface manifestation of igneous activity (Fig. 6.14). Because the composition and crystallization of lavas are closely related to those of intrusive rocks, both are treated in this chapter, although we shall not discuss the forms of volcanoes and the products of volcanic eruption until Chapter 15, as part of the general discussion of the processes that shape the surface of the earth.

Intrusive Igneous Rocks

Plutons. So far our discussion has centered on the crystallization of rock melts on or near the ground surface, or those produced artificially for laboratory experiments. We can presume that large masses of magma cool and solidify before they ever reach the surface. Large bodies of rock that are of igneous origin but that crystallize below the surface are called **plutons.** Many such rock bodies are exposed in mountain belts and in the stable interior shield areas of the continents, where ancient Precambrian plutons constitute a large proportion of all exposed rock. Magma reservoirs must exist today below all modern volcanoes, and evidence of cooling is found in many places where volcanic eruptions have not occurred for thousands or even millions of years.

Erosion over long periods of time has removed the surface veneer of volcanic rock and exposed the now-solidified, deeper-seated sources from which the volcanic rocks emanated. Because the level of erosion varies greatly, it is possible in some places to find offshoots and feeder systems leading from plutons toward the volcanoes that were above them. Not only are the roofs of large plutons sometimes exposed, but the deeper interior portions are also sometimes revealed. Thus, ample opportunity exists to study the once-buried magma chambers and their long-since frozen contents.

Igneous intrusions take many forms (Fig. 6.15 and Table 6.5). The largest of these intrusions and the ones to which many of the others are related are called **batholiths** (Fig. 6.16). These are arbitrarily

Figure 6.14
Volcanoes, such as this one in New Zealand, may form on the earth's surface over plutons if gases or magma rise to the surface. (National Publicity Studios of New Zealand)

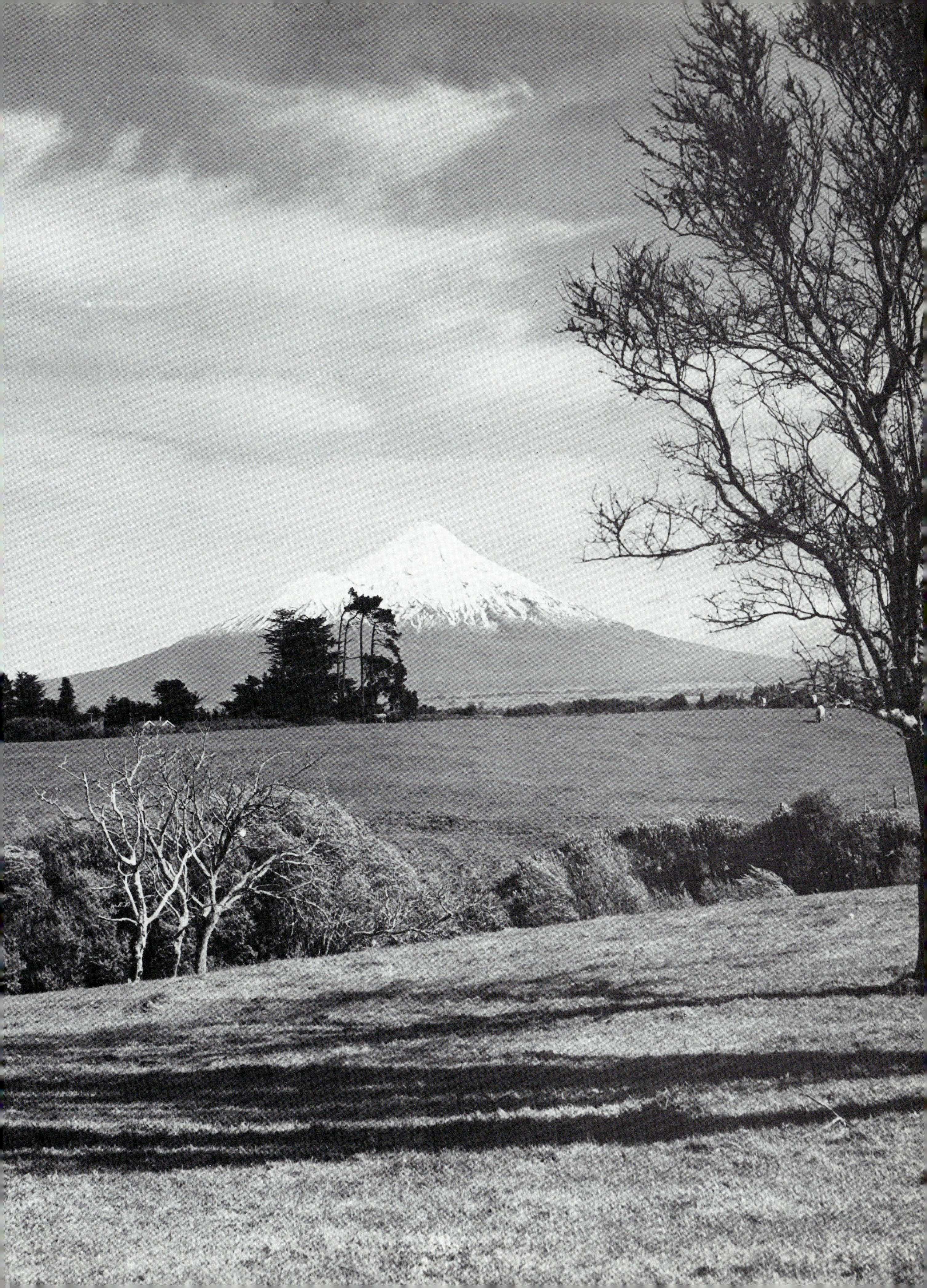

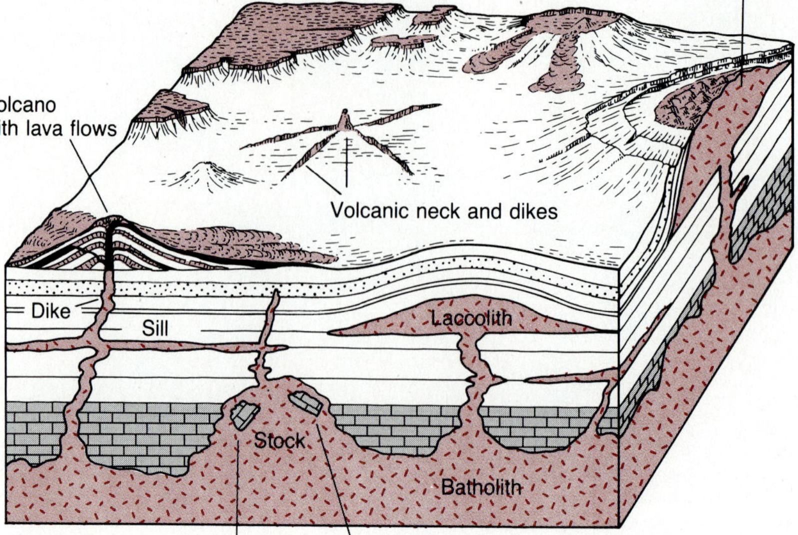

Figure 6.15
This schematic sketch shows most of the types of intrusions described in the text. In addition, common surface expressions of some of these are indicated.

Table 6.5 Classification of Common Intrusions Found in Continental Crust.

Type	Term	Characteristics
Concordant (injected along planes of stratification or layering).	Sill	Sheetlike body with large lateral extent relative to its thickness.
	Laccolith	A lens-shaped mass, convex on one or both sides, flattened in bedding plane of the invaded formation. The roof is arched over the laccolith.
	Lopolith	A lens-shaped mass whose thickness is approximately one-tenth to one-twentieth of its width. The top of the central portion is concave.
Discordant (injected across planes of stratification).	Dike	A body with parallel or nearly parallel walls that is narrow relative to its length.
	Dike swarm	Many more or less parallel dikes occurring together.
	Vein	A body that is less regularly defined than a dike.
	Ring dike	An arc-shaped to circular dike.
	Volcanic plug	Solid lava occupying a volcanic vent.
Subjacent masses (bodies of igneous rock that are not visibly floored). The mode of emplacement is often uncertain.	Batholith	A large body with surface area of 100 or more square kilometers.
	Stocks	Masses similar to batholiths but smaller in size.

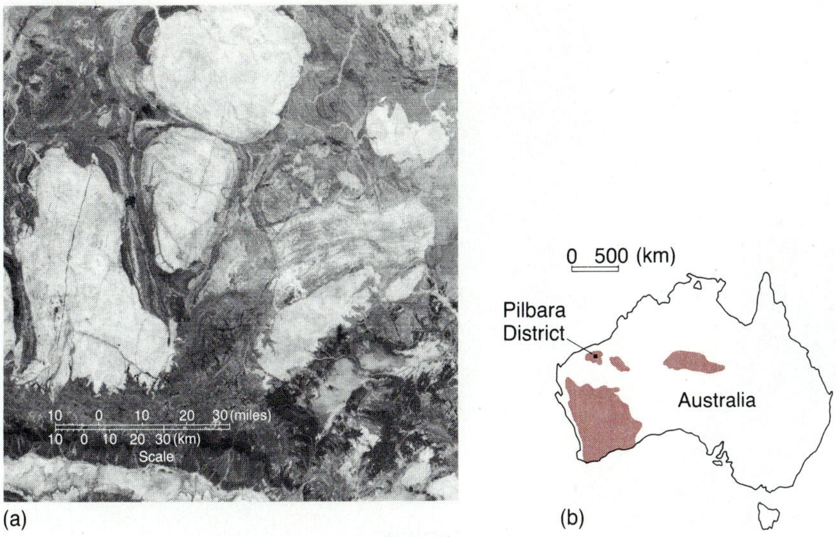

Figure 6.16
(a) This image, recorded by a Landstat satellite, depicts part of western Australia, showing the outline of several light-colored batholiths. Metamorphosed sedimentary rocks appear to wrap around the plutons. (b) Locator map.

defined as plutonic bodies having surface areas of at least 100 square kilometers, though usually they are much larger. Smaller intrusions, called **stocks,** are generally protrusions from the large batholiths. Ancient batholiths are exposed in both the vast stable areas of the world like central Canada and in mountain belts like the Appalachians and the Rockies. Although batholithic intrusions are common in both old and young mountains, they are not necessarily present in all mountain belts. Intrusions are rare in the young Alpine mountain system but are prominent in the mountain systems found around the Pacific. We will examine one of these latter areas, the Sierra Nevada batholith, in some detail because many of its features are also shared by other batholiths.

The Sierra Nevada Batholith: A Case Study. The name Sierra Nevada is applied to both the batholith itself and the mountains of eastern California and western Nevada (Fig. 6.17). This spectacular mountainous area, 700 km long by 100 to 150 km wide, is essentially a large block of the earth's crust that has been tilted up along faults on its eastern side. Geologically recent faults along which this movement has occurred are responsible for the precipitous drop from the high peaks of the mountains into the deep basins to the east.

The eastern side of the batholith has been cut and dropped down along these faults (Fig. 6.18). Thus, the faulted eastern edges of the batholith lie buried under recent fill deposits. However, the northern and western margin of the batholith and the rocks into which it was intruded are well exposed. Big masses of country rock, which were above the magma and thus formed its roof when it was intruded, are also intact in many places. Dissection of the uplifted fault block, first by streams and later by glaciers, has laid bare tremendous deep sections into the batholith such as that seen at Yosemite Valley (Fig. 6.19). Thus, exposures in many parts of the batholith are excellent.

The Sierra Nevada batholith is actually made up of a complex of smaller batholiths and stocks, which range in size from a square kilometer to more than a thousand square kilometers. The borders of these plutons are usually steep and smooth. Often where the plutons are close together they are separated from one another by thin walls of metamorphic country rock (as shown in Fig. 6.17*c*). Most of the plutons are composed of granitic rocks, but the composition ranges from granite to diorite, and individual plutons can be distinguished on the basis of their mineralogy and texture. In general, the granitic rocks are coarse-grained, but porphyries with big crystals of potash feldspar are common, as is the fine-grained rock along the borders and in the dikes extending away from the larger plutons. Pegmatites and quartz-rich veins extend as far as two miles into the surrounding country rock.

The evidence that these bodies were formed by intrusion of preexisting rock is clear. The contacts between the plutons and country rock are sharp, and the plutons cut across both the layering and other structures in the country rock. The igneous rock masses usually show fine-grained chilled borders;

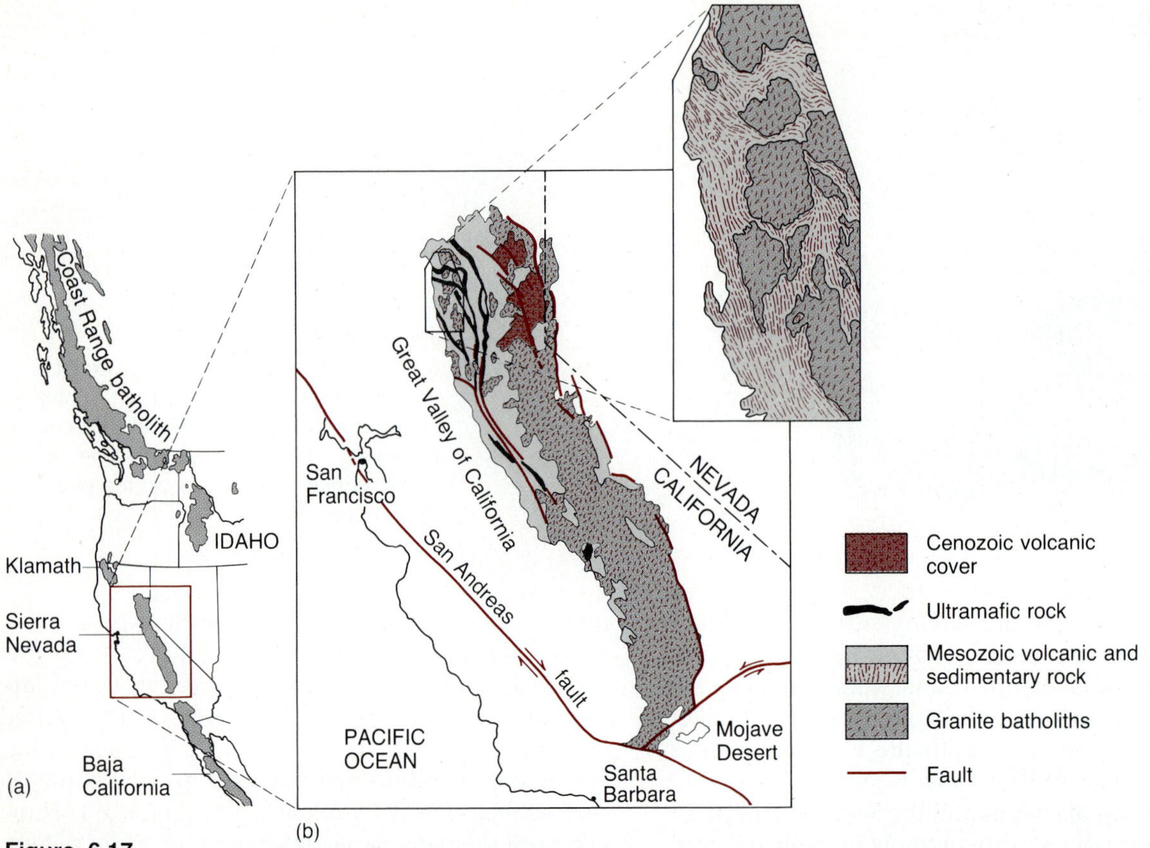

Figure 6.17
(a) This sketch of a portion of western North America shows the location of many of the large Mesozoic batholiths in the Cordilleran Mountain System. (b) General outline of the Sierra Nevada Mountains. Younger sediments fill the Great Valley of California and much of the region east of the high Sierras. (c) A more detailed map of part of the Sierra Nevada Mountains.

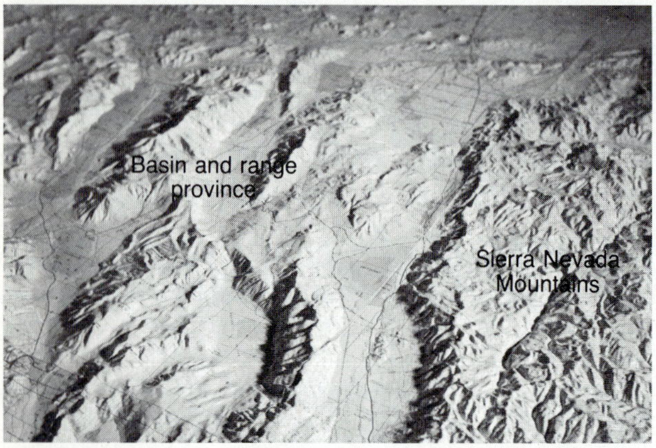

Figure 6.18
This view, looking south, shows the topography along the western side of the Sierra Nevada Mountains. High angle faults occur along the edge of the Sierras, separating them from the deep basin to the east.

dikes, sills, and tonguelike protrusions from the plutons extend into the country rock; metamorphic reactions due to heat and fluids can be seen in the country rock around the plutons; inclusions of pieces of country rock in the plutons are aligned as might be expected if they broke off or became engulfed in a flowing melt; and many of these pieces were partially melted and assimilated in the magma. As the magma intruded, the country rock was forced aside and arched upward to make room for the huge volume of granite. Although the contacts are generally sharp, a few show broad zones where the country rock and the melt mixed together during emplacement of the granite.

Many of the plutons have been dated by radiometric dating techniques. Most of those on the east side of the area yield dates from 170 to 210 million years ago; those in the western belt are 125 to 145 million years old; and a third group in the central part of the mountains yield dates of 80 to 90 million years ago. Thus, the igneous activity responsible for the Sierra Nevada batholith complex took place over a very long period of time. The coarse textures indicate that the plutons cooled and crystallized slowly and probably at considerable, though usually undetermined, depth.

The contact zones between batholiths and country rock are usually steep or slope away from the batholiths at depth. This, coupled with the fact that the bottoms of batholiths are almost never seen or at least identified, leads to the conclusion that batholiths become bigger at depth. This model presents a problem since virtually all of the batholiths on continents are granitic in composition, but all other lines of evidence indicate that the upper mantle is mafic (rich in magnesium and iron-bearing minerals), probably peridotite. If this is correct, the granitic batholiths must have floors somewhere in the continental crust. The problem is now at least partially resolved as a result of geophysical studies. Gravity surveys in general confirm that granitic plutons are less dense at depth than the surrounding rock. Geophysical studies in the western United States that cross the Sierra Nevada batholith indicate that the surface of the upper mantle is deeper below the batholith, leaving the crust thicker there (Fig. 6.20). According to this interpretation, the smaller plutons we see at the surface join at depth to become part of vast granitic batholiths that extend deep into the continental crust. The batholiths do not, however, extend into the mantle but taper at depth.

The fact that all granitic batholiths lie on continents certainly suggests a causal connection between the two. The bulk chemical composition of continental crust is similar to that of granite; consequently, it seems probable that modern granitic batholiths are produced by melting of part of the continental crust,

Santa Fe Railroad

Figure 6.19
The bare rock faces in Yosemite Park were cut into granites of the Sierra Nevada batholith by glaciers.

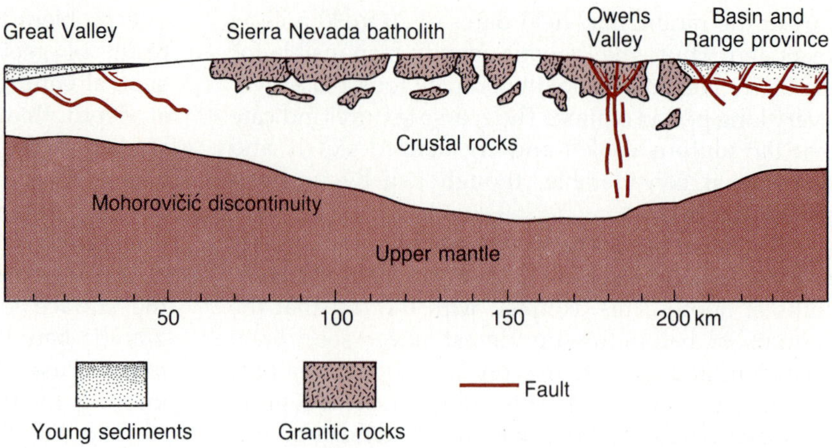

Figure 6.20
A cross section across the Sierra Nevada Mountains. This section extends into the upper mantle and shows the individual batholiths tapering off at depth in the upper part of the continental crust. Note the steeply inclined faults east of the Sierras and the low faults to the west.

as has been suggested for the Sierra Nevada batholith. The origin of the continental crust is more obscure, but most geologists now think it started to form during the early history of the earth. While the earth was largely molten, the less dense sialic minerals rose to the surface as they crystallized and may have formed islandlike masses of less-dense rock in a sea of mafic rock. This primitive crust has been modified and enlarged by addition of magmas at continental margins where, according to plate theory, subduction generates magmas that are added to the continental margins.

6.5 THE ORIGIN AND OCCURRENCE OF MAGMA

Most theories concerning the origin of the earth postulate that whether the earth originated as a mass of hot gaseous material or formed by the accretion of cool planetesimals, it was largely molten material at an early stage. The gases condensed to hot liquid in the first hypothesis. In the second, the planet became hot and melted as a result of the pressure, and heat was liberated first during the impact and gravitational collapse of the planetesimals and later by release of heat as a result of decay of radioactive isotopes. The liquid core and the concentric zonal character of the earth's interior, with more dense matter at the center, support the idea that the earth was once molten.

For many years scholars thought lavas reached the surface of the earth from a molten interior. But as we have seen, the outer 50 to 100 kilometers is now known to be solid. Most of the mantle is rigid enough to transmit the vibrations produced by earthquakes—something molten magma would not do. Where then and how is magma generated? It is easy enough to specify where on earth magma exists, for volcanic activity is a surface manifestation of magmas at depth. Most modern magmas can be accounted for in terms of plate theory—basaltic magmas are generated in oceanic crust both at the crest of **oceanic ridges** and at **"hot spots"** like that at Hawaii (Fig. 6.21). Andesitic and granitic magmas are generated over subduction zones in island arcs and under continental margins. Presumably, most magma is produced directly beneath the centers of volcanic activity. When rock melts, it is not only fluid but buoyant because the volume increases on melting. Thus, melts should tend to rise nearly vertically, and volcanoes should be located directly above places where magma is being generated by melting processes.

The Melting Process

The production of magma is governed by four primary variables: the composition of the rock to be melted, the temperature, the confining pressure (the pressure on the rock caused by the weight of the overlying rock) and the type of volatiles, if any, that are present. All minerals will begin to melt when the temperature reaches the **fusion point** of that mineral, that is, the temperature at which it melts for a given confining pressure. Some minerals melt at temperatures as low as 500 to 600 °C; others melt only above 1500 °C. The common minerals of which basalt is composed begin to melt in the range of 900 to 1000 °C at atmospheric pressure. The exact temperature depends on the amount of water that is present. Dry rock melts at higher temperatures than wet rock.

U.S.G.S.

Figure 6.21
Cracks in the surface of a lava lake in Hawaii show as bright white lines. Note the trees behind the fountain for scale.

Both temperature and confining pressure increase with depth in the earth. The increase in temperature is relatively rapid, but the increase in pressure raises the fusion temperature. Magma may be generated if either the temperature reaches the fusion point of the rock at its depth and pressure, or if the pressure on rock near its fusion point is reduced enough to achieve a lower fusion point.

The rate at which temperature normally increases with depth in the earth is calculated from measurements of heat flow at the surface, from well and mine records of temperature, and from geophysical models of the interior. Temperature increases with depth at a rate of about 25 to 30 °C per kilometer (about 1 °C per 30 meters) near the surface. This steep gradient suggests temperatures of 1,000 °C at a depth of only 40 kilometers. If the rate of increase of temperature with depth continued at this rate, all materials below 100 kilometers would be molten. Studies of vibrations generated during earthquakes and transmitted through the earth indicate that most of the mantle reacts as a rigid body and certainly cannot be totally molten at such shallow depths. Therefore, the geothermal gradient must lessen at depth. Our best estimates of temperature change with depth as is shown in Fig. 6.22. In contrast, pressure increases in a nearly linear fashion with depth.

Although the mantle is not molten, mantle rocks at depths of 100 kilometers are near their fusion temperatures, and it seems likely that much basaltic magma is generated at about that depth. Experimental melting studies demonstrate that granite begins to melt at considerably lower temperatures than basalt, so it could originate at shallower depths. But in neither the continents nor the ocean basins are crust temperatures high enough for granitic or basaltic magmas to form under normal conditions.

Magmas can be generated at shallow depths if temperatures become unusually high or if both high temperatures and volatiles are present. Several possible mechanisms have been suggested to cause localized increases in temperature. Heat liberated by radioactive decay, frictional heat accompanying movements in the earth, and heat rising from greater depths in the mantle are among the most likely sources of abnormally high temperatures in the crust. Melting is also favored for dry rocks in the crust by any process that reduces pressure. Stretching and thinning of the crust could have this effect.

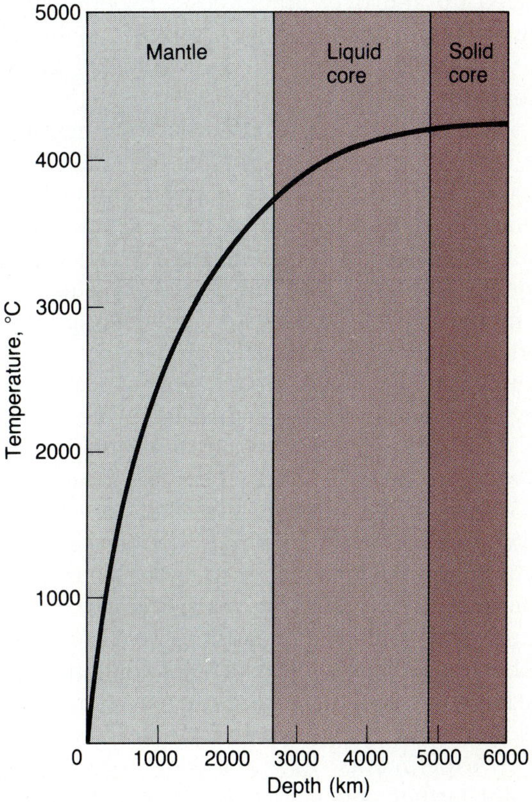

Figure 6.22
This graph shows the estimated temperature of rocks at various depths. The depth of the liquid core is indicated. Magmas responsible for volcanic activity occur at depths of no more than a few hundred kilometers.

Radioactive minerals occur in small quantities in many rocks in the earth's crust, but they are much more abundant in granitic rocks found on continents than in the basalts of the oceanic crust. Heat is liberated in radioactive decay processes, and because rock is such a poor conductor, the heat accumulates. It has been suggested that this source of heat may have been a factor in the melting of rock now found in the Sierra Nevada batholith. Concentrations of radioactive minerals may be sufficient to melt rock locally, but little is known about the amount or the distribution of radioactive minerals in mantle rocks, and it is from the upper mantle that most magma appears to originate.

Although frictional heating is often mentioned as a cause of magma generation, the evidence does not support this conclusion. Some heat may be generated where large lithospheric plates move past one another, but magma is not generated along major active faults like the San Andreas. Nor is it present along the big faults like those in the Alps where rock masses have moved up and over other rocks. Experiments involving the deformation of rock samples at high pressure also fail to indicate the generation of melt along breaks.

The rise of heat from greater depths in the mantle has long been an attractive alternative to geologists. Many have appealed to the idea of convection—the slow boiling motion in hot fluids that includes the rise of hot, less dense material and the fall of cooler, denser matter. Certainly, the temperatures postulated to exist in the earth's interior are high enough to make convection models plausible.

To this point we have been mainly concerned with ways of melting rocks by increasing the temperature or decreasing the pressure around them. But according to plate tectonic theory outlined in Chapter 3, slabs of the lithosphere sink or are forced down into the mantle in subduction zones. These slabs, which contain rocks formed at or near the surface, are depressed into the mantle where both temperature and confining pressures increase. Not only are these rocks composed in large part of minerals that have lower fusion temperatures than most mantle materials, but in their upper parts they are saturated with water. Experiments show that water lowers the fusion point farther and facilitates melting. For example, dry sialic rocks at atmosphere pressure (1 bar) begin to melt at about 960 °C, but if enough water is present to produce a water-saturated magma, melting will begin in these same rocks when temperature reaches 650 °C even at pressures of 4 kilobars. For this reason, we think large volumes of magma are generated from rock carried down in subduction zones. Just how magma generation and especially its geological setting are explained by the theory of plate tectonics will be discussed later, in Chapter 13, after we have studied the theory in more detail.

Clearly, magma is generated in several quite different geological settings. In some places heat and possibly magma generated deeper in the mantle rises into the lithosphere fusing the rocks there. Elsewhere rocks formed near the surface are lowered into the high-temperature region of the upper mantle until they melt. In both of these situations large volumes of rock are subjected to increasing temperature and, in the latter case, to increasing pressure as well. Long before melting begins, the rock is altered by increasing temperature and pressure. These changes consitute what we call metamorphism, the subject of the following chapter.

SUMMARY

Igneous rocks form by crystallization of minerals from a melt. Those formed well below the surface are intrusive; those formed at the surface are extrusive. The mineral composition of both types are similar—combinations of quartz, olivine, amphibole, pyroxene, feldspars, and micas—but they show differences in texture that depend on their conditions of cooling. Slow cooling allows time for crystal growth and so gives rise to the coarser-grained (phaneritic) rocks. Rapid cooling, common for all extrusions and some intrusions, produces either glasses (amorphous rocks) or fine-grained (aphanitic) rocks. These textural differences together with mineral composition are the basis for identifying and naming igneous rocks.

Experimental studies of cooling rock melts show that some minerals crystallize at much higher temperatures than others. After crystallization the mineral may react with the remaining melt to form new minerals. As crystallization and subsequent reaction of crystals and melt take place, the composition of the melt constantly changes, in turn affecting the composition of any new crystals that form.

A magma may form several different igneous rocks as a result of gravitational separation, based on mineral densities; or as a result of squeezing of liquid from a partially solidified mush; or by formation of two melt layers that are insoluble to each other.

Uplift and erosion have now exposed many plutons that crystallized below the surface, especially in mountain belts and in the continental shields. The Sierra Nevada batholith is one such formation.

Most igneous activity is associated with plate boundaries as marked by sites of volanic activity. At convergent zones magma is generated as rock is carried down into the mantle where the rise in temperature and the presence of water and carbon dioxide in the rock tend to melt it.

KEY TERMS

aphanitic	**intrusions**
batholith	**magma**
chilled margin	**oceanic ridge**
contact metamorphism	**phaneritic**
country rock	**phenocryst**
cross-cutting	**pluton**
dike	**sill**
dike swarm	**simple eutectic system**
extrusions	**stock**
fusion point	**vein**
gravitational separation	**xenoliths**
hot spot	

STUDY QUESTIONS

1. Compare the textures of igneous rocks with those of sedimentary rocks. Which textures are distinctive and which might be confused?
2. What differences might you expect to find in the rock texture if you could sample the same pluton at depths of 200, 2000, and 20,000 meters?
3. Summarize the ways a basalt magma might be differentiated to form a variety of igneous rocks.
4. Compare the typical intrusive rocks of oceanic ridges, continents, and island arcs.
5. A large proportion of the oldest rocks on earth are granitic intrusions. What do you infer about conditions on earth at that time?
6. Why can't basaltic rocks be produced by the differentiation of granitic magmas?
7. Describe the features you would expect to find in the border zone of a pluton.
8. Why do the moon and Mars lack granitic rocks?

REFERENCES

1. Nicholas Desmarest, *Mémoire sur l'origine et la nature du basalte à grandes colonnes, polygones, determinées par l'histoire naturelle de cette pierre, observée en Auvergne.* Mem. Acad. Royale des Sciences for 1771, pp. 705–775, Paris, 1774. Translated by Sir Andrew Geikie, in *The Founders of Geology.* New York: Dover, 1962, p. 150.
2. Abraham G. Werner, *Kurze Klassification und Beschreibung der Verschiedenen Gebirgsarten.* Dresden: 1787, p. 25. "Sent to press" in 1777 and printed by a friend. The statement is translated by Sir Andrew Geikie in *The Founder of Geology.* New York: Dover, 1962, p. 222.

SUGGESTED READINGS

Bowen, N. L., *The Evolution of Igneous Rocks.* New York: Dover, 1956.

Compton, R. R., "Trondhjemite Batholith near Bedwell Bar, California," *Geol. Soc. America Bull.*, 66 (1955), 9–44.

Geikie, A., *The Founders of Geology.* New York: Dover Publications, 1897.

Hamilton, W., and W. B. Myers, "The nature of batholiths," *U.S. Geol. Survey Prof. Paper 554-C,* (1967).

Hyndman, E. W., *Petrology of Igneous and Metamorphic Rocks.* New York: McGraw-Hill, 1972.

Wright, T. L., and R. T. Okamura, "Cooling and crystallization of tholeiitic basalt, 1965 Makaopuhi Lava Lake, Hawaii," *U.S. Geol. Survey Prof. Paper #1004,* (1977).

THE TRANSFORMATION OF ROCKS: METAMORPHISM

Chapter 7

The processes that lead to the formation of sedimentary deposits take place at the surface of the earth where they can be studied directly. Similarly, we can see volcanic rocks forming as lavas cool, and it is possible to trace volcanic flows back through dikes and feeders into the plutons from which they came. Yet, for a third class of rocks, the metamorphic rocks, the conditions of formation can only be observed through laboratory experiments or be inferred from the field relationships where metamorphic rocks occur.

Metamorphic rocks—those that have undergone a change of form—are found: surrounding igneous intrusions; in zones along which the crust has been so completely deformed that the rock texture is tremendously transformed, as in Fig. 7.1; and in areas where field evidence indicated that the rocks were once deeply buried. Generally, rocks in the last of these geologic settings also show signs of some deformation. These three environments suggest that metamorphic processes are closely related to three different sets of conditions: to the heat of igneous activity in the first case, to localized deformation of the crust in the second, and to large-scale regional subsidence and uplifts in the third. These conditions produce three types of change: **contact metamorphism, dynamic metamorphism,** and **regional metamorphism,** respectively.

In each of these types of metamorphism, the processes occur because the rock becomes unstable, either chemically or physically, as a result of changes in its environment. **Metamorphism** is formally defined as the mineralogical and physical adjustments that take place in rocks as a result of changes of environmental conditions. Many of these changes are caused by high temperatures; but if the rock melts, the resulting products are igneous, not metamorphic. The environmental changes are limited to those developing below the surface in order to eliminate any confusion with weathering processes, which represent the adjustment of rocks to surface conditions.

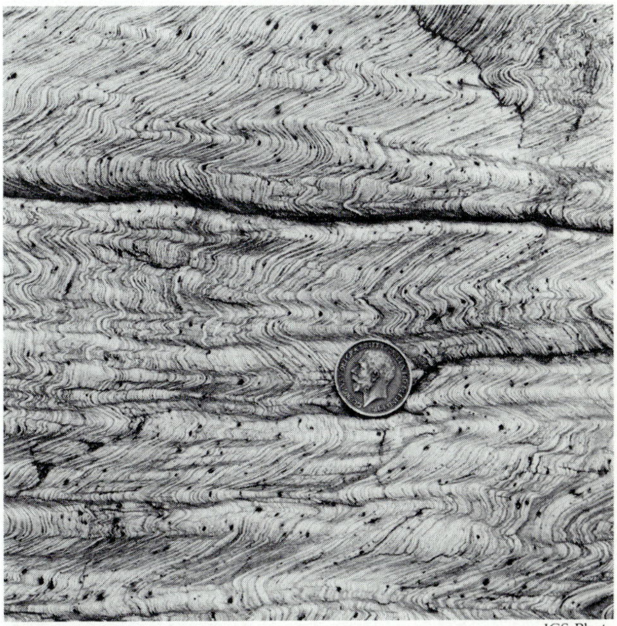

IGS Photo

Figure 7.1
The rock shown in this picture was originally a sedimentary rock. Now it shows the effects of strong pressure, and it is composed of minerals that formed during metamorphism.

7.1 COMMON METAMORPHIC MINERALS

Most of the rock-forming minerals described in Chapter 2 are found in metamorphic rocks. Quartz, orthoclase and plagioclase feldspars, pyroxenes, amphiboles, the micas, and calcite all occur in metamorphic rocks, but other minerals that are rarely, if ever, present in unmetamorphosed igneous rocks become major components of certain metamorphic rocks. Of these, the six minerals described here are especially important because they are diagnostic of certain conditions of metamorphism described in this chapter. All of them occur also in sediments derived from metamorphic rock by weathering and erosion.

Chlorite: $Mg_3Al(OH)_8AlSl_3O_{10}$. Crystals of chlorite are tabular, hexagonal, and often bent. Chlorite has one perfect cleavage and a hardness of $2\frac{1}{2}$, and is a type of mica. It is distinguished from other members of the mica group by its greenish color. It often occurs as a scaly coating on other minerals (see Fig. 7.2*a*), or as a fine groundness in the matrix of metamorphic rocks.

Biotite Mica: $K(Mg,Fe)_3(OH,F)_2AlSi_3O$. Members of the mica group form hexagonal crystals. Scalelike sheets formed as a result of mica's one perfect cleavage are the most characteristic occurrence in rocks. Biotite (Fig. 7.2*b*) is characteristically black, brownish, or blackish green; muscovite, another mica sometimes produced by metamorphism, is colorless, white, or yellowish. Like chlorite, biotite is rather soft, with a hardness of $2\frac{1}{2}$.

Garnet: $(Ca,Mg,Fe,Al)(SiO_4)$. Crystals of garnet are common and are cubes or dodecahedra, as shown in Fig. 7.2*c*. Garnet also occurs as masses of small crystals. Although the familiar deep red garnet colors are characteristic of some varieties, the mineral color is quite variable. Because it does not have true cleavage, its hardness ($6\frac{1}{2}$) is one of the best ways of distinguishing it from similar minerals.

Staurolite: $Fe_2Al_9O_6(SiO_4)_4(OH)_2$. The mineral staurolite is characterized by prismatic crystals. These are usually in the cross shape shown in Fig. 7.2*d*.

Kyanite: Al_2SiO_5. Long, bladed crystals like those in Fig. 7.2*e* that are curved and radially grouped are characteristic of kyanite. It is also distinguished by streaks or spots. The hardness varies with direction, being 4 in one direction and as much as 7 in the other.

Sillimanite: Al_2SiO_5. Sillimanite occurs as long, thin, needle like crystals (hardness $6\frac{1}{2}$) or as radiating fibrous masses in metamorphic rocks (Fig. 7.2*f*). A silky luster characterizes most specimens.

Serpentine: $(Mg,Fe)_3Si_2O_5(OH)_4$. Serpentine is a group of minerals characterized by a greasy luster, soapy feel, and conchoidal fracture. This group usually exhibits compact forms, but may be granular or fibrous. Most serpentines are green. Serpentine is formed by the alteration of magnesium-rich silicates such as olivine.

7.2 TEXTURE OF METAMORPHIC ROCKS

Like igneous rocks, metamorphic rocks have crystalline textures. None is composed of fragments, although relics of the original fragmental texture may be preserved in some metamorphosed sedimentary rocks. Two distinctly different types of crystalline textures are recognized, and these are used for purposes of classification. The first type is represented by three subtypes: the nearly parallel alignment of platy minerals, frequently micas; layers of minerals that may give the rock a banded appearance, resembling the layering of some sedimentary rocks; and minerals that have been flattened and drawn out as a result of deformation. Rocks that show any of these structural or mineral fabrics are said to be **foliated.** The second type, unfoliated metamorphic rocks, have a more randomly arranged mineral fabric. Representative examples of these textures are illustrated in Fig. 7.3. Using this brief description of texture and what we know of their mineral content, we can describe some of the most common metamorphic rocks.

Common Metamorphic Rocks

Slates are fine-grained perfectly foliated metamorphic rocks characterized by a tendency to break along nearly parallel foliations, called **slaty cleavage** (Fig. 7.4*a*). Slate is derived from shale, but in the process of metamorphism the clay minerals have been altered to small but aligned mica minerals. Individual minerals are too small to be identified in hand specimens.

Figure 7.2
The minerals illustrated here are common constituents of metamorphic rocks. Most of the minerals that occur in igneous rocks (described in Chapter 2) are also found in metamorphic rocks. Well-formed crystals of the types shown grow during metamorphism and are often visible. The minerals shown are: (a) chlorite; (b) mica; (c) garnet; (d) staurolite; (e) kyanite; and (f) sillimanite.

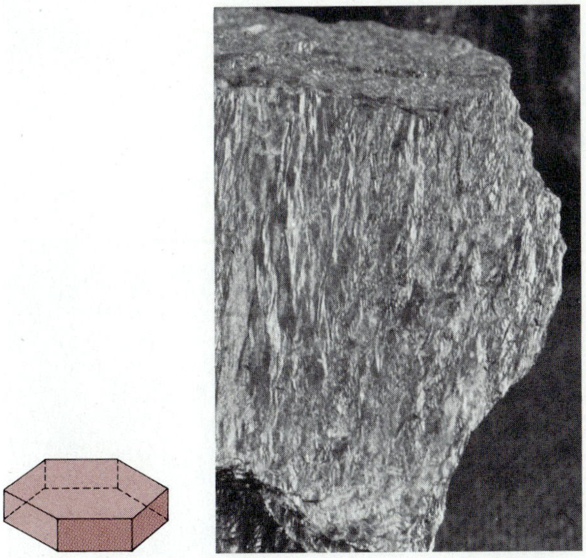
(a)

(b)

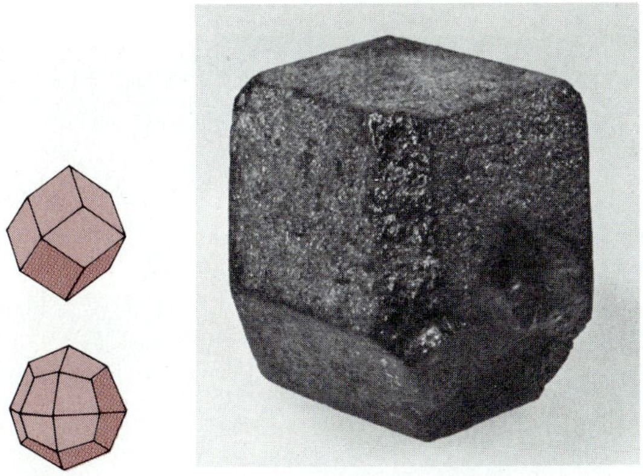
(c)

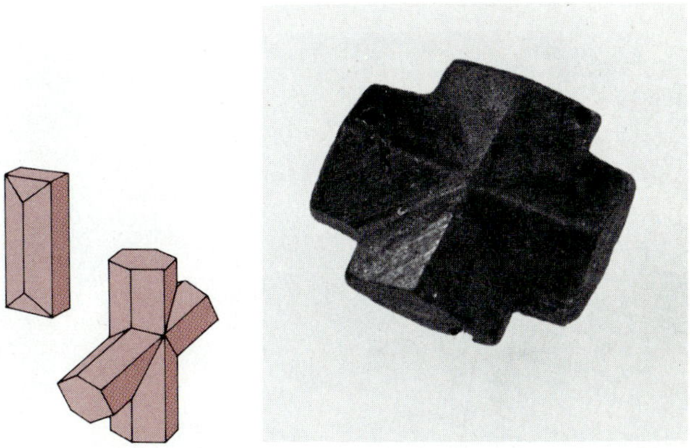
(d)

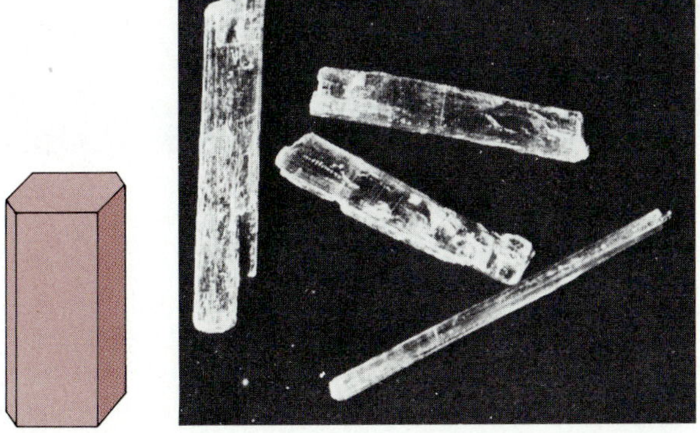
(e)

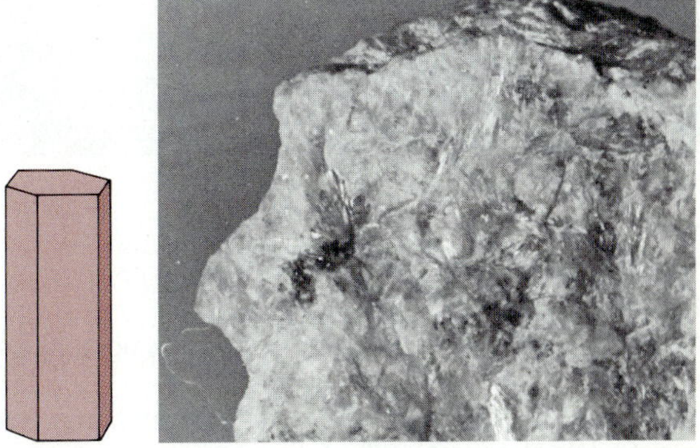
(f)

(a)

(b)

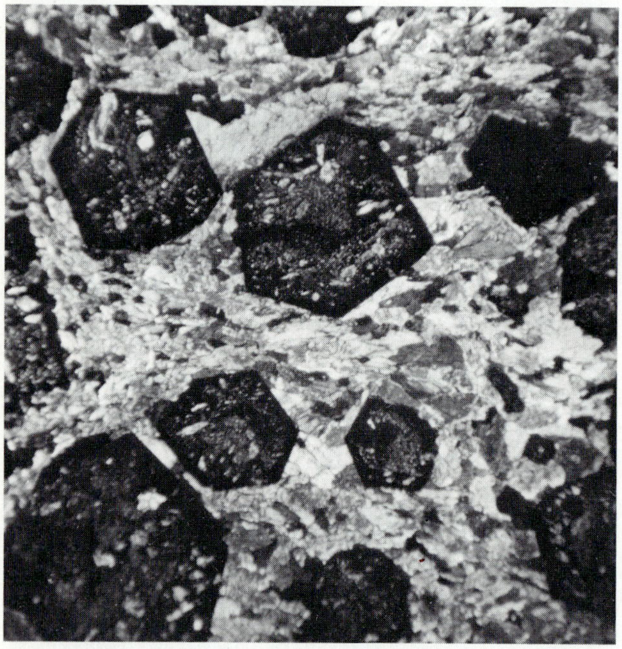

(c)

(d)

Figure 7.3
Metamorphic textures: (a) and (b) photomicrographs of rocks showing strongly developed foliation; (c) photomicrograph of an unfoliated rock showing perfectly formed garnet crystals; (d) a foliated texture, containing eyelike lense-shaped crystals, called augen. (5×)

Figure 7.4
The most common metamorphic rocks: (about $\frac{1}{4}$ true size): (a) slate; (b) phyllite; (c) schists; (d) gneiss; (e) marble; (f) quartzite; (g) hornfels; and (h) amphibolite.

Phyllites are fine-grained rocks identified by the lustrous, silky sheen (Fig 7.4*b*) caused by light reflected from the chlorite and muscovite micas of which they are composed. Quartz and albite are often present. The grain size of phyllites is larger than that of slates, but finer than that of schists. Phyllites are usually greenish or red and may show the initial stages of segregation of some mineral constituents into layers.

Schists are foliated rocks of medium to coarse crystalline texture (Fig. 7.4*c*). Unlike phyllites, most of the mineral constituents of schists are easily identified without the use of a microscope. Foliation is caused by the parallel or nearly parallel alignment of micaceous minerals. The most common minerals in schists are quartz, feldspars, and micas. If one of the constituents makes up 50 percent or more of the rock, its name is attached as a modifier (for example, mica schist, quartz schist, or hornblende schist). If no constituent comprises 50 percent, the names of the two most abundant constituents are used (for example, garnetiferous-mica schist).

Gneisses have a characteristic compositional layering that produces the banded appearance evident in Fig. 7.4*d*. Gneiss is medium to coarse grained. It usually contains quartz and feldspar interlayered with thin layers rich in hornblende or mica that induce places of weakness in the rock. The quartz, feldspar, and other constituents usually have interlocking boundaries. The layering is thought to develop as a result of segregation of mineral constituents during metamorphism. In some cases it corresponds to original bedding of sedimentary rocks.

Schists are distinguished from gneisses by the way they break. Schists break into thin slabs (1 to 10 milimeters thick) or into pencil-like columns, whereas gneisses break into much thicker slabs or even across the layering. This ease of breaking, caused by the presence of aligned crystal plates is called **schistosity** (Fig. 7.4*c*). Most gneisses have a much higher feldspar content than schists. In some classification systems, the amount of feldspar is used to distinguish schists from gneisses. Rocks with more than 20 percent feldspar are gneisses; those with less are schists.

Marble is formed by the metamorphism of calcite, limestone, or dolomite, and so is called their metamorphic equivalent. Marbles often contain micaceous impurities that give them a foliation. Marble's texture is usually that of an interlocking mosaic growth of calcite or dolomite crystals (Fig. 7.4*e*).

Quartzite, the metamorphic equivalent of quartz sandstones, is composed of about 80 percent or more quartz (Fig. 7.4*f*). Tightly cemented quartz sandstones may resemble the metamorphic form (or metaquartzites) in hand specimens, in that both often break across the sand grains of which they are composed. However, the grain boundaries of metaquartzites have usually grown together and become interlocked, and the rock often has some foliation.

Hornfelses are fine-grained, nonfoliated, dense, usually dark rocks (Fig. 7.4*g*) formed near the contacts of igneous intrusions. The minerals form a mosaic of interlocking grains. Hornfelses commonly break into splintery fragments that have translucent edges, like horn. The term is most frequently applied to baked shales.

Amphibolites (Fig. 7.4*h*) contain mostly plagioclase and hornblende, and often some biotite as well. The prismatic hornblende crystals lie in the plane of foliation, sometimes causing a planar or linear schistosity. Plagioclase too is often strongly aligned. Quartz-free amphibolite is usually derived from basaltic or mafic tuffs. The presence of quartz suggests that the amphibolite may be derived from sedimentary rocks.

7.3 NATURE OF METAMORPHIC CHANGES

Technically any rock that forms from a melt is igneous; so the changes that produce metamorphic rocks take place in solid rock. This is the formal distinction between igneous activity and metamorphic activity, but the two phenomena grade so imperceptibly into one another that it may be difficult to determine which name, igneous or metamorphic, is applicable. The problem centers on determining whether a rock melts at high temperatures. When metamorphism takes place at relatively low temperatures, its end products are less likely to be confused with those of igneous activity.

A new rock of different texture and often different mineralogical composition is produced through metamorphism. New minerals form by recombination of elements present, old minerals are recrystallized as larger crystals, new textures develop, and the internal structure and fabric of the rock change.

Metamorphic chemical reactions tend to establish an assemblage of minerals that is stable under the physical conditions at which the metamorphism takes place. When the reactions continue to completion, the resulting new rock may contain all new minerals. When equilibrium is not reached because metamorphic processes are interrupted, or when reactions fail to go to completion because they are too slow, some of the original minerals may remain. Plagioclase feldspar may show a compositional zoning, or crystals of one mineral may be partly altered to another mineral.

The tendency of metamorphism to reach an equilibrium condition can be demonstrated by comparing field observations and experimental evidence. In the field, progressive changes in mineral assemblages can be observed where an unaltered layer of rock is traced into zones of increasing metamorphism. In the laboratory, the original unaltered rock can be subjected to controlled conditions approximating those thought to be responsible for the metamorphism. If the same mineral assemblages observed in the field can be reproduced in the experiments, we can be fairly certain that the laboratory conditions approximate those under which metamorphism occurred.

Processes of Metamorphism

Field observations and studies of the rocks formed during metamorphism indicate that four processes taking place during metamorphism are responsible for most of the observed features of metamorphic rock texture, composition and structure. Two of these are mechanical effects: rock flowage and granulation; and two are chemical in nature: recrystallization and recrystallization with recombination. The particular geologic circumstance causing metamorphism to occur and the particular factors controlling the metamorphism determine which of the four processes are most effective. They often operate together.

Rock flowage, in the sense used here, is not the same as the viscous flow associated with fluids like water, but rather like the flow of glacier ice in which the ice remains solid while the movement occurs. Flowage of this type results from movements along planes of slip within the minerals making up the rock and from rotation and displacement of mineral grains relative to one another. Usually these are accompanied by recrystallization of minerals. Under extremely high confining pressure, rock can undergo large-scale deformation without developing cracks or being crushed (Fig. 7.5). It is most likely to take place at high temperatures, while the rock is deeply buried and under great pressure. Rock flowage that accompanies metamorphism may be accelerated by movement of solutions and migration of elements in the solids toward points of less stress.

Granulation is the crushing of rock under such conditions that no visible openings or spaces result. The stress acting on the rock must be great enough to exceed the crushing strength—the pressure that must be applied to crush the rock. Usually granulation is favored by intense and rapidly applied stress of short duration. It is most likely to occur in rocks that are composed of hard, brittle, insoluble minerals; otherwise the rock may flow or recrystallize. The rock

Figure 7.5
Tightly folded gneiss from the Montana Rockies. Note that this rock has folded without cracking. It yielded by processes called rock flowage by geologists.

produced by granulation is characterized by mineral grains that become flattened as though smeared between two hard surfaces, as shown in Fig. 7.6. The grains may be rotated as they break up, eventually forming a powdery substance resembling mortar.

Recrystallization, the complete reorganization of the ions and molecules of the original minerals in a rock, may take place when the rock is subjected to high temperature and pressure, and particularly when water is present. It is not necessary that any new elements be introduced or that any original elements be removed, although such additions and removals may also occur.

Recrystallization may occur in such a way that the original minerals grow into larger crystals or slightly different forms of the same minerals. But if several different minerals are present in a rock that is recrystallizing, the constituent elements may re-

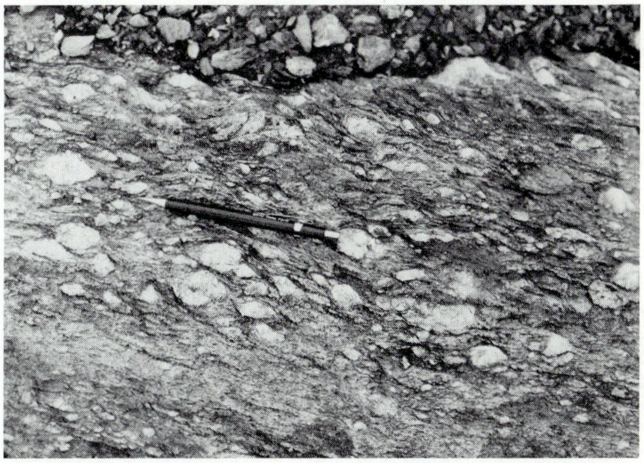

Figure 7.6
Strongly deformed rock. The original texture has been destroyed by the crushing and shearing of the rock.

combine to form new minerals. Unstable minerals will break down into their constituent parts; at the same time, these elements recombine during recrystallization to form new minerals that are stable. A reaction of this sort might be represented as follows: $AB + CD \rightarrow A^+, B^-, C^+$, and D^- ions that recombine to form $AD + BC$. This is represented schematically in Fig. 7.7. Under the metamorphic conditions, com-

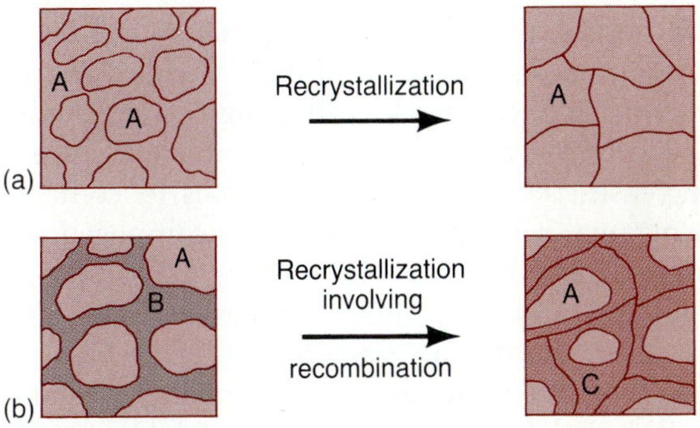

Figure 7.7
Recrystallization may occur with or without recombination of minerals in the rock. (a) A rock composed of mineral A and cemented with material with the same composition as A recrystallizes to a coarse grained metamorphic rock composed entirely of mineral A. (b) The original rock is composed of two minerals A and B. In this example recrystallization caused a recombination of A with B to produce mineral C. Small nuclei of mineral A remain because there was originally much more A than B; some of A was left after all of B was recombined. Actual crystals may be from a fraction of a millimeter to several centimeters across.

pounds AB and CD were unstable, whereas AD and BD are stable. These types of changes produce some minerals that are found only in metamorphic rocks.

Factors That Control Metamorphic Processes

The environmental conditions that prevail where metamophism is occurring and the original composition of the rock being altered determine which metamorphic processes actually take place. The temperature, the weight of the overlying rock (confining pressure), the amount and type of directed pressure (stress) present, and the presence of chemically active solutions are all important environmental variables. How long the processes act is also important in determining what the final product, the metamorphic rock, is. If the processes act over a long enough time, chemical changes may go to completion; and the rock will reach a new state of equilibrium with its environment. If time is not sufficient, the rock is incompletely metamorphosed. Fortunately, we find many such partially altered rocks, and study of these help us understand the reactions that were going on during metamorphism.

Original composition. The composition of the original rock limits the possible metamorphic reactions that can take place, because the new minerals formed by metamorphic reactions can only contain elements originally present or added during metamorphism. Some rocks can be altered to produce a wide variety of metamorphic rocks. Rocks with a high clay content are especially responsive to temperatures and pressure increases. For example, if it comes in contact with a magma, a shale that is initially composed of fine-grained quartz, mica, chlorite, hydrous aluminum silicates, and amorphous iron oxides may become a dense, hard, fine-grained hornfels, which is composed of the minerals quartz, andalusite, cordierite ($[Mg,Fe]_2Al_4Si_5O_{18}$), biotite, and feldspar. If the same shale is subjected to conditions of both high temperature and high confining pressure, it might be changed to a rock composed of quartz, muscovite, biotite, and garnet—a garnetiferous-mica schist. Under still other conditions of high directed pressure, the shale could be transformed into a slate, or, if recrystallized at great depth, a gneiss.

It may also be difficult to determine which of several possible original rock types was altered to produce a particular metamorphic rock. For example, amphibolite can be derived from such diverse rocks as basalt, gabbro, basaltic tuff, or an impure shaly dolomite, because all may have similar bulk chemical compositions.

Temperature. High temperature is an important factor in both mechanical and chemical processes of metamorphism for the following reasons:

1. It accelerates most chemical reactions.
2. It facilitates recrystallization and recombination by increasing molecular activity.
3. It favors dehydration (removal of water) and endothermic (heat-absorbing) reactions.
4. It makes most rocks more plastic and so accelerates rock flowage.

Heat is also important because it increases the chemical activity of water and other solutions.

Heat is the primary factor in the contact metamorphism that takes place close to igneous intrusions where the country rock may be baked without melting. In addition, experimental evidence indicates that temperatures in the range of 200 to 800 °C are common to regional metamorphism. Above 600 °C, granitic magmas may begin to form if water pressure is also high, and over 1000 °C basaltic lavas may be generated if water pressure is high.

Confining pressure. Confining pressure and **lithostatic pressure** are terms applied to the undirected pressure in rocks caused by the weight of the overlying rock. The pressure acts as a compression directed from all sides, like hydrostatic pressure in water. Lithostatic pressure increases rapidly with depth (Fig. 7.8), on the order of 250 to 300 bars per kilometer of depth (1 bar = 1 atmosphere or 14.7 pounds per square inch). Most observable metamorphic phenomena take place under lithostatic pressures in the range of 0 to 10 kilobars. Few of the rocks exposed at the surface of the earth have ever been at depths of much more than 10 to 15 kilometers.

Directed pressure. When rock is subjected to **directed pressure,** like that experienced by a rock pressed between the jaws of a vice, it either moves or changes shape (becomes distorted). If the magnitude of the directed pressure is low, the rock may behave simply as an elastic material that deforms but recovers when the pressure is released. Most rocks behave as brittle materials at low temperatures and low confining pressures, but as temperature and confining pressure are increased, rock undergoes a transition from brittle to plastic behavior. If the amount of directed pressure is high, and temperature and confining pressure are low, the crushing strength of the rock may be exceeded, and shearing, crushing, and granulation occur in the brittle rock. If conditions favor plastic behavior, directed pressure may induce rock flowage.

Directed pressure is also an important factor in bringing about recrystallization because solubility, the ease with which a substance dissolves, may increase with pressure. Granular rocks have points of greatest stress where the individual grains touch one another. The points of least stress are the pore spaces around the grains. Under stress the grains may go into solution at the points of contact; the material moves into the pores where it recrystallizes, giving rise to elongated minerals. In stratified rocks, the lines of easiest movement are parallel to the beds, not across them. Thus, when a layered rock is metamorphosed under pressure, bands of new minerals often form parallel to the old bedding surfaces.

Effects of chemically active fluids. Fluids may be extremely important in initiating chemical reactions and in providing a means by which the products of these reactions can move through rock. Many sedimentary rocks retain water from the time they were

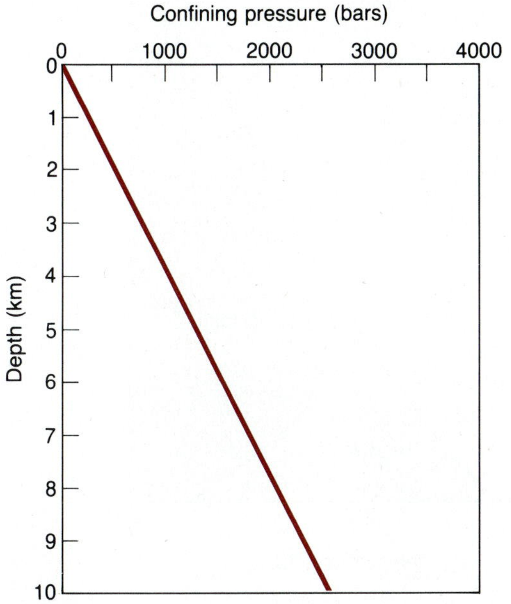

Figure 7.8
Pressure increases with depth in the earth. Because this pressure, called confining pressure, is caused by the weight of the overlying rock, it varies from one place to another. In most places the pressure increases at the rate of 250 to 300 bars per kilometer of depth.

deposited. Rock melts also introduce fluids into the surrounding country rock. Such fluids bring with them dissolved material that changes the net chemical composition of the country rock, increasing the chances for formation of new minerals. In a process of simultaneous solution and deposition, called **metasomatism,** the old minerals may be replaced by new minerals that are partly or wholly different in composition from the ones they replace.

Regardless of its origin, water facilitates chemical reactions and the temperature of the water affects the rate at which reactions proceed. For example, silica is almost insoluble in water at surface temperatures, but at elevated temperatures and with small quantities of alkalis present, significant amounts of silica will dissolve. As temperature rises, the mobility of water molecules is increased, and at 374 °C, the critical point is reached. Above this temperature the gaseous and liquid behavior of water merge. The resulting fluid is so active that it can penetrate even the most minute opening in rocks and crystals.

Because chemical reactions occur on surfaces, the surface area exposed to solutions is a factor in determining the rate of reactions between the solid and the solution. The total surface area of a given volume of rock increases as the rock is broken into smaller and smaller pieces. Thus, fine-grained materials dissolve faster than coarser ones, and reactions occur faster in finer materials. Consequently, rocks like shale, composed mainly of microscopic micaceous flakes, are more sensitive to metamorphic changes than are coarser sediments. Because of the way they formed, many sediments have extensive surfaces on which chemical reactions may take place, but even igneous and metamorphic rocks have some porosity due to fractures, cracks through crystals, and minute separations along grain boundaries. Reactions with active fluids occur on these surfaces.

Experimental Approaches to the Study of Metamorphism

Although duplicating the conditions under which most metamorphic processes operate poses many difficulties, the field of geology known as experimental petrology is highly developed and is the source of much of what we know about the complex chemical reactions that take place during metamorphism. Working with the high temperatures and pressures involved in metamorphism is a major difficulty, and the fact that rocks often consist of several minerals and elements further complicates the problem. Add

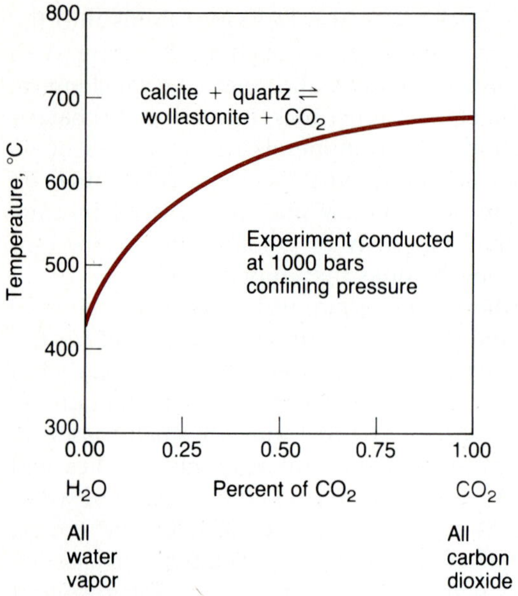

Figure 7.9
This diagram shows the conditions under which equilibrium exists between calcite, quartz, wollastonite, and CO_2 at 1000 bars pressure. Note that higher temperatures are needed for this reaction as the amount of CO_2 increases.

to this the important consideration of the amount of water present during these chemical reactions, and the task of duplicating nature is formidable indeed. Nevertheless, a few examples of the types of studies conducted will demonstrate the methods of experimental petrology and the significance of the results.

Many experiments are designed to find out what chemical reactions will take place between certain specified elements of compounds when they are subjected to a change in their environmental conditions. For example, the elements in clay will begin to react to form new compounds as the temperature is increased. As temperature continues to rise, the new compounds formed react to form still other compounds. Experiments can be used effectively to find the temperature at which each reaction occurs and the range of temperature over which each compound is stable. In a similar manner, the effects of confining pressure and other variables can be determined. For example, the graph in Fig. 7.9 shows the effect of increasing concentration of carbon dioxide on the temperature at which calcite will combine to form a new mineral, wollastonite ($CaSiO_3$), plus carbon dioxide (CO_2). We see that as the amount of CO_2 increases, much higher temperatures are needed to drive this equation. That certain reactions occur can be shown by examination of field and mineral relationships. For example, reactions are often incomplete, leaving a core of the original mineral within its alteration product. Or it is frequently possible to trace an unmetamorphosed rock, such as quartz sandstone cemented by calcite, to its contact with a pluton where the main contact mineral is a calcium silicate (wollastonite). Thus, we can be sure the quartz and calcite are reacting to form wollastonite. Experiments provide information about the conditions under which these reactions may occur.

A second type of experimental study is directed toward determining the temperature and pressure conditions under which metamorphic minerals are stable. An important example is that of the three key metamorphic minerals—andalusite, kyanite, and sillimanite. These minerals all have the same chemical composition, Al_2SiO_5, but because they differ in crystalline structure they have the different physical

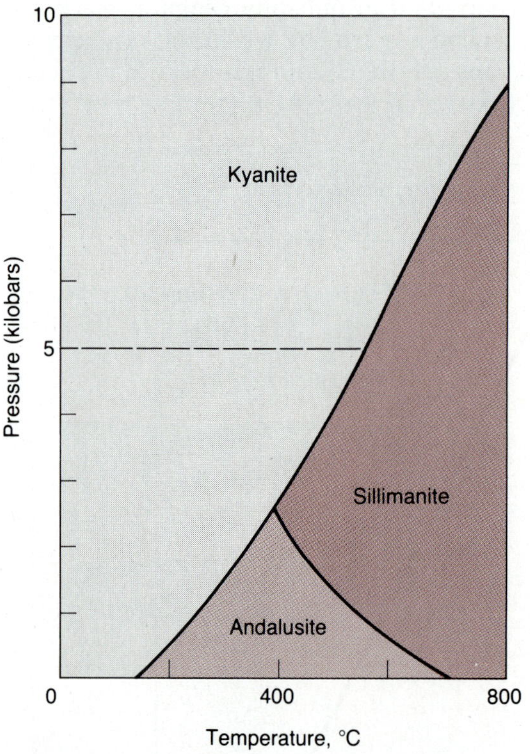

Figure 7.10
This diagram shows the combinations of pressure and temperature for which kyanite, andalusite, and sillimanite are stable. For example, if kyanite under 5 kilobars of pressure is subjected to steadily increasing temperature, it will change to sillimanite when a temperature of about 600 °C is reached.

properties described earlier. Which structure Al_2SiO_5 has is determined mainly by the temperature and pressure. Figure 7.10 summarizes the conditions at which each mineral is stable. For example, at the temperature and pressure corresponding to any point in the area marked kyanite, that mineral is the stable one. Clearly kyanite is the stable form at very high pressure and at moderate temperature, but at very high temperature and high pressure sillimanite is stable. Andalusite is favored by milder conditions—low pressure and moderate temperature. These minerals change from one structure to another in response to the changes of temperature and pressure that occur in rocks that sink deep into the crust.

The effect of water on reactions can also be explored in the laboratory. The role of water in metamorphic reactions depends on the confining pressure, water pressure, and the amount of water present, as well as on the temperature. The geologists who first studied and described the particular assemblage of minerals present in metamorphic rocks thought the differences they observed were related almost entirely to temperature. But experimental work revealed that the minerals thought to be typical of many different temperatures and pressures can be produced at an elevated but constant temperature simply by varing the amount of water present during the reactions. At lower water concentrations, minerals that ordinarily develop at higher temperatures come into existence. Thus either temperature or the amount of water vapor present may control which minerals evolve in many situations. Figure 7.11 summarizes the correlation between temperature of formation and water concentration for a series of minerals considered key to understanding the conditions of metamorphism.

Recognition of the role of water has led to extensive reevaluation of the kinds of reaction that take place during metamorphism. The chemical equilibrium of any reaction in which water appears as a reactant or a product depends on the **partial pressure** of water. Partial pressure of water is that part of the total vapor pressure that is due to water vapor. Other gases, such as CO_2, may also be present and contribute to the total vapor pressure. An example is seen in the following reaction:

muscovite + quartz $\rightleftharpoons$
K(orthoclase) feldspar + sillimanite + water

This reaction can proceed from left to right only so long as the vapor pressure of water remains low. If water vapor pressure rises, the reaction is driven right to left.

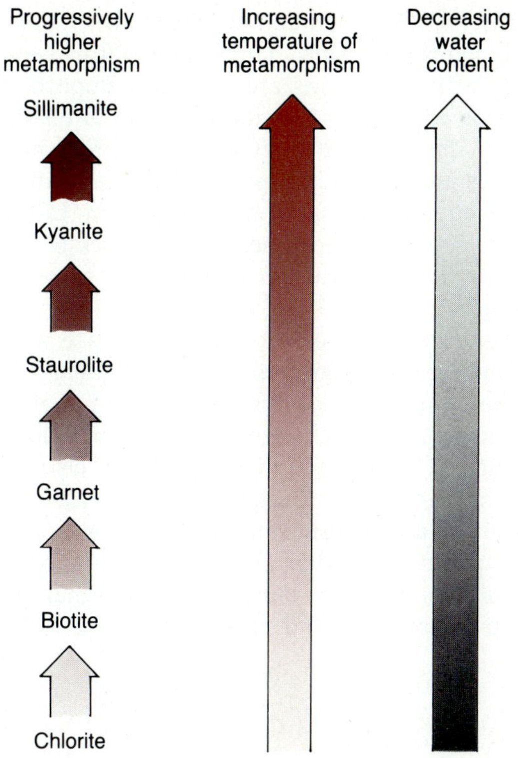

Figure 7.11
Key minerals in metamorphic rocks. When a rock containing many different elements is progressively altered by metamorphism, the intensity of that metamorphism is usually indicated by the presence of certain key minerals shown listed in order. At first, geologists thought these minerals indicated primarily progressively higher temperatures. Later experimental work showed that the amount of water present may also determine which minerals form. Thus higher minerals in the list may form at moderate temperatures if the rocks are dry.

Of course, the ultimate test of the value of laboratory experiments in studying the origin of rocks is how well they help us understand what we observe in the field.

7.4 FIELD OCCURRENCE OF METAMORPHIC ROCKS

The geological settings in which metamorphism occurs may be classed in three broad categories: contact metamorphism, dynamic metamorphism, and regional metamorphism. Contact metamorphism is, as we have seen, the alteration of the country rock around igneous intrusion. High temperature and chemically active solutions are the most important factors involved. Dynamic metamorphism is distinguished by the fact that it generally occurs under conditions of much greater pressure. Recrystallization usually accompanies dynamic metamorphism. In the absence of recrystallization the resulting rock is not classified as metamorphic no matter how severely deformed it is.

Temperature and pressure always increase with depth in the earth. Many rocks formed near the surface show the effects of this pressure and heat as they are buried. These changes due to burial alone may be referred to as **burial metamorphism.** Usually areas that have subsided enough to bring about burial metamorphism are also sites of enough other crustal activity to qualify as regional metamorphism—a combination of the effects of burial, rock deformation, elevated temperatures, and chemically active solutions.

Contact Metamorphism

The country rock around the margins of igneous intrusions is almost invariably altered, but the width and composition of the altered zone, known as a **metamorphic aureole,** varies considerably from place to place. It is not difficult to imagine some of the possible reasons for such variations. The temperature of the magma, the amount of volatiles present, the depth, and the length of time it took for the magma to cool are all likely causes of variations, but the most obvious and the easiest factor to evaluate is the effect of intrusions on country rocks of different compositions. To study this, we might examine the borders of granitic intrusions in country rock of relatively uniform composition.

The Onawa pluton in Maine, an intrusion of granodiorite in slates, is one such example. This intrusion, exposed as an oval stock about 70 km^2 in area, is surrounded by a series of concentric zones of altered country rock extending about one kilometer into the earlier metamorphosed slate. The mineralogy of the altered slate changes systematically with distance from the stock, as shown in Fig. 7.12.

The alteration in this contact metamorphic aureole is similar to that found in many other places where plutons have intruded into slates or shale. A typical feature of the least-metamorphosed slate is the appearance of spots of newly crystallized biotite and chlorite. Closer to the pluton new minerals not present in the slate—muscovite, andalusite, and quartz, for example—are found in a rock that is a schist rather than a slate. Near the contact, other minerals, such as cordierite, sillimanite, and orthoclase, may be present. The pluton is clearly a source of heat, and the alteration in the concentration zones in the country rock appears to be directly related to temperature. Country rock at the contact itself may be a dense, fine-grained, baked rock, a hornfels, but at greater distance, the chemical components of the slate have recrystallized to form larger crystals and different minerals from those originally in the slate.

Around plutons, dikes often are injected into the country rock, and some plutons contain large amounts of fluids that may invade and alter the country rock. This raises the question of whether the aureole is a result of recrystallization within the slate or the product of alteration due to reaction between fluids originally in the pluton with the slate. This question may be resolved by comparing the bulk chemical analyses (in which the quantity of each element in the rock is determined) of the slate and the rock in the aureole. Such comparisons often indicate that nothing has been added or removed during the metamorphism. In these cases, the alteration takes place in an essentially closed system and results from recrystallization and recombination of the elements originally present in the country rock.

Dynamic Metamorphism

Distortion of a rock may completely transform its texture, breaking mineral grains and causing other types of deformation in what is referred to as dynamic metamorphism. The best and most clear-cut examples of this are found in fault zones along which the rock masses on either side have been displaced, producing the kind of distortion known as **shear** (Fig. 7.13). The large-scale geometry of these zones is discussed in Chapter 10. Our concern here is with the processes that can so completely modify the texture and appearance of the rock within these zones. Deformation of this type occurs throughout the crust, but the effects vary greatly depending on the depth at which the deformation takes place.

All faults are zones where the rock has yielded to allow displacement of rock outside the zones, but the mechanism of this yielding appears to depend on the type of rock the fault crosses and the depth at which the movement occurs. Some fault zones contain rock that has been broken into angular fragments of varying size. The displacement has fractured and

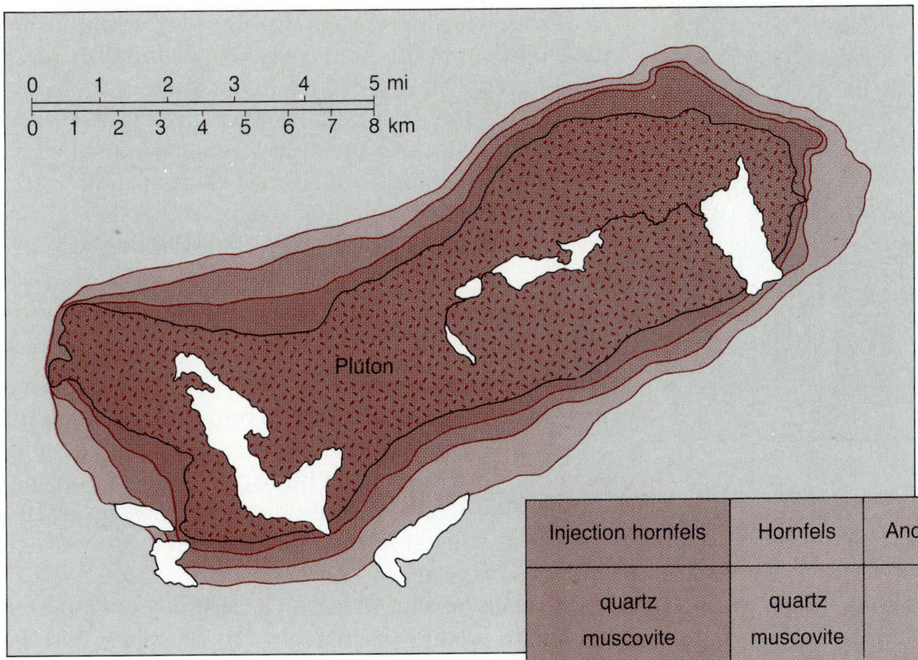

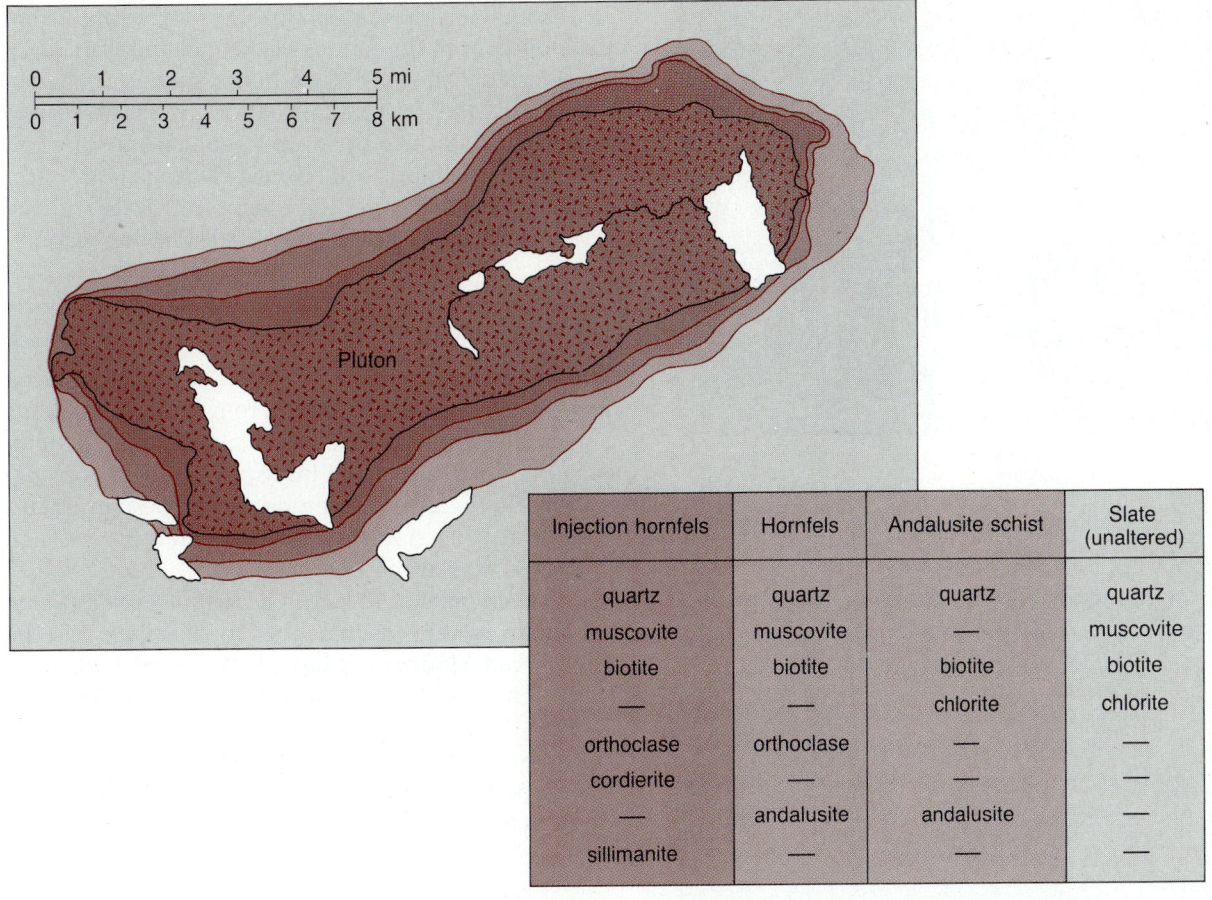

Injection hornfels	Hornfels	Andalusite schist	Slate (unaltered)
quartz	quartz	quartz	quartz
muscovite	muscovite	—	muscovite
biotite	biotite	biotite	biotite
—	—	chlorite	chlorite
orthoclase	orthoclase	—	—
cordierite	—	—	—
—	andalusite	andalusite	—
sillimanite	—	—	—

Figure 7.12
Map diagram of the metamorphic aureole around a pluton at Onawa, Maine. The table at the bottom shows the rock types and minerals found in each of the concentric zones around the pluton. The country rock, slate, is progressively altered toward the contact of the pluton.

rotated these bits, producing breccia. The breccia shown in Fig. 7.14*a* is a coarse one, but the process of breaking, rotation, and grinding may proceed until the rock is ground to a powdery rock flour. If the displacement occurs at greater depth, the rocks produced by this process are characterized by strongly aligned planar bands of fine layers of sheared and crushed rock (Fig. 7.14*b*). In extreme cases, zones of fine-grained, flinty rock called **mylonite** (Fig. 7.14*c*) are produced. These last two types differ from ground-up powder in that they have recrystallized during the deformation. The effects of the breaking up and shearing of the original rock can be seen in the deformed mineral grains, but the whole rock is solid and crystalline; the grains are interlocked, and often new crystals can be seen growing at the expense of the deformed grains.

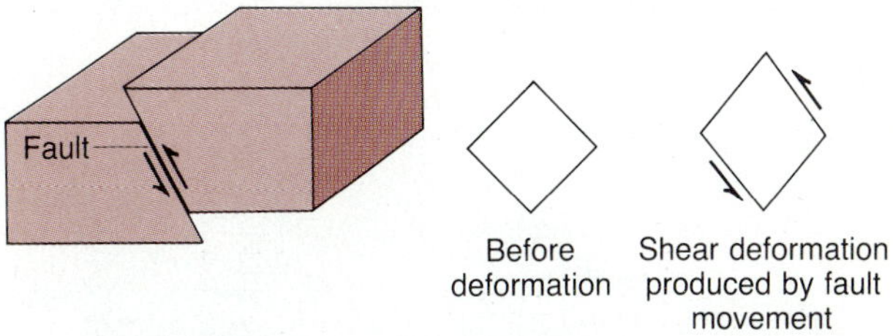

Figure 7.13
Movement along faults produces shear, a type of deformation. The arrows indicate the direction of motion. If these arrows were directly opposed to one another the material would be compressed as in a vise. In shear the movement and the forces causing it are not directly opposed. As a result, the rock is distorted as shown in the enlarged section. This movement tends to rotate, break, and smear the minerals in the rock. The fault could occur at scales ranging from microscopic to thousands of meters.

(a)

(b)

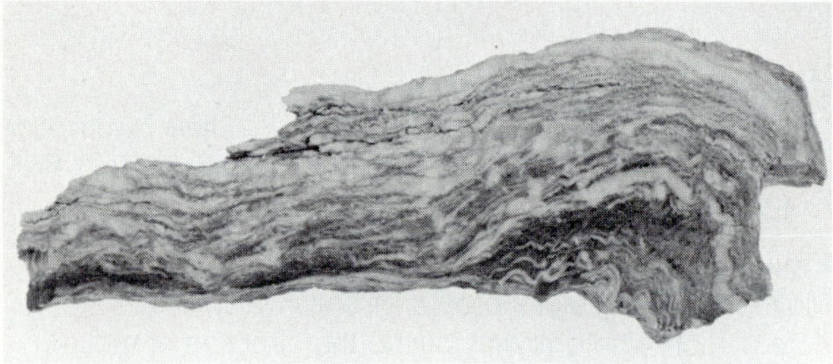

(c)

Figure 7.14
Textures caused by severe rock deformation. (a) A coarse breccia in quartzite. (b) A strongly sheared gneiss. The white streaks are feldspar. (c) Mylonite formed in a major fault.

Processes of deformation also play a role in the development of the fabrics of the metamorphic rocks outside fault zones. For example, slates are characterized by the presence of strongly aligned platy minerals that form in a plane perpendicular to directed pressure. This fabric, called **rock cleavage** (Fig. 7.15), is caused by the flattening of clay minerals and the crystallization of micaceous minerals in this plane.

Regional Metamorphism

The increase of temperature with depth is observed in deep mines and drill holes. From such observations and from theoretical calculations, models of the temperature structure of the lithosphere like the one we saw in Fig. 6.22 have been devised. These estimates suggest that even in the crust temperatures are high enough to cause changes.

Certain belts of the earth's crust, especially some mountain systems such as the Appalachians and the Cordilleran Mountains, have long histories of instability. They have been sites of deep subsidence, in which shallow-water sediment have accumulated to thicknesses of as much as 10,000 to 15,000 meters. These belts have also been sites of uplift, strong deformation, metamorphism, and concentrated ig-

IGS Photo

Figure 7.15
Slaty cleavage is well developed in this fold in Scotland. The outcrop is about 10 meters wide.

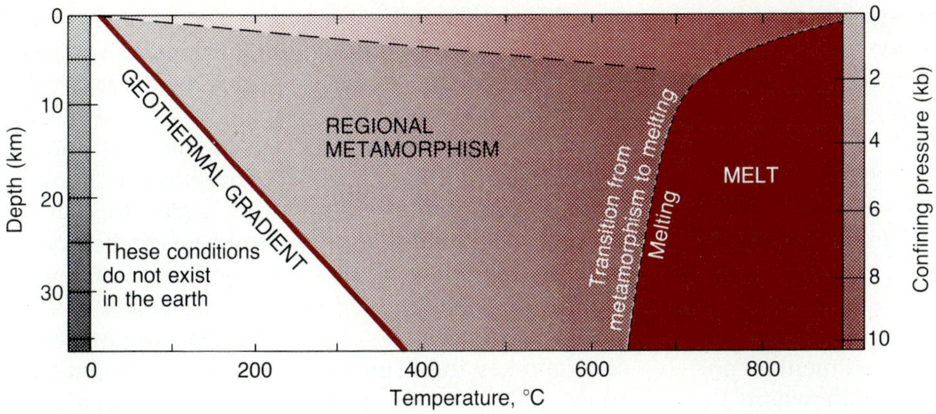

Figure 7.16
This figure illustrates the relationships among temperature, pressure, melting, and metamorphism. Temperature normally increases with depth as shown by the line labeled "geothermal gradient." Temperature–pressure combinations to the left of this line do not occur in the earth. However, higher temperatures do occur, leading to metamorphic reactions. At very high temperature (at right) melting begins. Contact metamorphism occurs adjacent to rock melts even at low temperatures found near the surface.

neous activity. The metamorphism in these belts may be due primarily to depth of burial in some places, but high temperatures, injections of magma, or strong directed stresses and combinations of these are more probable causes in large areas of these belts.

The common denominators of regional metamorphism are the increased confining pressures and elevated temperatures that are a consequence of depth of burial (Fig. 7.16). Metamorphic effects around the contacts of plutons rarely extend more than two or three kilometers beyond the contact itself, and the dynamic metamorphism of fault zones usually involves only a narrow zone along the fault, yet we find metamorphic rocks exposed over vast areas of the earth. Metamorphic rocks are major components of both the mountain belts and the Precambrian rocks that make up the shield areas of the continents (Fig. 7.17), areas that have been stable since the end of

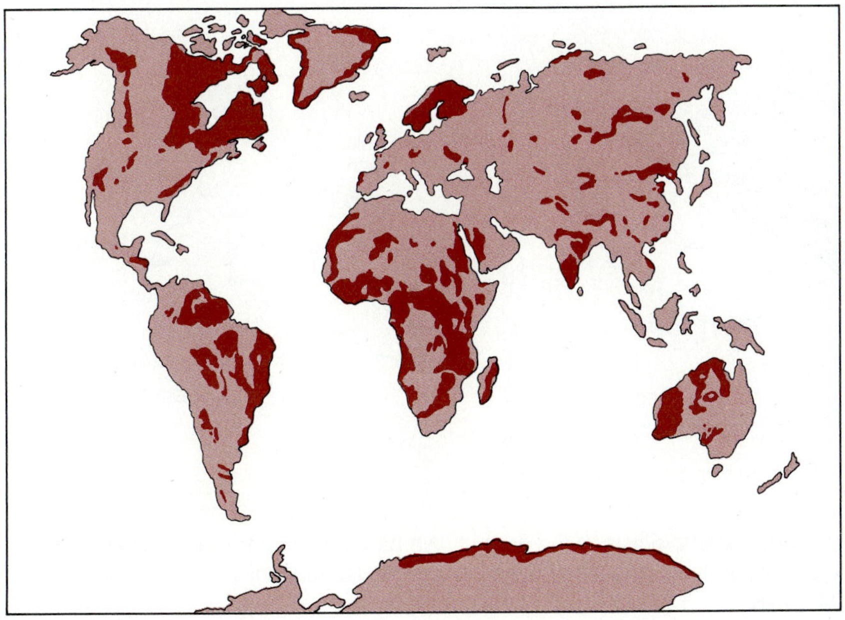

Figure 7.17
The areas shown in black identify areas where Precambrian metamorphic rock are exposed.

Precambrian time nearly 600 million years ago. Both of these regions have been deeply denuded as a result of erosion. The shields have long since had any older sedimentary veneer stripped from them, and the younger mountain belts are rapidly being laid bare by erosion even today. Both the shields and mountains are areas where rocks that were once much deeper in the crust can now be examined at the ground surface.

The most important insights into the character of regional metamorphism come from the areas where it is possible to trace slightly altered sedimentary and igneous rocks into metamorphic terrain where the sediments and igneous rocks become progressively altered. This gradual alteration shows up in progressive changes in mineral composition and texture.

The study of metamorphism in southern Scotland by George Barrow around the turn of the century is one of the classic analyses of progressive regional metamorphism. The region described by Barrow and later by others is located within a mountain system known as the Caledonides. This system, composed of the mountains that run through Norway and across Scotland and Northern Ireland, was uplifted and deformed in the early Paleozoic, in Ordovician time. The area of Barrow's study is between two major faults in Scotland, the Highland Boundary and the Great Glen (Fig. 7.18), each of which forms a sharp boundary abruptly cutting off the metamorphosed rocks. The rocks within this area were originally composed of fine-grained clay, coarse sandy sediments, and some carbonates. During his field studies of this region Barrow noticed that progressive changes in mineralogy and texture (from slates to coarse-grained sillimanite-garnet-mica schists) can be traced in rocks that have almost the same bulk chemical composition. He pointed out that the region can be subdivided into zones characterized by the presence of certain key **index minerals.** Barrow concluded that zones with the same index minerals had been subjected to similar conditions of metamorphism. This **concept of metamorphic zones,** as bodies of rock exposed during metamorphism to similar environmental conditions, is now widely used. Each zone represents a particular **grade of metamorphism,** a reflection of the intensity of metamorphism. Grade was originally thought to be caused mainly by the temperature to which the rocks were subjected, but we have seen in Section 7.3 that the amount of water also influences grade.

The zones recognized by Barrow in 1912 are listed below in order of ascending metamorphic grade. Some typical mineral assemblages found in each zone are also listed.

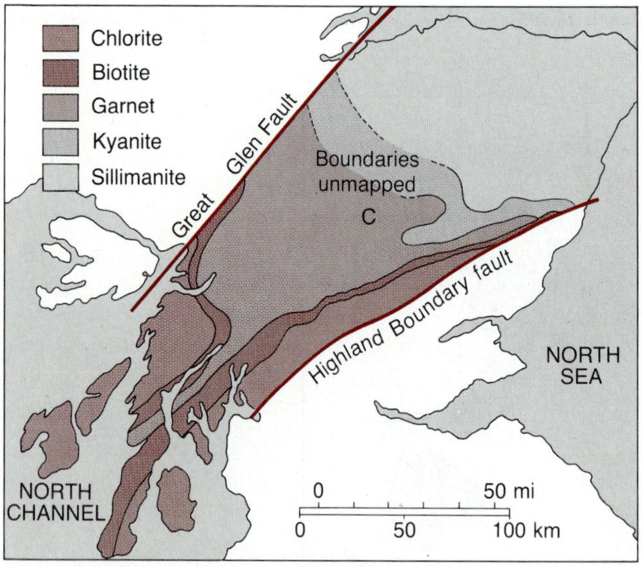

Figure 7.18
(a) Barrow made his classic study of metamorphism in the area shown here between the Great Glen and Highland Boundary faults. He recognized that certain key minerals can be used to indicate the intensity of metamorphic change. (b) Locator map.

1. **Chlorite zone.** The typical mineral assemblage is quartz, muscovite, albite, and chlorite.
2. **Biotite zone.** The appearance of biotite is characteristic.
3. **Garnet zone.** The typical mineral assemblage is quartz, muscovite, biotite, garnet, and albite.
4. **Staurolite zone.** The typical mineral assemblage is quartz, biotite, muscovite, plagioclase, garnet, and staurolite.
5. **Kyanite zone.** This typically contains quartz, biotite, muscovite, plagioclase, garnet, and kyanite.
6. **Sillimanite zone.** This typically contains quartz, biotite, muscovite, plagioclase, garnet, and sillimanite.

These zones are most easily demonstrated when the altered rocks are originally composed in large part of clay minerals. It is easy to see why—clay contains many elements that can combine during recrystallization to form a variety of minerals. Obviously, none of the index minerals could be formed from a pure quartz sandstone, a rhyolite, or in a pure limestone.

Barrow interpreted the metamorphic zones as giant metamorphic aureoles surrounding granitic rocks, but the situation is far more complex. More recent investigations indicate that the regional metamorphism in this part of the Scottish Highlands is related to the rise of a hot mixture of magma and country rock in large, deeply rooted folds. The metamorphism in many places here and elsewhere is the product of more than one heating and deformation.

According to the concept of progressive metamorphism, a rock could be metamorphosed to one grade during one event and later the progression could advance to a higher grade if higher temperatures were reached. It is also possible for the rock to regress—that is, for a mineral assemblage formed at higher temperature gradually to revert to those of a lower grade. The evidence for this is the discovery that some of the high-grade minerals are partially altered to lower-grade minerals, or that low-grade minerals have the shape of higher-grade minerals they have replaced.

Metamorphic Facies

The concept of metamorphic grade helps us recognize the progressive nature of metamorphic changes as temperature and pressure increase in regional metamorphism. The nature of this relationship between the composition of the original rock, environmental conditions, and the resulting metamorphic rock is further emphasized by a related concept—that of **metamorphic facies.** According to the facies concept, a particular mineral assemblage will result whenever rocks of the same bulk chemical composition are metamorphosed within a particular range of environmental conditions.

During metamorphism, some of the original minerals usually become unstable and are altered to produce a new group of minerals stable at the prevailing temperature, pressure, and other environmental conditions. If the metamorphic reactions proceed to completion or near-completion, the new minerals are approximately in an equilibrium state, and they tend to undergo no further change until the conditions are changed. Thus, the particular assemblage of minerals present in a metamorphic rock is an indication of the environmental conditions such as the confining pressure, water pressure, and temperature under which they formed.

At about the turn of this century, two now-famous geologists, V. M. Goldschmidt and P. Eskola, studied the mineral assemblages in metamorphic rocks in the regions of Oslo, Norway, and Orijarvi, Finland. Goldschmidt examined the contact zones around a number of plutons that had invaded a series of shaly, sandy, and calcareous sedimentary rocks. He found that the mineral assemblages in the contact aureoles surrounding these plutons were remarkably simple and similar in mineralogical composition, despite the great difference in bulk chemical composition of the country rock around different plutons. The rock in each metamorphic aureole was composed of only four or five minerals; but chemically equivalent pairs (pairs with the same kinds and amounts of elements) did not appear together in the same aureole. Eskola found the same effects in Finland and attributed the differences in mineralogy to differences in the physical conditions that had accompanied the metamorphism. These observations led Eskola (1915) to formulate the concept of facies which he later defined as follows (*1*): "A metamorphic facies comprises all rocks exhibiting a unique and characteristic correlation between bulk chemical and mineralogical composition."

According to this concept, the mineralogical composition of a metamorphic rock in a given metamorphic facies depends both on the chemical composition of the rock from which it was derived and on the environmental conditions under which metamorphism took place. The same group of minerals, representative of a facies, will always result if original rocks of the same bulk chemical composition are metamorphosed under the same environmental conditions, no matter what their original mineralogical composition was. A single facies can, however, consist of a number of different metamorphic rocks, each derived from original rocks of different chemical composition.

7.5 THE TRANSITION BETWEEN METAMORPHIC AND IGNEOUS ACTIVITY

The progressive nature of metamorphic changes as temperature and pressures rise is evident in the alterations that take place in the mineral assemblages of different metamorphic facies. These changes are progressive up to a certain limit. The limit is the point at which environmental conditions favor melting of the rock rather than further metamorphic alteration.

Metamorphic rocks subjected to temperatures and pressure greater than those that produce the amphibolite facies rocks begin to pass into a transition zone, often involving partial melting. So, it is not surprising that the boundary between these two major rock types is often vague. The most abundant of these highest-graded metamorphic rocks are the gneisses.

High-Grade Gneisses

Gneisses are characterized by banding or compositional layering. Typically, layers rich in quartz and feldspar are separated from layers rich in darker minerals such as biotite, hornblende, or pyroxene. Gneisses are a product of high-grade metamorphism. They often contain garnets, or higher-grade index minerals, and they represent a more complete reconstitution of the original rock than is found in slates, phyllites, or schists. Gneisses rarely show any original primary features of the sedimentary or igneous rock from which they are derived.

Not only have the minerals in gneisses formed by recrystallization, they have become segregated into layers of different composition. Several mechanisms may bring about this segregation:

1. A melt may be intruded parallel to original sedimentary layers or to an earlier structure, such as fractures or foliations.
2. A partial melt formed within the original rock may migrate locally into zones parallel to earlier bedding or structures.
3. Some beds in the original sediment may be enriched by the migration of ions of potassium, sodium, iron, or magnesium from adjacent layers.

The mineralogy of gneisses varies. Many are similar to granite and contain quartz, feldspar, and biotite; others are notable for the absence of biotite and other **hydrous** (water-bearing) minerals and for the presence of an **anhydrous** (waterless) pyroxene mineral (hypersthene). It seems probable that these latter gneisses, called **charnokites,** form under dry conditions, perhaps as the products of metamorphism deep in the crust. Many of them outcrop now in mountains, such as the Blue Ridge, where they could have been uplifted from great depths.

Mixed Igneous and Metamorphic Rock

As we have noted the distinction between metamorphism—a solid rock process—and igneous activity involving melts is unclear, because one grades into the other. At temperatures below the fusion point of all the minerals in a rock, the rock remains completely solid even though metamorphic reactions may occur within and between minerals in the rock. As temperatures rise, the activity of hot water and ionic solutions increases. Fig. 7.19 shows the types of rock that might form as temperature rises. At some point a few of the minerals begin to melt while most of the rock remains solid. At this time, the rock is a solid network of crystals filled with only pockets and pore spaces of melt. Figure 7.19(*a*) shows dark streaks of metamorphosed sedimentary rock invaded by a light-colored granitic rock. If the rock is compressed, the fluids may be squeezed off and form intrusive veins. The name **migmatite** is given to a mixed rock consisting of metamorphic rock invaded by veins that are usually granitic (Fig. 7.19, *b* and *c*). Some migmatites appear to form by the process just described. Others may result from the intrusion of melts from nearby magma into regionally metamorphosed country rock. As temperatures continue to rise, the molten fraction of the original rock increases and eventually the whole mass may become magma and later igneous rocks.

Origin of Granite

The origin of granite has been the subject of a controversy for many years. This controversy has been important in part because such large portions

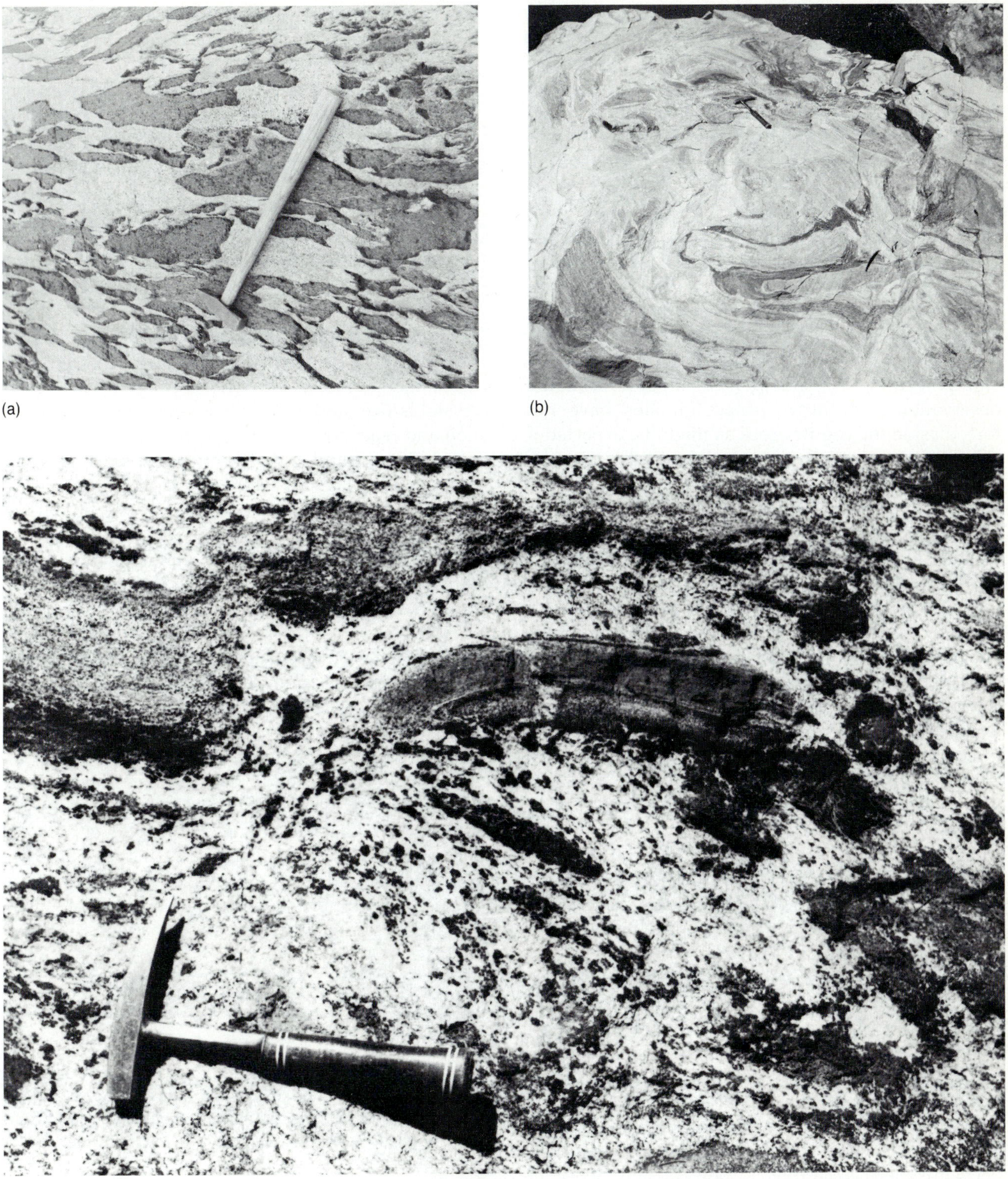

(a) (b) (c)

Figure 7.19
Rocks showing the transition from metamorphic to igneous types. (a) Dark gneisses invaded by light-colored granite. (b) Mixed gneisses with veins and masses of granitic composition. (c) Remnants of metamorphic rock in a locally formed pegmatite.

of the Precambrian shields are composed of granitic rocks. Some granites are clearly of igneous origin, as described in Chapter 6: they are surrounded by metamorphic aureoles; they cut across the younger country rock; and they contain blocks of the intruded country rock. But other granitic rocks are not so clearly intrusive: their borders are not sharp, and no aureole surrounds them. Instead, uniform granite passes into a foliated granite, and that gives way to banded gneisses. The foreign blocks are aligned so the bedding can be traced from one block to another—a relationship suggesting that the gneisses and granites may have been formed in place by metamorphic alteration.

The idea that granites might be formed by metamorphic processes also gained support from the observation that these suspect granites have not forced aside the country rock as might be expected if a huge volume of magma were forcefully injected. Yet another argument in favor of a metamorphic origin of granites came from the analyses of zircons in granites. Zircon is one of the first minerals to crystallize in granitic magma. Its crystals are usually perfectly shaped because they are free to grow in a melt. Thus, a typical igneous granite contains perfect crystals of zircon. Some granitic rocks, however, contain rounded zircons, and it is suggested that these, although originally derived from an igneous rock, have been recycled. They have been removed by weathering and erosion, rounded like quartz grains in sediment, and deposited to form part of a sedimentary rock that was later altered by metamorphism to a granite-like rock by a process known as **granitization.**

The idea that granite is formed by the transformation of other rock is not new; it was proposed by a group of French geologists in the nineteenth century and maintained and elaborated by the Finnish geologist Sederholm in the early part of this century. These geologists theorized that the rocks above granitic magmas may be transformed by emanations from the melt. One of the most persuasive early arguments of the transformists was that it offered a solution to the "room problem." It explained why the rocks surrounding gigantic masses of granitic rock are not arched, folded, or pushed aside as might be expected if the thousands of cubic kilometers of granite had been intruded. According to Bowen, only about 10 percent of a basaltic parent magma could differentiate to form granite; yet almost all of the plutons on continents are granitic in composition. If most granitic bodies are actually highly altered sedimentary rocks, it is not necessary to explain why we do not find the remaining 90 percent of the melt as basaltic or gabbroic plutons on the continents.

The field evidence we have described seems to support the transformists' view, but there remains a most serious problem for the transformists—how to explain the geochemical process that must take place for transformation of sediment to granite. The transformation is seen as taking place in solid rock by processes of ionic migration and reaction. The introduction of some ions is usually necessary because sediments do not often contain enough sodium to account for the quantity found in the feldspars of granite. Ions can move in minute amounts of solution through the pore spaces, around grains in the sediment, along grain boundaries, through spaces in the crystal lattice, and along crystal defects. Ionic migration and reaction would be greatly facilitated if water were present and if temperatures were high even though below fusion temperature. The transformist view is that, under proper conditions, sediment can be altered to form the crystalline rock, granite, by the progressive growth of quartz and feldspars at the expense of the clay, silt, sand, and feldspar fragments in the sediment.

Metamorphism has taken place and is continuing because the earth is a dynamic planet. The processes that promote metamorphic change come into play because magma is generated and because the crust of the earth is unstable. This instability involves both vertical movements, uplift, and subsidence, and lateral movement related to the shifting of lithospheric plates. In the next chapters we shall discuss the nature of the earth's interior, principles of rock deformation, and the causes of plate movement. With this background, we shall be able to examine in greater detail how magma generation and metamorphism are related to crustal dynamics.

SUMMARY

Metamorphism, the conversion by heat, pressure, and fluids of rock from one texture and mineral composition to another, occurs in three geological settings. Around igneous intrusions the heat of the magma bakes surrounding country rock and causes contact metamorphism. Where the crust has been deformed by strong direct pressure, rocks are transformed by dynamic metamorphism. Deep burial, with its heat and high confining pressure, especially when combined with crustal deformation, produces regional metamorphism. These conditions cause the rock conversion by four main processes, or combinations of them: rock flowage and granulation (physical changes) and recrystallization with or without accompanying recombination of the components (chemical changes). All the conversion processes are affected in some degree by the temperature and pressure of the environment and usually by the presence of chemically active fluids as well.

Metamorphic rocks are characterized by a set of minerals and by textures indicative of the conditions under which they formed and the rocks from which they formed. Geologists recognize a series of key metamorphic minerals that index the intensity, or grade, of metamorphism and therefore define metamorphic zones. According to the concept of metamorphic facies, rock of the same bulk chemical composition (but not necessarily the same mineral composition) will always form the same set of minerals when subjected to the same conditions of metamorphism.

As metamorphic conditions become more extreme, metamorphic rocks grade into mixed metamorphic and igneous rock (migmatites) and finally into igneous ones.

KEY TERMS

anhydrous
biotite zone
burial metamorphism
charnokite
chlorite zone
concept of metamorphic zones
confining pressure
contact metamorphism
directed pressure
dynamic metamorphism
foliation
garnet zone
grade of metamorphism
granitization
granulation
hydrous
index mineral
kyanite zone
lithostatic pressure
metamorphic aureole
metamorphic facies
metamorphism
metasomatism
partial pressure
recrystallization
regional metamorphism
rock cleavage
rock flowage
schistosity
shear
sillimanite zone
slaty cleavage
staurolite zone

STUDY QUESTIONS

1. What are the best ways to distinguish igneous rocks from metamorphic rocks?
2. Considering the conditions under which they originate, where would you expect to find metamorphic rocks exposed today?
3. What condition might prevent metamorphic reactions from going to completion?
4. If a chlorite grade metamorphic rock is subjected to temperatures of about 400 °C, what changes might you expect?
5. Why might a strongly foliated rock be a dangerous foundation rock for a dam?
6. Describe the transition from metamorphic to igneous activity.
7. Why is fine-grained rock more easily altered than coarse-grained rock?
8. What is metamorphic grade?
9. Can igneous rocks become metamorphosed?
10. Why are shales so sensitive to metamorphic changes?

REFERENCE

1. P. Eskola, "On the relation between chemical and mineralogical composition in the metamorphic rocks of the Orijärvi region," *Comm. Geol. Finlande Bul.* 44 (1915).

SUGGESTED READINGS

Barrow, G., "On the geology of Lower Deeside and the Southern Highland Border," *Proc. Geol. Assoc.*, 23 (1912), 274.

Miyashiro, A., *Metamorphism and Metamorphic Belts.* New York: John Wiley & Sons, 1973.

Philbrick, S. S., "The contact metamorphism of the Onawa pluton, Piscataquis County, Maine," *Am. J. Sci.*, 31 (1936), 1–40.

Turner, F. J., *Metamorphic Petrology, Mineralogical and Field Aspects.* New York: McGraw-Hill Book Co., 1968.

Vernon, R. H. *Metamorphic Processes.* New York: Wiley, 1975.

Winkler, H. G. F., *Petrogenesis of Metamorphic Rocks*, 5th ed. New York: Springer-Verlag, 1979.

THE DYNAMICS OF THE EARTH: THE INTERNAL PROCESSES

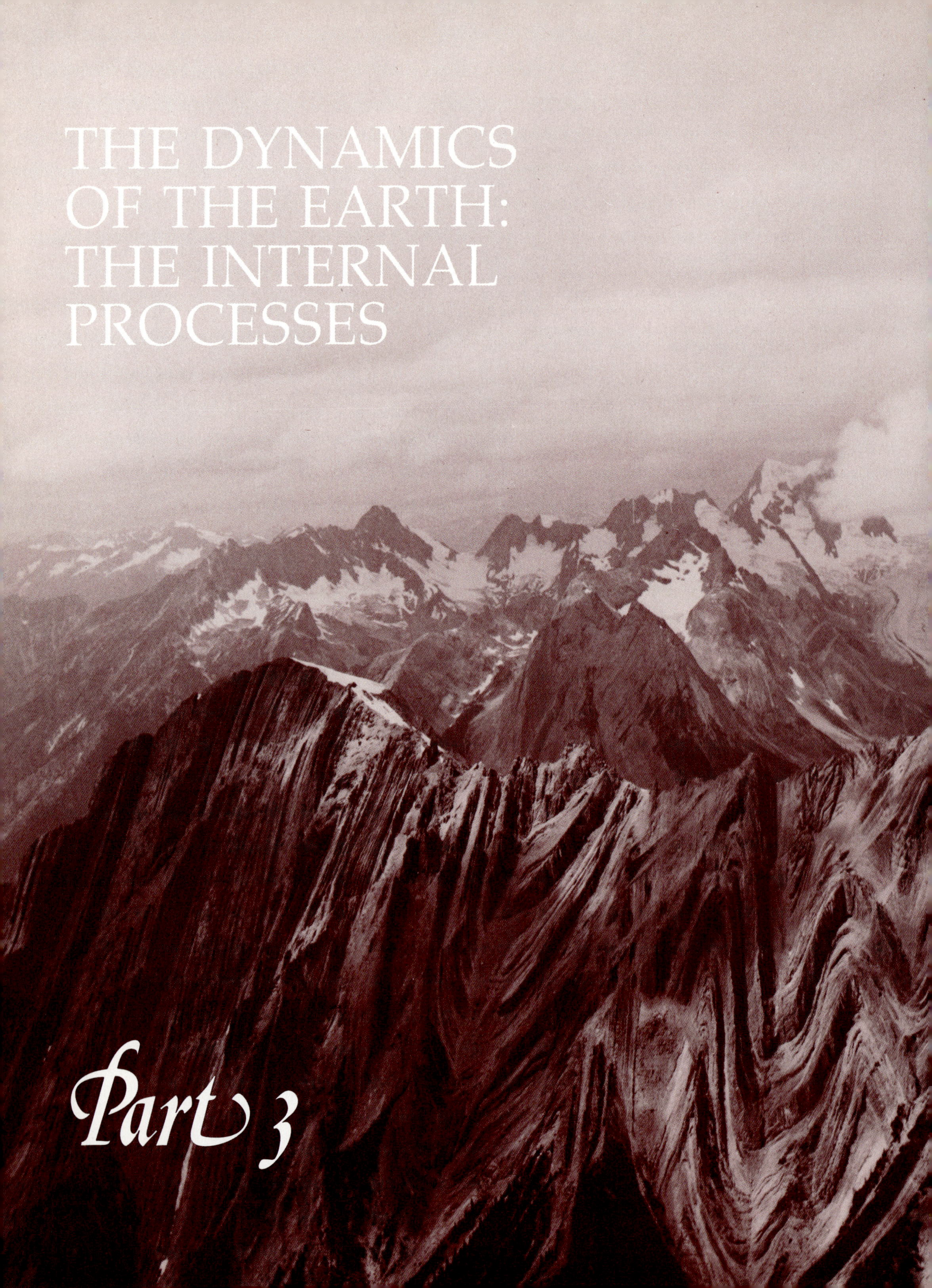

Part 3

Plate tectonic theory is often characterized as a revolution in scientific thought. Only a little more than two decades ago, most American geologists subscribed to the idea that continents and oceans were permanent features of the earth's crust. Today we find many types of evidence indicating not only that continents are moving, but that the floors of today's oceans are all young compared to the continents. One of the great attractions of plate tectonics is its power as a unifying concept. It provides a clear and simple explanation for the location of the zones of earthquakes and volcanoes, the generation of mountain belts at plate boundaries, the relationships between the various sedimentary rock types and their position on plates and the sites of origin for igneous and metamorphic rocks.

Part III is devoted mainly to the large-scale features of the earth: its interior, the structure of its continents and oceans, and the evidence that has led to our present understanding of the movements of the earth's crust.

EARTHQUAKES

Chapter 8

The vibration of the earth that accompanies an earthquake is one of the most terrifying natural phenomena known. We have good reason for wanting to understand this phenomenon and to find ways of reducing the loss of life and property damage caused by earthquakes.

From the geological perspective, earthquakes provide one of the most dramatic evidences of the instability of the earth's crust and are a logical starting point for any examination of the dynamics of the earth. Earthquakes are largely confined to relatively narrow zones in the lithosphere. These zones of high seismic activity are a key to identifying the boundaries of the major lithospheric plates. Recognizing this relationship better enables us to understand what causes earthquakes.

8.1 CAUSES OF EARTHQUAKES

Earthquakes occur from levels near the earth's surface to depths as great as 700 km. We can make direct observations of earthquakes that affect the surface. Most of the near surface earthquakes are closely associated with, and almost certainly caused by, abrupt movements along faults in the bedrock. The only other significant cause of near-surface earthquakes is volcanic activity, although minor tremors also accompany large landslides. The deeper earthquakes cannot be directly observed; and, as with other phenomena that take place deep in the earth, our working models of the processes involved are largely inferred from indirect observations.

The California earthquake of 1906, which destroyed much of San Francisco, is often taken by American geologists as the model of earthquakes caused by movement along a fault. This earthquake took place along the San Andreas fault, one of the largest in North America. The San Andreas fault is a part of a system that can be traced from the Gulf of California to Cape Mendocino in Northern California (Fig. 8.1). The fault is nearly vertical, and the land

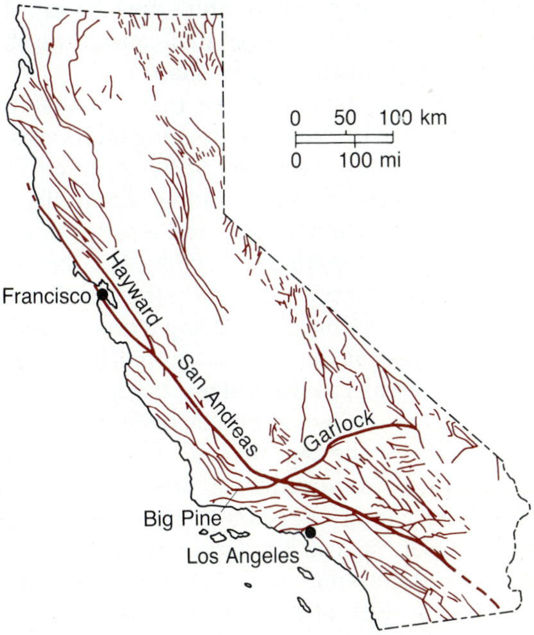

Figure 8.1
California is criss-crossed by an extensive system of faults, most of which are still alive.

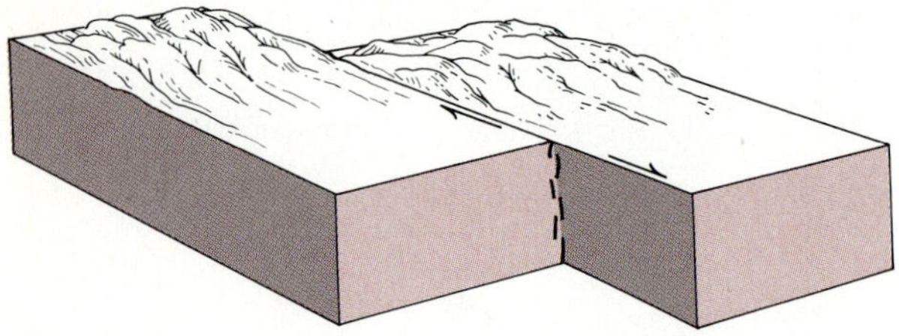

Figure 8.2
A steeply inclined break through a block along which a displacement, similar to that found on the San Andreas fault, has taken place.

on the west side of the fault is gradually moving north relative to the rest of the continent, as shown in the schematic diagram of Fig. 8.2. This movement is usually slow and causes distortion of the rock near the fault.

The movement along the San Andreas fault was first reported in 1906 by H. F. Reid, who made a number of precise surveys using markers located on either side of it. These repeated surveys revealed that lateral movements were taking place without causing earthquakes. Reid theorized that the changes in shape, called **strain,** caused by these movements finally became so great that the rock along the fault zone ruptured abruptly and the land on the west side moved north, as shown in Fig. 8.2. We may envision the rock masses as behaving elastically; that is, they were distorted or strained until they broke. Then with this release of the forces that caused the strain, they returned to their original shape.

Most geologists now agree that the shaking of the earth created by movements along faults is caused by a combination of vibration of the strained region when the release occurs, grating along the fault surface due to frictional drag, and a series of abrupt breaks along the length of the fault. As a break and movement occur at one point, the stress farther along the fault increases sharply, causing a break there and increasing stress at the next position. The result is a sequence of breaks moving along the fault.

This theory, known as the **elastic rebound theory,** is supported by the fact that most earthquakes are followed by a long series of aftershocks. During earthquakes, energy appears to be released from a series of closely spaced but abrupt events. The movement in the fault zone might be compared with the breaking of a block of concrete compressed in a vise until it cracks. From work done in recent years, we know that along some sections of the San Andreas fault very slow movements, called creep, occur almost continuously, releasing the strain as quickly as it builds up. Other sections, however, appear to be locked. These areas of strain buildup are sites of potentially dangerous earthquakes.

The mechanism of deeper earthquakes is more difficult to assess. These earthquakes occur at depths where both the great pressure of the overlying rock and the extremely high temperatures tend to cause the rock to flow like a hot plastic rather than rupture like a brittle, elastic solid as it does near the surface. In other words, the rock at these depths should yield before the strain can accumulate. The cause of deep earthquakes may involve abrupt slips along fault zones, but the movement is of a different kind. It does not involve elastic rebound, but is a phenomenon called **stick-slip**. The process may be somewhat like cutting a piece of asphalt into two parts and then pressing them back together while pushing them in opposite directions. At first, the friction between the blocks would stick them together, but eventually they would yield and move by sudden slips before once again sticking.

Earthquakes associated with volcanoes are much more localized and limited both in the extent of damage and in the intensity of wave motion produced than are those associated with faults. In some regions, particularly around the perimeter of the Pacific Ocean, volcanic activity and active faulting are closely associated. In earthquakes due solely to volcanic activity unassociated with faults, shocks may be produced by any of the following mechanisms: explosion of the volcano upon the release and expansion of gases and lavas; faulting within the volcano resulting from pressures in the chamber of molten rock; collapse of the center of the volcano into the space formed by the extrusion of gases and molten matter; and vibrations set up by the movement of magma toward the surface.

8.2 THE DISTRIBUTION OF EARTHQUAKES

The modern-day study of earthquakes, **seismology,** enjoys the advantage of having the results of many years of accumulated records from a large number of seismic observatories. A network of more than 100 stations located throughout the world have now been operating for more than a decade, and compilations of earthquake locations from these provide a good picture of the distribution of earthquakes in the earth's lithosphere, (Fig. 8.3).

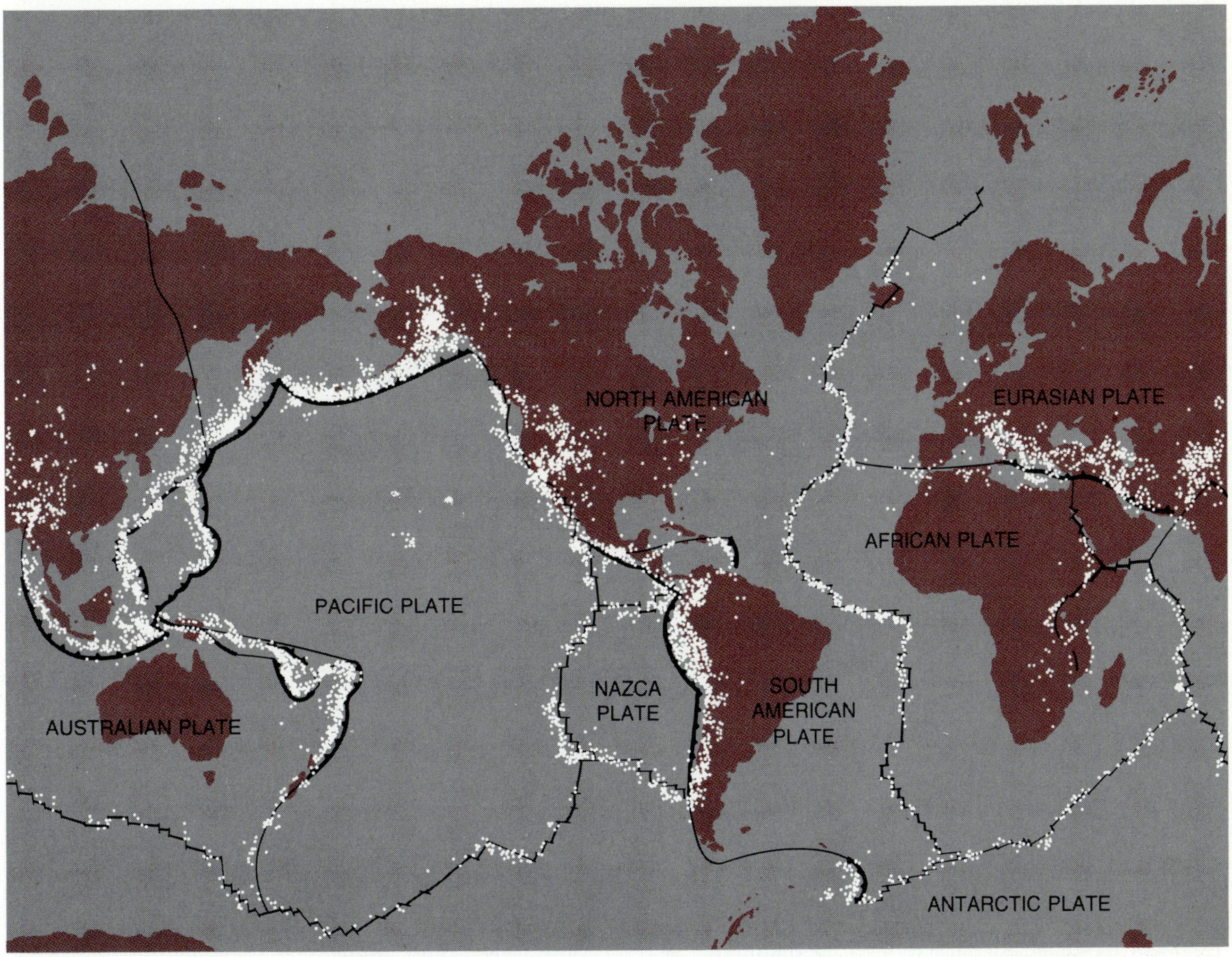

Figure 8.3
The dots mark the location of earthquakes. Notice how closely these coincide with the margins of crustal plates indicated by black lines.

The location of earthquakes is described in terms of two variables: the depth at which the shock waves originate, and the location on the ground surface directly above this point. The point of origin of the shock is called the **focus** of the earthquake. Its depth is referred to as **focal depth,** and the point on the ground above it is the earthquake's **epicenter.**

It is clear that most earthquakes are located along a number of long and relatively narrow belts separated by large areas in which few earthquakes occur. The most active belts include the island arc systems, the young and active mountain belts, and the oceanic ridge system. The relatively quiet areas are the continental interiors and the deep sea floor. Most earthquakes occur in the same zones where volcanoes are found (Fig. 8.4). They coincide most closely in the island arcs and along the crest of the oceanic ridge systems. A few volcanoes are also found within the young folded mountain belts. The main exception to this relationship is found in the ocean basins, especially in the Pacific, where chains of volcanic seamounts and some islands such as the Hawaiian Islands stand far removed from the main seismic zones.

The zones of high seismic activity are perhaps the best single guide to plate boundaries. The boundaries are of four main types:

1. *Divergent plate junctions* where spreading is taking place.

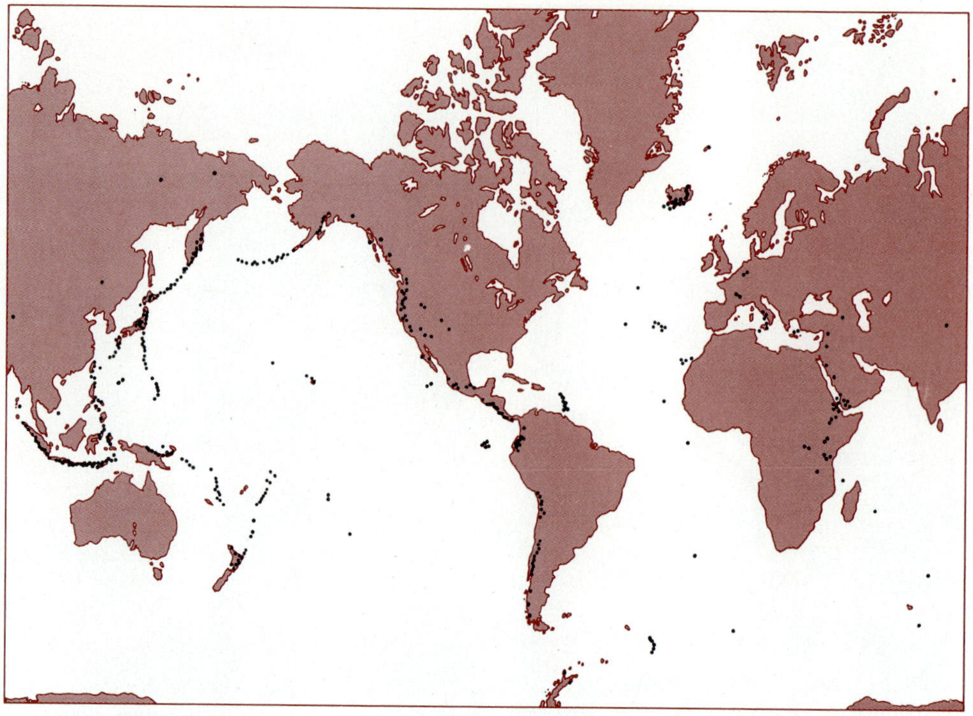

Figure 8.4
The dots mark the location of volcanic centers that have been active in historic times.

2. *Subduction zones* where plates are being consumed and lithospheric slabs are sinking into the mantle.
3. *Fault zones* along which plates are moving laterally relative to one another.
4. *Mountain belts* along which plates have collided.

Each of these is characterized by distinctive seismic activity.

The earthquakes along divergent junctions and fault boundaries are generally of shallow focus (less than 100 km deep). Most shallow-focus earthquakes and almost all deep-focus ones occur in mountain belts or in subduction zones, where some originate as deep as 600 to 700 km. Nevertheless, it should be clear from the case histories of both the San Andreas earthquake of 1906 and the Guatamela earthquake of 1976 that devastating shocks can result from movements on faults as well. Nor are generally stable areas completely exempt from earthquake activity as shown by the Charleston, South Carolina, earthquake of 1886, described later in this chapter.

8.3 EFFECTS OF EARTHQUAKES

Because earthquakes differ tremendously both in the amount of energy released and in the depth of focus, their effects vary greatly. In general, the greatest damage at the surface is caused by shallow-focus, high-energy earthquakes. Some of the extremely deep earthquakes may almost go unnoticed at the surface, even when they involve the release of great amounts of energy. Surface effects are also influenced by the type of material there. Damage to structures is always less when solid bedrock occurs at a shallow depth than it is when there is a considerable depth of unconsolidated material at the surface.

Interest in the effects of earthquakes preceded by many years the development of an extensive network of instruments, called **seismographs,** designed to measure earth vibrations. Often earthquakes occurred in areas far removed from scientists who might record their own observations. In order to achieve a relatively objective standard for comparing earthquake effects, an Italian seismologist named Mercalli developed a scaled set of observations that can be used to distinguish different levels of activity, referred to as **intensity**, during an earthquake. The observations are effects that most people could be expected to report objectively. Mercalli's intensity scale, first developed in 1902, has been modified, but it is still in use. The effects used to distinguish different intensities are given in Table 8.1.

Generally intensity decreases with distance from the focus. It also depends on the type of rock or sediment under the observer, on the focal depth, and on the amount of energy released at the focus. While

Table 8.1 The Modified Mercalli Scale.

I. Not felt, except by a few people under especially favorable conditions.

II. Felt by only a few people at rest. Delicately suspended objects may swing.

III. Felt quite noticeably under favorable circumstances. Parked automobiles may rock slightly.

IV. Felt by many or most people; some awakened. Dishes, windows, doors disturbed, walls cracked. Sensation like that of a heavy truck striking a building.

V. Some dishes, windows, and so on, broken; a few instances of cracked plaster; unstable objects moved.

VI. Felt by all; many frightened and run outdoors. Some heavy furniture moved; a few instances of fallen plaster or damaged chimneys.

VII. Everyone runs outdoors. Damage negligible in buildings of good design and construction; slight to moderate in well-built ordinary structures; some chimneys broken. Noticed by people driving cars.

VIII. Panel walls thrown out of frame structures. Fall of chimneys, factory stacks, columns, monuments, walls. Heavy furniture overturned. Sand and mud ejected in small amounts.

IX. Buildings shifted off foundations, ground cracked conspicuously. Underground pipes broken.

X. Ground badly cracked, rails bent, landslides considerable from river banks and steep slopes. Water splashed over banks.

XI. Few masonry structures remain standing. Bridges destroyed. Broad fissures in ground. Earth slumps and land slips in soft ground.

XII. Damage total. Waves seen on ground surface. Lines of sight and level distorted. Objects thrown into the air.

we often compare intensities of earthquakes, such a comparison does not allow us to compare the true scale of events.

With the development of sophisticated measuring devices, it became possible to devise a scale that allows us to compare the amount of energy released in different earthquakes. The scale now in use, the **magnitude** scale, was devised by Charles Richter, and it often bears his name. The magnitude of an earthquake is determined by measurements made on certain types of seismographs. Because it is a logarithmic scale, an increase of one on the scale means a tenfold increase in the amount of energy released. An earthquake of magnitude 8, for example, represents the release of 100 times as much energy as a magnitude 6 earthquake. The Richter scale is open-ended, but few earthquakes register above 8 on the scale. These are usually devastating earthquakes. In fact, the amount of energy even in low-magnitude shocks is enormous. A magnitude 5 earthquake is estimated to represent the release of about the same amount of energy as the explosion of 20,000 tons of TNT or a small atomic bomb detonated underground.

The effects of a few of the most carefully studied earthquakes are described in the following case histories.

8.4 CASE HISTORIES

The Chilean Earthquake of 1960

At 6:00 A.M. on May 21, 1960, a sharp earthquake was felt in the region around Concepción, Chile. This earthquake originated at a depth of 50 kilometers and was followed by small aftershocks all the next day. At 3:00 P.M. the second day, a second shock came, followed 15 minutes later by the most severe shock of what was to be a long period of violent earth movement; several of these shocks were rated as magnitude 7.5 to 8 earthquakes. The affected area was 150 kilometers wide and 1,600 kilometers long along the coast of Chile. People ran out of their houses during the first shock, and when the second shock came, waves were visible in the streets and ground. Trees and telephone poles swayed back and forth, loud noises came from the ground, cracks formed in the earth, and cars were tossed back and forth across the street, some falling into fissures. Almost all buildings and bridges in this region were damaged, many totally destroyed; and landslides caused by the shocks buried parts of a number of cities.

The strongest shocks were felt at Isla de Chiloe where trees were snapped off, the ground cracked open, highways were split, road fills slumped, and soils flowed like mud. After the main shocks, the sea began to retreat well below low tide and came back 20 minutes later as a large wave, a **tsunami** or **seismic sea wave** between 4 and 11 meters high, destroying parts of many towns and drowning 500 people in one town alone. These tsunami, like many others, were caused by subsidence of parts of the sea floor along the coast by as much as two meters in some places. When it occurs rapidly, a displacement of this type affects the water level as well as the sea floor,

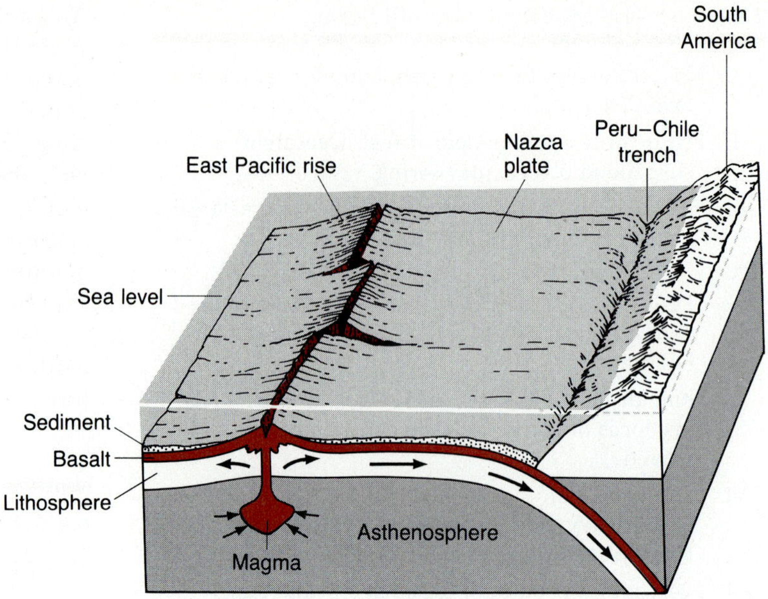

Figure 8.5
A model depicting the plate tectonic setting of the west coast of South America.

and sets up tsunami. The wave motion generated in Chile crossed the Pacific and destroyed towns in Hawaii and Japan.

A few days later, on May 24, an explosion occurred on the side of Mount Puyehye, one of the largest active volcanoes in Chile. Ash, steam, and finally lava poured from a fissure about 100 meters long. Aftershocks continued in Chile for several months and were responsible for billions of dollars in property damage and for the loss of more than 5000 lives in what may have been the most severe earthquake of this century.

This earthquake originated along the plate boundary between the Pacific Ocean and South American continent. This boundary is a subduction zone located where oceanic crust sinks and slides under the continent (Fig. 8.5). This movement between the two plates is responsible for high levels of seismic activity along much of the western coast of South America.

The Hebgen Lake, Montana, Earthquake of August 17, 1959

Near midnight on August 17, 1959, a magnitude 7 earthquake was felt by persons in nine states and three Canadian provinces (Fig. 8.6). Intensities of X were assigned near the epicenter on the basis of the extensive topographic changes that accompanied the earthquake, including the formation of escarpments up to six meters high (Fig. 8.7), warping of the ground surface, and a number of large landslides, among them the Madison Canyon landslide that will be described in Chapter 18 on slope stability.

In the immediate vicinity of the earthquake epicenter, people were thrown out of bed; the ground was set in violent vibration; mortar work broke; buildings were lifted and dropped back, tilted and knocked askew; brick chimneys in West Yellowstone

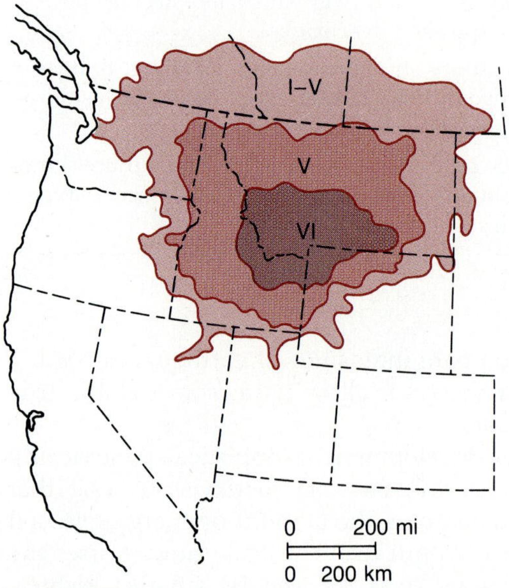

Figure 8.6
Map showing the intensities (I to VI) reported for the Hebgen Lake earthquake (small areas reported intensities as high as X).

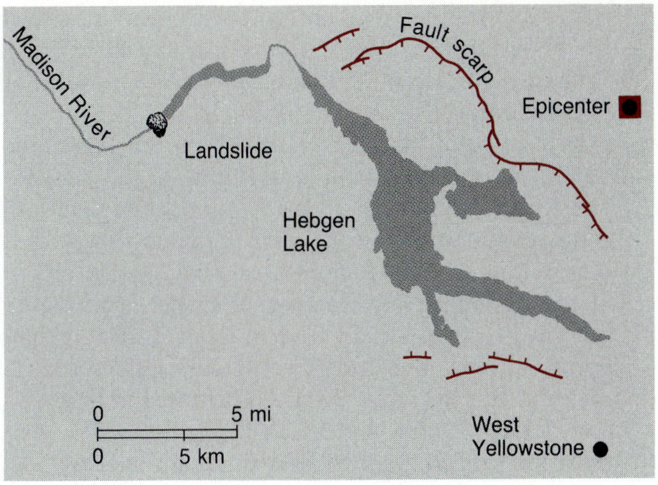

(a)

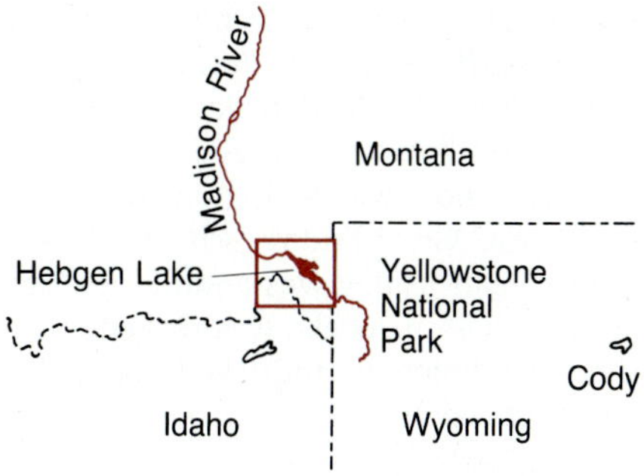

(b)

Figure 8.7
(a) Map of the Hebgen Lake area showing the main fault traces formed during the 1959 earthquake.
(b) Locator map.

were twisted on their foundations by ground motion; and bricks fell out of walls and chimneys as far as 100 kilometers from the epicenter. Rumbling and grinding sounds were heard as a roar near the epicenter, and some reported hearing the earth movements at distances of 80 kilometers. A long series of aftershocks followed the main shock but with a general decrease in frequency and severity.

Most of the geysers in Yellowstone National Park, which lay within the intensity VI zone (Mercalli scale) were unaffected, but dramatic changes did occur in the area around Hebgen Lake (Fig. 8.8). Among these were displacements in the ground surface, escarpments, which can be traced for a distance of nearly 32 kilometers across the area. Most of these escarpments occured in soil and **colluvium** (a mixture of soil, stream deposits, other coarse rock debris, and loose rock fragments) on the mountain flanks north of Hebgen Lake. Many originated when the colluvium slipped downslope over bedrock composed of metamorphic rock and consolidated sediment, as shown in Fig. 8.9. In some places these new scarps lie near the position of old faults, some dating back 70 million years to the Cretaceous Period when the modern Rocky Mountains were uplifted. Changes in the topography near the epicenter, caused by tilting of the region, were most dramatic along the shores of Hebgen Lake, one side of which was submerged and

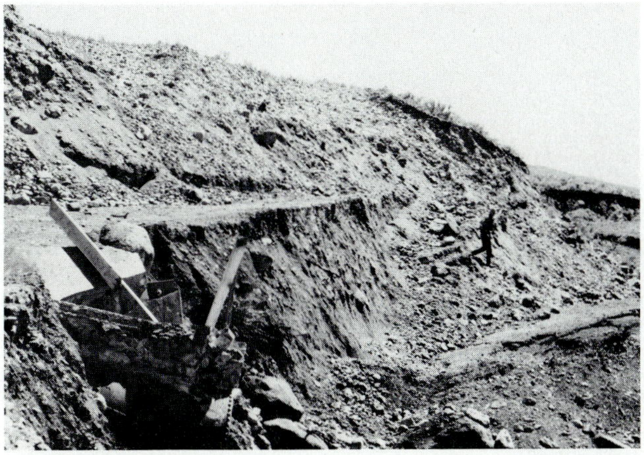

Figure 8.8
A fault scarp formed during the Montana earthquake of 1959.

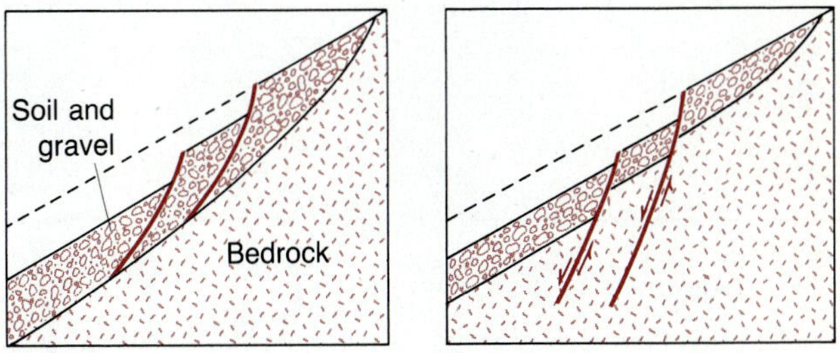

Figure 8.9
(a) Many of the fault scarps seen on the steep slopes around Hebgen Lake are due to slip of the soil. (b) Others may be due to steeply inclined faults that cut into bedrock.

the other raised so much that piers were lifted completely out of water (Fig. 8.10).

The Alaskan Earthquake: March 27, 1964

> One of the greatest geotectonic events of our time occurred in Southern Alaska late in the afternoon of March 27, 1964. Beneath a leaden sky, the chill of evening was just settling over the Alaskan countryside. Light snow was falling on some communities. It was Good Friday, schools were closed, and the business day was ending. Suddenly, without warning, half of Alaska was rocked and jarred by the most violent earthquake to occur in North America this century.

In these words, Wallace Hansen and Edwin Eckel begin their description of the effects of this earthquake in a comprehensive report published by the U.S. Geological survey (*1*). The effects of the Alaskan earthquake have been studied more intensively than those of any other. Much of the damage from this shock was caused by mass movement and tsunami, but the regional geologic setting and the large scale effects of this earthquake are of unusual interest. For one thing, the shocks that accompanied the Alaskan earthquake were of unusually high magnitude (estimates range from 8.3 to 8.7), of unusually long duration (ranging from 1 to 7 minutes), and particularly destructive. The epicenter has been located in the high mountainous region near the head of Prince William Sound, about 128 kilometers east-southeast of Anchorage (Fig. 8.11). Although the focus is estimated to have been between 20 and 50 kilometers deep, one should probably not think of this series of shocks as originating from a point source. A long series of aftershocks, following the main shock, were distributed over a region of about 260,000 square kilometers, generally located along the continental margin of the Aleutian trench, from Prince William Sound to Kodiak Island. Twenty-eight aftershocks of magnitude 6 or more occurred within the first day, and about 12,000 shocks of magnitude 3.5 were recorded within 70 days after the main shock.

This earthquake occurred within one of the most seismically active belts in the world, the nearly continuous belt that circles the Pacific. The seismicity of the Alaskan part of this belt is indicated by the density of epicenters of major earthquakes recorded between 1898 and 1961 (Fig. 8.12). Many of these earthquakes have been concentrated along the Aleutian Islands, an arcuate chain of active volcanic islands extending from the Alaskan peninsula to Kamchatka. Most of the epicenters lie north of the center of the deep Aleutian trench located south of the Aleutian Islands. Of the remaining epicenters, most are associated with the mountain system that includes the Alaska Range

U.S.G.S

Figure 8.10
As a result of faulting, the Hebgen Lake basin tilted, leaving these docks 1.8 meters out of water.

and extends down the western coast of Canada. These high mountains are the sites of present-day crustal deformation.

One of the most notable aspects of the Alaskan earthquake of March 27 was the vast area over which the level of the land was changed. This zone extends from Prince William Sound to Kodiak Island (Fig. 8.11). Benchmarks and triangulation stations provide the most accurate measurements of changes in level; but unfortunately these are widely scattered. Based on resurveys of highways and the displacement of marine organisms, such as barnacles, that live attached to the bottom in the intertidal zone along the coasts, geologists learned that a zone southeast of Kodiak Island was elevated, whereas a zone to the northwest was depressed. Maximum uplifts of 8 meters were found on Montague Island, and maximum depression of 2 meters was recorded in the Kenai Mountains. These movements are best described as a regional warping or tilting; only two faults were found, one of them on Montague Island. The deep-seated movement probably occurred on a fault that comes to the surface along the Aleutian trench and slopes back under the island arc. These movements occur between the oceanic (basaltic) and continental (granitic-andesitic) crust. The data support the idea of a relative movement between the oceanic and continental crust along a low dipping fault that lies under the volcanic islands.

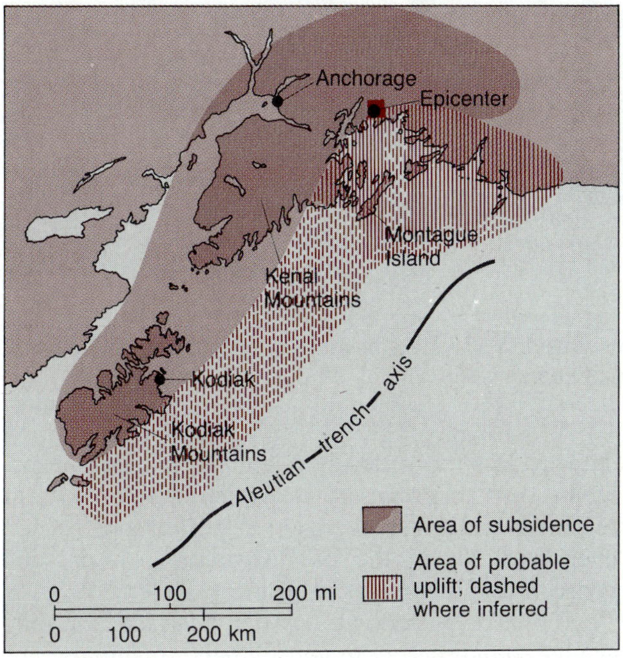

Figure 8.11
A large block of the crust north and west of the epicenter of the 1964 Anchorage, Alaska, earthquake subsided during the earthquake. Another large block was uplifted.

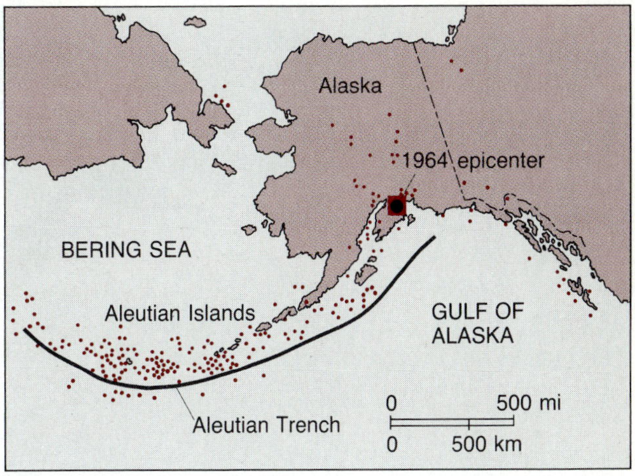

Figure 8.12
Map of Alaska and the Aleutians showing the epicenters of major earthquakes in the area during the period 1898 to 1961. The epicenter of the 1964 earthquake has been added.

The San Francisco Earthquake of 1906 and the San Fernando Earthquake of 1971

Frequent predictions of impending disastrous earthquakes in California and the grim reality of the 1971 San Fernando earthquake have caused attention to be focused on the seismicity of the West Coast of the United States.

The major earthquake that devastated San Francisco in 1906 resulted from movement on the San Andreas fault, one of a complex of faults that runs about parallel to the west coast from northern California to the Gulf of California (refer back to Fig. 8.1). This fault passes through the San Francisco area. Features on opposite sides of the San Andreas fault were displaced laterally as much as 4.5 to 6 meters in 1906, but very little vertical displacement occurred during that earthquake. Railroads, paths, orchards, and streams were displaced. Movements of the kind shown in Fig. 8.2 were recorded for over more than 300 kilometers along the fault; everywhere the west side moved north relative to the east side.

The earthquake caused extensive damage over a large part of California, but the greatest losses were in San Francisco and nearby towns where an estimated 500 to 700 people perished, and approximately

U.S.G.S

Figure 8.13
Lower Van Newman Dam, overlooking the heavily populated San Fernando Valley, was severely damaged and on the brink of catastrophic failure as a result of the San Fernando earthquake of February 9, 1971.

20 percent of the city was destroyed. Much of the damage was caused by fires that swept the city following the shock. Because the earthquake had completely disrupted water mains, there was no water available to fight the fires that broke out. They raged for days until they were finally extinguished by dynamiting the buildings in their path.

The San Francisco earthquake had a magnitude of 8.3 on the Richter Scale. Its intensity, however, varied greatly from place to place depending on the type of soil or bedrock in the area. The extent of damage was more closely related to foundation materials than to proximity to the fault. Intensities as high as IX and X were recorded in areas of construction over land fills in the Bay Area; in areas of solid bedrock, the intensities were much lower.

If this experience were not adequate warning against building on unconsolidated materials, the 1971 San Fernando earthquake in southern California should reinforce the message: again, construction located on weak materials suffered extensive damage. Buildings tipped over and collapsed, reservoir dams

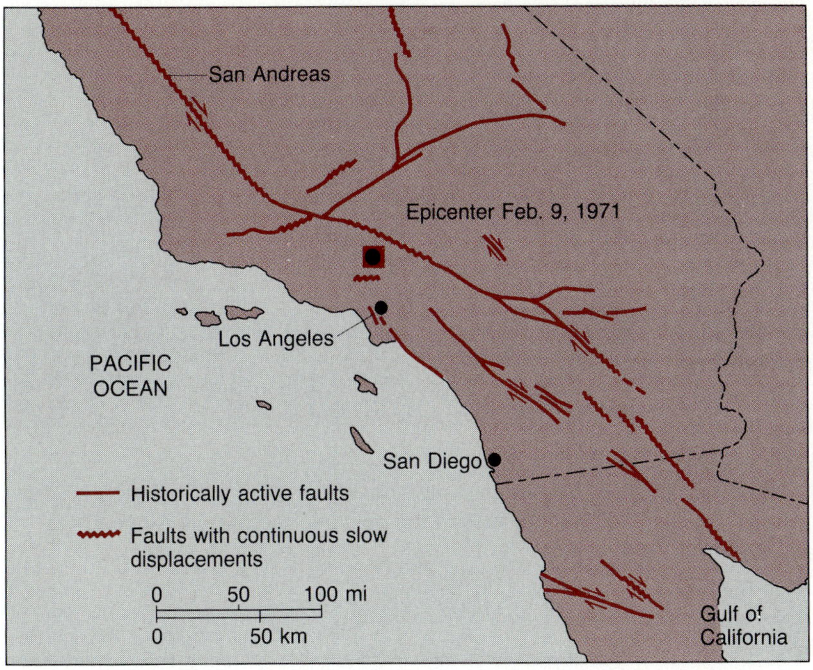

Figure 8.14
Historically active faults in southern California.

were damaged (Fig. 8.13), and overpasses fell, all despite the relatively low magnitude (6.6 on the Richter scale) of the 1971 earthquake.

The San Fernando earthquake resulted from movement on one of the many faults in southern California that are part of the system separating a strip of western California from the rest of North America (Fig. 8.14). This event emphasizes that severe earthquakes are not confined to the San Andreas fault alone. Sixty-four people were killed, and approximately half a billion dollars in property damage occurred in San Fernando and Los Angeles as a result of a moderate-sized earthquake.

Earthquakes Outside Areas of High Seismic Risk: New Madrid and Charleston

Most major earthquakes on continents are associated either with faults that break through the ground surface or with plate boundaries, but a number of large earthquakes have affected areas far removed from such boundary zones. One such earthquake affected the region around New Madrid, Missouri, in 1811, causing the Mississippi River to change course and creating a long lake, Reelfoot Lake, in the abandoned channel (Fig. 8.15). Another affected Charleston, South Carolina, in 1886. Both earthquakes occurred before modern seismic instruments were in use, but the intensity of the Charleston quake can be

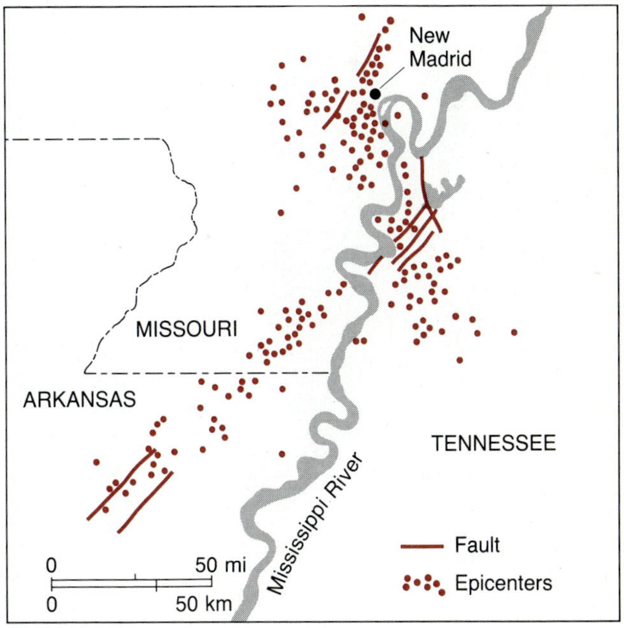

Figure 8.15
Setting of the New Madrid earthquake. The dots are earthquake epicenters.

judged from accounts of the damage. About 60 people were killed in the Charleston area, and the city fell in zone of intensity IX on the modified Mercalli scale. The movements were felt throughout most of the southeastern states, and rated an intensity V as far away as Chicago (Fig. 8.16).

It is not especially reassuring to note that seismic activity in the region continues today at a level higher than it was prior to 1886. A number of these earthquakes are located along a line that trends northwest from Charleston, South Carolina, as shown in Fig. 8.17. The eastern portion of South Carolina lies in the Atlantic coastal plain. The Cenozoic sediments exposed at the surface there are underlain by an igneous and metamorphic complex at depth. Although this basement is buried at a depth of about 900 meters at Charleston, it seems probable that the earthquakes originate as a result of movements on faults in this basement complex. Apparently, a number of steeply inclined faults have broken this basement into big blocks that are shifting. The faults may be ancient breaks formed as a result of continental drift when North America and Africa began to separate—a topic we will examine later.

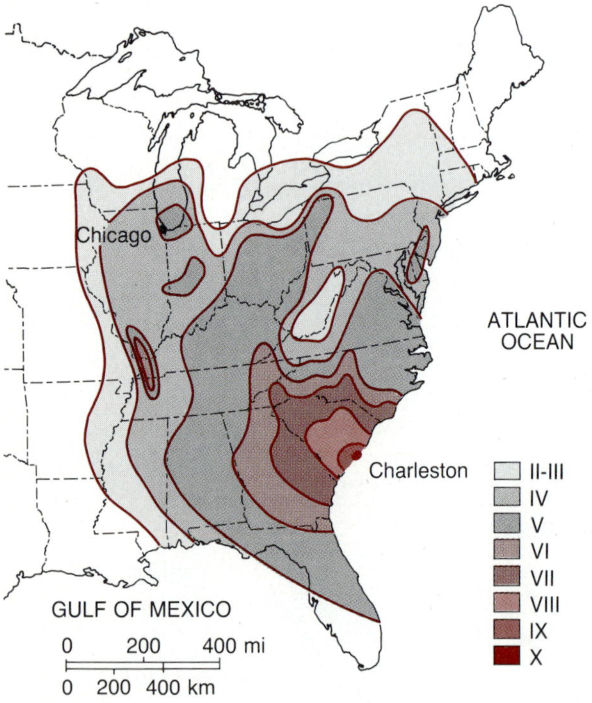

Figure 8.16
Map of the eastern United States, contoured to show the broad regional patterns of the reported intensities for the 1886 Charleston earthquake. Contoured intensity levels are color coded.

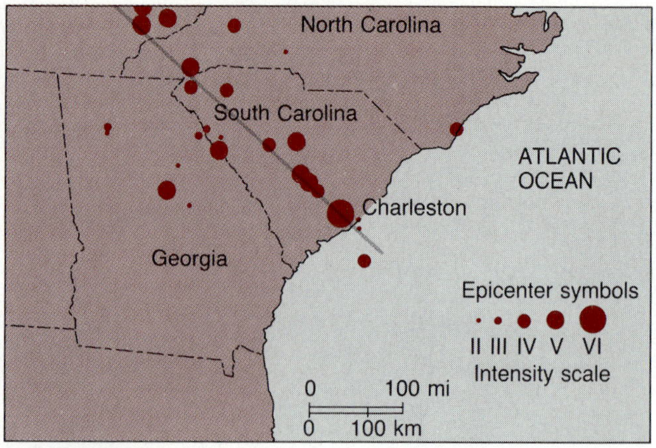

Figure 8.17
Earthquake epicenters for shocks during the period 1961 to 1975. Intensities in the epicenter area are indicated by symbols.

8.5 SEISMIC HAZARDS: PLANNING TO AVOID DISASTER

More people have been killed by earthquakes and earthquake-induced phenomena than by all other geological processes combined. Over the past thousand years seven earthquakes are believed to have had a death toll greater than 100,000. An estimated 800,000 people were killed in the 1556 Shensi, China, earthquake, which is the most disastrous earthquake on record; and approximately a quarter of a million people may have perished in the 1976 earthquake in China. By comparison, the death tolls of 5600 in Chile in 1960 and 114 in the 1964 Alaskan earthquake seem small.

Earthquakes pose dangers to people in many ways. The vibrations often cause masonry structures such as brick, block, and stone buildings to disintegrate (Fig. 8.18) and highway overpasses to collapse; landslides and avalanches are triggered; dams may fail; and power lines, gas lines, water lines, and transportation routes are disrupted at the very time emergency fire-fighting equipment and rescue equipment is desperately needed. Along the coast, tsunami may be generated, and in a few instances, volcanic eruptions may even be set off.

The potential loss of life and property is so great it would seem unlikely that human beings would choose to use land in proven earthquake areas; yet most of the population of California is located in such seismically active areas, and many of the most heavily

City of Los Angeles, Department of Building and Safety

Figure 8.18
Veterans Hospital rescue operations during the 1971 earthquake in Los Angeles.

populated areas of the world are located where these hazards exist. So, two quite different questions are raised: first, can we predict violent earthquakes or in some other way protect people in areas already developed; and secondly, what precautions should be taken to reduce land use in danger areas in the future?

A number of commonsense suggestions can be made to avoid earthquake damage in areas of known seismic activity. Dams, housing subdivisions, hospitals, schools, and other public buildings should not be built on faults or within fault zones (Fig. 8.19). Masonry structures should not be used in earthquake-prone areas. Reinforced concrete is much more stable, and wooden buildings are much safer than masonry structures. All major buildings should be built on bedrock foundations rather than on unconsolidated sediment or fill, because ground motion is almost invariable greater on the less-consolidated materials. Plans should be drawn up for the evacuation of people who live on filled or unconsolidated foundation materials. Special care should be taken to identify steep, potentially unstable slopes that might be sources of large landslides during earthquakes; and, of course, everyone who lives or builds in an earthquake-prone area should be aware of the potential hazards. Unfortunately, few of these precautions are taken in most seismically active areas.

Learning how to predict earthquakes is now one of the primary objectives of the field of seismology. A tremendous amount of time, energy, and money has been directed toward this goal in the last decade. We will look at some of the promising results of these studies in the next chapter after discussing the principles on which these methods are based.

SUMMARY

Most shallow-focus earthquakes occur as the result either of movement along a fault or of volcanic activity. According to the elastic rebound theory, in fault-generated earthquakes

Figure 8.19
Aerial view along the San Andreas fault. The housing subdivision has been built across a section of the fault that was displaced five feet in 1906.

some gradual movement takes place along a fault without causing tremors. This motion builds up strain in the rock across the fault until a sudden break and ground movement, the earthquake, occurs. Slippage at one point along the fault increases strain elsewhere, often giving rise to a cluster of aftershocks as strain is released in a series of lesser breaks.

Deeper earthquakes are thought also to involve movement along faults, but because the temperature and pressure at depth give the rock greater plasticity, the movement is stick-slip rather than elastic.

The location of earthquakes is described in terms of their focus, point of origin in the earth, and the location on the surface just above the focus, the epicenter. Most earthquake activity occurs in narrow belts that coincide with areas of volcanic activity and with the boundaries of lithospheric plates.

Two scales for expressing the size of earthquakes are in use. The Mercalli intensity scale reflects observed surface effects of the tremor and is given in Roman numerals. The Richter magnitude scale, a logarithmic scale reported in Arabic numerals, is based upon seismograph measurements of the earth's movement.

KEY TERMS

colluvium
elastic rebound theory
epicenter
focal depth
focus
intensity
magnitude
seismic sea wave
seismographs
seismology
stick-slip
strain
tsunami

STUDY QUESTIONS

1. What is the distinction between the focus and the epicenter of an earthquake?
2. What are the main causes of earthquakes?
3. What types of construction are most likely to survive the effects of an earthquake?
4. If you were making recommendations for the location of important buildings such as schools or hospitals in an area with high earthquake risk, what geological conditions would you consider about possible sites?
5. What is the distinction between earthquake intensity and magnitude?
6. What side effects often accompany earthquakes, as shown by the case histories you have studied?
7. Describe the geography of high seismic risk. Where are most earthquakes located?

REFERENCE

1. Wallace Hansen and Edwin B. Eckel. "A summary description of the Alaska earthquake—its setting and effect," *The Alaska Earthquake, March 27, 1964, Investigations and Reconstruction.* U.S. Geological Survey Professional Paper 541 (1966), *1*.

SUGGESTED READINGS

Bolt, B. A., *Earthquakes—A Primer.* San Francisco: W. H. Freeman Co., 1978.

Boore, D. M., "The motion of the ground in earthquakes," *Scientific American* offprint #928 (1977).

Eckel, E. B., "The Alaska earthquake, March 27, 1964: Lessons and conclusions," *U.S. Geological Survey Professional Paper 546* (1970).

Oakeschott, G. B., *Volcanoes and Earthquakes: Geologic Violence.* New York: McGraw-Hill Book Co., 1975.

Richter, C. F., *Elementary Seismology.* San Francisco: W. H. Freeman Co., 1958.

U.S. Geological Survey, "The San Fernando, California, earthquake of February 9, 1971," *U.S. Geological Survey Professional Paper 733* (1971).

THE INTERIOR OF THE EARTH

Chapter 9

Geologists can examine directly only the earth's outermost surficial parts; even the deepest oil wells penetrate a distance of only about ten kilometers (six miles) or approximately $\frac{1}{600}$ of the distance to the center of the earth. Uplift and erosion have exposed rocks that were once buried at depth, but probably none of the rocks presently exposed come from depths of more than a few tens of kilometers. What we know, therefore, about the earth's deep interior has been compiled from indirect examination of the interior using observations made at the surface and study of meteorites.

Early models of the earth clearly show that their designers were greatly impressed by volcanoes, for they envisioned a planet composed of a thin outer crust covering a fiery interior. This general notion found support both in the then-popular concept of Hades and in the various theories based on the hypothesis that the earth was once molten—an idea lending itself to the conclusion that the cool outer part of the planet represents a crust formed over a molten interior. As we will see, geophysical evidence shows that the earth's interior is solid halfway to the center.

The existence of a hot interior is confirmed by the temperatures encountered in deep mines and wells. These temperatures show an average increase of about 1 °C for every 30 meters depth. If projected to greater depth, this gradient predicts that at depths of only 30 kilometers temperatures reach 1000 °C, about the melting point of most rocks at ground surface.

Other important inferences about the interior are based on our knowledge of the earth's radius, mass, and density. We found in Chapter 3 that the density of the planet as a whole is 5.5 gm/cm^3, although the average density of rocks near the surface is only about 2.5 gm/cm^3. Therefore, the interior must contain enough high-density rock to compensate for the lower density near the surface. We can also infer that the mass of the earth is symmetrically distributed around its axis of rotation. Otherwise, the earth's movements, such as its rate of rotation, would be much more irregular. Such inferences provide some insight into the character of our planet, but our most important source of knowledge about this planet's interior is **seismology,** the study of seismic waves—vibrations caused by earthquakes.

Because the materials in the earth's interior are highly elastic, these vibrations penetrate and pass through earth's deepest parts. By analyzing the speed at which they travel, geologists have identified several major subdivisions of the planet's interior.

In this chapter we shall see how a model of the major features of the earth's interior has been developed using the analysis and interpretation of seismic waves, the earth's magnetic field, and meteorites. Similar methods will be used later to provide more details about the near-surface layers of the earth. Finally, we will see how seismic data contributed to the formulation of the plate tectonic theory.

9.1 PRINCIPLES OF THE SEISMIC METHODS

Generation of Seismic Waves

The term **seismic wave** is applied to all vibrations produced in rocks by earthquakes or generated by various types of artificial means such as the explosion of TNT. Most seismic waves generated at shallow depths along faults are caused by the elastic rebound of rocks deformed near the fault zone, as described in Chapter 8. A somewhat similar, but simpler, model can be envisioned for an underground explosion such as those associated with explosions set off in wells or with the explosive eruption of volcanoes.

Propagation of Seismic Waves

During an explosion, the force of the explosion exerts outward pressure on the enclosing rock. The rock is compressed as it is forced away from the center of the explosion, and this compression is transmitted through the rock as an expanding shell (Fig. 9.1). Because rock is highly elastic, it rebounds to its original shape and volume after the wave has passed, but the compression is transmitted out from the source. The leading edge of compression, called the **wave front,** is continually expanding as it moves farther from the source. Thus the energy associated with this wave is distributed over an expanding surface, and the amount of energy in any part of the wave front is continually diminishing. But the amount of energy released during many earthquakes is sufficient to send the seismic wave through the entire earth. It is often convenient to represent the movement of the wave by lines or arrows, called **rays,** indicating the direction of the movement of the wave. Both wave fronts and rays are illustrated in Fig. 9.2. Waves that are propagated throughout a three-dimensional medium like the earth are called **body waves.**

Experiments in the generation and transmission of seismic body waves through rock indicate that two quite different types of waves or elastic responses may occur when rock is suddenly deformed. One of these types, a **compressional wave,** is transmitted by the compression of the rock as previously described in Fig. 9.1. This type of wave motion is like the shock wave that is transmitted down the length of a bar when it is hit on the end (Fig. 9.3). This type of wave travels through liquids and fluids as well as through solids; in fact, sound waves are of this type. Because compressional waves travel faster than the other wave types, they always arrive first, and are also called the **primary** or **P waves.**

The second type of body wave is transmitted as an elastic response of the rock to distortion of its shape, as shown in Fig. 9.3. This type of movement involves changes of shape known as shear (Fig. 9.4). The movements involved in shear waves are perpendicular to the direction in which the wave front is moving. They may be thought of as the motion that is transmitted along a long bar if one end of the bar is hit from the side or twisted. The motion of a tight string that is plucked near one end is analogous. Waves can be seen to move along the string, but any

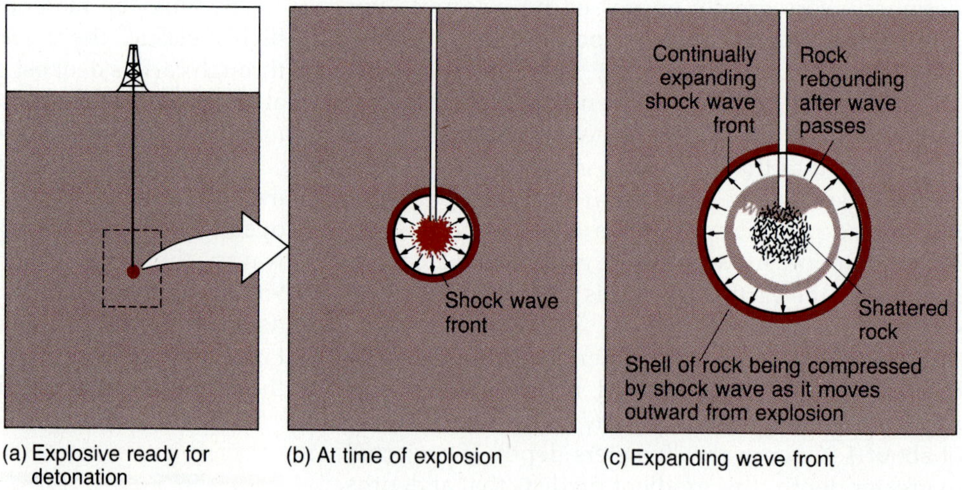

Figure 9.1
A shock wave like that generated by an explosion deep in the earth is simulated in these drawings. (a) As the explosion occurs, the rock surrounding it is compressed. (b) The wave then moves out as an expanding shell around the source. (c) As the wave passes, the compressed rock rebounds because it is elastic beyond the area of shattered rock at the site of the explosion.

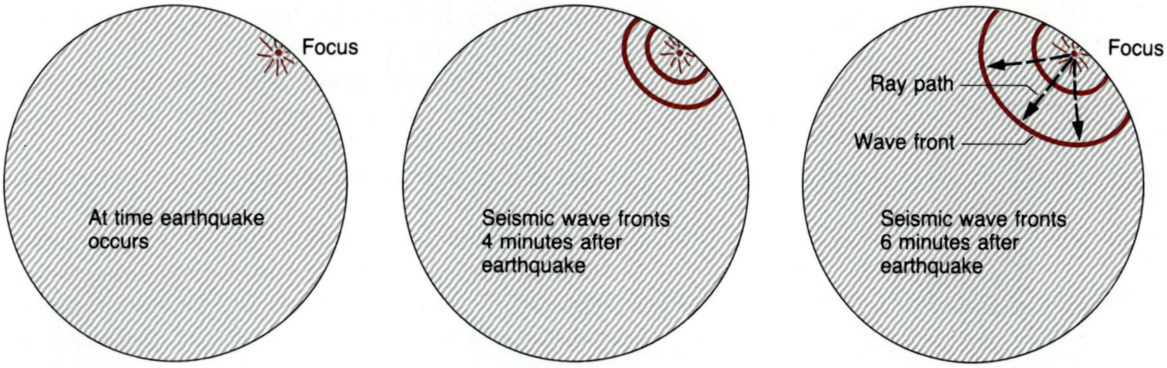

Figure 9.2
This cross section of the earth shows the propagation of wave fronts and a ray for body waves emanating from an earthquake focus. When an earthquake occurs, several different types of elastic waves are generated at the same time. At the instant of generation, all of the waves are together, but a few minutes later the P wave is ahead of the slower-moving S wave, and this separation grows with the passage of time.

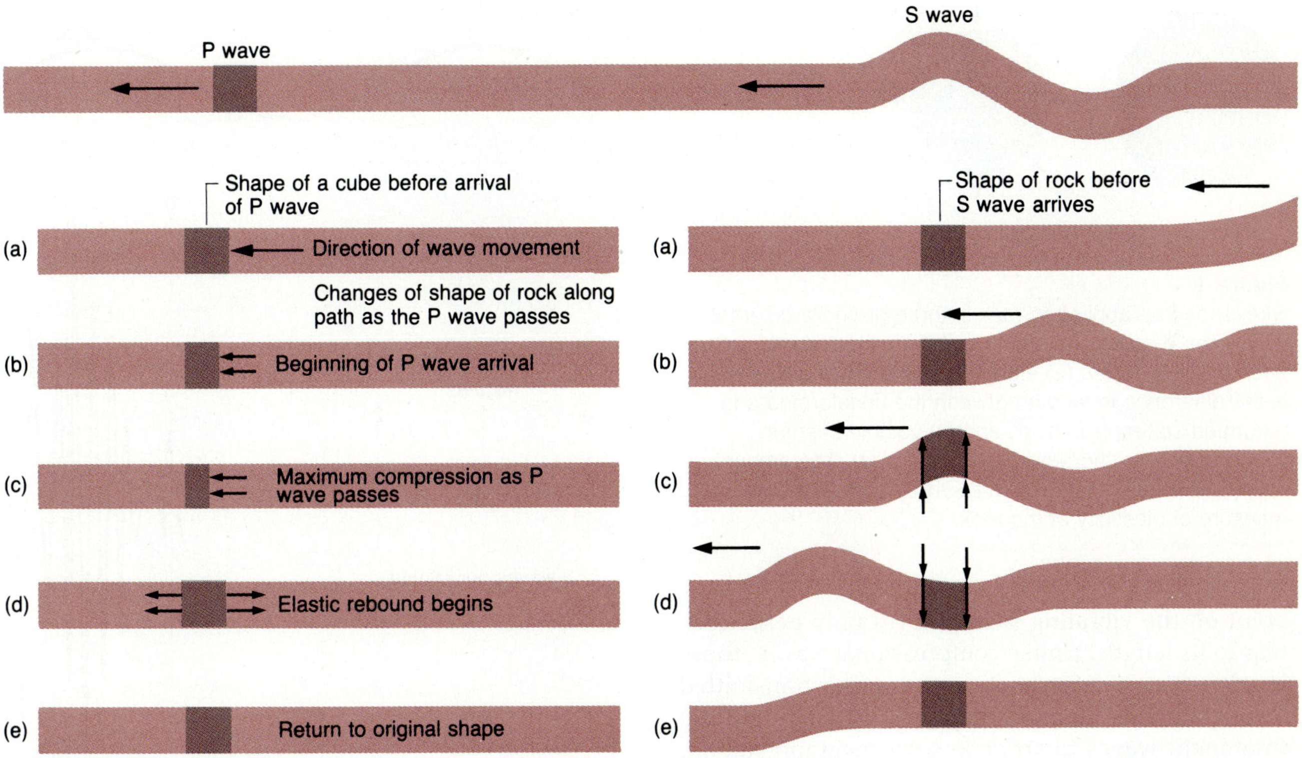

Figure 9.3
The rock along the ray path of the wave is first compressed when the P wave approaches as is shown in this diagram. Because rock is elastic, it returns to its original shape as the compression passes. When the S wave approaches, the shape of the rock is distorted, but again the rock returns to its original configuration after the passage of the vibration. The sequence of changes in shape which the rock undergoes with passage of P and S waves is represented in the sequential drawings (a–e).

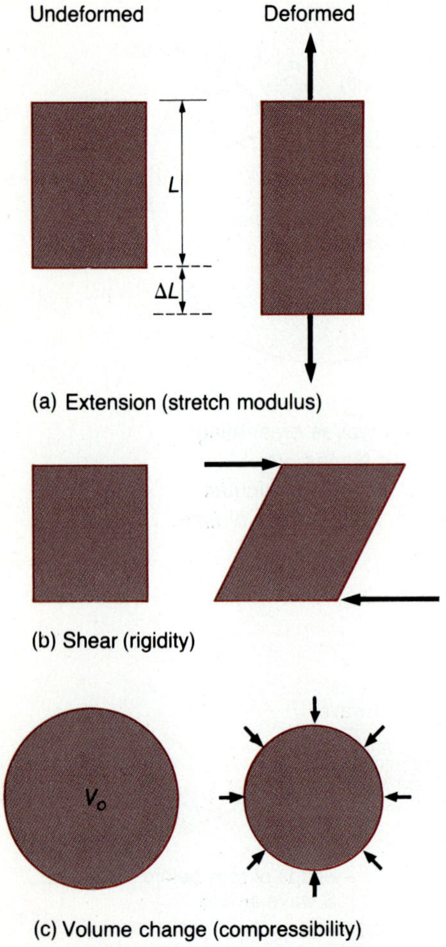

Figure 9.4
Stress may be applied to rock in three distinctly different ways. The resulting deformities of the rock is shown for each of these three: (a) elongation or stretching where ΔL is the difference in length between the undeformed and deformed states; (b) shear; and (c) volume change. Compression or shortening is similar to (a). The amount of change in shape or volume caused by a given stress is a measure of elasticity of the rock.

point on the vibrating string moves only perpendicular to its length. Unlike compressional waves, those involving a shearing motion cannot be transmitted through liquids. Because they are so often the second prominent waves to arrive at a seismograph station, they are referred to as **secondary waves** or **S waves.**

Seismic waves, like sound waves or light waves, are described in terms of two features, amplitude and frequency (Fig. 9.5). For seismic waves, the **amplitude** is the amount by which rock moves as the wave passes. It reflects, therefore, the intensity of earthquake effects. In Fig. 9.5(b) the background vibrations of the earth have very low amplitude at the left of the figure. The amplitude of the vibration is much greater for the waves shown at right in Fig. 9.5(b). The **frequency** is just what the name suggests—the number of vibrations per second that the wave causes at any point it affects. The **period** expresses the same information about a seismic wave; it is the length of duration (usually expressed in seconds) of a single vibration. New wave arrivals can be recognized on seismograms by abrupt changes in amplitude or frequency.

Surface Waves. Most earthquakes and explosions generate, in addition to body waves, wave motion that is confined to the earth's surface or near-surface regions. These **surface waves** are like the waves set up when a stone is dropped into a still body of water.

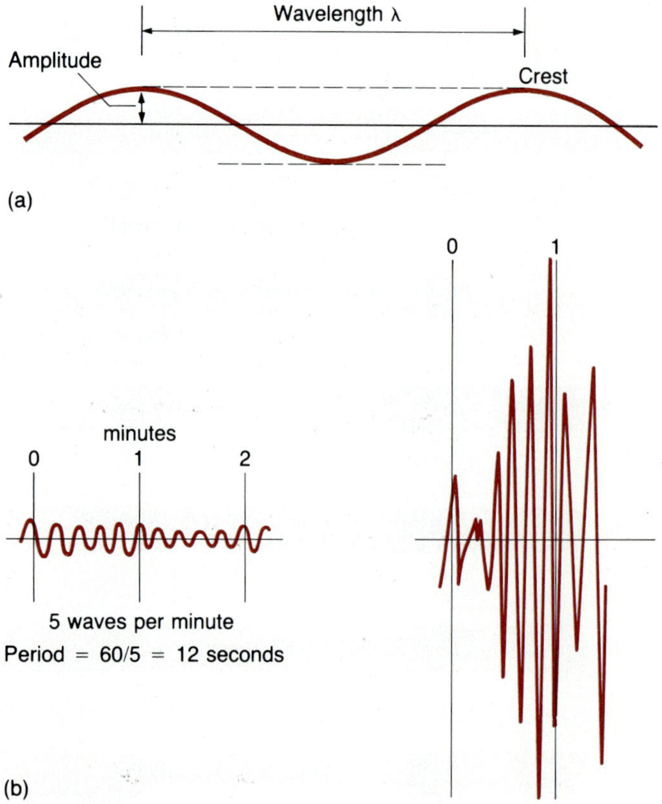

Figure 9.5
The ideal wave (a) is drawn with much greater wavelength than the waves traced from seismograms in (b). Note that the distance scale in (b) is marked off in minutes. This is done on seismograms because the wave motion is recorded as the waves arrive.

VELOCITY OF SEISMIC WAVES

The velocity with which seismic waves move through rock is determined by the elastic properties and the density of the rock. The elastic properties of all materials, including rocks, are described in terms of the extent to which the material changes shape or size in response to a given deforming stress (force applied per unit of area). For example, if a rock is deformed by stretching it, as in Fig. 9.4(a), the ratio of the force applied to the change in length is a measure of the elastic property called the **stretch modulus** (also Young's modulus). If a specimen is deformed by applying a shearing stress to it, as shown in Fig. 9.4(b), the ratio of the shearing stress to the amount it is distorted is called its **rigidity** (μ). A third elastic property, called **compressibility** (K), is a measure of how much the volume of a rock changes as the pressure on it is increased.

Theory and experiment confirm that the velocity of P and S waves in rock or other materials is governed by compressibility (K), rigidity (μ), and density (ρ) as shown below.

$$\text{Velocity of compressional waves, } V_P = \sqrt{\frac{K + \frac{4}{3}\mu}{\rho}}$$

$$\text{Velocity of shear waves, } V_S = \sqrt{\frac{\mu}{\rho}}$$

These formulas indicate that the velocity of a P wave is increased as the medium becomes more rigid and less compressible (that is, as K and μ increase), and that the velocity decreases as the medium becomes more dense. Actually, because more dense rocks have higher values of compressibility and rigidity, the velocities of seismic waves increase even in more dense rocks.

The velocity of an S wave increases with increased rigidity and decreases with increased density. But shear waves cannot travel through materials that lack rigidity, such as liquids. P waves do continue through liquids because their transmission depends on the compressibility of the medium as well as its rigidity. The formula for V_P shows that a P wave will have a lower velocity in a liquid than in most rigid rocks, but it will pass through.

The energy is distributed surficially over an ever-expanding circle or near-circle (Fig. 9.6).

The amount of displacement caused by the surface waves associated with earthquakes (their amplitude) is greater than that associated with body waves. Especially high surface waves are generated when the earthquake focus is near the surface of the earth (as in depths of 25 km or less). This surface motion turns statues on their bases and shakes buildings apart during major earthquakes. From the various ways objects on the ground are displaced, it is possible to recongnize two distinctive types of surface wave motions. One, called **Rayleigh waves,** involves a rolling motion of the ground similar to that found in waves in the ocean. The other, **Love waves,** involves a lateral or sideways displacement of the ground surface in the direction of motion of the wave front, as shown in Fig. 9.7.

Reflection and Refraction of Waves

When seismic waves are generated at the focus of an earthquake, the wave front, of the P wave for example, begins traveling out in all directions from the focus with equal velocity. The rays may thus be represented by straight lines. They will remain straight as long as

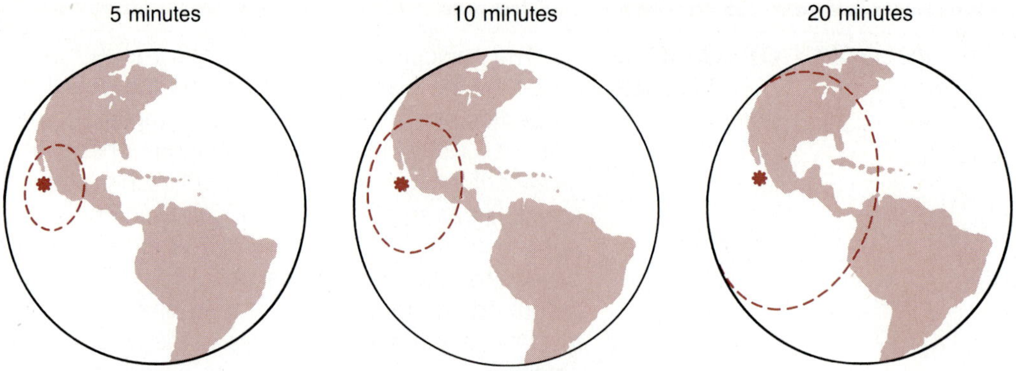

Figure 9.6
Spread of the surface waves 5, 10, and 20 minutes after an earthquake south of Baja peninsula. The wave front depicted here is the front of the fastest-moving surface waves.

the velocity is constant. This velocity is, as we have seen, closely related to the elastic properties of the rock. Thus, as the waves strike the boundary between layers of different rock types, the velocity changes abruptly.

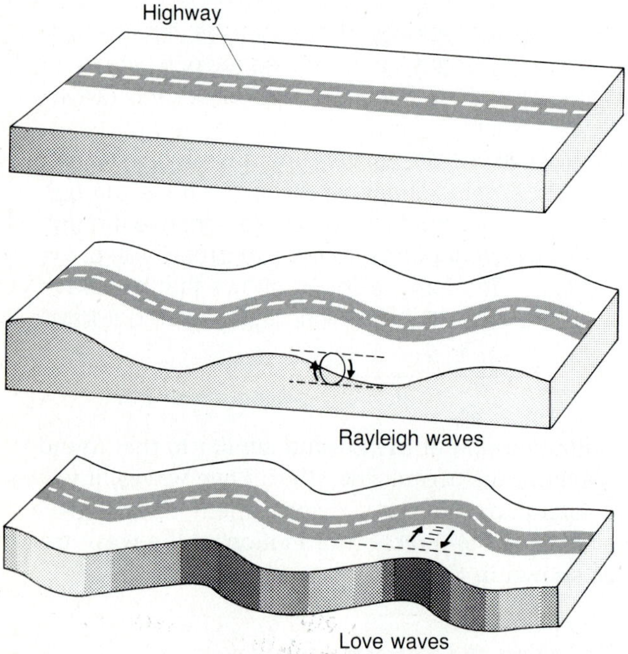

Figure 9.7
Surface waves. Two distinctly different types of waves are identified on the ground surface. Only rarely is the amplitude of the ground motion great enough to distort the surface as shown here, but movements of these types are associated with most shallow-focus earthquakes.

When waves reach a boundary between layers, part of the energy of the wave motion is **reflected.** The angle of reflection from the boundary is always equal to the angle of incidence of the ray, as shown in Fig. 9.8.

If the angle of incidence is not too great, some of the energy of the wave will also pass through the boundary and into the lower layer. Unless the wave hits the boundary perpendicularly, rays are bent as they pass into the next layer. This bending is called **refraction.** If the velocity of the seismic wave is greater in the second layer, the ray will be bent away from the perpendicular, as shown in Fig. 9.9. If the velocity

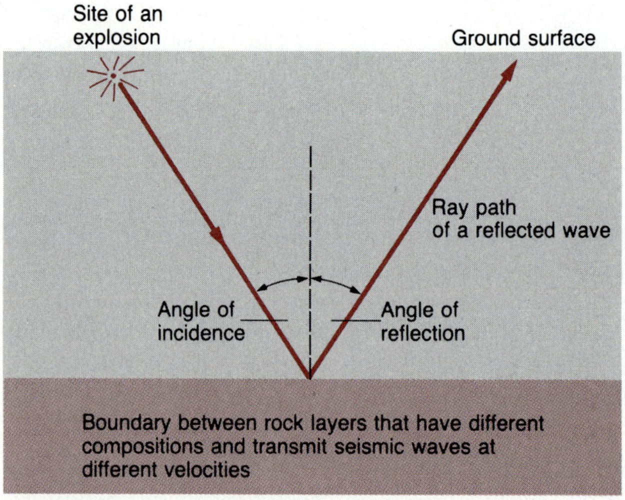

Figure 9.8
When a ray of a seismic wave is reflected from the boundary between two layers of rock, the angle of reflection is equal to the angle of incidence.

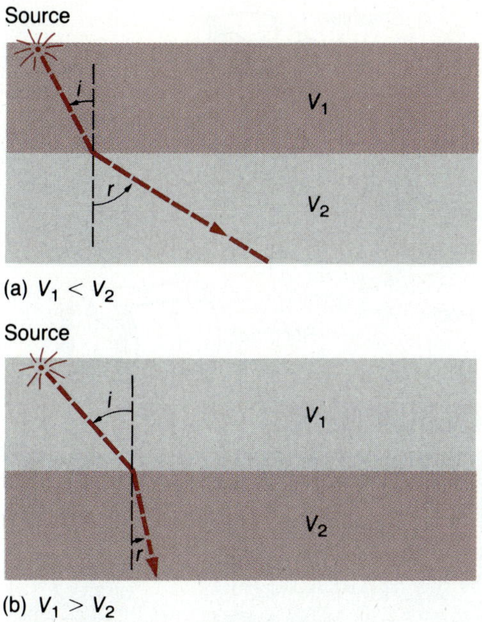

Figure 9.9
The refraction of seismic waves. (a) A layer with high seismic wave velocity lies under one with lower velocity. (b) The upper layer has a higher velocity. Note the difference in the way the ray bends on passing through the boundary in these two cases.

is lower, the ray will be bent to a steeper angle, or toward the perpendicular.

Detection of Seismic Waves

The ground motion caused by surface waves depends on how deep and how far away the focus of the generating earthquake is. Near the epicenter, this motion may be violent enough to turn buildings on their foundations or throw objects into the air, but its magnitude decreases rapidly away from the epicenter as the wave energy is spread over an ever-increasing surface area. At distances of a few hundred kilometers from the epicenter, the movement is rarely felt.

Delicately suspended objects, such as a chandelier, may be set in motion even when the earthquake is not strong enough to be felt. The chandelier's first apparent movement is actually due to the movement of the building around it. The suspended chandelier will remain relatively still at first because of its **inertia,** or tendency to remain still. As the earthquake vibration continues, the chandelier begins to swing like a pendulum, but its motion is still largely independent of the motion of the house and ground. This principle is used in the design of **seismographs,** instruments used to study ground motion.

Seismographs detect the amplitude and direction of ground movements. Most seismographs, such as those in Fig. 9.10, measure motion of the ground by recording the relative motion between the ground and the center of oscillation of some type of pendulum. The basic element in the design of most seismographs is a mass suspended by springs from a frame or support anchored to bedrock. The mass tends to remain still when the frame moves. In Fig. 9.10(*b*) a magnet is used for the mass, and a coil of wire fixed to the frame surrounds but does not touch the magnet. Thus the coil moves past the magnet when the ground is in motion. As the ground moves, the coils move through the magnetic field of the magnet, causing an electrical current to be generated in the coils. The magnitude of the current depends on how fast the ground (and therefore the coils) move, and over what distance. This current is amplified, fed into a recording device, and recorded either on magnetic tape or as a line on a piece of paper, called a seismogram (Fig. 9.11). This line (or trace) is straight when the ground is still, but it is deflected when the ground moves. The amount and direction of the deflection is a measure of the amount and direction of the ground movement. Seismographs of this kind are capable of magnifying ground motion 100,000 times or more, but they usually operate in a magnification range of only 500 to 20,000. Time marks imprinted at one-minute intervals on the record make it possible to determine the time of arrival of any particular ground motion at the seismograph station.

Since one instrument can detect motion in only a single direction, a seismograph station normally consists of at least three instruments. One of these is designed to detect vertical motion; the other two record movement in east–west and north–south directions. Other specialized seismographs are also used in the worldwide network of stations to detect waves that have different periods (time between passage of adjacent waves) or freqencies (number of waves per second) of movement.

The same physical principles used in the design of seismographs for detecting earthquake generated waves are also employed in small portable sensing devices called **geophones,** which are used to detect artificially produced seismic waves. These are used by oil companies and others to determine the shape and depth of underground rock layers in what has become an extremely important technique in exploration for oil.

(a)

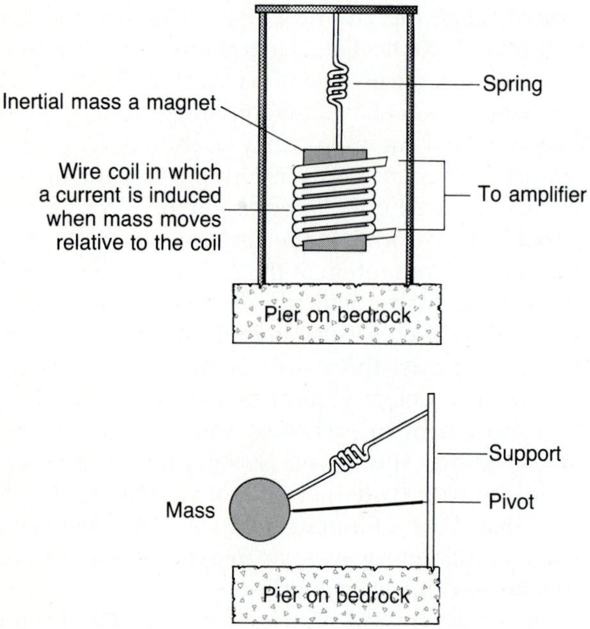

(b)

Figure 9.10
Seismographs, devices that detect seismic waves, use the principle of electromagnetic induction. (a) A seismograph designed to detect vertical movement. (b) Schematic drawings depict the methods used to detect vertical (top) and horizontal (bottom) movements.

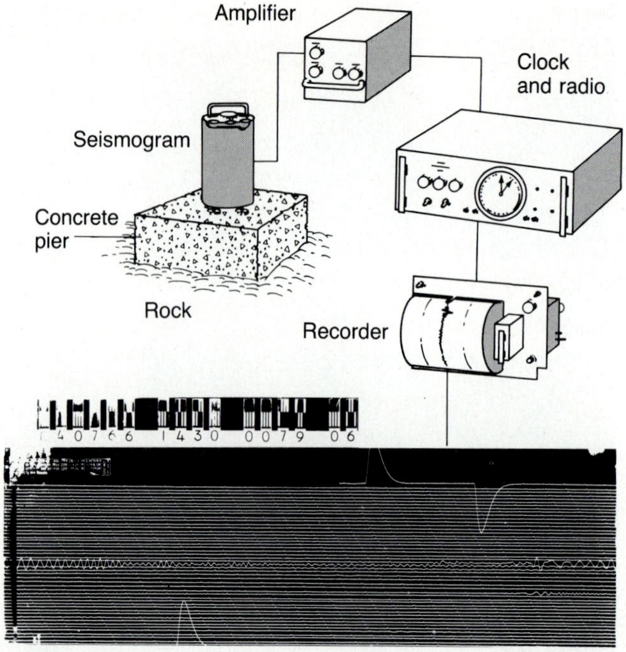

Figure 9.11
A schematic sketch of a seismograph showing the sensing device—the seismometer, an amplifier, a clock used to make time marks, a radio used to keep an accurate check on the clock, a recorder, and below, a record showing an earthquake.

9.2 SEISMIC EXPLORATION METHODS

Seismic waves can be generated in many ways, using everything from nuclear explosions to hitting the ground with a sledge hammer. Most modern exploration work involves use of truck-mounted devices that induce a vibration of known frequency in the ground. At sea, the explosion of a mixture of propane and oxygen detonated by a spark or a high-frequency sound signal is used. The seismic waves produced from these sources travel through water, soil, sediment, and rock just as earthquake-produced waves do. They differ, of course, in that they carry much less energy and cannot travel as far. Like other seismic waves, the velocity with which they are transmitted or propagated through the earth depends on the physical properties of the water, sediment, or rock through which they pass. These waves are also reflected and refracted (as shown in Fig. 9.8 and 9.9) when they encounter the abrupt changes in physical properties that generally characterize different layers of rock.

DETERMINING DEPTH BY SEISMIC REFLECTION

Determination of the depth from the ground surface to the top of a layer that reflects seismic waves is one of the primary objectives of seismic exploration. In the simplest case, the reflector is a horizontal layer, as shown in Fig. 9.12. Reflecting layers always have different elastic properties and seismic velocities from those of the overlying layer. Otherwise reflection will not occur.

In order to determine the depth (Z) to the reflection it is necessary to know or to measure the velocity (V_1) of seismic waves in the upper layer. The distance (AC) between the source of seismic waves (for example, an explosion or sonic wave produced by some type of vibrator) and the detector (geophone) is measured. By carefully timing the moment the waves are generated and when they arrive at the geophone, it is possible to measure the time required for the waves to travel from A to the reflector at point B, the point of reflection, and back to C.

The path ABC is the hypotenuse of two identical right triangles. The depth (Z) is one side of these right triangles and its length is given by the Pythagorean theorem

$$AB^2 = \left(\frac{AC}{2}\right)^2 + Z^2$$

Therefore

$$Z^2 = AB^2 - \left(\frac{AC}{2}\right)^2$$

$$Z = \sqrt{AB^2 - \left(\frac{AC}{2}\right)^2}$$

The distance AC is measured directly on the ground surface. The distance AB can be measured using the travel time for the seismic wave because $AB = \frac{1}{2}(V_1 \times \text{travel time})$.

$$Z = \sqrt{\frac{V_1T}{2} - \left(\frac{AC}{2}\right)^2}$$

These same principles explain the use of sonic waves to determine the depth of the sea floor in the ocean.

Reflection Methods

Seismic exploration methods use both the reflected and refracted waves that come back to the ground surface. This wave motion is detected by geophones, which record vibration of the ground. Figure 9.12 shows a simple diagrammatic layout for the use of geophones in a reflection seismic study. In this example a horizontal discontinuity at depth Z below the ground reflects waves from the shot point back to geophones at the ground surface along the paths shown.

The objective in exploration work is to determine the depth, Z. This can be determined simply if the following data are known: the distance from the shot point to the geophones; the time required for the wave to travel from the shot point down to the reflecting interface and back to the geophone; and the velocity of propagation of the wave in the layer. An example of this calculation is shown in the box on "Determining Depth by Seismic Reflection." Similar methods used at sea can produce continuous profiles along the track of the ship. An example of such a profile is shown in Fig. 9.13. Layers of sediment

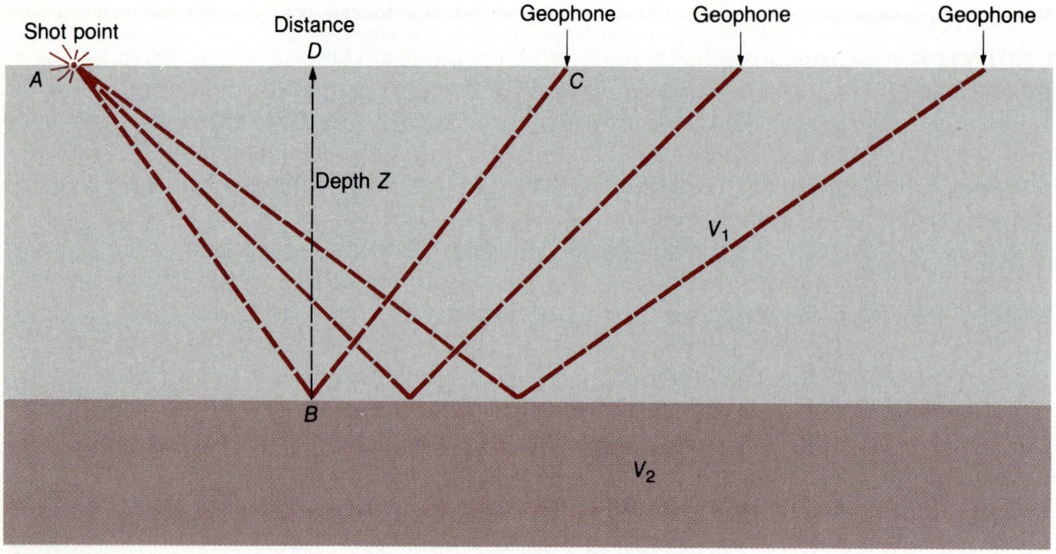

Figure 9.12
Outline of the method used to determine the depth to a reflecting boundary using the seismic reflection method. The boundary might be from a few meters to several thousand meters deep.

intruded by large pluglike masses of salt are revealed in this profile.

In practice, reflection seismology involves highly sophisticated techniques used to enhance the signals and to sort out reflections that are arriving back at the geophone from the many layers in the crust. The signals received at the geophones are recorded on magnetic tapes. These are manipulated by various computer techniques to strengthen them and make interpretation of the records easier.

Refraction Methods

Although part of the energy being propagated along any path will be reflected back toward the surface when the waves strike an interface, such as a rock

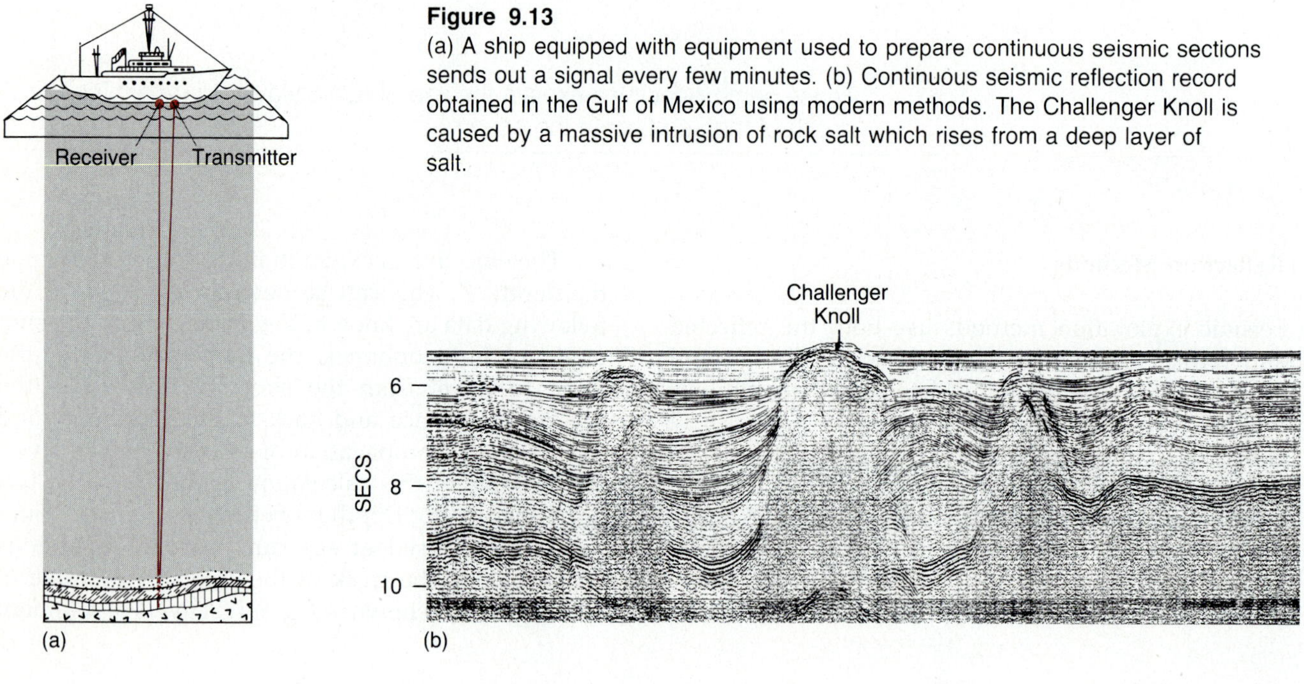

Figure 9.13
(a) A ship equipped with equipment used to prepare continuous seismic sections sends out a signal every few minutes. (b) Continuous seismic reflection record obtained in the Gulf of Mexico using modern methods. The Challenger Knoll is caused by a massive intrusion of rock salt which rises from a deep layer of salt.

layer, another part will be refracted on through the layer and into the one below. How much energy is reflected and how much transmitted depends in large part on the angle at which the wave in any given path strikes the interface. At low angles of incidence (the angle is measured from the perpendicular to the interface), much of the energy is refracted through the interface (Fig. 9.14), but at a certain angle, called the **critical angle,** the energy is refracted, so the energy of the wave travel is parallel to the discontinuity in the top of the lower layer. At all greater angles of incidence, total reflection of the energy occurs as along ray *C* in Fig. 9.14.

The waves that approach the boundary at the critical angle do not return in a direct path to the surface. As they pass through the interface, they are refracted and continue along the interface through the higher-velocity rock. As this wave motion moves along the boundary, part of the energy is refracted back through the interface and the upper layer to the surface. Because these waves must have approached the boundary at the critical angle, they will not return to the surface short of a certain distance. This critical distance from the shot point is determined by the thickness of the rock layer.

Although waves reflected from the discontinuity can arrive at geophones placed at any distance from the shot point, this is not the case with refracted waves. At some distance from the shot point, waves reach geophones on the surface faster than even the direct waves because they travel faster through the higher-velocity lower layer. This effect shows up clearly when the travel time of the first wave to arrive at each geophone is plotted against the distance traveled, as in Fig. 9.15. This plot of elapsed time versus distance from the shot to the geophones is ordinarily composed of two or more straight line segments. The slope of each segment gives the velocity of the seismic waves over part of the path. The first segment is the velocity of the direct waves, and the second is the velocity of the refracted wave in the second layer. Deeper layers may also be detected, and the depth to each layer can be determined from these data.

Although we have referred only to horizontal layers in the examples here, both reflection and

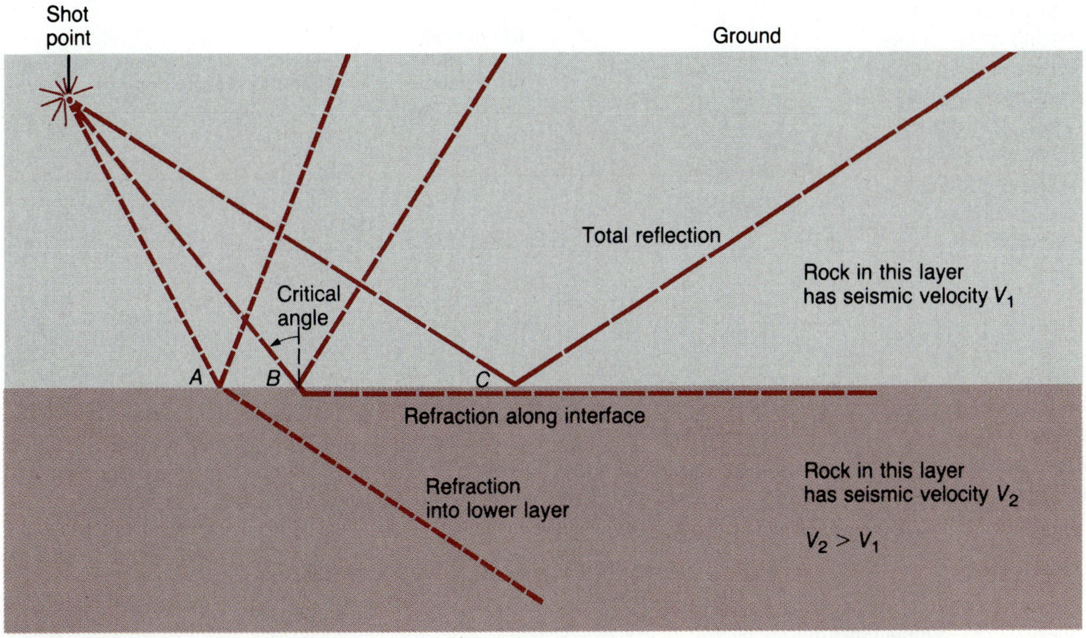

Figure 9.14
The three paths shown here illustrate the differences in what happens to wave energy that approaches a boundary at different angles of incidence. At point *A*, part of the energy is reflected and part is refracted into the lower layer at an angle. At *B*, part of the energy is reflected, and because this is the critical angle, part is refracted along the boundary. This will also be continuously bent back into the upper layer. At angles of incidence greater than the critical angle, as at point *C*, total reflection of energy occurs.

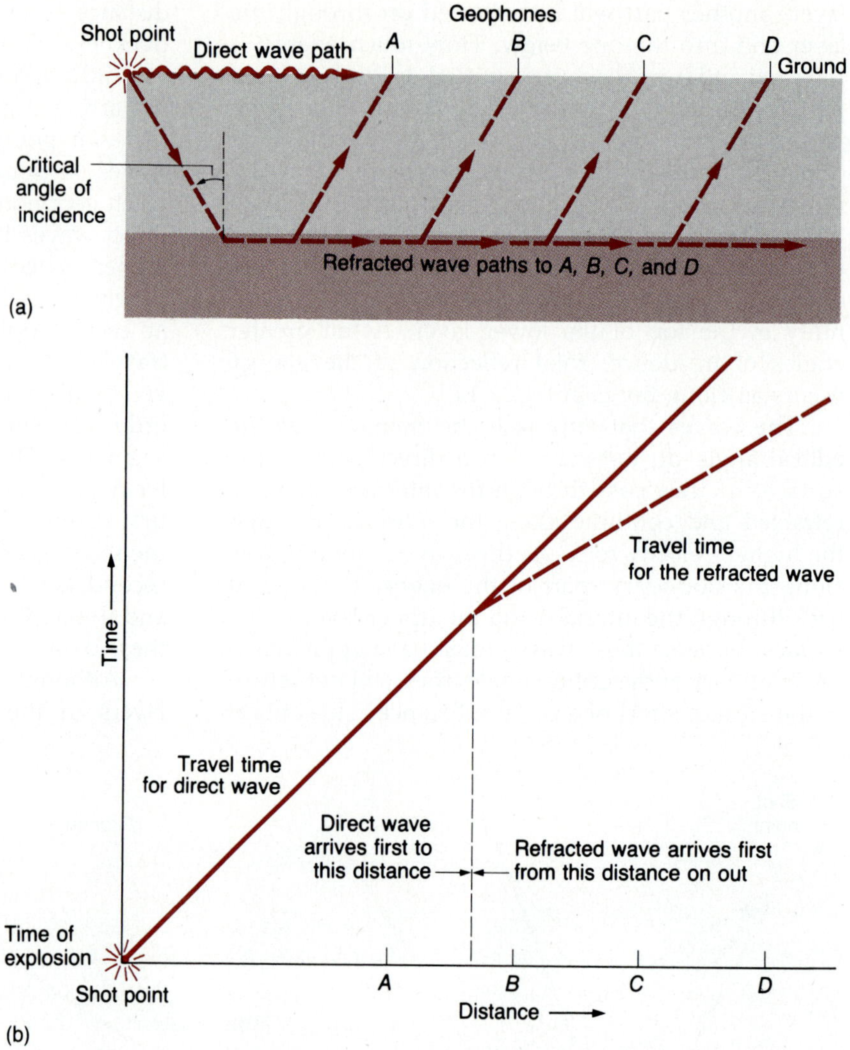

Figure 9.15
Schematic drawings of (a) a wave refracted along a boundary below and (b) the travel–time curve for waves reaching the geophones at distances *A–D.* Note the direct wave will reach the geophones before the refracted wave close to the shot. At greater distances the refracted wave will arrive first. Refraction methods can be used to determine depths to layers throughout the crust.

Figure 9.16
This seismogram of an earthquake originating in Japan was recorded at the University of Tasmania. The time marks below the seismogram trace indicated time elapsed after the earthquake. The first body wave (P) arrived in Tasmania about 13 minutes after the earthquake began. S waves arrived at 23 minutes, and surface waves began to arrive at 36 minutes. The arrivals of several other waves (SS, SSS) are discussed later.

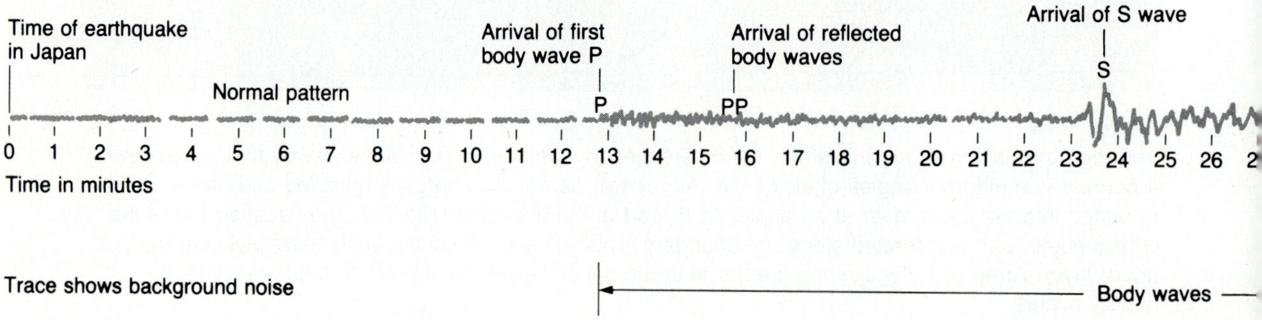

refraction methods can be applied to inclined layers as well. In such cases, the depth to the layers and the angle of inclination (also called dip) can be determined.

9.3 EARTHQUAKE SEISMOLOGY

Interpretation of Seismograms

Generally, in the absence of earthquake activity, the trace on a seismogram shows very slight ground movement called **microseisms.** These are caused by movements of trees in the wind, by waves breaking along the seashore, and even by trucks passing on nearby highways. When the first seismic wave from an earthquake arrives, the trace is immediately displaced, showing a sudden change in the direction of movement and greater aplitude of the trace's swing, as is indicated in Fig. 9.16.

Once ground motion starts, it continues, sometimes for several hours, and the succession of incoming waves generally shows a number of sharp breaks in trace amplitude and direction. Abrupt breaks in direction of movement of the trace are most prominent in the early part of the wave train. These early waves do not generally have high amplitude. The time of arrival of each wave at the station depends upon the path traveled by that wave and the velocity of the propagating wave. The first wave to arrive is the one that travels the most direct path with the highest velocity—usually the P wave. Other body waves that have traveled the same path as P but more slowly, or that have been reflected or refracted in route, follow the P wave. Finally, the ground is set in high-amplitude motion as the surface waves arrive. These waves are characterized by their high amplitude and regular patterns. Other body waves that may arrive after surface wave motion starts are difficult to identify because they are masked by the greater amplitude of the surface waves.

Constructing Travel–Time Curves for the Earth's Interior

Analysis of the time required for waves to travel from the focus of an earthquake to recording stations at various distances from the focus is the key to interpreting seismograms and to using them to study the interior of the earth. These data are plotted on a graph, called a **travel–time curve.** Time of arrival of seismic waves is plotted on the vertical axis, and the distance from the epicenter to the recording station is plotted on the horizontal axis. Distance is not expressed in meters or kilometers along the surface but rather in terms of the angle between the focus, the center of the earth, and the station. Thus, a station 180° away from the focus is directly on the opposite side of the earth (Fig. 9.17).

Because surface waves travel at the earth's surface and with constant velocity, we may anticipate that they will appear as straight lines starting at the point of origin of the travel–time graph. It is not so obvious what paths other wave arrivals on the seismogram may have taken inside the earth. Nor is it obvious that they travel with constant velocity in the interior. We can, however, follow in the footsteps of early seismologists and deduce the paths and velocities of various waves.

As a first step we might collect seismograms for a particular earthquake from stations at many different distances from the epicenter. If these records are laid out side by side it should be apparent that many of the same wave arrivals appear on adjacent records, although at later times for the more distant stations. Some arrivals, like those marked P, S, and surface

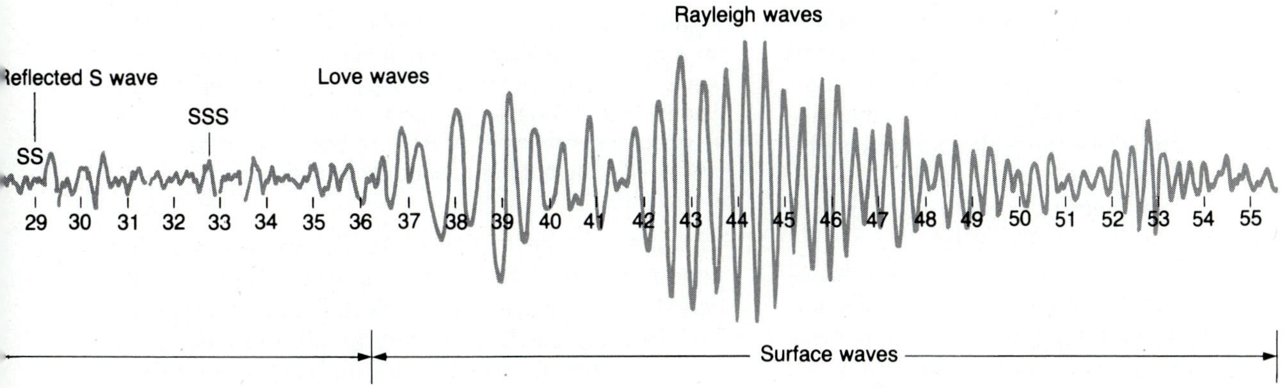

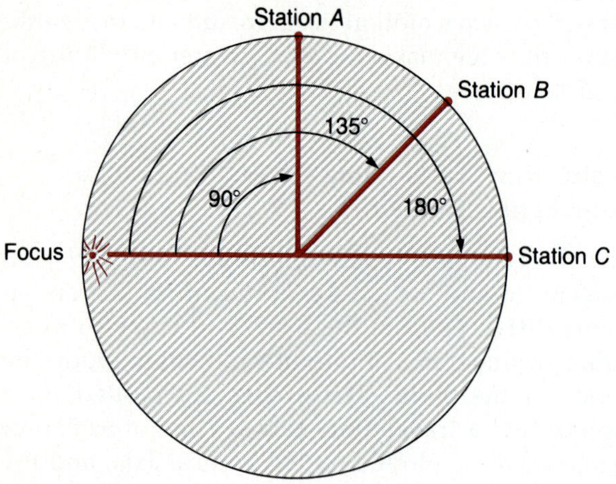

Figure 9.17
The distance from the focus of an earthquake to points on the surface is often given in terms of the angle between lines connecting the center of the earth to the focus and the point.

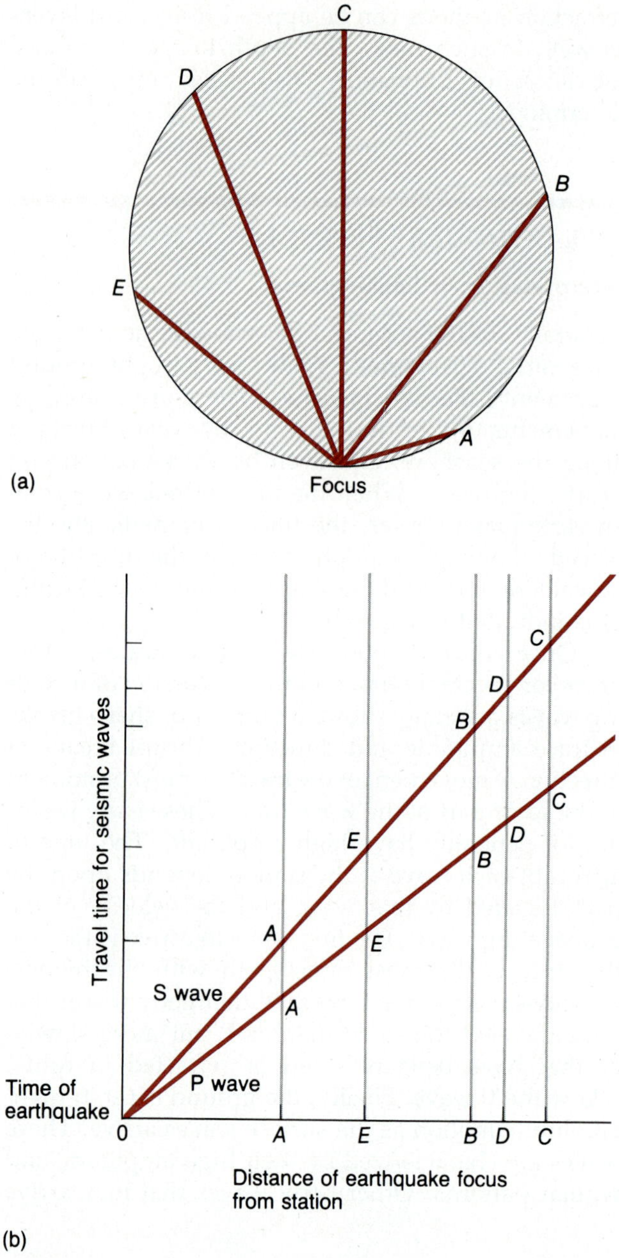

Figure 9.18
(a) A hypothetical model earth. If the velocity of seismic waves were uniform through the earth, a plot of the travel time for P and S waves traveling from a focus to seismograph stations at *A, B, C, D,* and *E* would be a straight line (b), but this is not the case, as is shown in Fig. 9.19.

waves on Fig. 9.16, are easy to identify. Others, such as those marked PP, PcP, and SS, (waves that have been reflected from the surface of the earth, such as PP and SS, or reflected from the core as PcP) are less obvious on some records than others.

If the time the earthquake occurred is known, we can plot the travel time for each wave against the distance of the station from the epicenter and construct a travel–time curve.

If the interior of the earth were uniform in composition and elasticity, the velocity of shock waves would also be uniform; the travel time for the first arrival (P wave) and for the S wave that travels the same path but more slowly would be a simple function of the distance from the focus to the station, as shown in Fig. 9.18; and all wave paths (rays) would be straight. However, the observed times of arrival for P and S show that as the distance becomes greater between the origin and observing station, the waves seem to have traveled faster. This is easily explained if the rigidity of rocks in the interior increases with depth, for as the waves penetrate deeper, their velocity increases. One consequence of the increase in velocity with depth is that the waves are gradually, but continuously, bent as they move into the interior. This refraction causes them to travel a curved path and come back to the surface (Fig. 9.19).

A number of other wave arrivals on the seismograms obviously follow different paths. The seismologist's job is to determine the paths they must have traveled in order to arrive at stations where they do. The waves identified as PP and SS on Fig. 9.16, for example, can be explained as P and S waves that penetrated into the earth, followed a curved path back to the surface, were reflected from the surface back into the earth, and were finally received at the station as they returned to the surface a second time. The path of such a wave is shown in Fig. 9.20.

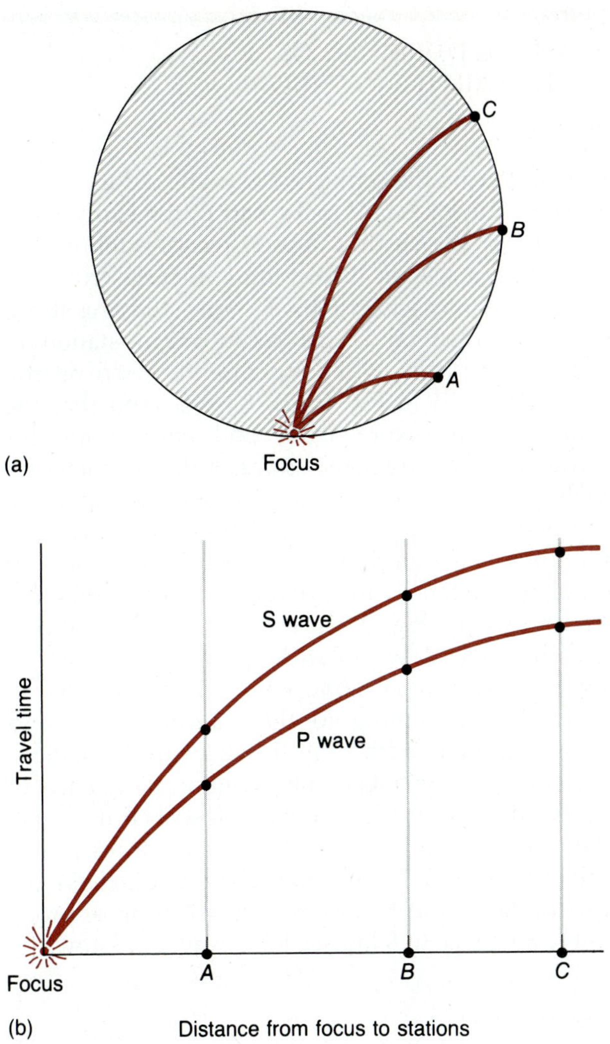

Figure 9.19
(a) Hypothetical earth model in which velocity of seismic waves increases with depth. This results in continuous bending of the waves and causes the travel time graph to be a curved line (b).

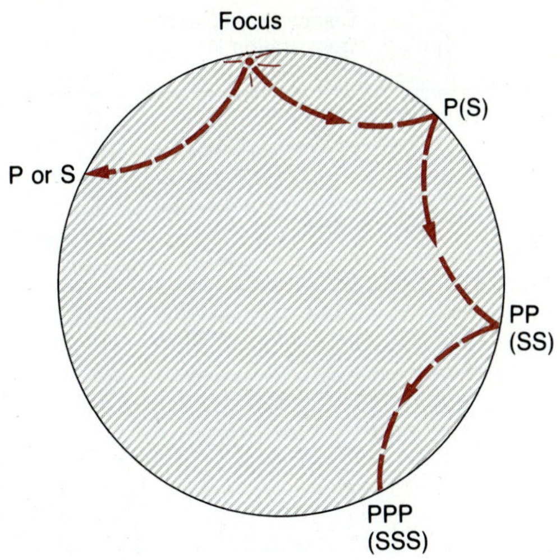

Figure 9.20
Paths followed by waves in the mantle. The paths to the right of the focus illustrate multiple reflections.

So far our analysis of travel times leads to the conclusion that the rigidity of the earth's interior increases with depth. With this much background, we can explore how earthquake epicenters are located. In later sections we shall see how the seismic evidence points to the existence of major divisions within the earth.

Locating Earthquake Epicenters

The position of the epicenter and the time of occurrence of an earthquake are first known only to those who live in the immediate area, and they are likely to be cut off from the outside by the disruption of lines of communication by the earthquake. The first indication of the earthquake at a station some distance away is the arrival of the wave train. If it is possible to identify the time of arrival of waves at several different stations, then it is a simple problem to determine the distance to the epicenter. P and S waves travel the same path but at different velocities. Thus, the distance to the epicenter is a function of the time lag between these two arrivals. This time lag tells us how far away the epicenter is from the station, but does not indicate the exact geographic location; the center could be anywhere on the circle with that radius. The precise location can be determined if the distance to the epicenter from three stations is known. The distance from each station to the epicenter is drawn on a map as a circle around the station. These three circles intersect in only one point—the epicenter (Fig. 9.21).

Earthquake Prediction

Various techniques have been tried in the effort to predict earthquake activity. Among them have been measurements of the magnetic field, surveys of the ground level across faults, surveys of line lengths in the vicinity of fault zones, and the measurement of accumulated strain. None of these has proven accurate enough to yield the type of information needed. In recent years, a more promising method has been developed. It involves monitoring the velocities of

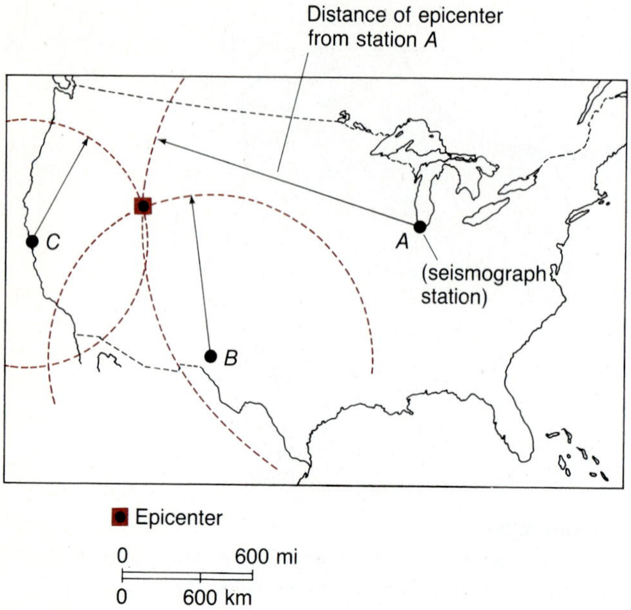

Figure 9.21
Location of an epicenter. Once the distance to the epicenter from each of three seismograph stations is known, it becomes possible to locate the epicenter by triangulation, as shown.

compressional and shear waves traveling across a fault zone. The ratio of the velocity of the compressional waves to the velocity of the shear waves changes in a systematic way before an earthquake. The process is thought to involve changes in the volume of the rock, the amount of water, and the degree of fracture of rock in the fault zone before an earthquake occurs. As strain accumulates in the fault zone, the rock approaches and reaches the stress level needed to initiate fracturing. Once fractures form, the rock begins to expand, and the velocity of the compressional (P) waves drop. Gradually, these pores fill with water, which lubricates the zone, reducing the friction and facilitating movement in the fault zone. As water content increases, compressional velocity begins to increase again. At this stage an earthquake is imminent. This technique is now being refined and applied to many fault zones, especially in the western United States. It requires careful and continuous measurement of seismic wave velocities for wave paths that cross major fault zones. Seismologists look for changes in velocity, which may signify that this sequence of events is in progress. It is hoped that the results will yield a method by which we may accurately predict earthquakes.

9.4 MAJOR DIVISIONS OF THE EARTH'S INTERIOR

The Crust

For many years the term *crust* was applied loosely to the outer consolidated, rocky part of the earth. Now the crust of the earth is more precisely defined on the basis of a seismic discontinuity first described by Andrija Mohorovicic in 1909. He was studying shock waves that arrived at his seismograph station at Zagreb, Yugoslavia. When the epicenters were nearby (for example, *A* in Fig. 9.22), he observed that the P-wave arrivals were sharp and abrupt. But for earthquakes beyond some critical distance (such as *B* in Fig. 9.22), the P waves started as a small-amplitude, long-period movement followed by an abrupt, higher-amplitude wave arrival. This led Mohorovičić to postulate that these early arrivals traveled two slightly different paths (Fig. 9.22): the first arrivals traveled from the focus down to an interface underlain by a higher-velocity layer along which the wave was refracted before coming to the station; the second followed a direct path from the focus to the station. This effect has since been observed in many parts of the world; and, in fact, we have described just such a boundary in the discussion of seismic wave refraction in Section 9.2. The interface identified on the basis of this abrupt velocity increase from approximately 6 km/sec to 8 km/sec for P waves is known as

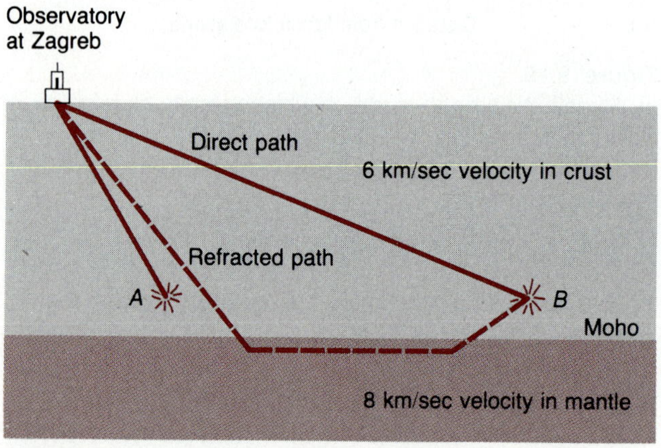

Figure 9.22
This schematic sketch shows the paths Mohorovičić postulated for seismic waves arriving at the Zagreb observatory. Those generated nearby follow path *A*; those originating farther away follow path *B*. The first arrivals from point *B* traveled the refracted path and arrived before those traveling the direct path from intermediate points.

the *Moho* or *M discontinuity*. Today it is recognized as defining the bottom of the earth's crust.

Under the oceans, the Moho occurs on average at depths of about 10 km, but under the continents it is much deeper—30 to 50 km (Fig. 9.23). Not only is there a great variation in the thickness of the crust under the continents and under the ocean, but its composition in the two regions also shows significant differences. The oceanic crust consists of a thin (0.5 km on the average) layer of sediment capping a much thicker one of basaltic lava flows and gabbro. In contrast, the sedimentary veneer on the continents is thin or absent in some places but thousands of meters thick in others. The thickest sections are in mountain belts, in larger basins, and along some continental margins. Usually, this continental sediment is underlain by granitic rocks. These are intrusive igneous rocks or the products of metamorphism of ancient sedimentary rocks.

The lower half of the continental crust is not often exposed for our inspection, but seismic studies initially conducted by Joseph Conrad about 1925 revealed that the lower part of the continental crust has higher seismic velocities than the upper granitic rocks, suggesting that these lower rocks are basaltic or gabbroic. Because of the difference in velocity, the boundary can be recognized seismically; it is known as the Conrad discontinuity.

The Mantle

Below the crust lies the earth's region known as the mantle. We have some direct evidence of the character of the uppermost part of the mantle, but the deeper parts are known only indirectly. The most significant subdivisions of the mantle occur in the depth range of 100 to 200 km where the velocity of seismic waves drops. We believe this low-velocity zone separates an outer solid and brittle part of the earth, including the crust and uppermost mantle, from a more plastic interior (Fig. 9.24).

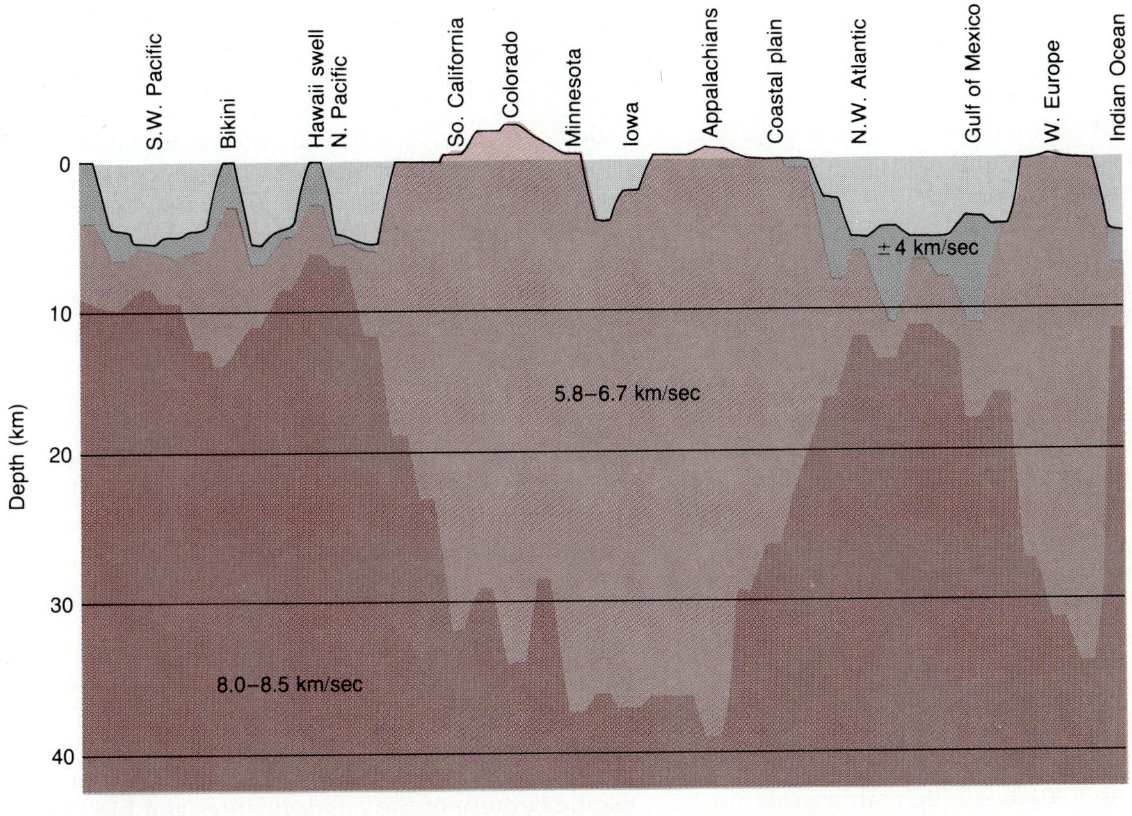

Figure 9.23
The crust of the earth varies greatly in thickness as shown here. Numbers indicate the seismic velocities observed at different depths. Note the low velocities near the surface and the high velocities (8 km/sec) just below Moho.

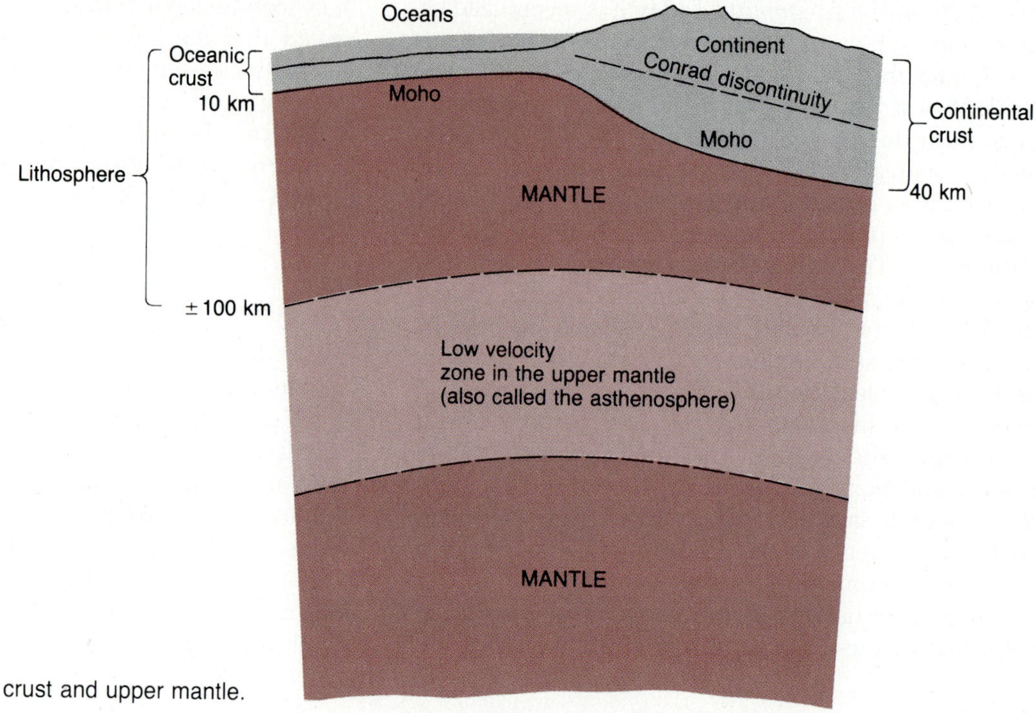

Figure 9.24
Major subdivisions of the crust and upper mantle.

The upper mantle. Most of what we know about the upper mantle is derived from seismic studies. In the few places where upper mantle rocks are thought to be exposed by upward faulting of these deep rocks along the edges of crustal plates, they are all rocks rich in magnesium and iron—rocks like dunite (all olivine), peridotite (olivine, pyroxene, and plagioclase), or their alteration products. For example, serpentine, $Mg_6(OH)_8Si_4O_{10}$, can be formed by chemical reaction between hot water and magnesium-rich minerals of ultramafic rocks. In most places, however, the upper mantle is too deep to be examined, and our knowledge of it is based mainly on the velocity of propagation of P and S waves in the upper mantle.

Velocity measurements in the upper mantle just below the Moho reveal considerable variation from place to palce, probably due to differences in the composition of the rocks there. Clearly the uppermost mantle is not uniform, but the differences within it are not nearly as great as those among crustal rocks. Differences in P-wave velocities in the upper mantle range from 7.8 to 8.3 km/sec; they are caused presumably by variations in mineral composition, temperature, or perhaps both. High temperatures change the elastic properties of the mantle rock reducing its rigidity and thus its ability to transmit seismic waves. This may explain why upper mantle rock has a lower velocity under such places as oceanic ridges, Yellowstone Park, the Columbia River Plateau, the island arcs, and other sites of recent volcanic activity. The higher velocities generally found under continental interiors suggest that mantle velocities are related in a direct way to the structure of the crust.

The lithosphere and low-velocity zone. Seismic studies clearly show that the mantle behaves rigidly in transmitting both compressional and shear seismic waves. This does not mean that all of it is strong, brittle, solid, and elastic like the rock we see on the surface of the earth. It could behave like a plastic material such as asphalt, which is rigid when subjected to rapid impulsive pressure like a seismic wave, but flows under small pressure applied over a long period of time.

We might expect such plastic behavior in the mantle because of the high pressures and high temperatures found there, and in Chapter 12 we shall see evidence of this plasticity, found by studying earth's gravity field. Thus, in terms of rigidity or strength, a distinction is evident between the lithosphere, which exhibits brittle behavior, great strength, and resistance to flow, and a lower zone, sometimes called the **asthenosphere,** in which high temperature and pressure promote plastic behavior and flowage.

The first seismological evidence of the depth of the boundary between the lithosphere and the asthenosphere came in 1954 with Beno Gutenberg's recognition of a **low-velocity layer** in the upper mantle at a depth of about 100 to 200 kilometers. Low-velocity layers are difficult to detect by seismic methods, because waves refracted into a low-velocity layer from above are bent downward so they do not return to the surface. Gutenberg's original evidence for the low-velocity layer was the existence of a shadow zone around earthquake epicenters. Within that region, the amplitudes of P waves coming to the surface at distances of 600 to 1,500 kilometers from an earthquake epicenter are much lower than the amplitudes of those P waves reaching the surface from distances beyond about 1,500 kilometers. Today additional evidence of other types confirms the presence of the low-velocity layer.

According to Gutenberg, the low-velocity zone is centered at a depth of about 150 kilometers, and it is about 150 kilometers thick; but the zone is neither sharply defined nor uniform in depth or thickness. Why the zone has been much easier to locate under oceans than it has under continents is still uncertain. Within this zone, seismic velocities are about 6 percent less than velocities just below the M discontinuity.

The composition of the material in the Gutenberg low-velocity zone may not be different from that of the mantle material above or below it; it may differ primarily in physical properties as a result of a particular combination of pressure and temperature conditions. Estimates of the temperature and pressure at that depth suggest that a mantle composed of such minerals as olivine, garnet, and pyroxene might begin to melt at the depth of the low-velocity zone. The velocity of seismic waves may be reduced because the mantle there contains some liquid. This liquid would substantially reduce rock strength, making the mantle plastic and causing it to behave like a pseudoviscous material such as asphalt or Silly Putty. Additional evidence about the character of the materials in the low-velocity layer comes from the observation that some lavas seem to originate at that level. The movement of molten material through conduits toward the surface generates shock waves; waves first appear within and just above the low-velocity zone, suggesting that the melt is forming there.

Mantle below the low-velocity zone. The composition of the deeper parts of the mantle is inferred from the velocity of seismic waves penetrating successively greater depths. The mantle is divided into upper and lower parts by a broad velocity transition zone located between 300 and 800 km depth, as shown in Figure 9.25. Even though the lower part contains much more of the mass of the earth than the upper mantle and crust combined, we know relatively little about it. We have no samples of the lower mantle and no hope of ever getting them; yet we are confident that the lower mantle is relatively solid because it transmits seismic shear waves. On the basis of what we know

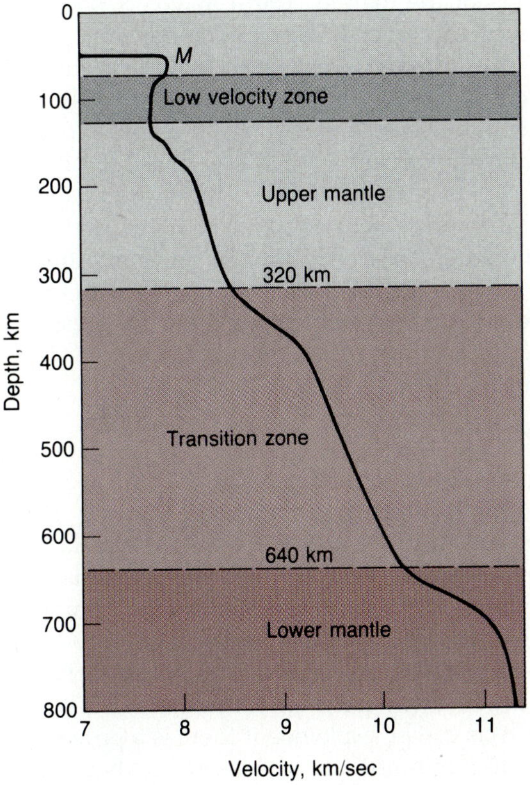

Figure 9.25
This curve shows the variation of P-wave velocity with depth in the upper 800 km of the earth's interior. The dashed lines at 320 and 640 km indicate approximate boundaries of a zone of transition. Above that zone, mantle velocities are around 8 km/sec. Below the transition zone, velocities in the mantle are over 11 km/sec.

about lower mantle seismic velocities and about composition and density of the upper mantle and of stony meteorites, we are confident that the lower mantle is composed of silicate minerals.

The most common minerals in both the stony meteorites—materials from which the inner planets possibly formed—and in the upper mantle are olivines and pyroxenes—both silicates containing iron and/or magnesium. We do not know what other minerals may be present or just how homogeneous the lower mantle is, but we expect some variation because the lower mantle does appear to be concentrically zoned—a conclusion based on the detection of several relatively minor seismic discontinuities that occur at the same depths everywhere.

Models of density variations in the mantle are based largely on seismic data. Recent models constructed to correspond to seismic velocity data include many density breaks and gradients in the upper 1000 kilometers of the mantle. These may correspond to changes in the crystal structure of the minerals at depth. Harold Jeffreys suggested that these changes in velocity might be due to a change in the crystal structure of olivine caused by high pressure at depth. For example, we know that one type of olivine, Fe_2SiO_4, transforms to another, the mineral spinel, with a 12 percent density increase at the pressures encountered at a depth of 120 kilometers. Another type of olivine, Mg_2SiO_4, will also transform to spinel with an 11 percent density increase at pressures predicted at a depth of about 400 kilometers.

The Core

The core–mantle boundary. Of the seismic discontinuities in the earth's interior, none is more fundamental than the one that marks the surface of the core. The presence of this core boundary is inferred from the observation that the amplitude of P and S waves begins to decrease markedly, or they are not recorded at all, at stations located between 103° and 143° away from an earthquake (Fig. 9.26). These effects can be explained if there is a prominent seismic discontinuity, the core—mantle boundary, located at a depth equal to about half the radius of the earth.

Stations located less than 103° away from an earthquake receive first arrivals of P waves that have traveled simple curved paths. Waves reaching a station exactly 103° away just barely pass by the core boundary and are tangent to it. Waves approaching at slightly steeper angles are reflected from the boundary. Waves that approach the boundary at still steeper angles strike it, are refracted into the core, and pass through part of the core. Any waves that pass through the core always travel with P-wave velocities. Apparently, no shear waves travel through the outer part of the core. The closest station that can receive a wave that has been refracted into the core is 143° away. This zone between 103° and 143° is known as the **shadow zone.**

If a core boundary of this kind is present, part of the energy reaching the core boundary should be reflected back toward the surface. Knowing the approximate wave velocity above the core from the velocity of the first arrivals of P and S waves, it should be possible to predict the time of arrival of a wave that travels from an earthquake focus to the core and is reflected from the core back to a seismograph less than 103° away. The waves that travel these paths, shown as PcP or ScS paths on Fig. 9.27, are frequently identified on seismograms at the time predicted. Thus, they offer an independent confirmation of the presence of the core.

The character of the core. The failure of S waves to reach stations located along paths that would carry them through the core is interpreted to mean that the core is incapable of transmitting the shear waves. P waves are transmitted through the core, but at a lower velocity than in the overlying mantle. These are the characteristics we would expect of waves being transmitted through a liquid.

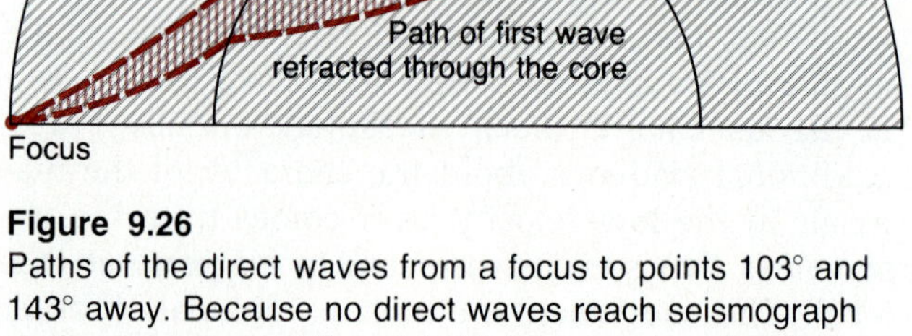

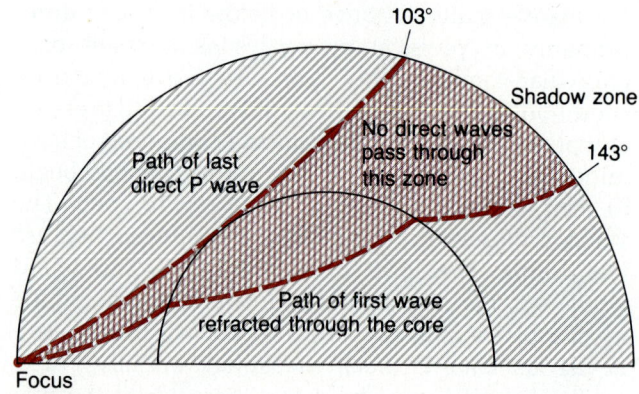

Figure 9.26
Paths of the direct waves from a focus to points 103° and 143° away. Because no direct waves reach seismograph stations between 103° and 143°, this zone is called the shadow zone.

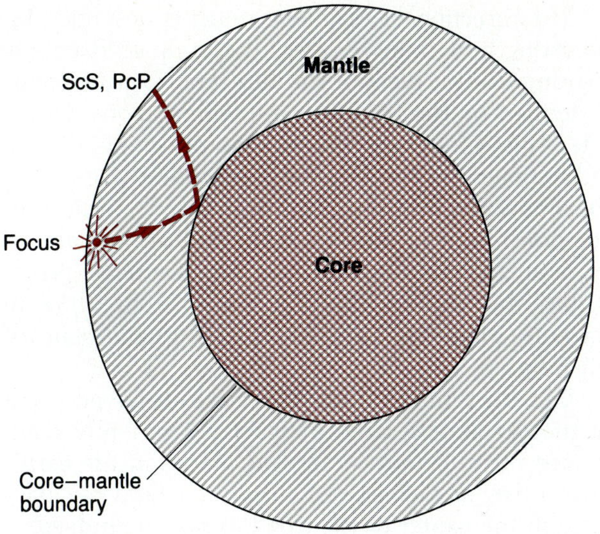

Figure 9.27
The wave paths shown here are those of a P or S wave that travels through the mantle to the core boundary. The waves are reflected from this boundary and return to the surface as P or S waves. The identification of these waves on seismograms confirms the existence of the core.

The outer surface of the core is nearly 3000 kilometers deep; so, our chances of ever obtaining a specimen are remote indeed, and it seems extremely unlikely that any of the rocks at the surface have come from the core. Yet we are highly confident that the core is dense (9 to 12 grams/cm^3), that the outer part of it is liquid, and that because it is capable of generating a magnetic field, it is probably a metal. Certainly the fact that has most impressed scientists concerned with this subject is that many meteorites are composed mainly of nickel and iron. Melts of these meteorites have properties similar to those of the earth's core. In addition, iron and several oxides are the only common minerals that are heavier than the silicates that compose the crust and mantle. The density of the core is so much greater than that of the crust and mantle that it is unlikely that silicate minerals are its main constituents.

The calculated density of the core is somewhat lower than that of iron alone. This suggests that something is probably mixed with the iron. Silicon might be present, or the core could consist of iron alloyed with other elements, possibly sulfur or hydrogen, which is thought to change to a metallic state with a sudden increase in density (0.4 to 0.8 g/cm^3) at the pressures found at a depth of 1600 kilometers.

The Inner Core

The core appears to be divided into two parts, an outer core and a small inner one. Evidence for a boundary between the two is found in the existence of wave arrivals that have apparently traveled the paths shown in Fig. 9.28. PKiKP is the path of a P wave that is first refracted into the outer core, then reflected from the inner core, and travels back to the surface. The inner core has a radius of 1300 kilometers and must be more rigid than the outer core in order to account for the higher velocities of waves, such as PKiKP (Fig. 9.28), that pass through it. This is interpreted to mean that the inner core is solid.

The composition of the inner core is in even greater doubt than that of the outer core. It may be a solid, as indicated by the increased velocity of P waves passing through it, but until we have a better understanding of the nature of the changes that can take place in materials under extreme pressures, some doubt will persist. Most scientists believe that the best model of the core at the present time is one characterized by a solid inner core composed of iron, nickel, and possibly other components such as sulfur,

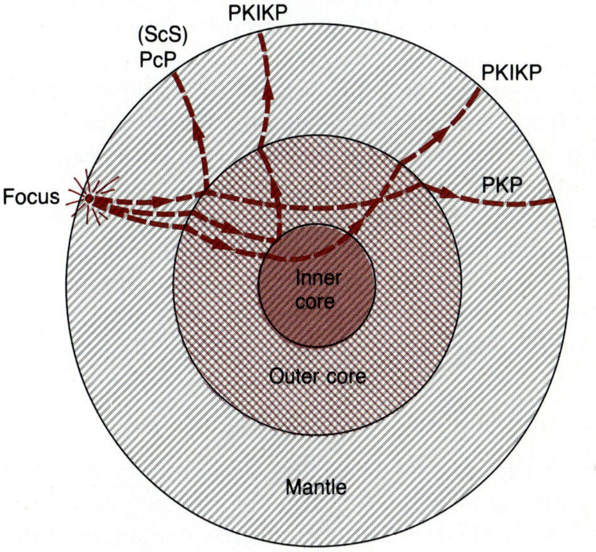

Figure 9.28
Paths of seismic waves that are reflected from the outer core (PcP) and inner core (PKIKP) and wave paths that penetrate and are refracted through the outer and inner core (PKP and PKIKP). Symbols used to represent wave paths include P or S to indicate compressional or shear wave paths; c indicates reflection from the outer core; K indicates a core path; and I indicates reflection from the inner core.

and a hot, liquid-iron outer core in which convection is taking place as a result of a high temperature difference between the inner core and the cooler core—mantle boundary. Evidence supporting this model of the core is also derived from the character of earth's magnetic field.

9.5 EARTH'S MAGNETIC FIELD

The Chinese knew of some magnetic phenomena as early as 3000 to 2000 B.C.; the Greeks knew of the magnetism of lodestone by the sixth century B.C.; and primitive compasses were in use by the Arabs and the Chinese by 1000 A.D. But it remained for the Englishman William Gilbert (1540–1603) to demonstrate that the earth as a whole behaves somewhat like a great bar magnet. Early observations had shown that several other materials, including steel, can be made into magnets if they are held close to lodestone. When a bar-shaped magnet of steel is placed in a pile of iron filings, the filings are attracted to the steel, and more filings are attracted to the ends of the bar than to the middle. The points that hold the most filings are called the poles of the magnet. If a steel needle is magnetized and carefully floated on the surface of a liquid it will rotate until one end, called by convention the south pole of the magnet, points north and the other end points south. This is the principle behind the compass. If two such magnetic needles or bar magnets are brought close together, opposite poles attact each other and similar poles repel each other with a force that is directly proportional to the product of their pole strengths and inversely proportional to the square of the distance between them.

Description of the Earth's Magnetic Field

The earth's magnetic field is described in terms of the force exerted on a magnet by the magnetic field, called the field strength, and the orientation of the field—that is, the direction in which a suspended magnetized needle, free to rotate in any direction, points. Such a needle will become aligned in the field. Close to the earth's surface the field is inclined, often at steep angles, (Fig. 9.29). Because it is difficult to suspend a needle in such a way that it remains horizontal, we customarily measure the orientation of the horizontal component of the field using a compass in which the needle is weighted so as to keep it horizontal.

The direction of magnetic north is described by indicating how many degrees the compass needle is pointing east or west of the direction to true north, as defined by the direction to the geographic North Pole. This difference between the directions of true and magnetic north is called **declination** (Fig. 9.30). The amount of this departure of magnetic north from true north varies considerably from place to place on earth and is very great near the magnetic poles. Within the United States it varies from 21°W in northern Maine to more than 22°E in the state of Washington.

Gilbert thought that the earth behaved as a magnet because a large body of permanently magnetized rock like lodestone was buried in the earth. Such a bar, as Gilbert visualized it, would not pass through the center of the earth, however. Indeed, it would miss the center by almost 960 kilometers. We now know that the north magnetic pole is not situated at the north geographic pole, on the axis of rotation, but rather in northern Canada at about latitude 75°N. Similarly, the south magnetic pole is located at latitude 68°S, south of Australia.

Not only are the magnetic poles far removed from the geographic poles, but the magnetic poles are known to shift in position, and the earth's magnetic field varies with time in both strength and orientation. The field is usually depicted by three maps: one showing lines that connect points on the surface where the inclination of the field is the same; a second showing lines of equal declination (Fig. 9.31); and a third showing field strength.

Although Gilbert was wrong in thinking that the earth's field was produced by buried magnets, we now know that rocks do differ in their magnetic properties. Some rocks contain large proportions of the mineral magnetite, and a great many other minerals are slightly magnetic. When sufficiently accurate maps of the earth's field are prepared, irregularities—magnetic anomalies—result from distribution of these rocks of differing magnetism. This fact has proved to be a most useful tool in exploration for economically important iron ore and other deposits containing magnetic minerals. When a measurement of the magnetic field is made, the instrument used is affected by a magnetic force field that includes various components: magnetic fields external to the earth, such as those of the sun; those arising from deep internal causes in the earth; and others caused by magnetic minerals in rocks of the crust. This last component of the field is of great importance to the geologist who seeks data on the location of mineral deposits.

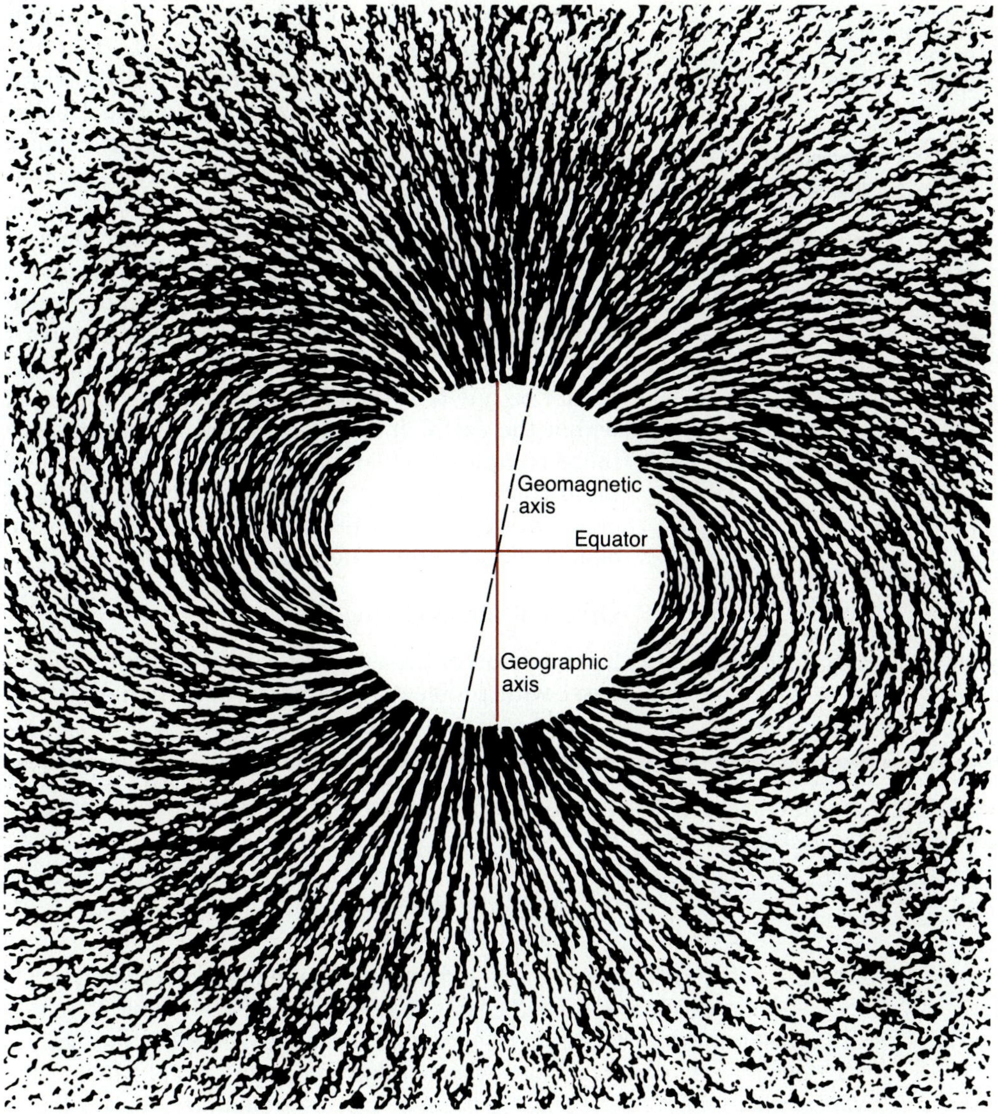

Figure 9.29
The magnetic field of the earth resembles the one shown here. The dark patterns show a pattern caused by the alignment of iron filings around a magnetized sphere. These bits of iron line up indicating the alignment of the magnetic force field. The geographic axis (axis of rotation) and a geomagnetic axis similar to that of the earth are shown.

Variations in the Field Through Time

Data collected over a period of nearly a century clearly show a number of different types of variation in the magnetic field. Some are periodic, varying with the time of day or the season of the year; these are related to motion of the earth and moon relative to the sun. The field strength undergoes strong and irregular fluctuations on "disturbed" days—days when the earth experiences magnetic storms related to sunspot activity. High-energy particles from the sun bombard the earth, sometimes extending the northern lights, normally observed only in high latitudes, to the middle and low latitudes.

Although most of the periodic changes observed in the magnetic field can be related to solar or other external causes, we are also well aware of other slow changes, called **secular changes.** These can be identified by analyzing the variation in declination, inclination, and intensity of the magnetic field over long periods of time. Even as early as 1692, the position

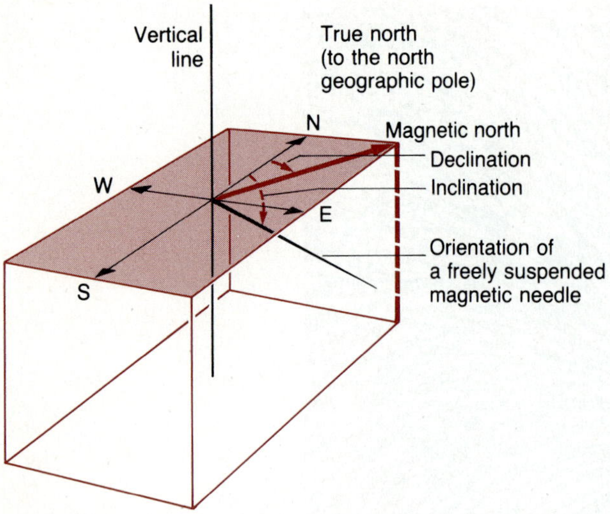

Figure 9.30
The orientation of the magnetic field would be shown by a magnetized needle suspended so it could rotate freely in space. We ordinarily measure the declination (angle between true and magnetic north) and inclination in order to describe this orientation.

of some areas of unusually high magnetic field strength were found to drift westward at a rate of about 0.5° per year. This can be confirmed by carefully plotting the position of the line of zero declination labeled "no variation" in Fig. 9.31 year after year. Similar analyses of the annual maps of declination, inclination, and field strength reveal that many complex changes are taking place in the field. These changes are not uniform over the surface of the earth, nor do they occur at a steady rate. These variations raise important questions about the cause of the earth's magnetic field. It seems clear that the magnetic field cannot be due to solid permanently magnetized rock within the earth. If it were, we should not observe these complex secular variations, which are much more compatible with ideas that the magnetic field arises from movements within a fluid—in this case almost certainly the outer core.

Origin of the Magnetic Field

Investigation of the origin of earth's magnetic field is beset with problems, for the earth's interior, where

MEASURING THE MAGNETIC FIELD

The force of attraction between two poles of strengths p_1 and p_2 separated by a distance d is given by

$$F = \frac{1}{\mu} \times \frac{p_1 p_2}{d^2}$$

where μ is a constant known as the **magnetic permeability** of the medium in which the poles are located; it has a value of 1 for air. A **unit pole** is of such strength that it will repel a similar pole of equal strength separated by 1 cm in a vacuum with a force of one dyne. The area around a magnet within which the influence of the magnet can be detected is called the **magnetic field.** The field for the earth is very great indeed, but for a small magnet the field loses strength rapidly with distance. A magnetic field at a point may be described in terms of the orientation a very small bar magnet at that point would assume if it were free to move and the intensity or magnetic strength of the field at that point. **Magnetic-field strength** (H) is the force (in dynes) that would be exerted on a unit pole at the point.

$$H = \frac{p}{\mu d^2}$$

The force is designated in terms of dynes (gram · cm/sec^2) per unit pole strength, or **oersteds.**

Field strength may be visualized in terms of lines of force in the field. The orientation of the lines indicates the direction of the magnetic field and the number of lines in a cross-sectional area indicates the field intensity. The number of lines of force per square centimeter is taken as equal to the number of dynes with which the field would act on a unit pole.

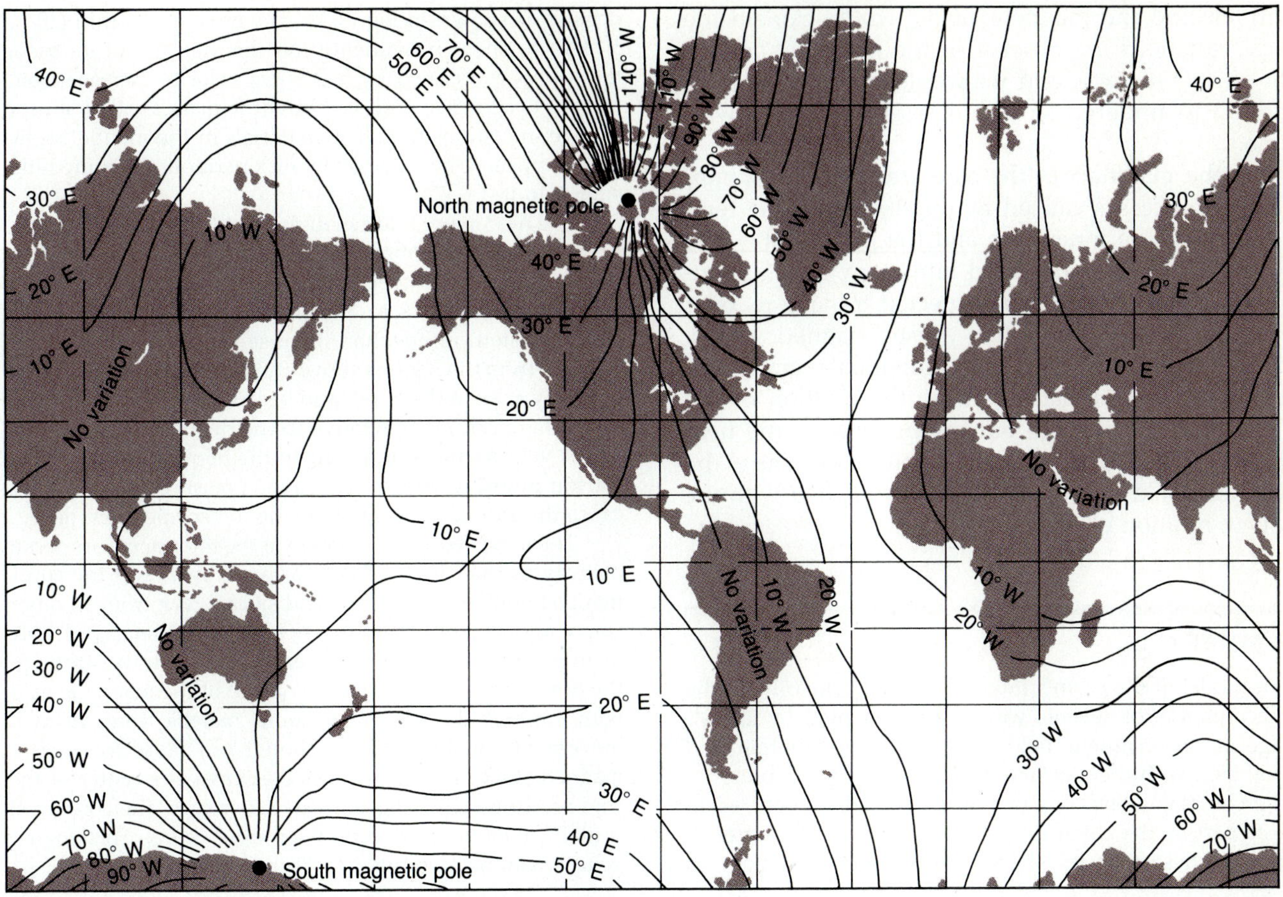

Figure 9.31
World map of magnetic declination; lines connect points of equal declination.

the field is thought to originate, is known only indirectly, and most conclusions regarding the composition, state of motion, and distribution of compositional variations are based on assumptions and theory. We can, however, dismiss the idea that the main field arises from permanent magnetization of rock in the crust or mantle because of variations in the field described in the last section.

Although many small anomalies in the field are clearly due to the distribution of iron-bearing minerals near the surface, intermediate scale anomalies that might be due to magnetic minerals in the mantle appear to be missing. This is no surprise, since rocks lose their permanent magnetization when heated above 580 °C, and temperatures should be high enough to cause such a loss at a depth of 25 km.

The most widely favored hypotheses for the origin of the magnetic field are those involving electromagnetic phenomena, such as the theory of the **self-exciting dynamo** first suggested for the sun by Joseph Larmor in 1919. (A dynamo is a machine used to convert mechanical into electrical energy.) According to this hypothesis, electric currents in the earth's core produce the magnetic field. Our picture of the core—its fluidity, its metallic composition, and its solid inner core, probably composed of ferromagnetic material—are all compatible with the existence of electric currents in the core. The largest anomalies, and the main part of the field, are universally attributed to the core.

The dynamo theory is certainly not proven, but several possible causes of electric currents in the core are recognized. Chemical and heat processes often generate currents at the surface of contact between materials of different electric properties. The boundaries between the mantle and core and between the outer and inner core would be likely places, and W. M. Elasser has suggested currents might be due to turbulent motion within the core. The presence of eddylike convection currents in the core, associated with electric and magnetic fields, would explain the local variations we observe in the magnetic field at the earth's surface, and we would expect such fluid motion to undergo slow progressive changes with time.

If the materials in the core are involved in convective processes caused by differences in the temperature of the inner core and the mantle, the core probably can be subdivided into many separate eddies. Each eddy might be expected to have its own magnetic field if convection causes magnetic fields. The magnetic fields caused by individual regions may not show up distinctly because the rotation of the earth tends to average out components of the field that are not aligned with the earth's axis. Thus, the magnetic field as a whole is crudely aligned with the axis of rotation of the earth.

SUMMARY

Our model of the earth's interior is based primarily on the interpretation of seismic waves supplemented by knowledge of the magnetic field and density distribution and combined with theories about the origin of meteorites.

The major divisions of the interior are recognized in large part by the changes in velocity of the body waves, P and S. These velocity shifts correlate with variations in density as shown in Fig. 9.32. The crust is too thin (10 to 50 km) to show prominently on this figure, but its base, the Moho, is defined by a P wave velocity increase from about 6 to 8 km/sec. This is thought to mark a rather abrupt change in composition. The materials in the mantle below Moho appear to be made of the rock peridotite or something similar to it.

The next major internal division, the low-velocity zone, separates the brittle materials of the lithosphere above from the weaker plastic materials in the zone and possibly below it. This zone appears to arise because the material in it is nearly molten, rather than because of differences in composition from the regions above and below it. Many magmas probably arise in the zone, and because of its weakness the lithosphere can move relative to the deeper mantle.

A velocity transition zone in the depth range of 300 to 700 km probably arises because the pressures at that depth cause the atomic packing of olivine to become more dense.

The most distinctive change in the interior occurs about halfway to the center of the earth. At this level the core–mantle boundary separates rigid silicate rock from an outer core composed of highly dense but liquid metal, probably mainly iron. The failure of shear waves to travel through the outer core and the variable nature of the magnetic field both support the idea of a liquid metallic core. In fact, movements in this core are thought to cause the earth's magnetic field. Because waves that cross the center of the core are faster than those in the outer core, the core must contain a high-speed central or inner core. This higher velocity can be accounted for if the inner core is solid.

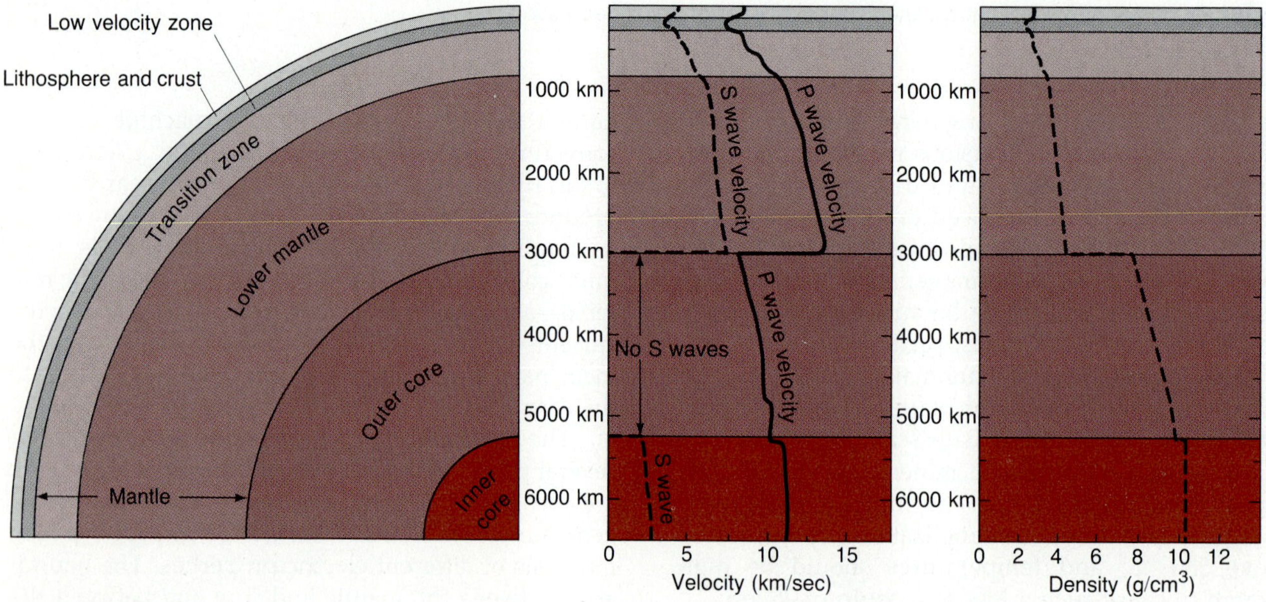

Figure 9.32
Summary of the physical model of the earth's interior. The section at left shows the depth of major internal divisions. At right, the variation of seismic wave (P and S) velocity and density are plotted against depth.

KEY TERMS

amplitude
asthenosphere
body waves
compressional wave
compressibility
critical angle
declination
frequency
geophones
inertia
Love waves
low-velocity zone
magnetic field
magnetic-field strengh
magnetic permeability
M discontinuity
microseism
Moho
oersted
period
P wave
primary wave
ray
Rayleigh wave
reflection
refraction
rigidity
secondary wave
secular change
seismic wave
seismograph
seismology
self-exciting dynamo
shadow zone
stretch modulus
surface waves
S wave
travel–time curve
unit magnetic pole
wave front

STUDY QUESTIONS

1. Which internal division of the earth differs from adjacent divisions in the behavior of the material rather than in its composition?
2. How is it possible to determine the time at which an earthquake takes place if no one is in the vicinity?
3. Why is it difficult to detect low-velocity layers in the earth?
4. Why do compasses used in the northern hemisphere have weights tied around the southern end of the needle?
5. How can we infer the composition and other characteristics of the earth's interior from analysis of the magnetic field?
6. Why should we be more concerned with the composition of the core and mantle than with the composition of the rocks exposed at the surface of the earth when we examine the earth as a planet?
7. Why aren't the same vibrations or shock waves generated during an earthquake detected at seismograph stations everywhere on the earth?
8. How could you design a mechanical device capable of detecting and recording movements such as a sudden shift in the position of the desk on which it is placed?
9. In what ways would the data we have about the interior of the earth be different if the core of the earth were composed of a material that had a density about the same as the mantle?
10. Geologists often refer to the outer shell of the earth as the crust or lithosphere. In what ways are these two divisions alike and in what ways do they differ?

SUGGESTED READINGS

Bingham, Roger, "Explorers of the earth within," *Science 80,* 1:6 (1980), *44–55.*

Bolt, Bruce A., "The fine structure of the earth's interior, "*Scientific American* offprint #906 (1973).

Bullen, K. E., "The interior of the earth," *Scientific American* offprint #804 (1955).

McKenzie, D. P., and Frank Richter, "Convection currents in the earth's mantle; "*Scientific American* offprint #921 (1976).

O'Nions, R. K.; P. J. Hamilton; and N. M. Evensen, "The chemical evolution of the earth's mantle," *Scientific American,* 242:5 (1980), *120–133.*

Richter, C. F., *Elementary Seismology.* San Francisco: W. H. Freeman Co., 1958.

Wyllie, Peter J., "The earth's mantle," *Scientific American* offprint #915 (1975).

ROCK STRUCTURE

Chapter 10

The brevity of recorded history, the lifespan of humans, and the slow rate of most processes of crustal movement obscure the constant changes taking place in the crust. Yet clear evidence remains that most of the earth's crust has been elevated and depressed at various times and that some parts of the crust have been intensely deformed. Perhaps the most convincing evidence of this crustal instability is the ground movement that accompanies large earthquakes. Other evidence of major crustal instability comes from analysis of the distorted shape and dislocation of rocks whose original configuration and position can be deduced from study of sea level and features known to be produced at or just below sea level, and from detailed surveying. In this chapter we shall see what information comes from each of these types of evidence.

Two essentially different types of deformation are found in the earth's crust. One type, characterized by broad uplifting or downwarping of large areas, is called **epeirogenic** movement. These movements may result in uplift or subsidence of segments of the crust. Strata are not dislocated or badly distorted by this type of deformation, although they may be tilted or warped.

The second type of deformation of the crust, called **orogeny,** applies to the processes of mountain-building. In orogeny, strata are bent as in Fig. 10.1; they may be uplifted to great heights, broken, and displaced laterally great distances. Large-scale regional uplift usually accompanies orogenic deformation, but the orogeny itself is most often confined to long, narrow belts of the crust, called mobile or orogenic belts. The central portions of these are generally metamorphosed, suggesting that orogeny involves thermal activity as well as uplift and shortening. It differs from epeirogeny in the intensity of the deformation and in the structural features produced; the two are caused by different although related processes.

U.S.G.S

Figure 10.1
Aerial photograph of the eastern margin of the Rocky Mountain Front Range west of Denver. Tilted layers of sedimentary rock form a prominent ridge.

10.1 EVIDENCE OF VERTICAL MOVEMENT OF THE CRUST

Changes in the Relative Level of Land and Sea

In the sixth century B.C., a large and rich city was built near the water's edge of the Bay of Taman on the northern shore of the Black Sea. The remains of this city, named Phanagoria, now being excavated, lie partially under water, suggesting that parts of this region have subsided relative to sea level since the city was built. Similar evidence is found near Naples, Italy, in the ancient Roman public market known as the Temple of Jupiter Serapis. Here marine gastropods bored holes in columns 5.4 meters above the floor of the temple, and the remains of some of these gastropods are still found in the holes. Evidently the shore was submerged beneath the sea after the Romans built the temple, and then later raised above sea level again where it stands today. Studies of crustal movements in the Bay of Naples demonstrate that the floor of the temple had been sinking at a rate of about 13 mm per year since 1538 when an eruption took place at Campi Flegrei ("fiery fields") near Naples. But from 1968 to the early 1970s rapid uplift amounting to more than 80 cm took place; and in 1969, new volcanic fissures opened, masonry walls were cracked, and parts of the area had to be evacuated.

Sea level is the level to which most crustal movements must be related. Not only is it the basis for comparison of modern altitudes; it is also useful for documenting past levels because the rock record often clearly shows whether an area was above, at, and below sea level. Marine sediments are deposited below sea level; wave action and reef growth are concentrated within a relatively narrow range of shallow water depth; and surface processes act above sea level. Thus, a reasonably close approximation of sea level is obtained where reefs, wave-cut terraces, or shallow-water marine sediments are found.

There is ample evidence of changing relative positions of sea level and land for many times in the past, but not all of them can be explained in terms of either epeirogenic or orogenic crustal movements. Fluctuations in sea level have also resulted from changes in the amounts of sea water as more or less water was stored as ice on the continents. The last glacial age was at its peak about 11,500 years ago. At that time sea level was more than a hundred meters below its present level (see Chapter 24). Enough ice remains today to raise sea level by at least 60 meters if it all melted. Many fluctuations in sea level have occurred as a result of glacial and interglacial periods during the past two million years.

Volcanic activity has also added water to the surface from the interior. How much water is added in this way is uncertain, but most geologists agree that an increase has taken place. Changes in the size and shape of the ocean basins also cause sea level to vary.

Any changes in the relative level of land and the sea must be carefully considered against this background of likely causes. It is not always possible to isolate the exact cause of these changes, but it is possible to demonstrate that some are caused by epeirogenic uplift. For example, many terraces originally formed by wave erosion at sea level now stand at elevations higher than the level at which the sea would stand if all the ice now stored in glaciers melted. A number of different levels can be identified in many different parts of the world. Examples can be seen along the coast of New Zealand (Fig. 10.2). Terraces of this type are also prominent along the western coast of North America, especially along the California coast near Los Angeles. Some terraces there stand 400 meters above present sea level.

Reefs also provide a good indication of former positions of the sea against the land. The top surface of coral reefs is usually only a few meters below water level. Yet reefs, such as those in Fig. 10.3, are found elevated above the sea. A few reef islands in Indonesia are found hundreds of meters above the sea. These are clearly high because the crust has risen. Elevated reefs closer to sea level and those found more than a few meters below the sea's surface may be related to worldwide changes in sea level caused by glaciation, but the higher lands must be due to regional uplift along the coast.

Downward movements of the sea floor are indicated by flat-topped submerged mountains called **guyots** (Fig. 10.4), first discovered in the Pacific basin by H. H. Hess in the 1940s. Dredging across the top of these mountains yields rounded pebbles and cobbles of basalt. This supports the idea that the mountains were originally volcanoes that were eroded by streams and wave action until their tops were nearly flat and close to sea level. Reefs of the Cretaceous and Tertiary ages are found on the tops and around the edges of some guyots. Reef-forming organisms now live only within the upper 100 meters of the sea, yet these reefs are submerged to depths of one to two kilometers. Either the mountains must have been

New Zealand Publicity Service

(a) (b)

Figure 10.2
(a) Aerial view of the sea coast of the South Island, New Zealand, showing elevated marine terraces. (b) Locator map.

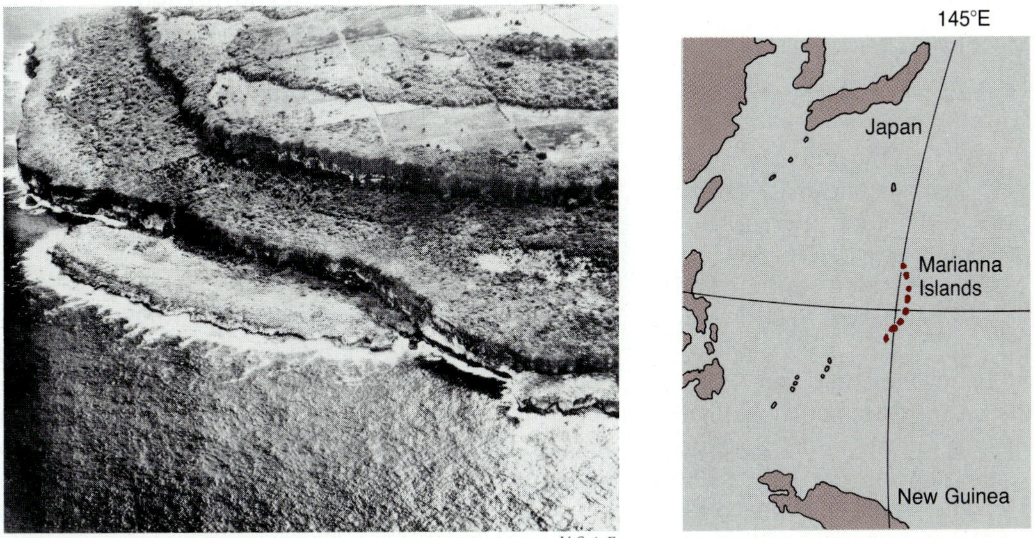

U.S.A.F.

(a) (b)

Figure 10.3
(a) Uplifted coral terraces. Unlike the wave-cut terraces, these were formed by the growth of reefs to a position just below sea level. (b) Locator map.

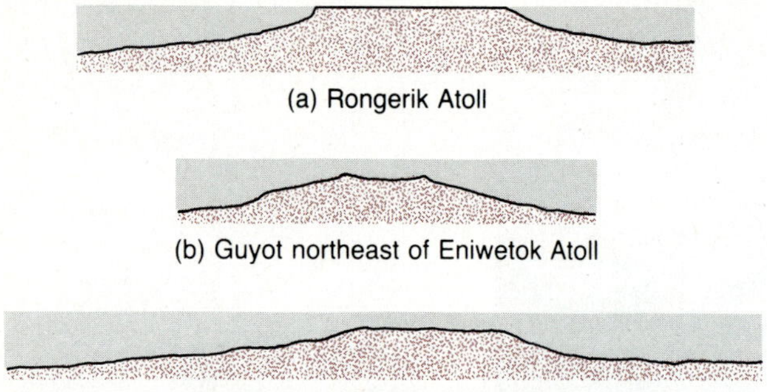

Figure 10.4
Profiles of Rongerik Island and two guyots. The similarity of form suggest that guyots are islands that have been submerged.

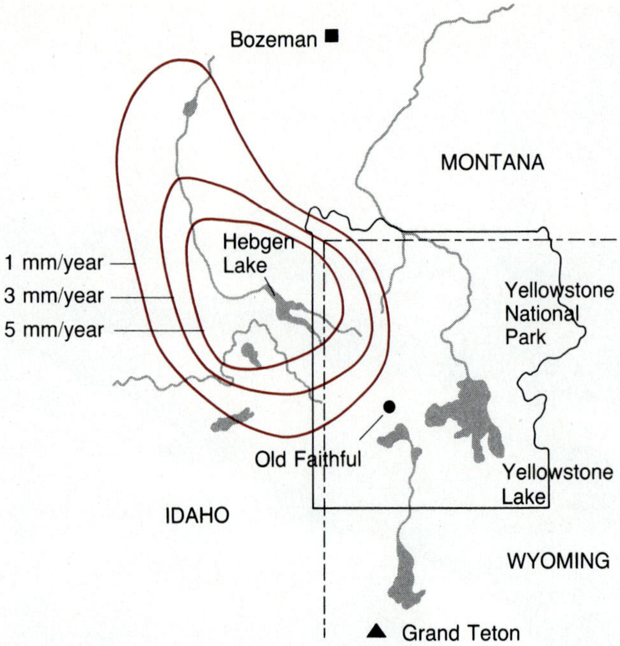

Figure 10.5
The region around Hebgen Lake west of Yellowstone National Park is rising. The lines connect points that are rising at about the same rate. Hebgen Lake is the site of one of the major earthquakes affecting the United States.

drowned as the result of a phenomenal rise of sea level, or the sea floor has subsided.

Results of drilling and seismic work on a few islands in the southwestern Pacific bear on this problem. On Eniwetok, drillings showed coral to a depth of 1400 meters; 1000 meters of coral were found at Funafuti. The ages of the coral can be identified from their fossils. The fossil corals of particular ages occur at different depths under the various islands, indicating that changes of sea level are not sufficient to account for the elevations. The crust must also have subsided.

Detection of Modern Movements by Leveling

Although evidence of large-scale movement of the crust had been recognized by Aristotle, one of the first efforts actually to measure the rates of such movements came in 1731. At that time marks were made at sea level on rock cliffs along the coast of Sweden. These marks are now more than two meters above sea level. Such a rise is extremely rapid. The region would be uplifted 1.5 kilometers in less than 150,000 years at this rate.

Modern techniques of leveling and surveying make possible such precise measurements that we can detect minute changes of elevation and position by carefully resurveying the same area. The simplest method involves setting up a series of survey markers, such as concrete posts or markers on solid rock or on concrete floors. Repeated careful surveying between these markers reveals any change in the relative elevation or position of the markers. For example, resurveying in the area at Yellowstone Park reveals uplifts in some areas of more than 5 mm per year. The most rapid rate of uplift is in the area affected by the earthquake described in Section 8.4. These movements may be related to rise of magma (Fig. 10.5). The relative movements between stations ultimately must be related to stations that are stable in order to determine which markers actually moved and how much they moved. If the network of stations is large enough so that some of the stations are located in areas that undergo no change, then the regional patterns of movement can be determined.

If the aim of the surveying is to determine horizontal movements as well as elevation changes, a method known as triangulation is used. A base line is carefully laid out between two points. The length of this line is measured precisely. A new triangulation station, visible from both ends of the base line, is located by taking a bearing (reading the compass direction) from each end of the base line to the new point, which is located where these two bearing lines cross. A network of stations, such as that shown in Fig. 10.6, is set up in this fashion. Any horizontal movements that occur within the network between resurveys show up as changes in the bearings between stations. A combination of leveling and triangulation makes it possible to determine both vertical and horizontal movements within the network.

Evidence of Crustal Deformation from the Rock Record

A striking feature of the view from the rim of the Grand Canyon (refer back to Fig. 5.24) is the great lateral extent of layer on layer of horizontal strata. Some of these layers can be traced for hundreds of kilometers through the region of the great plateau into which the Colorado River has cut its course. The path descending into the canyon leads the hiker down through more than a thousand meters of horizontal strata. But before the inner gorge is reached, this pattern changes—the beds below a certain level are tilted, as shown in Fig. 10.7. When traced upward, each of these angled beds is abruptly cut off, and the first horizontal layer lies across them. Careful inspection of the contact between the tilted older layers and the horizontal layers reveals that it is irregular, and at the boundary there are gravel beds now consolidated to conglomerate. Intervals like this one are found in many places, typically containing gravel beds; somtimes channels can be identified, and occasionally ancient soil is seen.

It seems clear that the break between the tilted strata and the flat strata was once the ground surface. The beds below must have been tilted earlier, before the flat beds were deposited. Erosion by streams leveled the upturned beds; producing a nearly flat surface. The area was later inundated as younger sediment was laid down almost horizontally. In some cases, the horizontal beds, as well as the deformed beds, are composed of marine sediment. This configuration, where an erosion surface separates deformed beds below from less deformed layers above, is known as **angular unconformity.** It clearly represents a significant amount of crustal deformation including in the case of the Grand Canyon: uplift, faulting, and

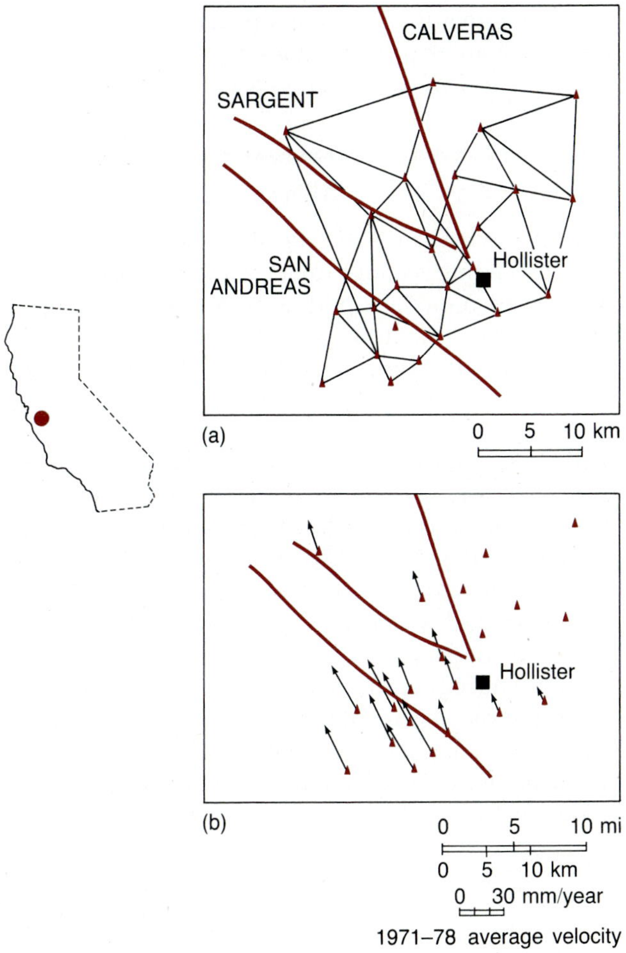

Figure 10.6
A network of surveying stations set up near Hollister, California, has been used for about 50 years to measure movements across several major faults. (a) The triangulation stations. (b) The arrows indicate the direction and relative movement of triangulation stations during the period from 1971–1978. Measurements reveal that the area west of the faults is moving northwest relative to those east of the faults.

John Shelton

Figure 10.7
Angular unconformity near the bottom of the Grand Canyon.

tilting of the older beds; erosion; subsidence as the younger horizontal beds were deposited; uplift of the region to its present elevation; and cutting of the canyon by the Colorado River.

Stratified rocks may show two other types of breaks also caused by erosion. One, called a **disconformity,** occurs when sedimentation is interrupted by uplift (without other deformation) and erosion followed by subsidence and continued sedimentation. An example of the other type of stratigraphic break occurs deep in the Grand Canyon. Here an erosion surface separates tilted rocks above from highly contorted and metamorphosed schist and intrusions below. A major break in the history of the region must be represented by this surface, a **nonconformity.** The early metamorphism, igneous activity, and deformation took place deep in the crust. Uplift and extensive erosion must have preceded the deposition of sediment on top of it.

All unconformities mark intervals of erosion preceded by regional uplift and usually followed by subsidence. The time of the uplift can be dated, at least approximately, as younger than the youngest rock below the unconformity and older than the oldest rock above it.

Streams flowing out of high mountains are almost invariably loaded with boulders and finer gravel. Even at great distances from the high elevations, the Platte, the Arkansas, the Rio Grande, and other streams that flow away from the Rocky Mountains contain heavy loads of sand and gravel. The quantity of debris being removed in any year is impressive, but consider what this implies if extended over millions of years. As the mountains erode, their volume is reduced; most of that rock is converted to sediment, and a large part of that sediment may be deposited in the low areas or in nearby seas on the flanks of the mountains. Sedimentary accumulations of this type lie in the Great Plains east of the Rockies.

Huge accumulations of sedimentary rocks are formed in this way, often surviving long after the mountains from which they were derived are eroded to a much lower level. The Catskill Mountains of New York are an excellent example. The Catskills are composed of sediments laid down to the west of high mountains that stood in New England about 300 million years ago. Streams carrying gravel flowed west depositing large fan-shaped bodies of clastic rock. Gravels in the east grade into finer sand deposits farther west, and these grade eventually into marine deposits containing marine fossils (Fig. 10.8). The bodies are referred to as **clastic wedges,** because the sediments, mainly fragmental rocks, thin away from the source areas producing a deposit that is wedge-shaped when viewed in cross section.

10.2 LATERAL MOVEMENTS IN THE CRUST

Most of the evidence of crustal instability discussed in the preceding sections can be explained as resulting from broad regional uplifting and downwarping of the crust that constitutes epeirogeny. This frequently occurs without strong distortion of the rock masses involved. Other lines of evidence point to strong deformation of crustal rock in lateral, horizontal, or nearly horizontal directions. Deformations of this type are best displayed in the orogenic belts—in the mountain systems of the world. There we find layers of rock distorted and bent to form what are called **folds,** and broken and displaced along narrow zones of movement, called **faults.** Both of these features occur in a great range of sizes. The largest faults, such as the San Andreas fault, described in Chapter 8, show horizontal displacements amounting to hundreds of kilometers. Faults and other features commonly formed by crustal deformation are described in the following sections of this chapter. Other types of evidence

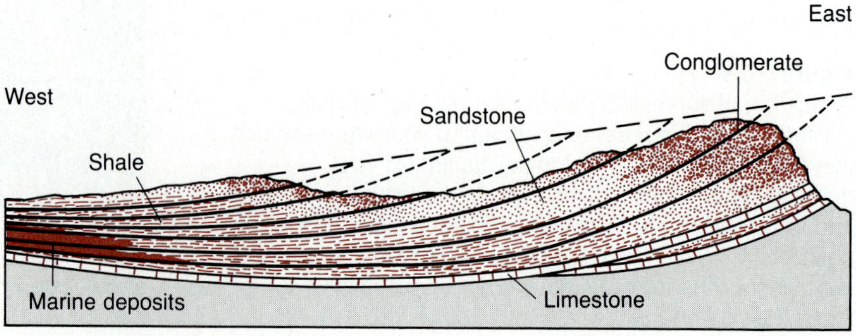

Figure 10.8
Cross section across the Catskill Mountains showing the thinning beds toward the west. Coarse conglomerate redbeds in the east and near the top of the mountains indicate the rise of high mountains in the east.

indicating large-scale lateral movements in the crust and even of entire continents will be presented in Chapter 12 in the discussion of continental drift.

Stress

Structural features such as faults and folds develop in response to directed pressure or stress, the force applied across an area that arises in the crust whenever the pressures are greater in some directions than they are in others. We saw some of the effects of directed pressure in discussing metamorphism (Chapter 7) and the causes of these imbalances will be explored in Chapter 12. For the moment, we will simply acknowledge that unbalanced stresses do exist and that they may be applied in several different ways, as illustrated in Fig. 10.9. If two opposed stresses act toward each other, the material subjected to this pressure is said to be compressed, and usually the material is shortened. This situation is analogous to what happens to a block compressed in a vise.

If the stresses are directed away from one another, the material is extended or stretched, and the stresses are said to represent tension. If pressures are applied so they are oppositely directed but not aligned, the material is sheared. All three of these methods of stress application cause deformation or distortion of the material. Such changes in shape are called **strains.** One other important reaction to the application of stress may occur—the stressed material may move.

Most of the strains that occur when rocks are deformed are permanent; so, we can study them long after the deforming stresses are released. The most common strains are such structural features as fractures (cracks or breaks in rock), faults, and folds.

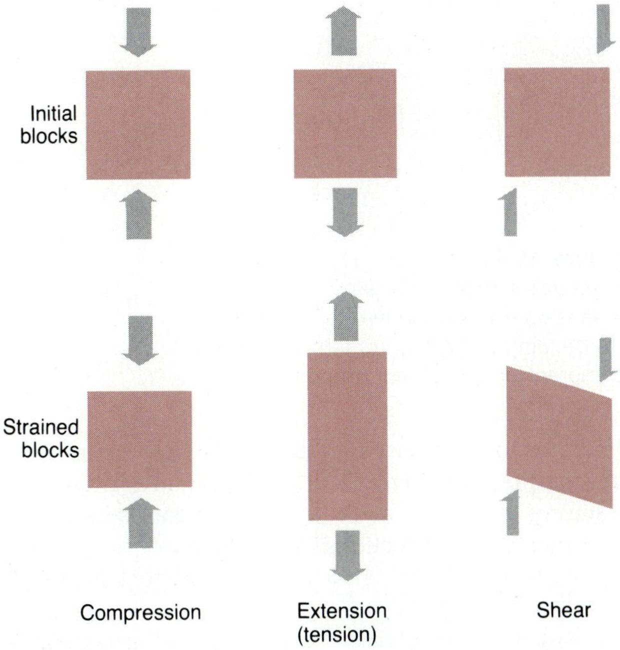

Figure 10.9
Stress and strain. Force is being applied as shown by the arrows to a uniform block of material. Compressive stress results in shortening at left. The material is extended by tensile stress in the middle. Shear stress results in distortion at right.

The Geometry of Rock Bodies

In order to communicate a relatively exact description of the configurations of deformed rocks, geologists have developed a nomenclature for the features commonly found in the crust. The most general of these terms, strike and dip, describe the orientation in space of any plane such as a sedimentary layer, a fracture, a fault, or other plane feature. The **strike** of a plane is the compass direction of a horizontal line on the plane (see Fig. 10.10). This direction is expressed in terms of the angle between the line and north (for example, N20°E). The **dip** of a plane is the angle of inclination of the plane, measured from the horizon to the plane and at right angles to the strike, as shown in Fig. 10.10. Linear features are described in a somewhat similar way. A line may be defined by the bearing of its horizontal projection, and the angle, called its **plunge,** between the line and its horizontal projection. Each of the main structural features of the crust has its own set of specific descriptive terms, as we shall see in the next section.

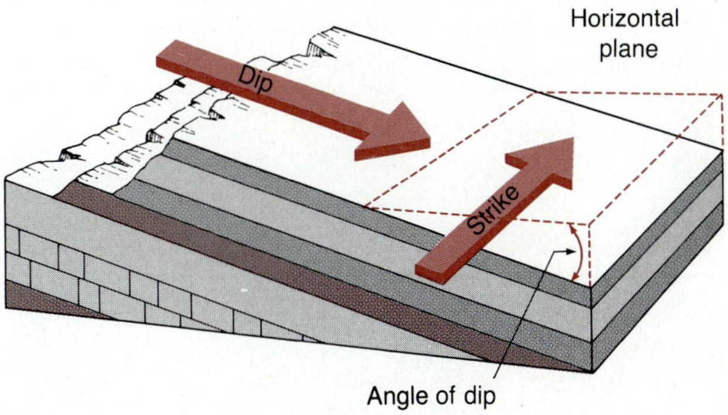

Figure 10.10
The strike and dip of a bed are indicated on this block diagram. The direction of dip is at right angles to the strike direction.

Fractures

Almost all consolidated rock and many unconsolidated sediments are criss-crossed by cracks called fractures (Fig. 10.11). They form when the stresses on the rock exceed its strength. Because most rock near the surface of the earth is brittle, movements in the crust easily crack it. Relatively slight warping of the crust is sufficient to form fractures. Some fractures are caused by the contraction that results from the drying out of sediment or from the cooling of magma. Others are clearly related to larger structural features such as folds and faults.

Directed stress causes two general classes of fractures: those that result from **extension** and those that result from **shearing.** Stretching or extension of rock may occur in a number of different ways, as shown in Fig. 10.12. For example, the upper surface of a layer that is bent to form a fold or a dome is stretched (Fig. 10.12*b*). Rock may also extend in one direction if it is compressed in another. This condition can be duplicated in a simple experiment with clay cakes, wax, or even with brittle materials. If a block of brittle material is compressed in a vise, the first fractures to form are generally parallel to the direction of compression (Fig. 10.13). These are extension fractures generated as the material expands in one of the unconfined directions. Most such extension fractures are open because they form as the rock is pulled apart.

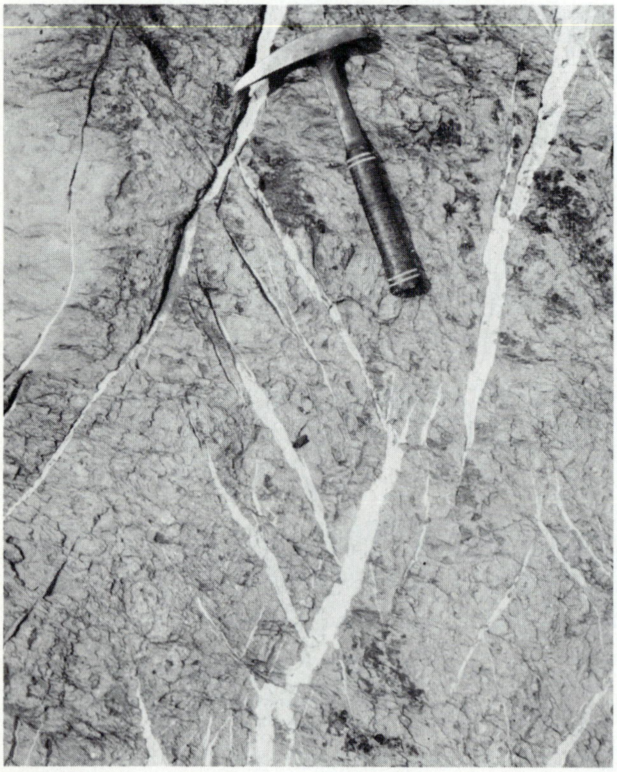

Figure 10.11
Fractures filled with quartz veins.

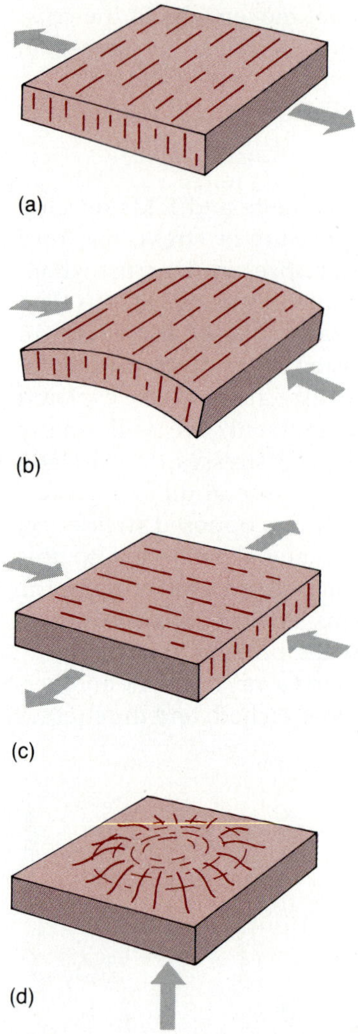

Figure 10.12
Fractures formed as a result of extension: (a) tension in one direction; (b) bending or folding; (c) tension and compression combined; (d) doming.

Shear fractures develop when a rock breaks as a result of shear displacement (Fig. 10.14). Shear fractures may also be formed by an experiment similar to the one described above, if the material compressed in the vise is not brittle. In such an experiment, shear fractures form, oriented between 30° and 40° to the axis of the vise. A shear displacement develops as

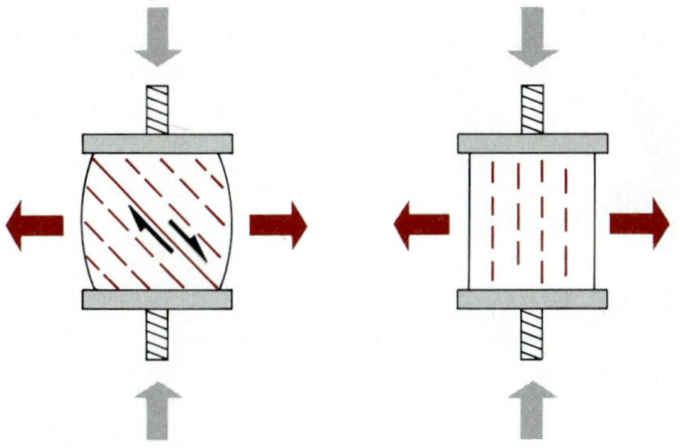

(a) (b)

Figure 10.13
(a) Shear fractures formed in material that is not brittle and (b) extension fractures formed in brittle material in response to compression in a vise.

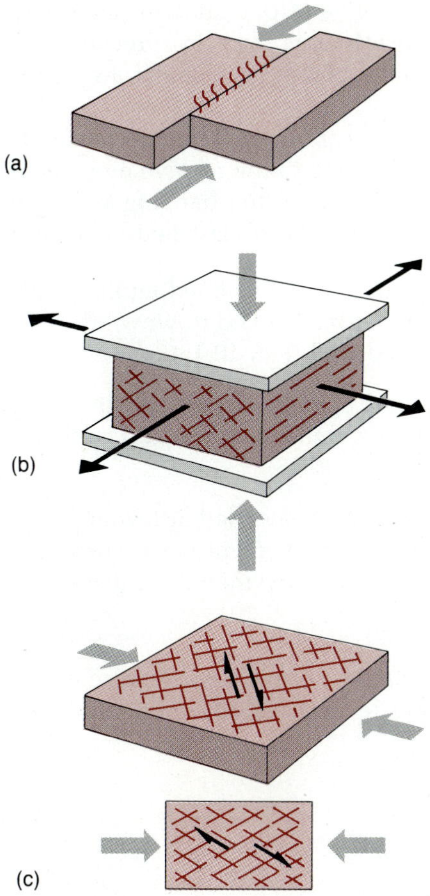

Figure 10.14
Fractures formed as a result of shear: (a) fractures formed along a line of shear; (b) fractures resulting from vertical compression; (c) fractures resulting from horizontal compression.

deformation continues. Often these fractures occur as a set of two shear fractures that intersect at angles of approximately 60°, as shown in Fig. 10.14.

Because rock loses cohesion across fractures, they weaken the rock. Because fractures usually occur as parallel sets, the rock mass may be weak in some directions but strong in others. In the design of dam or building foundations or in the construction of tunnels, the existence of fractures as well as their orientations and spacings are crucial factors. Weakness due to fracture often plays an important role in the failure of steep slopes in solid rock. Fractures are also important because they create almost the only void space or porosity in many igneous and metamorphic rocks. Water frequently occupies the space in fractures, and even oil and gas are produced from fractures in a few fields.

Faults

Description and nomenclature. Although we discussed some effects of movement along faults in Chapter 8, we did not consider at that time their variety, geometry, or classifications. Faults are zones of shearing across which displacement of the rock has occurred. Some faults are well-defined narrow zones, almost planes, along which movement has occurred. More often, however, the fault is a zone of some width within which rock is broken and crushed as a result of the shearing movement. Rock in the zone may be a coarse breccia of angular fragments in a ground-up matrix. Sometimes crushing is so complete that the breccia is reduced to a powder called **gouge.** If this is recrystallized, it forms a porcelainlike solid called mylonite. Because of the shearing movement, rocks in fault zones frequently exhibit scratch marks called **slickensides** caused by the scraping of rocks against one another.

The fault plane or zone divides the rocks of the crust into blocks, one on either side of the fault. Some faults are vertical, but more frequently the fault zone is inclined so that one block lies over the other. In this case the upper block is referred to as the **hanging wall block,** and the lower block is called the **footwall block** (Fig. 10.15). These are terms first coined by miners who mined metal deposits that were concentrated in fault zones. In the mine they walked on the lower block and the upper block formed the ceiling of the mine.

These terms, hanging wall and footwall, are used in the classification of faults. When it is possible to determine the direction and amount of movement

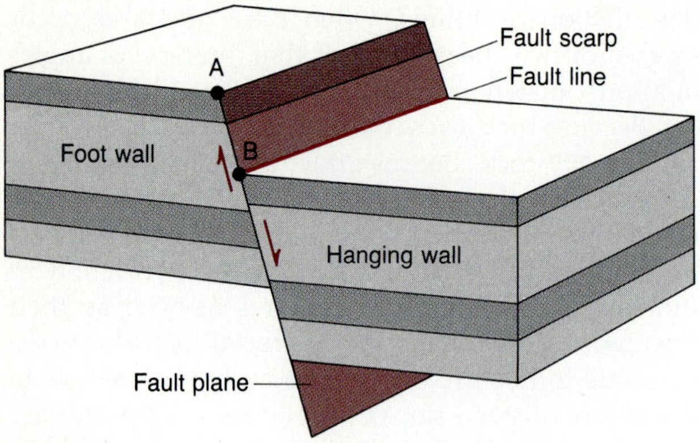

Figure 10.15
Block diagram of a steeply dipping normal fault. AB is the net slip of the fault.

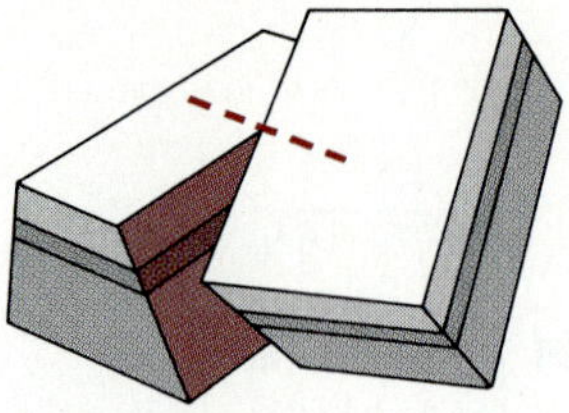

Figure 10.17
Block diagram showing a rotational movement between the blocks.

which has taken place on a fault, what is known as the **slip classification** may be used. The **net slip** or **displacement** on a fault is the straight-line distance, after movement, between two points that were originally adjacent on opposite sides of the fault, as shown in Fig. 10.15.

Linear features, such as a stream channel or the line of intersection of two dikes or a dike and a bed, that are cut by a fault usually provide the best guides to net slip. The orientation of slickensides in the fault plane may also indicate the direction of slip.

In the slip classification, faults are grouped according to the relationship of the direction of net slip to the strike or dip of the fault zone, as shown in Fig. 10.16. In a **strike-slip fault,** the direction of net slip is horizontal along the strike of the fault (Fig. 10.16*b*). If the net slip is in the direction of the dip, the fault is a **dip-slip fault.** If the slip is oblique across the fault plane, the fault is said to be **oblique slip.** These terms, strike-slip, dip-slip, and oblique-slip, can be applied if the blocks move in essentially straight lines relative to one another. Sometimes the fault blocks are rotated relative to one another, as in Fig. 10.17, so that displacements vary in direction and amount from place to place. Such faults are called **rotational faults.**

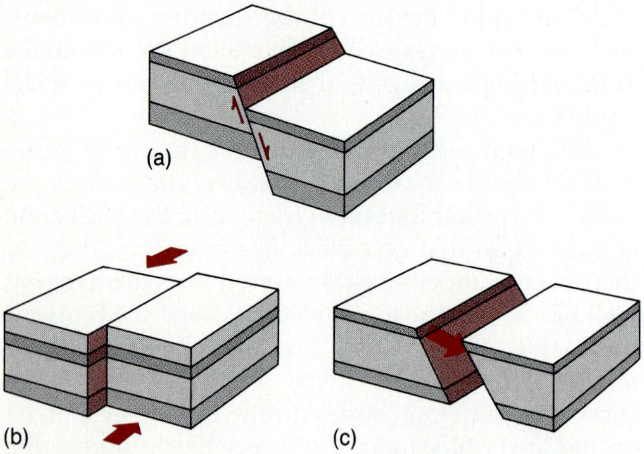

Figure 10.16
Block diagrams illustrating the movements involved in (a) dip-slip (also a normal fault); (b) strike-slip; and (c) oblique slip (also a normal fault).

When the net slip cannot be determined, faults may be classified according to the relative movement that has taken place between the hanging wall and foot wall. On this basis, faults are classified as follows:

Normal faults are steeply inclined faults, along which the hanging has moved downward relative to the footwall (Fig. 10.15).

Reverse faults are steeply inclined faults, along which the hanging wall has moved up relative to the footwall (Fig. 10.18).

Thrust faults are faults inclined at low angle, along which reverse movement has occurred. Many thrusts involve movements parallel to bedding.

Occurrence. High-angle normal and reverse faults are found both inside and outside orogenic belts.

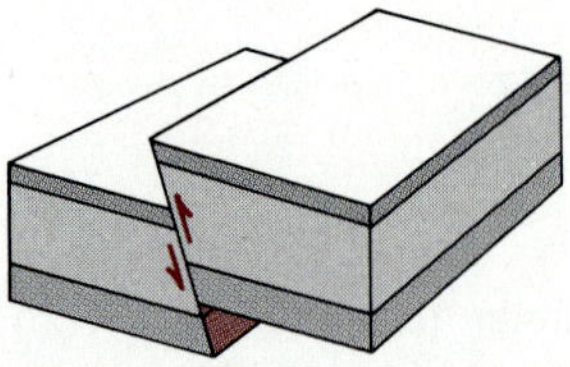

Figure 10.18
Block diagram of a reverse fault.

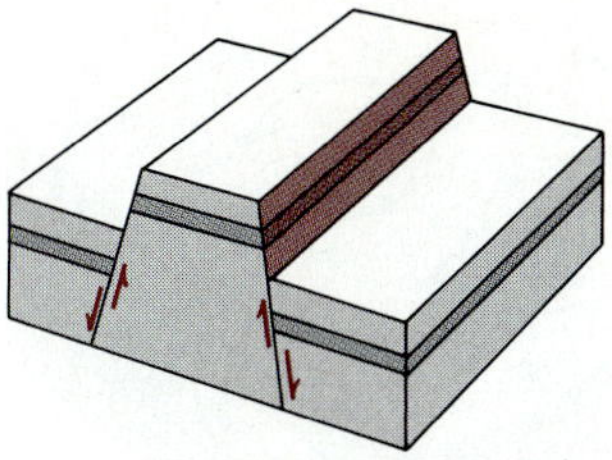

Figure 10.19
Block diagram of a horst.

Many of the high mountains of the western United States owe their elevations to uplifts on steeply dipping faults. The Sierra Nevada Range of California and the Tetons of Wyoming are both fault-block mountains. They are essentially large crustal blocks tilted up along one side along steeply dipping faults. Many other mountains in the basin and range province (parts of California, Nevada, Arizona, and Utah) are blocks bounded on two sides by steep normal faults. Uplifted blocks with this geometry are called **horst** (Fig. 10.19). Often these are separated by depressed areas, called **grabens,** which are also bounded by high-angle normal faults, as shown in Fig. 10.20. The typical graben is a straight-edged, gravelike depression bounded on each side by one high-angle fault or a series of step-like faults. The geometry of grabens, (Fig. 10.21), suggests that normal faults, grabens, and horsts are all caused locally by arching of layers or by actual stretching of the crust over a large area.

Not all grabens are of the grand scale described in the western states. Smaller grabens occur in sedimentary layers that are stretched over rising dome-shaped masses of salt in the gulf coast states of Texas, Louisiana, and Mississippi. Salt domes from one to ten kilometers in diameter are rising from depth. These salt plugs pierce the sediments as they move upward, but the strata above them are forced into domed shapes. At this stage normal faults and grabens are created.

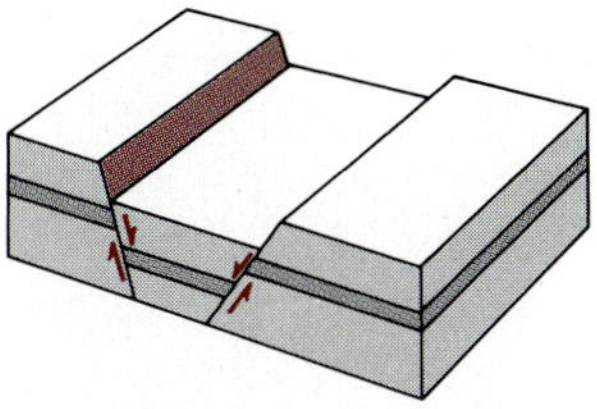

Figure 10.20
Block diagram of a graben.

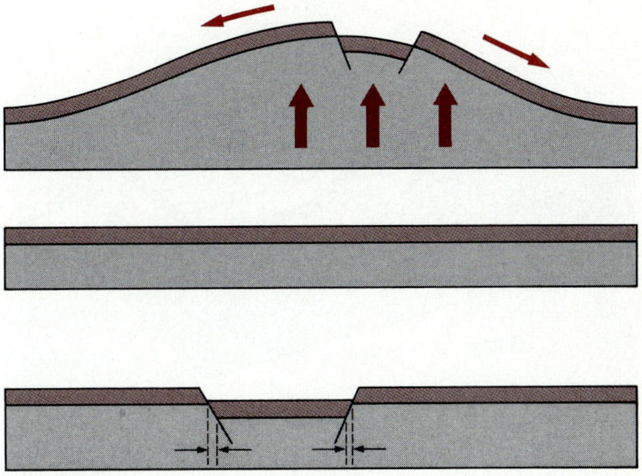

Figure 10.21
Schematic cross section showing an underformed layer, center; graben formed over a dome, top; graben formed as a result of horizontal extension, bottom.

Strike-slip faults—those that are nearly vertical and along which movement occurs parallel to their strike-slip—are the longest faults and have the greatest displacement. Along the San Andreas fault, for example, the displacement is believed to be more than three hundred kilometers. A long, narrow section of North America is being shifted north relative to the rest of the continent along this fault. Other comparable faults are found in several parts of the world. We will see later that some of these adjoin and separate major plates of the lithosphere.

Other strike-slip faults are much more limited, but most of them are nearly vertical, and, for this reason, they show up on maps and photographs as almost straight lines.

The last major group of faults we will consider are the thrusts (Fig. 10.22). Like others faults, thrusts occur on many scales. The largest, involving the movement of slabs of rock hundreds of kilometers long, are invariably in orogenic belts. The fault that forms the lower boundary of such slabs may be inclined, but often is nearly horizontal. When sedimentary rocks compose the slab, the fault often coincides with a bedding plane for great distances.

Because the slab of rock that is moved over thrust faults is usually thin compared with its breadth, the slab may be referred to as a **thrust sheet** or called by the French name **nappe.** When thrust sheets are exposed near the surface of the ground, they are

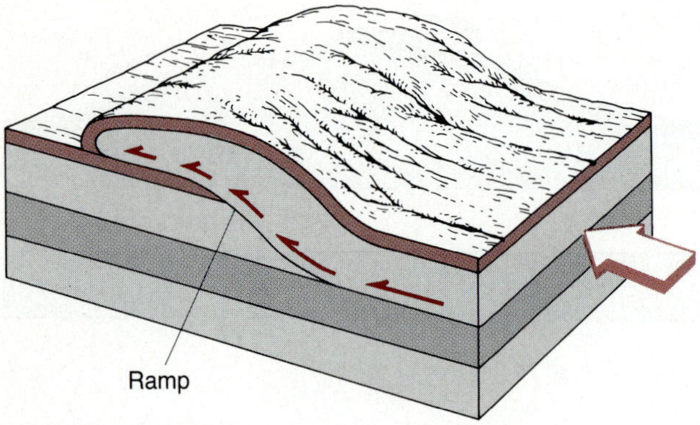

Figure 10.22
Many thrust faults like the one in this diagram are oriented parallel to bedding, except at a few places where the fault cuts across beds to another bedding plane.

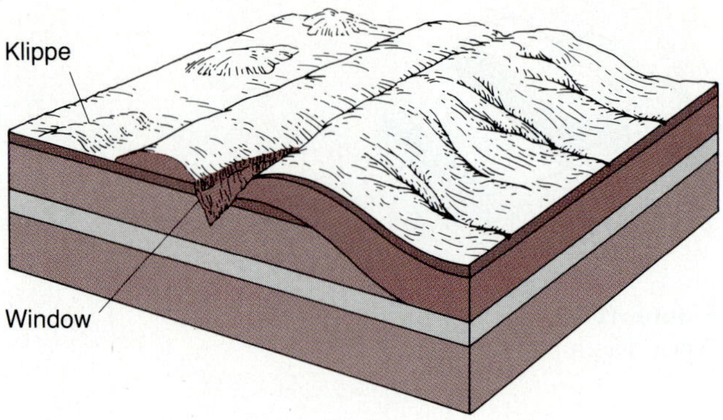

Figure 10.23
The sheet of rock carried above the thrust may be eroded, leaving isolated pieces of the thrust sheet, called klippes, and holes through the sheet, called windows.

eroded. This erosion may eventually cut some areas of the sheet off from the rest of the sheets, forming isolated pieces called **klippes.** The term **window** (or fenster) is applied where erosion cuts through a sheet, exposing rocks beneath the sheet in such a way that rocks beneath the thrust are surrounded by rocks above the fault (Fig. 10.23). Nappes are characteristic features of the Alps and of most other major mountain systems of the world. Many of the nappes have become detached from their original sites and have moved great distances.

The origin of low-angle thrusts of the sort described here is often difficult to determine. The traditional view is that they originate as a result of compression that shortens the crust and causes the overthrusting. But some thrusts appear to result from the sliding of sedimentary layers downslope along inclined bedding planes (Fig. 10.24). This sliding

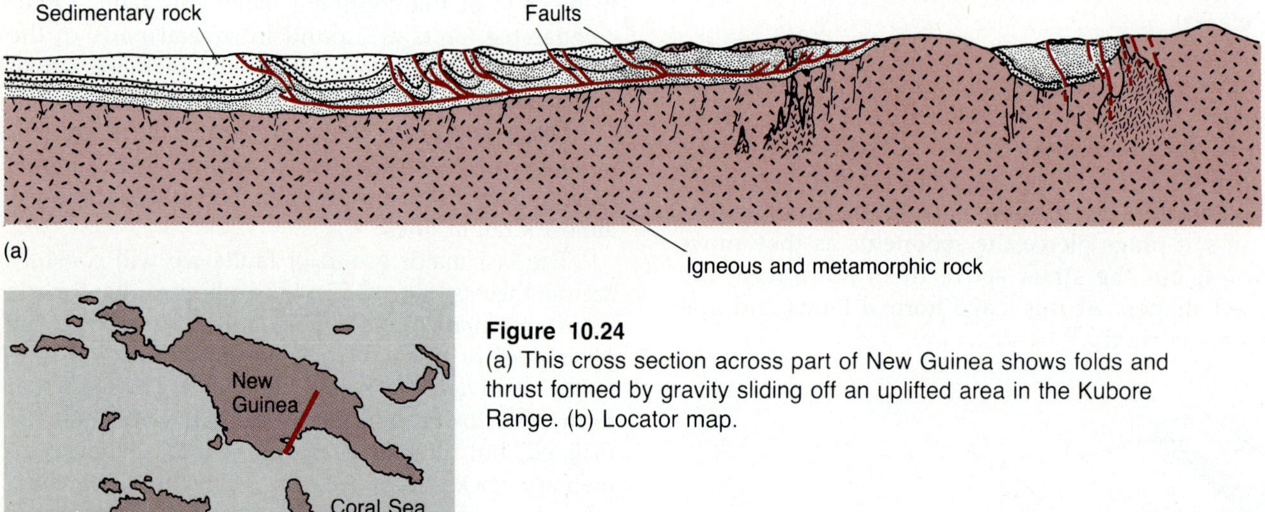

(a)

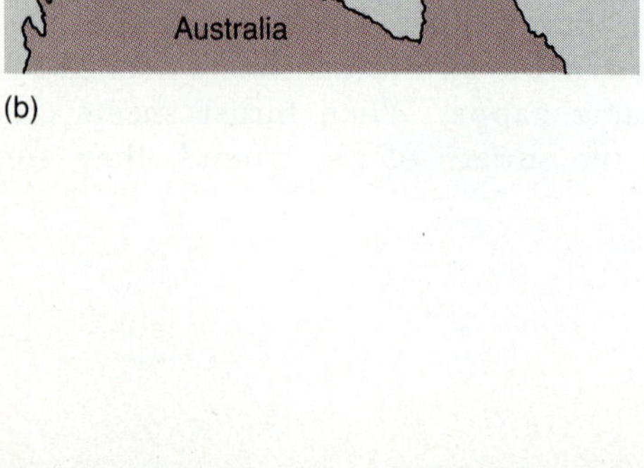

(b)

Figure 10.24
(a) This cross section across part of New Guinea shows folds and thrust formed by gravity sliding off an uplifted area in the Kubore Range. (b) Locator map.

under the force of gravity is made possible only by reducing frictional drag along the plane in which sliding takes place. Lubrication by water can reduce the friction—especially where a porous and permeable layer below the zone of movement is saturated with water under pressure. This water pressure buoys up the overlying rock and makes it much more likely to slide down low slopes.

Whatever the ultimate causes of the movement exhibited by thrust faults, it often transports large masses of rock more or less horizontally. The crust behaves as though it is being compressed horizontally, usually resulting in the bending or folding of rocks as well as thrusting.

Folds

Description and nomenclature. Folds are bends in rock that form in response to directional forces. Unless the rock contains some marker, the undeformed shape of which is known, it is not possible to recognize that the rock has been folded. Fortunately, many sedimentary and metamorphic rocks originally contain layers of different composition or texture. Most strata are originally formed horizontally, and this provides the reference plane we use to recognize most folds.

Usually, these reference planes are bent into forms that can be described in terms of relatively simple geometrical shapes. Figure 10.25 summarizes the terminology we use to describe parts of a fold. If a fold has a reasonably regular geometrical form, its shape is further defined by describing the relationship between the two limbs. If the limbs, the sides of a fold, are almost perfect mirror images of one another, the fold is said to be **symmetrical.** If the limbs are different from one another in shape or angle of dip, the fold is said to be **assymetrical.** The plane through the high or low point of a fold that divides the fold most nearly in half is called its **axial plane.** The orientations of the limbs and axial plane are used to indicate the amount and direction of tilt of the fold. In this regard, folds vary from being upright (axial plane vertical) to recumbent (overturned and lying down with the axial plane horizontal), as indicated in Fig. 10.26. A complete progression, like the one illustrated here, is rarely found intact. Usually, thrust

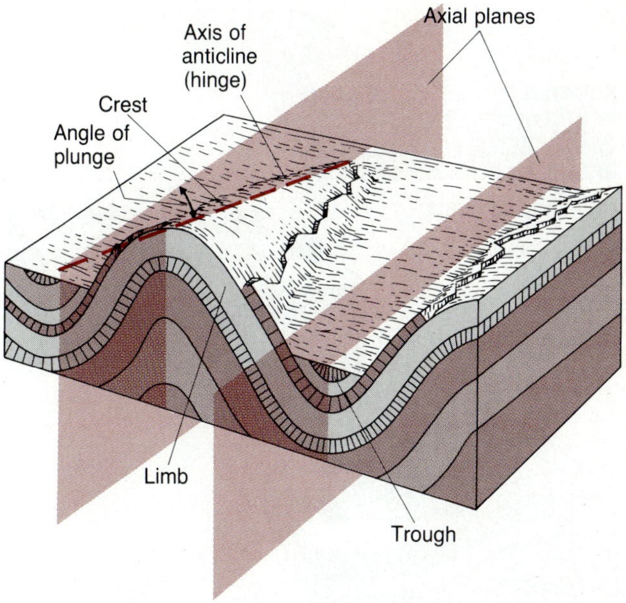

Figure 10.25
This upward fold or anticline shows the chief features of a fold. The *limbs* are the flanks or sides of a fold; the *crest* is the highest part of a fold, and, unseen here, the *trough,* the lowest part of a fold. An imaginary surface that approximately bisects the fold into symmetrical halves is called its *axial plane.* The imaginary line formed by the intersection of the axial plane with a bedding surface is the *axis* or *hinge* of the fold. The hinge is located where the curvature of the bed is greatest.

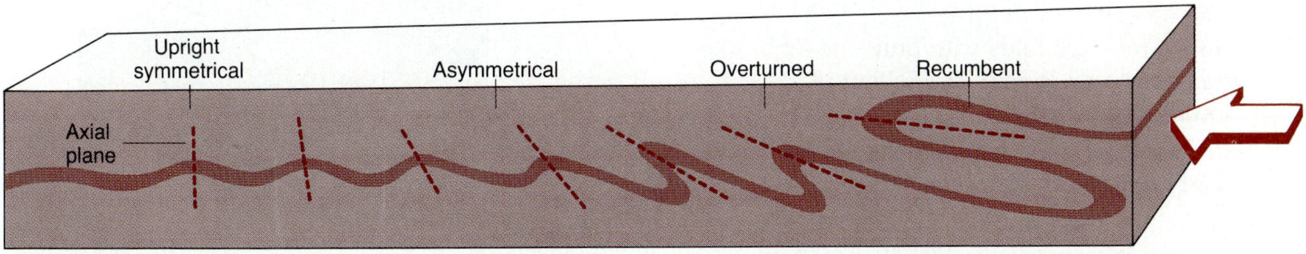

Figure 10.26
Schematic diagram showing variations in the inclination and symmetry of folds.

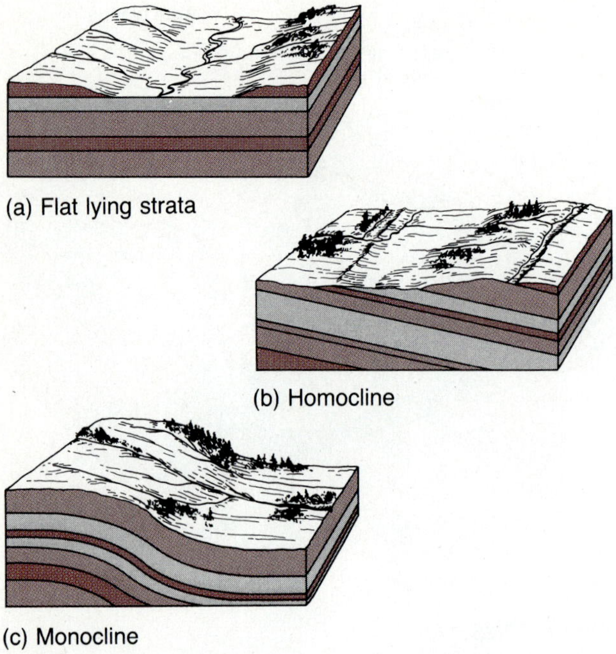

Figure 10.27
Block diagrams showing (a) flatlying strata; (b) a homocline; (c) a monocline. Note that although both homoclines and monoclines have only one limb, the slope of the monocline is not uniform, distinguishing it from a homocline.

faults occur within the series, especially after folds begin to be strongly overturned or recumbent. In a progression like the one in Fig. 10.26, the amount of deformation increases from left to right.

Fold classification. Two systems of fold classification are used here. One is based entirely on the geometry of the fold, the other on the behavior of the rock during folding. On the basis of geometry, folds are classified as follows:

Homoclines are tilted beds with only one, fairly uniform slope (Fig. 10.27). One limb of a large fold may be described as homoclinal.

Monoclines are folds with only one limb, like the one shown in Fig. 10.27(*c*) but of nonuniform slope.

Structural terraces are folds that produce a flat, nearly horizontal surface.

Anticlines and **synclines** are upfolds and downfolds respectively (Fig. 10.28). In an anticline, the limbs dip away from the fold axis. Older beds are located in the core of an anticline; thus, if the crest of an anticline is eroded away, older rocks are exposed. In a syncline, the limbs dip toward the axis, and younger beds are exposed along the trough of a syncline.

A **dome** is a circular or elliptical uplift. Older beds lie in the center, as shown in Fig. 10.29.

A **basin** is the corresponding circular or elliptical downwarp, with younger beds in the center.

The second system of fold classification is based on mechanical considerations of what happened to the rock during the folding process. On this basis folds may be classed as flexural or passive. **Flexural folds** are those in which slip occurs between the layers as folding takes place. The layering exercises an active control on the deformation, and the resulting folds represent a true bending of layers. Flexural slip is analogous to the movement involved between cards in a flexed deck of cards (Figs. 10.30 and 7.5). Where flow or slip crosses the layer boundaries, the fold is classed as **passive.** The layering exercises little or no control on the deformation. Layer boundaries serve merely as markers.

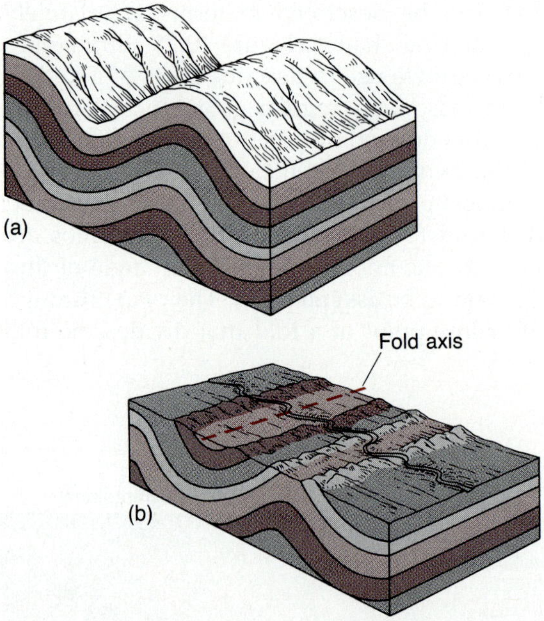

Figure 10.28
Block diagrams illustrating (a) an anticline and (b) the appearance of folded strata on the ground surface after erosion has leveled the region.

U.S.A.F

(a)

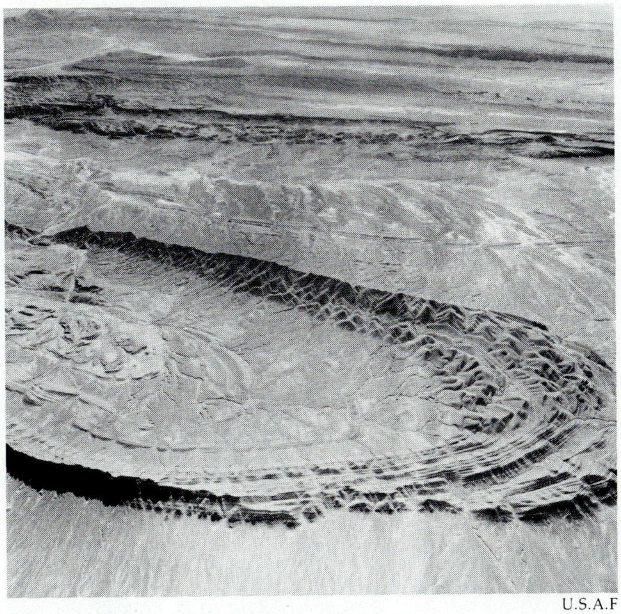

U.S.A.F

(b)

Figure 10.29
(a) A structural dome. The top of this dome has been planned off by long, continued erosion. The oldest rocks are exposed at the center, and succesively younger units are exposed away from the center. The distance across this dome is several kilometers. (b) A structural basin. Ridges are exposed around the margins of the structure. The youngest units are exposed near the center of the structure.

From these descriptions, it should be clear that passive folds are likely to form in rocks that are uniform, for example, in thick shale deposits or in rocks being metamorphosed at high pressure and temperature. Flexural folds usually result when layers of greatly different physical properties are folded. Folding of interbedded sandstone and shale will fold by flexural slip.

Occurrence. Folds occur in many different geological settings, both within and outside of orogenic belts. Those outside orogenic belts are usually broad warps, domes, or basins, often so large they can only be identified by mapping whole regions. Many examples of these features are found in the central United States. The axis of a broad archlike uplift passes through Lexington, Kentucky, and Cincinnatti, Ohio; most of Michigan and southern Illinois are situated in large structural basins. The Ozark Mountains are at the center of a broad dome. These features appear to be caused by regional uplift and downwarping rather than by strong horizontally direct stress.

When a layer of rock is subjected to strong horizontal pressures it is compressed and shortened, as shown in Figure 10.31. This effect leads geologists to conclude that great amounts of shortening take place across most orogenic belts. For example, geologists estimate that across the Appalachians the crust has been shortened by more than 100 km.

Folds formed as the result of directed stress are characteristic of the orogenic belts. For this reason, orogenic belts are often referred to as **folded mountain belts.** Most orogenic belts consist of a central core region in which metamorphic rock predominates. The flanks of this core are composed of sedimentary rocks deformed into long sinuous belts of folds such as the one in the Atlas Mountains of Morocco (Fig. 10.32), and in the Appalachians. The fold belt in the Appalachians varies from a few tens to a couple of hundred kilometers in width and can be traced continuously from Alabama to New York. The belt consists of a succession of anticlines and synclines, some of which are more than a hundred kilometers long. Much smaller folds occur in some rock types and in the metamorphic core of the orogenic belt. Folds of this type are illustrated and discussed in Chapter 7.

The complex fold patterns seen in metamorphic rocks (Fig. 7.19) demonstrate that rocks are not always as brittle as they seem. In order to understand rock behavior, we must consider the physical conditions under which they are deformed.

Figure 10.30
Slip between layers accompanied the folding that caused this larger fold exposed in the Canadian Rockies. (Photo by E. L. Fitzgerald.)

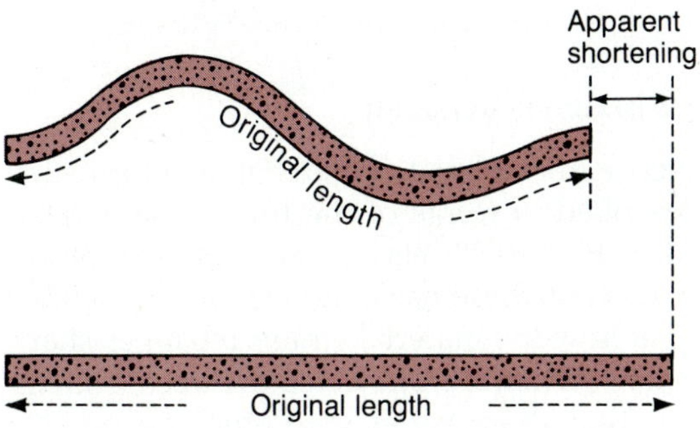

Figure 10.31
Schematic cross sections showing unfolded layer, below, and a fold and local shortening, above.

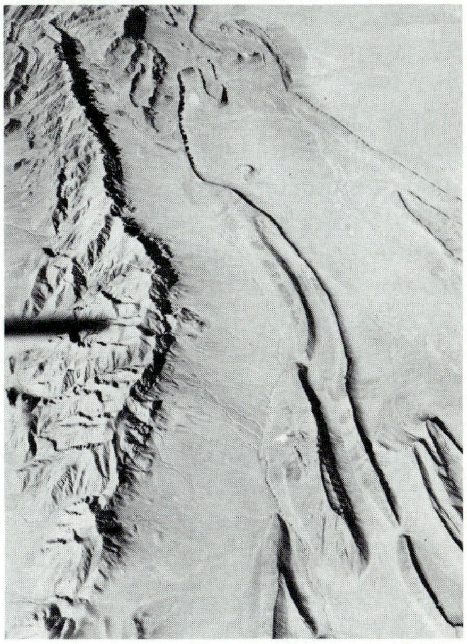
U.S.A.F.

(a)

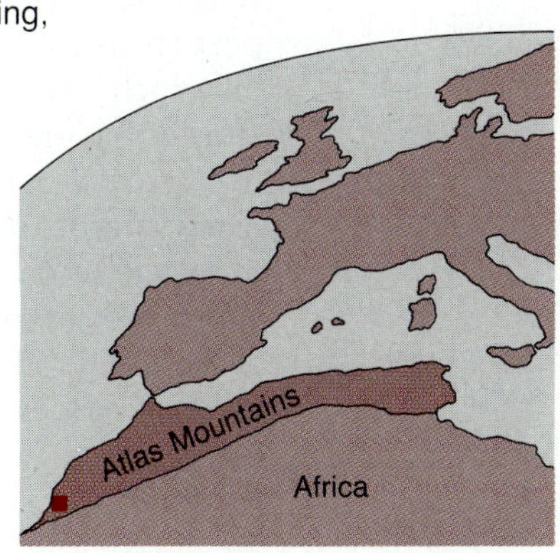

(b)

Figure 10.32
(a) A belt of folded mountains (Atlas Mountains) extends across the northwestern corner of Africa (see locator map, b). Deformation and uplift of the crust of the earth is one of the main factors in maintaining parts of the crust at high elevations. Already these mountains have been deeply eroded. Sediments removed from the mountains lie on the flanks of the ridges.

10.3 MECHANICS OF ROCK DEFORMATION

Conditions That Affect Rock Behavior

Rocks like granite, gneiss, sandstone, quartzite, and limestone appear to be hard, durable materials capable of sustaining great loads. They can, however, be broken when high stresses are applied rapidly; and when they do break it is usually in the manner of brittle materials. The development of fractures in slightly deformed rocks is compatible with these observations. Fractures result from the failure of rock when stresses exceed the rock strength. Yet, beds of all types of sedimentary and metamorphic rocks can be found folded within mountain belts, and often the folded rocks show no evidence of fracturing—they behave as plastic materials. Thus, rocks that are brittle under some conditions are plastic under others, and it is not difficult to identify some of the reasons for this difference. Some are related to the nature of the rock itself; others are caused by the physical conditions under which the rock is deformed and the length of time involved.

The behavior of a rock reflects in a large degree the behavior of its component minerals. Minerals vary greatly in strength and in the way they yield to pressure. Some deform under much less directed stresses than others. A number of minerals, like calcite, yield by microscopic slips along directions of weak bonding in the crystal structure; others, like quartz, do not yield in this way. The behavior of clay is quite different from sand because clay minerals can be squeezed and flattened under conditions that either would not deform or simply would crack quartz grains.

The confining pressure produced by the weight of the overlying rock increases with depth. Based on the occurrence of deformed rocks, it appears rocks are more likely to undergo deformation without fracturing when they are deformed at great depth than when near the surface. Experiments confirm this observation.

Temperature increases with depth, and this, too, promotes greater yielding before fracture. Many metamorphic rocks, formed when temperatures reach high levels, are highly deformed. The presence of solutions in pore spaces also promotes movement of materials. Solutions facilitate movement of ions and lubricate mechanical movements, such as grain rotation and slip on fractures.

Time. The effect of time in rock deformation is the hardest factor to evaluate. In classical ruins, as well as in old graveyards, we find monuments made of marble slabs supported only at the ends. Often the center of a slab that has been standing for hundreds of years has sagged enough to be easily measured. When these marble slabs were originally cut, they were smooth and flat, so the sags must have formed in the time since the slabs were mounted. Here, then, is evidence that solid rock can deform under a confining pressure of one atmosphere with nothing more than the force of gravity acting as a stress, if the time involved is great enough. This suggests that the contortions in the earth's crust may result from small directed stresses applied over long periods of time.

Deformation is time-dependent even among materials with which we are familiar, such as asphalt, beeswax, and Silly Putty. These materials are brittle under sudden deformation, but they flow like liquids under low stresses if given sufficient time. The history of mountain-building suggests that most rock deformation takes place over long periods of time. Millions of years may be involved in the formation of some of the structural features now exposed in mountain belts—plenty of time for even the slowest processes of deformation to produce impressive effects.

Models of Material Behavior

The difference in response of material to deformation can be described in terms of the three basic models illustrated in Fig. 10.33: elastic, viscous, and plastic behavior. Each of these basic models can be defined according to how the material strains (changes shape or dimensions) when it is subjected to directed stress, and to how that strain varies with time.

Elastic model. The ideal elastic material is one in which the volume or length changes in direct proportion to the amount of stress applied; that is, one in which the strain is directly proportional to the stress. The graph of the relationship is a straight line up to some particular stress (Fig. 10.33*a*), called the **elastic limit.** Once the elastic limit has been passed, the material no longer behaves as a purely elastic substance. A spring is an excellent mechanical model of an elastic material. Strain remains constant for a given stress level over any amount of time, provided the stress is not great enough to stretch the spring permanently.

Viscous model. The ideal viscous material is one in which strain takes place continuously for any amount of applied stress. Most liquids exhibit this behavior. No increase in stress is needed to cause strain to

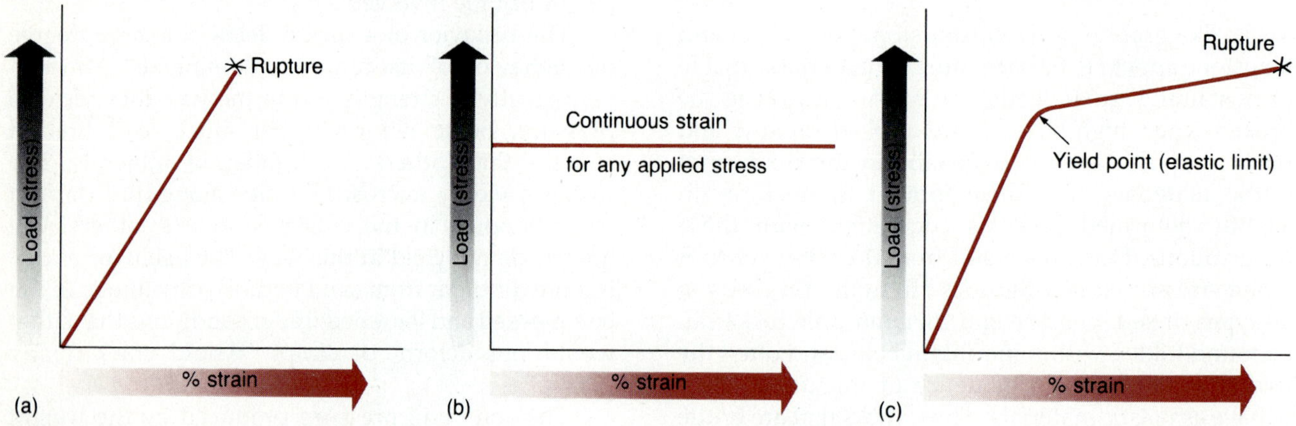

Figure 10.33
Graphs showing the relationship between the amount of stress (load applied) and the resulting strain for (a) ideal elastic; (b) viscous; and (c) plastic materials.

increase; strain continually increases with time at any applied stress.

Plastic model. The ideal plastic material is one in which behavior is initially elastic. As the amount of stress applied is increased, the amount of strain increases proportionally. This continues up to a certain stress level, called the elastic limit or **yield point,** beyond which the material strains continuously with little or no additional stress. The behavior may be likened to that of a block of clay. When small weights are added to the clay, strain occurs, but deformation stops after each addition until the weight added produces a stress greater than the elastic limit. Once this weight exceeds the yield point, the clay starts to flow and deforms continuously.

Materials of all three kinds are well known. In order to see how closely rock behavior fits these models, experiments have been conducted in which various rocks are deformed under different kinds of conditions.

Experimental Study of Rock Deformation

Modern equipment makes it possible to subject specimens to high temperature and confining pressure at the same time that a directed stress is applied. Most experiments in the investigation of rock behavior are conducted on small specimens, typically about 2.5 cm in diameter and several centimeters long. The specimen is inserted in a heavy steel cylinder in which a hole slightly larger than the specimen has been bored. Confining pressure is built up within an oil that surrounds the specimen, and the specimen is then deformed by means of directed stresses applied by a piston for a short span of time. Both compression and extension experiments can be performed in this way.

The ideal experimental procedure involves controlling all variables that may influence the rock's behavior, holding all but one constant. The effects of varying that one are then measured. Most experiments with rocks are conducted to evaluate the effects on rock deformation of composition, temperature, confining pressure, amount of applied stress, and time. In typical expriments, the confining pressure and temperature are held constant and the specimen is compressed. The amount of shortening that results from each increase in the amount of directed pressure is measured. The results of these experiments are plotted on graphs like those in Figs. 10.34, 10.35, and 10.36. Each dot corresponds to a particular observation. Only after many observations is it possible to draw the continuous curves. Because these curves show the amount of strain (shortening) produced by any amount of directed stress, they are called stress–strain curves.

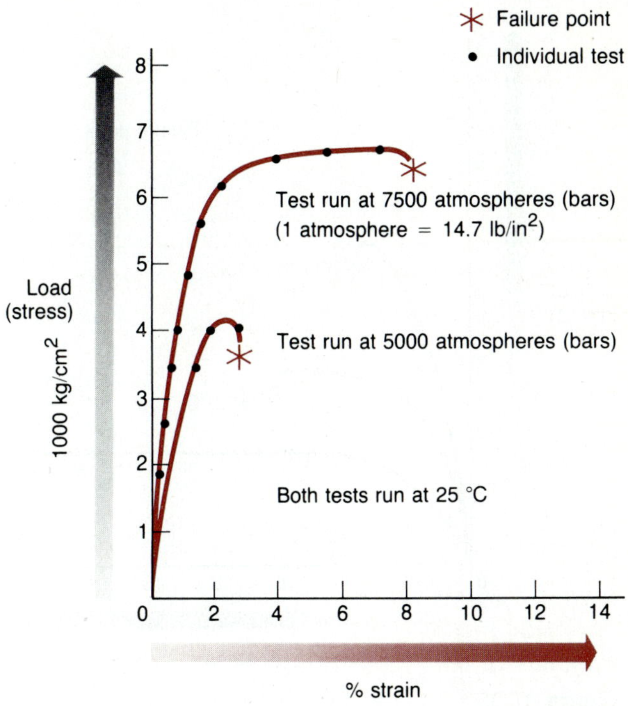

Figure 10.34
Stress–strain curves obtained by deforming a limestone. The temperature was held constant at 25 °C for both tests and the confining pressure was varied—5000 bars for one test and 7500 bars for the other. The specimen broke at the point marked ×.

By conducting stress–strain tests at a number of different confining pressures or temperatures, we can compare the curves and evaluate the effects of these variables on the rock's behavior. Stress–strain curves obtained by deforming a limestone at different temperatures and confining pressures are shown in Figs. 10.34 and 10.35. Under most conditions the limestone behaves initially as an elastic material, but as the stresses are increased, its behavior changes, and it becomes more like that of the plastic model. This kind of behavior, typical of most rocks, is best described as a combination of the ideal models. Rocks are, therefore, often termed **elasticoviscous** materials, showing elastic behavior under some conditions and plastic or fluid behavior under others.

The results of two series of tests designed to evaluate the effects of confining pressure are recorded in Fig. 10.34. During the first series of tests, the confining pressure was held at 5000 atmospheres and

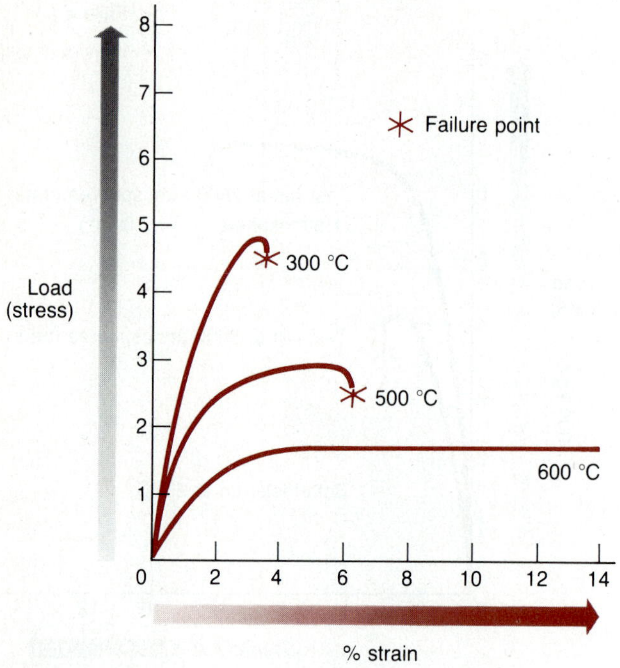

Figure 10.35
The effects of temperature are illustrated by their stress–strain curves. In each case limestone was deformed at a constant pressure. The specimen failed at the point marked ×.

temperature was held constant at 25 °C. Confining pressure was raised to 7500 atmospheres in the second series of experiments (run on another specimen of the same rock). Notice that in both series of tests, the amount of strain is almost directly proportional to the amount of stress up to a certain level. This is indicated by the straight line on the graph. At higher pressures (where the linear relationship ceases to exist), the specimen shows large amounts of strain for small increases of stress. Finally, the specimen breaks. The higher confining pressure caused the limestone to sustain much greater strain before failure, and a higher stress was needed to cause a given amount of strain at higher confining pressures.

The effects of temperature are evaluated for the same limestone in Fig. 10.35. At high temperatures, much greater amounts of strain are possible before failure of the specimen, and less stress is needed to produce a large strain. This relationship holds for most rock types.

Because rock deformation is thought to take place very slowly, experiments designed to evaluate the effects of time on rock behavior are especially interesting. In these experiments environmental factors, such as temperature and confining pressure, are held constant. Directed stress is applied and also held constant. Then strain is measured over a relatively long period of time. Experiments may be continued for weeks or even months, but such times are still very short compared with the time involved in rock deformation in the crust. These experiments indicate that most rocks do not yield or yield only very slowly when they are subjected to low confining pressures and temperatures. Under such conditions they are indeed "solid as a rock." But if these same rocks are deeply buried and subjected to the high confining pressure and high temperature, they will yield slowly and continuously even under low applied stresses, a type of deformation called **creep.** This type of deformation is probably responsible for much of the complex folding we see in metamorphic and in some sedimentary rocks.

Knowledge of the behavior of the rocks under varying temperature and confining pressures is crucial to understanding the mechanisms of crustal movements and the formation of mountains. The background information in this chapter on structures produced when crustal rocks are subjected to directed stress and on the conditions under which rock deformation occurs will be useful throughout the following chapters. In describing the continents and the ocean basins, we shall see how and where these features occur in the major subdivisions of the crust.

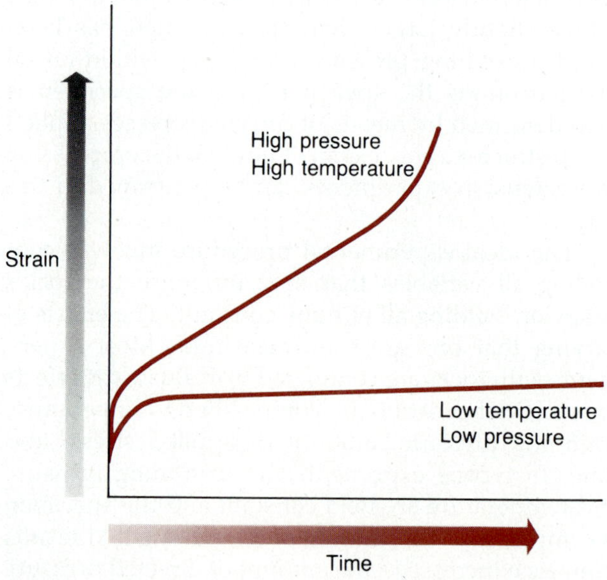

Figure 10.36
The effects of confining pressure and temperature on the variation of strain with time when rocks are deformed.

SUMMARY

The rock record of the earth's crust shows large-scale deformations of two types: epeirogenic movement, a broad lifting or subsidence that leaves the strata relatively undeformed; and orogenic movement, the movement of crust that forms mountains with intense folding and faulting of underlying rock.

The existence of such features as marine sedimentary rock and wave-cut terraces well above sea level show clearly that parts of the crust have been uplifted. Parallel evidence shows that other parts of the crust have subsided relative to sea level. Although some changes in elevation above sea level can be attributed to changes in the sea itself, others are much too large to have come about in this way. Today vertical movement is measured by the technique of periodically resurveying markers mounted securely in the crust. Evidence of past deformation comes from observation of rock layers. The unconformities within a series of sedimentary layers points to a period of uplift and erosion followed by subsidence and renewed deposition of sediment. The details of the rock below the unconformity spell out more of the region's history.

Stress within the crust causes horizontal movement, resulting in folds and faults. Faults are classfied according to their direction of net slip as strike-slip, dip-slip, or oblique-slip. On the basis of the relative movement of the blocks they are normal, reverse, or thrust faults. In combinations, faults give rise to uplifted areas called horsts and dropped ones, grabens, both of which are features of the mountains of the western United States.

Folds represent a different response of the crust to compression. They may be broad and gradual upward or downward warps of the crust or the more intensely compressed and deformed features typified by the fold belt of the Applachians.

Just what happens to rock under the intense pressure that accompanies folding and faulting depends on a variety of factors: the mineral composition of the rock, the confining pressure, the temperature, and the time over which the stress acts. The different possible types of behavior are summarized in the elastic, viscous, and plastic models. Laboratory experiments on rock samples help determine which model best describes the behavior of a rock to a given set of conditions.

KEY TERMS

angular unconformity
anticline
assymetrical
axial plane
basin
clastic wedge
creep
dip
dip-slip fault
disconformity
displacement
dome
elastic limit
elasticoviscous
epeirogenic
extension
fault
flexural fold
fold
folded mountain belts
footwall block
gouge
graben
guyots
hanging wall block
homocline
horst
klippe
monocline
nappe
net slip
nonconformity
normal faults
oblique slip
orogeny
passive fold
plunge
reverse fault
rotational faults
shearing
slickensides
slip classification
strain
strike-slip fault
structural terrace
symmetrical
syncline
thrust fault
thrust sheet
window
yield point

STUDY QUESTIONS

1. Under what conditions do rocks yield like plastic materials?
2. What evidence might you look for to indicate crustal movements in the area where you live or go to school?
3. What relationship exists between the way stress is applied to rocks and the types of fractures that form when the rock is brittle?
4. Under what conditions do normal faults usually occur?
5. Explain how you could detect slow uplift if a very large region is affected.
6. In what ways would you expect the fault zone of a normal fault to differ from the fault zone of a thrust fault?
7. In what way does a passive fold differ from a flexural fold?
8. What types of folds would you expect to find in rocks metamorphosed to a high grade at the time of folding?

SUGGESTED READINGS

Dennis, John G., *Structural Geology.* New York: The Ronald Press Company, 1972.

Spencer, Edgar W., *Introduction to the Structure of the Earth.* New York: McGraw-Hill Company, 1977.

STRUCTURE OF THE CONTINENTS AND OCEAN BASINS

Chapter 11

Having described many of the small-scale structural features of the earth's crust in the preceding chapter, we can now examine the large-scale geological settings in which these features occur. As we saw in Chapter 3, the earth's crust consists of two major crustal elements, continents and ocean basins. These may be subdivided into a number of smaller elements on the basis of their topography, the composition of the rocks that compose them, and their internal structure. Thus, continents consist of mountain systems (Fig. 11.1) and continental shields. The ocean basins are subdivided into abyssal plains and oceanic ridges. And the transition zones between continents and ocean basins are of three types. These margins around most of the Atlantic and others similar to them are designated Atlantic-type continental margins. A second type is characterized by trenches and arcuate chains of volcanic islands, the island arcs. And the third type, like the margin of western North and South America, consists of young mountain systems now being formed.

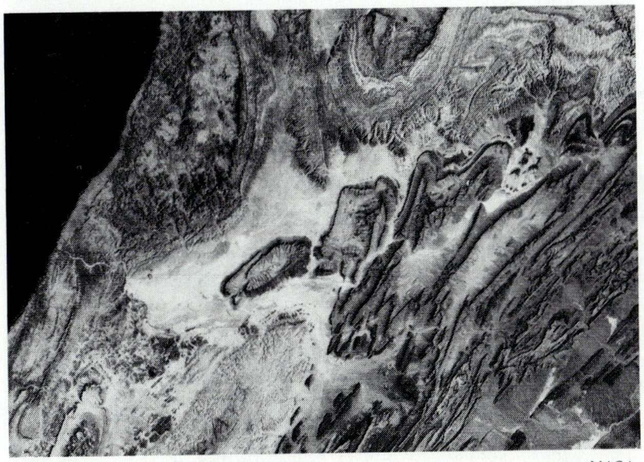

NASA

Figure 11.1
This Landstat image shows part of the folded mountain belt that stretches across the northwestern edge of the African continent. An oblique photograph of part of this mountain belt is shown in Fig. 10.32. Mountains like this are important parts of the continents.

With the background acquired through the study of earth materials, geophysical methods, and structural features, we can now reexamine these major subdivisions of the crust. This will pave the way for better understanding the evolution of thought about how these major features of the earth form and how they change through time. In this way, the material of this chapter will prepare us for a discussion of crustal dynamics in the next two chapters.

11.1 MAJOR FEATURES OF THE CONTINENTAL CRUST

Geologists had been impressed by the great difference in the average elevation of continents (800 m) and depth of the ocean basins (4 km) long before the seismic methods devised by Mohorovičić and others proved that continents and oceans are also drastically different in their deeper structure and composition. Continental crust averages 40 km in thickness, and even greater thicknesses are found under the high mountain systems. In contrast, the oceanic crust averages only 10 km thick, including the water. Both continents and oceans bear a veneer of sediment and sedimentary rock that covers most of their surfaces.

This veneer tends to be much thicker on some parts of the continents, especially in the mountain systems, than it is on the shields and platforms or in the oceans. Beneath this sedimentary cover, continents are composed of igneous and metamorphic complexes. Vast volumes of these crystalline rocks are granitic in composition, many of Precambrian age. In contrast, the floor of the ocean basins is largely basaltic lava flows. These are thin compared with the granitic rocks of the continents, and none is more than a couple of hundred million years old—young by comparison with the granitic rocks of the continental crust, many of which are more than two billion years old.

Shields

Every continent is comprised of one or more shield areas (Fig. 11.2). These are now and have long been the most stable portions of the earth's crust. During most of the 600 million years since the Precambrian era, they have been remarkably free of volcanic and tectonic activity. Even earthquakes are relatively rare in the shields. Because of this stability and the long periods during which they have been subjected to erosion, the shields are generally regions of low relief.

Despite their present stability, the rocks of the shields bear the imprint of an earlier history of geological activity. Some parts of shields are probably residual masses of the crust that originally formed as the earth cooled. If so, they were greatly modified during Precambrian time by volcanic and tectonic activity. Ancient, strongly folded, and faulted metasedimentary rocks, with structural features like those found in younger mountain systems, underlie large parts of most shields (Fig. 11.3). These have been

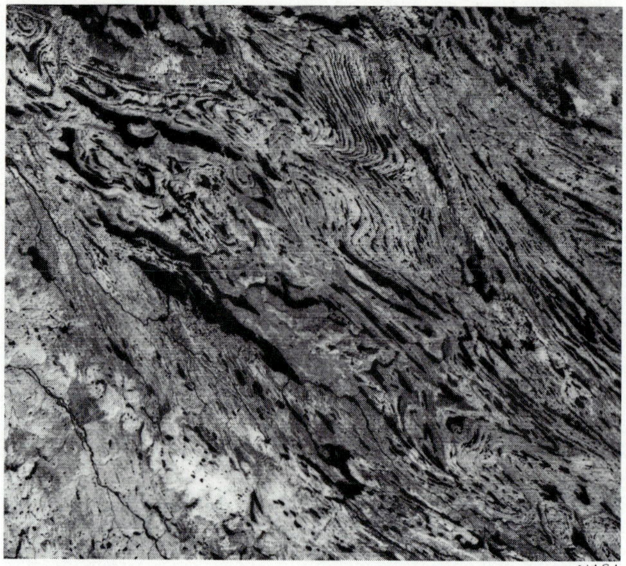

NASA

Figure 11.3
This area of Labrador in northeastern Canada is underlain by ancient metamorphic rocks that were strongly deformed more than a billion years ago. Although the area has little relief, the folds are still clearly visible in this satellite image. The area shown is about 100 km across.

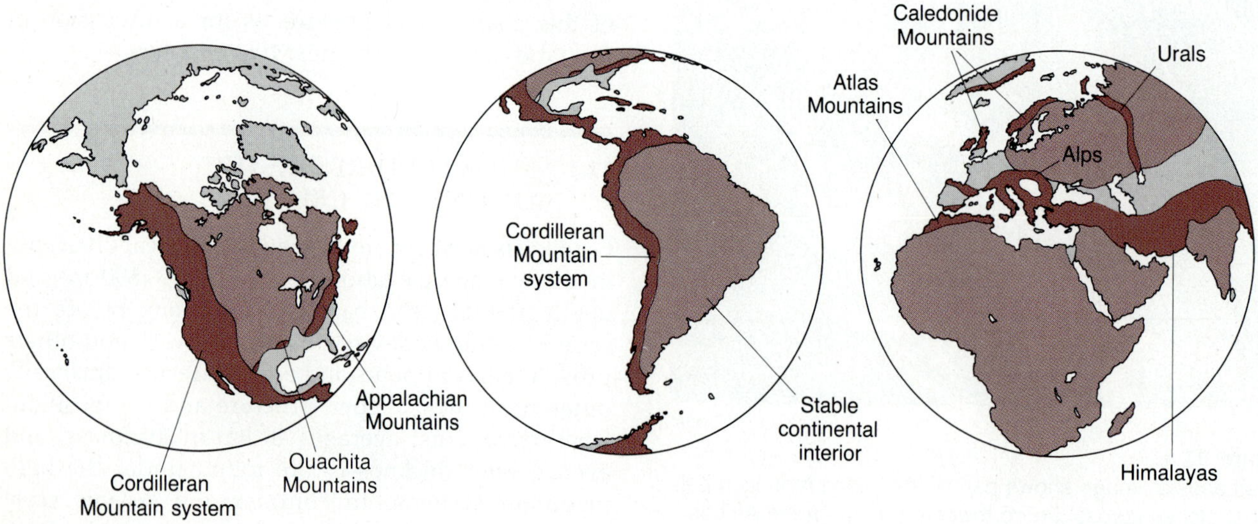

Figure 11.2
Continents consist mainly of relatively young mountain systems and older stable shields and platform areas shown here.

invaded by huge quantities of granitic magmas, and, in places, vast areas were covered in Precambrian time by the outpouring of basaltic lavas. Before the end of the Precambrian era, the shields had become consolidated and stable components of the crust, destined to remain largely passive over the millions of years that followed.

Although the shields seem to have been immune, since Precambrian time, to the type of folding that has affected the young mountain belts, some shield areas have been broken apart by extension that produced systems of grabens. The deep valleys of East Africa are excellent examples of this splitting of a shield. The edges of the shields have also become involved in mountain-building in places. Huge blocks of the edge of the shield of North America, known as the Canadian shield, are uplifted in the Rocky Mountains. Because of this involvement in deformation, they are no longer part of the shield, but belong instead to the mountain systems.

Mountain Systems

Most of us are accustomed to thinking of mountains as more or less isolated peaks or as places where the land stands much higher than its surroundings, but a mountain system is both larger and more complex geologically than individual mountains or even mountain ranges. Individual mountains may result from the buildup of lavas and ash, as has happened in Hawaii, where Mauna Loa rises nearly 10,000 meters from the sea floor of the Pacific basin. Others, such as the Sierra Nevadas, described earlier in Section 6.4 and shown in Fig. 11.4, may be thought of as large blocks of the crust which have been uplifted and tilted along faults creating **fault-block mountains.** Still other mountains are essentially erosional remnants of uplifted areas. A **mountain system,** however, may consist of all these and other types of individual mountains and ranges. Mountain systems, or **orogenic belts** (orogens), as they are called by geologists, are the result of strong deformation of the crust during uplift, usually folding and faulting, and sometimes the formation of plutons and volcanoes. These deformations affect thick sections through the crust and leave permanent scars that can be recognized long after the uplifted mountains are eroded away. It is much easier to study the younger and still-high mountains where rocks are exposed and deformation is continuing.

Most of the modern high mountains of the world lie within one of two major mountain systems: the Cordilleran system, which extends along the western margin of North and South America; and the Alpine system, which extends from Africa through southern Europe and across southern Asia (Fig. 11.2). These are the two highest and youngest orogenic belts on earth; both have many peaks over 5000 meters high, and both have been formed in their present configuration during the Cenozoic era (the last 70 million years). Both of these systems include many major mountain ranges. The Andes, the Rocky Mountains, the Sierra Nevadas, and the Coast ranges are all part of the Cordilleran system. The Alps, the Pyrenees, the Hindu Kush, and the Himalayas all belong to the same system, and the structure of each of these component ranges is in itself highly complex.

An older mountain system, the Appalachians, now eroded and reduced in elevation, lies along the eastern margin of the North American continent.

U.S.G.S

Figure 11.4
The Sierra Mountains, seen here from the east, are bounded by a high-angle fault along which the mountains have been uplifted like a huge block tilted up on one side.

Although orogeny there has long since ceased, many of the structures formed more than 300 million years ago are now clearly exposed, allowing us to compare this older mountain belt with its younger couterparts in other parts of the world.

The long, relatively narrow zones of strongly deformed continental crust in orogenic belts generally lie along or close to continental margins. The modern exceptions to this, such as the Himalayas, seem always to mark zones that were continental margins before the mountains formed. Proof is found in the rocks that make up these mountains. Without exception, the rocks in mountain belts include large volumes of marine sedimentary rocks deposited along continental margins. The thickness of these deposits is sometimes tremendous. Nearly 10,000 meters of marine sedimentary rocks lie in the Appalachian Mountains, and similar thicknesses occur locally in many other mountain belts.

These piles of sedimentary rocks are deformed into long systems of folds. Often these folds are closely associated with large thrust faults, and, typically, the folded and thrust-faulted sedimentary rocks lie on the flanks of great uplifts in which igneous and metamorphic rocks are exposed. In most mountain systems, the areas of greatest uplift are composed of Precambrian plutonic and metamorphic rocks, the younger rocks having been stripped off by erosion.

Each of the shields and mountain belts in the world is distinctive. Thus, we are afforded a rich variety of structural as well as topographic forms to study. In the following section, we will briefly consider the large-scale structural features of the North American continent as representative examples of continental structural features.

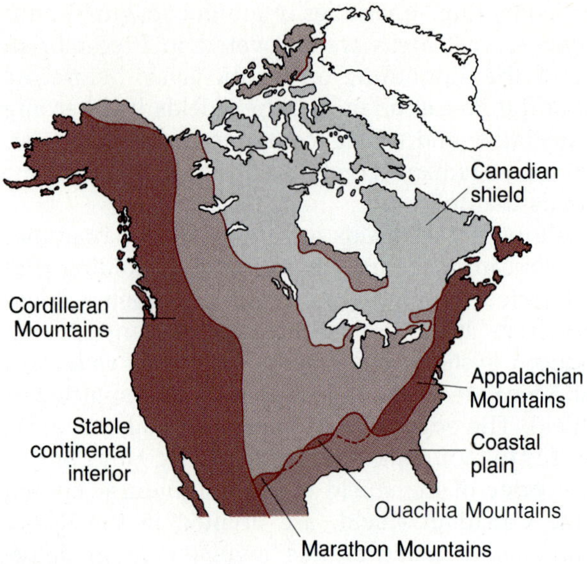

Figure 11.5
The major structural divisions of North America.

The Structural Framework of North America

The North American continent is composed of a large central shield area, the Canadian shield, bordered on its southern and western edges by a broad, stable platform (Fig. 11.5). This stable central core of the continent is bounded by mountain systems. On the east, it is flanked by the Appalachian Mountains. Remnants of this mountain system occur along the southern edge of the platform in the Ouachita and Marathon Mountains. A much younger system of mountains, collectively the Cordilleran Mountains, lies to the west of the central platform.

Deformation is still active in the western part of the Cordilleran Mountains, along the western coast of North America. This has recently found expression in the earthquakes and volcanic activity along the coast. In contrast, the eastern margin of North America has been stable for a long time—nearly 200 million years. As a consequence of this long stable condition, the oceanward side of the Appalachians has been eroded to a low level and young undeformed sediments now lie undisturbed along most of the Atlantic and Gulf coasts in a flat coastal plain.

The stable continental interior. The interior of the continent, comprised of the Canadian shield and adjacent platform, is a vast region of subdued topography. The shield is composed of the eroded remains of ancient continental crust and mountain systems that were consolidated and welded together in Precambrian time. This basement underlies the platform as well. The entire region has been quiet, having experienced very little volcanic activity, seismic activity, or mountain-building since the Precambrian era.

Much of this region was, however, covered by seas during early and middle Paleozoic time; some of the sediment deposited at that time remains as a veneer across the platform area. For reasons that are still not clear, several large structural depressions, or basins, formed in the platform area and became sites of thick deposits of sediment. An especially prominent one is located under the state of Michigan, where several thousand meters of sediment accumulated. Others are shown in Fig. 11.6. During the middle part of the Paleozoic era, the water-covered area of

Figure 11.6
The stable shield and platform of North America is bounded by orogenic belts on both sides. Igneous and metamorphic rocks with complex structure are exposed in the shield. The platform area includes a number of long basins and domes. Shading indicates the thickest accumulations of sedimentary rocks on the platform.

the Michigan basin was encircled by reefs, and the areas behind these reefs contain salt deposited as the water evaporated. These salt deposits are important economically, and some of the reefs now contain oil and gas.

Most of the vast region from the Rocky Mountain front in Colorado to the Appalachian Mountains are covered by flat layers of Paleozoic sedimentary rock. These ancient marine deposits are exposed at the ground surface east of the Mississippi River. In the Great Plains these rocks of marine origin are covered by younger sediments of the Rocky Mountains, spread across the plains by streams flowing eastward.

The Appalachian Mountain system. The name Appalachian Mountains, or more correctly Appalachian Orogen, is applied not only to the mountains in eastern North America, but to all of the region underlain by rocks that were deformed during a long period of crustal unrest in the Paleozoic era. The Appalachian Orogen extends all the way from Newfoundland, where the rocks and structural features can be studied in the wave-cut cliffs along the coast, to Georgia and Alabama (Fig. 11.7). At its southern end, the strongly deformed rocks of the orogen pass under and are covered by Cenozoic sediments of the coastal plain, forming one of the most pronounced angular unconformities on the North American continent (Fig. 11.8). Oddly, neither the structural features of the orogen, the folds and thrust faults, nor the igneous and metamorphic complexes show signs of dying out at either end of the system. At the northern end, they pass out to sea and, if continental drift reconstructions are correct, they are continuous with similar orogenic belts in Greenland, Ireland, Scotland, and Norway. At the southern end of the Appalachians, the deformed rocks are covered by a thick sedimentary veneer. It is difficult to trace them under the coastal plain, but rocks of similar age, also deformed in the Paleozoic era, and in somewhat the same style, emerge from beneath the Gulf coastal plain to form the Ouachita Mountains in Oklahoma and Arkansas and the Marathon Mountains in western Texas.

Although the details of structure, rock type, and history vary greatly from place to place in the Appalachians, several distinctive parts of the mountain system can be identified and traced for great distances. The most prominent subdivisions are: a long, narrow zone of folded and thrust-faulted sedimentary rocks called the **foreland fold belt,** (including the Taconic, Allegheny, Ouachita, Marathon, and other mountains); another long, narrow zone characterized by Precambrian basement ridges (including the Blue Ridge, Reading, Green, and Long Mountains); and a region largely underlain by Paleozoic igneous and metamorphic complexes (the Piedmont and New England highlands, and central and eastern Newfoundland).

The **foreland** of a mountain system is the region on the flanks of the more highly uplifted central core of the system; it is a transition zone between the undeformed region outside the mountains (for ex-

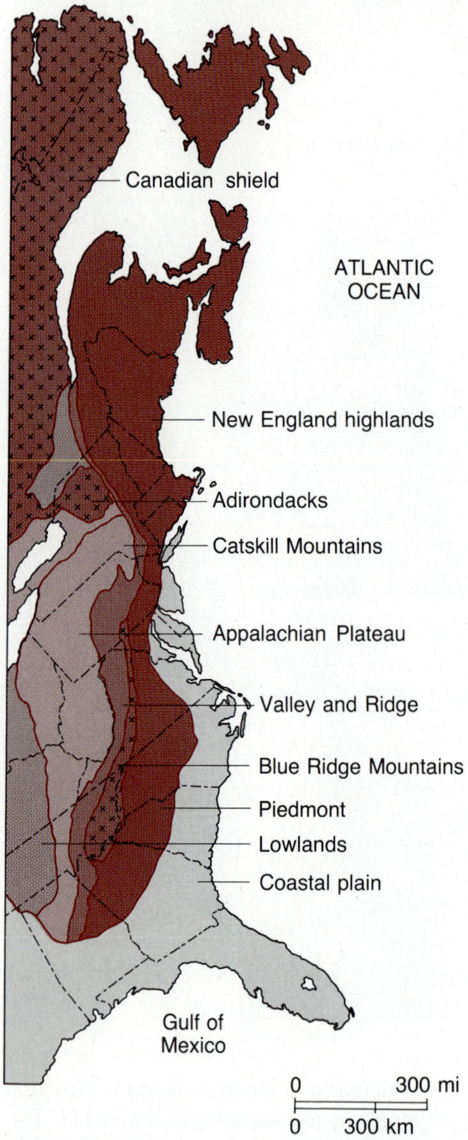

Figure 11.7
This map of eastern North America shows the geographic distribution of the major subdivisions of the southern and central parts of the Appalachian orogen.

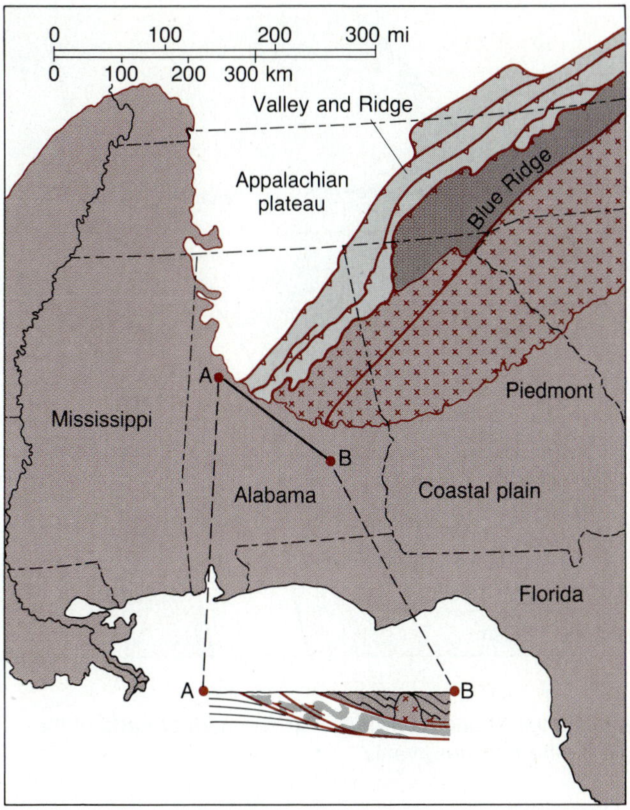

Figure 11.8
At their southern end, the folds, thrusts, and metamorphic complexes of the Appalachians are covered by younger and undeformed sediments of the Coastal Plain. This prominent angular unconformity is also present, as shown in Fig. 11.5, in the Ouachita and Marathon Mountains.

ample, the continental interior) and the zone where the most intensely deformed and highly metamorphosed rocks are exposed. The foreland of the Appalachians is a broad belt in the southern part of the Appalachians and a narrow belt in the north. The rocks in it are sedimentary deposits originally laid down over a 300-million-year period in seas that covered the region in the first half of the Paleozoic era. These sedimentary rocks are thick (about 10,000 meters) in the eastern part of the foreland, but they are thinner toward the continental interior.

Although they were originally laid down as nearly horizontal layers, the rocks of the foreland are now deformed into long narrow folds. These folds resemble the folds you might cause in a table cloth by shoving it across the table. This is suggested in the satellite image of part of the folded Appalachians shown in Fig. 11.9. The folds are broad, open, and widely spaced at the outer edge of the foreland. Toward the interior of the foreland, they are tight, assymetric, overturned, and broken by faults. Most of these faults are thrusts that originate along bedding planes. But as movement proceded, the fault cut up through the sediments, generally carrying older rocks up onto younger units, as shown in Fig. 11.10.

At the inner edge of the foreland in the southern Appalachian Mountains, Precambrian igneous and metamorphic rocks, some from deep in the continental crust, have been thrust up onto the folded rocks of the foreland. These uplifted, crystalline rocks form the Blue Ridge Mountains. It is as though a gigantic block of continental crust has been forced upward and to the northwest. Because this folding and thrusting affected late Paleozoic rocks in the southern Appalachians, it appears that mountain-building in this area took place near the end of the Paleozoic era. But much older mountain-building (Devonian and Ordovician) had taken place both in the northern Appalachians and in the region east of the Blue Ridge.

Evidence for these older episodes of crustal deformation is seen best in New England and in eastern Canada. Devonian mountain-building in the region of New England, eastern New York, and Pennsylvania created high mountains. These uplifted areas became the source of sediments now found in the Catskill Mountains (Fig. 11.7). Ironically, the sediments deposited on the flanks of this uplift remain, but the original mountains have been eroded to a much lower level. The Taconic Mountain of New York and Vermont were formed even earlier. The rocks in the Taconic Mountains constitute a huge mass of material that appears to have originated as a submarine slide formed as the sedimentary cover slipped and slid toward the west off an uplifted region in eastern New England in the Ordovician period, about 450 million years ago. During this same period, large masses of gabbro, seprentine, and pillow basalts, thought to be slabs of oceanic crust, were forced up onto the continental margin in a number of places. Unusually good exposures occur in Newfoundland. Another, similar slab may be represented by the gabbros located in Baltimore.

What we know of the structure and history of the Appalachians suggests several important generalizations about the mountain-building process in this system. First, the intensity of deformation and the amount of uplift is greatest in the east, toward the continental margin; it becomes less toward the continental interior. Second, movements during the deformations proceded from east to west and are of a

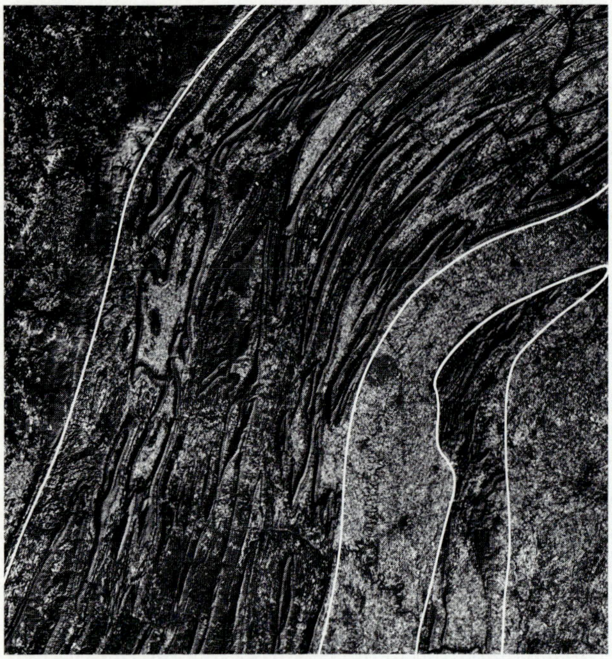

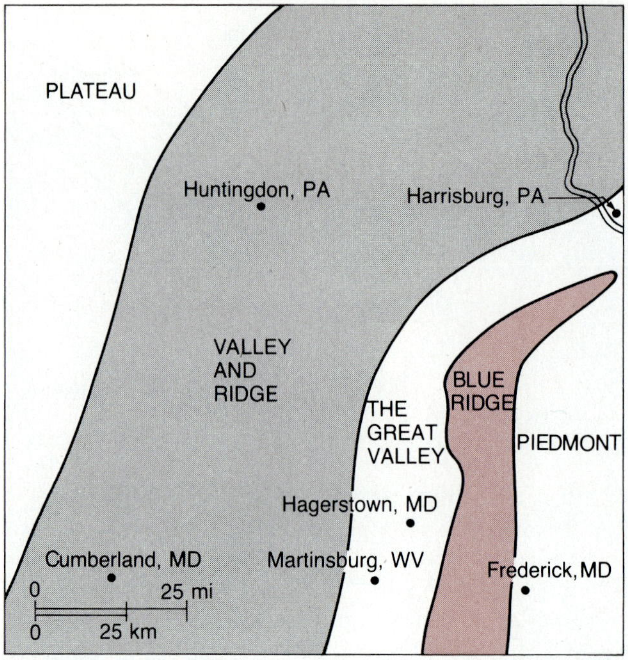

NASA

Figure 11.9
This satellite image shows parts of the Appalachian plateau, folds in the valley and ridge province, the northern end of the Blue Ridge, and part of the Piedmont. Most of the area shown is in southern Pennsylvania.

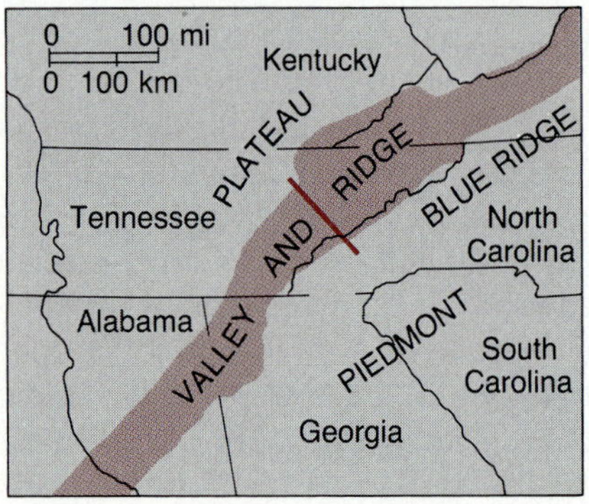

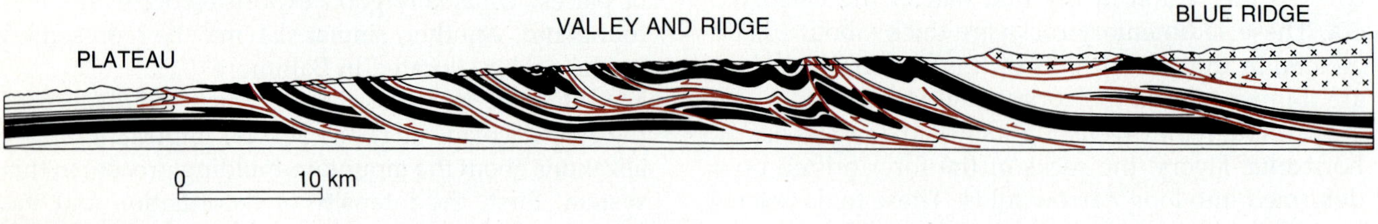

Figure 11.10
(a) A structural cross section across the southern Appalachian Valley and Ridge province in central Tennessee. In this interpretation, the Precambrian basement is nearly flat beneath the folded and thrust-faulted Paleozoic sedimentary rocks. (b) Locator map.

type that might be produced by a tremendous lateral compression or shove. Third, deformation was concentrated in several major episodes, all of which occurred during the Paleozoic Era.

The Coastal Plain. A region of low and subdued topography, the Coastal Plain (Fig. 11.11), lies along most of the eastern and southern margins of North America. This band of relatively flat ground separates the Atlantic Ocean and the Gulf of Mexico from the Appalachian and Ouachita Mountains and from parts of the platform.

The Atlantic and Gulf Coastal Plains are distinguished not only by low topography but also by the young and almost flat-lying sedimentary strata that lie under them. These marine strata were deposited across the edge of the continent beginning about the end of the Mesozoic era, approximately 70 million years ago. By that time, the Appalachian and Ouachita Mountains, which had been uplifted to form high mountains earlier, had eroded to much lower levels. The sea slowly advanced across at least the eastern part of the former mountains, burying them under young marine sediment. These sediments were later

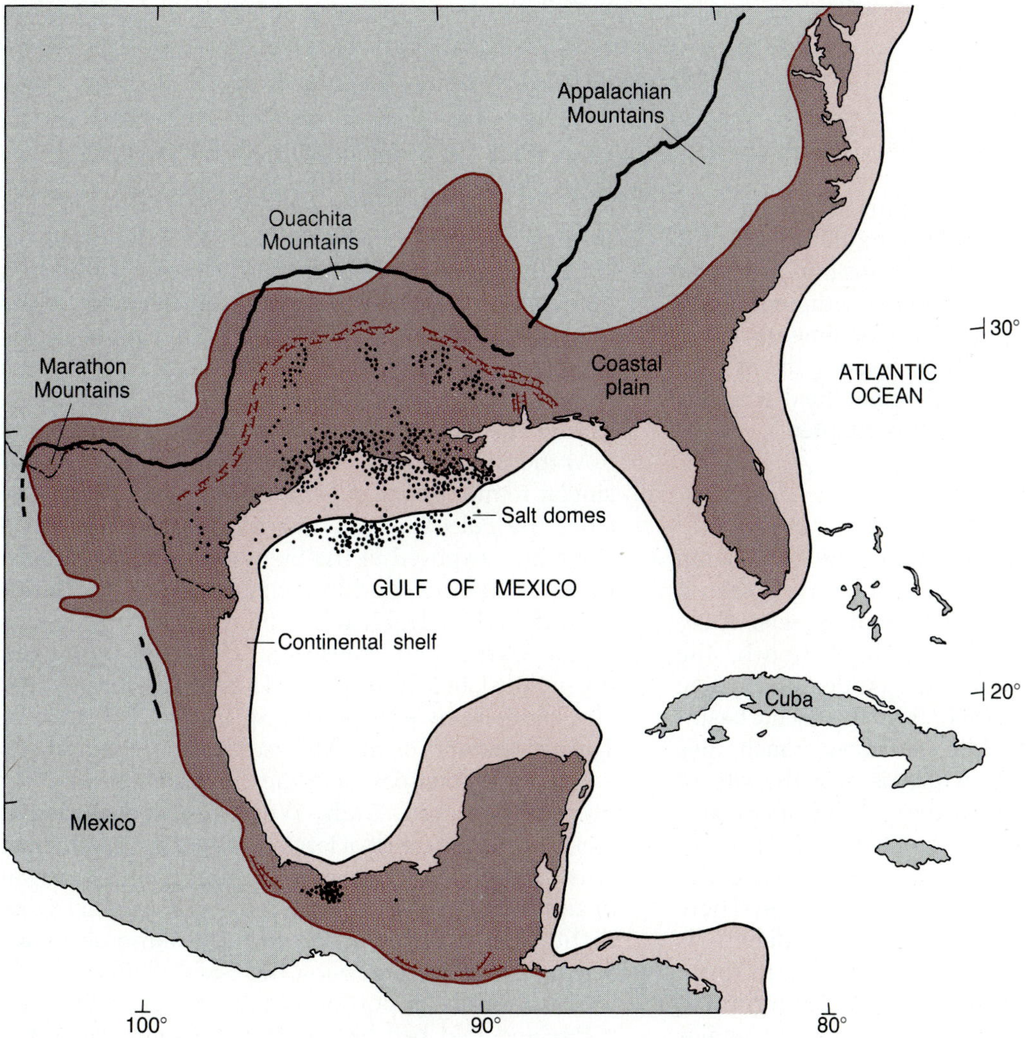

Figure 11.11
Regional sketch map of the Atlantic and Gulf Coastal Plains. A major zone of normal faults and a number of buried salt bodies, called salt domes, are indicated. The borders are shown approximately.

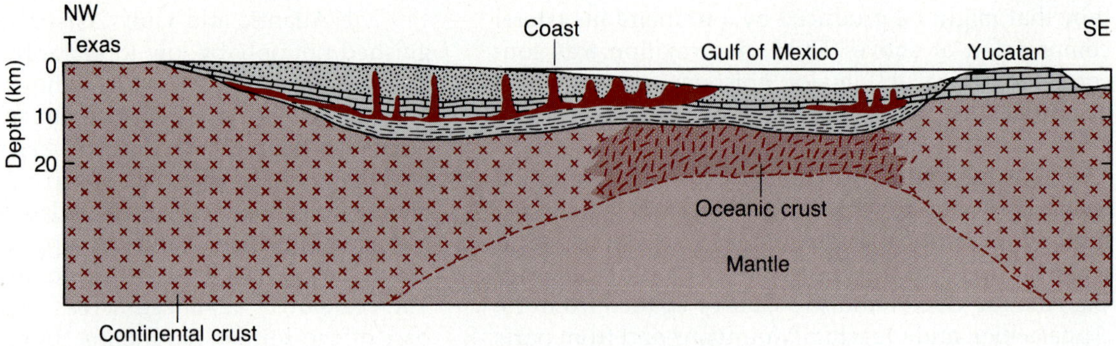

Figure 11.12
This schematic cross section across the Gulf of Mexico shows the thick sediment under the Gulf Coastal Plain and continental margin. A thick layer of salt occurs in this sediment. The salt has been squeezed to form long carrot-shaped intrusions into the overlying sediment.

exposed as the seas retreated from the edge of the continent.

Coastal plain sediments are thinnest at their inner edge. They are inclined and become thicker toward the Atlantic and Gulf. As they continue into the ocean and Gulf, they form thick prisms of sediment along the continental margin (Fig. 11.12). Sediments under the coastal plain are especially thick along the margin of the Gulf of Mexico, as we will see in the discussion of continental margins later in this chapter.

The Cordilleran Mountain system. The Cordilleran Mountain system, (Figs. 11.2 and 11.13) forms a relatively narrow but long belt that is seismically active throughout its entire length. A trench, the Peru–Chile trench, separates the Andes and active andesite volcanoes found there from the Pacific basin. Another trench, the Middle Americas trench, lies west of Central America, which is also the site of active andesite volcanism. Farther north from Mexico to Alaska, neither volcanoes nor trenches are present along the continental margin, except in the Cascades Mountains of the northwestern United States. There volcanoes are present, but too much sediment is being deposited off the coast for a trench to form. The two long sections of the continental margin that lack trenches and volcanoes are marked by major faults, the San Andreas and the Fairweather faults. These are nearly vertical strike-slip faults, along which the western side is moving north relative to the eastern side.

The processes of orogeny in the Cordilleran system started several hundred million years ago, in the Paleozoic era, but the best-known part of this history involves the period from the middle of the Mesozoic era to the present, a time span of nearly a hundred million years. During that time, the character of the region we know as the western United States has changed drastically. Most of the region west of the present Rocky Mountain ranges was oceanic crust during the Paleozoic era. The region along what is now the west coast of the United States was probably similar to modern island arcs during most of the time since the Paleozoic era. The huge batholithic intrusions now exposed in the Sierra Nevada Mountains, in Baja California, and in the Coast Ranges of British Columbia (Fig. 11.13) were the magmas from which volcanoes rose in the Mesozoic. These were emplaced during the late Mesozoic era. They have been uplifted and laid bare by erosion during subsequent periods. Toward the end of the Mesozoic era, the mountains we know as the Rockies in the United States (in New Mexico, Colorado, Utah, Wyoming, and Montana) were formed by uplift of large blocks of the continental crust, mainly along steeply dipping faults. Renewal of uplift in the last two million years produced the high mountains that make up the striking peaks we now encounter on crossing the Great Plains.

The Canadian portion of the Rockies is different; a foreland fold belt like that in the southern Appalachians forms its eastern edge. The sedimentary veneer that covers the ancient Precambrian rocks has been folded into a long chain of anticlines and synclines broken in places by low, dipping thrust

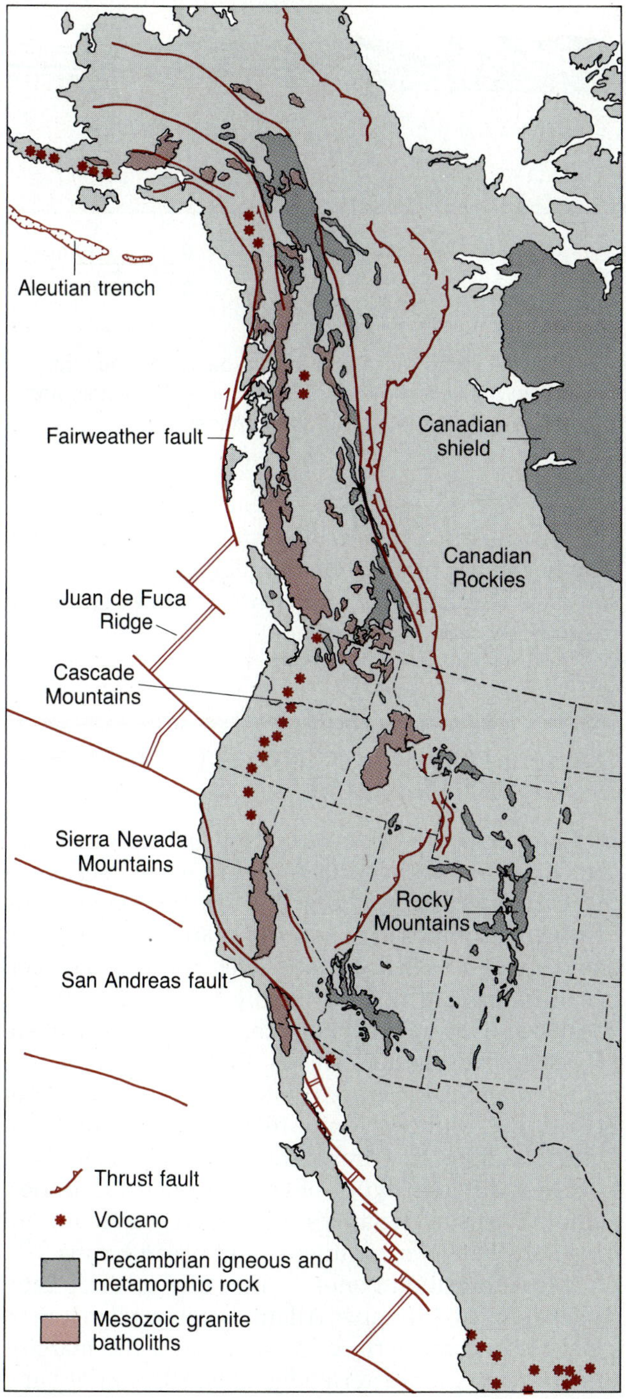

Figure 11.13
A regional sketch map of the North American section of the Cordilleran Mountain system, showing major structural features and the location of the largest volcanic and igneous rock bodies.

faults over which the sheets of sediment have been moved toward the continental interior (Fig. 11.14). This fold belt can be traced into the United States where it crops out in Montana, western Wyoming, and western Utah, but it has been broken up by younger block-faulting and is not nearly as continuous or distinct as its Canadian counterpart.

At the present time, the area between the Rocky Mountains and the west coast is especially complex because much of the region has been affected by faulting. Many of the high mountains are uplifted fault blocks. These are separated by grabens forming a region known as the Basin and Range Province. During this period of block-faulting, the Sierra Nevada Mountains have been uplifted along high-angle normal faults on their eastern side.

Farther north, a vast region is covered by basalt lava flows, which poured out of fissures and covered large parts of Oregon, Washington, and Idaho with lava to a depth of more than a thousand meters. Seismic studies indicate as much as 18 kilometers of basalt may be present in some places. Most of this lava was extruded during the Tertiary era, but the volcanoes of the Cascade Mountains are still active.

Along the west coast and west of the Sierra Nevada Mountains, we now find younger rocks that are extremely deformed and contain siliceous deposits with radiolarians and other deep-sea sediments. These rocks also contain serpentine and peridotites of the types found in the mantle. The combination of minerals and fossils suggests that some of the material in this deformed zone has been uplifted from a former deep-sea trench.

It is interesting to compare the Cordilleran system of North America with the much older and stabilized Appalachian Mountains. Both systems are located along continental margins; both contain fold and thrust zones, the foreland fold belts, which appear to have been formed by lateral compression. In both foreland belts, movements proceeded toward the continental interior. Deformations have taken place over long time spans in both systems, and igneous activity has emplaced huge batholiths at times but not continuously. The two systems also differ in several remarkable respects. There is no evidence of major strike-slip faulting in the Appalachians, nor have the folds and thrusts been broken up by high-angle normal faults in the Appalachians as they have in the Basin and Range region. Finally, we see no counterpart in the Appalachians of the basement uplifts that constitute the Rocky Mountains in the west.

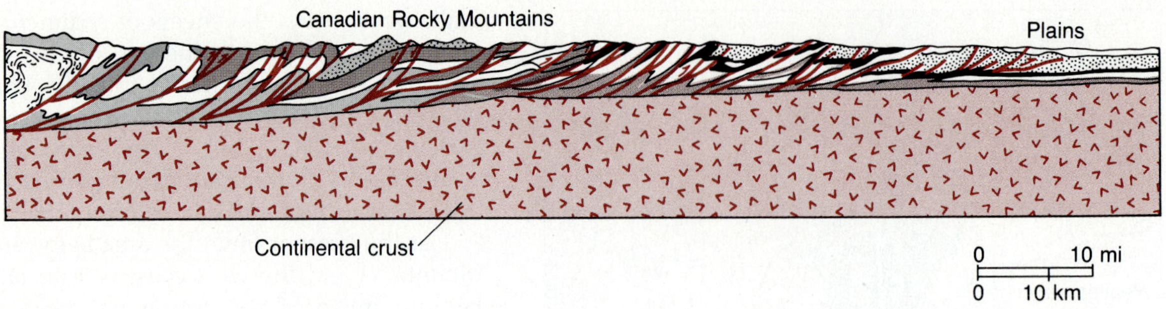

Figure 11.14
Cross section across the Canadian Rocky Mountains. The Precambrian rocks that form a basement beneath this part of the mountains are uplifted as blocklike mountains in the U.S. Rockies. The covering Paleozoic and Mesozoic sedimentary rocks are complexly folded and thrust toward the interior of the continent.

It is clear that mountain systems are not all alike. One of the important jobs for geologists is to understand and explain how and why they differ and what features they bear in common.

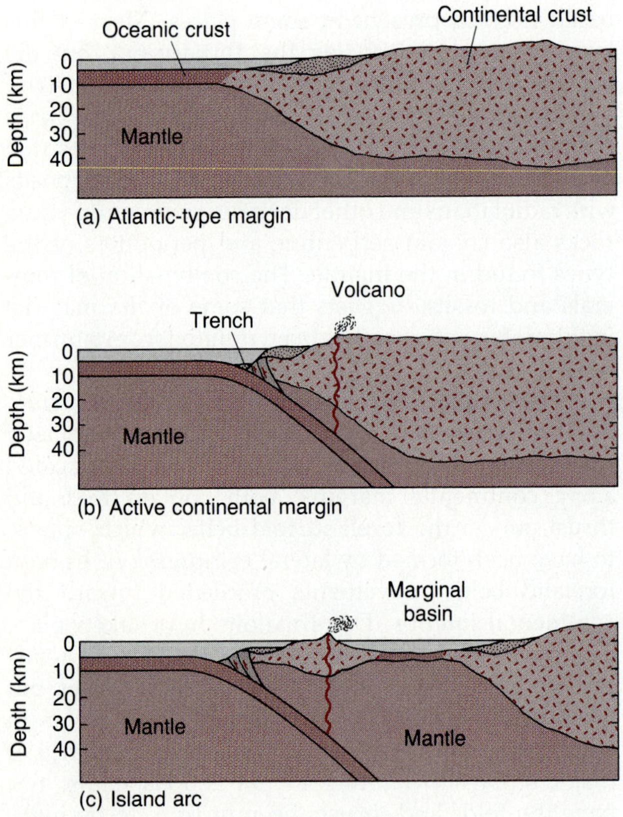

Figure 11.15
Most continental margins can be classified as one of the three types shown here. The coast of California, which is described in Chapter 13, is a notable exception.

11.2 THE TRANSITION FROM CONTINENTS TO OCEAN BASINS

In Chapter 3 we reviewed some of the geological characteristics of the largest geographical features of the earth's crust. We saw that on the average the continents composed of rocks of granitic composition stand about 0.8 km above sea level, and that the deep ocean basins are floored mainly with much denser basalts and lie nearly 4 km below sea level. Geophysical investigation shows that the base of the continental crust averages 40 km depth, while under oceans the Moho is only 10 km below sea level. Understanding this transition zone between these two very different types of crust is critical to understanding crustal dynamics—the movements and interactions between continents and ocean basins.

Most transition zones fit into one of three categories (Fig. 11.15): quiet Atlantic-type margins; marginal zones characterized by island arcs; and margins like those on the western edge of North America and South America, where young mountains are now forming.

Our knowledge of these transition zones, the continental margins, has expanded greatly in the last two decades. The application of geophysical techniques, such as those described in Chapter 9, has been an especially important source of new data about both continental margins and the ocean basins. Drilling has also supplied new data. Some of this drilling is done as oil and gas exploration on the continental shelf, but most of the drilling in deeper water has been carried out as part of a program known as the Deep Sea Drilling Project (DSDP). This is a joint research program sponsored by many oceanographic research institutions and supported by the National Science Foundation. Although this project has involved drilling in many parts of the ocean, we shall consider here only results obtained in a few localities. The first of these is along the eastern margin of North America.

Atlantic-type Continental Margins

We saw in Chapter 3 that the continental margin along the east coast of the United States is in many respects typical of margins around most of the Atlantic, Arctic, Indian, and Antarctic Oceans. Except in a few sections, these margins are stable. They are seismically and volcanically inactive. They are not flanked by deep-sea trenches. The margin is broad for long distances along them, and they are characterized by the types of sedimentation described for North America's margin in Section 5.3.

The sea floor along a typical Atlantic-type margin, such as that off the United States (Figs. 11.16 and 11.17), gradually drops off from the shore to the deepest part of the ocean in a systematic way. The shallow, gently sloping surface of the **continental shelf** lies just offshore. The continental shelf is less than 180 meters deep and averages 30 to 40 meters in depth. The bottom drops in depth less than one meter for every 1000 meters out from shore (a slope of 1:1000). The shelf is very narrow off southern Florida, but extends more than 300 kilometers out into the Atlantic farther north.

At the outer edge of the shelf, the slope of the sea floor steepens to 1:40 or more (3 to 6°). This next, steeper-sloping province, the **continental slope,** continues out into water depths of 1 to 3 kilometers. Its landward edge is usually marked by a relatively sharp break in slope, but the base is usually less definite.

The next deeper submarine province, the **continental rise,** is characterized by slopes of 1:100 to 1:700. It can generally be recognized on submarine profiles because it is much less steep than the continental slope. The outer edge of the continental rise often passes without an abrupt change in slope onto the nearly flat deep-sea floor of the abyssal plains.

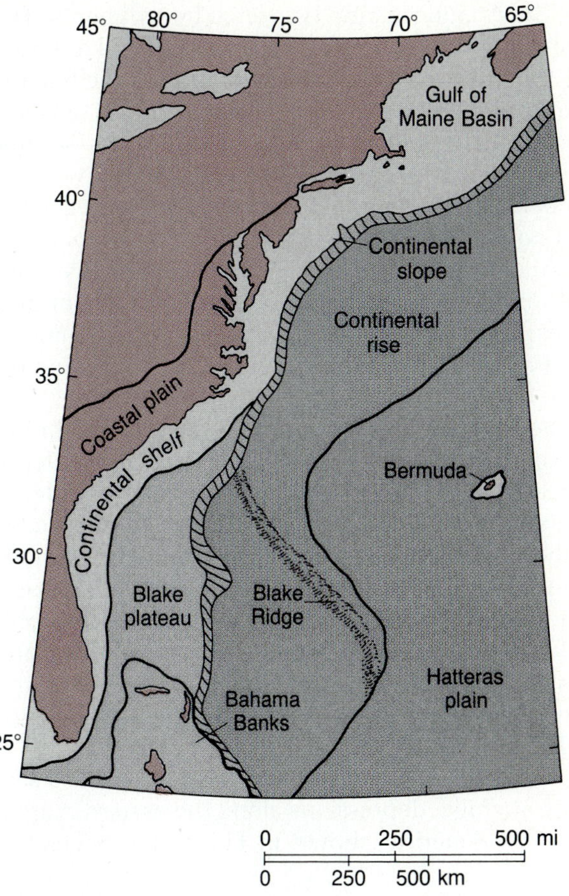

Figure 11.16
Major divisions of the continental margin and ocean basin east of the United States.

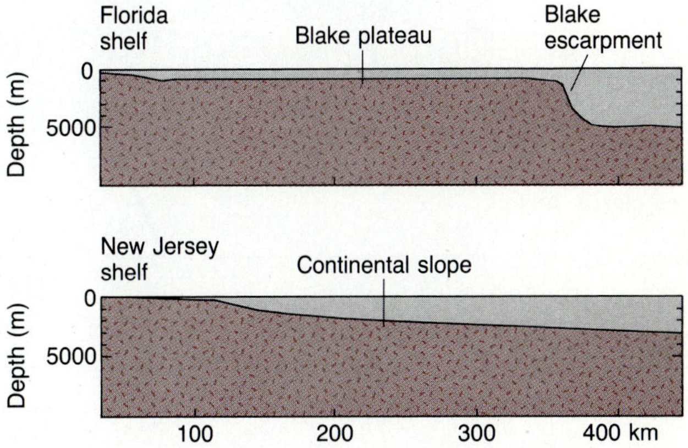

Figure 11.17
Profiles across the continental margin of the eastern United States are typical of those found on Atlantic-type margins throughout the world.

The Atlantic margin is covered by sediments. The younger and shallower of these lie as a blanket across the top surface of the continental shelf. Sediments also cover the slope and rise. These shallower deposits are especially well revealed by continuous seismic profiling, such as that shown in Fig. 11.18.

Seismic studies and a few deep wells reveal what we know of the structure beneath the sea floor along the Atlantic margin. The Cenozoic sedimentary rocks that underlie the Atlantic Coastal Plain continue beneath the continental shelf (Fig. 11.19). Some of them crop out on the steeper slopes of the continental slope and on fault scarps.

Seismic studies also show the deep structure of the continental margin. The Moho rises from a depth of nearly 40 kilometers under the continent to a depth of no more than 10 kilometers under the outer edge of the continental rise. The crust above the Moho is complex in this transition zone. The crystalline basement rocks, consisting of granites and metamorphic rocks in the continental interior, become thinner, and they are overlain by a thick prism of sediment and sedimentary rock under the continental margin (Figs. 11.19 and 11.20).

This accumulation of sedimentary rocks fills several troughlike depressions along the eastern margin of North America, shown in Fig. 11.21. Several of

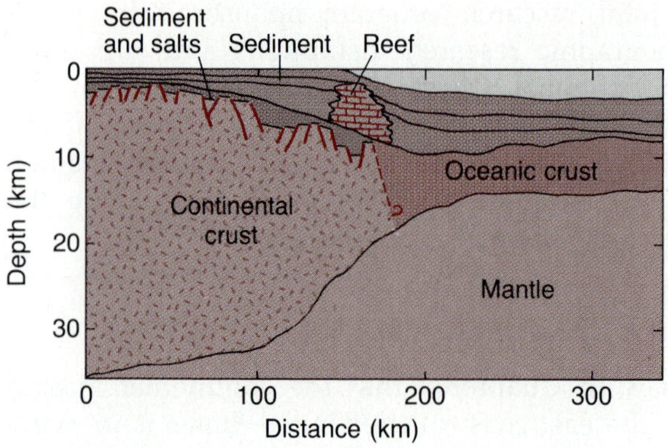

Figure 11.19
This schematic cross section drawn across the continental margin near Cape Cod depicts the young sediments that cover both the continental and oceanic crust along the margin. Block-faulting is present in the continental crust, and a reef buildup may have determined where the edge of the continental shelf is located. Compare this section with the one farther south in Fig. 11.20.

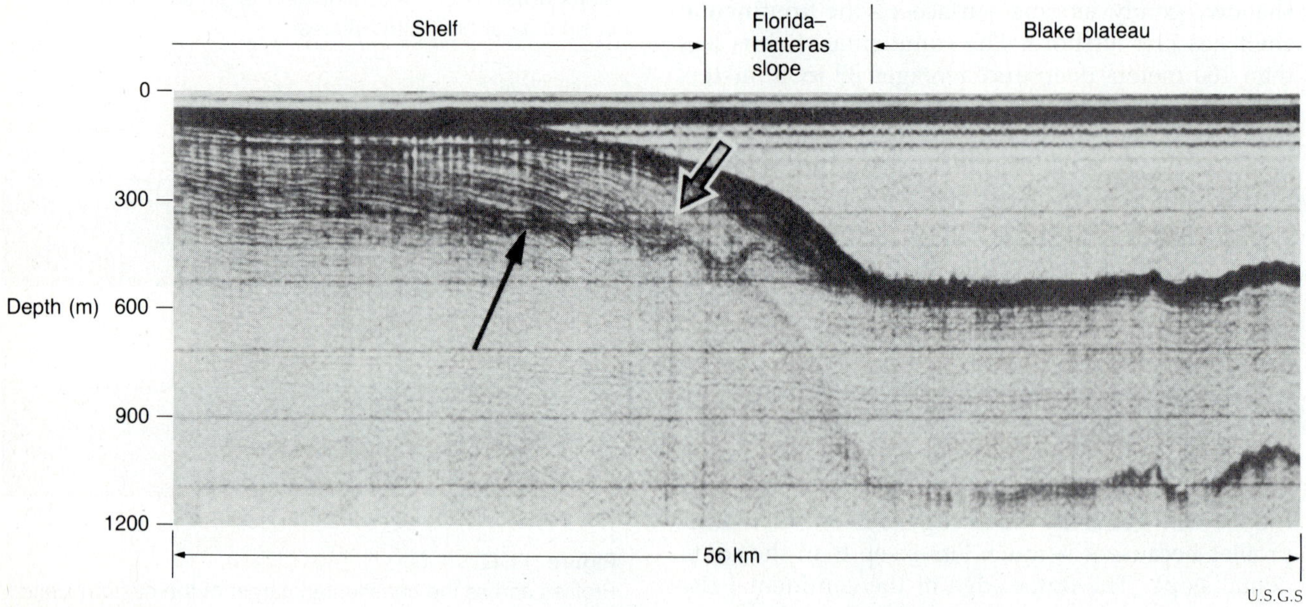

Figure 11.18
This seismic profile reveals details of the shallow structure under the continental shelf and Blake Plateau off the coast of Florida.

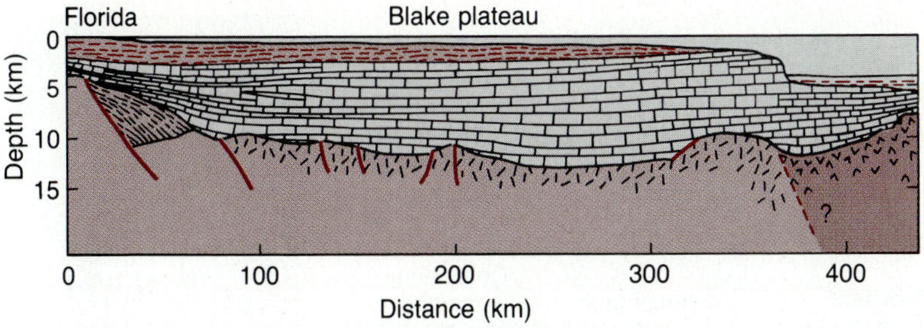

Figure 11.20
This schematic cross section across the Blake Plateau illustrates how flat-lying the younger sediments under the plateau are. Steep faults mark the outer edge of the plateau.

these troughs, especially the Baltimore Canyon trough are of special interest because they are being explored for oil and gas. Shows of hydrocarbons have been discovered in the Baltimore Canyon trough, but no major discoveries of gas or oil have yet been made. The deepest part of these troughs lies under the continental slope. The structure of these troughs is still uncertain, partly because they are so deep that no drilling has yet penetrated them. They may be long synclines parallel to the continental margin, or, more probably, grabens bounded on both sides by high-angle normal faults, like the ones shown beneath the continental shelf in Fig. 11.20. Many sections across the continental margin show reef buildup beneath the outer edge of the shelf. The reefs are

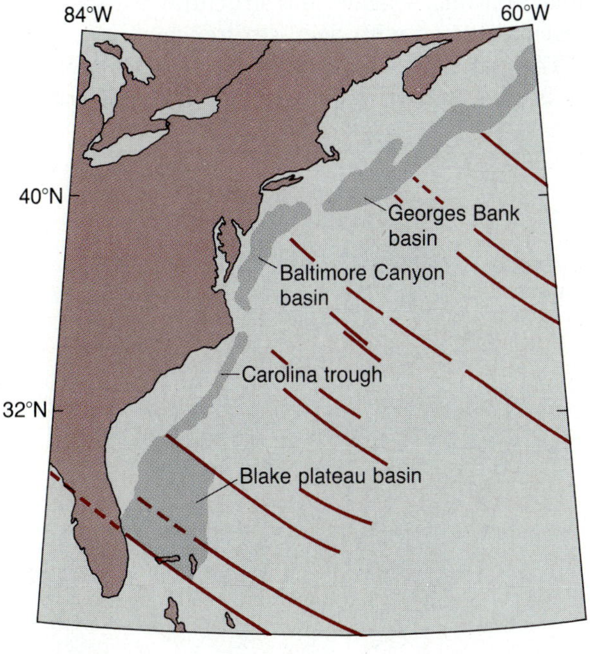

Figure 11.21
Thick accumulations of Mesozoic-age sediment lie in large basins (in color) beneath the continental margin of North America.

now dead and buried, but they formed sediment traps along the margin of the continent for millions of years in the Mesozoic era.

In some places along the edge of the continent, normal faults have formed undersea cliffs. The escarpment on the east side of the Blake plateau (Fig. 11.20) appears to be a high-angle fault in some places. The ocean basin has moved down relative to the continental side. Other normal faults occur on the continent in the Appalachian Mountains. These faults formed at the end of the Paleozoic era while the mountains were high. The resulting down-faulted basins were filled by sediment eroded from the Appalachians during Triassic time. The deeper parts of the troughs along the continental margin and under the continental slope formed at the same time. Normal faults that are generally shown on the side toward the ocean basin are also well known on the European side of the Atlantic. Figure 11.22 shows the prominent block fault pattern found in the subsurface along the continental margin west of France.

The normal faults on the Atlantic margin formed after the Paleozoic mountain-building and appear to be the result of extensional deformation on the continental margins. They clearly derive from deformation of a different type from that responsible for the mountain-building. Slight movements continue on some of the normal faults along the Atlantic margin, and these are a concern in the location of nuclear reactors. But compared with the island arcs and areas of active mountain-building, the Atlantic margins are very quiet.

Volcanic Island Arcs

As we have seen in earlier chapters, the arcuate island chains are extraordinary parts of the earth's crust in several respects. Most modern andesite volcanoes are located in these islands. The island arcs also lie in the most seismically active parts of the crust. Not only are earthquakes frequent in the arcs, but almost

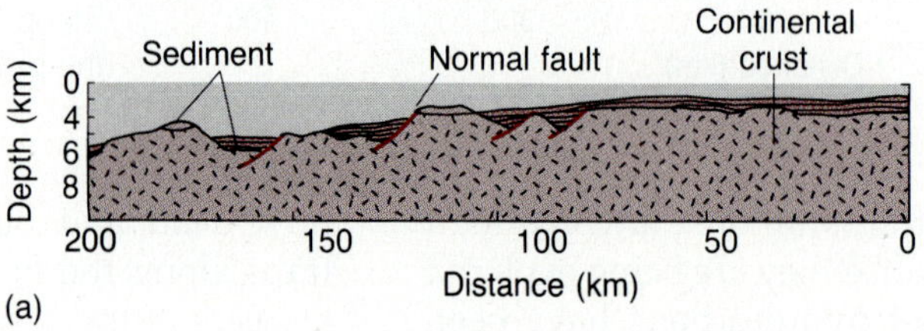

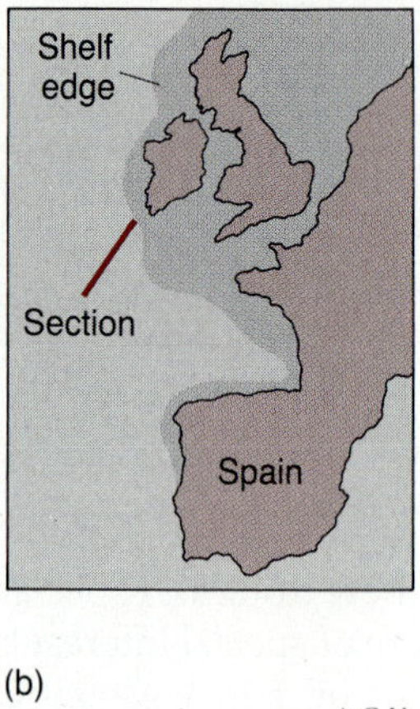

Figure 11.22
(a) This cross section of the European continental margin reveals normal faults in the subsurface, similar to those found along the North American continental margin.
(b) Locator map.

A.G.U.

all deep and most of the highest-magnitude earthquakes occur here as well. In addition, the deepest depressions in the crust, the deep-sea trenches, lie in these arcs. They are clearly one of the most active parts of the crust.

Figure 11.23 shows the location of the island arcs and their relation to the oceanic ridges and continents. Most of the island arcs lie along the margin of a continent and present a consistent pattern of features. The typical island arc consists of a chain of active andesite volcanoes forming a curve convex toward the ocean. A deep-sea trench, sometimes 10 kilometers or more deep, lies on the oceanward side of the arc between the abyssal plains of the ocean and the volcanoes. A long, narrow sedimentary basin lies between the trench and the volcanoes. Behind the arc, large, relatively shallow seas separate the volcanic chain from the continent. This sea usually has one or more arc-shaped submerged ridges parallel to the island arc.

Not all chains of volcanic islands exhibit this neat pattern. The most notable exceptions occur in the southwest Pacific, in the Caribbean and the Scotia Arcs (Figs. 11.24 and 11.25), and along the coast of South and Central America (Fig. 11.24). These areas are grouped with the island arcs because they share similar volcanic, seismic, and structural histories, and all are associated with deep-sea trenches.

The pattern of earthquake foci in the volcanic arcs is one of their most distinctive characteristics.

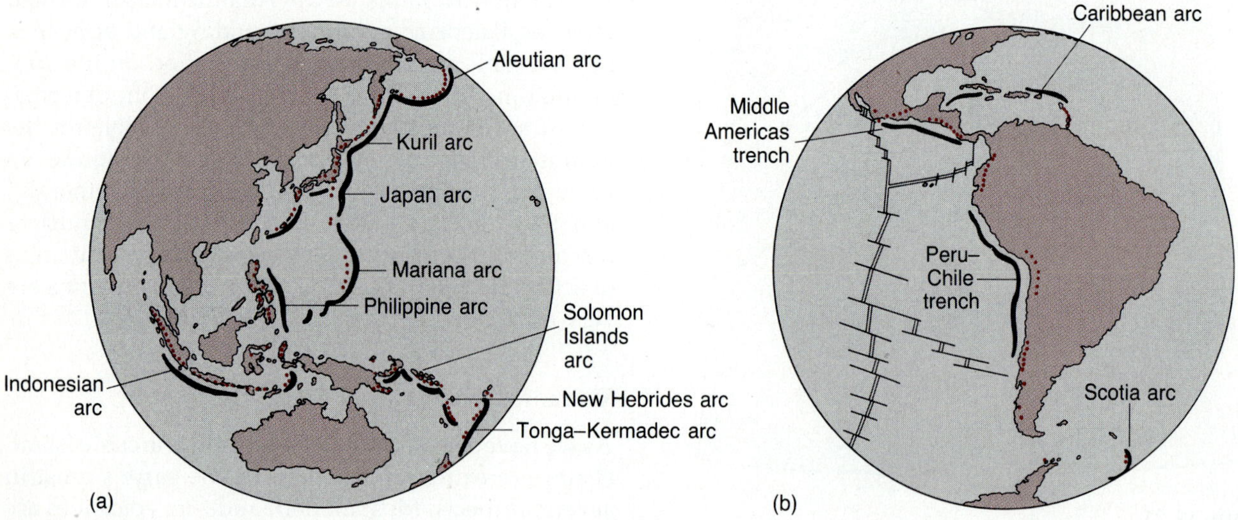

Figure 11.23
The relationship of the island arcs and trenches to the oceanic ridges and adjacent continents is shown in these sketch maps.

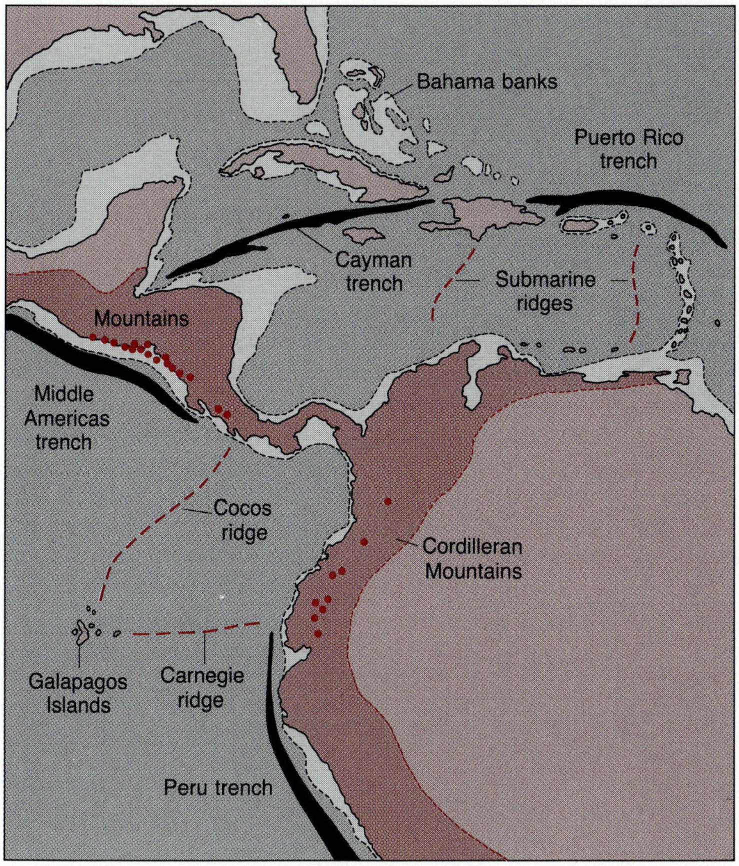

Figure 11.24
Sketch map of the Caribbean Islands and portions of the west coast of Central and South America.

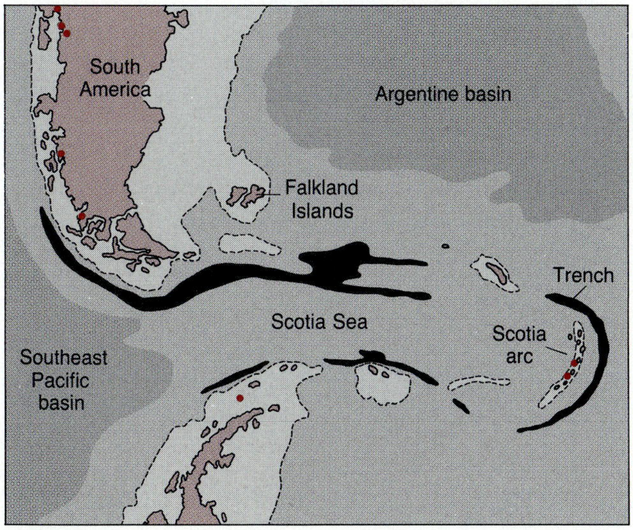

Figure 11.25
Geographical setting of the Scotia Arc.

Shallow-focus (25 km or less) earthquakes tend to occur in well-defined inclined zones called Benioff or seismic shear zones, which extend from the trench under the volcanic arc. In oceanic settings this zone extends deep under the marginal basins. The deepest earthquakes occur at depths of 600 to 700 kilometers.

All seismic shear zones are inclined from trenches under volcanic arcs, but not all have the same angle of inclination. The seismic shear zone below the Aleutian arc (Fig. 11.26), is typical of many, but others are inclined at lower or steeper angles.

The composition of the lavas erupted in the arcs is another of their most characteristic features. Lavas of varied composition are erupted, including some rhyolites and basalts, but the most typical lava is intermediate in composition, andesite. Thus a sharp distinction exists between lavas erupted in arcs and those that make up the oceanic crust and flow from oceanic volcanoes.

Volcanic arcs vary greatly in structure. In many, especially the arcs composed of volcanic islands like the Aleutians, few structural features are evident in the recent volcanoes. More complex structures occur in the large land areas of Japan, Java, and Sumatra, and along the coast of South America. Block-faulting, with development of grabens and horsts aligned parallel with the arc, is a common feature, as are broad arches and synclines.

Seismic profiles provide valuable clues to the shallow crustal structure in submerged areas. These profiles show that the crust in the abyssal plains near trenches is like that in other parts of the deep-sea floor. Close to the trench, the layers in the crust bend down and appear to pass beneath the inner wall of the trench. These sediments are often broken by steep, normal faults, probably formed by the bending of the crust. The floor of the trench usually contains some sediment accumulation; and, as we saw in Chapter 5, in some places the trench is literally filled.

The inner wall of the trench appears to be composed of marine sediments that are folded and thrust-faulted in such a way that layers are repeated, as shown in Fig. 11.27. This is interpreted as resulting from the scraping off of sediment and basalt from the sea floor as it slides down into the mantle under the volcanic arc. From the trench to the volcanic arc, a terrace is usually present. Seismic profiles here show undisturbed sediment beneath the terrace. Closer to the volcanoes lava flows and layers of volcanic ash predominate.

The marginal basins that separate many island arcs from continents, as the Marianas are separated from the Philippines, receive sediment both from the continent and the volcanic island. The crustal structure of most such basins is more like that of the ocean floor than of the continents. In some it appears that new sea floor has been created within this basin. This is certainly true of the sea between New Zealand and Australia and of the Philippine sea. In these cases, lava must be finding its way to the surface along zones similar to those of the oceanic ridge system.

Evidence supporting the interpretation of oceanic

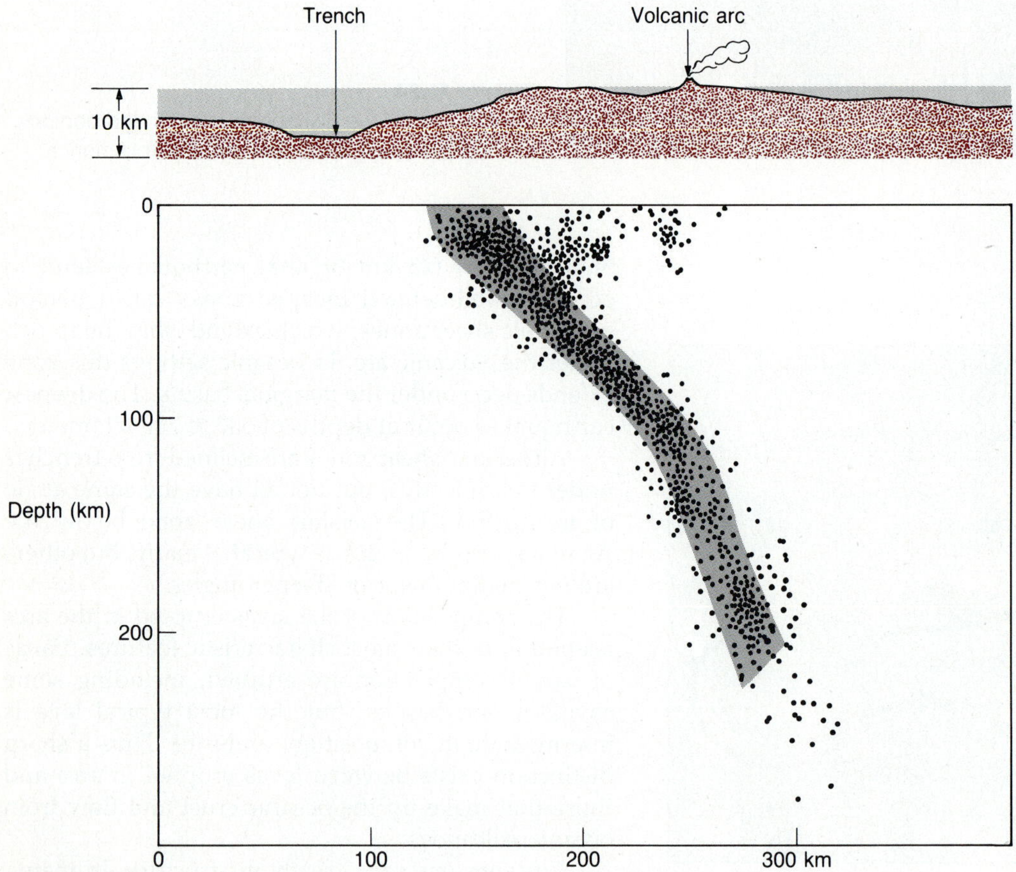

Figure 11.26
(a) A profile across the Aleutian Arc. (b) The focal locations of earthquakes that have occurred beneath the arc. Foci have been projected into a single plan from along the arc.

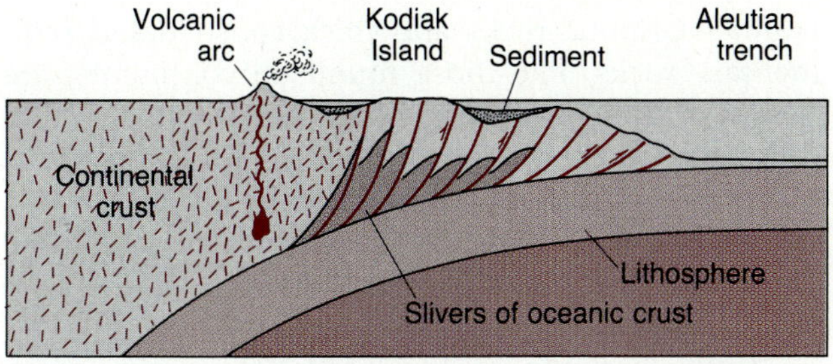

Figure 11.27
Seismic studies of the area between trenches and volcanic islands often reveal the presence of fault slices of sediment and oceanic crust. A model of this structure from the Aleutian Islands is depicted here.

ridges as spreading centers and the volcanic arcs as subduction zones is a main subject of Chapter 12.

11.3 THE OCEAN BASINS

The ocean basins themselves present a somewhat less complicated geological picture than do the Atlantic-type continental margins or the island arcs. However, they do have a far more varied and complex topography, structure, and history than geologists originally suspected. Four major types of geological features are present in the oceans: the oceanic ridge system; faults that cut across the ridge; abyssal plains; and more or less isolated volcanoes.

Oceanic Ridge System

In 1854, Matthew Fontaine Maury published the first map of the North Atlantic Ocean, on which he indicated the presence of a ridge in mid-ocean. We have now traced this ridge 72,000 kilometers into all oceans of the world (see map on inside front cover), and we know that it has rugged topography along much of its length. Off the flanks of the ridge the irregular topography is buried by sediment.

Detailed studies have been made of the oceanic ridge system in a number of localities. Notable among these are studies in the area just south of Iceland (along the Reykjanes Ridge), on the Mid-Atlantic Ridge due west of Gibraltar, and an area near the Galapagos Islands rift (Fig. 11.24). Many of the figures in this section are derived from these studies.

Because the Mid-Atlantic Ridge passes directly through Iceland, we can expect a close relationship between the geology of Iceland and that of the ridge. Iceland (Fig. 11.28) is covered by volcanic rocks of basaltic composition. Many of these are lava flows that rise to the surface along faults and spread out laterally as sheets of lava. The faults are normal faults, many of which have been active in recent times. Epicenters of shallow earthquakes lie close to these faults. The faults define a system of grabens that cross the island in the same direction as the mid-Atlantic ridge.

A number of wells have been drilled on Iceland, some to depths of two to three kilometers. These wells penetrate one layer of basalt after another, interrupted only by an occassional layer of volcanic ash and marine sediment interpreted as graben filling during extension of the area. Thus Iceland appears to be underlain by a thick pile of basaltic lavas. The bottom layer of its structure is a nearly horizontal lava flow broken by normal faults caused by the extension of the area in an east-west direction.

A similar pattern seems to hold for the submerged portions of the oceanic ridge system. Profiles across much of the ridge system—especially the Atlantic and Indian Ocean parts—reveal a mountainous topography on the flanks of the ridge and a valley, the median valley, along its crest (see Fig. 3.11). A median valley is also found along an offshoot of the East Pacific Rise, the Galapagos rift. The median valleys are the sites of much shallow earthquake activity, and of the most recent volcanic activity, as is evident in the map of the Galapagos rift shown in Fig. 11.29.

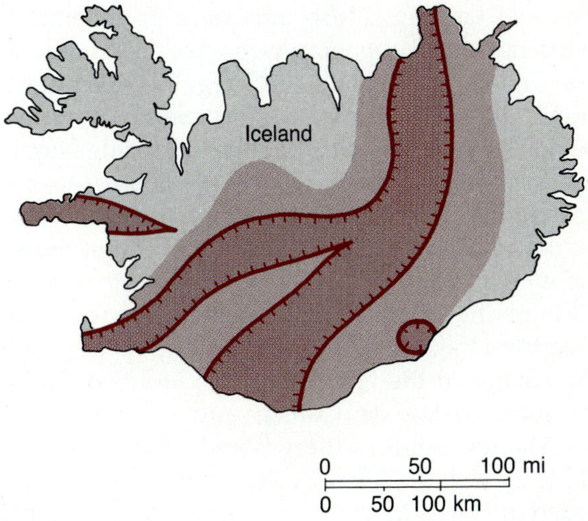

Figure 11.28
Iceland is composed primarily of oceanic basalts. The barbed lines, probable positions of important normal faults, outline the areas underlain by the youngest basalts (7 million years old or less). The darker shading indicates exposed lava flows, 7 to 3.1 million years old. Still older flows lie farther out to the east and west.

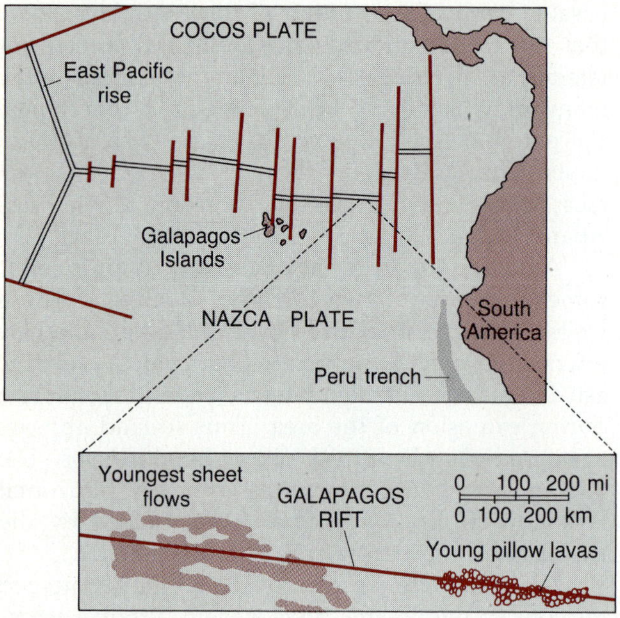

Figure 11.29
The Galapagos Islands are located near a junction between the East Pacific Rise and a spreading center which extends toward South America. Part of this is called the Galapagos rift. Mapping has demonstrated that the youngest lavas in the area lie close to the spreading center.

There basaltic lava is apparently rising to the surface along steep normal faults, just as it is in Iceland.

Normal faulting is also important in generating the structures and topography on the flanks of the ridges. Figure 11.30 shows the irregular topography encountered across the ridge crest. Seismic profiles show that high-angle faults parallel to the ridge occur far out on the flanks of the ridge. These too are normal faults. Most face in toward the ridge and show the same type of extensional movement found in the median valley.

Where the oceanic ridge rises high enough to emerge from the ocean, it forms volcanic islands such as the Azores in the North Atlantic, and São Paulo, Ascensión, Tristan da Cunha, and Gough in the South Atlantic. Some of these islands, the Azores for example, have been sites of volcanic eruptions within recorded history. Most of the lavas are of basaltic composition, as are the igneous and volcanic rocks dredged and recovered by drilling on the ridges. Although basalt predominates, several other rock types, including dunite (100 percent olivine, gabbro, and serpentine) have been found. Significantly, all of these are rich in magnesium and iron, and all are the rock types we associate with the deep crust and mantle. Granitic rocks and thick consolidated sedimentary rocks, like those found on continents, are almost completely absent in the ocean basins. A thin sediment cover may be present on the ridge crest, but bare volcanic rock is often exposed on the sea floor along the ridges. Thicker sediment (about 0.5 km) occurs in the flanks of the ridges and in the deep abyssal plains.

Cross sections of the crust of oceanic ridges prepared from seismic and gravity surveys usually reveal a crust composed of 0.5 kilometer or less of sediment, underlain by about 3 kilometers of rock with seismic velocity of 5 km/sec (approximately that of basalt) and deep substratum of velocity 7.3 km/sec.

The most significant distinction between this and sections of the deep oceanic crust is the low-velocity (7.3 km/sec) zone found under the ridge crest. This velocity is intermediate between that ordinarily found in the deep crustal rocks of the oceans (6.7 km/sec) and those of the upper mantle (8.1 km/sec). The material with 7.3 km/sec velocity is probably serpentine, or it may be a zone where the mantle and crust are mixed as a result of volcanism and igneous intrusions.

The accumulation of evidence from drilling and seismic studies supports the idea that the oceanic ridge system is formed by upwarping of the sea floor caused by the creation and rise of magmas in the mantle. These magmas find surface expression in the

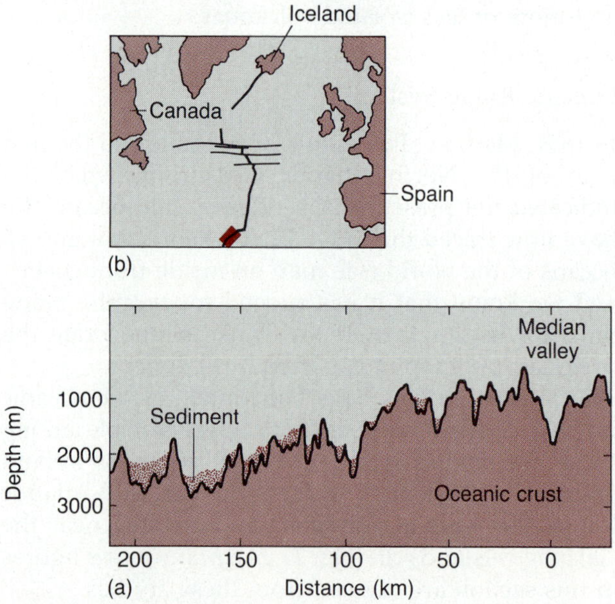

Figure 11.30
(a) A cross section across the Mid-Atlantic Ridge west of Gibraltar. Note the great vertical exaggeration.
(b) Locator map.

volcanic islands, and they are the source of the pillow-shaped masses of lava, called pillow basalts, found in the upper part of the crust under the ridge crest. Perhaps the most impressive conclusion from what we know of the structure of the crust under these ridges is that the mountains under the sea are strikingly different in structure, composition, and origin from those on the continents.

Transverse Faults in Oceanic Crust

In many places the oceanic ridge system is cut by major transverse faults like those in the Mid-Atlantic Ridge (Fig. 11.31). Similar faults are found in the sea floor west of North America (Fig. 11.13). These zones were first identified by the sharp topographic breaks seen along them. Some are marked by steep escarpments across which the sea floor lies at different depths. Differences in elevation of half a kilometer are found in places. Others, like the profile across the Romanche fracture zone profiled in Fig. 11.32, form deep valleys where they cut across the ridge. The relief across the Romanche zone is nearly 3000 meters in this section—twice that of the Grand Canyon.

That these fault zones are active is shown by the concentration of earthquakes along them. As we will see in the following chapters, these are now recognized as a special variety of strike-slip fault caused by the spreading of the oceanic ridge.

Because the valley walls along these fault zones are deep, they provide an unusual opportunity to sample rocks deep in the oceanic ridge. Of the rocks recovered by dredging along the walls of the Romanche zone, much of the deepest material is peridotite altered to serpentine or metamorphosed gabbro; basalt and some sediment were recovered at shallower depths. These samples confirm the model of the structure of the ridge originally interpreted from measurements of seismic velocity.

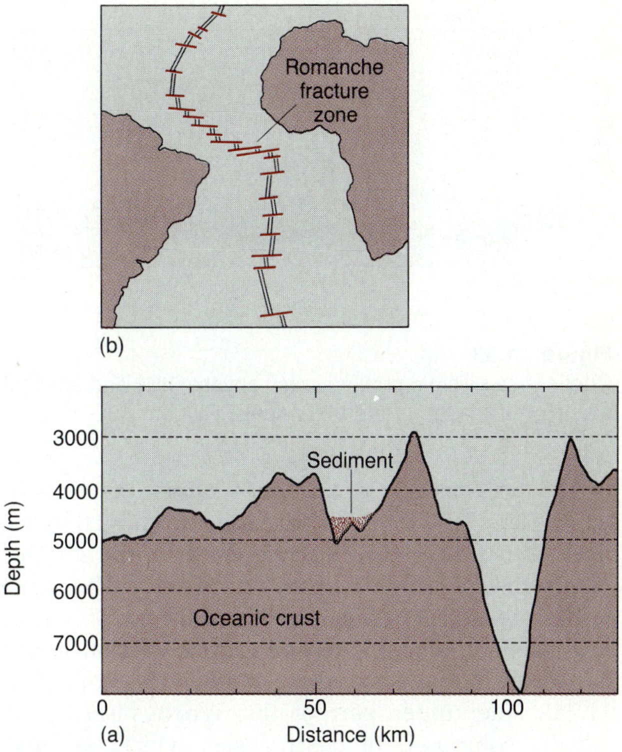

Figure 11.32
(a) Profile across one of the transverse faults, the Romanche fracture zone, which cuts the oceanic ridge system in many places. (b) Locator map.

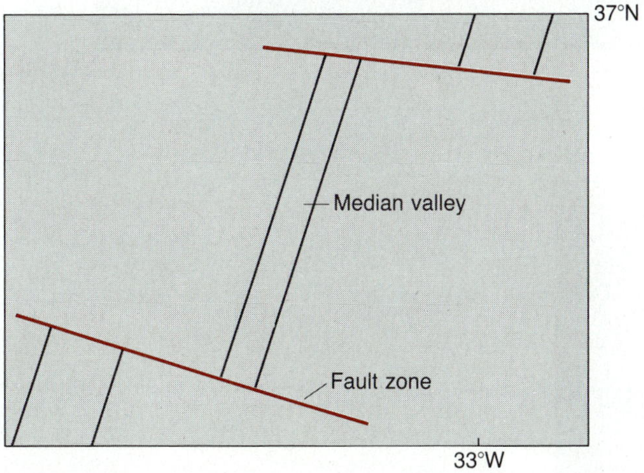

Figure 11.31
Sketch map of the study area on the mid-Atlantic ridge showing the ridge crest and the transverse zones along which the ridge is offset. The transverse zones are about 50 km apart.

Abyssal Plains

Seismic profiles reveal a thin veneer of sediment in depressions near the crest of the Mid-Atlantic Ridge. Thicker accumulations floor low areas on the flanks of the ridge, and this blanket of sediment becomes heavier as it is traced into the flattest areas of the ocean, the abyssal plains. These plains lie at depths of about five to six kilometers in most oceans. Based on the seismic records, we know that the plains are flat because they were formed by the deposition of sediment, settling of material from surface waters, and turbidity currents.

According to seismic evidence, the irregular block-faulted form of the volcanic rocks seen on the flanks

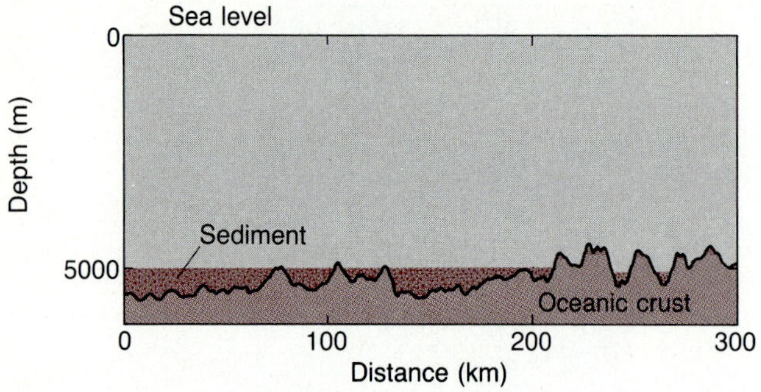

Figure 11.33
Continuous seismic profile across abyssal hills and abyssal plains on the edge of the Mid-Atlantic Ridge.

of the oceanic ridges continues under the abyssal plains (Fig. 11.33). Seismic records also show that the deep structure of the oceanic crust in the abyssal plains resembles that under the ridges. The crust under the ocean basins consists of 0.5 to 1.0 km of unconsolidated sediment and sedimentary rock and 3 to 4 km of igneous and metamorphosed rock (Fig. 11.34). The upper part of this igneous section is usually composed of basalt lavas, which, because they were extruded under water, have a characteristic pillow shape. Gabbro and metagabbro or serpentine (gabbro or basalt altered by hot water) usually occur beneath the pillow lavas. The solid oceanic crust is thus only about 4 to 5 kilometers thick. This places the bottom of the oceanic crust at a depth of about 8 to 10 kilometers below sea level. This type of typical oceanic crust underlies the abyssal plains and much of the continental rise where the sediment cover thickens toward land. It extends close to most deep-sea trenches and up onto the flanks of oceanic ridges.

In Chapters 10 and 11, we have studied the types of structural features found in the earth's crust. Chapter 10 provided a definition of most of the common features and some insights to the physical conditions that give rise to and affect the behavior of rock masses. The geological setting in which these features occur is set forth in this chapter. We are now prepared to consider in the following two chapters some of the ideas that have been advanced to explain how these features, both large and small, have come into existence and why they are found in some parts of the crust but not in others.

SUMMARY

The earth's crust is subdivided broadly into the continents and the ocean basins, with the transitional zone of continental margins between them. Within each of these major divisions we can distinguish geological regions. The continents all contain shields, broad, tectonically quiet, and

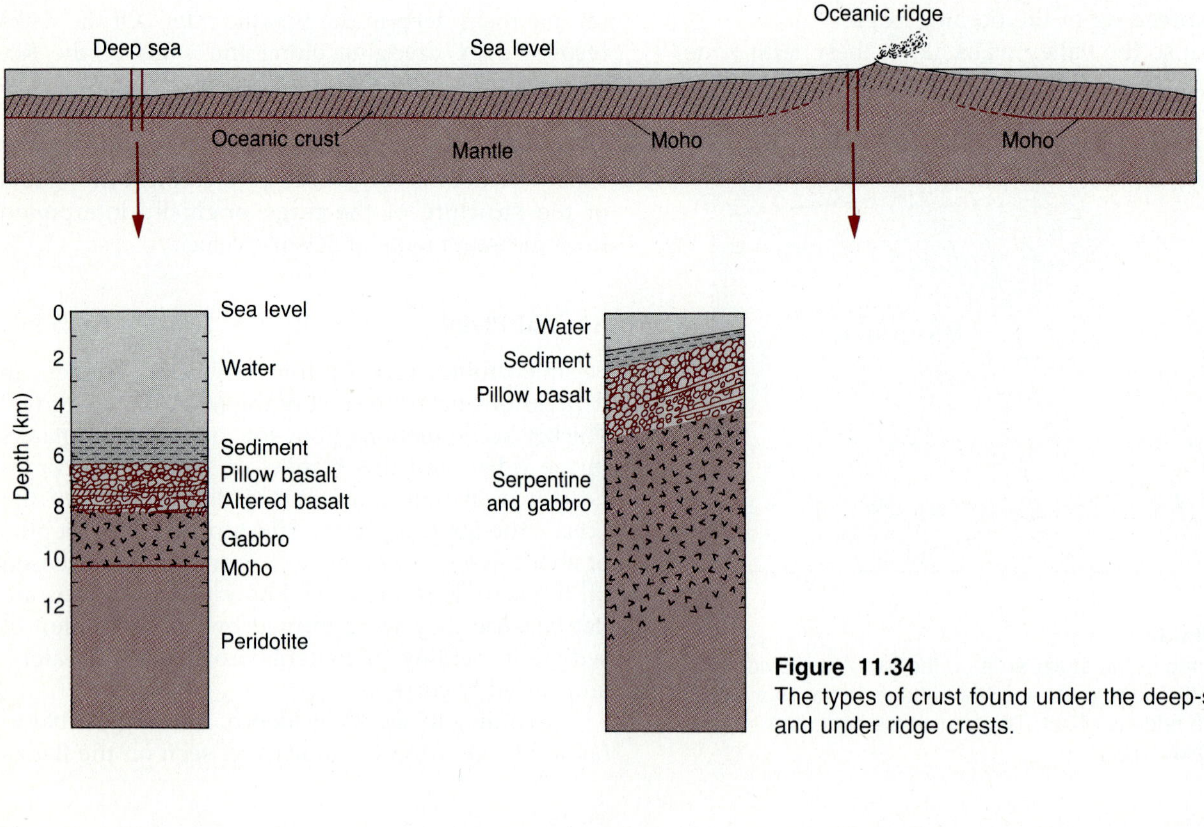

Figure 11.34
The types of crust found under the deep-sea floor and under ridge crests.

relatively flat areas where any deformed underlying rock is ancient. Mountain systems, by contrast, either are or have been tectonically active. Orogenic belts are formed by strong deformation of the crust involving folding, faulting, uplift, and sometimes igneous activity. Most young mountains belong to one of two systems, the Cordilleran or the Alpine. Older mountains, such as the Appalachians, show signs of similar deformation, but erosion has lowered their topography. The Appalachians today contain remnants of a long zone of folded and thrust-faulted sedimentary rock, the foreland belt, running parallel to a zone of Precambrian rock masses and a region of Paleozoic igneous and metamorphic rock. East and south of the Appalachians is a low, flat coastal plain underlain by flat-lying sedimentary rock.

The younger Cordilleran system remains seismically active today, with a chain of andesitic volcanoes running parallel to deep-sea trenches along much of Central and South America. The volcanically active Cascades probably also parallel a now-filled trench.

The transitional zones between ocean and continent are of three sorts. The first is typified by the west coast of the Americas where mountain building is ongoing. The second, the Atlantic-type margins, are quiet, with neither seismic nor volcanic activity. The sea floor drops off gradually, passing from continental shelf to continental slope and then the flatter continental rise. All are blanketed by sediment broken occasionally by fault-formed escarpments and troughs. The third type of continental margins, island arcs, contrast sharply with Atlantic-type margins. The island arcs are generally both volcanically and seismically active, the foci of earthquakes defining a seismic shear zone that slopes downward under the island. On the ocean side, a deep-sea trench separates the islands from the ocean's abyssal plains. Behind the island arc a shallow sea separates the chain and the continent—sometimes marked by one or more submerged ridges parallel to the island arc.

The main features of the ocean basins are the mid-ocean ridges and the abyssal plains. The ridge is a submerged volcanic mountain system of basaltic composition, split by a median valley running along its crest. Normal faults bound the median valley and mark the flanks of the ridge as well. Over the ridge is a layer of sediment, thin near its center and thickening toward the abyssal plains, all over a layer of basaltic lava and serpentine. The abyssal plains themselves are low (5 to 6 km), flat sedimentary regions over a deep structure much like that of the ridges. Within the ocean basins there are also a few volcanic islands that are neither part of the mid-ocean ridges nor of island arcs. Many more such basaltic mountains lie below the ocean surface.

KEY TERMS

continental rise
continental shelf
continental slope
fault-block mountains
foreland
foreland fold belt
mountain system
orogenic belts

STUDY QUESTIONS

1. If the shields were sites of orogenic belts early in their history, how would they differ now from young orogenic belts?
2. What features characterize young orogenic belts? How do they differ from older inactive orogenic belts?
3. Geologists have long debated the relative importance in orogeny of vertical and horizontal movements. What evidence can you cite to prove that both types of movement are involved?
4. Where does the sediment which occurs on and under continental shelves originate?
5. Based on the cross sections of the North American continental margin in Figs. 11.19 and 11.20, what features have acted in the past to trap this sediment?
6. Summarize the various lines of evidence that indicate that the oceanic ridges are located where the lithosphere is being extended.
7. Why are the abyssal plains so flat?
8. How does the continental interior (the platform) differ from the coastal plain? Be sure to describe what lies beneath the rocks in each of these areas.
9. Why are so many volcanoes and earthquakes located in island arcs?
10. Compare the Appalachian-Ouachita mountain belt with the Cordilleran system. In what ways are they alike? How do they differ? Include mention of the location, age, and types of structural features found in each. Be sure to examine the sketch maps in Figs. 11.7 and 11.13.

SUGGESTED READINGS

Heezen, Bruce, and I. D. MacGregor, "The evolution of the Pacific," *Scientific American*, 229:5(1973), *102–106* and *109–112*.

Johnson, H., and B. L. Smith, *The Megatectonics of Continents and Oceans*. New Brunswick, N.J.: Rutgers University Press, 1970.

King, P. B., *The Evolution of North America*. Princeton, N.J.: Princeton University Press, 1977.

Nairn, A. E. M., and F. G. Stehli, *The Ocean Basins and Margins. Volume 1: The South Atlantic*. New York: Plenum Press, 1973.

Windley, Brian F., *The Evolving Continents*. London: John Wiley & Sons, 1977.

EVOLUTION OF PLATE TECTONICS

Chapter 12

How permanent are the continents and ocean basins? Has the present distribution of continents and ocean basins persisted since the continents first formed; or are the ocean basins, as some of the evidence suggests, much younger? Have the continents grown by the welding on of successive orogenic belts, by **continental accretion?** Or did the continents originate as one or more continental plates that subsequently were rifted apart, moved, and modified by orogeny to their present form, as suggested by proponents of the theory of continental drift? What causes mountains to form? Has the size of the earth changed during its history? If so, is it contracting or expanding? These questions have long been asked by geologists, but only in the last two decades have discoveries been made that led to widely acceptable and satisfactory explanations.

Speculations and hypotheses about the origin of the earth's large-scale features preceded the emergence of geology as a science. Of the many proposed over the centuries, several hypotheses stand out as major unifying concepts in the study of the large-scale features of the earth's crust: the hypothesis that the earth is cooling and contracting, or, alternatively, that it is expanding; the hypothesis of continental drift; the idea that there are convection currents in the earth's interior; the hypothesis that the sea floor is spreading; and the theory of plate tectonics. These are ideas to which many other hypotheses and data can and must be related. The way geologists use multiple working hypotheses (described in Chapter 1) to better understand the earth is illustrated by the evolution of tectonic hypotheses.

12.1 FORERUNNERS OF MODERN TECTONIC THOUGHT

Theory of the Contracting Earth

Lord Kelvin, who viewed the earth as a heat engine that is slowly running down, was the first scientist to propose that the earth is cooling. He observed that great quantities of energy are expended in volcanic activity, crustal deformation, and in seismic activity and faulting. The source of this energy is not immediately apparent, although it must come from the earth's interior. According to the dynamic encounter hypothesis (see Chapter 4), the earth originated as a mass of hot gases derived from the sun as a result of a near-collision with another star or a comet. The gases condensed, forming magma; then on further cooling, a crust formed over the magma. Kelvin believed the earth's interior to be composed of liquid left from this original magma. According to Kelvin, heat lost from the cooling of the interior is the source of the energy expended in volcanism and orogeny. Elie de Beaumont (1829) saw in this cooling a simple way to account for the folded and thrust-faulted mountain belts formed by strong lateral compression in the crust. According to this hypothesis, the volume of the earth's interior decreases as it cools; and the outer crust is forced to accommodate itself to a shrinking internal sphere by wrinkling or folding of the crust.

The idea that the orogenic belts originated as a result of crustal compression, caused by contraction of the earth, was the most popular theory of mountain-building until the advent of convection hypothesis in the 1930s. But the fundamental assumption of the theory—that the earth is cooling and contracting—is no longer accepted. The discovery of large quantities of radioactive minerals in the crust, which release heat as they decay, makes it doubtful that the earth is cooling. No clear evidence can be found that proves that the earth has contracted, and some evidence, such as the grabens along the oceanic ridge system, even suggests that it is expanding.

The Theory of Isostasy

Neither George Everest nor any member of the team that went with him to India in 1850 to measure the length of a one-degree meridian arc realized that a discrepancy in their measurements would stimulate inquiries that would lead to the development of one of our most important concepts about the nature of the lithosphere. The Everest survey used two techniques to measure the difference in latitude between two cities located just south of the Himalayas. One method involved careful measurement of the north–south distance between the two cities by use of ground measurements and conventional surveying techniques. The other method involved determining the difference in the angle to a star measured from each city. As the work progressed, Everest discovered that the difference in latitude determined by these two methods was much greater than expected. The error amounted to $5\frac{1}{4}$ seconds of latitude in a distance of only 600 kilometers. He could find no error in either set of data to explain a discrepancy of this size. These data quickly attracted the interest of a number of scientists, including a clergyman named J.H. Pratt in Calcutta, who concluded that the error had been introduced because the leveling system, a plumb bob (a weight suspended on a string), used to set up the telescope for the astronomical observations had been thrown off by the force of gravitational attraction between suspended weight and the mass of the Himalayas. In order to prove this point, he calculated what the deflection should be at each of the two stations, using his estimate of the mass of the nearby mountains. To his surprise, his calculations indicated that the two readings should have been off nearly 16 seconds instead of $5\frac{1}{4}$ seconds. The mountains did not cause nearly as much deflection of a plumb bob as they should; this led Pratt to think that part of the attraction must be compensated for within or under the mountains by the presence of a mass of less dense rock.

Pratt was not the first to reach this conclusion. Pierre Bouguer, a member of a French expedition sent to Peru in 1749 to establish the length of a meridian arc there, reported that the Andes Mountains have less mass than had been expected, but he did not bother to explain this finding, and the idea received little further attention until the Everest expedition. Much earlier, in the sixteenth century, Leonardo da Vinci commented on the question of why mountains stand so much higher than surrounding lands, and he too concluded that mountains stand high because they are composed of less dense matter than the plains.

We now have a much more accurate picture of the great differences in elevation that exist on the surface of the earth than either Leonardo or Bouguer had. Great differences in level between the highest mountains and the ocean floor have been sustained over the time span of human observations; they indicate that either the earth's lithosphere is rigid and strong enough to support these differences in mass indefinitely, or that, if the earth's interior is fluid, the weight in these different crustal elements is buoyed up in some way. Of course, there would be no problem in explaining great relief on the surface of the earth if the earth were solid or if the outer shell of the earth were rigid enough to sustain the weight of mountains. But we would in that case face other problems—such as how the mountains are uplifted, or how they are maintained high in the face of unrelenting erosion over billions of years. In the times of Leonardo and Pratt, the notion of a rigid earth was rejected because the interior of the earth was thought to be molten throughout; it is rejected today because seismic studies indicate the existence of a plastic layer, the Gutenberg low-velocity zone, the asthenosphere (see Section 9.4), beneath the lithosphere.

Even accepting the existence of a plastic or fluid zone at depth, there are still several possible explanations for the variation of density in the overlying rocks. Pratt concluded that high land areas, like mountains, are caused by the expansion of rock that was originally deep and of uniformly high density. He envisioned a model in which the density of rock is uniform below a certain depth but decreases with elevation above that depth, as shown in Fig. 12.1(*a*). This model allows a mountain range to have lower density than rock beneath adjacent plains, accounting for the observed deflection of the plumb bob. Consequently, the various parts of the earth that now stand at different elevations do so because they are underlain by rock of different densities—the higher the elevation, the lower the density. His calculations indicated that surficial differences in gravity would be evened out at a depth of about 150 kilometers, and this, according to Pratt, is the depth at which compensation occurs. Below that, the density should be nearly uniform. The total mass of each of the main units of the lithosphere, such as mountain ranges, shields, or ocean basins, should be nearly the same, although the average density of the rock in these units might differ.

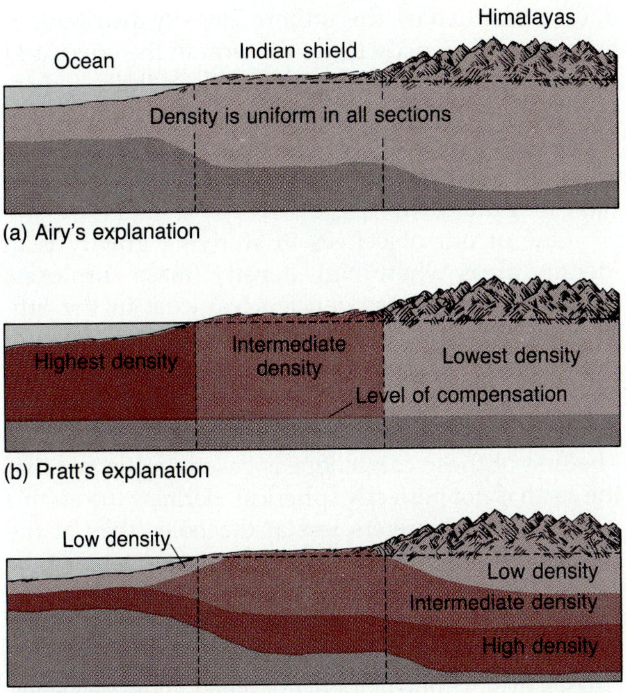

Figure 12.1
(a) Pratt's theory of isostasy. Pratt recognized that the crust of the earth is not made up of materials of uniform density. He visualized isostatic equilibrium as being the result of density variations. He postulated that elevation of an area is an indication of crustal density—the lower elevations correspond to higher density crust. Furthermore, he postulated a depth or level of compensation at which the effects of different densities above were balanced. The horizontal distance across these sections is several hundred kilometers. (b) Airy's theory of isostasy illustrates another way isostatic equilibrium might be obtained if the crust of the earth consisted of columns of rock of different heights with essentially uniform density. Airy would account for the mountains by the presence of deep low density roots of crustal material extending down into the high-density mantle. Because the mantle rock can flow, the crust may be thought of as floating on it. (c) Heiskanen's theory of isostasy. He combines the assumptions of both Pratt and Airy. According to his theory, density varies both between columns and within each column. The higher densities are represented by the shorter columns and toward the bottom of each column. This theory accounts for the roots of mountains, and for the variations in density in different parts of the crust.

Within a few months of the time Pratt submitted his paper to the Royal Society of London, another scientist, G.B. Airy, submitted a paper (1855) on the same subject. Airy had a different idea about the mechanism of density distribution. He suggested that the density of various segments of the lithosphere is uniform, and that they are of lower density than the underlying material. Thus, he believed the mass deficiency in mountains is best explained if it is assumed that the mountains have low-density, root-like projections beneath them that protrude down into a denser substance, as shown in Fig. 12.1. He used the analogy of a number of floating blocks of wood of different size (Fig. 12.2). (Remember that an object submerged in fluid is buoyed up by a force equal to the weight of the fluid it displaces.) The block that extends deepest into the water also has its top surface farthest out of the water. Thus originated the idea that mountains have roots.

Airy's and Pratt's hypotheses were proposed long before we had the results of seismic exploration of the crust and upper mantle. They even preceded the availability of much information about the density of various crustal elements. We discussed many of the relevant discoveries in earlier chapters:

1. Continents are composed of less dense rock than the basalt of the sea floor.
2. Continents are somewhat stratified according to density—that is, less-dense sedimentary rocks predominate at high elevation. They are under-

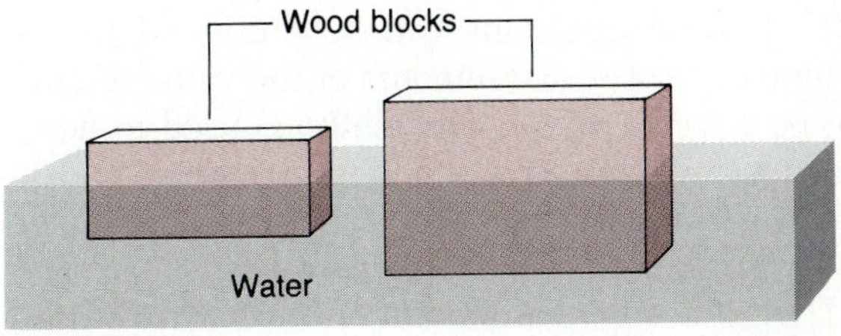

Figure 12.2
Blocks of wood of different sizes but the same density floating in water are analogous to Airy's hypothesis of isostasy.

lain by denser granite; and, at depth, the lower part of the crust is probably still-denser gabbro.

3. Moho is much deeper under continents (40 kilometers) than it is under ocean basins (10 km), and is even deeper under mountains under most continental crust.
4. Both continental and oceanic crust rest on solid and rigid high-density peridotite in the uppermost mantle.
5. A weak plastic zone (the low-velocity zone) lies at greater depth (100 to 200 km).

Neither the Pratt nor the Airy model fits what we now know about the lithosphere. A better model would incorporate elements of both. Both mountains and continents do have roots as Airy suggested by his model, but the average densities of different crustal elements vary as Pratt proposed. Actually, density also varies with elevation in different parts of the lithosphere. A model that includes these three features, and thus comes closer to fitting our current knowledge, was proposed by a Finnish geophysicist, Heiskanen. Airy's, Pratt's, and Heiskanen's models are compared in Fig. 12.1.

All three of these models have one thing in common. They all suggest that mass near the surface of the earth is essentially floating on the material below. The outer part is said to be in a state of **flotational equilibrium,** an idea that was expressed by C.E. Dutton in 1880 as the theory of **isostasy.**

It follows from the notion of isostasy that areas of less-dense rock stand higher than areas of high-density rock. In addition to this, we know that some sections of the earth's crust are rising, and others are subsiding. Thus, important questions for us are: how close is the earth to being isostatically balanced, at what level does this equilibrium exist, which vertical movements are related to isostatic adjustments, and how fast do these adjustments occur?

Measuring gravity to test the theory of isostasy. Some of these questions can be answered by the ingenious use of measurements of the value of gravitational acceleration, g. The methods used to determine g using pendulums and gravimeters are described in Section 3.2. It would be useful to review that section before proceeding here.

The value of g at any point on the earth's surface results from the attraction of the mass of all matter near the measuring instrument. The largest of these is the mass of the earth itself. Taken as a whole, the effect of the mass of the earth on the measuring instrument is the same as if it were concentrated at the center of the earth.

The value of g would be the same everywhere if the earth were perfectly spherical and uniform in density (or if the density were uniformly distributed in homogeneous concentric shells). The deep interior does appear to have this uniform density distribution, but we know density is not uniform in the crust, and seismic velocities in the upper mantle indicate density and compositional variations there as well. These near-surface variations in density should be reflected in variations in the value of g measured at different places on the earth.

One of our objectives in studying gravity is to identify areas where high density (mass surpluses) or low density (mass deficiencies) exist in the lithosphere. In order to find these areas, we follow a procedure that includes first measuring the local value of g and then taking into account any already-known factors that would affect the gravitational attraction. The first of these corrections is based on the fact that the earth is not perfectly spherical. Because the earth's radius is 21 kilometers less at the poles than at the equator, readings of g at high latitudes tend to be higher than those at low latitude.

The effect of elevation above sea level on the reading must also be removed. The value of g decreases by 0.3 milligal for every meter above sea level. This effect, the **free air correction,** depends on differences in the distance to the center of the earth at various places. But there is also additional mass between the point of measurement and the center. The attraction of the rock between sea level and the elevation at which the reading is made is removed. This correction, called the **Bouguer correction,** is about 1 milligal per meter of elevation.

Once these three corrections are made, the value g would have on an ideal homogeneous earth, is subtracted from the adjusted reading. Any resulting difference is called a **Bouguer gravity anomaly.** A negative value signifies a mass deficiency in the area where the reading was taken; a positive value signifies a mass surplus. In general, mountains and continents are associated with negative Bouguer anomalies. The deep-sea floor and a few areas like volcanic centers usually have positive anomalies.

Gravitational analysis has been extended to identify areas that are not isostatically balanced. This is done by estimating the value of g at successively greater depths in the earth. For each calculation, the change in elevation and the effect of the mass in the slab of rock above this elevation are removed. The results of this type of calculation indicate that very few anomalies exist at a depth of about 113 km. Any remaining anomalies at this depth are called **isostatic anomalies.**

Isostatic anomalies apparently vanish because of the low-velocity zone. This zone is the most probable place for flotational equilibrium of the lithosphere to exist, because the material in this zone is not rigid. Presumably, it can flow like a plastic.

Isostatic adjustments. The theory of isostasy provides the best explanation for the cause of many vertical uplifts of the earth's crust, which is known to vary in thickness, composition, and density, and is constantly undergoing change through the shifting of loads of rock on the surface as a result of erosion, transportation, and deposition. Over long periods of time, these processes shift mass mainly from the high mountains to the lower parts of the surface. In addition, the distribution of mass is changed by the igneous activity that builds volcanoes and lava plateaus and emplaces plutons, and by the orogenic processes that form mountains.

According to the theory of isostasy, over long periods of time, loads on the crust should be approximately balanced. If mass is removed from one area on the earth's surface, equilibrium is upset, and mantle rock, behaving like a plastic, flows and exerts a pressure on the crust, tending to elevate the area from which mass has been removed and to restore equilibrium.

Glacial rebound. The most dramatic demonstration of isostasy comes from areas that were covered by thick masses of ice during the last major Ice Age (Fig. 12.3). Thickness of continental glaciers exceeded a thousand meters in central Canada and in Scandinavia before the climate warmed, and the glaciers began to recede about eleven thousand years ago. If isostatic theory is correct, the extra weight of the ice in these areas depressed the lithosphere as the ice accumulated, and the lithosphere should rise or rebound as the ice melted. How quickly these adjustments occur depends on how rapidly the ice accumulated and melted and the rate at which material in the mantle can move to make the necessary adjustments.

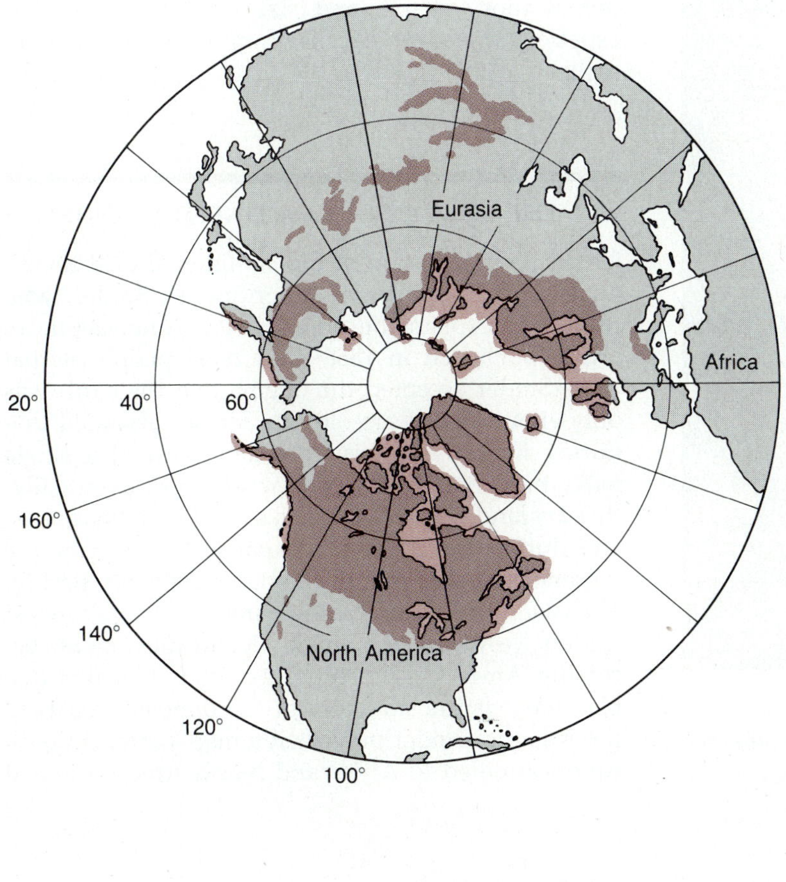

Figure 12.3
Polar projection of the northern hemisphere showing areas covered by ice during the last glacial maximum.

Repeated surveys of the level of the land along railroads from southern to central Canada show that the area centered on Hudson Bay is indeed rising, and gravity surveys reveal negative isostatic gravity anomalies in these areas. Other evidence, for example the existence of shoreline features (such as beaches) that are now elevated and warped upward the Hudson Bay region, proves that this region is today much higher than it was immediately following the melting of the glaciers. The fact that this region is still rising proves that the rate of isostatic adjustments is slow, requiring thousands of years in some cases. We might envision the movements in the mantle below the lithosphere as being a slow creeplike flow, perhaps analogous to the creep of a plastic under pressure.

Uplifted beaches and other shoreline features along the coast of Scandinavia, including marks left on rocks where boats were docked by early seafarers, have been dated and used to define the rate of uplift of Scandinavia. The results, in Fig. 12.4, show the rates are greatest in central Scandinavia where the greatest thicknesses of ice accumulated. This, also, is an area of negative isostatic gravity anomalies.

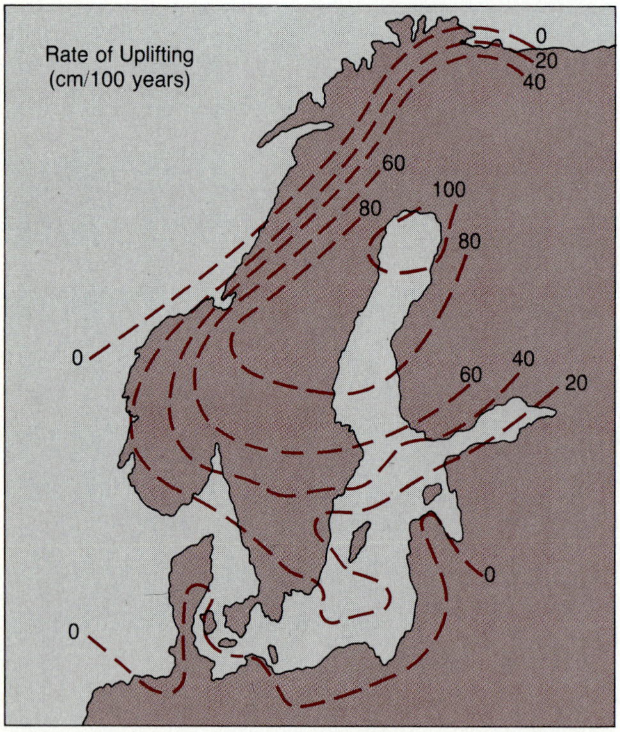

Figure 12.4
Rate of uplift in Scandinavia. This map shows the present rates of uplift in centimeters per 100 years.

Later in this chapter and in Chapter 13 we shall see that the theory of isostasy plays an important role in modern theories of continental drift, plate tectonics, and mountain-building. It is also critical to the next theory we will consider, that of convection currents. It did not, however, lead to the development of the theory of continental drift. For many decades, the two theories coexisted but were not clearly related.

The Convection Current Hypothesis

Geologists have postulated for decades that there is convection of the hot materials in the earth's interior, caused by the temperature difference between the surface and the core. Hot plastic material deep in the interior rises toward the surface as matter cooled near the surface sinks. This movement presumably set up large cells in which movement is directed upward on one side, down on the other. Convection cannot exist in the crust because the rock there is too rigid, but within the mantle, where high temperature and confining pressure are thought to make the material plastic, convection is possible. It is also possible in the fluid outer core.

The nature of the earth's magnetic field and temperature gradients within the earth strongly support the idea that convection exists in the earth, but the location, shape, and size of convection cells, especially the deeper cells, cannot be detected directly.

12.2 THE THEORY OF CONTINENTAL DRIFT

Credit for conception of the continental drift theory appears to belong to an American, A. Snider, who used it to explain the origin of the Americas in an article published in 1858. Like most people at that time, Snider accepted the dogma that the earth was only about 6000 years old. He further assumed that during its early history, dry land formed a single mass hardened by underground fires. Eventually, this dry land was split by cracks lined with volcanoes, and during Noah's Flood, expansion caused by the volcanoes further split the land along a major fissure. The island of Atlantis, accompanied by its volcanoes, drifted westward in a single violent movement, becoming America. The proof he offered of this displacement is the similarity of the opposite coasts of the Atlantic. Snider published a map showing South America joined to Africa and North America joined to Europe—a reconstruction he made to explain the similarities of fossil plants found in Carboniferous coals of both Europe and North America.

The modern debate over continental drift started with papers published by F.B. Taylor in 1910 and by Alfred Wegener in 1915, and since that time the question of the permanence of continents and ocean basins has been intensely argued. Wegener advanced the idea of continental drift to explain anomalous distribution of ancient climatic belts—for example, the observation of fossils of warm-climate plants in coal deposits near the poles. Evidence of a late Paleozoic glaciation on parts of widely separated continents proved to be one of the most convincing lines of evidence of continental drift. This glaciation affected parts of Antarctica, South America, Africa, India, and Australia, as shown in Fig. 12.5. Many aspects of this glaciation support the idea that these continents were close together during the growth and decay of the ice sheets. The same fossil flora, known as the *Glossopteris* flora, occurs in sediments deposited close to the ice on all these continents. Rocks apparently picked up and carried by the glaciers in Africa are now found in South America. Grooves cut in rock by advancing ice show parallel directions in South America and Africa but only if in a predrift reconstruction of the southern hemisphere. In addition to these matches, it would also be much easier to explain glaciation if all these areas were close to the pole. He attempted to fit all continents into a single protocontinent called **Pangaea** (Fig. 12.6); the close puzzlelike fit of the coastal outlines on either side of the Atlantic remains one of the convincing lines of evidence for continental drift. Other predrift reconstructions have favored the existence of two primitive continents—**Gondwanaland,** which included Africa, South America, India, Australia, and Antarctica; and **Laurasia,** which was composed of the continental masses in the Northern Hemisphere.

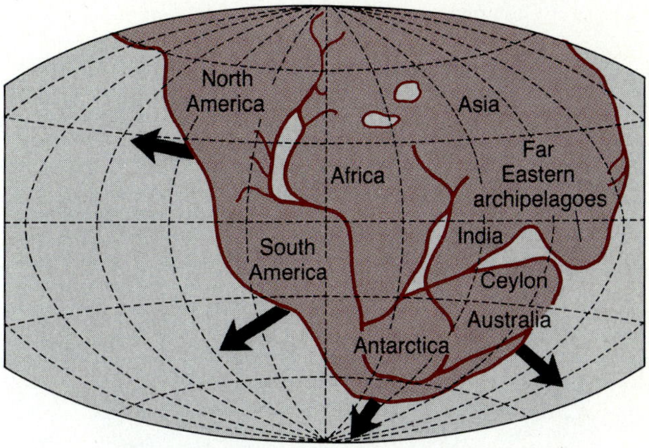

Figure 12.6
Hypothetical arrangement of the world's continents in a single supercontinent called Pangaea. From a single mass like this the continents are thought to have drifted apart.

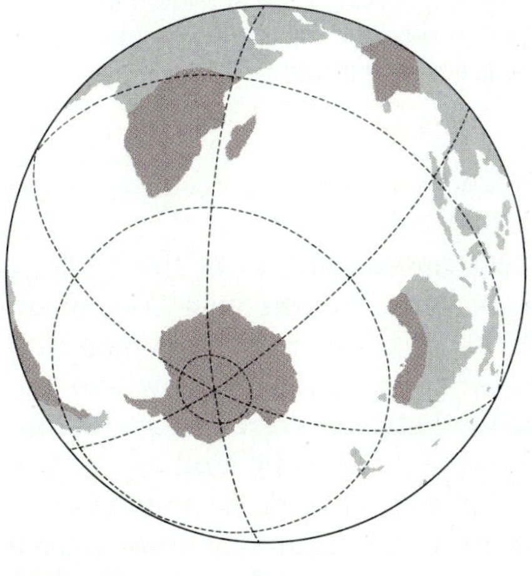

Figure 12.5
Ancient climatic zones were one of the main interests of Wegener that led him to postulate Pangaea. This map shows the present distribution of the land areas that were glaciated toward the end of the Paleozoic era. Note that all of these areas would be close together and close to the south pole in the predrift reconstructions.

The Fit between Africa and South America

Many attempts have been made to reconstruct the continents as they were before drift took place. Generally those who try to work out the puzzle attempt to match the shapes of the continental margins, and most efforts have met with only limited success. It would be rather unlikely for the lines along which the original continental plate(s) broke to be faithfully preserved during the millions of years of separation and drifting and through all subsequent changes. In fact, it is remarkable that the fit between South America and Africa is as close at it is. S. W. Carey, an Australian geologist, first demonstrated the exactness of this fit in 1958 (Fig. 12.7), by using maps drawn on a spherical surface. He found that the best fit is obtained when the continents are connected along a line 2000 meters below sea level. This fit is so perfect that it stood as one of the most convincing lines of evidence throughout the long debate over the theory of continental drift.

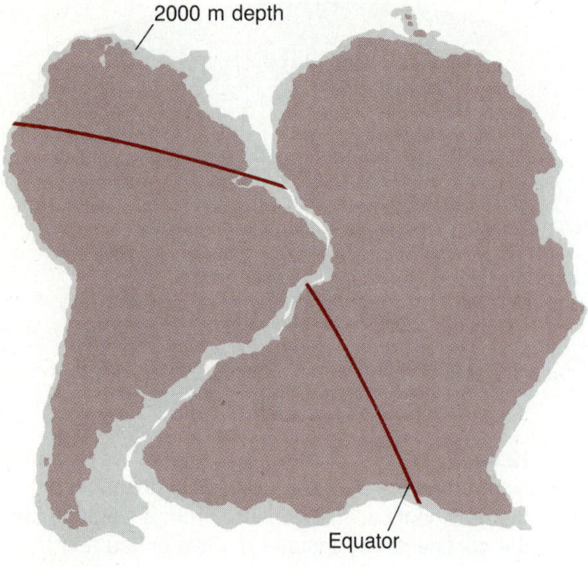

Figure 12.7
Fit between South America and Africa at a depth of 2000 meters, as constructed by S. W. Carey. The present-day equator is shown for reference.

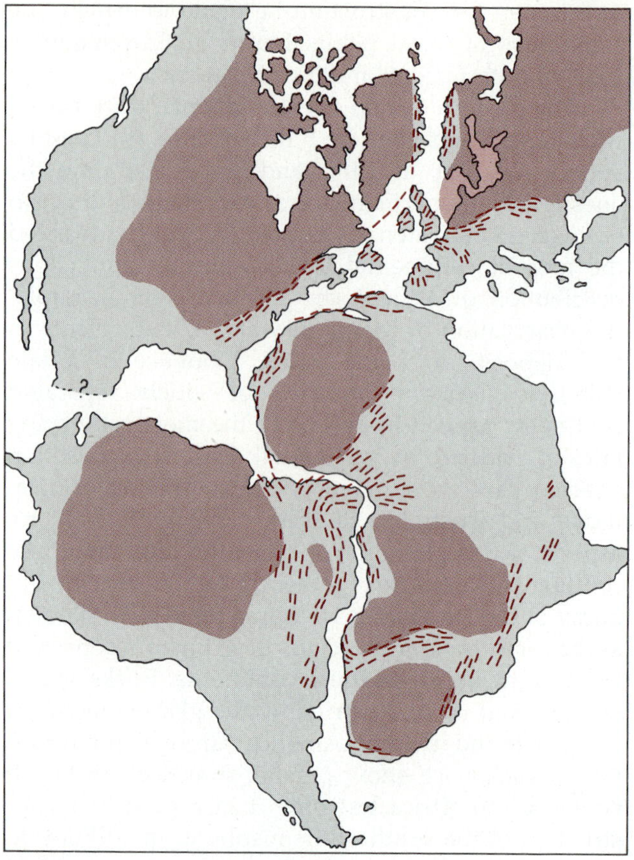

Figure 12.8
This reconstruction of continents to their predrift position is based on a computer program developed by Sir Edward Bullard. The approximate position of the shields and platforms (dark gray) and orogenic belts (red lines) close to the line of fit are also indicated.

Good fits are also found across the Gulf of Aden, the Red Sea, the Gulf of California, between Africa and Madagascar, and between Africa and India. Perhaps the reason for the good fit between Africa and South America is that the break between these two occurred through a Precambrian shield in which the older orogenic belts had been completely welded together by metamorphism. In contrast, complex Paleozoic mountain belts of North America, England, and eastern Europe lie along the line of rifting farther north (Fig. 12.8).

Many geologists have worked at matching such geologic features as rock types, age provinces, and fold belts on one continent with those of another, and these efforts have often been unsuccessful—a fact that is not too surprising, for if the fit should be made at the 2000 meter depth, the geologic provinces on land were originally separated by the width of the continental shelves including part of the continental slopes on either side of the ocean. Thus, only approximate fits can be obtained. For example, the Appalachian Mountains, whose northern extremity is at Newfoundland, may be connected with the Caledonide Mountains of Ireland, Scotland, and Norway. Restoration by closing the North Atlantic leaves a covered shelf 320 kilometers wide between Newfoundland and Ireland on which no rock exposures can be seen.

One of the most careful studies of the geology across the Atlantic was made in the mid-1960s when P.M. Hurley and others compared the geology on either side of the Atlantic at the point where the predrift reconstruction between West Africa and Brazil, based on topography, places the land areas close together, as shown in Fig. 12.9. A large number of samples were taken on each continent from coastal localities. The results of this study show a close correspondence between structural trends and ages when the continents are restored to their predrift positions. The geology in these areas coincides so closely that the match left little doubt in most geologists' minds that the two continents were part of a

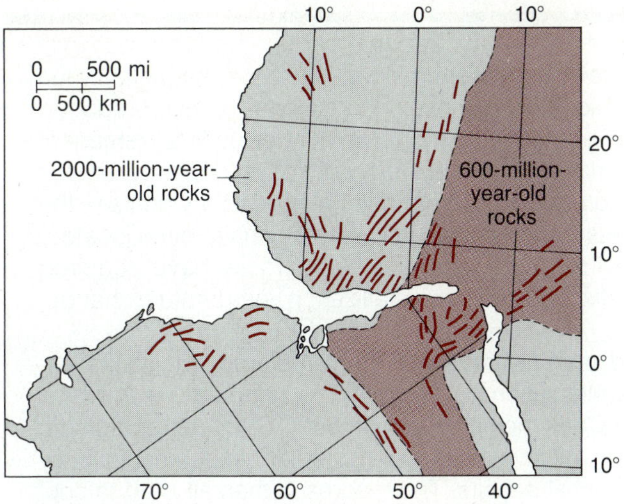

Figure 12.9
Two shield areas adjoin one another along this section of the predrift reconstruction between Africa and South America. Some of the structural features in the shields, such as fold trends and areas containing rocks of similar ages, match in this fit.

single continent when the rocks exposed there were metamorphosed and deformed.

Despite the evidence of the fit of the continents in Precambrian time and the continuity of the fossil record across the oceans, most American geologists remained unconvinced of continental drift—largely because there was no mechanism to explain it. Interest in continental drift renewed in the late 1950s with the analysis of new evidence of a surprising kind—permanant magnetism in rock.

When an iron bar is cooled or even stored in a magnetic field it becomes a magnet itself. Geologists have learned that some rocks also acquire magnetic properties as they gradually cool in earth's magnetic fields (see the box on "Remanent Magnetization"). In effect, such rocks indicate the direction of the magnetic field at the time they chilled.

Polar Wandering: Revitalization of Continental Drift Theory

The development of techniques for determining the orientation and strength of the magnetic field from rocks makes it possible to determine where the poles were located at times in the geologic past. Many studies have been made of these magnetic characteristics for rocks of various ages and for rocks of the same age located in diffrent areas. The results of these studies indicate that the position of the poles relative to continents has not remained constant through time. Studies in North America indicate that the position of the north magnetic pole has shifted along the line indicated in Fig. 12.10 from a Precambrian location in the central Pacific to a Triassic position in Siberia, and finally to its present location. This indicates a shifting of the pole in relation to modern geographical distribution of land.

Plotting paleomagnetic data from Europe on the same diagram shows a similar shift of poles, but the two lines do not coincide as they should if the geographic positions of Europe and North America had remained fixed relative to one another. The two polar wandering paths do coincide, however, if we shift the geographic positions of Europe and North America back to the suggested predrift position shown in Fig. 12.8. Studies of the same type from other continents also reveal shifting of continents in relation to one another.

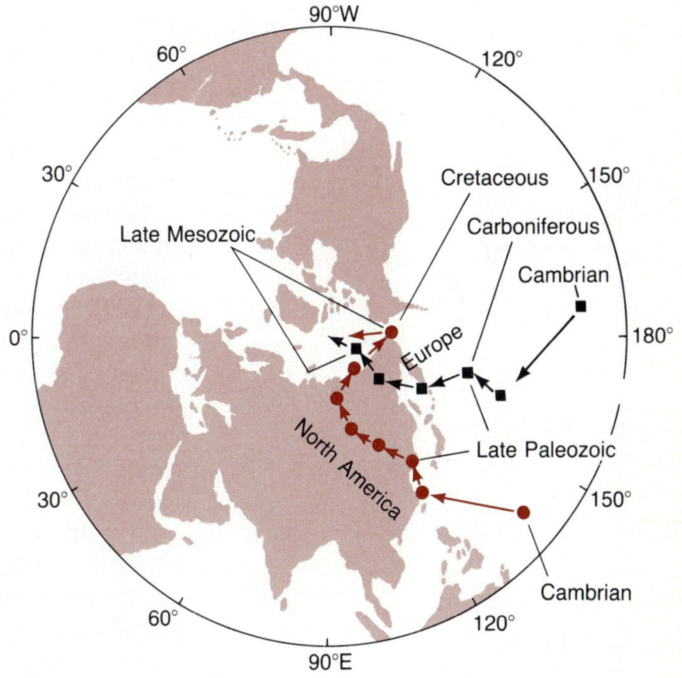

Figure 12.10
The two heavy lines show the position of the magnetic north pole as determined in Europe (squares) and in North America (circles) at various times in the geological past. These apparent polar wandering paths are not identical, because North America and Europe have drifted apart.

REMANENT MAGNETIZATION

Many minerals, especially iron-bearing ones, become magnetic when placed in a magnetic field. Most of those minerals retain a much-reduced magnetism when removed from an external field, referred to as **remanent magnetization.** The remanent magnetization of rocks depends on the quantity and composition of the ferromagnetic minerals they contain, the environment (heat, pressure, chemical) in which they have been located, and the history of environmental changes. Most rocks have acquired their permanent magnetization during crystallization, temperature change, or deposition.

Most igneous and many high-temperature metamorphic rocks acquired permanent magnetization on cooling. The temperature at which this takes place, the **Curie point,** is about 550 °C for ferromagnetic minerals. Magnetization of lavas has been studied experimentally by reheating volcanic rock until it melts. The magma is then allowed to cool in a magnetic field of strength similar to that of the earth. The magnetic properties of the lava are observed as temperature decreases. The intensity of magnetization in the specimen increases as temperatures are lowered through the Curie point. When the cooling lava reaches normal temperatures, the magnetization is permanent and is unaffected by changes in the external magnetic field.

Magnetization of sediments was first noted in studies of clays deposited in glacial lakes. The clays exhibit remanent magnetization, ascribed to the preferred orientation of iron oxide grains within the earth's magnetic field that existed at the time of deposition. The grains behave much like small compass needles. The theory is easily tested by dispersing the clay in water and allowing redeposition in the laboratory. When this is done, the clays become magnetized in the same general direction as the external field surrounding them when they are deposited.

Red sandstones and siltstones are often strongly magnetized and sometimes show a marked magnetic stability. When the direction of magnetization of Precambrian pebbles cemented in a much younger (Triassic) conglomerate was measured, the directions were found to be random. Thus the ancient remanent magnetization had been stable for millions of years through processes of erosion, deposition, compaction, and cementation. Similarly, it has been found that the direction of magnetization of some folded sandstones does not agree with that of unfolded rocks of the same age until the directions are rotated by the amount of the folding (Fig. 12.11).

In order to determine the strength and orientation of the magnetic field for any particular time in the geological past, called paleomagnetism, it is necessary first to find rocks of that age that possess a remanent magnetization imparted to them at that time. Next, the orientation of the remanent magnetization of the sample is determined. If we have been careful to observe the orientation of the sample in the rock, the field in the sample can be related to that in the rock outcrop (usually a core drill is used to obtain samples). The orientation (the direction to the north pole) should be consistent within an undeformed region.

Although the earth's magnetic poles do not now coincide with the poles of rotation (the geographic poles), paleomagnetic studies indicate that the magnetic poles have remained close to the poles of rotation, at least during the last couple of million years. The observed polar wandering of earlier periods is not attributed by geologists to movement of the magnetic poles relative to the poles of rotation. It is thought, instead, to be the result of movements of the earth's crust relative to the deeper interior of the earth where the magnetic field originates.

The determination of polar wandering paths in the 1950s was one of the most important factors in rekindling interest in the theory of continental drift among American geologists. Here was an independent line of evidence of drift, and though it did not resolve the problem of drift mechanics, it did stimulate many earth scientists to look for new ways to evaluate the drift hypothesis. One approach was to take a more careful look at the fit between continents and especially at the one between Africa and South America, described in Fig. 12.9.

A mechanism for continental drift finally emerged in the 1960s, beginning with the idea that new sea floor is being created along the oceanic ridges.

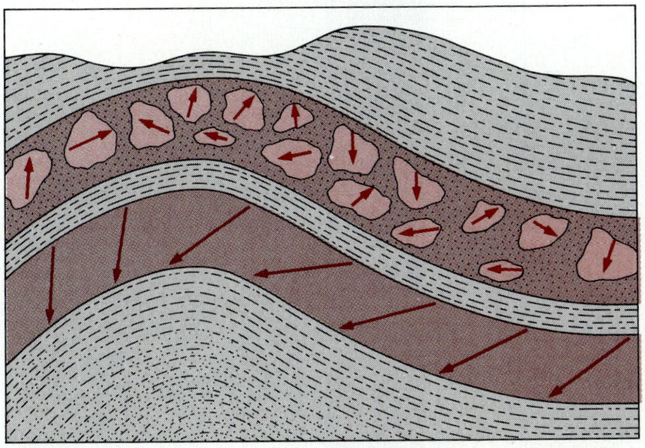

Figure 12.11
The arrows in this figure indicate the orientation of the remanent magnetization. Each rock in the conglomerate bed has a different orientation. Thus, we may infer the magnetization was acquired before the conglomerate formed. In contrast, the lower bed contains a systematically changing orientation clearly related to the folding. All of those arrows are parallel if the bed is unfolded. This magnetization was acquired after the bed was deposited but before the folding.

12.3 SEA-FLOOR SPREADING

The concept that both continents and ocean basins are fixed in position and that they have been essentially permanent features of the lithosphere was popular among North American geologists until the late 1950s, when the preponderance of accumulated geological and geophysical evidence clearly favored the theory of continental drift. One of the principle arguments against drift was the lack of a mechanism by which the continents could be moved. The continental crust is over 40 kilometers thick, and the oceanic crust is composed largely of basalt. How could a continent possibly move through a rock as strong as basalt? This mechanical difficulty was finally resolved when a mechanism known as sea-floor spreading emerged in the early 1960s.

The 1950s saw great achievement in the study of the sea floor and of the earth's crust and upper mantle. The convection current hypothesis, greatly strengthened by heat-flow measurements over ridges, came to be widely accepted as explaining the process by which trenches, mountains, and oceanic ridges are formed. Knowledge of the composition and thickness of the crust of continents, continental margins, the sea floor of the deep basins, island arcs, and oceanic ridges had been obtained from gravity and seismic studies, and these results were being incorporated in our model of the crust.

The concept of sea-floor spreading first found expression in print in papers by H.H. Hess (1961) and R.S. Dietz (1961), both of whom suggested that continental drift is driven by convection currents operating in the mantle. These currents rise under the oceanic ridge system and then spread laterally from the ridge, carrying the oceanic crust with them (Fig. 12.12). This spreading causes the sea floor to split and spread apart at the ridge crest. In response to this spreading, fault blocks, especially dropped grabens, form median valleys on the ridge crest. Magma rises from below the ridge crest, comes to the surface along the opening fissures, and erupts as submarine lava flows and, in places, as volcanic centers such as those at Iceland and in the Azores. This near-surface magma and upward-directed convection currents account for the high heat-flow values observed over the ridges.

The Gutenberg low-velocity zone in the mantle, which forms the base of the lithosphere at a depth

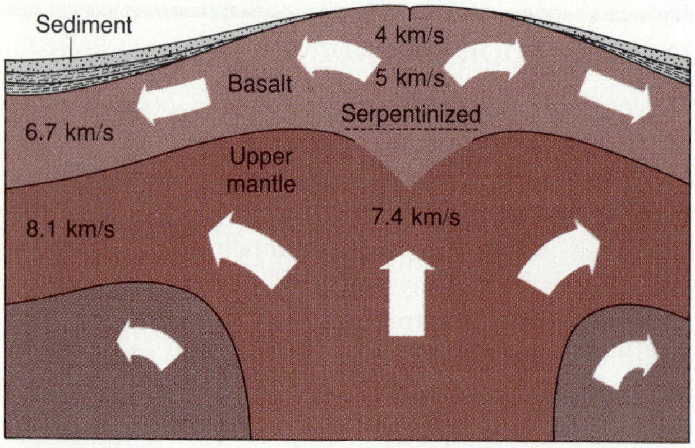

Figure 12.12
Convection currents rising beneath oceanic ridges were suggested by Harry Hess as the reason for sea-floor spreading.

of 100 to 200 kilometers, is an important feature of this model. Partial melts and high temperature make the material in this zone more plastic than that in the upper mantle and crust. This, then, is a likely place for the lithosphere to become detached from and move relative to the mantle. From this, we should expect that the continents move, not through the sea floor, but with it, riding passively on the lithosphere with parts of the oceanic crust. Thus, the mechanical problem that had so perplexed American geologists was resolved.

The theory of sea floor spreading quickly became a central focus for many facets of geology—among them studies of crustal structure, of land form, of rock origin, and of the history of ocean basins. The theory was put on an even firmer footing as a result of research designed to test its validity by studying magnetic data collected along the mid-Atlantic ridge.

Reversals of the Earth's Magnetic Poles

The remanent magnetization in a rock specifies the directions to the magnetic north and south poles at the time the rock acquired its remanent magnetization. As data from remanent magnetization accumulated, a surprising result appeared: the polarity of some of the specimens from particular rock masses is the reverse of the present magnetic field; they

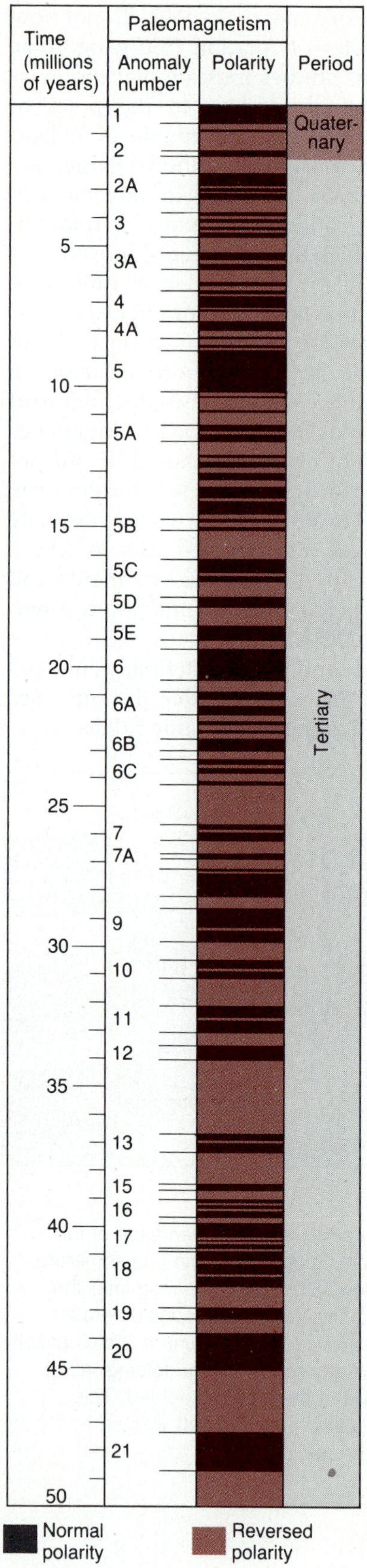

Figure 12.13
The times during which the magnetic field was normal and reversed have now been determined with enough accuracy to allow construction of the geomagnetic time scale shown here.

indicate that the north magnetic pole was in the Southern Hemisphere when the rock was magnetized. **Magnetic reversals** of this sort have been found in Tertiary lavas of the Massif Central of France; the Columbia plateau of the northwestern United States; Iceland; Japan; and other places in all parts of the world. Reversals have also been found in sediments of most geologic ages. Two explanations appear possible: either the earth's field has periodically reversed, or some process produces a reversal in rocks that is independent of the geomagnetic field. The first of these hypotheses is almost universally accepted as the best explanation for most of the observed reversals.

Analysis of the magnetization direction of dikes and the immediately surrounding country rock that was heated above the Curie point during intrusion of the dikes indicates that both the dikes and the altered country rock have the same direction and polarity. If real reversals of the geomagnetic field do not occur, we might expect that the dikes and country rock would, at least occasionally, have opposite directions of magnetization. Actually, both have identical direction regardless of whether it is the same or the reverse of the present geomagnetic field, and about half the time it turns out that both are the reverse of the present field.

The earth's field has apparently undergone frequent reversals. The times of these reversals can be established when radiometric ages as well as polarity can be determined for rock samples. These data make it possible to construct a geomagnetic time scale. The scale for the last 50 million years is illustrated in Fig. 12.13. As might be expected, details of the scale have changed as more and more data have been obtained, but the main features seem now to be well established. Long-term weakening of the present geomagnetic field has been well documented. The field may weaken gradually until it fades out and reappears reversed, gaining in strength to another maximum. If the field did disappear, the middle latitudes of the earth's surface would be subjected to a much greater bombardment of particles from the sun, which are normally funneled into the polar regions.

Magnetic Anomalies: A Key Test of Sea-Floor Spreading

When American geologists Raff and Mason measured the earth's magnetic field off the west coast of North America in the late 1950s, they found surprisingly great variations in its strength. Plotted and contoured on a map, the values of field strength showed a pattern of long linear belts of alternating low and high magnetic field strength (Fig. 12.14). The bands of unusually high or low field strength became recognized as **magnetic anomalies,** abnormally high or low values.

In 1963, F. J. Vine and D. H. Matthews suggested that these variations in intensity are caused by changes in the strength and polarity of the earth's magnetic field that had become "fossilized" in the oceanic crust. As new crust forms along the crest of the oceanic ridges, it cools, passing through the Curie point (temperature of about 550 °C), at which the magnetic-field strength and polarity are "frozen" in iron-bearing minerals as remanent magnetism. According to this interpretation, the numerous belts of unusually high field strength observed over and parallel to the oceanic ridges indicate the location of oceanic crust (mainly composed of basalt) that was intruded and cooled through the Curie point at times

Figure 12.14
This pattern was obtained when anomalies in the strength of the magnetic field off the western coast of North America were colored. All black areas have unusually high magnetic field strengths. The red areas have unusually low magnetic field strengths. This study was one of the first in which this type of alternating banded anomalies was discovered.

when the polarity of the magnetic field was the same as it is today—so the remanent magnetism adds to the present field. The belts of negative anomalies, low field strength, mark intervals of reversed polarity. Vine and Matthews reasoned that if this is the correct explanation of the magnetic anomalies, then a belt of high values should be found over the ridge crest where new basalts are now cooling; this high would coincide with the present interval of normal polarity. In addition, if the sea-floor spreading theory is correct, then belts of high and low magnetic anomalies should run parallel with the ridge crest, and they should be more or less symmetrically disposed on either side of the ridge crest. The ridge crest should be immediately flanked by belts of low field-strength value, caused by the presence of underlying basalt formed during the last period of reversed polarity; and successive anomaly belts out away from the ridge crest should correspond to oceanic crust formed during each successive change in polarity. The magnetic fields over the Indian Ocean Ridge and the Mid-Atlantic Ridge just south of Iceland both show a pattern of magnetic anomalies that fits Vine and Matthews's model (Fig. 12.15). These patterns have proven to be a powerful tool for interpreting the history of the ocean basins.

The anomalies can be correlated with reversals found in rocks on continents that can be dated by radiometric methods (Fig. 12.13). Thus, the anomaly belts provide a way to determine the age of almost any part of the sea floor, and they make it possible for us to measure the rate at which spreading is taking place. Using dated samples from land areas, it is possible to construct a record of ancient reversals—essentially a **paleomagnetic time scale.** By numbering the anomaly stripes corresponding to polarity changes out from a spreading ridge crest, we can determine the age of a particular strip from the geomagnetic time scale. The distance of an anomaly belt of a known age from the crest of the ridge is used to calculate the rate of spreading. Plots of anomalies against distance (Fig. 12.16) indicate that the sea floor in the Atlantic has been spreading at a

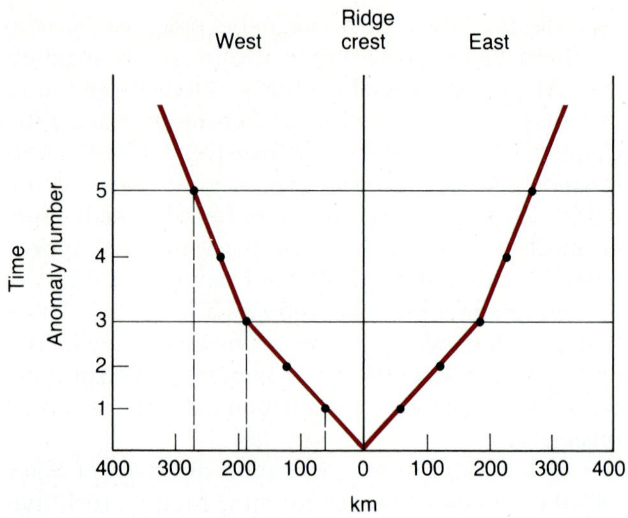

Figure 12.16
Each clearly recognizable band of magnetic anomalies on the flank of ocanic ridges is assigned a number. These numbers correspond to a particular time interval in the geologic past. Therefore, a plot of anomaly number against distance of the anomaly from the ridge crest allows us to determine spreading rates. In the example shown here, spreading between anomaly 3 and 5 was much slower than that between anomaly 1 and 3.

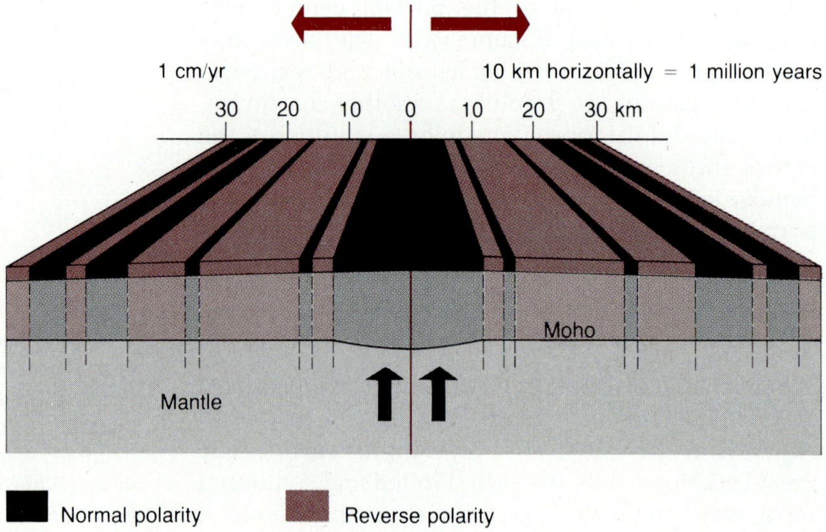

Figure 12.15
The origin of a banded magnetic anomaly pattern across a ridge crest is suggested in this schematic drawing. Sea floor at the ridge crest is polarized in the same direction as the present (normal) magnetic field. The flanking negative anomalies are caused by reversed polarity of the remanent magnetization of sea floor, formerly solidified at the ridge crest when the field was reversed.

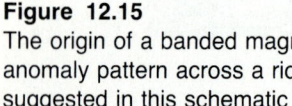

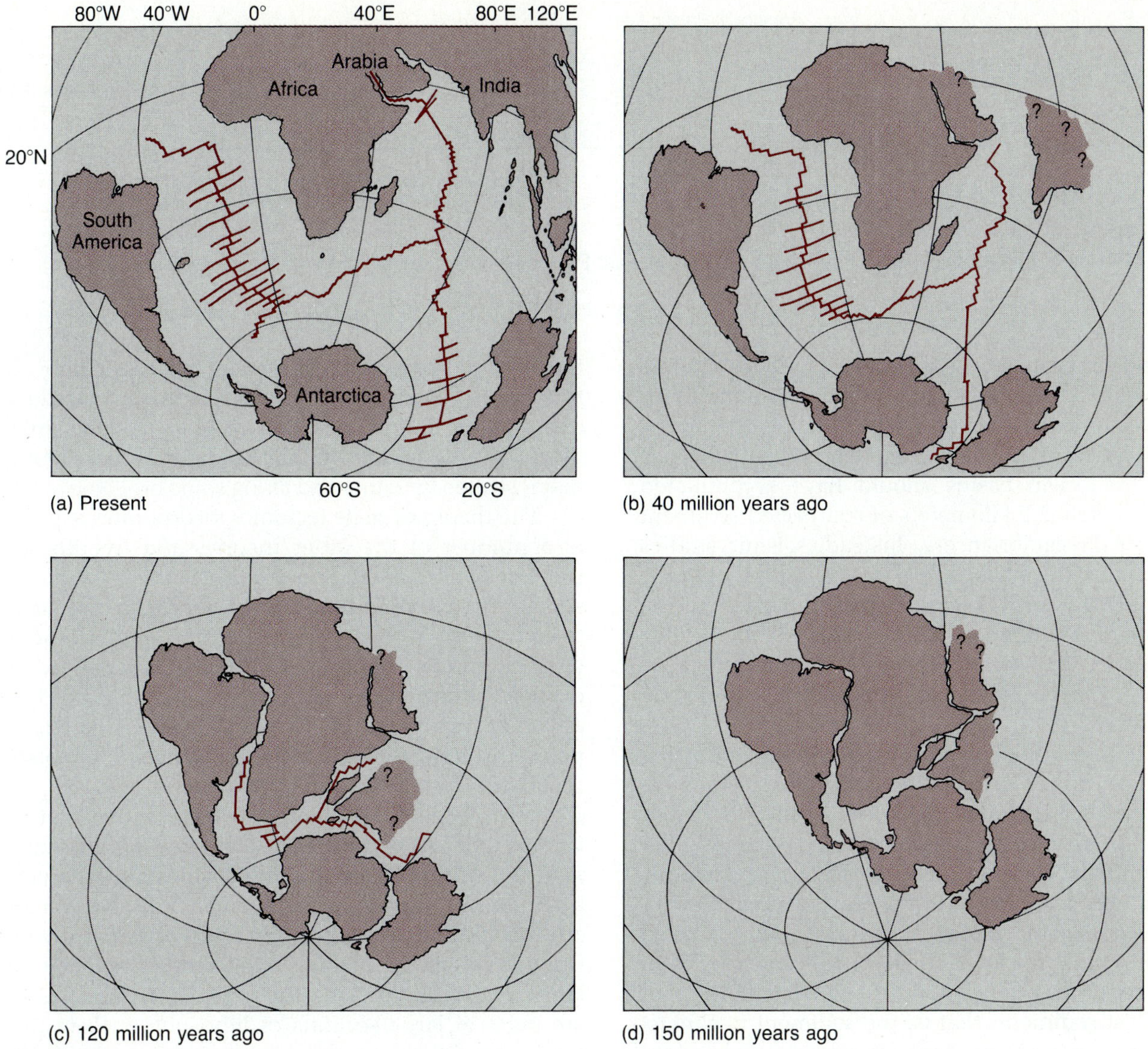

Figure 12.17
Maps showing the breakup of Pangaea and opening of the southern oceans. These maps were constructed by restoring the magnetic anomaly strips to the positions they would have had at the times indicated. (a) The present, shown for reference; (b) 40 million years ago; (c) 120 million years ago; (d) 150 million years ago.

rate of between 2 and 4 centimeters per year; spreading rates of 8 to 12 centimeters per year have been calculated for parts of the East Pacific Rise.

Some of the anomaly belts can be traced along much of the length of the oceanic ridge system. Once enough measurements are obtained to make possible tracing of the anomalies, we can restore the continents to their positions at the time a particular anomaly formed. To accomplish this restoration we could cut out the portion of a map between any given magnetic anomaly on either side of a spreading ridge and rejoin the two halves of the map. For example, Fig. 12.17 shows the southern oceans restored to the shapes they had at three times in the geologic past. The present configuration is also shown for comparison.

Confirmation of Sea-Floor Spreading from Sediment Distribution

Rates of modern sedimentation in the oceans are calculated by measuring the thickness of sediment between two layers that can be dated. These meas-

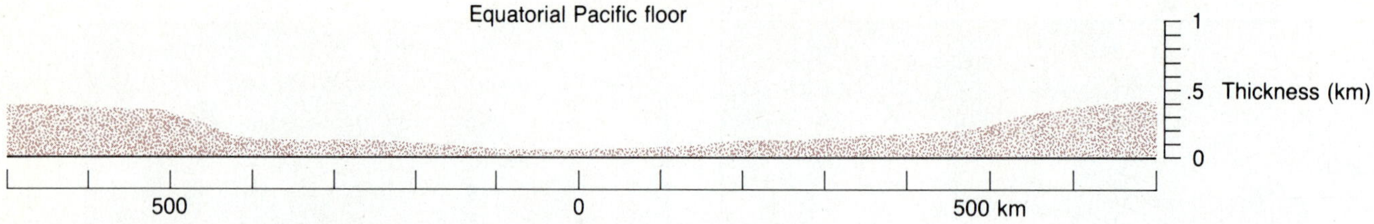

Figure 12.18
Variation in the observed thickness of sediment across the East Pacific Rise is depicted schematically.

urements indicate that sediment accumulates at a rate of between 5 millimeters and 2 centimeters per thousand years. If the smaller figure, which is the rate of accumulation of red clay in the deep ocean basins, is used, the ocean basins should have accumulated approximately 8.5 kilometers of compacted sediment since the Precambrian era. Instead, seismic studies indicate that the average thickness of sediment in the deep basins is only 1.3 kilometers. Long before the concept of sea-floor spreading was advanced, this discrepancy led some geologists to wonder if the oceans could be much younger than was generally thought.

If new sea floor is actually being formed at the mid-ocean ridges and moving away from it, then we would expect sediment thickness to increase away from the crest of the mid-ocean ridge, and older sediment to be found at progressively greater distances from it. Sediment thickness measured in many places by seismic methods confirms the predicted distribution of thickness (Fig. 12.18), and results of deep-sea drilling have now proven that the age of the oldest sediment, that is, the sediment in contact with the underlying basalt, is progressively older away from the ridge crest.

12.4 THE PLATE TECTONICS SYNTHESIS

The concepts of sea-floor spreading and continental drift have now been incorporated into a more comprehensive theory known as plate tectonics. According to the theory of **plate tectonics**, the dynamics of the lithosphere can be analyzed in terms of the interactions of a number of large, thin, relatively brittle lithospheric plates, which can move over a weak, plastic layer in the upper mantle, the asthenosphere. These plates interact with one another along their boundaries. One type of boundary is formed by the oceanic ridge system, where new sea floor is formed as plates move apart. A second boundary exists in the island arcs, where old sea floor sinks deep into the mantle. Other boundaries exist along major strike-slip faults and along some orogenic belts.

The theory of plate tectonics incorporates aspects of a number of the other theories and hypotheses previously discussed. For example, the location of spreading centers and subduction zones can be related to convection in the mantle. The concept of sea-floor spreading is an integral part of plate theory, used to explain the structure and history of the sea floor. The apparent movement of the position of the magnetic poles through geologic time is explained by plate theory in terms of the varied movement of the different plates. Presumably, the plates move relative to magnetic poles that themselves are fairly permanent in position. Plate theory provides a mechanical basis for the theory of continental drift. No longer are the continents thought of as great rafts moving through a sea of oceanic crust—they are parts of larger plates that include the oceans. The continents are parts of large sections of lithosphere. Plate tectonics provides a basis for explaining much of our current and rapidly expanding knowledge of the sea floor, and it has been employed successfully to explain the location of volcanoes, earthquakes, and some of the most perplexing features of orogenic belts, as we will see in Chapter 13.

Plate Boundaries

The long, narrow zones of recent structural, seismic, and volcanic activity define the boundaries between modern plates. (See inside front cover.) Plots of seismic activity alone, like that in Fig. 8.3, provide a reasonably accurate outline of plate boundaries, but when these seismically active belts are compared with crustal structure and other geological and geophysical characteristics, it is apparent that a number of distinctively different types of plate boundaries exist. Oceanic ridges, such as the Mid-Atlantic Ridge; island arcs; the young and active folded mountain belts; and a number of major fault zones comprise the plate boundaries.

The oceanic ridges, as outlined by the symmetrical bands of magnetic anomalies, define boundaries where new sea floor is forming and where sea-floor spreading is taking place. The island arcs and some of the young orogenic belts, especially those along continental margins, are sites where the margins of the spreading plates are consumed as they sink into the mantle to become heated and eventually reassimilated into the mantle. These zones where the plate edges are consumed and destroyed are known as **subduction zones.** There the oceanic plate subsides into the mantle, carrying with it the magnetic anomaly stripes imprinted on the oceanic crust as it formed. Because continents ride passively on lithospheric plates that include sea floor, we might expect that a continent would eventually reach a subduction zone, and, indeed, it appears that this has happened. However, the low density of continental crust precludes its sinking into the high-density mantle. When a continent reaches a subduction zone located along the margin of another continent or an island arc, the edges of the two continents are deformed and uplifted in the collision, forming mountains. Much, but not all, of crustal dynamics is best understood in terms of what happens at plate boundaries.

Junctions between spreading plates. The topographic highs of ocean ridges form where upwelling of molten mantle material lifts the oceanic crust. As we saw in Chapter 11, volcanic islands, like Iceland, the Azores, and the Galapagos Islands, located along the oceanic ridges confirm the existence of melts beneath the ridges. Volcanoes there erupt basaltic lava derived from materials that make up the mantle.

In recent years, outcrops near the ridge crest have been inspected from deep-diving submersibles. These explorations reveal that large areas of the sea floor are covered by lava flows, many of which contain pillow-like masses of basalt that characteristically form when extrusions of lava occur under water. The discovery of open fissures along the ridges confirms that extension is taking place, that the ridges are being pulled apart. Such zones of extension are undoubtedly the places where magma is most likely to rise to the surface.

The rocks found along the actively spreading ridges are primarily basalt with a thin sedimentary veneer, but serpentine, gabbro, and a few blocks of peridotite have also been found. The more mafic of these rocks are probably derived directly from the lower part of the oceanic crust or the upper mantle, and serpentine is produced by chemical reactions between the peridotite and hot water. Serpentine is sometimes found in the deep valleys that cut across the oceanic ridge, and it is probable that large quantities of serpentine underlie the ridge crest. The relatively low density of serpentine would account for both the negative gravity anomalies that occur over the ridges and for the abnormally low seismic velocities found beneath the ridges. The density of serpentine and other hydrated mantle rock explains why the ridges are topographically high—the low-density materials tend to be buoyed up as predicted by the theory of isostasy.

As the sea floor spreads away from the ridge crest, sediment accumulates, and the newly formed oceanic crust is buried. The blanket of sediment insulates the crust from water above, preventing heat loss from the hot rocks below. The resulting heat buildup causes the hydrated crust to lose water. As dehydration takes place, the density of the rock increases, and the heavier crust gradually sinks. In this fashion, the gradual lowering of elevation away from the ridge and toward the abyssal plains can be explained as an isostatic adjustment caused by variations in density of the oceanic lithosphere; any upward movement due to convection would also cease on the flanks of the ridge.

Rifting of continents. At the present time, divergent junctions pass under continents at the Aden Sea (Fig. 12.19) and in the Gulf of California (Fig. 12.20). In both places the continental crust is apparently being split and is spreading apart. The Gulf of Aden, the Red Sea, the system of grabens known as the East Africa rift system, and the Rio Grande rift in the western United States all appear to be products of extension involving continental crust. The relative movement of Africa away from the Arabian peninsula may well represent an early stage of the type of breakup and separation that took place between the continents on either side of the Atlantic in the early part of the Mesozoic era. These continents moved apart as the ocean gradually grew by addition of new crust at the spreading ridge, but each continent underwent little or no movement relative to the

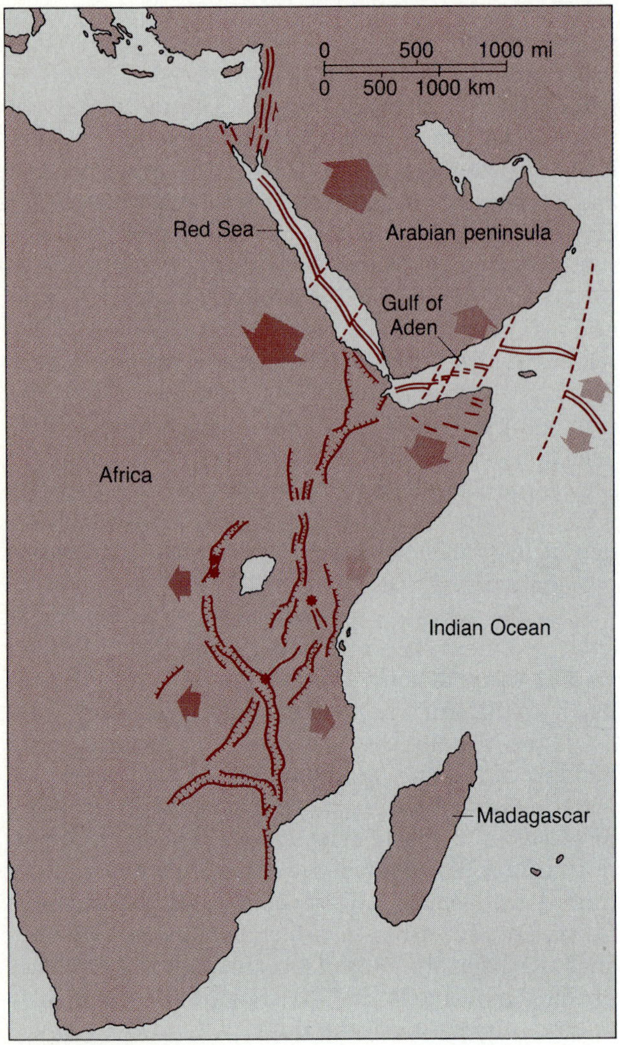

Figure 12.19
East Africa rift system. The Red Sea, Gulf of Aden, and the extensive system of fault blocks in eastern Africa are all caused by extension of the crust in this region. The arrows indicate the local relative directions of extension.

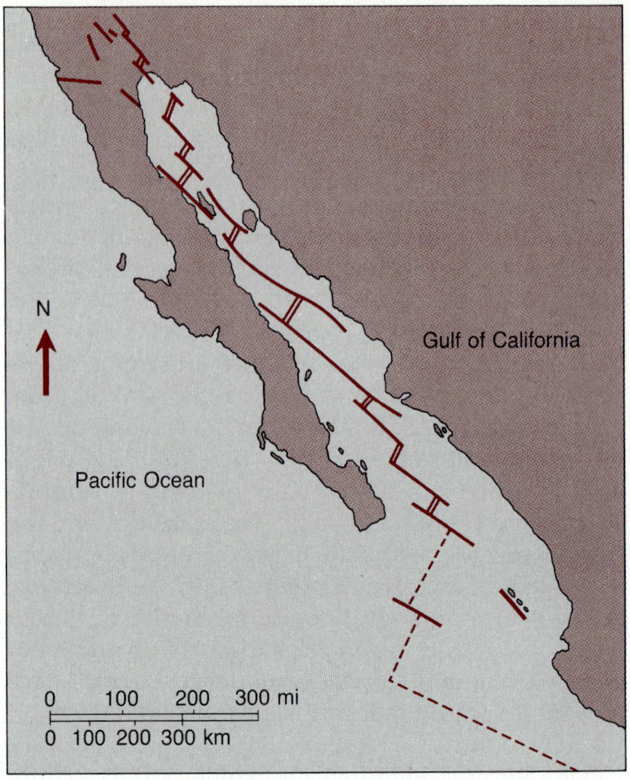

Figure 2.20
The Gulf of California has originated as a result of the movement of Baja and the land west of the San Andreas fault to the northwest. This movement has taken place as a result of spreading along the short ridges shown in the Gulf. This ridge is cut by numerous northwest trending transform faults.

mantle immediately under it or to adjacent sea floor that is part of the same plate. This accounts for the quiet tectonic character of the Atlantic-type continental margins found around most of the border of the Atlantic Ocean, the exceptions being in the Caribbean and Scotia Island arcs.

Fault boundaries. Major faults that extend deep into the lithosphere and along which great apparent displacement has occurred are found within both continental and oceanic crust. In some places these faults mark boundaries where one plate slips laterally past an adjacent plate without vertical displacement, what we called strike-slip movement in Chapter 10. The San Andreas fault of California, the Alpine fault of New Zealand, and the Great Glen fault of Scotland are among the most famous faults of this type. Faults like these form some plate boundaries.

Studies of the ocean ridges revealed deep valleys that cut across and apparently offset the ridges in the way that strike-slip faults would. The magnetic anomaly stripes are also offset across these valleys. J. Tuzo Wilson was the first geologist to recognize what the consequences would be if a vertical fault cut across an oceanic-ridge spreading center. Strike-slip faults are usually thought of as the type of movement that

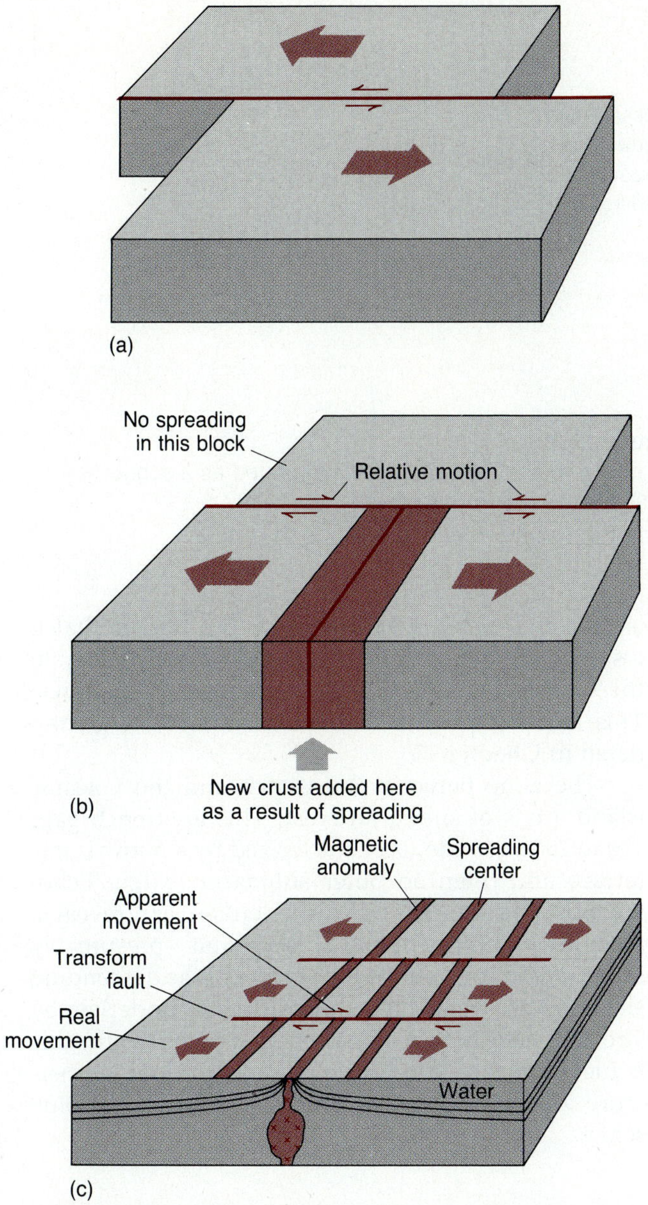

Figure 12.21
(a) The conventional strike-slip fault. (b) This same type of movement is seen in the transform fault at a spreading center, but notice that the relative movement is reversed across the spreading center. (c) Because of this, the offset of magnetic anomalies along a ridge crest appears to indicate a sense of motion opposite to that which is actually taking place.

occurs if two rigid blocks are moved by one another, but a second motion is involved if the fault cuts an ocean ridge center: movement in opposite directions across the spreading center. Movement along faults that cut across spreading centers therefore occurs in two opposite directions as illustrated in Fig. 12.21. Wilson coined the term **transform fault** for this special variety of strike-slip faults. It now appears probable that many of the great strike-slip faults, including the San Andreas, are of this type.

Transform faults, like all strike-slip faults, are nearly vertical, and they are caused by horizontal movement of the rocks on either side; but the relative movement across the fault is reversed from what it appears to be along the segment between the offset ridge crest. Transform faults cutting spreading ridges may separate parts of the ridge that are spreading at different rates or places where the direction of spreading is unequal; or they may simply mark places where the position of the spreading center is offset. Transform faults may also connect spreading centers with subduction zones, but most of them cut across oceanic ridges and are located within plates.

Subduction zones. The evolution of plate theory coincided in time with many of the new discoveries about the structure of the sea floor. This is evident in the interpretation of island arcs as zones of subduction. The geologic setting of island arcs had been known for decades. Deep trenches usually occur in the sea floor on the oceanward side of the volcanic arc (see Section 11.2). Both an outer nonvolcanic ridge and an active inner volcanic arc (Fig. 12.22) are usually present, and the volcanic extrusions are andesitic. The earthquake foci located under the arcs lie in and define a planar fault or shear zone that comes to the surface in the trenches and dips at an angle of about 50°. Such planes are referred to as **seismic shear zones** or **Benioff zones**.

Among the studies in the 1950s and early 1960s that set the stage for the subduction hypothesis was the discovery of abnormally low heat-flow values through the oceanic crust in the trenches. This was interpreted as indicating that cool rock may be moving down under the trenches. High heat-flow values in the vicinity of the volcanic part of the arc are accounted for by the ascent of magma.

An especially important study was made by Jack Oliver and Bryan Isaacs in the early 1960s, on the behavior of seismic surface waves under island arcs. Surface waves are propagated within channel-like

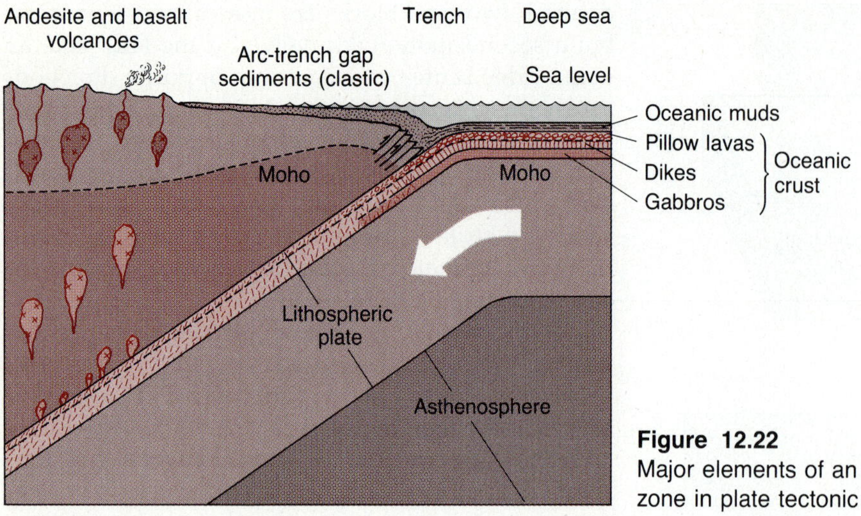

Figure 12.22
Major elements of an island arc interpreted as a subduction zone in plate tectonic theory.

paths formed by the surface of the lithosphere and the top of the low-velocity zone. Surface waves die out (attenuate) more quickly where they pass through plastic material, like that of the low-velocity zone, than when they move through the more rigid lithosphere. To study the propagation of surface waves across island arcs, a network of seismograph stations was set up to receive seismic waves coming from earthquakes in the Benioff zone under the Tonga arc north of New Zealand. Analysis of the records demonstrated that a thick slab of material, similar in attenuation characteristics to the oceanic lithosphere, lies immediately beneath the seismic shear zone. Oliver and Isaacs recognized that this might well be a slab of lithosphere sinking deep into the mantle under the island arc. The slab becomes less distinct at depth and probably does not exist much below the depth of the deepest earthquakes (about 700 km), which result from frictional effects between the sinking slab and adjacent overlying mantle.

Because temperature in the mantle increases with depth, the descending lithosphere is gradually heated as it goes down. The initial effect of this heat is thermal regional metamorphism, which progressively alters the sediments and lava flows that make up the slab or oceanic crust. Sediments dragged down into the subduction zone are affected by the temperature more readily than are the igneous rocks, which originally crystallized at high temperatures. Eventually all of the descending slab is affected by the high temperature. The slab begins to melt at depth and is assimilated into the mantle to such a degree that it loses its identity. Some of this melt finds its way to the surface and erupts, forming andesite volcanoes. This melting process will be examined in greater detail in Chapter 13.

The zone between the trench and the volcanic island arc is of special interest. This **arc-trench gap,** 100 to 200 km wide, is characterized by a nonvolcanic terrace and often an outer submarine ridge. Echo-sounding techniques in the Aleutian arc reveal a complexly deformed pile of sediment—presumably sediment scraped off the top surface of the descending slab. The arc-trench gap should thus be underlain by strongly deformed rock. At some greater depth, this is metamorphosed under conditions of low temperature and high stress to produce a rock called blue schist.

12.5 MECHANISMS OF PLATE MOVEMENTS

Evidence that the earth's crust can be subdivided into lithospheric plates, and that many features of the crust, including seismicity, volcanism, and mountain-building, can be explained in terms of plate movements is well established. But geologists have yet to put the causes of these movements on as firm a footing.

Two very different types of movements involve plates—vertical movements, both up and down; and lateral movements. The theory of isostasy explains many of the vertical movements. These are due to instability related to the relative density of the lithosphere and the asthenosphere, on which the lithosphere "floats." Estimates of the density of the asthenosphere are based on observed seismic wave velocities, temperatures, and compositions of the materials that make up the mantle. These estimates indicate that the density of the asthenosphere is close to that of the lithosphere. Some studies indicate that the asthenosphere is marginally less dense than the overlying lithosphere. A less-dense asthenosphere overlain by a denser lithosphere is, of course, highly unstable. The denser layer will tend to sink into the less dense layer—as apparently happens at subduction zones. Not all of the lithosphere can sink. The continental crust is significantly less dense than either the oceanic crust or the asthenosphere. Consequently, continents can not be subducted.

Sinking of parts of the lithosphere at subduction zones would not only account for vertical movements in these zones, but would also exert pull on the entire plate. Thus, the plate is drawn toward the subduction zone as the slab sinks.

At the oceanic ridges, warm, less dense lithosphere has been elevated. Evidence of this elevation is seen in the Moho, which is inclined away from ridge crests. Thus, slabs of lithosphere are high at ridge crests and low in subduction zones. The slabs are slightly inclined, and they are underlain by a weak plastic material over which they may move easily. In this view, gravity acting on the inclined slab is the driving force that causes slabs of lithosphere to move.

The role of convection is still unclear. Slow convection is almost certainly taking place in the asthenosphere, but whether the convection actually moves the lithosphere, as Vening Meinesz suggested, is unclear.

The Rock Cycle

We have reviewed many aspects of the operation of the rock cycle in plate theory (Fig. 12.23). New igneous rock materials are introduced in and on the lithosphere at oceanic ridges, where magma from the mantle rises and creates new sea floor which spreads away from the crest. As the sea floor moves, it becomes the site for deposition of organic sediment as well as that deposited by turbidity currents. When the sea floor reaches a subduction zone, new sediment and volcanic ash may be added to it before it descends. During subduction, the rocks in the slab are metamorphosed first to blue schist and eventually to other

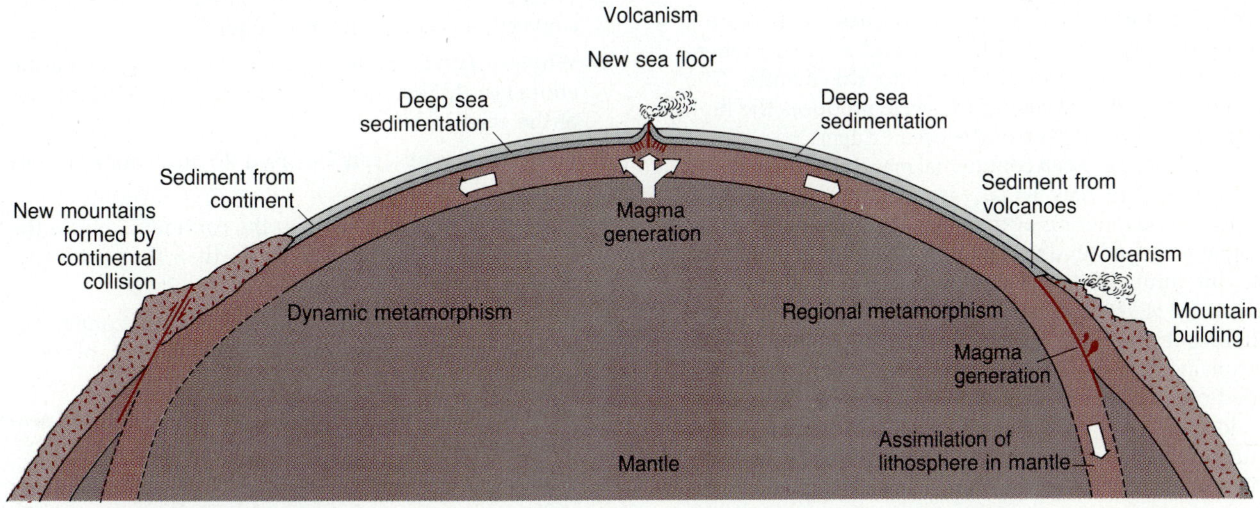

Figure 12.23
A schematic summary of geologic processes related to plate tectonics and the rock cycle.

higher temperature and pressure forms, until andesitic magmas are produced. These magmas rise and create the high andesitic volcanoes behind the zones of subduction.

Two especially important links in the rock cycle are yet to be clarified—one is the plate mechanism by which larger volumes of rock are uplifted to form mountains, and the other is the process by which these uplifted rock masses are broken down, forming sediment that is ultimately carried into the sea. These parts of the cycle are topics for later discussion.

SUMMARY

The modern theory of plate tectonics has its roots in earlier theories, each of which has contributed concepts to our present view of crustal dynamics. The theory of isostasy rests upon differences in earth's gravitational field that accompany crustal features such as mountains, ocean basins, and deep-sea trenches. Analyses of these gravitational variations suggest that the density of continental crust is less than that of oceanic crust and that the Mohorovičić discontinuity is much deeper under the continents than under the ocean basins. According to the theory of isostasy, the continents are buoyed up in the plastic asthenosphere by their relatively less dense roots. The vertical rebound of regions once weighted down by continental glaciers gives evidence of the operation of isostasy.

The convection current hypothesis holds that at least part of the mantle is plastic enough to support convection cells between the very hot core and the cooler crust. According to this theory, deep-sea trenches are associated with the descending part of the cell and are areas where the crust is being carried down into the mantle. The corresponding upward half of the cycle produces the high topography and heat flow of the ocean ridges.

The close fit between continental margins, the similarity in fossil record of now widely separated continents, and the apparent wandering of the magnetic poles as recorded in magnetic rock all tend to support the theory of continental drift. But until the late 1950s, geologists could offer no mechanism to explain the enormous displacements of continents through the rigid lithosphere. Sea-floor spreading explains that movement. If new sea floor is formed at the mid-ocean ridges and spreads laterally away from them, it could carry with it existing sea floor and the continents floating in the dense lithosphere. The observation of bands of reversed remanent magnetism running parallel to the spreading centers and symmetrically across the ridge tends to confirm this model of crustal movement, and even offers a mechanism for determining its speed. Evidence from the increasing age of sediment on the flanks of the ridge and the ocean floor strengthen our confidence in the theory.

Plate tectonics is the synthesis of ideas from the theories of isostasy, convection, continental drift, and sea-floor spreading. According to plate tectonic theory, new crust is generated at the ocean ridges, and old crust is carried down into the mantle at subduction zones—regions of earthquake and andesitic volcanic activity. Some combination of convection, gravitation, and isostasy drive the movement of the crust over the weaker, plastic asthenosphere.

KEY WORD LIST

arc-trench gap
Benioff zones
Bouguer correction
Bouguer gravity anomaly
continental accretion
Curie point
flotational equilibrium
free air correction
Gondwanaland
isostasy
isostatic anomalies
Laurasia
magnetic anomalies
magnetic reversals
paleomagnetic time scale
Pangaea
plate tectonics
remanent magnetization
seismic shear zones
subduction zones
transform fault

STUDY QUESTIONS

1. What discoveries caused the revival of interest in continental drift in the 1960s?
2. Why did the idea that the earth is contracting lose favor?
3. What causes convection in the earth, and why doesn't convection occur in the lithosphere?
4. Why can polar wandering be explained by having the entire lithosphere move as a unit relative to the interior of the earth?
5. What rock types can be used to find ancient pole positions?
6. Suggest several reasons why the continents across the South Atlantic fit together better than they do across the North Atlantic.
7. Why was the discovery of the low velocity zone in the mantle so important in the development of plate tectonics?
8. What information is needed to determine rates of sea-floor spreading?
9. What are the objections to the notion that continents move over or through the mantle and oceanic crust?
10. How does movement along a transform fault differ from that on other types of strike-slip faults?

SUGGESTED READINGS

Hurley, Patrick, "The confirmation of continental drift," *Scientific American 218* (1968), *52–64*.

Sullivan, Walter, *Continents in Motion*. New York: McGraw-Hill, 1974.

Takeuchi, W. S., S. Uyeda, and H. Kanamori, *Debate About the Earth*. San Francisco: Freeman and Cooper, 1967.

Totten, S. M., "Frank B. Taylor, plate tectonics and continental drift," *Journal of Geological Education* 29:5 (1981), *212–220*.

Uyeda, S., *A New View of the Earth*. San Francisco: W. H. Freeman, 1971.

Wegener, Alfred, *The Origin of Continents and Oceans*. New York: Dover, 1966.

CRUSTAL DYNAMICS: MOUNTAINS, MAGMA GENERATION, AND METAMORPHISM

Chapter 13

Plate tectonic theory has provided a new, more lucid, and all-encompassing framework in which to view the processes that account for igneous activity, metamorphism, and mountain-building. For more than two decades geologists have sought to explain exactly how these processes are related to plate movements. These efforts have been successful, even though many questions remain to be resolved.

We shall begin by examining the composition and geological settings of modern igneous rocks. In particular, we shall look at igneous activity in spreading centers and subduction zones. As we examine the processes acting in subduction zones, it will become obvious that these must also be the location of most modern metamorphism. Next, we will turn to mountains where igneous and metamorphic rocks commonly abound. And, finally, we will see how plate theory can be applied to explain the origin of the Appalachians and the mountains of western North America.

13.1 MAGMA GENERATION

Igneous Provinces

Active volcanoes (Fig. 13.1) and recent volcanic deposits are our best indication of the places where magmas have been generated in the lithosphere in recent geological times. These fall into several distinctly different tectonic situations, or provinces. One province lies along the oceanic ridges, where new sea floor is being formed and where spreading and divergence of plates is taking place. A second province corresponds to the island arcs and young folded mountain belts where sea floor is being subducted. A third group of volcanic centers, known as hot spots, is not obviously connected with plate generation or destruction, but occurs well within plate boundaries. Hot spots are represented by sites such as the Hawaiian Islands, in the middle of an ocean basin; and Yellowstone National Park, in the interior of a continent.

Among the older sites of igneous activity, the Sierra Nevada batholith is an example of a type of igneous province that is described as a continental-plutonic province. This intrusion occurred in a mountain system, but its precise relationship to plate boundaries is still uncertain. The continents also

National Publicity Studios, New Zealand

Figure 13.1
Active volcanoes like this one in New Zealand are our most direct evidence of where magma is now being generated.

contain large complexes composed of ancient granitic rocks—the shield areas. The rocks in the shields are so ancient, and the configurations of these rock bodies are so complicated, that the application of plate tectonic theory to this province and to the events of the early history of the earth is unresolved.

Chemical Composition and Igneous Province

One relatively simple way of comparing the rocks of different igneous provinces is to examine the bulk chemical composition of the main rock types found in them. Representative chemical analysis of rocks from the Mid-Atlantic Ridge, Hawaii, the Tonga Island arc, and Sierra Nevada batholith are tabulated in Table 13.1. The contrast between the Sierra Nevada batholith and the other provinces is most striking, especially in the content of silicon, iron, magnesium, calcium, and potassium. Clearly, the magma sources are dramatically different. The rocks of the island arc, the oceanic ridges, and the Hawaiian Islands could all have been derived from partial melting of the iron- and magnesium-rich lower part of the crust or more likely from the upper mantle. The Sierra Nevada rocks, however, are very rich in silica and were probably derived, at least in part, by partial melting of ancient continental crust that already had a high silica content. This difference is confirmed by other types of geochemical analyses. However, the geochemical data given in Table 13.1 show that basalts at ridge crests and oceanic islands are very similar. It is clear that distinctions between the rocks found at the spreading centers and those formed at oceanic islands must be sought in characteristics other than the bulk chemical composition of their lavas.

Generation of Magma at Spreading Centers

Oceanic ridges are the sites of the greatest sustained injection of magma and extrusion of the largest volume of lava on earth. The entire lower part of the oceanic crust has been created here; and it is here that new sea floor is now being created—sea floor that is composed mainly of basaltic lava flows and the feeder dike systems along which these magmas rise to the surface. The mid-ocean ridges are composed almost entirely of igneous materials covered by only a trace of sediment at their crest and thicker sediment on their flanks.

Geophysical studies also reveal that the Mohorovičić discontinuity rises toward the ridge crest, suggesting that the ridges are high because the mantle under them is arched upward. This arching is presumed to occur because the ridges mark places were huge convective currents in the mantle rise close to the surface, bending the overlying rock upward, stretching it and breaking it apart. Measurements clearly show that heat flow from the interior is especially high at the ridges. That magma is generated under ridge crests is hardly surprising when we consider that not only is heat rising from deep in the mantle, but that also the arching of the lithosphere in those areas causes fissures to form, thus reducing pressure—an effect we considered earlier, in Section 6.5.

Compositionally, all the lavas dredged from ocean ridges are basalts—usually olivine basalts, containing augite. All of these basalts are thought to be products of partial melting of materials in the upper mantle. Partial, rather than complete, melting is the preferred assumption because mantle rocks, such as peridotite, contain much more iron and magnesium and less

Table 13.1 Chemical Analyses of Representative Rocks from Some Important Geological Provinces.

	Continent (Sierra Nevada Batholith)		Island Arc (Tonga)		Ocean Ridge (Mid-Atlantic)		Mid-ocean Island (Hawaii)	
SiO_2	67.0		55.4		49.0		50.0	
Al_2O_3	15.0		15.6		16.0		13.9	
Fe_2O_3	1.5	} 4	3.3	} 10.5	2.7	} 10.5	1.0	} 10.8
FeO	2.5		7.2		7.8		9.8	
MgO	1.7		4.8		6.4		7.1	
CaO	3.8		9.8		10.5		11.3	
Na_2O	3.2		1.8		3.0		1.5	
K_2O	3.8		0.4		0.1		0.5	

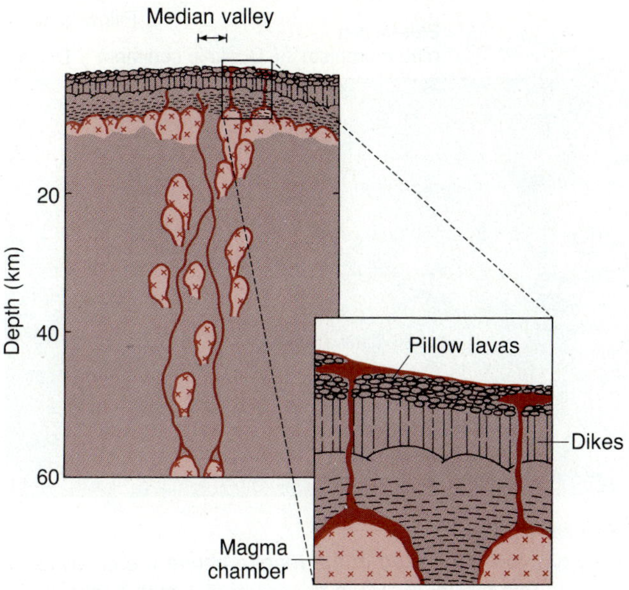

Figure 13.2
Magma chambers under oceanic ridges. The deep magmas become differentiated as they rise in the upper mantle. Olivine crystallizes early and sinks out of the magma. Some of the residues rise and are extruded as pillow lavas along the ridge.

silicon than does basalt. Potassium and titanium are also low in abundance in the ocean ridge basalts, and calcium is high.

A few volcanoes penetrate the surface of the ocean to form islands along the oceanic ridge system, but most igneous activity takes place under water. Lavas apparently reach the top of the ridge in most places by moving up along an extensive network of feeder dikes (Fig. 13.2). These dikes are envisioned as occupying fracture and fault zones parallel to the ridge crest, formed by the pulling apart of the crust at the ridge crest.

When magmas reach the crest of the ridge, some flow out as fissure-type eruptions, the lava pouring out along the fault zone. Others are concentrated more or less around an eruptive center as though the magma is moving up along a pipelike feeder and out of a single central vent. Eruptions of both kinds can be studied in Iceland, the Azores, and at a number of other centers along the ridge crest, where piles of lava have built up above sea level. The lava that erupts under water forms a characteristic type of lava formation called **pillow lava** (Fig. 13.3). Older examples of these lavas have been examined on land, where they have been elevated and exposed; but in recent years, their formation has been studied by volcanologists who put on diving gear and photographed the edge of an active underwater flow. Lava is chilled immediately when it is extruded under water. For this reason, flows, like those we see on the ground surface, cannot form. Instead, a crack or hole in the edge of a flow develops, and a bag-shaped mass of lava, usually no more than a meter across, oozes out. It is a red hot mass when it first touches water, but its surface is chilled immediately and turns dark. This outer chilled shell is glassy, but the pillow is still molten just a few centimeters inside. Pillows may break away from the edge of the flow and roll downslope, or they may join a pile of these masses, resembling blobs of toothpaste, built up along the edge of the flow. When the pillow has solidified, it generally has a glassy outer rim containing gas bubbles and a somewhat coarser interior. The spaces between the pillows are filled with mixtures of sediment and glass fragments. If this is the mechanism of crust formation at spreading centers, as is now thought, much of the crust under the sea floor must have formed as a thick layer of pillow lava and dikes feeding these flows.

Generation of Magma in Subduction Zones

The lithospheric plate approaching a subduction zone generally consists of a veneer of water-saturated sediment several hundred meters thick, underlain by a pile of pillow lava and basalt dikes a kilometer or more thick. These lie on top of upper mantle rock of

Figure 13.3
"Toothpaste" lava photographed by cameras aboard a submersible.

peridotite and gabbro. The sediment cover generally includes red clay and the siliceous remains of radiolarians and diatoms. Near the trench formed by the descending plate, other sediments derived from nearby land may be added to the sediment cover. These sediments are likely to include volcanic ash of andesitic composition derived from volcanoes in the island arc, mixed with other land-derived sand, silt, and clay. These components—the sediment veneer, the basalt lavas, and the upper mantle—comprise the slab of rock that moves into and down the subduction zone into the mantle.

Although andesite is the characteristic type of lava erupted from volcanoes located over subduction zones, basalts and rhyolites are sometimes present in quantities equal to that of the andesites. The composition of the lavas and plutons over most modern subduction zones is relatively well known. Basalts, similar to those of the islands over the mid-oceanic ridge, occur with andesite where the subduction zone is located within the oceanic crust. But rhyolite and other quartz-rich magmas are generated along with andesite and some basalt where subduction occurs beneath continental crust, as it does in the northern part of the Kuril Islands or in the eastern part of the Aleutians.

Variations in magma types also occur across most island arcs. In Japan, for example, basalts, like those of the oceanic islands and spreading centers, are abundant along the front of the arc but decrease toward the back of the arc, where andesites become more abundant.

Two models are now used to explain the origin of the magmas in subduction zones. In the first one, the temperature in the descending slab of oceanic lithosphere increases with depth, as does the pressure of the increasingly thick overlying rock. These changes in temperature and pressure cause the materials in the slab to be metamorphosed. At some depth, the fusion point of the slab materials is reached, and parts of the slab may melt (Fig. 13.4). The melt is likely to be composed of the minerals that made up the basalt and sedimentary cover of the sea floor. If the sediments and basalt melt and mix, they form a magma of composition intermediate between those of granite and gabbro.

According to the second model, magma may also be generated by the partial melting of peridotite in the mantle over the sinking slab as a result of the infusion of water into it. Because a volume increase accompanies melting, any magma produced has a lower density than the surrounding mantle, and it

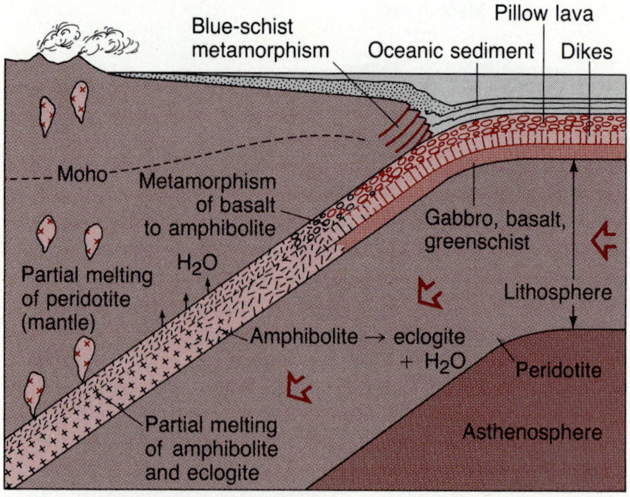

Figure 13.4
Magmas are generated in the mantle above the downgoing slab. In this model, water is driven off of a slab composed of serpentine and amphibolite. This causes partial melting of mantle rock, which rises and becomes differentiated to form the types of basalts found in island arcs.

tends to rise along with the hot solutions formed by heating of the water-saturated rocks in the upper part of the slab. Experimental studies have clearly established that the addition of small amounts of water into such rocks lowers their fusion point and results in the generation of partial melts of intermediate (andesitic) composition.

An independent line of evidence supports the idea that magmas erupted over subduction zones originate within or close to the descending slab. This evidence derives from the observation that the horizontal distance between the trench, where oceanic crust starts sinking, and the line of volcanoes in the volcanic arc is less in subduction zones where the descending slab is sinking at a steep angle than it is where the slab is more gently inclined (Fig. 13.5). This is exactly what we would expect if the fusion temperature is reached at about the same depth in different subduction zones, and if, the magmas are generated within or adjacent to the sinking slab.

13.2 METAMORPHISM IN SUBDUCTION ZONES

Although metamorphism may occur in other geological settings as well, subduction zones are ideal localities for all three types of metamorphism described in Chapter 7. Contact metamorphic aureoles surely surround the large plutons generated along ridges and in subduction zones. Rocks caught between colliding continents or in the seismic shear zones are subjected to great directed stress sufficient to break down the more brittle minerals and promote recrystallization or the other types of internal adjustments that characterize dynamic metamorphism. And regional metamorphic effects must occur as the entire descending slab, formed near the surface, sinks slowly to greater depths where the materials in it are subjected to steadily increasing confining pressure and higher and higher temperatures, until they melt.

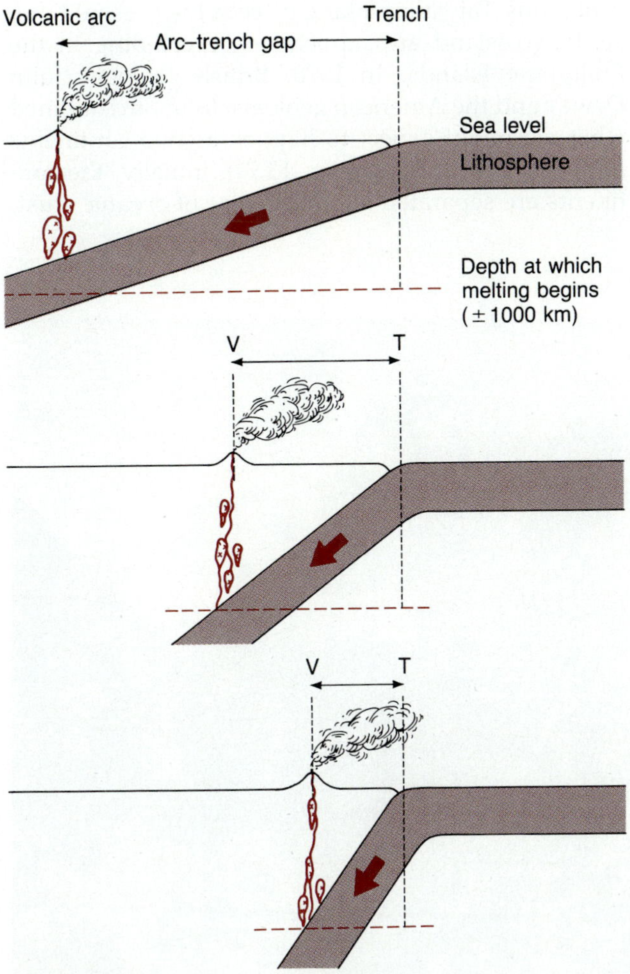

Figure 13.5
The horizontal distance between the trench and the volcanic islands depends on how steeply the downgoing slab is inclined, because melting and magma generation are caused by the increasing temperature depth.

Long before rock in the slab melts, it will have been greatly altered as a result of recrystallization, recombination of constitutents, and other pressure-induced changes. Among these, one has special significance. The common minerals of basalt and gabbro that make up a large part of the oceanic crust are pyroxene, olivine, and plagioclase. These recombine at depth to yield a rock called **eclogite,** which is rich in garnet and quartz (see Fig. 13.4). This transformation is accompanied by a change in density from about 3 gm/cm^3 for gabbro to 3.5 gm/cm^3 for eclogite. The increase in density that accompanies the transformation to eclogite is probably an important factor in the sinking of the slab.

In many subduction zones, some of the accumulated sediment is scraped off the descending slab against the inner face of the trench wall. The pile of sediment formed is subjected to great lateral pressure as more and more sediment is driven into it. This compaction is believed responsible for the formation of an unusual metamorphic rock, **blue schist,** which contains minerals that develop under experimental conditions of high directed stress but relatively low confining pressures (equivalent to shallow depths). Blue schist is found in many zones now thought to have been subduction zones. This interpretation is strengthened by the presence in these zones of trench sediments and even ultramafic rocks, such as serpentine, peridotite, and eclogite, which occur in the upper mantle. These slices of sea floor are occasionally faulted up into the overlying sediment as a result of horizontal compression in the subduction zones. This upward movement of oceanic crust is called **obduction.**

As we will see in the following sections, sediments carried down and metamorphosed in subduction zones may be caught between colliding crustal fragments, such as continents, and forced back up. Thus, many mountain systems contain regionally metamorphosed sedimentary cores.

13.3 MOUNTAIN-BUILDING

> It is the study of mountains which above all else can quicken the progress of the theory of the earth or geology. The plains are uniform, and allow the rocks to be seen only where these have been excavated by running water or by man. The high mountains, on the other hand, infinitely varied in their composition as in their forms, present gigantic natural sections wherein the order, the postion, the direction, the

thickness and the nature of the different formations which they are composed, as well as the fissures which traverse them, can be seen with the greatest clearness and at one view. . . . I desire that we should never lose sight of the great masses, and that we should always make a knowledge of the great objects and their relations, our aim in studying their small parts.

With these words, H. B. de Saussure (*1*), a famous Swiss geologist, in 1796 described a fascination with mountains that has been shared by many geologists through the years (Fig. 13.6). No theory of global tectonics could possibly gain widespread acceptance unless it provided a satisfactory explanation for what we know about mountains. Plate theory not only does this, but has proven flexible enough to explain the diversity of forms found in different mountain systems.

The Collision Model of Mountain-Building

Plate tectonics revitalized the old theory of continental drift and with it an explanation for the origin of mountains, an origin based on the idea that mountains form along the line of collision between two continents. In developing the continental drift theory, Wegener thought of those collisions as taking place when continents drifted through the sea floor and encountered one another. According to plate theory, continents approach one another as the oceanic crust between them is subducted along the edge of one or both continental masses.

Collisions in plate theory are not confined to the continent–continent type; they might also occur between a continent and an island arc, as shown in Fig. 13.7, or even between two island arcs. The boundary between India and Tibet, marked by the Himalayan Mountains, is thought to represent a continental collision of the kind diagrammed in Fig. 13.8. A collision between mainland China and the island arc containing Taiwan is taking place at the present time. And two island arcs appear to be colliding in the Philippine Islands. In 1970, British geologist John Dewey and the American geologist John Bird outlined what we might expect to happen as two continents approach one another (Fig. 13.7*a*). Initially, the continents are separated by an expanse of oceanic crust,

Figure 13.6
A group of geologists studying the structure of the Swiss Alps. The pronounced change in color of the rocks on the ridge marks the location of a major thrust fault. The rocks on top of the ridge have been moved from the south to this position.

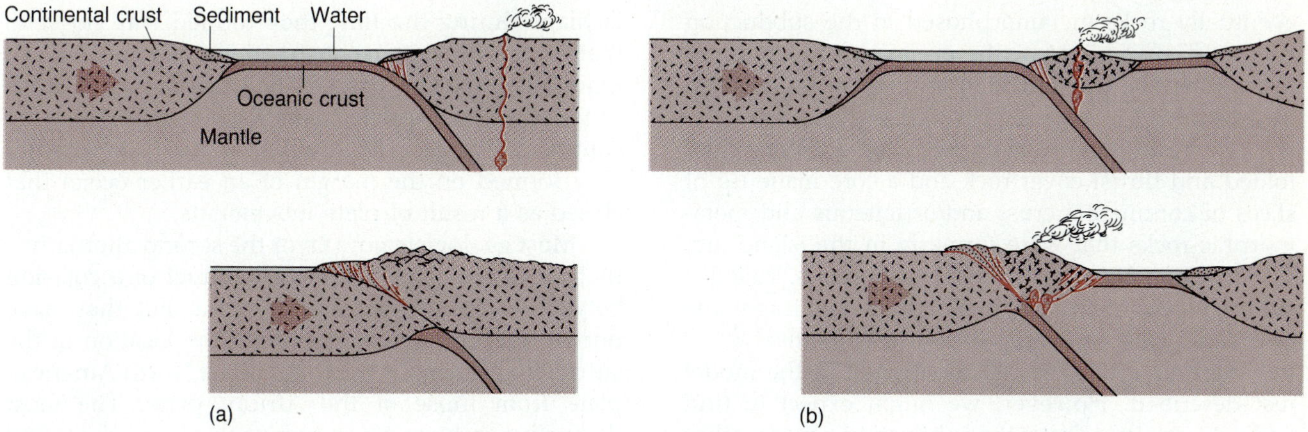

Figure 13.7
Mountain-building is often caused by the interaction of plates along their boundaries. Two types of collisions are illustrated here: (a) continent–continent collision; (b) a continent–island arc collision.

and a subduction zone must exist along the margin of one of the continents. According to this model, the continental margin that includes the subduction zone is the site of a trench and a seismic shear zone. The trench receives both deep-sea sediment, including radiolarian oozes, and sediment washed in from the nearby continental margin. The oceanic crust between the continents is passive and receives normal deep-sea sediment, consisting of deep-sea oozes and red clay. The approaching continent has a passive Atlantic-type margin consisting of carbonate banks, deltaic deposits, and sands, silts, and clay. Based on present rates of sea-floor spreading, we may assume that the two continental masses are approaching each other at a rate of 4 to 12 cm/year.

As the continents come closer together, the sediment built up on the margin of the approaching continent begins to flood into the trench, filling it up. The quantity of sediment on the sea floor is more than can be carried down the subduction zone. At this time, the trench disappears and the sedimentary cover begins to be compressed between the two continents.

Compression of the sedimentary cover forces it into a system of folds and lifts it up. The gigantic mass of the approaching continent can not be subducted because of its low density, but large slivers of the sea floor and the edge of the continent may be sheared off and forced upward into the folded sedimentary rocks above. As compression continues, the edge of one continent may be tucked under the edge of the other, creating a very thick crustal section of the type now found under the Himalayas.

As uplift of the rocks caught between the two continents proceeds, deeper rocks are progressively uplifted. Sediments are at the top of this pile, but

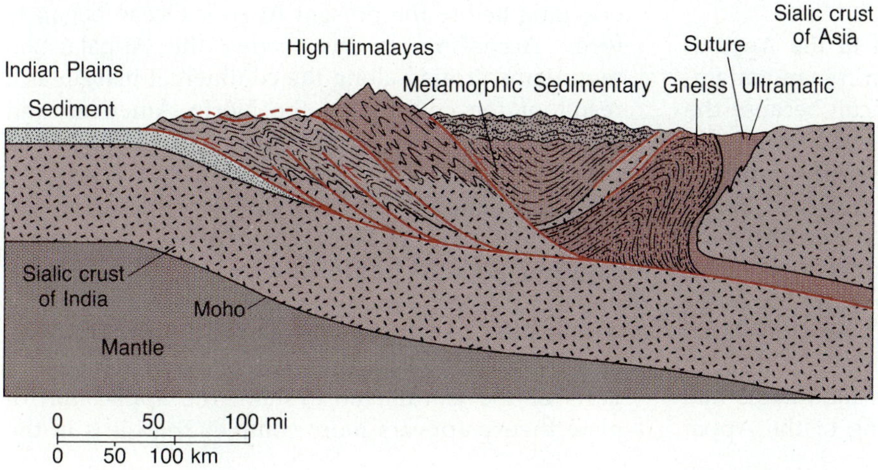

Figure 13.8
A schematic cross section across the Himalayas showing an inferred deep crustal structure and some details of the near-surface geology, including the suture zone between the Asian and Indian plates.

eventually rocks metamorphosed in the subduction zone and even slices of the oceanic crust or upper mantle may be uplifted high enough to be exposed by erosion.

The resulting orogenic belt has a foreland of folded and thrust cover rock and a core made up of slices of continental crust and/or igneous and metamorphic rocks that were formerly in the island arc. The junction between the two continents, called a **suture,** may contain slices of blue schist and fragments of oceanic crust or even pieces of the mantle.

Volcanic activity is not mentioned in the model just described. However, we might expect to find andesite volcanoes over the subduction zone in other cases. The pronounced volcanic activity in Central America and in the Andes seems clearly related to subduction. In the Andes, subduction occurs along the edge of a continent. Volcanic activity is also an important element in the eastern Mediterranean, where it appears that the African plate is approaching Europe.

Collision between an island arc and a continent would resemble a continent–continent collision in many respects. However, the sediments involved would include more of volcanic origin; and we might expect to see more igneous and metamorphic rocks in the resulting mountains.

Collision models are being applied to most modern mountain systems with some degree of success. The Himalayas (Fig. 13.8) are attributed to a collision between Asian and Indian continents, and the Alpine chains are thought to arise from collisions between Europe and Africa. Both the Appalachian and Cordilleran mountain systems are now being explained by collision models. However, both of these major North American mountain systems require additional explanations as well.

Application of the collision model to the Appalachians. Unraveling the mechanism of mountain-building in the Appalachians is difficult because the mountains were formed such a long time ago—more than 200 million years ago. Consequently, much of the original mountain mass has been eroded away, and the eastern part of the system is covered by younger undeformed sediments in the coastal plain. Even more importantly, the geographic setting of the system has changed. In Chapter 12, we considered evidence that the Atlantic Ocean began to form in Triassic time, after the Appalachian mountains had formed. Thus, the geographic setting of the Appalachians during the time they formed was closer to that in the predrift reconstructions of Fig. 12.17(*d*), than to that we know today. This raises the questions of whether the Appalachians formed within a large continent (Pangaea) before it split apart, or whether they formed on the margin of an earlier ocean that closed as a result of plate movements.

Most geologists now favor the second alternative, that the Appalachians are the product of a collision between North America and Africa, but they have not yet reached agreement about the location of the suture that separates the rocks of the North American plate from those of the African plate. The most distinctive features of a suture zone—blue schists and slices of sea floor—are not clearly evident throughout the belt but rocks that may have been part of the oceanic crust have been identified at a number of localities. The best example is in Newfoundland, where large masses of mafic and ultramafic rock thought to be pieces of the sea floor have been thrust up and over Precambrian continental basement exposed in the Long Mountains of western Newfoundland (Fig. 13.9 and 13.10). East of this, a highly complex zone of volcanic, plutonic, and sedimentary rocks appear to be the remains of an island arc system now plastered on the edge of the continent. Many of the rocks just east of these remnants are intensely sheared, as might be expected in a suture zone. Still farther east, we find slightly deformed rocks thought to be a part of the African plate.

Farther to the south, in the Taconic Mountains, a gigantic submarine landslide mass slid into place about the same time the slices of oceanic crust were thrust westward in Newfoundland. Still farther south, a large mass of gabbro near Baltimore is thought by some to be another slice of ancient sea floor of an ocean named **Iapetus** that existed during the Paleozoic era, long before the present Atlantic Ocean began to form. According to this model, the Appalachian mountains formed along the continental margin as a result of the collision of the North American and African plates. For a long time at the end of the Paleozoic era, the plates were joined. When the modern Atlantic Ocean began to form, this large plate split, leaving part of the original African plate on the eastern margin of North America.

Application of plate theory to the Cordilleran system. Although mountain-building is still going on in parts of the Cordilleran system, the application of plate theory appears more complex than it is in the

Appalachians. Perhaps this is because the Cordilleran system is a younger, better-exposed, and much longer orogenic belt than the Appalachians. The collision model applied to the Appalachians does not fit what we see in the Cordilleran system for at least two important reasons. First, no other continents that could have been involved in a collision with the Americas are located along or west of the continental margin. And second, an unusual relationship exists between the western margins of the Americas and the East Pacific Rise, the only major spreading ridge in the Pacific Ocean.

At the present time, new sea floor created along the east Pacific rise and moving eastward passes into a subduction zone marked by the Peru–Chile and Middle Americas trenches (Fig. 13.11). A line of active andesitic volcanoes in Central America and in the Andes is fed by magmas generated in this subduction zone.

At its northern end, the east Pacific rise can be traced closer and closer to the coast and into the Gulf of California. What happens to it farther north is uncertain. Some geologists think the spreading ridge continues northward under the North American continent; other geologists think the ridge is cut off at the northern end of the Gulf where the San Andreas fault cuts across it. The Juan de Fuca ridge in the northern Pacific may be the continuation of this spreading ridge. It is thus connected to the east Pacific rise by the San Andreas and Mendocino faults, as shown in Fig. 13.11.

Presumably, oceanic crust generated along the Juan de Fuca ridge is being subducted along the coast

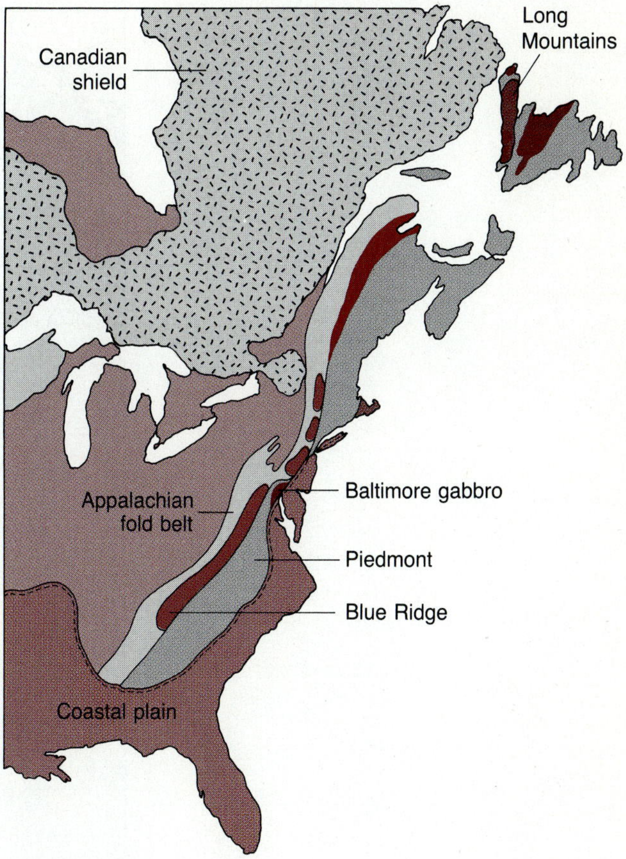

Figure 13.9
The areas shown in color on this map indicate the location of the largest masses of rock thought to have been portions of the oceanic crust of an ocean, the Iapetus Ocean, that predated the Atlantic. The location of Iapetus sea floor in Canada is based on work by Harold Williams.

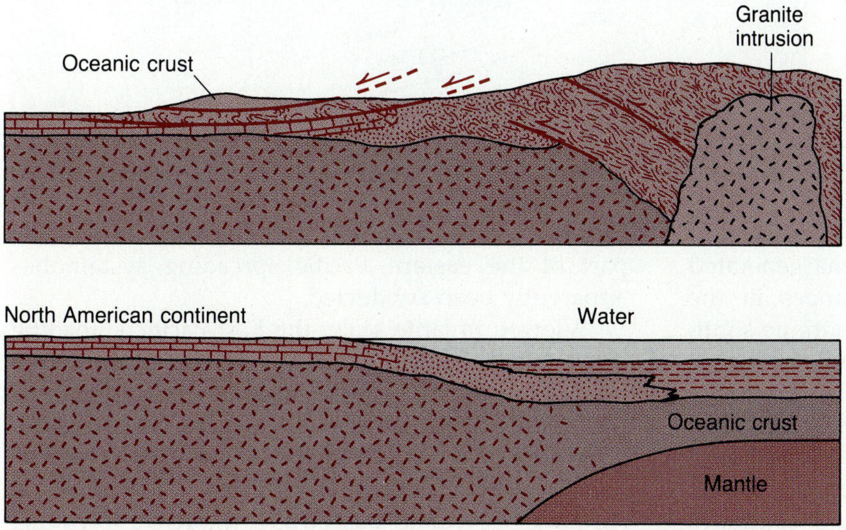

Figure 13.10
Cross sections depicting the evolution of western Newfoundland in the early Paleozoic era. In (b), deformation culminates in the emplacement of sea floor on the continent.

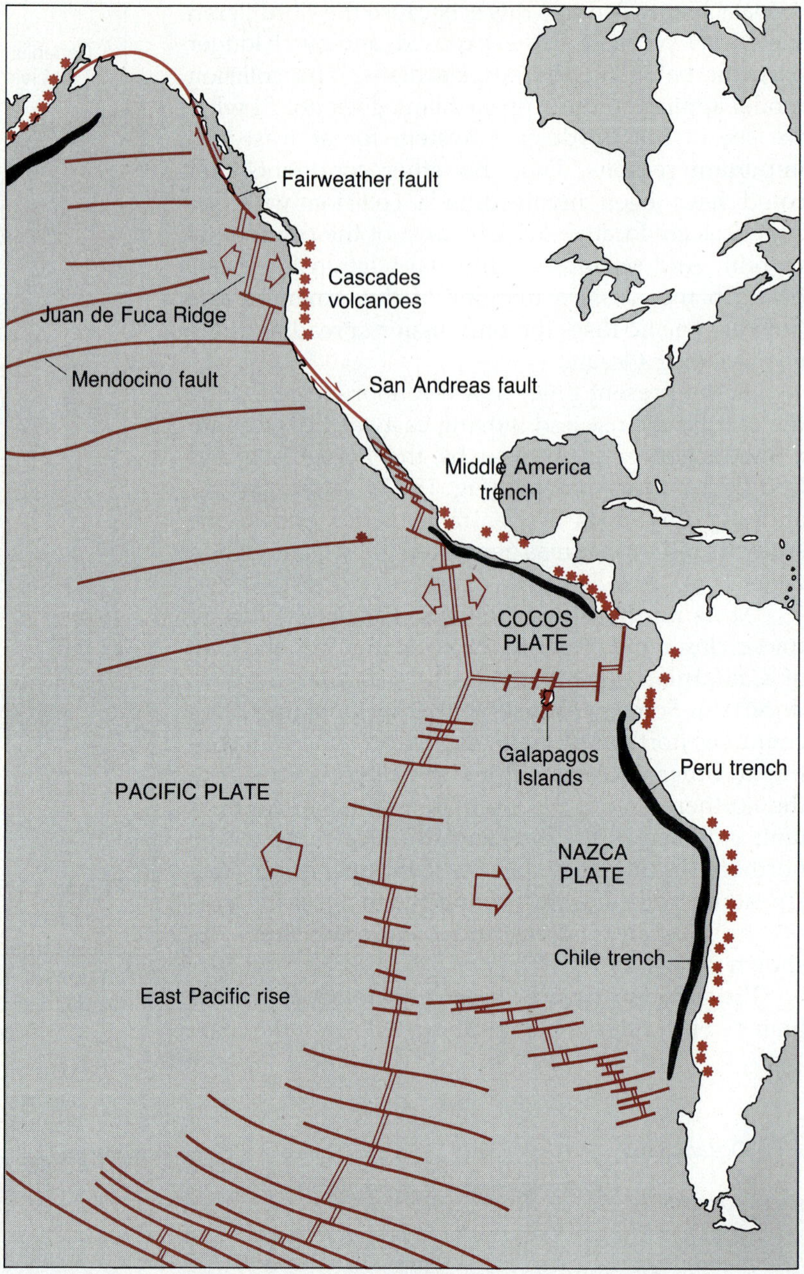

Figure 13.11
Most of the active volcanoes of North, Central, and South America are fed by lavas generated where subduction zones pass under the continental margins.

of Washington and Oregon, and magma generated from this slab feeds the active volcanoes in the Cascade range. The Cascades do not continue south of the Mendocino fault, nor do they extend northward along the Canadian coasts. The reason for this is found in the structure of the Juan de Fuca ridge, which extends to the northeast and ends abruptly at another major strike-slip fault along the coast of British Columbia. This fault, the Fairweather fault, continues northward to the Aleutian trench, where the northern part of the eastern Pacific spreading system has apparently been subducted.

Viewed in large scale, the East Pacific spreading ridge system is apparently being overridden by a westward-moving North American plate (remember that North America is moving westward as the Atlantic Ocean continues to widen). The San Andreas and Fairweather faults have formed where North America has actually passed over the active spreading ridge. In this interpretation, these faults are a special type of transform fault. The evolution of this fault system during the Cenozoic era is depicted in Fig. 13.12.

According to this interpretation, a subduction zone was located along the west coast of the United States sixty million years ago. Oceanic crust generated along the East Pacific Rise was subducted in this zone. At this time, North America was moving westward as the Atlantic Ocean grew larger. About forty million years ago the west coast reached the crest of the East Pacific Rise. The western part of California and Baja California subsequently moved over the crest of the rise. At the present time, this strip of land west of the San Andreas fault and the Gulf of California are moving northwest in the same direction as that part of the Pacific plate west of the East Pacific Rise. Many of the more recent features of the North America Cordillera can be related to the interaction of a continental margin bordered by a subduction zone and a spreading ridge as the two collide.

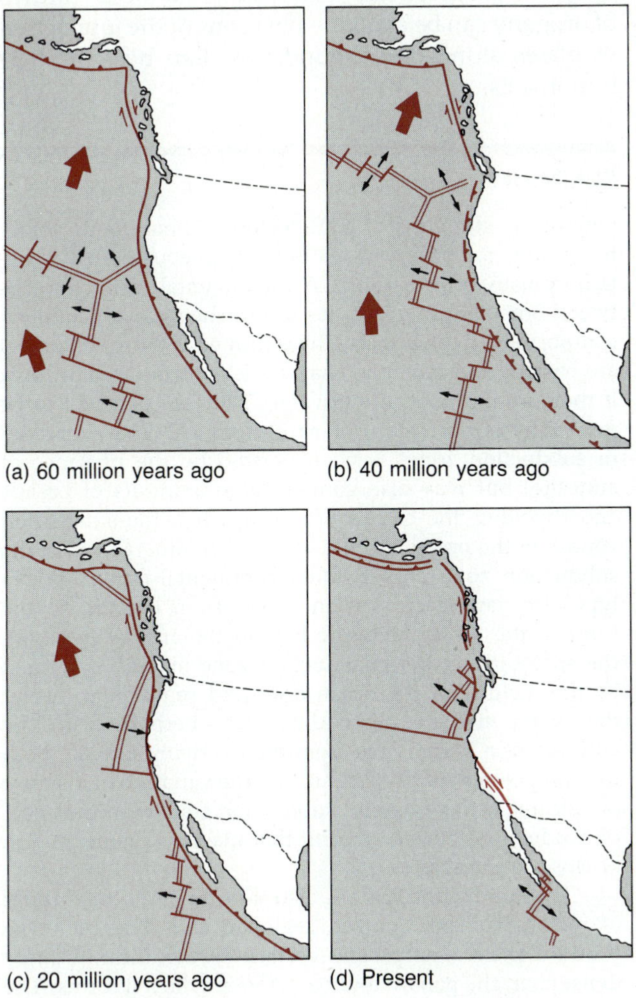

Figure 13.12
Evolution of fault systems, subduction zones, and spreading ridges along the western margin of North America over the last 60 million years. (a) A triple spreading junction is shown (b and c) migrating into the Aleutian subduction zone. At the same time, North America is moving westward, and a subduction zone off the southwestern margin of the continent is slowly transformed (d) into a strike-slip fault as the coast moves over the spreading ridge.

Many of the older features of the North American Cordillera are now interpreted as results of the collision of the continent with islands and island arcs now incorporated in the continent. Other features, especially the igneous intrusions, were caused by magma generated along an eastward-dipping subduction zone.

Dewey and Bird also postulatred a model for mountain-building in the absence of collision. They referred to this as a model of Cordilleran-type orogeny. Although collisions seem to play an increasingly important role in North American interpretations, their model, depicted in Fig. 13.13. may be applicable to the Andes and parts of western North America. The initial condition is that a subduction zone is located along the margin of a continent. Sandstones, carbonates, and shales are deposited at the edge of the continent, but these grade into sediments rich in volcanic ash derived from an island arc off the coast and near the subduction zone. As subduction goes on, magma is produced. The sediments in the subduction zone yield water, which rises into the mantle over the descending slab, facilitating the melting of the mantle there and causing alteration of the ultramafic mantle rocks to large quantities of serpentine.

The serpentinized rock and magmas generated within and above the slab form a large mobile core that begins to rise both because it is less dense than the surrounding rocks and because it is plastic. As it rises, some magmas find their way to the surface and others crystallize as plutons at depth, but the whole region is uplifted. Sediments eroded from this uplifted area flood into the adjacent waters as silts and sands, where they are deposited by turbidity currents. Because erosion and sedimentation are fast, these sediments often contain minerals like feldspar. When the slopes on the flanks of the uplifted core zone become steep enough, large sheets of sediment and even rocks that were formerly deep break off and slide down into the sedimentary basins.

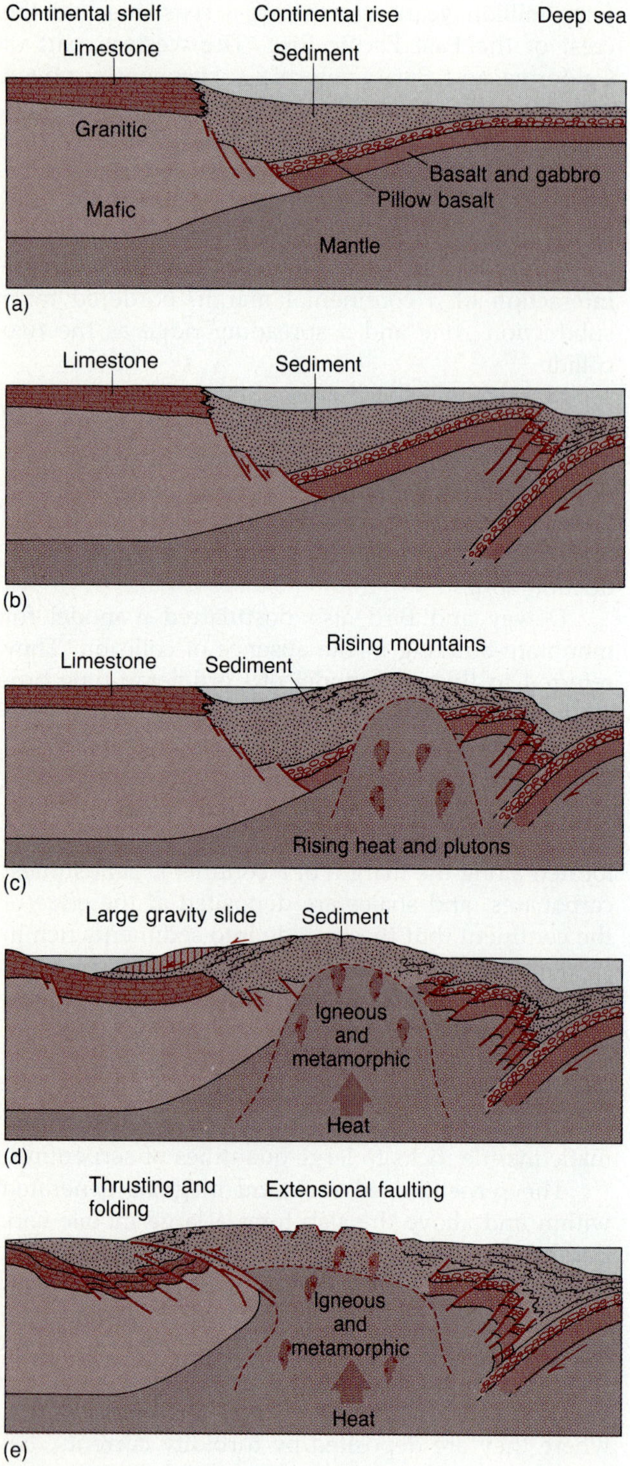

Figure 13.13
These cross sections illustrate the sequence of changes that may occur as a subduction zone is uplifted to form a mountain system. These types of changes have been postulated for the evolution of western North America.

The rise of the mobile core beneath the mountains brings with it rocks that have been metamorphosed in and above the subduction zone. This rising material exerts lateral pressure on the rocks into which it rises, lifting and pushing them aside. This lateral spreading results in the extension and rifting of the high-level rocks, creating block fault mountains, such as those in Nevada.

It would certainly be unfair to suggest that all the problems of mountain-building have been resolved, but great progress has been made, and more features of orogeny can be explained in terms of the interaction of plates along their boundaries than by any other hypothesis.

SUMMARY

One of the strengths of plate tectonics theory is its ability to account for observed features of igneous activity, metamorphism, and orogeny. A fundamental distinction between igneous provinces lies in the chemical and mineral composition of their rocks. Lavas formed at the ocean ridges are basaltic and have the characteristic form of pillow lava if they are extruded underwater. They are believed to be formed by partial melting of mantle rocks. Magma generated in subduction zones is characterized by the presence of andesite, but may also contain large amounts of basalts and rhyolites, the basalts occurring when the subduction zone is in the ocean and the quartz-rich minerals when the subduction zone passes under continental crust. Magma type also varies across island arcs—from basaltic on the front of the arc to andesitic behind the arc, farther from the spreading center. Subduction zone magma is formed by the melting of the ocean floor and sedimentary cover that make up the descending slab—perhaps with the addition of a partial melt from the overlying mantle. Both the composition of the magma and the variation in distance of volcanic activity behind subduction zone with the angle of subduction tend to confirm that magma is generated in or close to the slab.

The subduction zone is also the site of all three types of metamorphism: contact, regional, and dynamic. One product of metamorphism, eclogite, is significantly more dense than the gabbro and basalt from which it forms. The increase in density of the slab that occurs as eclogite forms is believed to contribute to the sinking of the slab. Sediment on the slab may be metamorphosed to blue schist, composed of minerals that develop under high directional stress at low confining pressure.

According to plate tectonics, great mountain-building episodes occur along subduction zones, especially when two continents or a continent and an island arc approach each other. The continental margin that includes the subduction zone becomes the site of a trench, a seismic shear zone, and possibly andesitic volcanoes fed by magma generated in the descending slab. Sediment caught in the trench between the slabs eventually fills the depression and begins to be compressed, folded, and raised as subduction continues. As one continental slab begins to override the other, the continental crust becomes extremely thick. The Himalayas are believed to have been created by such a collision between India and Asia.

The Appalachians are believed to have formed in a similar collision between North America and Africa. Much of the evidence of such a past has been removed by erosion over the last 200 million years, but remnants of sea floor found in Newfoundland and Maryland tend to confirm it.

The Cordilleran system did not form by continental collision, but probably did involve the addition of island arcs on the western edge of the continent. Along the west coast of North America, the continental plate is overriding the east Pacific ridge where crust is being generated. The San Andreas and Fairweather faults are believed to have formed where the continent has actually passed over the spreading zone. In Central and South America, the Cordillera may have been formed when water-rich sediments were carried down into the mantle by the subducting plate. The presence of water facilitates melting of the mantle rock, and so helps generate a large mobile core capable of moving upward to raise the mountain regions to the east of the subduction zone, pushing aside the rock into which it intrudes. The upward pressure on the rocks is believed responsible for the block-fault mountains of the basin and range.

KEY TERMS

blue schist
eclogite
Iapetus
obduction
oceanic ridges
pillow lava
suture

STUDY QUESTIONS

1. How do magmas generated under oceanic ridges differ from those generated in and above subduction zones?
2. How might you recognize the site of an ancient and now-inactive subduction zone in rocks of the continental crust?
3. In what ways does a "hot spot" differ from a divergent plate junction? Cite two examples of hot spots to illustrate your answer.
4. What inference about the environment in which extrusion took place can you make from the presence of pillow lava?
5. What is eclogite? From what rocks is it formed, and why is it so significant in plate tectonic theory?
6. What causes plates to move?
7. In what ways would the composition of mountains formed by a continent–island arc collision differ from those formed by a continent–continent collision?
8. What evidence supports the idea that the Appalachian Mountains formed as a result of a collision?
9. On the basis of what you know about the Cordilleran system, why would you conclude that not all mountains are formed by collision?
10. What changes in the position of the land, major faults, subduction, and volcanic activity would you predict will occur along the west coast of North America over the next 20 to 30 million years?

REFERENCE

1. H. B. de Saussure. *Voyages dans les Alpes*, vol. 3, 1796, as quoted in Sir Andrew Geikie, *The Founders of Geology*. New York: Dover, 1905.

SUGGESTED READINGS

Bird, John M., ed., *Plate Tectonics*. Washington, D.C.: American Geophysical Union, 1980.

Cloud, Preston, "Beyond plate tectonics," *American Scientist* 68 (1980), *381*.

Condie, Kent C., *Plate Tectonics and Crustal Evolution*. New York: Pergamon Press, 1976.

Cox, Allan, *Plate Tectonics and Geomagnetic Reversals*. San Francisco: W. H. Freeman and Company, 1973.

Marvin, Ursula B., *Continental Drift*. Washington, D.C.: Smithsonian Institution Press, 1976.

Sullivan, Walter, *Continents in Motion, The New Earth Debate*. New York: McGraw-Hill Book Company, 1974.

U. S. Geodynamics Committee, *U. S. Program for the Geodynamics Project, Scope and Objectives*. Washington, D. C.: National Academy of Sciences, 1973.

Wyllie, P. J., *The Way the Earth Works*. New York: Wiley, 1976.

THE SURFICIAL PROCESSES

Part 4

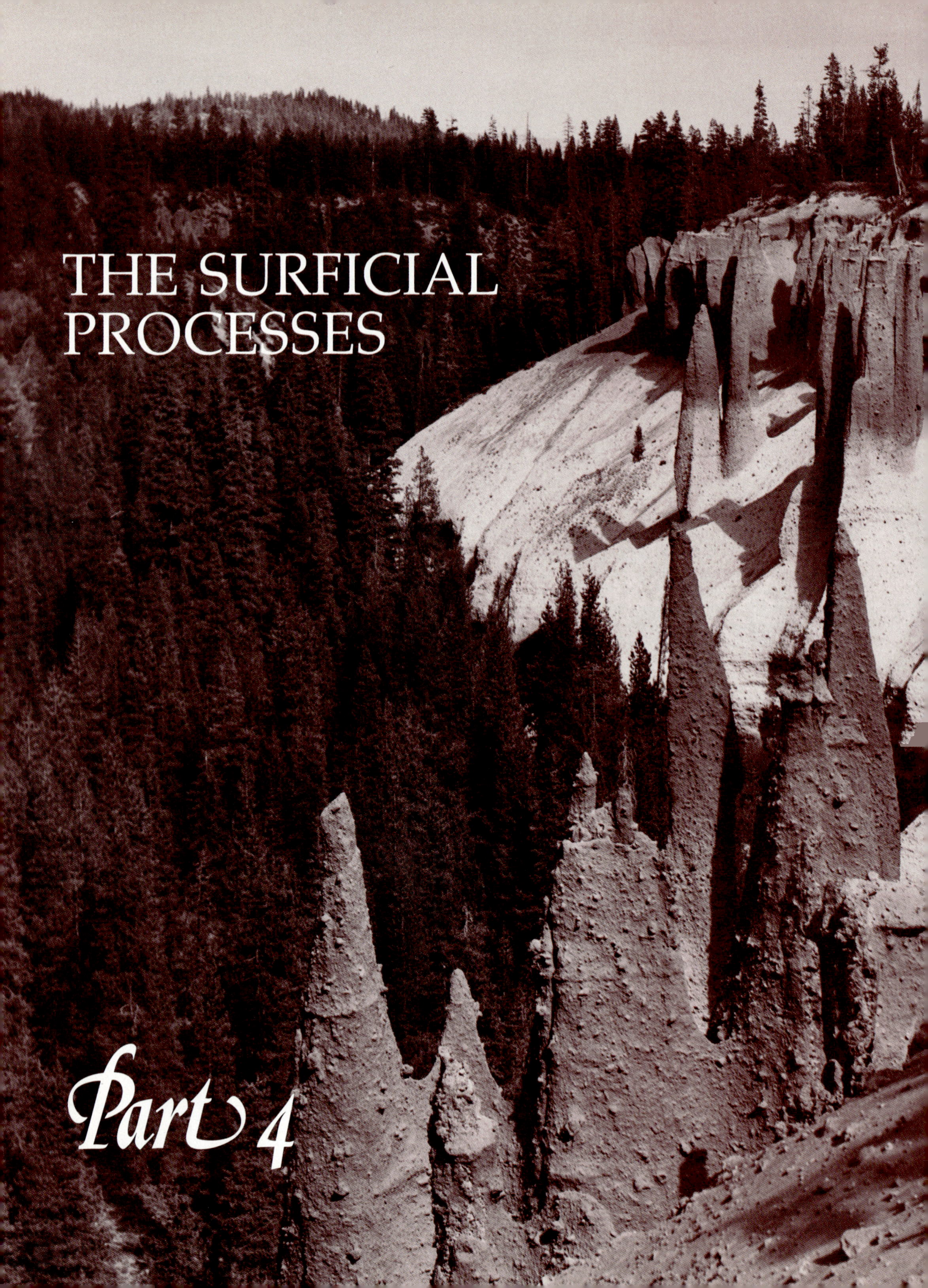

Humans and their activities come in contact, and often into conflict, with natural processes at the earth's surface. Often we have spread our cities and developments across the land without regard for the suitability of the land for our purposes, and we have frequently paid a high price in lives and property damage as a result of volcanic eruptions, earthquakes, hurricane damage, flooding, and slope instability. As we learn more about the natural processes that shape the surface, we should also gain a better understanding of how to live in harmony with these processes. In many cases, this means avoiding the conflict; in others it means designing our developments in ways that are compatable with the environment.

Part IV is concerned primarily with understanding the operation of surficial processes. As we consider these, we should continually be alert to the implications of these processes for the human environment.

THE DEVELOPMENT OF LANDFORMS

Chapter 14

In this part of the text we shall discuss the processes that act on the surface of the earth, shaping the land and dramatically influencing our environment. Some of the most violent of these are floods, volcanic eruptions (Fig. 14.1), earthquakes, landslides, and storm waves, all of which can have disastrous effects on people and property. For example, in August 1959 an earthquake in central Montana triggered a gigantic landslide in which 27 million cubic meters of rock moved half a mile in the span of five minutes. In 1960, violent earthquakes in Chile caused landslides that cost more than 5000 lives and extensive property damage over thousands of square kilometers. Tsunami, giant waves caused by these earthquakes, crashed on shores surrounding the Pacific Ocean, resulting in millions of dollars in damage as far away as Japan. Hurricanes annually batter the southeastern United States and the Caribbean islands, creating severe wave and current action, and sometimes accompanying rains cause vast flooding inland, as happened along the Rio Grande valley in 1967 and in Virginia in 1969. In 1980, the explosion of Mount St. Helens removed the top of the volcano and spread ash over much of the northwestern United States.

Rapid changes of this sort are catastrophic in character, and hundreds of them occur each year. Although these catastrophes are important in modifying land forms, most sculpture of the earth's crust takes place through processes acting so gradually that we rarely even think of them as significant. The slow, downhill movement of soil, the bits of dislodged rock that fall downslope, the gradual solution of limestone by rainwater percolating through it, the movement of grains of sand by the wind, and the constant transport of sediment in streams over long periods of time remove more material than the catastrophic events. We are conscious of the muddy water in streams after heavy rainfalls. We take note of the shifts in sand bars along the coast. We see glaciers, covered by rock fragments, slowly pushing their way

National Archives

Figure 14.1
Paracutin, a volcano located in southwestern Mexico, grew from a small cone to a volcano nearly 1000 meters high in a few months. The ground is covered by ash and dust.

down mountain valleys. But we do not always perceive that this is part of a large-scale pattern by which the land forms so familiar to us were brought into existence and by which they will one day vanish from the face of the earth.

The study of the form of the earth's surface is called **geomorphology.** Because of the important role they play in shaping the form of the land, streams, underground water, wind, glaciers, waves, currents, and gravity-induced downslope movement of material are known as **geomorphic agents.** This chapter provides a brief overview of the framework within which these agents operate. The following chapters are devoted to exploring these surficial processes in hopes of gaining an appreciation both of the complexity of the natural forces at work and of the way they affect people's lives.

14.1 GRADATION

The surface of the earth is under constant attack by the constituents of the atmosphere and by the geomorphic agents. The long-term effect, to level the land to a uniform shape or grade, is referred to as **gradation.** This is accomplished by a combination of **degradation**—lowering of the surface—and **aggradation**—building up of sedimentary deposits. The pattern is one of lowering of the earth's surface in one place by shifting of sediments, and raising the surface elsewhere by redeposition. Degradation is accomplished through processes known as weathering, denudation, and erosion. **Weathering** includes all those physical, chemical, and biological processes that cause rocks and minerals exposed at or near the ground surface to disintegrate or decompose. The stripping away of vegetation, soil, or unconsolidated sediment to expose underlying bedrock is known as **denudation. Erosion** includes all processes by which earth materials are loosened and moved from place to place. The term is often used to include both weathering and transportation, but many geologists prefer to restrict its use to processes involving movement of materials.

Movement of materials is accomplished by water, glacial ice, downslope movement of surficial materials, and wind. It is almost always directed from higher to lower elevations on land, and it is driven primarily by the force of gravity, although other forces are at work as well. Material is both moved by the geomorphic agents and deposited from them. The sand, silt, and gravel in streams reside in the stream channel much of the time. Likewise, sediments settle out of the wind, and glaciers form deposits that reveal their mode of origin to those who know how to recognize them. This building up of the surface through deposition (Fig. 14.2) is the aspect of gradation known as aggradation.

The long-range tendency of these processes is toward a leveling of the irregularities of the earth—a cutting down of the high places and a filling in of the lower ones. To estimate the rate of lowering, geologists determine the quantity of material moved out of an area by the major streams. By carefully measuring the quantity of sediment that is suspended in a stream plus the amount of material that is dissolved in the water, they can calculate the approximate quantity of material being removed by that stream. By this method, it is possible to show that the elevation of the land surface of the United States is being lowered at an average rate of between 3 and 7 cm/year. High areas are reduced in elevation much more rapidly than areas of low relief, but at this average rate of removel of materials (primarily by stream action) from the continents, almost all materials above sea level would be removed in a period of 20 to 40 million years. The earth is estimated to be at least 4.5 billion years old; thus, enough time has elapsed for the surface to have been eroded flat many times over.

Because the surface of the land is nearly a kilometer above sea level on average, it is obvious that other processes more than counterbalance the destructive tendency of degradation. These are the constructive forces of mountain-building and other movements of the crust referred to as **tectonic activity** (or **diastrophism**) and igneous activity, including volcanism. Both igneous and tectonic activity move material from lower to higher positions. Thus, the particular configuration of the surface of the crust at any given time—the present, for instance—represents a sort of balance between the forces of upheaval in the crust and the forces of erosion, denudation, and degradation.

The Three Phases of Gradation

The gradation of the land surface usually consists of three distinct phases:

1. The decay of the surface layers of rock or sediment.

Figure 14.2
These sand and gravel deposits along the Toklat River near Mount McKinley in Alaska illustrate the temporary building up of the surface, or aggradation.

2. The erosion of the weathering products.
3. The deposition of the transported materials elsewhere, usually at lower levels.

Erosion is facilitated by both chemical decomposition and mechanical disintegration of rocks. All three phases are controlled in large part by climate.

The products of weathering are usually unconsolidated, granular materials easily washed downslope by surface runoff of rain, a process called **surface wash.** These materials also tend to move gradually by rolling, sliding, and slipping downslope under the force of gravity. Eventually, they come in contact with running water or glacial ice and are carried along until either the velocity of the water becomes too low to move them or the ice melts. They are then deposited and eventually may become consolidated, compacted, and cemented into sedimentary rocks. Although some sediments may be deposited on land, the transitional environments of the continental shelves and slopes receive much larger quantities of sediment.

14.2 GEOMORPHIC AGENTS

The surface of the land is shaped by a number of quite different natural processes related to the action of streams, wind, waves, currents, glaciers, movement of water underground, and the gravitational pull exerted on all surficial materials. Distinct processes are associated with each of these geomorphic agents, and certain characteristic landforms are produced by both the erosion and the deposition associated with each of these agents.

Of the geomorphic agents, mass movement due to gravity is the most widespread (Fig. 14.3). It is an effective process in all climates and plays an important role in conjunction with each of the other agents. Gravity is important in moving materials downslope into streams where current action carries it along; it brings about subsidence and collapse of caverns eroded by groundwater solution; it accounts for the debris found on top of valley glaciers; it acts with the

(a)

(b) U.S.G.S.

Figure 14.3
Downslope movement of loose materials on the earth's surface takes place everywhere. (a) The lower slopes of the peaks in the Spanish Peaks area, Montana, are covered by cone-shaped piles of rock fragments that have broken off the steep slopes. (b) Rock of the steep slope of this valley located west of Yellowstone Park was dislodged during an earthquake in 1959, and a gigantic landslide followed, which dammed the Madison River.

wind to shape sand dunes; and it operates in the oceans as well as on land.

Streams are present in most areas outside the polar regions, but their role as a geomorphic agent varies considerably from arid to humid climates (Fig. 14.4). The runoff of water after rain has important effects in both climates, although streams in arid climates are often intermittent and the drainage pattern more poorly developed. Where rainfall is abundant, streams are both more numerous and larger, and the drainage system they form is a more prominent part of the landscape.

Glaciers exist both in polar regions and in the high mountains, even in equatorial latitudes (Fig.

(a) IGS Photo

(b) New Zealand Publicity Service

Figure 14.4
Streams are important geomorphic agents in most parts of the world. (a) The V-shaped valley and smooth slopes are characteristic features formed in areas that are eroded by streams. (b) When streams reach lower slopes, or when the amount of sediment being transported is excessive, streams often assume the braided form seen here on the Waitaki River in New Zealand.

Royal Canadian Air Force

(a)

Figure 14.5
(a) Glaciers have been an important geomorphic agent in mountains and at high latitudes, as in northern Canada.
(b) Brady Glacier in the Fairweather Range, Alaska.

U.S.G.S.

(b)

14.5). The present distribution of glaciers, however, is a poor indication of their importance in shaping the land surface. Glaciers were much more widespread and active only a few thousand years ago. About 11,000 to 12,000 years ago, sheets of ice over a thousand meters thick advanced far south into the United States and covered northern Europe and Siberia. The courses of the Missouri and Ohio Rivers are located approximately along the former edge of the ice sheet. So little time has elapsed since the glaciers covered these areas that the landforms found there still bear the imprint of the glacial processes. Most of the topography of central Canada and other formerly ice-covered areas reflects the smoothing effects caused by the movement of ice sheets of continental dimensions, and the deposits from these sheets cover vast areas. Most of the mountains in middle and high latitudes have been eroded by glaciers that occupied former stream valleys. The deposits from these glaciers are often prominent features in the landscape of the lowlands below the mountains.

Wind is an important agent of gradation in areas where winds are strong or persistent and especially where supplies of fine sediment are abundant (Fig. 14.6). Wind is an important agent along most coasts. Fine sand is blown from the beach face to build up ridges and sand dunes just inland from the shoreline.

(a) U.S.D.A.

Figure 14.6
Wind shapes the land in deserts and along many coasts. (a) A dust storm approaches over a New Mexico town in this picture. (b) Air view of the sand dunes in the western Sahara Desert. The distance across the area shown is several kilometers. (c) The entire area shown here is covered by sand being blown from left to right. A number of people can be seen on the left in the distance. These dunes are located in Sand Dunes National Monument, Colorado.

(b) U.S.A.F.

(c)

Wind is also effective in desert regions where the vegetative cover is thin and products of weathering and erosion are exposed and unprotected by plant cover, root systems, or soil moisture. Dust blown from the deserts of North Africa can be identified in Europe. The wind also serves an important function in dispersing ash and dust from volcanic eruptions. Following major eruptions, deposits of ash can be traced thousands of kilometers from their source. In 1980, ash from Mount St. Helens fell hundreds of kilometers from its source, across many of the plains states. Traces of ash fell on the east coast.

Part of the water that falls on and moves across the ground surface infiltrates the ground, becoming part of a system of underground water moving through pore spaces in sediments and through fractures in consolidated bedrock. This is an important source of water, and may even become an important factor in shaping the ground surface in areas underlain by soluble rock such as limestone. The moving groundwater gradually dissolves and carries off mineral matter. This process may eventually form large underground caves. As these are enlarged by removal of subsurface rock, the overlying roof becomes unstable and may subside or collapse into the cavity. In areas underlain by limestone, landforms may be strongly influenced by the process of solution and collapse, which eventually produces a landscape characterized by streams that disappear abruptly into the ground, by caves and natural tunnels, and by enclosed depressions called **sinks.**

The landforms of coastal areas and the shallow parts of the continental shelves are determined in large part by the effects of wave and current action (Fig. 14.7). Currents move sediment along the shoreline and out into deeper water. How crucial this supply of sediment is becomes apparent when the currents that move along shore are deflected by barriers constructed out into the ocean. Deprived of current-carried sediment, beaches are eroded more rapidly and may even disappear completely. Where the supply of sand is sufficient, wave action is important in building beaches and in changing them from one season to the next. Where beaches are not present, waves break directly against the land, cutting cliffs into the bedrock and constructing terraces offshore.

Not all coastal areas are shaped primarily by wave and current action. Some have forms inherited from earlier times, when glaciers or streams were more

National Park Service

(a)

(c)

IGS Photo

(b)

Figure 14.7
Waves and currents act on the materials along coasts, producing a rich variety of landforms. (a) Waves break along the Maine coast. Such waves are capable of moving vast quantities of sediment and causing coastal erosion. (b) Sea cliffs produced by wave action along the coast of Wales, in Great Britain. (c) Beach deposits built by wave and current action along the coast of North Carolina.

Montana Travel Promotion Bureau

Figure 14.8
The spectacular peak shown here, Going to the Sun Mountain, in Glacier National Park, Montana, is but a small part of a vast mountain system in western North America. These rocks exposed in these mountains have been uplifted as a result of deformation of the earth's crust.

active; some are shaped by volcanoes; and others are strongly influenced by organic deposits, such as coral reefs.

14.3 PROCESSES THAT BUILD UP THE LAND

If the calculated rates of denudation are correct, all land areas should have been eroded to sea level long ago by the various geomorphic agents. Clearly, other processes are acting to maintain elevation of the land—to rejuvenate the continents. Evidence of rejuvenation abounds. Igneous activity, especially as expressed by volcanism, is the more obvious of the processes that build up land. Volcanoes build up great piles of lava, ash, and other fragmental materials on continents and on the sea floor, as in Hawaii. Both high peaks, such as the peaks in the Cascade Ranges, and vast lava flows, as in the Columbia River plateau, are results of these processes. No less important is deep-seated igneous activity in which large bodies of magma invade older rocks, engulfing them or uplifting and forcing them aside. As these intrusions crystallize, they become welded to, and part of, the continents, adding their bulk to the land.

Although uplift of the land is often associated with igneous activity, it also occurs without accompanying volcanism. In the high mountain belts (Fig. 14.8), for example, we often find marine sediments and underlying crystalline rocks bent and folded, broken and displaced by faults, and uplifted—in some cases to thousands of meters above sea level.

Geologists now believe that the greatest of the uplifts associated with mountain belts are caused by the interaction of lithospheric plates along their margins, interactions of the kind outlined in Chapter 13. Where continents collide, the edges of two plates may buckle, and fault slices of one plate may be forced up onto or under the other plate. Both of these processes cause thickening of the continental crust and uplift of mountains.

Once the continental crust is thickened, it usually stays elevated even though the uplifted mountainous portion of the continent is subjected to more rapid denudation and erosion than lower regions; the less dense continental crust tends to float on the denser rock of the mantle (see Section 12.1). As the high peaks are eroded, the crust beneath them rises to maintain flotational equilibrium of the type discussed in Section 12.1.

The configuration of the surface of the land at any given time represents a balance between the forces of upheaval in the crust and the forces of erosion, denudation, and degradation. Although processes of degradation are constant and effective, continents have persisted as land areas throughout all of geologic time for which a rock record is available. Their shapes, sizes, and positions have, nevertheless, changed.

14.4 FACTORS IN LANDFORM DEVELOPMENT

As we study the shape of the land surface, it will become clear that the forms are not just a random assortment, but are strongly influenced by conditions that are easily recognized. We can predict in a general way the landforms that are likely to be present in an area if we know what these conditions are. The most important factors in determining land forms are:

1. The climate, with its direct influence on geomorphic agents.
2. The structure of the underlying bedrock, a product of the tectonic setting of the area.
3. The length of time particular geomorphic agents have been operating.

Bedrock Structure

If the earth were composed of a homogeneous material, without internal variations in shape or resistance to weathering and erosion, the form of the land surface would be due largely to the geomorphic agents in operation. However, rocks differ greatly in their susceptibility to weathering and erosion; and the thickness, form, and distribution of rock masses is quite variable. Both the external shape of a rock mass and its internal fabric are referred to as **rock structure.** Some of these structures, such as the layering of a sedimentary rock or the shape of an intrusion, originated when the rock was formed. Others result from subsequent deformation by tectonic activity. The original rock may be fractured; parts of it may be displaced along faults; or the originally flat-lying layers may be folded. All of these introduce a variety of forms that may eventually be reflected in the shape of the land (Fig. 14.9).

Sedimentary rock masses illustrate the importance of rock structure especially well because the layers are so often different in their resistance to weathering and erosion. With few exceptions, the large blanketlike deposits of sediment, such as those on the continental shelf, are laid down on nearly flat surfaces. This has been true of the sedimentary rock bodies that cover most of the continents as well. They were deposited in shallow water like that of the present continental shelf. Consequently, the initial form of most sediments is that of a relatively thin layer of great lateral extent. Some strata cover hundreds of thousands of square kilometers of surface area. The layers thin out toward the former shore, and they often change composition laterally. Such layers, with only slight inclination, are found throughout much of the Coastal Plain of the United States and in the Great Plains. In mountainous regions, similar layers are deformed; the layers are tilted, folded, offset along faults, and even overturned. (The most common types of structural features are reviewed in Fig. 14.10; a more complete description is given in Chapter 10.) These structural features are usually reflected in the topography that develops, especially if streams, wind, or wave action are the effective geomorphic agents.

Tectonic Setting

The **tectonic setting** of a region is both the type of large-scale structural features present there and the tectonic activity going on in the area. The types of

U.S.A.F.

Figure 14.9
Deformation and uplift of the crust of the earth is one of the main factors in maintaining parts of the crust at a high level. These mountains, the Atlas Mountains of North Africa, have been deeply eroded. Some of the materials removed from the mountains lie on the flanks of the ridges.

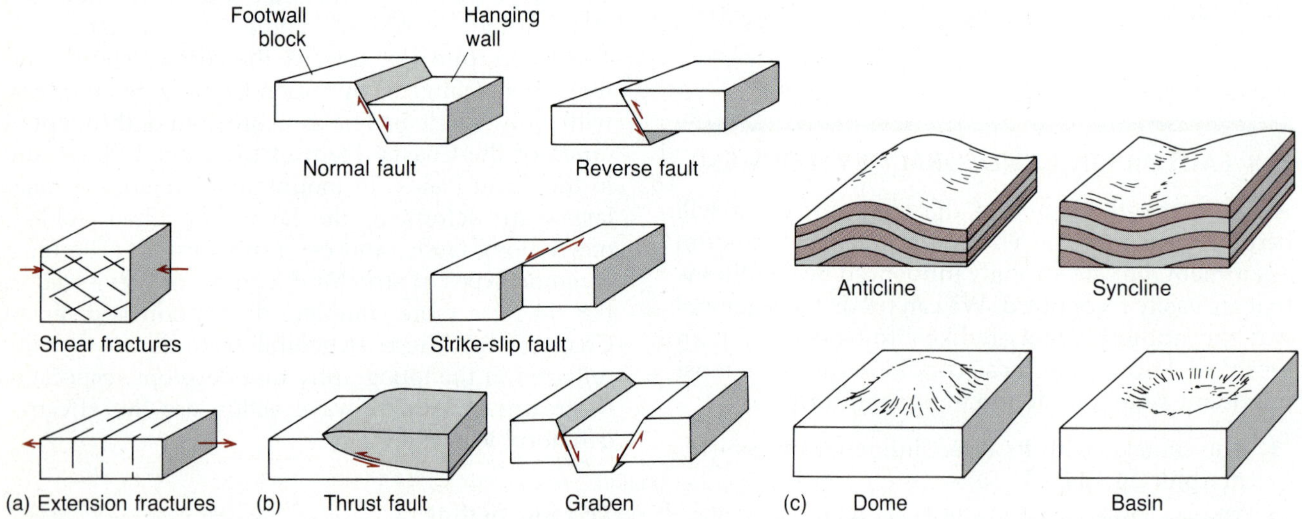

Figure 14.10
Some of the most commonly encountered structural features are depicted here. Refer to Chapter 10 for a more complete discussion. These features may occur on many scales.

structural features are directly related to the history of an area; they may be caused by modern active tectonism, or they may be a product of ancient tectonic activity. For example, the folds that are so prominent in the topography of the Appalachian region formed more than 200 million years ago, while many features in the landscape of western North America are due in part to the effects of movements still taking place.

Tectonic activity can take many forms. At one extreme are stable areas where the earth's crust is neither rising nor subsiding, almost no displacement is taking place along faults, and volcanism is absent. Vast areas of the central and eastern parts of the United States fall into this category. Much of central Canada is almost this stable, although a very slow elevation of the region is taking place where the crust was once depressed by thick ice.

The western United States exhibits a much greater variety of types of instability. Some areas, notably in the northwestern states, are sites of volcanic activity, which clearly can change landforms rapidly. Many areas are affected by active faults and crustal uplift. Sharp escarpments often mark the location of active faults (see Fig. 8.8). Movements on faults may disrupt drainage, even cause natural dams to form lakes.

Where uplift is taking place, the form of the land may reflect this upward movement. In New Zealand, originally flat surfaces formed along the coast are being deformed into broad, archlike upwarps. Of course, where uplift occurs, the forces of degradation immediately go to work. Consequently, areas of active uplift are being eroded down as they form. How high they become depends on how fast they are uplifted relative to the rate at which they are eroded, and an important factor in the rate of wearing down is the resistance of the uplifted material to erosion.

Resistance to Weathering and Erosion

The ease with which sediments and rocks are broken down, decomposed, and moved depends on many factors. Their composition, texture, degree of consolidation or crystallization, and the types of climatic and erosive agents to which they are subjected are all important. Climatic influences are reflected in the susceptibility of certain minerals and rocks to weathering. Quartz and quartz-rich rocks, for example, are highly resistant to chemical weathering. As a result, sandstones, quartzites, and rhyolites tend to weather and erode much more slowly in humid climates than do limestones, shales, or rocks rich in feldspar, olivine, or pyroxene. For this reason, quartzites often form prominent ridges in areas with humid climates.

The degree of consolidation of a rock body or a sedimentary deposit has a pronounced effect on how easily it can be eroded, especially by water and wind. Wind is rarely capable of moving fragments larger than coarse sand size (2 to 4 mm), but if unconsolidated sediment of smaller size is present on the ground surface, the wind may be effective in moving fragments or even in picking them up and removing them, as it does dust. Since loose sediment is also more likely to be moved by surface wash or in streams, unconsolidated sediments are more easily eroded than are compacted and cemented sedimentary rocks. Rocks, like quartzite, that are tightly cemented with strong insoluble cement are especiallly durable. Thus, in any area where rocks of different types are exposed, some types are more quickly broken down and removed than are others. These differences in resistance produce the effects called **differential erosion** (Fig. 14.11).

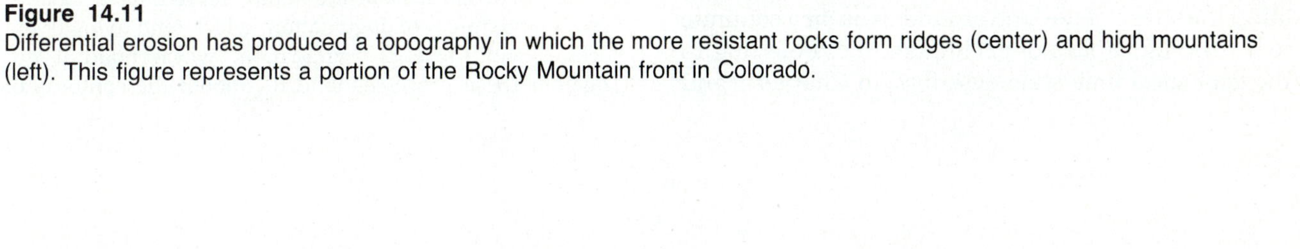

Figure 14.11
Differential erosion has produced a topography in which the more resistant rocks form ridges (center) and high mountains (left). This figure represents a portion of the Rocky Mountain front in Colorado.

Climate and Landform Development

Climate, the long-term averages of weather conditions, is an important factor in landform development, because the climate determines which of the weathering and geomorphic agents are likely to operate and to be effective in a region. These effects are most obvious if we compare regions with climatic extremes, such as those of the polar, desert, and tropical humid regions. Temperature conditions not only affect the speed of chemical reactions involved in the weathering process, but they also determine whether the precipitation is in the form of rain, ice, or snow. The amount and kind of precipitation determines a host of conditions, such as vegetation, soil type, soil moisture, the type of chemical weathering processes that will occur, and even more importantly whether glaciers, streams, or wind will prevail as important geomorphic agents.

Base Level

There is an elevation below which the surface of the land cannot be lowered by geomorphic agents acting on the land. John Wesley Powell, the one-armed Civil War veteran who made the first trip by boat through the Grand Canyon, conceived the idea of such a **base level.** He was undoubtedly impressed by the abrupt change that occurs in the Colorado River where it flows out of the canyon. In the narrow canyon, the water moves into high velocity through a series of rapids. As it emerges, the slope of the stream becomes much less, rapids disappear, and the river flows across a region of low relief near sea level. Powell recognized that the river had cut the canyon, and he could see that the river could not cut the canyon deeper as it approached sea level. Clearly the sea is the ultimate base level for geomorphic agents acting on land.

Locally, lakes and enclosed or land-locked depressions may act in a similar way and limit downward erosion of local areas. However, in the context of geologic time, these must be viewed as **temporary base levels.**

The Time Factor

It is difficult for us to realize how drastically the surficial processes we witness from day to day will affect the shape of the land around us as they continue to act through spans of geological time. Referring to the geological time scale described in Chapter 5 and shown on the inside back cover, we can see that at present rates of erosion all of the land above sea level could have been removed twenty times over since the beginning of the Paleozoic era! Although rocks from all four geologic eras crop out on the ground surface, most landforms are no more than a few million years old. It is true that many of the forms we see reflect structural features formed much earlier, but in most instances, the part of the structure exposed today lay buried in the crust until Cenozoic time. Thus, the folds we now see in the Ouachita and Appalachian Mountains are old. They range in age from 200 to 500 million years of age, but the parts of the folds now exposed to view could not have been seen at the start of the Cenozoic era, 70 million years ago. Indeed, some of these folds were almost certainly buried under much younger sedimentary rocks 70 million years ago.

Much of our landscape has come into existence in the last two million years, a time called the Pleistocene and best known as the "Ice Ages." Ice still shapes the land in Antarctica and Greenland and in some high mountains. Earlier sheets of ice several thousand meters thick repeatedly covered much of the latitudes above 40° north, and glaciers occupied mountain valleys all over the earth. These glaciers were responsible for shaping most of the land forms we see in mountains and at high latitudes. Their effects are seen in regions far beyond the edge of the ice. Thus, much of our landscape is geologically new, and we can see it is changing even today.

By studying the forms that emerge as the various geomorphic agents act, geologists have learned to recognize progressive alteration of the land. Given the structure of the bedrock, the tectonic setting, the climate, the resistance of the rock, and the geomorphic agents at work, we can predict the types of landforms that should emerge as the area is progressively eroded toward base level.

In the following chapters, we will study the operation of the various geomorphic agents and see how much we can infer about the history of a region by studying the form of its land.

SUMMARY

The form of earth's surface today represents a balance between the forces of degradation, which tend to lower the surface, and those of aggradation, which build it up. Together these processes tend to smooth the contours of the earth, wearing away the high places and filling in the low ones. Degradation is accomplished by weathering, denudation, and erosion, including transportation of material by water, wind, glacial ice, or the force of gravity. Whatever is transported must come to rest somewhere. Deposition in streams and oceans, the settling out of windborne particles, and the deposits left by glaciers all contribute to aggradation. In each of these cases, it is gravity that ultimately brings a particle to its resting place. Over the long range, deeper tectonic processes are also at work, building the main structure on which these forces act.

The geomorphic agents at work degrading the surface include streams, wind, waves, ocean currents, and glaciers. Where the underlying rock is soluble, underground water may form subterranean caves that eventually collapse to form sinks. Just what landforms develop as a result of the combination of degradation and aggradation depends on the set of conditions that prevail in a region. Among the determining factors are climate, the structure of the underlying bedrock (a product of the region's tectonic setting), and the length of time that geomorphic agents have been at work there.

KEY TERMS

aggradation
base level
climate
degradation
denudation
diastrophism
differential erosion
erosion
geomorphic agents
geomorphology
gradation
rock structure
sink
surface wash
tectonic activity
tectonic setting
temporary base levels
weathering

STUDY QUESTIONS

1. What geomorphic agents operate where you live?
2. What evidence, if any, suggests that bedrock structure has influenced the origin of the landforms of your area?
3. Distinguish among the terms *weathering, denudation,* and *erosion.*
4. What would you have to do to determine the amount of material removed from the nearest major river basin during a year's time?
5. How do various geomorphic agents differ in the way they move materials?

SUGGESTED READINGS

Bloom, A. L., *Geomorphology.* Englewood Cliffs, N.J.: Prentice-Hall, 1978.

Bradshaw, M. J.; A. J. Abbot; and A. P. Gelsthorpe, *The Earth's Changing Surface.* New York: John Wiley & Sons, 1978.

Cooke, R. U., and J. C. Doornkamp, *Geomorphology in Environmental Management.* Oxford: Clarendon Press, 1977.

Howard, A. D., and I. Remson, *Geology in Environmental Planning.* New York: McGraw-Hill, 1978.

Ollier, C. D., *Tectonics and Landforms.* London: Longman, 1981.

Weyman, Darrell, and Valerie Weyman, *Landscape Processes, An Introduction to Geomorphology.* London: George Allen & Unwin, 1977.

VOLCANIC FORMS AND PROCESSES

Chapter 15

The explosive eruption of Mount St. Helens in 1980 made us acutely aware of the force of volcanic activity and of its tremendous capacity to shape the surface of the land. Because volcanoes have such distinctive forms and because volcanism is one of the two important processes raising and maintaining the land areas above sea level, volcanism is in one sense a surficial process. It should be clear, however, that volcanism is the surface manifestation of deep-seated igneous activity. Both aspects of volcanic activity are important to our understanding the origin of materials. In addition, volcanic activity is one of the dramatic proofs that earth is a dynamic planet. As we saw in Chapters 12 and 13, volcanic processes are clearly related to the movements of the lithospheric plates.

Both emplacement of masses of molten rock below the surface and the eruption of molten materials above ground play important roles in forming the landscape. Both of these processes build up the land—volcanoes add new rock directly on the surface, and emplacement of magma at depth often lifts up the surface from below. Distinctive landforms are created by each of the various types of volcanic activity, but an intrusion is usually expressed in the landscape as a result of differential erosion that etches out its shape and internal structure as the overlying rock is gradually removed. Because igneous rock generally has more resistance to weathering and erosion than does sedimentary rock, the form of an igneous intrusion in sediments or sedimentary rocks is often gradually exposed as denudation proceeds.

15.1 GEOLOGIC SETTING OF VOLCANIC ACTIVITY

Early students of volcanism thought that the earth's molten interior was covered by a thin crust through which fiery masses from below erupted. Through seismology, we know today that the earth's interior is largely rigid to a depth of about 2900 kilometers. Because of this rigidity, it is highly unlikely that magmas are coming from such a depth. The modern view is that most magmas are generated in the upper part of the mantle, at depths ranging from 50 to 100 kilometers.

Both the location and composition of erupted magmas of most volcanoes can be explained in terms of plate tectonics. As soon as volcanologists began to draw maps showing the location of volcanoes, they were impressed by the distribution of active and recently active volcanic centers. As early as 1903, a distinct compositional contrast between volcanoes in the central part of the Pacific basin, which erupt basalt (50 to 55 percent SiO_2; see Section 6.2), and those around the margin, which erupt andesite (55 to 65 percent SiO_2) was recognized, and a boundary known as the **andesite line** was drawn.

Most andesite volcanoes are interpreted as forming over subduction zones, such as those of the island arcs that bound the western Pacific, the so-called "ring of fire." Basalt volcanoes occur along the crest of the oceanic ridges, where sea-floor spreading is taking place, and along some faults and fracture zones in the oceanic crust. Other volcanic centers, including the famous ones at Hawaii and Yellowstone Park, do not lie in either of these types of plate boundaries. They are thought to be situated above places where hot plumes rise in the mantle. The heat from these hot spots produces magmas that rise and cause volcanic activity. Mantle plumes rising under oceanic crust produce basalt; those rising under continental crust or derived from sinking slabs of the lithosphere produce both andesitic and basaltic lavas.

15.2 ERUPTIONS

Once generated, magma moves in response to stresses in the surrounding rock and as a result of the volume increase that accompanies the melting process. The expanding body of molten matter is forced upward along any existing fractures and faults, or it pushes the overlying strata upward into domes through which it may eventually break along newly formed cracks.

When magma finds its way to the surface, a volcanic eruption of some type takes place. If great quantities of gases are held under pressure, the eruption may be explosive, but if gases are not present or if they have escaped en route to the surface, the eruption is usually relatively quiet and may consist mainly of the extrusion of lava. Lava, gases, and fragments of the rocks through which the magma has moved constitute the principal products of volcanism. Magmas do not always break through to the surface. Many of them cool and crystallize beneath the ground surface, giving rise to the great variety of

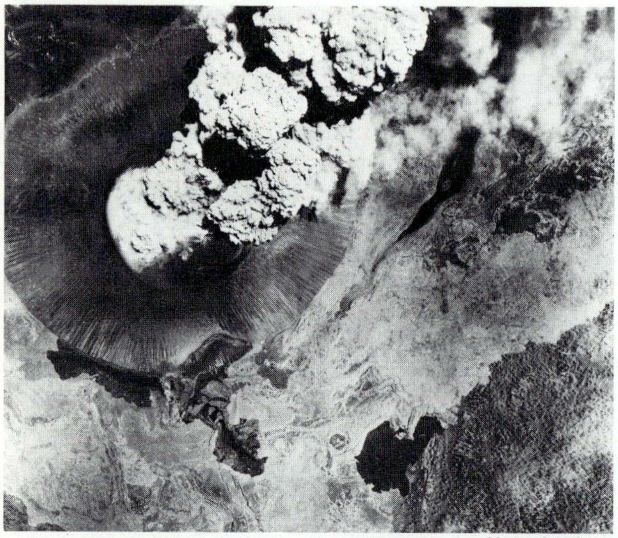

(a) National Archives

(b) National Archives

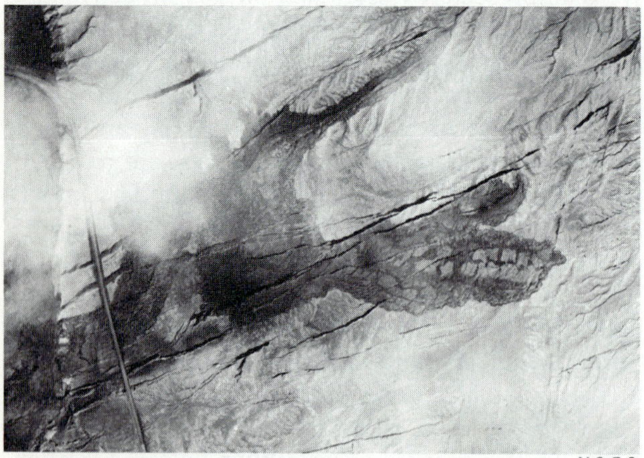

(c) U.S.G.S.

(d) U.S.G.S.

Figure 15.1
(a and b) Central-vent eruptions, such as the one shown here at Paracutin volcano in Mexico, derive their name from the eruption of material at a single point. (c and d) In contrast, lava and gases rise to the surface along a line in fissure eruptions such as the ones shown here in this vertical photograph of the flanks of Mauna Loa in Hawaii. Note the highway in (c) for scale. The long straight lines are the fissures, and dark lava is seen along two of the fissures.

igneous rock types described in Chapter 6. Those that do reach the surface may break through along long surface cracks causing **fissure eruptions,** or if the magmas move up mainly along a pipelike opening, they give rise to **central-vent eruptions** (Fig. 15.1). Fissure eruptions are long flows or clusters of volcanic centers along the fissure; central-vent eruptions produce the more isolated peaks that we generally think of as volcanoes.

15.3 PRODUCTS OF VOLCANIC ERUPTIONS AND THEIR FORMS

Lava

Although we can learn much about lavas by studying the solidified igneous rocks they form, eruption of lavas provides an opportunity to examine the melts directly while they are still liquid. Because they are extremely hot, usually over 1000 °C, and because volcanoes are unpredictable, collecting samples of lava is a pastime not without hazards, and a few volcanologists have paid the supreme price for their curiosity. Fortunately, some volcanoes, especially those in the middle of oceans, are less prone to violent eruption and are more accessible for study. Samples of melts and solidified lava have been collected from these volcanoes in many different locations, and representative chemical compositions of a few of them, expressed as percentages of oxides, are compared in Fig. 15.2. The figures shown for Hawaii were obtained from chemical analysis of samples of melt collected during an eruption. Average figures for andesite and rhyolite from other parts of the world are shown for comparison. The significant chemical differences between these three most common types of extrusive rock clearly indicate that lavas are not all alike chemically. Similar variations in the composition of their plutonic equivalents, gabbro, diorite, and granite, clearly indicate that magmas vary greatly in composition. Less dramatic changes in composition can be found in the lavas erupted from a single volcanic center. Knowledge of the cause of these

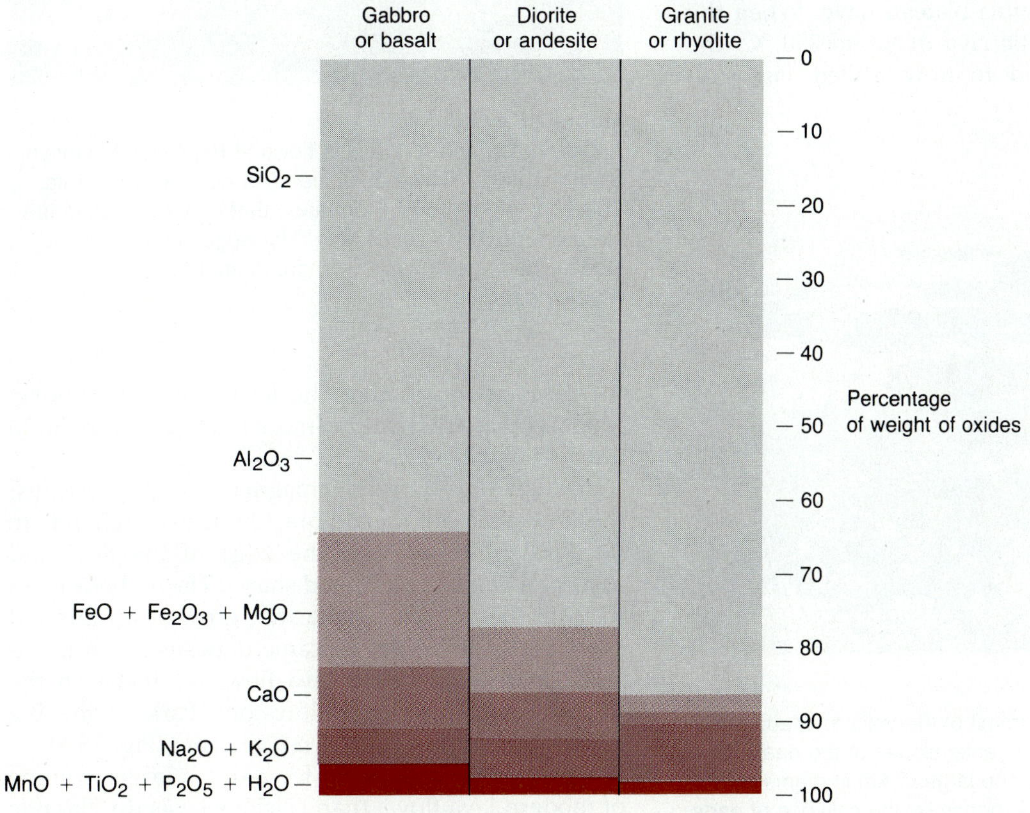

Figure 15.2
The relative abundance of the common rock-forming elements in representative samples of basalt, andesite, and rhyolite.

variations is one of the important objectives of igneous petrology.

The composition of the melt, its temperature, and the amount of dissolved water and gas it contains all influence its fluidity. In general, high temperature, high water content, high content of volatiles, and low silica content all favor high fluidity. The temperatures of lavas are usually measured optically based on their color alone, making it possible to determine temperature from a relatively safe distance of several meters from the melt. The orange, golden yellow, and white colors of most lavas signify temperatures in the range of 900 to 1200 °C. At these temperatures, basaltic lava is highly fluid and flows like a stream, following valleys in the topography and forming lava lakes as enclosed depressions fill. The vast plateau surface in the northwestern United States, the Columbia plateau, was built up in this way, as was the gently sloping, shield-shaped surface of Mauna Loa on the island of Hawaii (Fig. 15.3). The lava in most lava plateaus and shield volcanoes is basaltic in composition.

Not all flows are fluid enough to spread out the way those in the Columbia plateau have. When the temperature of a basaltic lava drops to 750 °C, the resistance of the liquid to flow (called viscosity) increases so much that the lava will barely move. Magmas that are rhyolitic (more than 70 percent SiO_2) are also highly viscous.

U.S.G.S.

Figure 15.3
This oblique view of the summit of the volcano Mauna Loa, in Hawaii, shows the broad gentle slopes of the peak. The summit itself is marked by one large (2 km in diameter) and several smaller craters formed by the collapse of parts of the summit over areas from which lava has been extruded.

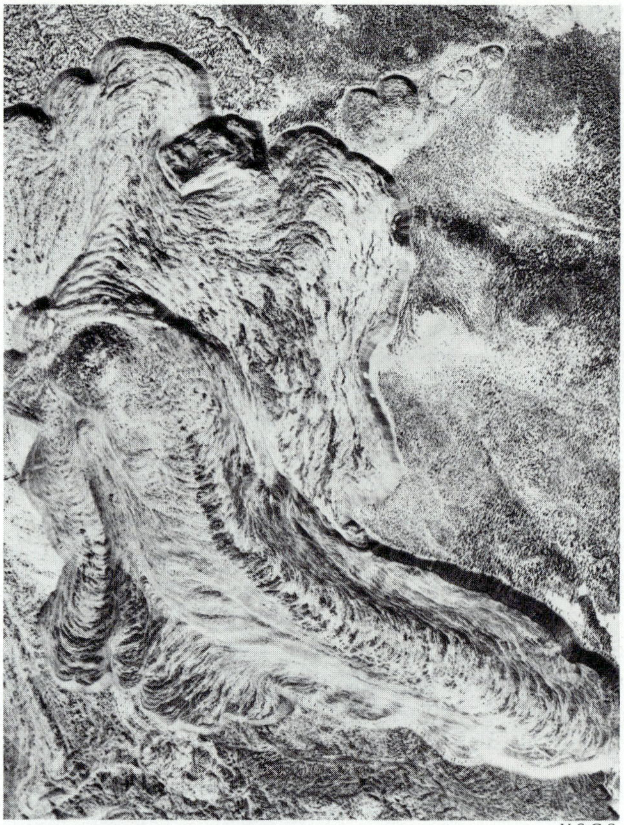
U.S.G.S.

Figure 15.4
Aerial photograph of the lava flows at Big Glass Mountain, Medicine Lake Highlands volcano, in northern California. This flow is one type of formation that may result from the extrusion of very viscous lava. The edges of this flow are steep. The large flow shown here is about one-half kilometer across.

When the extruding magma is highly viscous, the flow does not spread quickly. It usually tends to form steep slopes near the edge of the flow and assume a bulbous or domed shape. The bulbous mass may fill the crater of a volcano, grow as an injected mass under the older volcanic deposits as an intrusion, or develop into a lava flow like those on the Pitchstone plateau in Yellowstone Park or the Big Glass Mountain in northern California (Fig. 15.4).

Basaltic lavas account for a much larger volume of modern lava flows than silicious lavas do. Basaltic lava flows can usually be classed as one of two distinctively different types originally recognized in

U.S.G.S.

(a)

National Archives

(b)

Figure 15.5
(a) The surface of this fresh lava flow in Hawaii shows structure typical of ropy or pahoehoe lava. The hat provides a scale for this flow, but notice how similar the surface of this flow is to that of the much larger flow in Figure 15.4. (b) The smoldering front of a blocky or aa lava flow that formed during the eruption of Paracutin volcano in Mexico is shown here. Such blocks are usually from 10 to 30 cm across.

Hawaii—*pahoehoe* or ropy lava (Fig. 15.5), and *aa* or blocky lava. Ropy lava, characterized by its ropelike form, is formed as the outer surface of a thin sheet of lava cools and begins to crystallize. The cooled surface wrinkles as it is pushed forward and slowly twists into the ropy form. Blocky lava usually forms when a thick crust develops over the surface of a slowly moving lava flow. This crust, usually composed of a dark, fine-grained rock made porous by holes formed by escaping gases, breaks up as the lava continues to flow. Because the liquid center of the flow moves, the crust is dragged with it. Blocks fall forward off the steep front of the flow and are buried as the flow advances, but, if the rate of advance of the lava increases, the flow may break out and proceed as a river of lava.

Magma generally solidifies in the feeder systems of fissures under volcanoes, causing the formation of dikes, which are occasionally exposed by later erosion. Some pipes are also etched out by erosion, as Devil's Tower, Wyoming, and Shiprock, New Mexico. These towering columns of basalt are all that remains of once-grand volcanoes. The pipe and several dikes that radiate out from it can be seen in the air view of Shiprock in Fig. 15.6. A series of sketches shows how such a feature evolves as a volcano is eroded.

Pyroclastics

The fragmental or **pyroclastic** materials erupted from volcanoes include pieces of rock torn from the sides of the pipe or feeder system through which the lava moved, pieces of the cone, and pieces of lava that cooled and at least partially solidified in the air. The largest masses of rock blown out of volcanoes, called **blocks,** may weigh tons and sometimes are moved considerable distances by the force of the eruption. For example, a 2000-kilogram block was blown a distance of about 3.2 kilometers during an explosion of the Italian volcano Stromboli. Even larger blocks, weighing tens of tons, have been moved short distances. Most pyroclastics, however, are smaller fragments such as **bombs** (32 millimeters in diameter or larger), **lapilli** (4 to 32 millimeters in diameter), **ash**

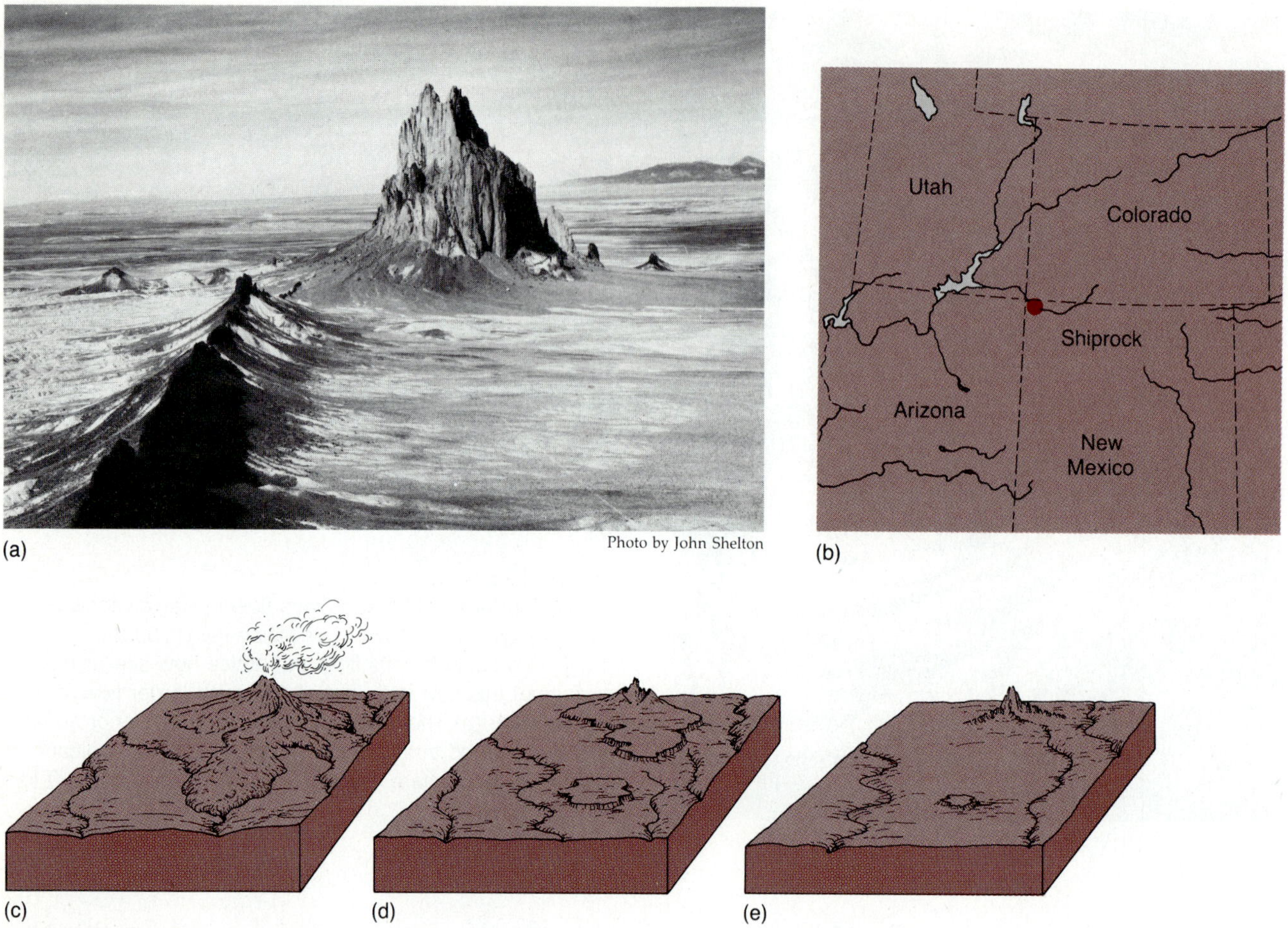

Figure 15.6
Shiprock, New Mexico. (a) The high central mountain is what remains of lava that solidified in the pipe of an ancient volcano. Several basaltic dikes form ridges which radiate out from this center. (b) Locator map. (c, d, and e) Hypothetical stages in the creation and destruction of a volcano leading to the formation of a mountain like Shiprock.

($\frac{1}{16}$ to 4 millimeters in diameter), and **dust** (less than $\frac{1}{16}$ millimeter in diameter). Bombs and lapilli are formed from bits of lava blown into the air, where the outer part is chilled and partially solidified before the fragment hits the ground. They usually have an oval center, but most are turned in the air, producing twisted ends (Fig. 15.7). Finer fragments may be blown great distances, and volcanic dust can be moved hundreds, even thousands of kilometers. Prevailing winds carried particles of ash (Fig. 15.8) from Mount St. Helens hundreds of kilometers. Dust from Italian volcanoes can be found in North Africa, and some dust may even reach the upper atmosphere and circulate around the earth, causing brilliant sunsets. If a strong wind is blowing, the small fragments may be sorted. That is, the finest fraction is carried far away, while the coarser materials fall near the vent.

Many smaller volcanic peaks are made up largely of pyroclastics, often mainly cinders, which tend to form steep-sided cone-shaped masses like those of Paracutin, shown in Fig. 15.1 (*a* and *b*). Because the fragments are irregular in size and shape, they can stand up on steep slopes, usually between 30 to 40°. As the cone is built, an inverted cone-shaped depression, or **crater,** normally develops around the central vent, from which materials are ejected. Materials that do not clear the rim of the crater are

U.S.G.S.

Figure 15.7
Pyroclastic fragments. These are small pieces of lava, in this case scoria, that solidified in the air. The largest is about 3 cm long.

E. A. Vincent

Figure 15.8
This scanning electron micrograph shows a sample of ash from Mount St. Helens, collected in Montana. It is seen enlarged about 3000× here.

funneled back into the vent. Excellent examples of this type of cinder cone can be seen in Craters of the Moon Park in Idaho and in Oregon, Arizona, and Mexico. They may be found also in Hawaii, Iceland, New Zealand, and other places.

Many volcanoes, circular in plan and concave upward in profile (Fig. 15.9) are called **composite cones** because they are formed of a mixture of lava and pyroclastics. Many of the most famous volcanoes, such as Mount Fujiama in Japan, are of this type. Extrusion of lava may alternate with ejection of pyroclastics. Often, lavas form a lava lake in the crater before spilling over the rim and flowing down the sides of the cone. Sometimes lava fills the feeder system and forms a lake. This causes pressure to build up in the volcano and this pressure may be

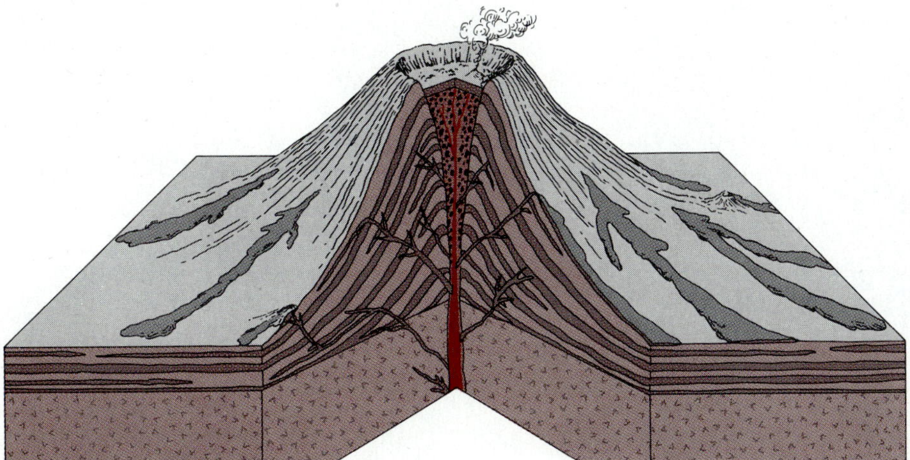

Figure 15.9
Composite volcanoes are composed of mixtures of lava and cinder. The steep slopes are usually due to the pileup of cinders. Lavas are shown breaking through on the flanks of this cone.

Figure 15.10
Long, tongue-shaped lava flows are clearly evident in this view of a snow-covered Mount St. Helens. This picture was taken in April 1980, after activity started but before a violent eruption destroyed the top of the volcano. (U.S.G.S. photo.)

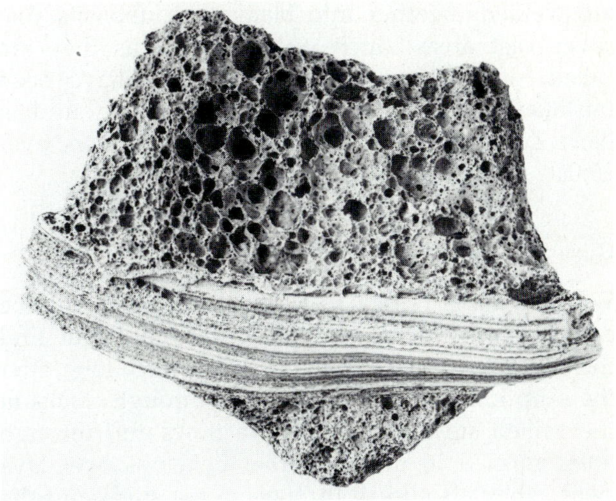

U.S.G.S.

Figure 15.11
A fragment of scoria showing holes formed by escaping gases. This specimen is about 10 cm wide.

released by flank eruptions. Lava flows tend to expand the base of composite cones (Fig. 15.10), whereas the pyroclastics build the rims higher, leading to the characteristic profiles.

A light-colored, porous, glassy rock called **pumice** is formed when viscous rhyolitic lava contains large quantities of gas. Often pumice is formed when gases that have been prevented from escaping underground are suddenly released in a violent eruption. As a fragment enters the air, the outside pressure on the fragment is much reduced, causing gases in the lava to expand and form bubbles that grow to form cells separated by membranelike walls of glass. The city of Pompeii was buried under pumice and ash when Vesuvius erupted in A.D. 79.

Scoria is another vesicular volcanic rock formed by the expansion of gases in lava as it is ejected, but scoria is heavier and darker in color because of its more mafic (higher magnesium and iron) composition (Fig. 15.11). If the fragments of lava are hot enough to remain plastic when they hit the ground, they may weld together to produce mounds or towerlike structures called **spatter cones** (Fig. 15.12), or layers called welded tuff, described below.

Particularly disastrous and violent eruptions occur when the ejected mass is composed of incandescent clouds of gases and lava droplets, usually lava rich in silica. The gas in these clouds of red-hot dust and ash lubricates the solid particles and enables them to flow as rapidly as 100 km/hr. The intensely hot clouds spread quickly; and as the material, initially carried upward in the air, condenses and falls, it forms a deposit in which the ash and glassy particles

U.S.G.S.

Figure 15.12
Spatter cones composed of scoria formed on lava flows in Hawaii. Note the man standing beside the cone in the foreground.

are welded together into blanketlike deposits that cover vast areas. Such **welded tuffs,** as they are called, often resemble lava flows and have been mistaken for them. A notable example is found in New Zealand where one such deposit covers over 20,000 square kilometers.

Gases

Volcanic gases (Table 15.1) escape in great volume during most eruptions, and they continue to find their way from the magma to the surface long after the eruption stops. Gases emerge through cracks in the chilled surface layer of lava flows and through holes formed in the crust that develops over lava lakes. They also break through to the surface in the central vent of cinder cones after making their way upward through the ash and cinder that fill the conduits. Gases are particularly abundant where explosion or collapse of a volcano has fractured the rock over the magma (Fig. 15.13).

Table 15.1 Composition of Volcanic Gases from Kilauea Volcano, Hawaii. (Data gathered by the U.S. Geological Survey.)

Constituent	Percent of total volume
H_2O	67.7
CO_2	12.7
N_2	7.65
SO_2	7.03
SO_3	1.86
S_2	1.04
H_2	.75
CO	.67
Cl_2	.41
A	.20

Water vapor is the most abundant volcanic emanation (Table 15.1), but carbon dioxide, hydrogen sulfide, and sulfuric and hydrochloric acids may also rise from small vents called **fumaroles.** The ground

New Zealand Publicity Service

Figure 15.13
White Island, Bay of Plenty, New Zealand. The continued degassing follows an explosion that destroyed the central part of the cone.

National Park Service

Figure 15.14
Crater Lake, in southern Oregon, formed in a depression left by the collapse of a composite cone.

around these fumaroles is usually yellow from sulfur crystallization. Sometimes sulfur and sulfur-bearing minerals, such as pyrite crystals, also form an encrustation around the vent of a fumarole.

Many hot springs and all geysers (discussed more fully in Section 21.4) are thought to arise from the heating of groundwater in the rocks over and around magmas or cooling igneous rocks. Groundwater is heated to form steam in geysers, but the violent expulsions of steam and water that characterize geysers are relatively rare compared with the number of active volcanoes.

Calderas

The eruption of large quantities of lava or pyroclastic debris from any type of volcano may leave an empty chamber under the volcano. Such voids may be created by the expulsion of material driven by gas pressures released during eruption, or by the contraction of the magma as it cools and crystallizes. The loose volcanic debris and lava that make up the volcano are rarely strong enough to stand up over a shallow void, and usually subside or collapse into the magma chamber, creating an enclosed or partially enclosed depression called a **caldera,** (Fig. 15.14). Similar calderas are occasionally created as the result of a violent explosion of pent-up gases beneath the volcano. Explosions violent enough to destroy a cone completely are rare, but a number of cases have been recorded, and two of these are described in the following section on the eruptions of Krakatoa and Tamboro. Explosive development of a caldera leaves a deposit of the materials that made up the cone. Heavier fragments fall, of course, nearby, but finer ash and dust may be deposited hundreds or even thousands of kilometers away.

Crater Lake, Oregon (Fig. 15.14 and 15.15), is located in one of the best known calderas in the United States. This one, like many others, resulted from collapse of the cone following a major eruption. The sequence of events that is thought to have led to the present form are depicted in the sketches of volcanologist Howell Williams in Fig. 15.15. One of the last events in this sequence was the formation of a small cinder cone in the caldera and filling of the caldera by water, leaving the cone as a small island, called Wizard Island, in the lake.

Another of the famous calderas in the United States is that of Kilauea volcano in Hawaii (Fig. 15.16). This caldera is especially well exposed because activity at this center is continuing, and there is little vegetative cover. A deep pitlike opening can be seen within a large area that has subsided. This pit has been the site of lava lakes described later in this chapter. The geological sketch map of this caldera clearly shows that the cliffs around the caldera are high-angle faults. These ring the caldera and form a system of concentric scarps along which the central sections have dropped.

One of the most remarkable series of photographs taken of the planet Mars reveals a volcano, Olympus

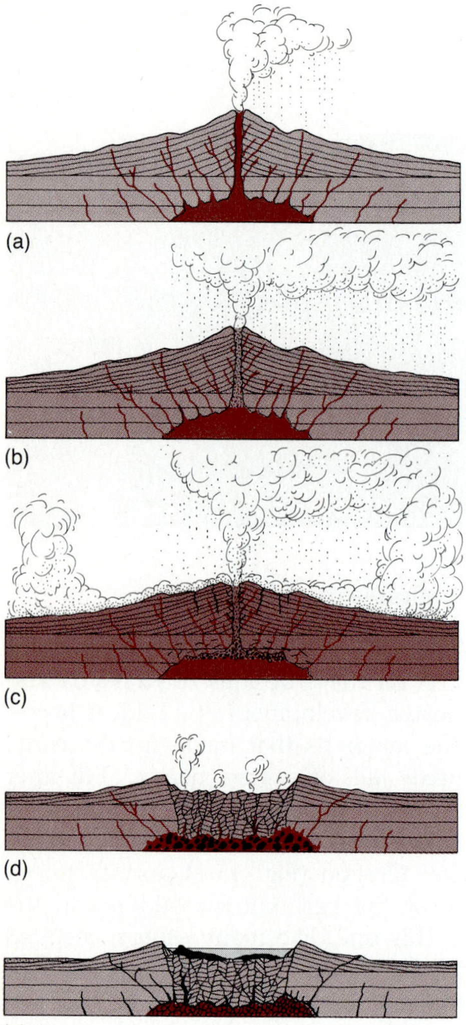

Figure 15.15
The evolution of Crater Lake. (a) Beginning of the great eruption. (b) Eruptions increase in violence, and showers of pumice become heavier. Lava level sinks in the conduit. (c) Climax of activity: glowing avalanches of pumice and scoria sweep down the mountainsides; magma reservoir is rapidly drained. (d) Top of volcano collapses into the caldron. (e) Crater Lake today, showing Wizard Island. Magma in the reservoir is largely if not entirely solidified.

Mons (Figs. 15.17 and 15.18), which has a caldera very similar to that at Kilauea. Olympus Mons is the largest volcano known on Mars; it is about as wide as the state of New Mexico (600 km) and 30,000 meters high—about three times the height of Mount Everest. Volcanic activity is one of the many ways Mars resembles Earth.

U.S.G.S.

Figure 15.16
The crater of Kilauea volcano. Part of the shield volcano has collapsed and subsided. Mauna Loa is visible at top of picture.

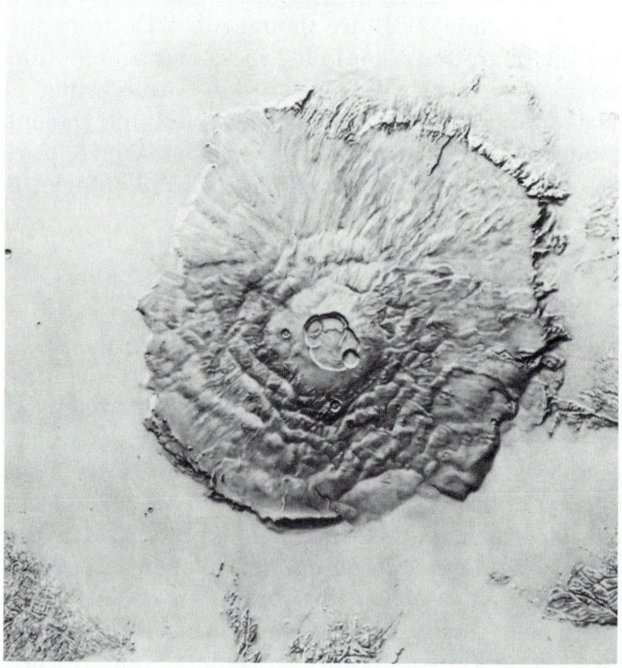

NASA

Figure 15.17
This composite photograph shows the huge volcano Olympus Mons on Mars. This volcano is more than 20 kilometers high and 80 kilometers across. By comparison, Mauna Loa rises about 10 kilometers from the sea floor.

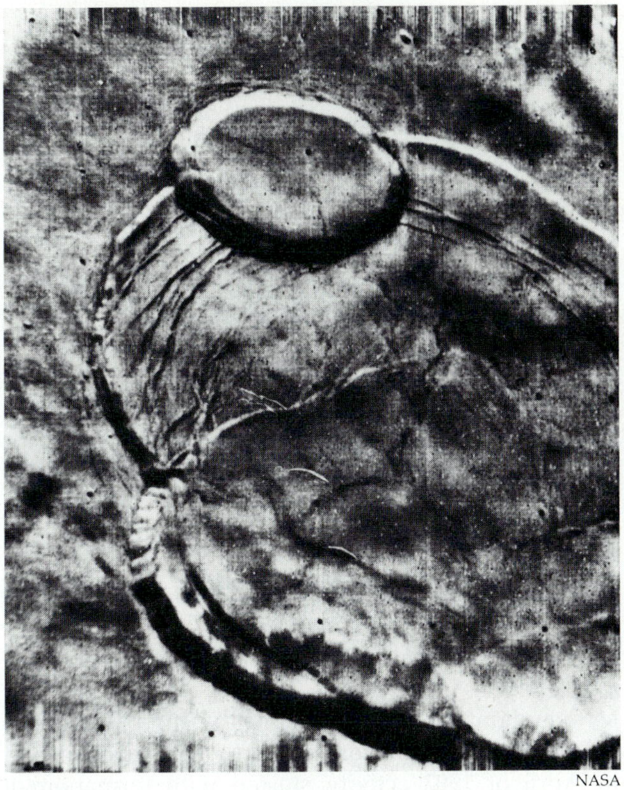

Figure 15.18
View of the calderas at the summit of Olympus Mons. Superimposed collapse structures record a series of volcanic events. This caldera is very similar to the one at Kilauea, except this one is 30 kilometers wide, nearly 15 times the size of Kilauea's caldera.

15.4 CASE HISTORIES

The following case histories illustrate the variety of types of activity at volcanic centers. The passive activity at Hawaii is typical of many oceanic island volcanoes, while the explosive eruptions of Krakatoa and Pelée are characteristic of the often disastrous eruption of the andesite volcanoes situated over subduction zones in the island arcs. Stromboli, located in an orogenic belt between the converging continental plates of Africa and Europe, lies in yet a third type of tectonic setting.

The Explosive Eruptions of Krakatoa and Tamboro

The most violent explosion of recorded history occurred on a small island in the East Indies in 1815 when the volcano Tamboro blew about 200 cubic kilometers of the earth's crust into the air. The debris created by this tremendous explosion covered islands for hundreds of kilometers around. Explosions of this sort are rare, and most of them occur in volcanoes that have long been dormant. The explosion releases pressures built up beneath the volcano's vent in the pipe, the connection between the volcano and the magma, or within the magma itself. Magmas contain large quantities of compressible gases. Pressure may build up until it exceeds the strength of the material blocking escape of the gases through the pipe. If the gases can be released slowly, violent explosion is unlikely; if not, the pressure builds up and may reach a level sufficient to blow the volcano apart.

One Sunday afternoon in August of 1883, a few mild explosions rocked the island of Krakatoa, located in the Sunda Strait between Java and Sumatra. The next morning the cone of this volcano was ripped apart by an explosion that sent more than 4 cubic kilometers of rock into the air. A cloud of dust, gases, and debris rose nearly 27 kilometers into the atmosphere. Heavier debris fell back to earth nearby, but smaller particles of dust were caught in upper air currents and blown around the earth. For the next two years, before this great quantity of dust settled, it colored sunsets around the world. Krakatoa and a neighboring island were literally blown to bits. A sounding over the site of the original peak of the volcano indicated that elevation, formerly 792.5 meters above sea level, was 304.8 meters below sea level. Although few people lived on Krakatoa, thousands were killed in the lowlands of the southwest Pacific islands by the **tsunami,** or **seismic sea wave,** generated by the awesome eruption.

The site where Krakatoa once stood had previously been occupied by other volcanoes. A fringe of islands still marks the rim of a former volcano that collapsed or exploded before recorded history (Fig. 15.19), and now a new volcano has been built up near the site of the 1883 eruption.

Stromboli

Stromboli, located at the northern end of the Aeolian Islands in the Mediterranean Sea, typifies the type of volcanic activity called **strombolian.** Because Stromboli is so consistently active and because a red glow follows each explosion, it has been called the "lighthouse of the Mediterranean." Its activity has been persistent, moderate, and uniform since the earliest records were made, at about the time of Aristotle.

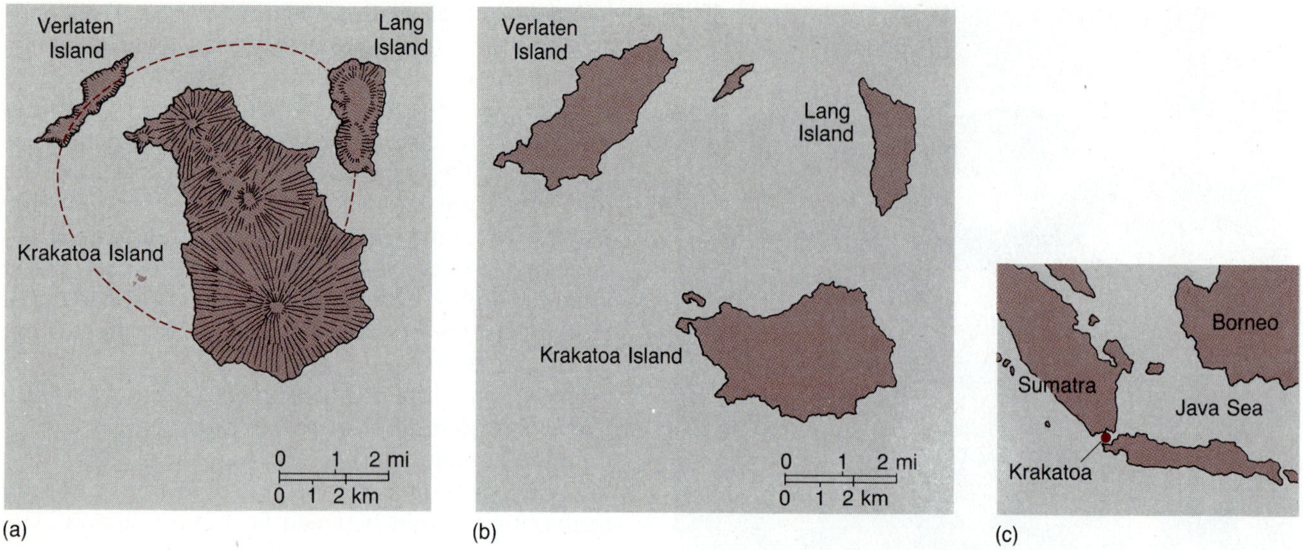

Figure 15.19
Krakatoa is located on a site of repeated explosive eruptions. (a) Note the outline of an earlier caldera (colored line) and the outline of the islands before the eruption of 1883. (b) Outlines of the islands after the 1883 eruption. (c) Locator map.

Stromboli has a near-perfect composite cone that rises about 3500 meters from the sea floor and stands 915 meters above sea level. The crater has steep-sided walls around most of its perimeter, but one side is lower than the others, probably because it was destroyed in a prehistoric eruption. Now when eruptions take place, lava and ash flow down a steep slope on this breached side of the crater. Large blocks fallen from the surrounding cliffs protrude from the crater floor. The crater has three vents in which lava rises and falls by as much as six meters. During a typical eruption, large bubbles form in the rising lava. When the bubbles burst clouds of steam form and glowing fragments are thrown into the air; then the lava subsides. At other times basaltic lava flows almost continuously, cascading down the slope toward the sea. This type of activity prevails as long as the vents remain open, but when the vents become temporarily blocked, more violent explosions occur, breaking up the material in the vent and producing ash-laden clouds.

Mount Pelée

Mount Pelée, situated on the island of Martinique in the West Indies, is famous for an eruption that took place in 1902. This eruption was so extraordinary and its effects so disastrous that the name **peléan** is used to describe this particular type of violent eruption (Fig. 15.20). Mount Pelée was known to have been active, but eruptions were rare and never caused serious damage, so residents of the city of St. Pierre below the mountain were not seriously alarmed when in April of 1902 activity started with formation of fumaroles in the upper valley of the volcano. A few weeks later, ash began to be ejected and residents smelled sulfur gases; and late in April explosions began, and increasing amounts of ash fell on St. Pierre. The population of the island became uneasy, but officials assured them that there was no danger, and troops were used to prevent people from leaving the city because an election was in progress. By the morning of May 8, activity had subsided and only a thin cloud of smoke was rising from the crater when suddenly, at about 8:00 A.M., the volcano exploded in four great bursts. A huge black cloud of incandescent gases and ash, called a **nuée ardente,** spread out from the volcano and flowed down the slope of the 1340-meter-high mountain, englufing the city and port in two minutes. The cloud, consisting of hot gases and dust, wiped out the entire population of the city, estimated to be about 30,000, in a matter of minutes. The only survivors were people on two of the eighteen ships in the harbor, a prisoner confined in a dungeon, and a cobbler in his shop.

Passive Volcanic Activity of the Hawaiian Islands

The Hawaiian Islands form a chain of islands extending nearly 2560 kilometers along a northwest-southeast line in the mid-Pacific. They are composed almost

(a)

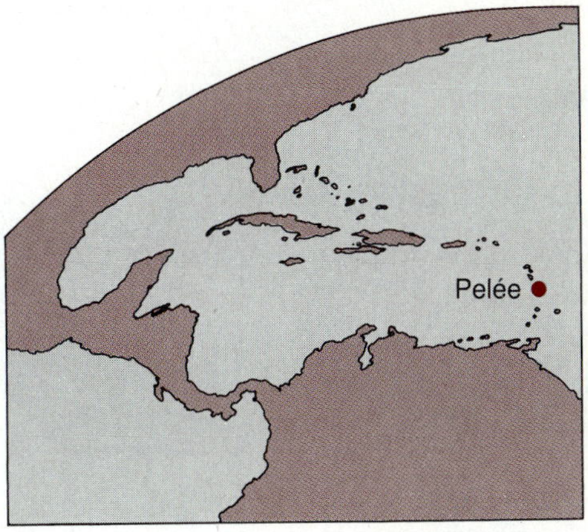

(b)

Figure 15.20
(a) This view of the city of St. Pierre shows the devastation that followed the eruption of Mount Pelée (in the background) in 1902. (b) Locator map.

entirely of volcanic rocks and the sediments derived from them (Fig. 15.21). The islands rise approximately 4800 meters from the sea floor, and the highest peak, Mauna Loa, reaches 9725 meters above the sea floor. This enormous pile of lava and ash has accumulated during thousands of flows, which continue today much as they have for millions of years. The highly fluid lava flows have formed volcanoes that have the profile of a broad shield—hence the name *shield volcano* (Fig. 15.3). Movements of magma in the islands may be traced by sensitive seismographs that record earth tremors produced by the movement of magma through the feeder system that carries it to the surface. Magma apparently originates at a depth of at least 60 kilom-

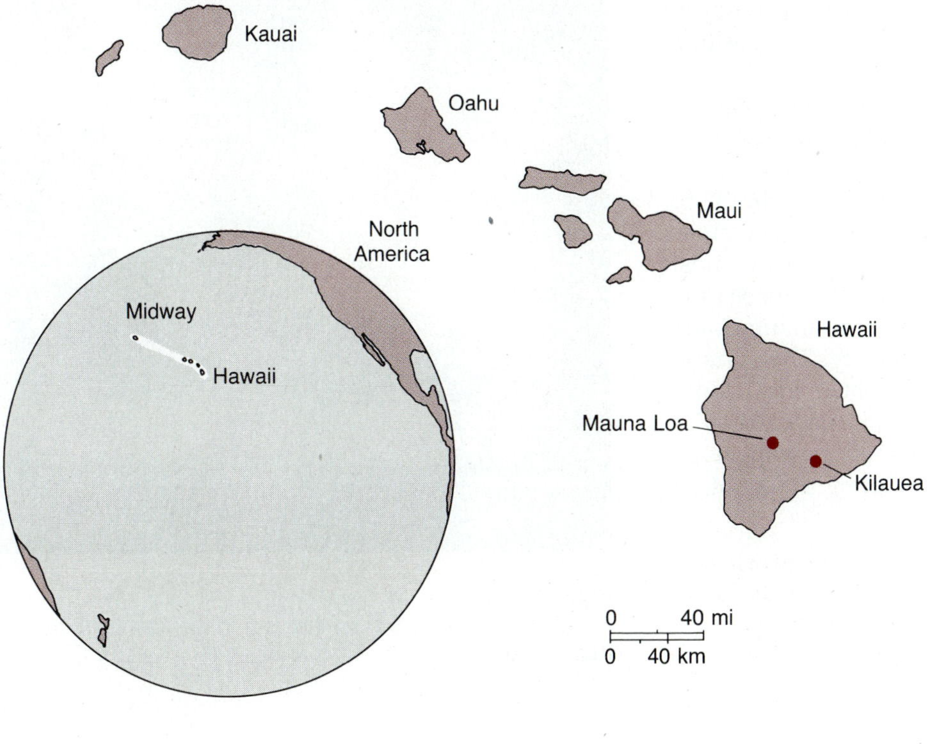

Figure 15.21
The Hawaiian Islands are the main surface expression of a long submarine ridge formed by volcanic activity. Midway Island lies on this same 2000-kilometer-long ridge.

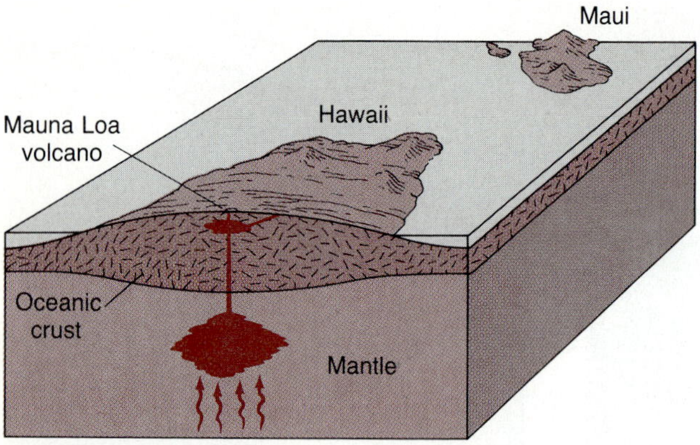

Figure 15.22
Magma generated in the mantle beneath Hawaii rises into a chamber under the active centers and breaks through to the surface in the calderas or along fissures on the flanks of the volcano.

eters in the mantle (Fig. 15.22), rises because of its low density, and flows out on the ground surface at temperatures of about 1200 °C. Explosive eruptions are rare; usually magma flows through central vents or along fissures that open in the flanks of the volcanoes. When these lavas consolidate, they form olivine basalt composed of plagioclase, pyroxenes, well-shaped olivine crystals, and minor amounts of magnetite. The parent rock from which all the Hawaiian lavas are derived is oceanic crust or upper-mantle material, thought to be relatively uniform in composition. The observed differences in composition of lavas are explained in terms of processes acting within the magma chamber or during the magma's ascent. The results of one such study are described in Chapter 6.

Because activity there is so quiet, the U.S. Geological Survey maintains a volcano observatory on the rim of the crater of Kilauea, one of the most active volcanic centers located on the flank of Mauna Loa. Studies at this observatory provide an unusually complete record of the volcanic activity. The volcano observatory has made it possible to obtain much more complete descriptions of activity in Hawaii than is ordinarily available. Eaton and Murata describe what was recorded at the observatory during the eruption at Kilauea in 1959–1960 (*1*):

> The first suspicious sign appeared during September 1959, when a series of very shallow, tiny earthquakes began recording on the North Pit seismograph . . . By the first of November, quakes of this swarm exceeded 1000 per day. . . . A hurried remeasurement of tilting [tiltmeters are sensitive leveling devices used to detect changes in the slope of the volcano sides] at bases around the caldera . . . revealed that dramatic changes were in progress: the summit of Kilauea was swelling at least three times faster than during previous months. In mid-afternoon on 14 November earthquakes emanating from the caldera suddenly increased about tenfold in number and intensity. At frequent intervals during the next five hours the entire summit region shuddered as earthquakes marked the rending of the crust by the eruptive fissure splitting toward the surface. Then, at 8:08 P.M., the lava broke through in a half-mile-long fissure about halfway up the south wall of Kilauea Iki crater, just east of Kilauea caldera. Abruptly the swarm of earthquakes stopped, and seismographs around the caldera began to record the strong harmonic tremor characteristic of lava outpouring from Hawaiian volcanoes.
>
> During the next 24 hours the erupting fissure gradually shortened until only one fountain remained active [Fig. 15.23]. But then the rate of lava outpouring, which had decreased as the erupting

U.S.G.S.

Figure 15.23
Fountains of lava rise from the vent in the lava lake in Kilauea Iki.

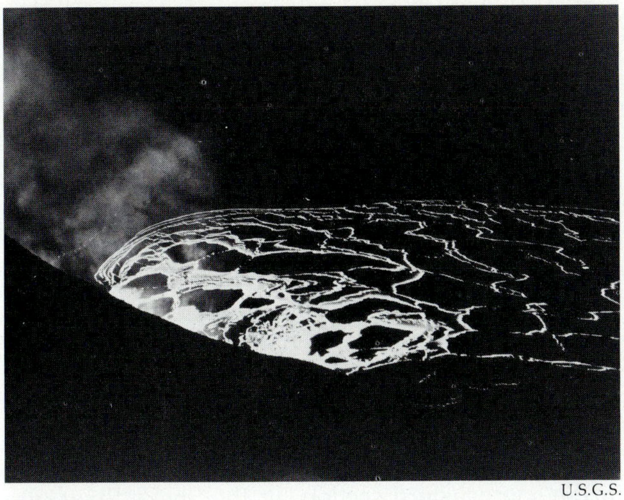

U.S.G.S.

Figure 15.24
This nighttime view of the lava lake at Kilauea Iki, Hawaii, shows lava pouring back into the feeder vent from which it came. The thin cool crust on the lava lake is black. Light areas show exposed red hot lava. The lake is several hundred meters wide.

fissure shortened, began to increase again, and it continued to increase steadily until the fountain died out suddenly on 21 November. The 40 million cubic yards of lava poured into Kilauea Iki crater filled it to a depth of 335 feet, slightly above the level of the vent [Fig. 15.24]. . . .

During the following three weeks 14 more eruptive phases of shorter and shorter duration but with increasingly vigorous fountaining took place at the Kilauea Iki vent. The highest fountain was measured during the 15th phase, on 19 December, when a column of incandescent gas-inflated lava jetted to 1900 feet, by far the greatest fountain height yet measured in Hawaii. At its highest stand, at the end of the eighth phase, the lava pond was 414 feet deep and contained 58 million cubic yards of lava. At the end of each phase the fountain died abruptly, and from the 2nd to the 16th phase, a mighty river of lava surged back down the vent as soon as the fountaining stopped. . . .

Tiltmeters around Kilauea caldera showed that the volcano was swelling rapidly as phase after phase of the eruption delivered lava to the surface and then swallowed it up again. When surface activity ceased at Kilauea Iki on 21 December, far more lava was stored in the shallow reservoir beneath the caldera than when the eruption began. It appeared that Kilauea was in unstable state and that further activity was very likely.

During the last week of December a swarm of small earthquakes began to record on the seismographs at Kapoho. . . . The source of these earthquakes was soon traced to the east rift zone of Kilauea, about 25 miles east of the caldera. . . . The magma that inflated the summit region most probably exerted pressure on the plastic core of the rift zone, and earthquakes revealed where the rift zone yielded and where dikes began to extend toward the surface. . . .

On 13 January the village of Kapoho was rocked by frequent, very shallow earthquakes, and by nightfall a graben [a grave-like depression bounded by faults] 0.5 mile wide and 2 miles long that contained about half of the town had subsided several feet. At 7:30 P.M. the earthquake swarm gave way to harmonic tremor, and the flank eruption broke out along a fissure 0.75 mile long near the center of the subsiding graben, a few hundred yards north of Kapoho and nearly 30 miles east of the summit of Kilauea.

During the next five weeks nearly 160 million cubic yards of lava poured out of the vent north of Kapoho and reshaped the topography of the eastern tip of Hawaii [Fig. 15.25].

On 17 January, only four days after the flank eruptions began, the summit of Kilauea began to subside precipitously as lava began to drain from beneath the caldera and to move through the rift zone toward the Kapoho vent.

Activity at Kilauea continued until April.

For many years the volcanic activity in Hawaii was explained as resulting from mantle magmas rising along a crack or fault in the oceanic crust. This idea

U.S.G.S.

Figure 15.25
Lava flows into the sea, creating columns of steam, during the 1960 eruption in Hawaii.

was appealing because the volcanic centers of the Hawaiian Island chain are aligned along a straight, narrow zone. The main problem with this hypothesis is that only one center, that on the island of Hawaii, is active. Later, as more age determinations were made it became clear that volcanic materials increase in age from east to west in the chain, the oldest lying under Midway. With the advent of plate theory a new explanation was proposed—that the volcanic activity is taking place over a rising hot plume located in the mantle. The plume or hot spot remains essentially fixed in position as the lithosphere moves over it. Thus the Hawaiian ridge is aligned in the direction the plate has moved over the hot spot. This hypothesis has now been applied to other volcanic centers as well.

Volcanic Provinces of the Northwestern United States

The northwestern states have been the site of exceptionally extensive volcanic activity during the last 70 million years, a period of time geologists call the Cenozoic era (Fig. 15.26). A chain of some of the most beautifully formed composite cones on earth, the Cascade Mountains, runs north–south along the coast from Lassen Peak in California to north of Mount Rainier near Seattle. Lassen volcano erupted from 1914–1917, Mount Baker heated up in 1978, Mount St. Helens became active in 1980, and signs of incipient activity are present at both Mount Hood and Mount Rainier. Many eruptions have taken place in the Cascades in recent geological times, and we can expect that activity will continue in the future. The Cascades stand at the western edge of the vast Columbia Plateau, a region of a quarter of a million square kilometers, built up by the extrusion of lavas, much of it to a depth of several kilometers. Some of the most recent activity in this region has taken place in the southeastern section, in the Snake River plateau, and the easternmost section in and around Yellowstone Park.

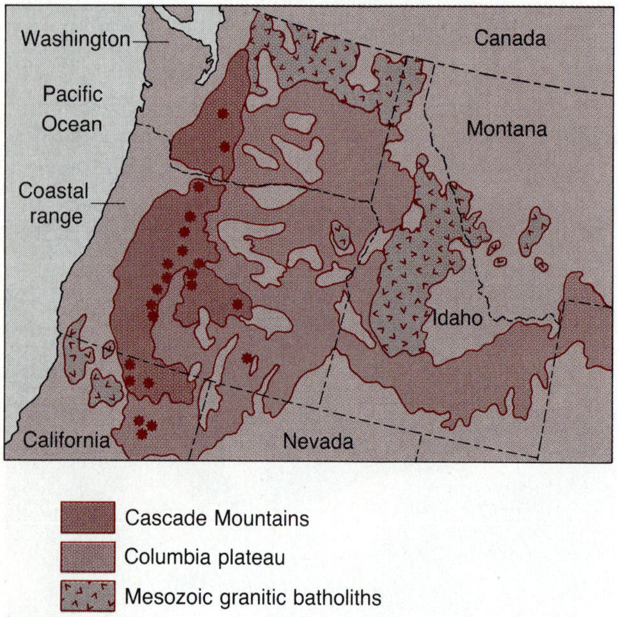

Figure 15.26
Volcanic and igneous rocks are exposed over a vast area in the northwestern United States. The batholiths shown here are older features (Mesozoic age) now exposed as a result of erosion. Younger lavas are present in the Columbia Plateau and Cascades. The white areas are underlain by other types of rocks.

The Columbia Plateau has been the site of volcanic activity for many millions of years. The oldest extensive lavas are of Eocene age (38 to 53 million years ago), but most of this area is covered by younger flows of Miocene age (7 to 27 million years old); and Pleistocene (the last 2 million years) volcanics cover most of the Cascade region, the Snake River, and Yellowstone areas. Despite their great age, the older flows remain nearly horizontal over vast areas and are folded or broken by faults only locally. The thickness of individual flows is generally only about 5 meters, but some flows are more than 100 m thick, and the total thickness of the flows as measured in many river canyons amounts to two to four kilometers. The total volume of lava exceeds 417,000 cubic kilometers. The extent of the flows, some of which can be traced over 160 kilometers, their thickness, and their general uniformity of composition and horizontality indicate that they were produced by fissure eruptions of great quantities of highly fluid lava. The flows filled topographic lows and extended toward the edge of the large regional basin in which they lie as flow after flow erupted. Subsidence of the surface after the extrusion of such large quantities of lava must account for some of the folding and faulting seen in the flows, but much of this deformation is probably due to the continuing tectonic activity along the western margin of North America.

Cascade Mountains. The region west of the Columbia Plateau was uplifted to a broad and long arch several million years ago. Some of the oldest lava flows were folded into long upfolds at that time, and these older flows and deposits were covered by the volcanic

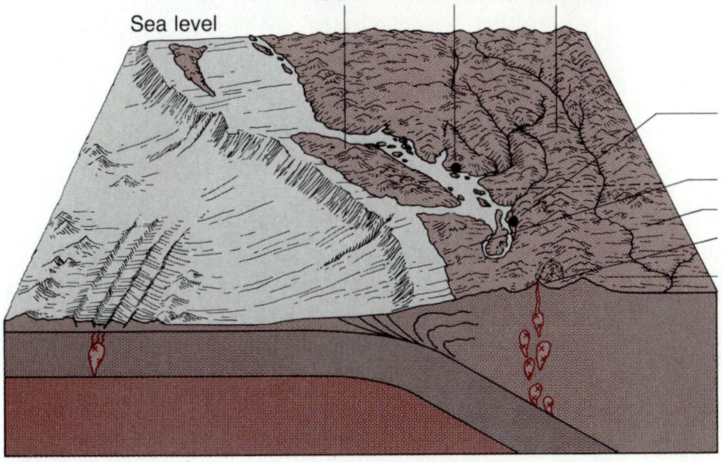

Figure 15.27
The volcanoes in the Cascade Range are probably fed by magmas generated at depth along a slab of lithosphere that is sinking just west of the coast of Washington and Oregon.

materials that make up the large composite cones known as Mount Hood, Mount Lassen, Crater Lake, Mount Adams, and Mount Rainier. These cones are so perfectly shaped (Fig. 15.9) that they cannot be very old. A number of explanations for this volcanic activity have been offered, but the one in favor at this time is that the volcanoes are caused by the partial melting of a lithospheric slab that is moving down into the mantle along the coast of Washington and Oregon (Fig. 15.27). This slab originates along the Juan de Fuca ridge, spreads east, and descends along the coast; but the site of subduction is not marked by a trench because the depression has been filled with sediment transported into the sea by the Columbia River.

Mount St. Helens, Washington. Studies of the lava flows and ash deposits around Mount St. Helens (Fig. 15.10) helped geologists identify it as the volcano most likely to erupt in the Cascades long before that eruption started in March 1980. Its eruption in April 1857 was but the last of a long series of eruptions at about 100- to 150-year intervals over the past few thousand years. These eruptions, once started, tended to continue off and on for a number of years; the duration of some earlier eruptive phases is estimated at 20 to 30 years.

Events leading up to the explosion of May 18, 1980, provide an unusual opportunity to trace the evolution of an eruption. These events, as reported by the U.S. Geological Survey, can be briefly outlined as follows:

March 20–27: Seismic activity began in the vicinity of Mount St. Helens with a magnitude 4.1 earthquake 5 km below the surface. This initial shock was followed by a week of increasing seismic activity. By March 24, low-magnitude earthquakes were occurring every few minutes.

March 27: Eruptive activity began with an explosion audible 15 km away. A cloud of steam and ash rose from the summit. When the air cleared, a new crater about 70 meters in diameter could be seen inside the large summit crater.

March 28: A second explosion the next day sent a cloud 2 km high, and an avalanche of ash flowed down the northwest flank. Observers found two arc-shaped fissures, one 1.5 and the other 5 km long, running east–west across the summit. By dawn, a dense ash cloud rose 3 km above the summit, and blocks were being ejected. Small mudflows formed by meltwater and ash moved down the northeast flank. Water levels in Swift Reservoir were lowered more than nine meters to accommodate possible melt waters.

March 29: A new crater formed next to the first one, and in both, blue flames could be seen.

March 30: A morning explosion generated a great anvil-shaped cloud of ash, which fell to earth as far away as Bend, Oregon, 250 km south. Six more explosions followed. Three earthquakes of magnitude 4.5 to 4.7 and only one kilometer deep took place on March 30 and April 1.

April 1: Ash from a large explosion on this day was collected in Spokane, Washington, 500 km to the east.

April 2: Ash and steam eruptions continued about once every three hours. During this period, the two new craters merged to form a larger one about 350 meters across. For the next three weeks, minor eruptions of ash and steam continued but with diminishing intensity. Sometimes large blocks of ice that had fallen into the crater were ejected. Tiltmeters installed on the flanks of the mountain showed erratic ground movements. Generally, uplift preceded the minor eruptions taking place at this time.

April 22: Explosive activity ceased. Up to this time, no new magma could be identified in the ejecta. Steam continued to rise from the new craters until early May. During this period, comparison of a 1952 topographic map with new maps prepared from photographs taken in early April revealed that the upper flank of the volcano had been uplifted. Careful comparisons indicated that an area 4 km^2 had risen about 25 meters, and local uplift of as much as 100 meters was found. Careful leveling surveys prompted by this observation revealed that one area, near a place called Goat Rocks, rose 6 meters between April 24 and 29. The two fissures identified earlier could now be seen to define a graben across the summit area (Fig. 15.28). The central block had dropped by 150 to 200 meters relative to the adjacent ones. The north flank of the volcano was rising and moving northward. An upward movement of 75 meters and lateral movement of 100 meters north was determined by April 29.

April 30: The Spirit Lake area was closed because of the danger of major ice avalanches from glaciers near the summit that had been severely broken up by the uplift of the north flank. The bulge on the north flank continued to grow in the next weeks, and steam was emitted, but otherwise the volcano remained in a dormant state.

May 18: At 8:23 A.M., a tremendous explosion heard 350 km away ripped the north face of the volcano apart and sent a cloud of ash 15 km into the atmosphere. In a remarkable series of photographs, the north flank bulge could be seen to grow rapidly upward and outward. The over-steepened slopes began to slump down the side of the volcano, but before this was complete, an explosion blew away much of this side of the volcano, leaving the top 250 meters lower than it had been. A magnitude 5 earthquake took place seconds before this eruption, but its relationship to the eruption is uncertain. Three distinct phases of the eruption have been identified: (1) a directed blast that blew hot ash out of the north flank of the cone and blew down trees up to 24 km away from the summit (Figs. 15.29 and 15.30); (2) a pyroclastic flow and landslide that carried material from the north flank across the lower slopes and 28 km down the Toutle River, burying it to a depth of 60 meters in places; (3) a pumice-pyroclastic flow funneled through the break in the north flank. This material represented the first fresh magma erupted. Ash

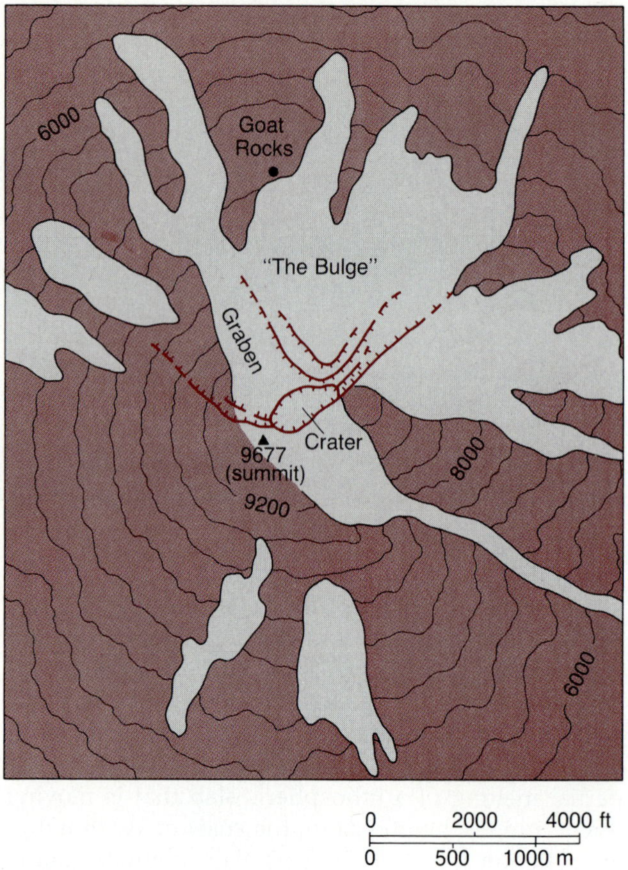

Figure 15.28
The crater, graben, and bulging area between the crater and Goat Rocks looked like this on May 4, 1980. Elevations and contours are in feet.

U.S.G.S.

Figure 15.29
A huge cloud of ash bellows from the crater after the May 18, 1980, blast destroyed the north side of the summit of Mount St. Helens.

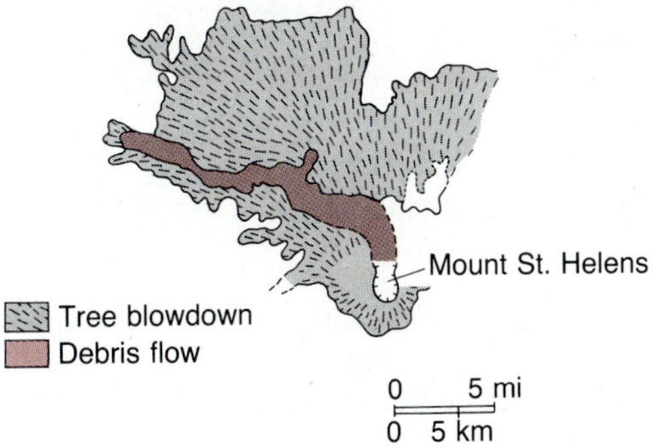

Figure 15.30
Sketch map of Mount St. Helens, prepared by the U.S. Geological Survey. The area affected by the May 18, 1980, directed blast is delineated by tree blowdown. The debris flow shown is in the valley of the north fork of the Toutle River.

from the major eruption on May 18 was carried mainly to the east by prevailing winds. At a distance of 140 km downwind, 10 to 12 cm of ash fell. It reduced visibility to a few meters in Spokane, 500 km northeast of the summit, and up to 4 cm of ash was reported in western Montana. Eruption of ash and steam continued for four days. At the end of that time the toll was 21 known dead, 68 missing, and hundreds of millions of dollars in property damage. Activity at Mount St. Helens was continuing as this was written in 1982, and, if its past history is an accurate guide, it may still be active as you read this.

Yellowstone Park. Yellowstone Park is located east of the great lava fields of the Columbia River Plateau (see Fig. 15.26), but unlike the basaltic lavas there, it is composed mainly of volcanic rocks of intermediate composition with large quantities of rhyolite. Volcanic activity had continued there for many millions of years. Eruptive centers of Eocene age (40 million years old) from which basalt and andesite were extruded have been found in the high mountains both east and west of Yellowstone Park. Much of the material erupted here was ejected as broken-up blocks, breccias with mixed sizes of material, some of which spread as huge mudflows; other violent eruptions spread welded tuffs of rhyolite across the countryside. All of these older volcanic rocks have been faulted and displaced, sometimes thousands of meters, as a result of block faulting during the last few million years.

In more recent volcanic activity, fluid basaltic lavas poured over the Snake River plateau during the Ice Ages; but for some reason, flows of the same age on the Yellowstone Plateau are mainly rhyolitic, and they are surrounded by slightly older, welded rhyolite tuff. Some of these rhyolites cover glacial deposits, showing that this volcanism took place during or after the advance of glaciers into the region. The welded tuffs may have been formed during a particularly violent eruption, but the rhyolite flows were viscous masses erupted from fissure zones along the crest of the Yellowstone Plateau. These flows can be traced nearly 50 kilometers along former river valleys, and they are estimated to contain about 400 cubic kilometers of material. The surfaces of these flows are marked by distinctive concentric ridges like those seen in Fig. 15.4, and their leading edges, still well preserved and moderately steep, indicate how viscous the lava was. Some of these flows, those located on the Pitchstone Plateau, are very fresh in appearance. Rhyolite flows in the Yellowstone area occupy a collapsed caldera approximately 48 kilometers in diameter that formed about 2 million years ago when the roof of the older lava and tuff accumulations subsided or collapsed into the underlying magma chamber (2). The caldera now has been filled by more recent rhyolite flows, tuffs, and minor amounts of basalt, but the cooling rock melt under Yellowstone Park is responsible for the high thermal gradient (over a hundred times normal) and geyser activity which still occur there (see Section 21.4).

The fact that volcanic activity in the Yellowstone area is geologically young raises the question of whether the area could become active again. Three similar cycles of volcanism have occured there in the past 2 million years. The last one started 1.2 million years ago with the formation of ring-shaped surface

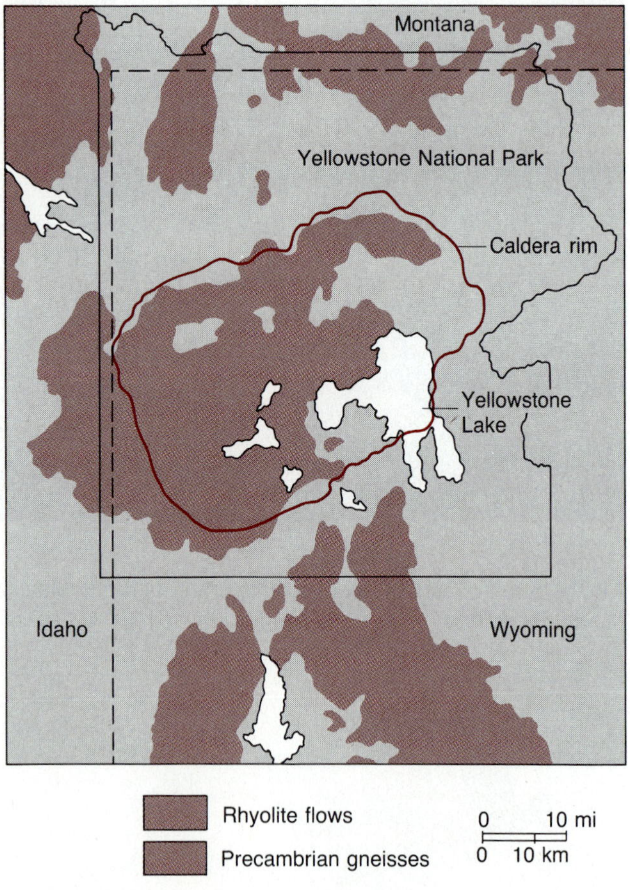

Figure 15.31
A large caldera is located in the central part of Yellowstone Park. Much of it is filled with young lavas, shown in color. The central part of this caldera has been rising since 1923. Old gneisses crop out in mountains around the volcanic areas.

fracture zones over and around two magma chambers at depth. This led eventually to an explosive eruption of rhyolite, about 600,000 years ago, and then collapse of the surface to form two large calderas. Recent leveling studies in the park revealed that part of the Yellowstone caldera is rising at a rate of about 14 mm/yr. This is a relatively rapid rate of uplift, probably caused by the movement of magma in the underlying magma chamber (Fig. 15.31).

The causes of volcanic activity at localities like Yellowstone or the San Juan Mountain region of Colorado, which are far removed geographically from plate margins, are intriguing mysteries. The possibility that they are located over mantle hot spots is one of the most appealing explanations. The magmas are clearly too rich in silicon, aluminum, and other components concentrated in continental crust to be derived directly from the mantle, but heat and solutions rising in the mantle could cause melting or partial melting of the overlying continental crust to form the volcanic materials.

15.5 VOLCANIC HAZARDS

Clearly, volcanoes constitute a serious hazard to human life, a hazard that is widespread especially around the perimeter of the Pacific. What may not be so obvious is that the effects of a volcano may extend far beyond its immediate vicinity and may be more varied than we might at first expect. The most obvious and immediate hazard is posed by the blast effects of volcanic explosions, by the falling debris, and by hot ash and gases. Great areas may be affected by these. The extent of some old volcanic deposits provide an indication. The incandescent gases and nearly molten ash and pumice from past eruptions have in some cases covered hundreds of thousands of square kilometers. Deposits of welded tuffs cover thousands of square kilometers on the north island of New Zealand, and similar deposits of rhyolitic material are found over large areas in the southwestern United States. It is hard to conceive the toll such an eruption, covering several states with incandescent gases, would take if it occurred today. Unfortunately, we have little assurance that one will not. Volcanic explosions and eruptions of incandescent clouds pose grave immediate threats to those in the areas affected, but the more usual eruptions of ash and gas may pose a variety of other dangers.

Great clouds of water vapor are often present over erupting volcanoes. This water usually condenses as it rises, creating tremendous rainfalls near the volcano. The rainwater, mixed with the ash on the flanks of the volcano, may produce debris slides or mudflows that extend far beyond the volcano. The size of these flows is impressive. Figure 15.32 shows the areas covered by two flows, one 500 years old, the other 5000 years old, in the vicinity of Mount Rainier. Similar flows moved down the slopes of Mount St. Helens.

Major ash falls may temporarily dam streams, which later flood as the dams are topped and washed away by the water. Lava flows may also create temporary dams. People living near an erupting

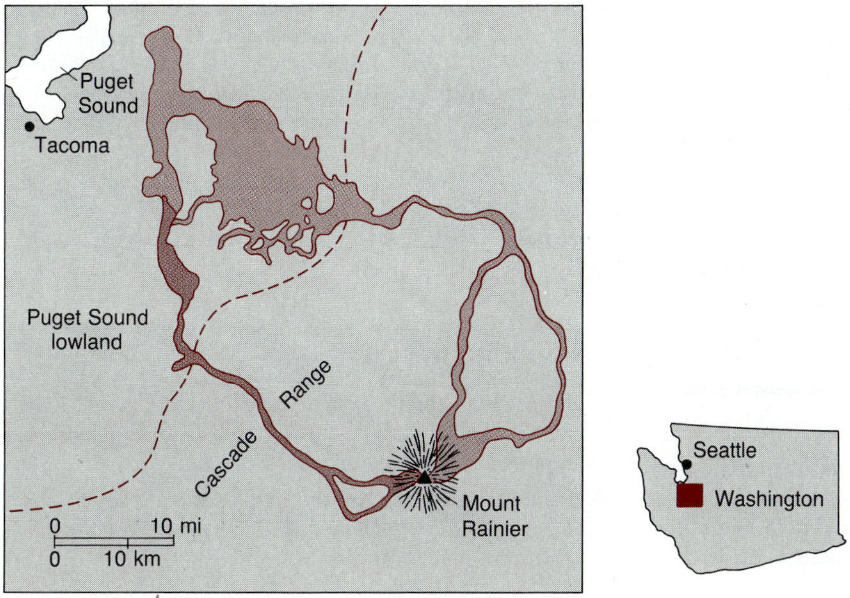

Figure 15.32
Map of Mount Rainier and vicinity, showing extent of the Osceola mudflow in the White River valley and the Electron mudflow (pattern) in the Puyallup River valley.

volcano might have to cope not only with the ash and lava at the volcanic center, but with landslides, floods, and torrential rains, as well as blocked roads and failure of communications. People living close to sea level may be affected by volcanic eruptions occurring thousands of kilometers away. Major explosive eruptions in the ocean often displace enough water to set up tsunami. These waves travel rapidly out from their source area and can cross the Pacific in a day, causing huge waves to wash up on the coast all around the ocean. In extreme cases, tsunami may knock down houses and trees tens of meters above sea level. How high they reach depends on the size of the wave and especially on the slope of the coast and the sea floor off the coast.

Volcanic eruptions may cause long-term as well as immediate health hazards. Much concern has been expressed about the hazards posed by breathing the dust-filled air. People in large portions of the northwestern states were exposed to volcanic dust following the eruption of Mount St. Helens, and studies are in progress to evaluate this danger.

Volcanoes certainly warrant our careful attention for the hazard they present. They should be monitored for seismic tremors that indicate magma movement, for changes of slope due to swelling of the summit as magma moves up, and for evidence of heat, which can now be monitored readily by satellite photography.

It is clear now that volcanic activity has been a part of the history not only of earth but of other bodies in the solar system as well. Apparently, melts can be generated in a number of ways, for the tectonic setting of volcanism on each planet may be quite different. Many volcanoes on earth are associated with plate boundaries; those on the moon appear to be related to impact basins. On Io, one of the moons of Jupiter, volcanism is thought to be related to a tidal bulge or to radiation. Volcanic activity on Mars is related to tectonic activity. In all cases, however, volcanic activity has shaped the surface of the planet creating distinctive forms that are easy to recognize and that hold valuable clues to tectonics and geologic history.

SUMMARY

Igneous activity, one of the two processes that tend to raise the surface of the land, acts through the eruption of lava from volcanoes and through the emplacement of magma below the surface. In either case, the molten material involved is believed to be generated in the upper mantle, 50 to 100 km below the surface. Differences in the magma's composition correlate closely with the location of igneous activity—andesitic composition being found along subduction zones and basaltic at spreading zones or hot spots.

The composition of the melt, its temperature, and the amount of its dissolved water and gases affect its fluidity and the type of eruption it is likely to produce. Lavas that reach the surface with a high content of trapped gas are likely to erupt violently as the pressure on the lava is released; otherwise the lava may flow out relatively quietly. The formation built by an eruption depends on the viscosity of the melt. High-temperature basaltic lavas, being highly fluid, form wide, low flows of shield volcanoes and lava plateaus. Cooler or more silicic lavas are far more viscous and build up steeper slopes.

An explosive eruption hurls into the air a variety of pyroclastic materials ranging in size from huge blocks to fine ash and dust, which are dropped in sorted deposits at varying distances from the volcano. Such eruptions can produce characteristic steep-sloped cinder cones. If both lavas and pyroclastics are erupted the volcano will form a composite cone, of which the concave upward profile of Mount Fuji is typical.

Eruptions of gas-rich lavas often form the vesicular rocks, pumice and scoria, sometimes as mounds called spatter cones.

Many volcanic centers are marked by large depressions, calderas, formed either by an explosive eruption or by the collapse of the summit after the magma chamber below it has emptied.

KEY TERMS

andesite line
ash
blocks
bombs
caldera
central-vent eruption
composite cones
crater
dust
fissure eruption
fumaroles
lapilli
nuée ardente
Peléan eruption
pumice
pyroclastic
scoria
seismic sea wave
shield volcano
spatter cones
Strombolian eruption
tsunami
welded tuff

STUDY QUESTIONS

1. What characteristics of the eruptions at Hawaii, Pelée, and Stromboli allow us to distinguish them as different types of volcanic eruptions?
2. Examine and compare the maps showing the location of most active volcanoes (Fig. 8.4), earthquakes (Fig. 8.3), and plate boundaries.

3. If you live in an area near active or dormant volcanoes, what types of precautions would you like to see the local government take to insure public safety from volcanic eruption?
4. What are the most important determinants of lava viscosity? How does viscosity affect the form of volcanic features?
5. Heavy rains often accompany volcanic eruptions. What types of deposits would you expect to form as a result?
6. What causes the violent explosions that occur at some volcanoes?
7. What evidence related to the shape of a volcano is a good indication that a historically inactive volcano has been active in the geologically recent past?
8. What geological evidence found in the vicinity would indicate that the caldera of a volcano was caused by explosion rather than collapse of the summit?

REFERENCES

1. J.P. Eaton and K.J. Murata, "How volcanoes grow," *Science* 132:3432 (1960), *925–938*. Copyright 1960 by the American Association for the Advancement of Science.
2. W.B. Hamilton, "Late Cenozoic tectonics and volcanism of the Yellowstone region: Wyoming, Montana, and Idaho," in *West Yellowstone—Earthquake Area.* Billings (Mont.) Geological Society, 11th Annual Field Conference Guidebook (1960), *92–105*.

SUGGESTED READINGS

Bullard, Fred, M., *Volcanoes of the Earth.* Austin: University of Texas Press, 1976.

Decker, Robert, and Barbara Decker, *Volcanoes.* San Francisco: W. H. Freeman and Company, 1981.

Green, Jack, and N. M. Short, eds., *Volcanic Landforms and Surface Features.* New York: Springer-Verlag, 1971.

Harris, S. L., *Fire and Ice.* Seattle: The Mountaineers, Pacific Search Press, 1980.

Herbert, D., and F. Bardossi, *Kilauea: Case History of a Volcano.* New York: Harper & Row, 1968.

Hyman, Randall, "Fire in Iceland," *Science 81* (1981), 62–28.

Williams, Howell, and A. R. McBirney, *Volcanology.* San Francisco: Freeman, Cooper and Co., 1979.

ROCK STRUCTURE AND LANDFORM DEVELOPMENT

Chapter 16

Structural features in the bedrock clearly control the shape of the land over vast areas of the continents, and they may be equally important along oceanic ridge crests where oceanic crust is not buried by sediment. The forms of stratified rocks, the shapes of plutons, and the fractures and faults that cut them may be prominent in the landscape. Occasionally, landforms directly reflect a structure as it comes into existence. The scarp formed by displacement along a fault is an example (Fig. 16.1). Newly formed folds, developing so quickly (by geologic standards) that the ground surface domes or arches over the rising fold, are also found in some modern orogenic belts. But most structural features develop so slowly, and at such depth in the earth, that we generally see them only after the shape of the affected rock is exposed at the ground surface by denudation and erosion of overlying rock. Such features are reflected in the topography mainly as a result of differences is the susceptibility of the bedrock to weathering and to the action of the geomorphic agents. Streams, groundwater, and wind tend to carve out these differences much more effectively than does mass movement, wave action, or glaciation. Also, the relationship of bedrock structure to landform is much clearer in arid climates because bedrock there is better exposed; plant cover is sparse.

Removal of the soil cover and the etching out of structural form are unlikely unless the area has been elevated relative to the regional base level. Thus, most of the best examples of structurally controlled landforms occur in mountains or elevated regions. There streams steadily dissect the land and bring rocks of different resistance into relief. But as erosion continues, the high areas are gradually lowered. Eventually, the relief is reduced, and the entire area approaches a uniform level even if the rocks involved

Figure 16.1
The cracks in the ground surface formed only a few days before this picture was taken. These breaks occurred during an earthquake in Montana in 1959.

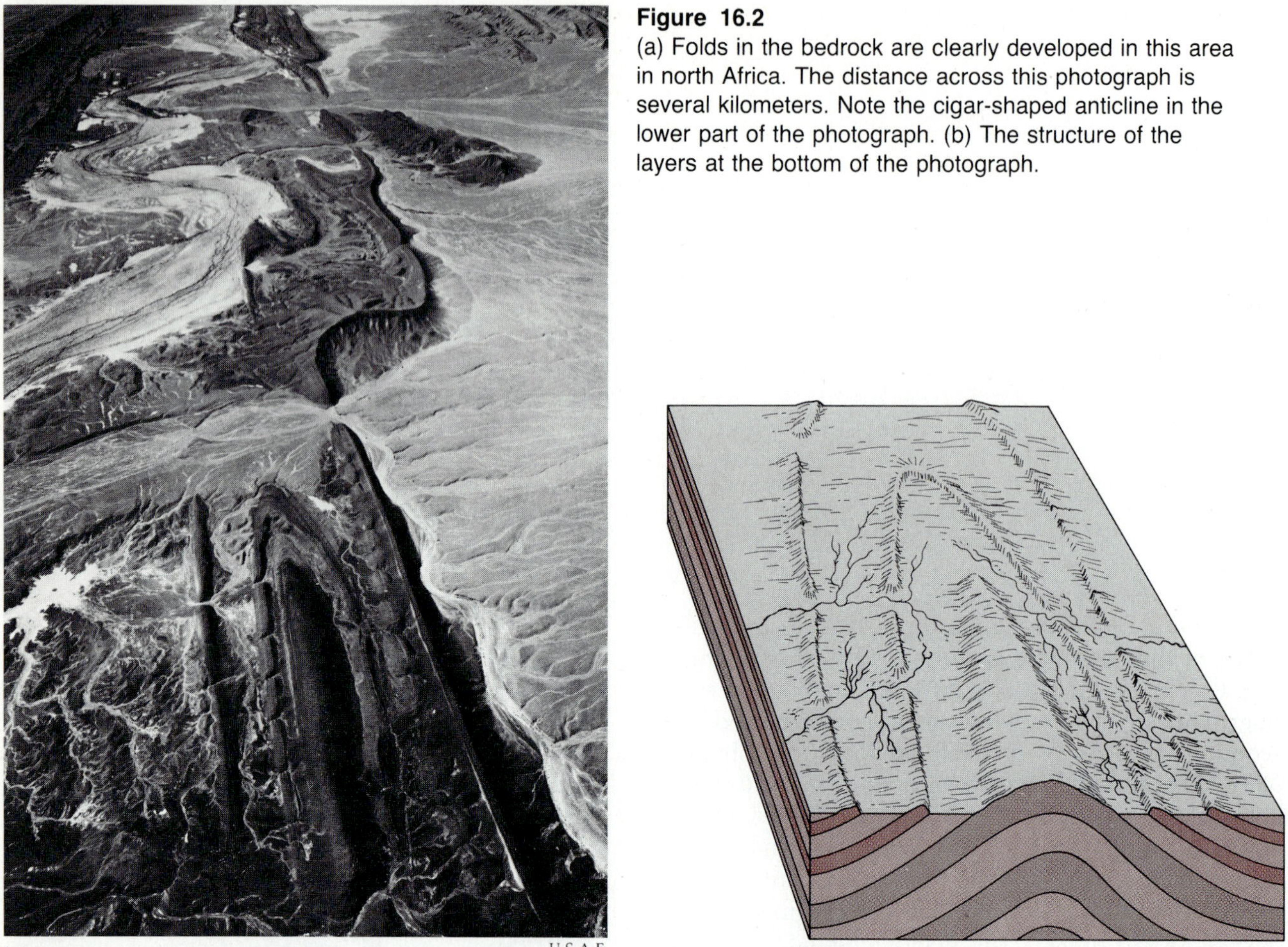

(a) (b)

Figure 16.2
(a) Folds in the bedrock are clearly developed in this area in north Africa. The distance across this photograph is several kilometers. Note the cigar-shaped anticline in the lower part of the photograph. (b) The structure of the layers at the bottom of the photograph.

differ greatly in their resistance to erosion. In arid regions, especially, the rock structure may be clearly visible in the outcrop pattern, as can be seen in Fig. 16.2, even though the relief has been reduced.

In this chapter, we shall look in some detail at the Colorado plateau, a section of the central Appalachians, and the Mississippi valley, which provide good representative examples of the relationship between the structure of the bedrock and landforms. Later, in Chapters 19 and 20, we shall follow up this discussion by examining the relationship between streams and the form of the land in these regions.

16.1 LANDFORMS ALONG FAULT ZONES

Active faults frequently show up in the topography because they cause abrupt offsets of cultural and physiographic features. The earthquake at Hebgen Lake in Montana in 1959 (Fig. 16.1) was caused by movement on a fault. Displacements of the ground surface created scarps over a meter high along the faults. Vertical offsets of as much as six or seven meters occurred where the fault crossed steep slopes. The great displacements on the valley sides are probably due in part to downslope movements of the loose soil, as well as to movements in the underlying bedrock, as was discussed in Chapter 8.

The San Andreas fault, the best-known active fault in America, is representative of major strike-slip faults throughout the world. It can be easily recognized on aerial photographs like that in Fig. 16.3, because it shows up as a nearly straight feature cutting across and offsetting a variety of other features. Movements of the fault caused an essentially sidewise horizontal displacement of the two sides. Features commonly seen along the San Andreas and other strike-slip faults are shown in Fig. 16.4.

(a)

(b)

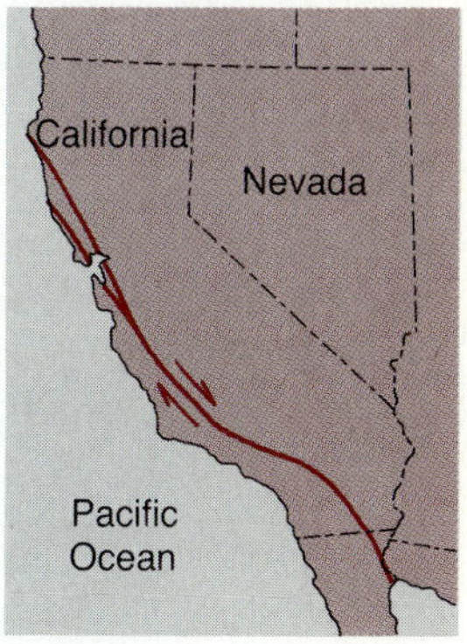

(c)

Figure 16.3
Air views of the San Andreas fault, California. (a) The fault zone occupies the narrow valley flanked by ridges in this photograph. The distance across the photograph is several kilometers. (b) Stream offset about one-half kilometer by the San Andreas fault. A highway above the fault in this view provides a scale. (c) Locator map.

Movement along the San Andreas fault has continued intermittently throughout the Cenozoic era (the last 70 million years). The total displacement of the two sides is enormous—on the order of several hundred kilometers. Consequently, totally different types of bedrock, structural features, and landforms are juxtaposed across the sharply defined zone of movement. Modern streams are offset, as are dikes, plutons, and other rock bodies cut by the fault. Along the San Andreas fault, the continuation of an offset feature is always found displaced to the right as one faces the fault.

The fault zone is straight because it is nearly vertical, but it is not a single continuous line all the way from the Gulf of California to Cape Mendocino. Straight-line segments of the fault zone end, but invariably another segment picks up, usually parallel to and overlapping the last. The zone of intense

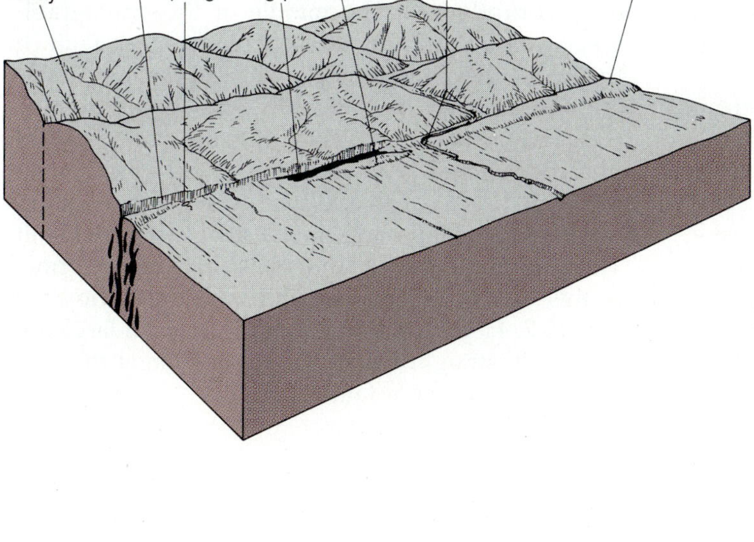

Figure 16.4
Block diagram showing newly developed representative geologic features along recently active strike-slip faults. Sag ponds are small bodies of water that accumulate in localized depressions along faults.

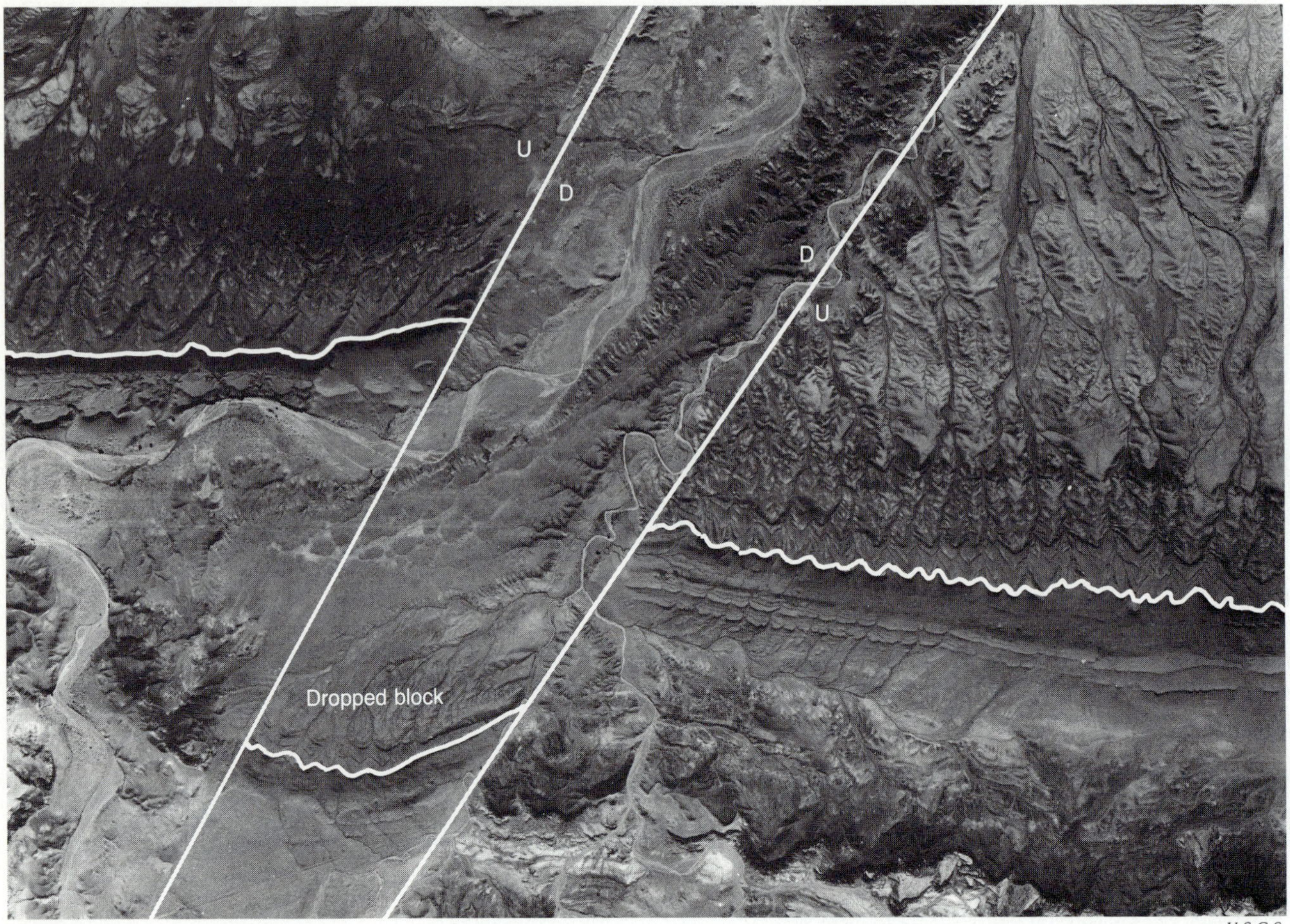

U.S.G.S.

Figure 16.5
Vertical aerial photograph of part of the San Rafael swell, Utah, showing two faults (identified by lines) that cut and offset a sequence of sedimentary layers that dip at a low angle toward the top of the photograph. Displacement on the fault is largely vertical not strike-slip as might first seem likely. Scale is about 1:24,000.

deformation in which the bedrock is sheared and torn apart is often more than a kilometer wide. Small linear ponds, called **sag ponds,** frequently appear in the zone, and parallel ridges commonly develop in the rock on one or both sides of the fault zone.

Inactive faults may or may not be prominent in the topography. They are most likely to be well-developed features when the fault is a high-angle one, when the fault zone contains a wide band of deformed rocks that are more susceptible to erosion than surrounding rocks, or when rocks of markedly different resistance type or structure are located across the fault zone. Inactive faults are usually recognized because they interrupt, cut off, and displace what would otherwise be a continuous stratum or structure. For example, notice in Fig. 16.5, that a long east–west ridge characterized by distinctive topography ends abruptly near the middle of the photograph, where it is cut and the strata are offset by a fault. In this case, the affected rocks are sedimentary layers and the faults are normal faults. In a second example (Fig. 16.6), the bedrock is the igneous and metamorphic rock of the Canadian shield, and the escarpment developed as a result of differences in the resistance of the rocks juxtaposed across a steeply inclined fault. The magnitude of the landforms that can be formed in this way is evident in a third example, shown in Fig. 16.7. Here in the Canadian Rockies, a large and unusually straight valley has formed along the trace (outcrop) of a major fault.

Royal Canadian Air Force

Figure 16.6
Oblique aerial view of a valley developed along an ancient, inactive fault that cuts across Precambrian rocks of the Canadian shield. The distance across the photograph is several kilometers.

Royal Canadian Air Force

Figure 16.7
The broad but straight valley that can be traced hundreds of kilometers through the mountains of British Columbia is known as the Rocky Mountain trench. It marks the trace of a major fault zone.

16.2 THE EFFECTS OF FRACTURES AND JOINTS

Like faults, both fractures and the large breaks known as **joints** develop from shear or extensional failure of the rock (Section 10.2 and Fig. 10.14); they differ from faults in that there is no displacement across the break. Nevertheless, fractures and joints often play an important role in landform development, because they usually occur as sets of parallel cracks that subdivide the rock into blocks. Weathering occurs along the fracture surfaces of the blocks, and the blocks may be dislodged by frost action, by being frozen into glacier ice, or by downslope movements.

Joints tend to produce angular topographic features where they control the shape of the land. This is clear in the vertical aerial photograph of an area in the Colorado Plateau shown in Fig. 16.8. Joints show up as especially well-developed linear features on the left side of the picture. Erosion and removal of the jointed sandstone has produced cliffs with prominent V shapes as seen in this vertical view.

Extensional (stretching) joints often form rectangular blocks. This may show up in drainage patterns, at a large-scale, or as block-shaped outcrops at a small scale. Because the rock is pulled apart when it fails in extension, these joints are sometimes intruded by

Photo by John Shelton

Figure 16.9
This sharp ridge stands out in the topography because the center of it is a resistant dike.

(a) U.S.G.S.

Figure 16.8
(a) The influence of the two sets (directions) of joints in the development of stream-cut valleys in a light-colored sandstone layer is apparent in this photograph. This photograph was taken looking down vertically on the east side of the San Rafael swell in Utah. Scale is about 1 : 50,000. (b) Locator map.

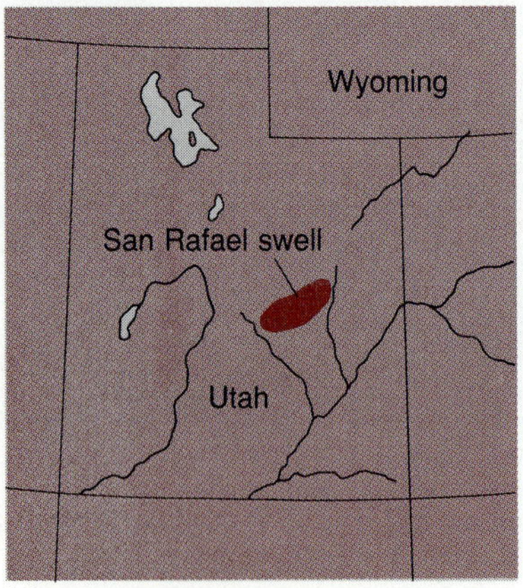

(b)

dikes. If the dike rock is more resistant to weathering than the country rock, the dike may become a sharp ridge (Fig. 16.9).

16.3 LANDFORMS IN REGIONS OF FLAT-LYING AND GENTLY DIPPING STRATA

Although the structure of flat-lying or gently dipping strata is the simplest we find, the topography developed on such rocks is quite varied. Where strata are flat, the same rock is often exposed over a large area. The uniformity of rock type and rock structure generally results in the development of a large area of more or less uniform topography. These areas often differ in the degree of consolidation of the exposed rocks and in the elevation of the area above base level. As examples of areas with flat or gently dipping strata, we will consider parts of the Coastal Plain along the Mississippi valley (Figs. 16.10 and 16.11), the Appalachian Plateau (Figs. 16.12, 16.13, and 16.14), and the Colorado Plateau (Fig. 16.15). Each of these areas is underlain by gently dipping strata, but they differ markedly in the resistance of the sedimentary rocks to erosion and in the relief of the area above base level.

The Mississippi River drains much of the vast continental interior of North America. The upper part of the valley is located on undeformed Paleozoic age (see inside cover) sedimentary rocks, part of the sedimentary veneer that covers the continental interior. South of Illinois and Missouri, the river flows across the Coastal Plain on nearly horizontal unconsolidated sediments and consolidated sedimentary rocks of Cenozoic age (Fig. 16.11). Rocks of this age and of similar structure cover a large region in and around the Gulf of Mexico and along the Atlantic coast. The strata are usually almost horizontal, although inclinations (dips) of a few degrees are common (Fig. 16.11).

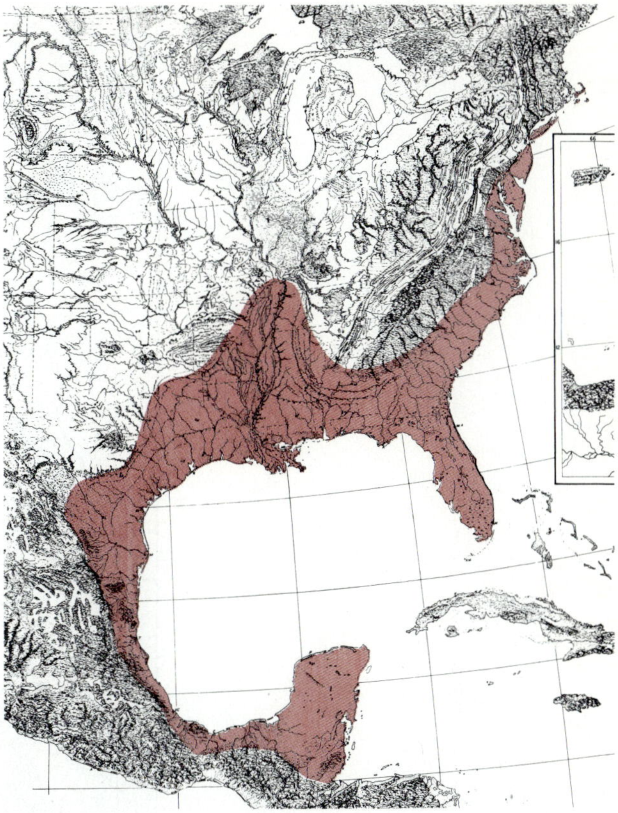

Figure 16.10
The Coastal Plain extends from Cape Cod into Mexico. A large embayment occurs along the Mississippi River valley. The entire Coastal Plain province was under water about 70 million years ago but has slowly emerged.

The Appalachian Plateau (Fig. 16.12), an area that extends from Tennessee to New York, is underlain by upper Paleozoic sedimentary rocks made up mostly of consolidated sandstone and interbedded shales.

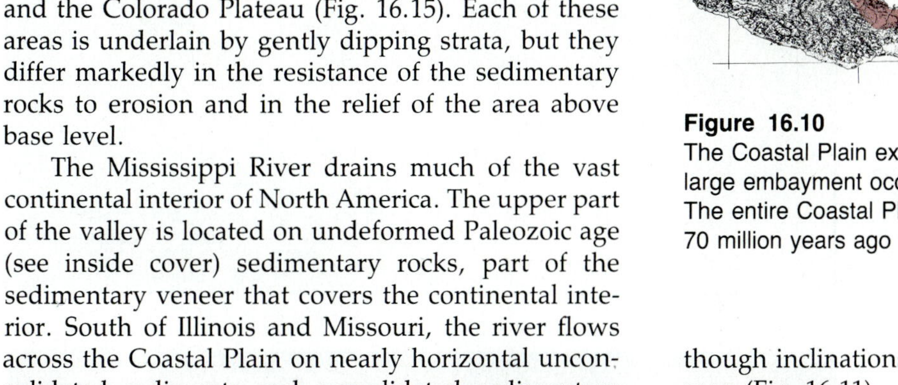

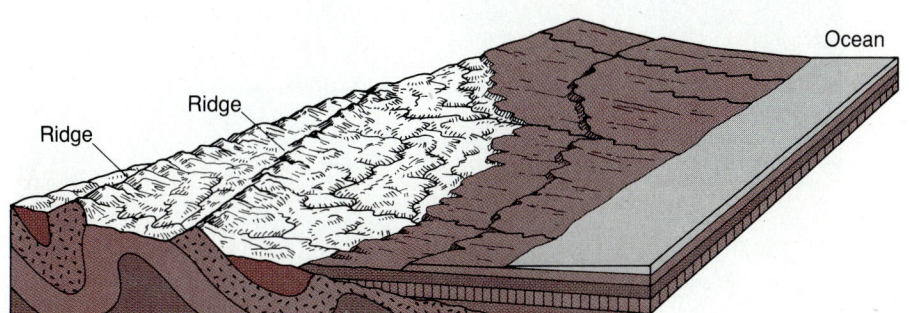

Figure 16.11
The Coastal Plain is underlain by nearly flat-lying sedimentary rocks, which cover older and strongly deformed rocks that are exposed at the surface in the Appalachian, Ouachita, and Marathon Mountains.

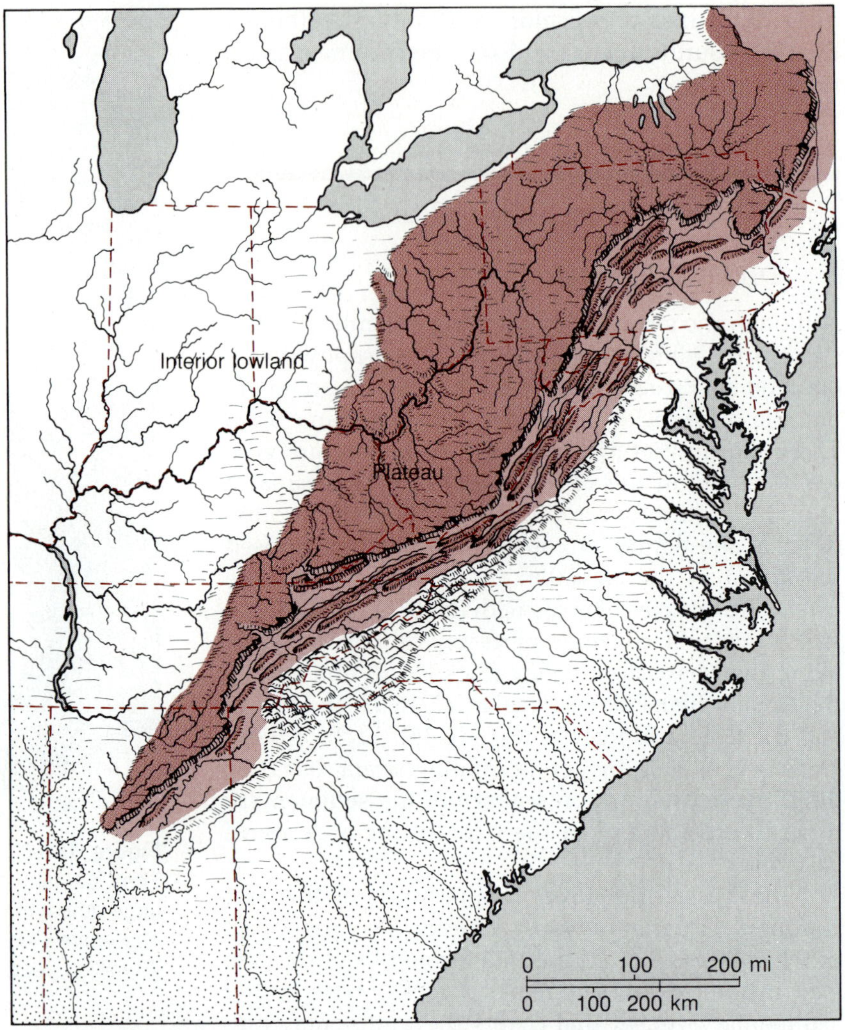

Figure 16.12
The Appalachian Plateau is a vast region of flat-lying rocks in the western part of the Appalachian Mountains. It includes many small subdivisions, some of which are indicated.

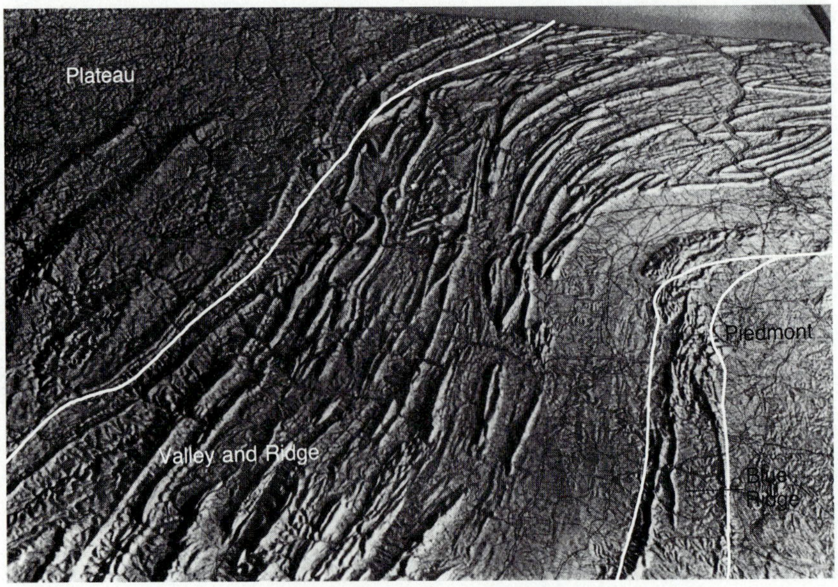

Figure 16.13
Oblique view of the Appalachian Plateau (left), the structural front, and the fold belt of the Appalachians' Valley and Ridge province.

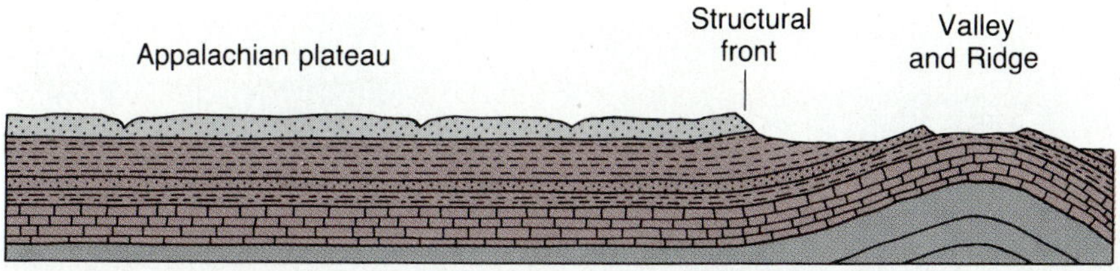

Figure 16.14
The rocks exposed near the surface in the Appalachian Plateau are nearly horizontal. These contrast sharply with the folded rocks of the Valley and Ridge farther east.

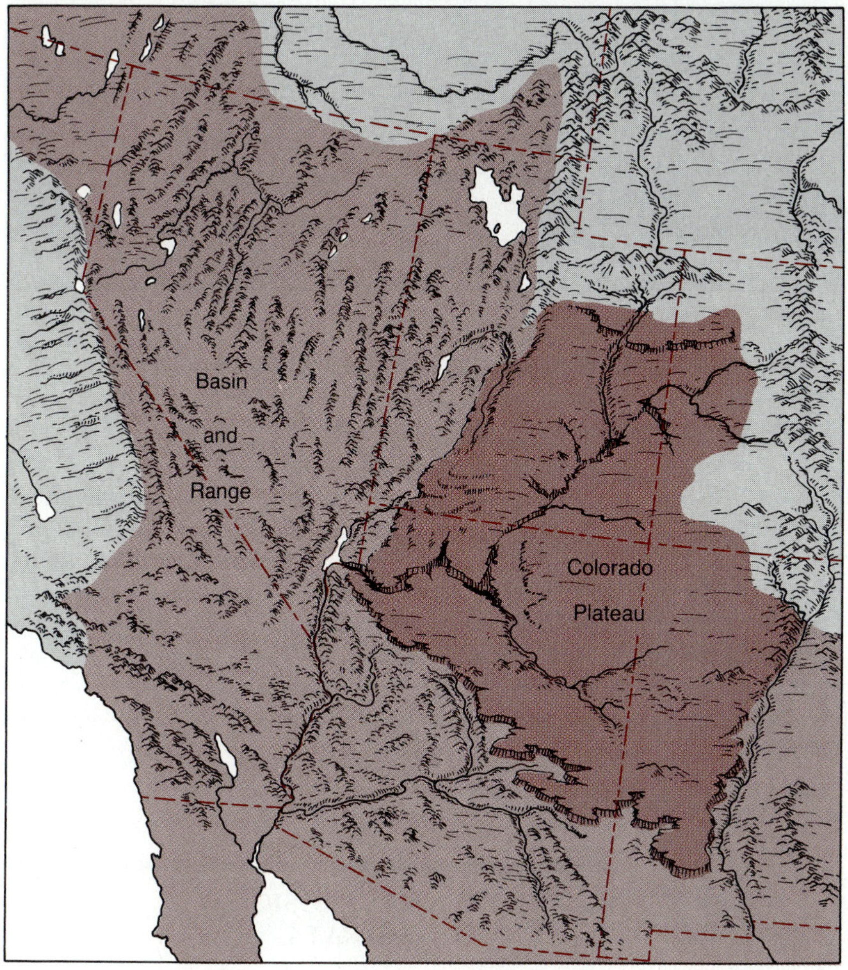

Figure 16.15
The Colorado Plateau is surrounded by complexly deformed rocks of the Rocky Mountains, but the layers are flat or gently arched across much of the plateau.

Strata are flat-lying over most of the region although broad arches and folds are present, and more numerous, toward the east. The plateau is distinctly different from the strongly folded rocks of the Appalachians (Figs. 16.13 and 16.14).

The Colorado plateau, famous for its many spectacular gorges, including the Grand Canyon, is a roughly rectangular shaped area in Arizona, New Mexico, Utah, and Colorado (Fig. 16.15). The exposed rocks are mainly Mesozoic and Cenozoic consolidated sedimentary rocks, although some volcanic centers are present. The strata display flexures, broad archlike bends, and are broken by high-angle faults, especially around the edge of the plateau; but over most of the region the layering is nearly flat (Fig. 16.16).

The flat surfaces, as seen from ground level, are the most striking similarity among these regions, and in all of them the structure of the underlying bedrock is the principal cause of this flatness. These examples differ in that the flat surfaces in the Appalachian and

Figure 16.16
This photograph, taken in the Colorado Plateau just upstream of the Grand Canyon, depicts the flat strata of the area and the abrupt, steep canyon walls cut into these strata.

Colorado plateaus are located at high elevations. Also these surfaces were produced by long-term denudation and erosion, which stripped away considerable thicknesses of overlying rock. The rocks in the Mississippi valley lie nearer sea level and are younger than those in the other two examples. Also the level of the ground surface has been reduced much less. It differs too in that sediment is today being deposited, and new strata are forming as the Mississippi delta is enlarged and extended into the gulf.

Because erosion is fast in easily removed rocks, like shale, and slow in more resistant rocks, like sandstone, resistant rock layers tend to form the flat exposed surfaces of the plateaus.

Why then isn't the unconsolidated rock in the Coastal Plain also stripped away rapidly? The answer lies largely in the low elevation of the regional base level and in the extremely low relief in the Mississippi Valley. In many places the Colorado River flows at a level more than a thousand meters below the plateau surface. The tributaries of the Ohio that drain most of the Appalachain Plateau are hundreds of meters below that plateau surface. But the Mississippi River passes through a region that has much lower relief—often a few tens of meters—and the river channel itself is only about 150 meters above sea level where it enters the Coastal Plain 600 kilometers from the gulf. Because processes do not reduce regions below their base level (see Section 14.4), the potential relief in these three areas is quite different. The deep canyons cut into the Colorado Plateau, at least in part due to its low base level, are familiar to most students. The Appalachian Plateau is also deeply dissected. And because much more rain falls in the humid region of the Appalachian Plateau than in the arid Colorado Plateau, the drainage system is much better developed (Fig. 16.12)—more streams are present, and they are closer together.

16.4 DIFFERENTIAL EROSION OF FOLDED STRATA

Fold forms are most clearly reflected in the shape of the land when:

1. The folds are large.
2. The folded rocks consist of interlayered units of greatly different resistance to weathering and erosion.
3. Relief is moderate.
4. The land has been subjected to erosion long enough for the structure to be etched out.

These optimum conditions are found in the central Appalachians, on the flanks of the Rocky Mountains, and in the less strongly deformed Colorado Plateau.

As we saw in Chapter 13, the central and southern parts of the Appalachian Mountains can be subdivided into several long physiographic divisions. The Appalachian Plateau is the westernmost of these. The eastern edge of the plateau is defined in many places by a pronounced change in both landform and bedrock structure. This structural front separates flat-lying rocks in the plateau from rocks of similar age and rock type that have been folded into long belts of anticlines and synclines (Figs. 16.12 and 16.13). These folds have been etched out as very long valleys and ridges, giving the area the name Valley and Ridge province. This province provides a classic example of the long orderly folds so often found developed in the sedimentary cover in major mountain systems. Farther east, old igneous and metamorphic rocks crop out in the Blue Ridge and Piedmont. These rocks are also strongly deformed, but because the folds are smaller in scale and the rocks more uniform in their resistance to erosion, their structure is not reflected in the topography.

Fold belts structurally similar to the Valley and Ridge province occur in the Alps, Iran, North Africa, and in the Rocky Mountains. In the Rockies, this belt is most continuous along the eastern edge of the Canadian Rockies, and it can be traced into Montana and Wyoming. But unlike the Valley and Ridge, the fold belt of the Rockies has been partially covered by younger volcanic rocks, and it has been broken and displaced by younger faults. Over most of its length it is so severely cut and broken by these younger structural features that the landforms developed on it do not resemble those in the Appalachians.

In both the Appalachians and the Rockies, degradation has gone on for many millions of years, and it is perhaps easier to understand how the features there originate if we examine an area, such as the Zagros Mountains in Iran, where deformation is much younger. Part of the Zagros Mountains is shown in the satellite photograph of Fig. 16.17. In general, the high topography there is formed of domes or anticlines. The deformation is so recent that the contours of the land directly reflect the immediately underlying rock structures. However, erosion has proceeded far enough to make its future course clear. Streams flowing off domes, such as the one labeled *A* in Fig. 16.17, show a well-developed radial pattern. These streams are clearly removing rock from the crest of the dome. Similar effects are seen on the limbs of many anticlines (*C, D,* and *E*). The crustal zones have been cut into, deeply exposing more easily eroded rock, which shows up as scars on the otherwise symmetical shapes of the land.

It is not surprising that the crest and steep flanks of these domes and anticlines show the effects of the attack of erosion more distinctly than synclines. Erosion by streams and downslope movement of materials are most effective on steep slopes. Thus, material on steep slopes is removed more rapidly than that on low slopes, other factors, especially bedrock resistance, being equal. But once erosion has broken through the crest of the folds and the topographic

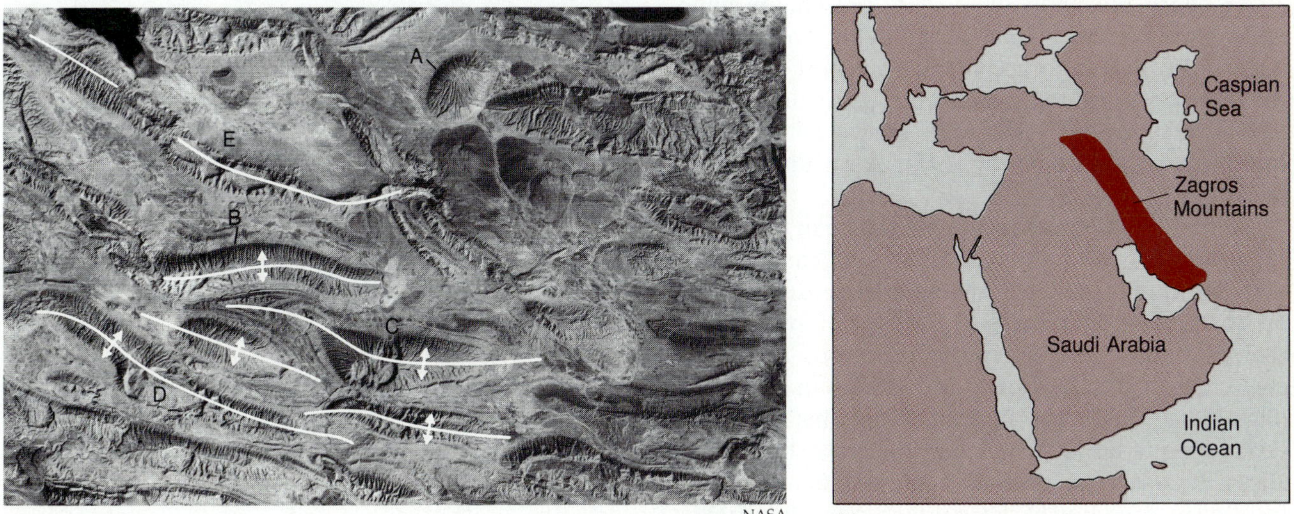

(a) (b)

Figure 16.17
(a) Satellite view of the folded rocks in the Zagros Range, Iran. Axes of folds are indicated. (b) Locator map.

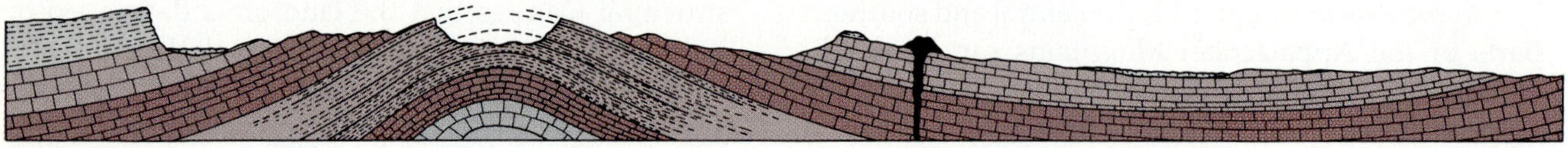

Figure 16.18
Cross section prepared across a fold in Wyoming, showing the relationship between rock structure and topography. Note the small valley developed in the top of the anticline. This type of feature is referred to as an inversion of topography.

level of the original high ground reduced, the shape of the ground surface looks less and less like the structural form of the bedrock. We see these stages in the Appalachians and in the Rockies.

The folds of the Appalachian Valley and Ridge involve strata that include both resistant sandstones and quartzites and less resistant shales and limestones. The ridges in the Valley and Ridge are held up most frequently by quartzites, some in layers more than 100 meters thick. These are both underlain and overlain by much thicker layers of shale and limestone, which lie in the valleys on either side. Presumably, these folds once resembled those we now see in Iran. The anticlines at that stage formed ridges. But as the resistant beds on the crest of an anticline were eroded through, older, less-resistant beds were exposed. Because these could be eroded rapidly, they were lowered and became valleys. Thus, anticlines do not always form ridges, nor do synclines always form valleys. In fact, the reverse is more often the case in the Valley and Ridge, a situation sometimes referred to as **inverted relief.**

A slight inversion of relief is present in some folds in the Rockies, as is depicted in the cross section (Fig. 16.18) drawn to accompany a report of one of the first surveys of the western territories directed by F.V. Hayden in 1877. The folds shown lie in a large intermountain basin south of the Wind River range in Wyoming.

Clearly differences in the resistance of the strata to weathering and erosion, acting in combination with the configuration of the bedrocks, exert important controlling influences on the type of topography that develops. This is especially obvious along the eastern front of the Rocky Mountains in Colorado (Fig. 16.19). Here the highest mountains are composed of igneous and metamorphic rocks, but differences in resistance of shales and sandstones have resulted in formation of sharp ridges, called **hogbacks,** where these strata have been turned up and eroded along the mountain front.

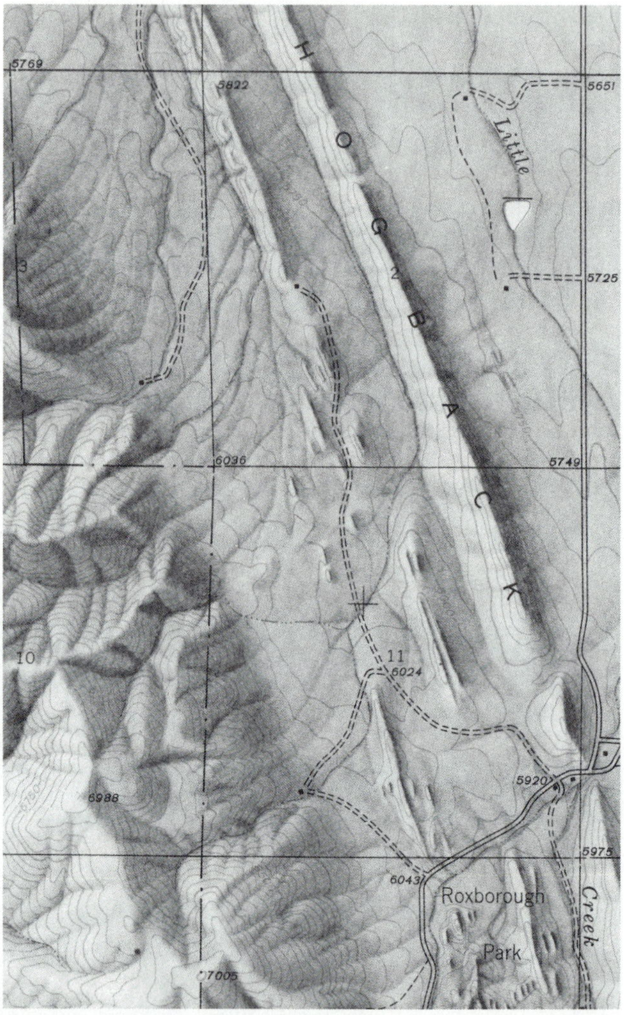

Figure 16.19
The long ridge labeled Hogback is caused by the upturned edge of a resistant sandstone along the eastern front of the Rocky Mountains south of Denver. The grid lines are one mile apart.

Structure of the bedrock is but one of several important factors that influence how landscape looks and the way it evolves. But in areas of moderate and high relief, the structure of the rock is usually a key to understanding why the land is shaped as it is. Thus, uplift of large segments of the crust and orogenic deformation leave an imprint that is reflected both in the internal structure of the crust and in the form of the surface. The role structural features play will become even more evident as we examine the operation of the geomorphic agents in the following chapters.

SUMMARY

The structure of the underlying bedrock exerts an important influence on the surface landforms in an area. Although deformation of the bedrock sometimes has an immediate surface effect—as in formation of a fault scarp—more often structural features are revealed on the surface only as differing rates of erosion bring them into relief.

Along fault zones, both physiographic features and constructed ones (such as roadways and fencelines) may show the lateral or vertical offsets. Ridges and stream beds may also show the extent of movement across a fault as do many features along the San Andreas fault. Inactive faults may be marked by a band of deformed rock along the fault zone that has markedly lower resistance to erosion than rock on either side of it. Sometimes movement along a fault juxtaposes very different rock types. This may cause sharp contrast in landforms across a fault.

Fractures and joints in bedrock may affect surface landforms even if no movement occurs along them. They increase the susceptibility of the rock to weathering and erosion, sometimes forming cliffs, characteristic rectangular drainage patterns, or other features.

Where the underlying strata are flat-lying or gently dipping, the topography that evolves is affected by the elevation of the surface above base level and by variability in the resistance of the strata to erosion. The Mississippi valley, for example, is a very low-lying flat region, already close to base level. The Colorado Plateau, on the other hand, lies far above base level and has, as a result, spectacular deep canyons. The Appalachian Plateau exemplifies intermediate topography influenced largely by the bedrock carved by an extensive drainage system that developed as a result of the region's climate.

New fold belts may clearly show ridges and valleys formed by anticlines and synclines. Older belts reflect ridges and valleys formed by the differential erosion of hard and soft layers in the underlying strata. Erosion may even continue to produce an inverted relief in which the synclines remain as ridges while the anticlines have been worn into valleys.

KEY TERMS

hogbacks	**joints**
inverted relief	**sag ponds**

STUDY QUESTIONS

1. What features of faults make them likely to stand out in the topography?
2. Why are so many flat-topped mountains capped by a resistant rock layer?
3. Why do shales and other unresistant rocks sometimes occur as hills or at high elevations?
4. If the Colorado Plateau, at an average elevation of 1500 meters, were being reduced in elevation at the average rate from the continent ($\pm$ 2 cm/year), how long would it be before the plateau would stand close to sea level?
5. Why don't we find deep canyons or valleys similar to those in the Appalachian and Colorado Plateaus in the Gulf Coastal Plain?
6. What changes would be necessary to create deep canyons in the Coastal Plain?
7. What is the bedrock structure in the area where you live or go to school?
8. Do you know of places in your area where structural features are reflected in the shape of the land?
9. What changes would you expect to find in the Appalachian Plateau if the climate became arid and remained arid for a few thousand years? Explain why you reached this conclusion.
10. Why don't the highest mountains generally have landforms that reflect bedrock structure?

SUGGESTED READINGS

Garner, H. F., *The Origin of Landscapes.* New York: Oxford University Press, 1974.

Hunt, Charles B., *Physiography of the United States.* San Francisco: W. H. Freeman and Company, 1967.

Ollier, Cliff, *Tectonics and Landforms.* London: Longman, 1981.

WEATHERING AND LANDFORM DEVELOPMENT

Chapter 17

The breakdown of rock at the earth's surface, weathering, is largely the result of interaction between the atmosphere, water, and rock minerals. It is not surprising, therefore, that both the **weather,** the physical condition of the atmosphere at any particular time, and **climate,** the long-term averages of weather, are important factors in controlling the biological and geological processes that bring about weathering. Climatic conditions affect both the chemical decomposition and physical disintegration that occur where earth's inorganic materials come in contact with the atmosphere and biological processes. Climate, more than any other factor, controls which of the geomorphic agents—streams, wind, groundwater, or ice—shape the land. Indirectly, climate also influences the action of waves, currents, and even the process of downslope movement of surficial materials.

The combination of heat and moisture, the two primary factors in climate, determines whether a land area is desert, rain forest, or the site of accumulation of snow and glacial ice (Fig. 17.1). They are the deciding factors in how the land is shaped by water. They govern the sculpturing by continuous or intermittent flow of surface runoff and stream action or by the movement of glacier ice. They also determine if the ground is subject to the heaving action of freezing water, or the slow, but continuous, leaching action of warm water as it percolates through the soil and into the underlying rock.

Dramatic differences in how geomorphic agents operate occur in areas of climatic extremes. For example, in a desert area, the entire annual rainfall of a few centimeters may fall in one or two downpours;

National Park Service

(a)

(b)

Figure 17.1
(a) The land is shaped by effects of freezing water and ice in frigid areas such as Alaska. (b) Moisture is scarce in desert areas such as Arizona.

in a hot, humid region subjected to several hundred centimeters of annual rainfall, the moisture may come as an almost continuous light rain or mist. At the cold extreme, almost all precipitation falls as snow or sleet, and the amount of snowfall exceeds the amount melted during warm periods. The following chapters are concerned with the processes by which the various geomorphic agents shape the land. Climate, in conjunction with the structure and composition of the bedrock and the tectonic setting, establishes the framework within which these agents act. The tremendous impact of changes in climate will become apparent in Chapter 24 when we examine the effects of the glacial advances that have distinguished the last two million years of earth history.

Scientists' understanding of the various natural factors that contribute to climatic changes is steadily improving. In the process we have learned that we ourselves may be influencing earth's present and future climate. Later in the chapter we will take a look at some of the potential and sometimes alarming consequences of those changes.

17.1 WEATHER AND CLIMATE

Weather is described in terms of temperature, moisture, types of clouds present, wind direction and speed, and other measurable physical conditions of the atmosphere. These conditions are constantly changing; some last a few hours or a few days, others are seasonal. They can change rapidly, and often vary considerably from place to place even within a relatively small area. This is especially evident in the mountains and is dramatically illustrated in the Cascades and in the Sierra Nevada Mountains, where large amounts of rain and snow fall on the western slopes, while areas a few miles to the east remain dry.

The climate of an area is described by the long-term average of the charateristics of the weather. The average annual temperature and rainfall provide some indication of this. Regions with average temperatures below zero are quite different from those with averages above 20 °C. A much more accurate picture of climate emerges, however, if we know not only the averages but also the range of temperature and rainfall during each month or season.

Because weather is usually recorded continuously at weather stations, the data for any time interval can be compiled. Monthly, seasonal, and annual data for the United States is tabulated in an atlas published by the Environmental Science Services Administration. If you want to know the average annual snowfall, the average date of the last frost, the prevailing wind direction during any month, the amount and distribution of rainfall during the year, or similar information for any area of the United States, it is available in this atlas.

Climate and Landforms

Climates are grouped into three very broad categories—humid, arid, and frigid. Figure 17.2 shows the climatic regions of North America. Frigid areas are those in which most precipitation is in the form of ice or snow, and glaciation is the primary geomorphic agent. The modern frigid areas include the polar and most subpolar regions and many of the high mountains in all parts of the world. The landforms in all of these regions have been shaped by glaciers during the Pleistocene era. Distinctive glacial features dominate the topography even in areas from which the ice has long since departed. Melting of the ice started about 11,000 to 12,000 years ago and signaled the beginning of new climatic regimes for most of the glaciated regions. Some eventually became deserts, but most now have humid climates. Consequently, the landforms first shaped by glacial erosion and deposition are being modified by the more recent effects of running water and wind.

Several distinctly different arid climates occur on earth. The frigid or polar climate is arid in that precipitation is sparse, but because the operation of geomorphic agents in the polar climates is so unlike those in other arid regions, they should be considered separately. Most of the world's largest deserts occur in the horse latitudes (30° to 35° North or South latitude) where cool, dry air descends from high altitude. The deserts of North America, the Middle East, China, South Africa, Australia, and South America all fall in this zone. Annual temperatures in these regions average around 8 °C, but they may exceed 32 °C in summer. The rainfall, averaging less than 25 cm (10 inches) per year, falls mainly in the winter season. Drought conditions are common, and no rain at all falls during some years.

Humid climates are widely distributed in both hemispheres. Lands that lie along the equator, such as the Amazon basin, Indonesia, and central Africa, are both warm and wet most of the year. The

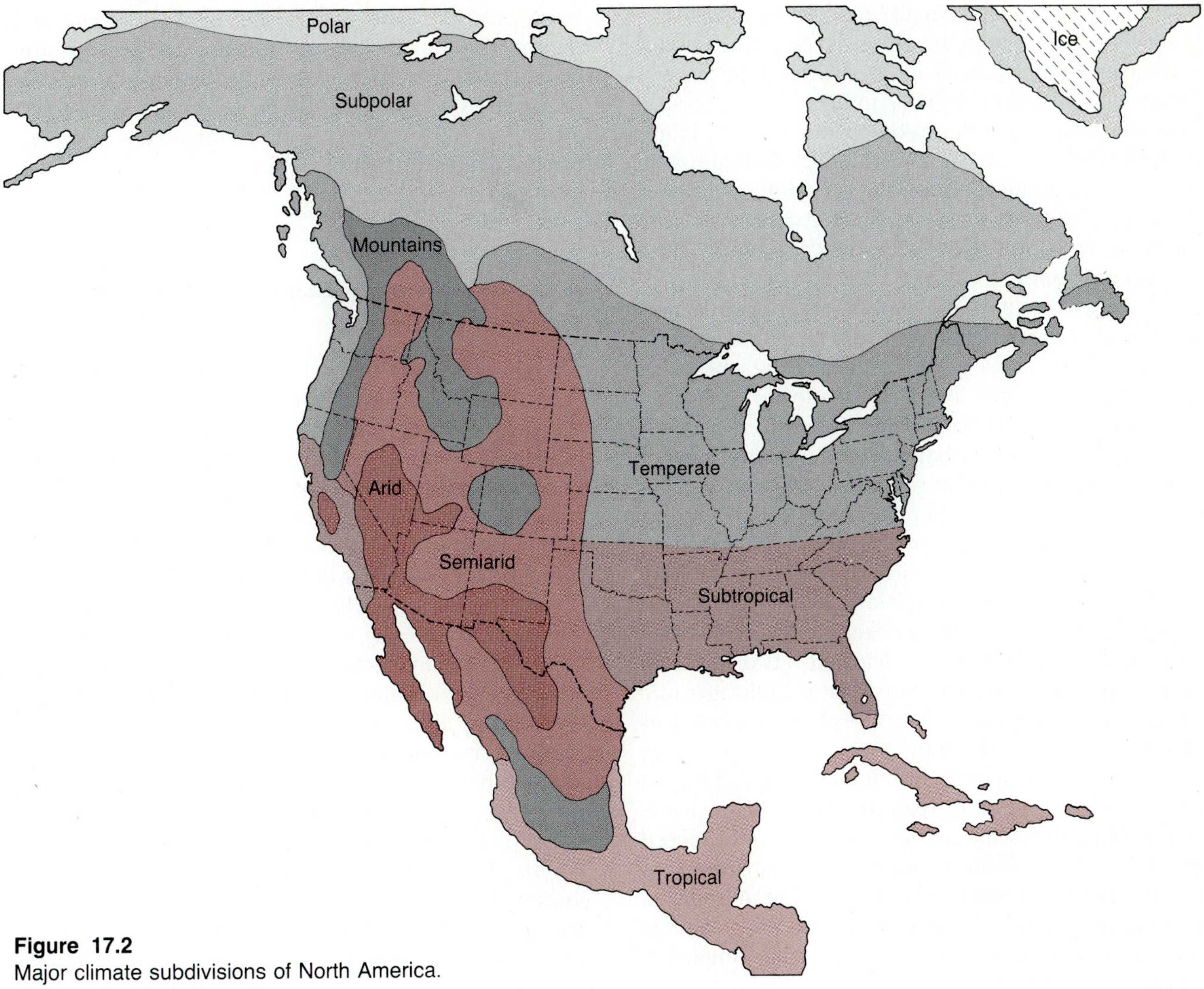

Figure 17.2
Major climate subdivisions of North America.

bordering regions to the north and south are warm most of the year, but rainfall is largely restricted to one season. The trade winds keep the areas relatively dry at other times. Still farther to the north and south, much greater variation is found in the middle latitudes. High rainfall occurs in some coastal areas, where moist air moving in from the ocean provides a ready source of water, and mountains along these coasts may be especially wet. The high seasonal rainfall of monsoons occurs in areas where major pressure systems cause the winds to bring moist air from the oceans inland during certain times of the year. In India, for example, low pressure systems develop over the continent in the summer when hot air over the land rises to high altitude. This draws in moist air from adjacent oceans toward these low-pressure areas. When the moving air encounters mountains, it rises and is cooled, and rain results. The southern flanks of the Himalayas are drenched by hundreds of centimeters of rain every summer for this reason.

An interesting contrast is found in the climates of the southern United States (Fig. 17.2). The entire region is located in a belt where dry air descends to the surface from high altitude and might, therefore, be expected to have a desert climate. The southwestern states are arid or semiarid, as might be expected, a condition enhanced by the coastal mountains of California. The Southeast, however, is humid, with rainfall generally exceeding 100 cm per year. The main reasons for this lie in the proximity of the warm waters of the Gulf of Mexico and the regional pattern of atmospheric circulation. Low-pressure systems containing moist air often develop in the Gulf and are swept northward and eastward in the general pattern of circulation. In the summer, a high-pressure system situated over Bermuda produces a circulation pattern that also carries air from the Gulf into the continental interior. Because the prevailing winds are from west to east across most of the country, the moist air from the Gulf rarely reaches west of the Mississippi valley.

Weathering, mass movement, and stream action are the main geomorphic factors at work in both humid and arid climates. Why then are the landforms of the two different? The basic distinction between arid and humid areas lies in the difference in rainfall. Not only is there much more water in the humid climatic regions, but that rainfall tends to be distributed more evenly throughout the year.

The effects on the earth's surface of abundant moisture are many—chemical weathering is more pronounced, unstable minerals are more completely decayed, weathering extends deeper into the ground, and soluble constituents are leached out of the soil. If the bedrock is soluble more water infiltrates into the ground, and solution effects are prominent. Streams are larger and more numerous, floods are more frequent, and the level of sustained flow much higher than in arid or semiarid regions. All of these factors suggest that denudation and erosion should proceed more rapidly in humid regions. But moisture also stimulates the growth of plants, which tend to retard erosion and stabilize the surface. We will examine the typical land forms of humid and arid climates in Chapters 20 and 22.

Arid climates receive relatively little rainfall, but because the total often falls within a few hours during torrential storms, its effect may be dramatic. Streams rise rapidly, causing floods; and great quantities of silt and soil are washed across the low slopes into drainage channels and downstream. So, despite the low total rainfall, its effectiveness as an agent of erosion is great, especially where vegetation is sparse.

The high rate of evaporation and the great depth of groundwater in deserts cause the amount of water in streams to decrease downstream. Thus, the streams often do not flow out of the desert into the ocean. Some deserts have no external drainage. Others, like most of the southwestern United States, are drained by only a single major stream, in this case the Colorado River, which has its headwaters and the source of its water beyond the desert.

In conclusion, climate determines which weathering processes and geomorphic agents operate in a region. These effects are surficial, modifying the bedrock that has been shaped by volcanic or tectonic processes described in earlier chapters. Once the forms shaped by these internal forces are exposed, the initial step in their modification is usually caused by weathering.

17.2 WEATHERING

The changes that take place in rocks and minerals exposed to conditions at the surface of the earth are the effects of weathering. The chemical and physical processes involved result in chemical decomposition and/or physical disintegration of the rocks and minerals. Weathering effects vary greatly from place to place, because neither the environment nor the type of material exposed at the surface is uniform.

A description of local climate is one of the best ways to characterize an environment. We have seen in the preceding section some of the general underlying causes of climatic conditions. But in order to understand the effects of weathering at a specific locality it is important to draw a more precise picture of its climate. Even within very short distances there may be significant differences in climate, giving rise to microclimates. This is especially true in mountains where climate varies rapidly with altitude as the temperature decreases. Strong local contrasts also usually exist between sunny and shady slopes or between slopes subjected to persistent wind and those in a wind shadow. Below the timber line, local variations show up dramatically in the types of plants present or in their abundance. Usually changes can also be clearly seen in the type of soil present.

Chemical Weathering

The susceptibility of a rock to chemical decomposition depends largely on the composition of the minerals in the rock and on the climate. Especially significant are the amount of moisture present and the temperature. Surprisingly few of the atmosphere's constituents (Table 17.1) are actively involved in chemical weathering. Of the dry gases, only oxygen and carbon dioxide are important agents of weathering. Water is by far the most important component of the atmosphere as far as weathering is concerned. The effectiveness of these chemical agents depends on the composition of the rock, the size of the particles that

Table 17.1 The Gases Present in Dry Air.*

Element	% by Volume	Important Weathering Processes
Nitrogen (N_2)	78.08	none
Oxygen (O_2)	20.95	oxidation
Argon (A)	0.93	none
Carbon dioxide (CO_2)	0.03	carbonation
Neon (Ne)	0.002	none
Helium (He)	0.0005	none
Methane (CH_4)	0.0001	none
Hydrogen (H_2)	0.00005	none

* In some locations, sulfur and nitrous oxide from burning of hydrocarbons or smelting of sulfide minerals are important causes of acid rain.

make it up, and the ease with which solutions move through the rock. Quartz, for example, is almost insoluble in rainwater, and is unaffected by carbon dioxide and other constituents of the atmosphere. Thus, rocks composed primarily of quartz decompose slowly. Most other minerals, though, are susceptible to water and show signs of reaction within a few years of exposure to the atmosphere. Five processes—solution, hydration (absorption of water), hydrolysis (reaction with water), oxidation (reaction with oxygen), and carbonation (reactions involving carbon dioxide)—are responsible for most chemical weathering.

Although most silicate minerals dissolve slowly in pure water, over the long periods of geological time **solution** plays an important part in the breakdown of rocks. Salt, gypsum, and, to a lesser extent, carbonate minerals dissolve relatively rapidly. In arid climates, salt and gypsum are often found in the soil, where they are dissolved at depth, siphoned to the surface as a result of evaporation, and accumulate in the top part of the soil as the water carrying them evaporates. In contrast, salts are rare in the soils of humid climates, where they are carried away by surface and subsurface movement of water.

Hydration and hydrolysis both involve the reaction of water with minerals. **Hydration** is the absorption and combination of water with other compounds; **hydrolysis** is the name given to reactions between water and other compounds. Hydrolysis is particularly important because it is the process by which most silicate minerals (such as feldspars, the most abundant minerals in igneous and metamorphic rocks) are chemically weathered. The exact mechanism of feldspar weathering is not yet adequately defined, but the general reaction is as follows:

$$\text{orthoclase feldspar} + \text{water} \rightleftharpoons \text{potassium} + \text{clay (kaolinite)} + \text{silicic acid}$$

$$4KAlSi_3O_8 + 22H_2O \rightleftharpoons 4K^+ + 40H^- + Al_4Si_4O_{10}(OH)_8 + 8H_4SiO_4$$

In the laboratory this reaction occurs at a noticeable rate only at temperatures over 200 °C. It would take place quite slowly at surface temperatures, and probably occurs as a series of steps rather than the single reaction written. It may be that aluminum (Al) and silica (Si) are separated from the feldspar as very fine (colloidal) particles that later combine to form clay minerals (Fig. 17.3). The silica that is set free goes into solution as silicic acid (H_4SiO_4), which, if present in high concentration, may result in the deposition of colloidal or amorphous (noncrystalline) SiO_2. The

Figure 17.3
Although this rock looks like the granitic gneiss from which it was altered, it is now soft because the feldspars are largely changed to clay.

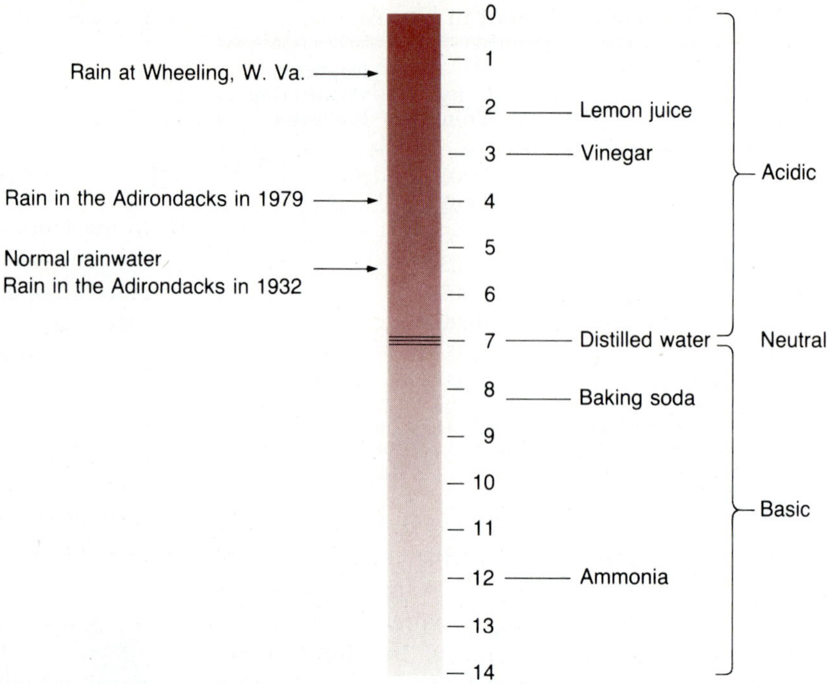

Figure 17.4
pH scale showing the acidity of some common substances and rainwater.

silica ultimately may go into the formation of clay; it may be taken up by plants; it may be recrystallized into small quartz crystals; or it may be deposited as chert or some other form of silica.

Reactions of carbon dioxide and acid rain. Only in recent years has the acidity of the rain attracted widespread attention and concern. But rainwater is naturally slightly acid. Chemists express the acidity of solutions in terms of a **pH scale,** which ranges from 0 to 14. A pH rating of 7 indicates a neutral condition; lower numbers correspond to increasing acidity, which depends on the concentration of hydrogen ions in the solution. The pH scale increases logarithmically. Thus, each increase of one in the pH corresponds to a tenfold increase in hydrogen ion concentration.

The pH values for some common substances and rainwaters are shown in Fig. 17.4. Normal rainwater has a pH of about 5.5. This acidity can be traced to the presence of carbon dioxide in the atmosphere, which combines with water to form carbonic acid (H_2CO_3). Carbonic acid reacts with the copper in copper roofs to produce their characteristic green copper carbonate coating. Carbonic acid is, however, much more important geologically in the weathering of carbonate minerals, such as calcite and dolomite, since both minerals are dissolved by acids. **Carbonation,** the solution of calcite and limestone by rainwater, can be summarized as follows.

Pure water and carbon dioxide combine to form carbonic acid.

$$H_2O + CO_2 \rightleftharpoons H_2CO_3$$

This breaks down to form hydrogen and biocarbonate ions in solution.

$$H_2CO_3 \rightleftharpoons H^+ + HCO_3^-$$

If this carbonic acid solution comes in contact with calcite, some of the hydrogen ions combine with the carbonate $(CO_3)^{2-}$ ions to form bicarbonate. Calcium ions are left in the solution.

$$CaCO_3 + H^+ \rightleftharpoons Ca^{2+} + HCO_3^-$$

The two above reactions can be summarized as follows:

$$CaCO_3 + H_2CO_3 \rightleftharpoons Ca^{2+} + 2HCO_3^-$$

These reactions are reversible. When the reaction proceeds from left to right, calcium carbonate is dissolved; when it proceeds from right to left, calcium carbonate is deposited. Thus, the equation helps us understand the solution of limestone as well as its deposition in the sea and in caves. Other acids as well as carbonic acid can dissolve carbonates. Natural waters may contain humic acids derived from decaying organic matter, and nitrogen acids formed by bacterial action and decay of organic matter. Sulfuric acid, which is very strong, can be found naturally in areas around volcanoes, where sulfide minerals are oxidized, or downwind of industralized areas where sulfur-bearing coal is burned. This last source of acidity is primarily responsible for acid rain.

Rainwater has almost certainly been slightly acid for millions and probably hundreds of millions of years. This acidity is responsible for the development of caves and the shaping of the land by groundwater, which we will explore in Chapter 20. The concern today about acid rain is prompted by a surprising increase in the acidity of rain and lake waters. Highly acid rain was first noticed in Scandinavia about twenty years ago, when it was connected with a decline in fish populations in lakes. Ten years later, the acidity had been traced to emissions of sulfur dioxide and nitrogen oxides. The likely sources include power plants burning coal, oil, and gas; iron and copper smelters; and automobiles, in addition to natural sources such as volcanoes.

Only in recent years have scientists realized how widespread the abnormally acid rains have become. The trend presumably started with the industrial revolution in Europe, but it has clearly spread around the world, as is shown by the change in the pH of the waters in lakes in the Adirondacks over a forty-year period (Fig. 17.4). Even waters in high lakes in the Rockies are becoming more acid, possibly as a result of emissions from jets and from industries as far away as Japan.

Several important problems are associated with acid rain. Many of the Scandinavian and Adirondack lakes no longer support fish life. Salamanders, frogs, even bacteria have died off in some lakes. Because limestone tends to neutralize acid, lakes located in areas where no limestone is present show the fastest increase in pH. Where limestone is present, its solution is accelerated, and lime is rapidly removed from the soil, increasing the need for lime fertilizer for some crops. Other crops, including tomatoes and corn, may benefit from the increased acidity.

Many of the beautiful and intricately carved statues and other stonework carved from limestone and marble have been destroyed or severely damaged by acid rain. The details of the carving are first softened and then gradually dissolved away, leaving rounded, almost featureless forms. These effects are seen throughout Europe, but are most pronounced in places like Venice, located close to industrial centers. The same effect is seen on many old marble gravestones (Fig. 17.5) in humid climates. The reaction of acid rain with metals is a potentially serious health problem because metal ions may be freed and con-

(a)

(b)

Figure 17.5
Compare the weathering effects in a humid climate on these tombstones, which were cut about the same time; (a) is composed of marble, (b) of granite.

taminate drinking water supplies. Higher-than-normal levels of both iron and lead are being detected in the water of the Adirondack lakes.

The outlook for solving the problems of acid rain are not encouraging. We are turning increasingly to greater utilization of coal, which contributes both CO_2 and SO_3 to the atmosphere. The high stacks used to protect localities from industrial emissions only serve to put these gases higher into the atmosphere, where they are spread more widely. The eventual solution probably will rest in development of a technology for removing acidic oxides from stacks before they enter the atmosphere.

Oxidation. Up until about two billion years ago, oxidation was not an important weathering process; for until then the quantity of oxygen in the atmosphere was too small to cause oxidation of iron minerals. The oxygen content since that time has been maintained both by photosynthesis and by breakdown of water vapor by solar radiation. **Oxidation** is the chemical reaction in which oxygen combines with other elements and compounds. Oxygen has a particular affinity for sulfur, manganese, and iron compounds. Thus, it is not surprising that minerals containing these elements are among the most commonly oxidized natural materials. The common rock-forming minerals—pyroxenes, hornblende, biotite, and olivine—all of which contain iron and are subject to oxidation, are described in detail in Section 2.5. As the rock containing these minerals is oxidized, iron is converted to ferric oxide (hematite, Fe_2O_3) or to ferric hydroxide (limonite). This is accompanied by a color change in the minerals and in the soils developed from them. Red, yellow, or brown are typical colors produced in this transformation. Because heat and water promote oxidation, it takes place most readily in warm, moist climates. Soils in these climates are often bright red or yellow, or shades of these colors.

Susceptibility of rocks to chemical decomposition. The texture, structure, and composition of rocks cause varying susceptibility to weathering. Because chemical reactions take place on surfaces, fine-grained rocks or intensively fractured rocks open to solutions are more easily broken down than those that are coarse-grained and impervious. Fractures, bedding planes, pore spaces, and even intergranular contacts all provide lines along which solutions can move.

Rock composition is of particular importance in chemical weathering (Fig. 17.6). Some minerals react slowly with the gases and solutions normally found in the atmosphere and in the ground; others break down rapidly in their presence. Of the most common rock-forming minerals, quartz is virtually unaffected by chemical weathering; salt, gypsum, and calcite are soluble; and feldspar, amphiboles, pyroxenes, olivines, and micas undergo complex chemical reactions with atmospheric constituents. In general, the minerals that crystallize at the highest temperatures are most susceptible to weathering.

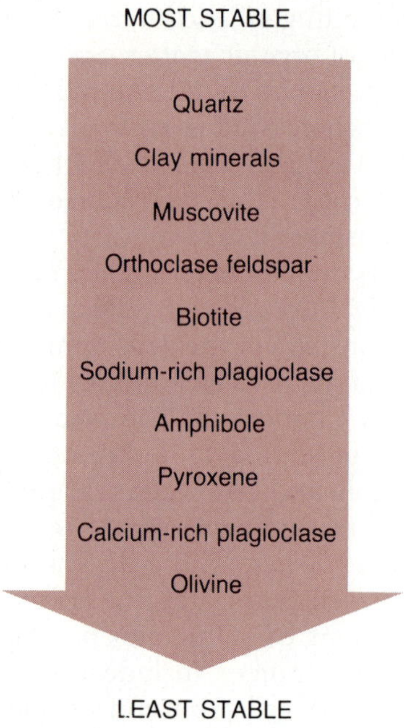

Figure 17.6
Relative susceptibility of some common minerals to weathering.

Animals and Plants as Weathering Agents

Worms and other burrowing organisms tunnel through the soil, loosening it and providing passageways for atmospheric gases and solutions. Bacteria play an important role in breaking down organic compounds, in the process liberating humic acids. Several marine invertebrates, especially gastropods and echinoids, bore into the rock, producing holes or pits in which they live, and thereby contribute to weathering in intertidal zones. Human activity is particularly important in influencing the rates of weathering through agricultural and construction practices that accelerate the erosion of the soil and affect the capacity of the soil to hold water.

Plant growth affects rock weathering in a number of ways. Certain crops remove particular chemical constituents from the soil, changing the chemical balance and accelerating the chemical reactions that are taking place in the soil and underlying rock. Roots force their way into the soil, opening it and leaving a cavity when the plant dies. Sometimes roots even grow down along fractures in solid rock, forcing the cracks open and extending them farther into the rock mass. But, on the other hand as the system of roots becomes interwoven, they form a mat of root matter that helps protect the underlying soil and rock from erosion. Plants intercept rain, reducing its impact on the ground surface. They also take up water, reducing the amount of moisture that reaches the ground, especially where rainfall is of short duration.

Some plants called **accumulator plants,** absorb unusually large amounts of certain elements from the soil. For example, horsetail, corn, palm trees, bamboo, reeds, and other grasses take up and store large amounts of silica. Silicon dioxide comprises about 77 percent of the inorganic matter in giant reeds, 47 percent in blue grass, and 70 percent in China laurel and palms. The significance of this high mineral content becomes apparent if we consider the amount of silica that can be stored in plants in tropical forests, where many of these accumulator plants are abundant. T.S. Lovering (*1*) estimates that such a forest would take 0.4 ton of silica per acre per year out of the soil. If the underlying rock is basalt, containing about 49 percent SiO_2, the basalt covering one acre only one foot deep would contain about 2000 tons of silica—an amount that could be removed in about 5000 years. One of the results of this process is the formation of tropical soils that are notably deficient in silica. Of course, not all silica removed by plants is lost, because the silica is returned to the ground surface when a plant dies and decays, but this silica is vulnerable to rapid erosion resulting from stream action. With the surficial removal of silica, the soil is left deficient in silica and relatively enriched by other constituents.

Processes of Mechanical Disintegration

The most forceful and dramatic agent of mechanical disintegration is the impact of falling meteorites, which may crush or melt the spot on which they land. Early in earth's history, meteorite impact was one of the most important processes causing breakdown of rocks, but it is no longer a significant factor, nor has it been for many millions of years. Other mechanical processes, notably freezing of water, insolation (exposure to the sun's rays), crystal growth, unloading (removal of weight), and, increasingly, human activities are responsible for the disintegration of surficial rocks.

Freezing and thawing water is effective in breaking rocks because, when it crystallizes, water expands with a 9 percent increase in volume. When water is frozen in an enclosed vessel, pressures up to 2100 kilograms per square centimeter are developed at temperatures of −22 °C. Such pressures far exceed the strength of even the hardest and most durable rocks. The development of this pressure is an important weathering process because rainwater seeps into fractures in rock and freezes. Even though most fractures are open at the top, the initial freezing partially seals that opening and enough pressures may develop to spread the fracture apart and extend it deeper (Fig. 17.7). The effects of freezing and thawing are cumulative; it is the often repeated thaw-and-freeze cycle operating year after year, that makes these processes so effective.

Frost action. Frost heaving is a process that disrupts the soil or other loose sediment, lifting it as layers of ice form in the soil. Once freezing begins in the soil, water is pulled toward the surface of the ice by capillary action along intergranular spaces in the soil and rock. When crystallization takes place, a slight amount of heat is liberated, keeping a film of water on the ice. Because this film is connected to water in pore spaces in the rock, the process continues. Water freezes on the ice surface; the ice slowly expands, and more water is drawn into the freezing surface. Layers of considerable thickness may eventually form, as is shown in Fig. 17.8. As the ice thickens, overlying soil is forced ever upward, disrupted, and broken up.

Even where frost heaving occurs only rarely, or to a limited extent, it loosens soil and allows air to pass into it. This process may be healthful for the soil, but it may also be destructive, since it can displace building foundations or loosen and uplift soil on steep slopes making it susceptible to downhill creep or erosion. Where the soil is cultivated, rock fragments forced up through the soil by frost heaving must be removed from the plowed fields year after year. Frost heaving of stones begins when water immediately beneath them cools to the freezing point and ice begins to form in the small capillary-sized opening under the stone. As the ice builds up, the stone is forced upward through the soil, until it eventually comes out on the surface.

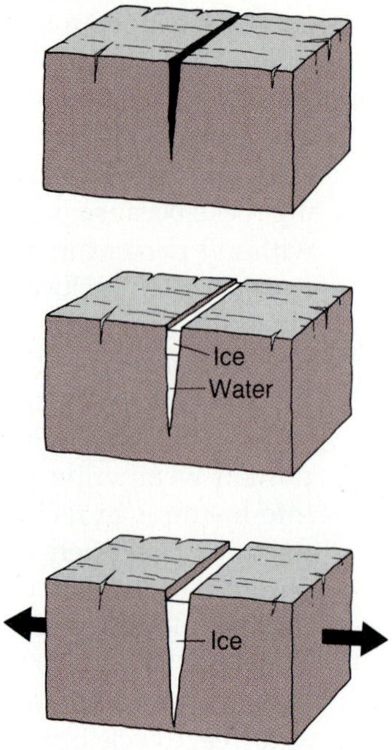

(a)

(b)

Figure 17.7
(a) Expansion caused by freezing of water in a fracture often forces the fracture open and extends it deeper into the rock. (b) These meter-long blocks of granite have been forced apart by frost wedging, through the process described in (a).

Crystal growth. Water in the ground dissolves limestone and other soluble mineral constituents. Evaporation at the ground surface takes place and may draw these solutions back to the surface through the capillary-sized openings in the soil, much as a wick carries fluid. As the water evaporates, the material in solution precipitates as crystals or crystalline aggregates in the soil. This **crystal growth** exerts pressure on the soil, pressure easily sufficient to break up soil and sometimes even rock. Some sulfates, phosphates, nitrates, gypsum, calcite, alum, epsom salts, and saltpeter can grow in this manner.

Unloading. Joints or fractures, called **sheeting,** may develop in bedrock as the confining pressure is released by removal of the weight of the overlying rock, or **unloading.** These joints form nearly parallel to the surface, as shown in Fig. 17.9. The process of unloading takes place most frequently in quarries and in places where glaciers have removed the load from massive rocks like granite. The fractures formed in this manner become planes of weakness in the rock, and they are often open, allowing solutions to penetrate and weather the rock by chemical or frost action.

Figure 17.8
Ice layers like this one formed in mud near Fairbanks, Alaska, loosen soil and make it more susceptible to erosion.

Figure 17.9
A thin sheet of granite formed and arched upward after overlying granite had been removed in the Mount Airy Quarry, North Carolina.

Insolation. Daily variation in temperature is one of the most universal processes, but its importance in bringing about mechanical breakdown of rocks is not yet clear. Daily variations of 20 °C (about 36 °F) or more may take place in desert regions where exceedingly high temperatures reached during the day are followed by rapid cooling at night. When solids are heated, they expand, and when they cool, they contract; but the amount of expansion and contraction is very small. The question is whether or not repeated expansion and contraction will, over a long period of time, lead to mechanical breakdown of a rock. In laboratory experiments, repeated heating and cooling of rocks through normal ranges of temperature have failed to produce mechanical disintegration even in experiments continued for years. Any importance of thermal expansion and contraction in nature is due to the continuation of this process over much longer periods of time than experiments. It seems more probable that chemical effects in combination with thermal effects may be much more significant. For example, slight differential expansion of minerals in a rock may allow gradual breakdown between the grains, and once grains are loosened, solutions can penetrate and bring about chemical decomposition. The development of scalelike sheets, called **exfoliation** (Fig. 17.10), on exposed rocks may result in part from the effects of solar radiation or **insolation,** but similar scales are also found on rocks that are protected from sunlight. These scales are more likely to be caused by unloading where they are developed over large flat surfaces or by chemical weathering along fractures and bedding surfaces. The spherical surfaces in Fig. 17.11 are formed by chemical weathering.

U.S.G.S.

Figure 17.10
Half Dome in Yosemite National Park, California. Note the disintegration of the rock. Exfoliation and frost action are the dominant processes promoting the disintegration now. Unloading due to glacial erosion has also been an important factor causing the exfoliation.

Figure 17.11
The rounded solid mass shown here is the result of chemical weathering along bedding and fractures in uniformly inclined beds of shale that outcrop in western Virginia. This is called spheroidal weathering.

17.3 SOIL

Engineers define **soil** as all unconsolidated material above bedrock. For agricultural purposes, it is defined as the natural medium at the surface of the earth capable of supporting plant growth. For most geologists, soil is the name applied to the unconsolidated matter formed by the decay of rock and generally altered by organic processes. In any case, its importance is clear: we depend on soils for our food supply, for most of our building materials, and for many other products. Most buildings, highways, dams, canals, and other large-scale structures are built on or in soil. Consequently, soil properties directly affect the stability of these structures.

The mechanical properties of soil are especially important from an engineering point of view. Engineers are concerned with the load-bearing strength of soil, with its potential for swelling or collapsing when wet, or with its corrosive characteristics and their effect on buried metal, such as cables.

Geologists are primarily concerned with the processes by which soil originates, and of these, the various weathering phenomena are the most important. Since soil develops largely by the weathering of bedrock (Fig. 17.12), it might seem probable that each type of bedrock would produce a certain type of soil. Although bedrock composition does strongly influence the resulting soil type, many different types of soil may be produced from a single kind of rock. Other factors are also important—among these, cli-

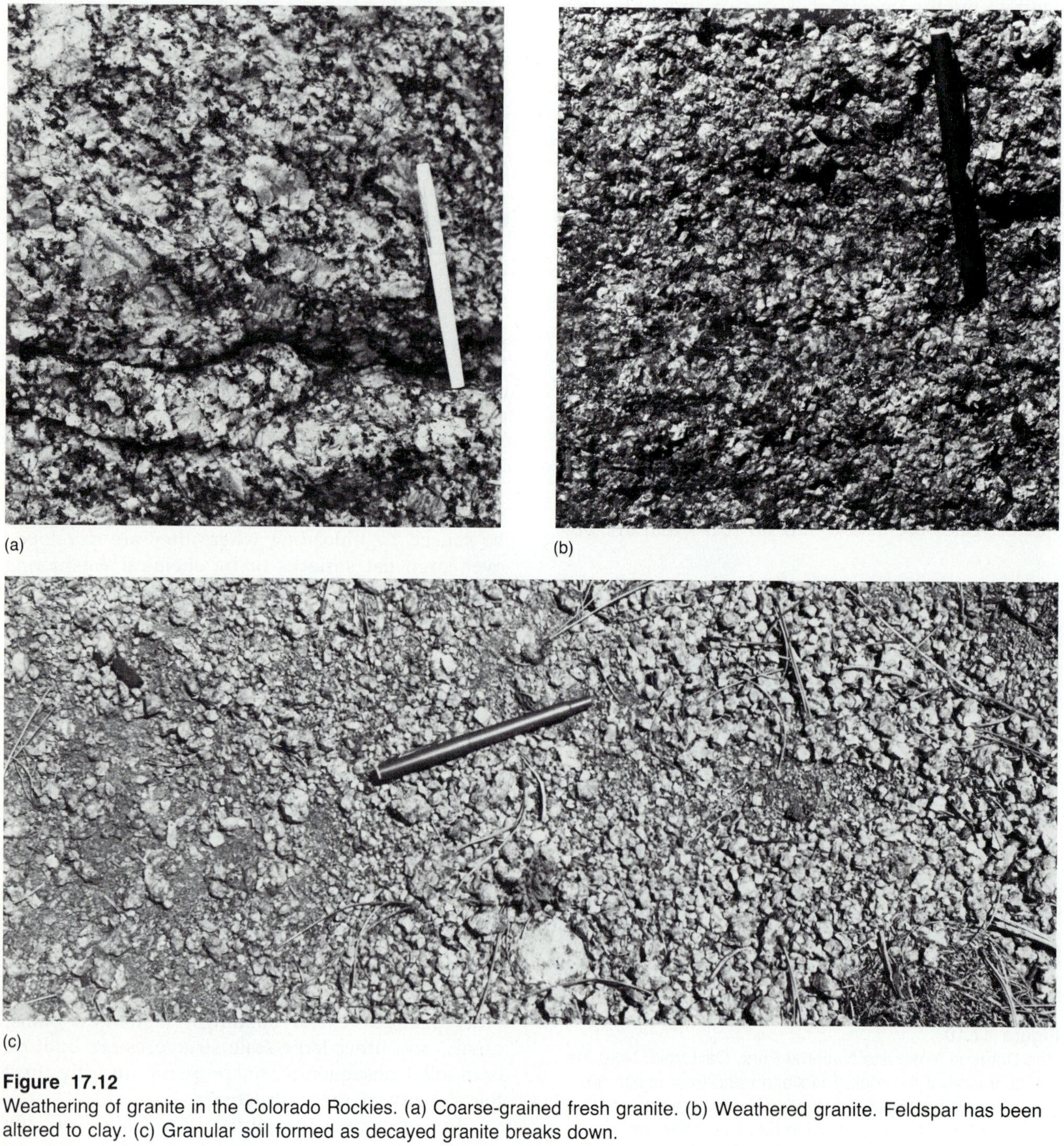

(a) (b) (c)

Figure 17.12
Weathering of granite in the Colorado Rockies. (a) Coarse-grained fresh granite. (b) Weathered granite. Feldspar has been altered to clay. (c) Granular soil formed as decayed granite breaks down.

mate, the slope of the surface on which the soil develops, the length of time involved, and various biological factors.

Soil produced by the weathering of pure quartz sandstone or pure limestone will obviously have a much more restricted chemical composition than one produced from a shale or granite that contains several mineral constituents. The climate and slope help determine which constituents remain in the soil. If the soil is well-drained and situated in a climate characterized by warm and abundant rain, the soluble constituents of the soil will be leached and carried away. In an arid climate, soluble constituents may actually become concentrated in the upper part of the soil by the wicklike mechanism described in Section 17.1. Climate is also an important determinant of which plants grow and how dense the vegetation becomes. These factors in turn affect the soil composition, since dead plant matter is a prime component of the upper part of the soil.

Through careful study, the processes governing fertility and the formation and destruction of soil can be determined. Understanding the processes at work provides the basis for conservation of soil and perpetuation of its agricultural value. But in areas where the study of soil has not been given careful attention, soil has often been depleted of its nutrients, subjected to erosion, and sometimes gradually abandoned as a spent natural resource.

A Typical Soil Profile

Well-defined zones, called *A, B,* and *C* horizons, are often seen in cuts made through soil to the underlying rock. Figure 17.13 shows the result of such a cut in the soil of the coastal plain in Mississippi. The *C* horizon contains weathered parent material, such as loose and partly decayed rock, from which the overlying soil is derived. Soils may, of course, develop on rocks that have been transported by streams, glaciers, or landslides from their place of origin. Consequently, the rock in the *C* horizon may not be related to the bedrock below.

The *B* horizon (Fig. 17.14), which is commonly called the **subsoil,** contains fewer organisms than the overlying layers, but more than the *C* horizon. Weathering has reduced all rock to fine material in the *B* horizon, and it usually contains more clay than either the *A* or *C* layers. For this reason, it is harder when dry and stickier when wet than either of the other two. The *B* horizon is the site of accumulation of suspended materials from the overlying soil. In humid

Figure 17.13
The top part of this exposure of sediment is dark because plant matter is abundant in the *A* horizon of the soil. This thin layer of soil is developed on coastal plain sediment.

U.S.D.A.

Figure 17.14
This soil profile has a distinct zonal appearance. Plant matter is seen in the *A* horizon. The subsoil, *B*, contains clay and silt. These horizons are developed in a wind-blown deposit underlain in the horizon marked $11C_2$ by sand and gravel. Depth is shown in inches.

and tropical climates, iron and aluminum hydroxide tend to be concentrated in the *B* horizon, whereas in arid climates, calcium is concentrated in this layer.

The top layer of soil, the *A* horizon, is the site of maximum organic activity and the place where chemical breakdown of mineral matter, solution, and leaching occur. From the top down, through an idealized *A* horizon, we might find leaves and loose organic debris that are largely undecomposed; then matted and partially decomposed organic material; next a mixture of organic and mineral matter; and finally a transition into clayey subsoil.

Soils vary both in type and in the degree of development of the three horizons described above. Some soils show no zones at all, others have a zone missing, and still others have subzones of the general zones outlined. The length of time weathering has taken place has an important influence on the development of the profile. A newly deposited ash, stream, or glacial deposit may show no zonal development. Texture and structure of soil are also highly variable and depend largely on the relative amounts of organic matter, sand, silt, and clay present. These occur in all degrees of mixture.

Soil Chemistry

Chemical decomposition of a rock's constituent minerals is of particular importance in soils. Igneous and metamorphic rocks (granite, basalt, gneiss, and schists) are composed primarily of quartz, feldspar, micas, pyroxene, amphibole, and olivine. Quartz tends to be stable, but the others break down into clay minerals, iron compounds, silicon dioxide, and other compounds in solution. Thus, of the over one hundred elements, only a few are common in soils, although many more may be present in trace amounts. Over nearly three-quarters of the land surface, sedimentary rocks form the underlying rock beneath soil. Common sediments are clay, sand (particularly quartz sand), lime, mud, and mixtures of these occurring in varying degrees of consolidation. Of the various minerals present, clay minerals are of particular significance, not only because they contain a number of different elements but because their size allows them to be more readily involved than other minerals in chemical reactions.

Typical soil contains clay colloids, fine mineral fragments, organic matter, and some moisture. The mineral fragments undergo chemical decomposition very slowly; some dissolve into their ionic constituents in the soil moisture. Some of the ions stay in solution, but because clay colloids are negatively charged they can hold some of the positive ions, such as hydrogen, calcium, and sodium, on their surface. Smaller quantities of aluminum, magnesium, and potassium are also held in this way. The concentration of hydrogen ions in a solution determines the acidity of the soil, which ranges from pH 4 (fairly acid) to 10 (basic or alkaline). The most prevalent ions in acid soils are hydrogen and aluminum; calcium and magnesium are common in neutral soils (pH 6 or 7); and sodium is abundant in soils that are alkaline. Each type of plant grows best in soils of a particular acidity.

The chemistry of soils and soil solutions is complex, but it may be viewed simply in terms of an equilibrium condition. Positive ions (such as H^+, Ca^{2+}, and Na^+) are exchanged between the solid and liquid phases in the soil. An equilibrium condition is established in which the concentration of ions in soil water is maintained. As ions are removed from the soil by plants for use as food, they are replaced by ions liberated from minerals, organic matter, and clay colloid surfaces. Continued application of lime is necessary to keep many acid soils neutral, the condition under which most plants thrive. When hydrogen ions in an acid soil solution are neutralized by lime, more hydrogen ions are released (or exchanged for calcium ions) from the surfaces of the colloids. The properties of soils that control their ability both to exchange ions and to maintain a supply of needed ions determine their desirability as a medium for plants growth.

In most soils the total supply of plant nutrients is high compared with the need of plants supported by it, and most of the ions return to the soil as decaying organic matter when the plants die. But if the plants are crops, the soil loses these constituents because the crops are removed; and the soil may become depleted in some of the elements needed for plant growth. The same thing happens when plants are burned off, a common practice in the tropics, and to an even greater extent if the ashes are washed away. For this reason, it is often desirable to add certain fertilizers, rotate crops, or plow under crop stubble to replenish soil.

SUMMARY

All materials at the earth's surface are subject to the combined effect of chemical, physical, and biological processes that break down and/or erode rocks and minerals. Climate is of overriding importance in determining the course of weathering and erosion. Three general climate categories are recognized: humid, arid, and frigid. The combinations of moisture and temperature that characterize each climate determine which geomorphic agents prevail in a region and thus also the surface landforms.

The weathering of rocks and minerals involves physical and chemical interaction with water and atmospheric gases. The chemical composition of a rock, the size of its particles and the ease with which solutions can move through it determine its susceptibility to weathering. Chemical weathering involves five principle processes: solution by water, hydration, hydrolysis, oxidation, and reactions with carbon dioxide solutions. The last of these accounts for much of the weathering of carbonate minerals. Biological agents also contribute to weathering processes by both chemical and physical mechanisms. Plants, for example, may affect the pH and chemical composition of soil, thereby accelerating chemical weathering. They may also contribute to physical weathering as their roots enlarge cracks in rock. On the other hand, plant cover generally slows denudation of rock by holding surface soil in place.

Water is the most important agent of physical weathering. Freezing water in rock fractures can eventually split rocks. Frost action may force stones to the surface and make soils more susceptible to downhill creep and erosion. As a solvent, water can lead to formation of mineral crystals that, like ice, can enlarge cracks in rock.

The weathering of rock, combined with biological activity, forms the soil that covers most of the continents. In addition to the nature of the bedrock from which a soil is derived, the factors of climate, surface slope, time of development, and biological activity all affect the nature of a soil. In general, soils consist of three zones. The *A* horizon is uppermost and richest in organic materials mixed with sand, silt, and clay. The *B* horizon or subsoil is less rich in biological activity and richer in clay and other components leached from the *A* horizon. The *C* horizon contains weathered fragments of the underlying rock.

KEY TERMS

accumulator plants	**oxidation**
carbonation	**permafrost**
climate	**pH scale**
crystal growth	**sheeting**
exfoliation	**soil**
frost heaving	**solution**
hydration	**subsoil**
hydrolysis	**unloading**
insolation	**weather**

STUDY QUESTIONS

1. Prepare a chart of igneous and metamorphic rocks, showing probable susceptibility to weathering in a humid climate.
2. How could you use a knowledge of weathering if you were involved in locating industrial plants?
3. Outline the criteria by which you could differentiate a transported soil from a soil formed in place.
4. What weathering processes are most active in the area where you live? What environmental conditions favor each?
5. Why are clay minerals so important in soil-chemistry studies?
6. Explain differences that may be found in the soils developed on rocks of granite composition as a result of variations in climate.
7. What field evidence would prove that feldspar does weather to form clay minerals?
8. What factors cause soils developed on the same type of bedrock to differ from one place to another?
9. Describe the soil profile at a local construction site or road cut.

REFERENCE

1. T. S. Lovering, "Accumulator plants and rock weathering," *Science* 128:3321 (1958), *416–417.*

SUGGESTED READINGS

Birkeland, P. W., *Pedology, Weathering and Geomorphology.* New York: Oxford University Press, 1974.

Bloom, A. L., "Rock weathering," in *Geomorphology: A Systematic Analysis of Late Cenozoic Landforms.* Englewood Cliffs, N.J.: Prentice-Hall, 1978.

Carroll, D., *Rock Weathering.* New York: Plenum Press, 1970.

Hunt, C. B., *Geology of Soils: Their Evolution, Classification and Uses.* San Francisco: W. H. Freeman, 1972.

Likens, G. E., and F. H. Bormann, "Acid rain: A serious regional environmental problem," *Science* 184 (1974), *1176–1179.*

Paton, T. R., *Formation of Soil Material.* Boston: Geo. Allen & Unwin, 1978.

UNSTABLE SLOPES

Chapter 18

Following an earthquake in 1959, more than 27 million cubic meters of rock and soil slid off the side of the Madison River Canyon just west of Yellowstone Park. Only rarely are the downslope movements of the materials exposed near the surface as dramatic as the Madison slide, but downhill movement of rock, sediment, and soil caused by the force of gravity is the most universal of all processes of erosion. Downslope movement of materials, sometimes referred to as **mass movement** or **mass wasting,** acting in combination with transportation by running water, glaciers, wave action, wind, groundwater, and sea currents is responsible for most erosion. The principles governing mass movement are simple, but the variety of types of movement, material moved, and forms assumed by these masses is great. Gravity, which is the main driving force in all these processes, is most effective in moving materials that are unstable, such as loose or water-saturated soil, or masses of rock and soil resting on surfaces over which they might slide. All such movements take place more rapidly on steep slopes than on low slopes; and they occur more easily in water-saturated materials than in dry materials (Fig. 18.1). But evidence of such movements is found under all climatic conditions, and they are important both on land and in the sea. Mass movement is significant on the moon, too, where neither atmosphere nor water is found; and landforms attributed to it are seen on all the planets we have inspected closely.

Slope instability occurs naturally, but is often induced in rock and other surficial materials by our own modifications of the surface of the earth. As our use of land has intensified, it has become increasingly difficult to avoid using areas where potential natural instability exists—areas where, for examples, floods, wave action, or earthquakes (Fig. 18.2) may trigger disastrous mass movements. In addition, human activity has become a chief cause of mass movements during the construction of building foundations, dams, reservoirs, bridge abutments, tunnels, highways, and canals. Costly and sometimes even disastrous results have followed where the dangers of potential mass

U.S. Forest Service

Figure 18.1
Heavy rains often cause loose soil and sediment to flow and slide down steep slopes. Mudslides similar to the ones shown here were especially severe in the San Francisco area in January 1982.

Figure 18.2
Collapse of this hillside and highway was triggered by the 1959 earthquake west of Yellowstone Park.

movement have not been fully recognized or efforts to meet the danger have been inadequate.

In areas of steep slopes and high relief, surficial deposits are often at or near their natural angle of repose. Such deposits are just barely stable and undergo frequent adjustments of slope by sliding or other types of downslope movement. When construction projects encroach on such areas, the results are often costly. Because most roads and other construction are located in valleys rather than on ridges, the chances are high that they may undercut a marginally stable deposit. When a cut is made into a loose deposit, sliding may be immediate; when the deposit is marginally stable, the deposit may stand until it is disturbed somehow—perhaps by an earthquake (Fig. 18.2) or by saturation of the deposit with water. Much of this damage can be avoided if the mechanisms that cause instability are understood, and if proper precautions are taken in design and construction of projects that involve modifying the land.

18.1 MATERIALS INVOLVED IN SURFACE INSTABILITY

All of the earth materials discussed in earlier chapters may become involved in movement on slopes or in subsidence and collapse. Because slope stability is critically important in many construction projects, it is one of the subjects in which civil engineers and geologists share an interest. Some of the terms used in the following sections are based on the broad categories of materials defined in civil engineering.

Rock is defined to include both solid unbroken materials and consolidated fragmental materials.

Earth is the general term used to describe disintegrated rocks and loosely consolidated sediments.

Soil is the product of rock disintegration and decomposition by weathering, modified by geological agents and capable of supporting plant life.

Mud is the mixture of water and finer-sized particles of earth and soil.

Debris is the general term applied to mixtures of rock, soil, plant matter, or mud.

Engineers do not ordinarily make distinctions among the different types of unconsolidated materials. They use the terms soil and earth to include all unconsolidated materials, such as those found in sand dunes, beaches, mud flats, river beds, deltas, and many other types of deposits. Geologists usually describe materials in terms of their mode of origin as well as their texture and composition. In the following discussions of principles, we will use the engineering terminology but we shall consider the nature of the deposits more carefully in case studies.

18.2 PRINCIPLES GOVERNING MASS MOVEMENTS

The force of gravitational attraction, acting throughout the materials of the earth's crust and directed toward the center of the earth, drives mass movement. The force of gravitational attraction is nearly uniform on surficial materials, yet some materials are so situated that they become unstable and move downslope, subside, or collapse, while others remain stable. Chief among the factors that influence stability are the slope and planes along which gravity acts; the nature of the material (particularly the presence of clays); the presence of water, ice, compressed gases, or steam in the material; and the presence or absence of stabilizing ground cover.

The Effects of Gravity on Masses on an Inclined Plane

The effect of slope on the downhill movement of solids can be shown graphically in terms of the forces that act on a solid block placed on an inclined plane (Fig. 18.3). The plane may be thought of as the surface of the ground, but the principles apply equally well to the downslope movement on any plane that provides a solid unyielding base over which the material above slides.

The weight of the block, which is the gravitational pull exerted by the earth on the block, is directed vertically down as shown in the figure. This attraction may be resolved into two components—one acting parallel to the inclined plane and the other perpendicular to the plane. If the block is not moving, the component of force directed down the plane is counterbalanced by frictional resistance to movement along the contact between the plane and the block. If the slope is steepened or the frictional forces between the plane and the block are reduced, the block may become unstable and slide down the plane, as shown in Fig. 18.3(*b*).

In nature, slope can be increased by erosion or uplift; or human activity may cause the increase by bulldozing or blasting away material. The friction between the block and the plane depends on the nature of the surface, the presence of lubricants, and the weight of the block. The slope required for sliding is steep when the friction is great, but movements can occur on low slopes when frictional resistance is low.

The plane along which mass movement occurs may be a fractured surface, a fault, a bedding plane, or the surface of the ground. One of the most common and potentially dangerous situations is that where steeply inclined strata or fractures dip in the same direction as the slope of the ground surface and are undercut in the valley by streams, glaciers, or highway construction. When undercutting penetrates massive rocks and cuts into shales or other rocks with less resistance to shearing, failure is likely because most strata are already fractured, permitting separation as part of the mass slides downslope. Water usually enters the zone of movement through fractures and lubricates the surface below the slide. The effect of water is particularly great if the slide zone contains

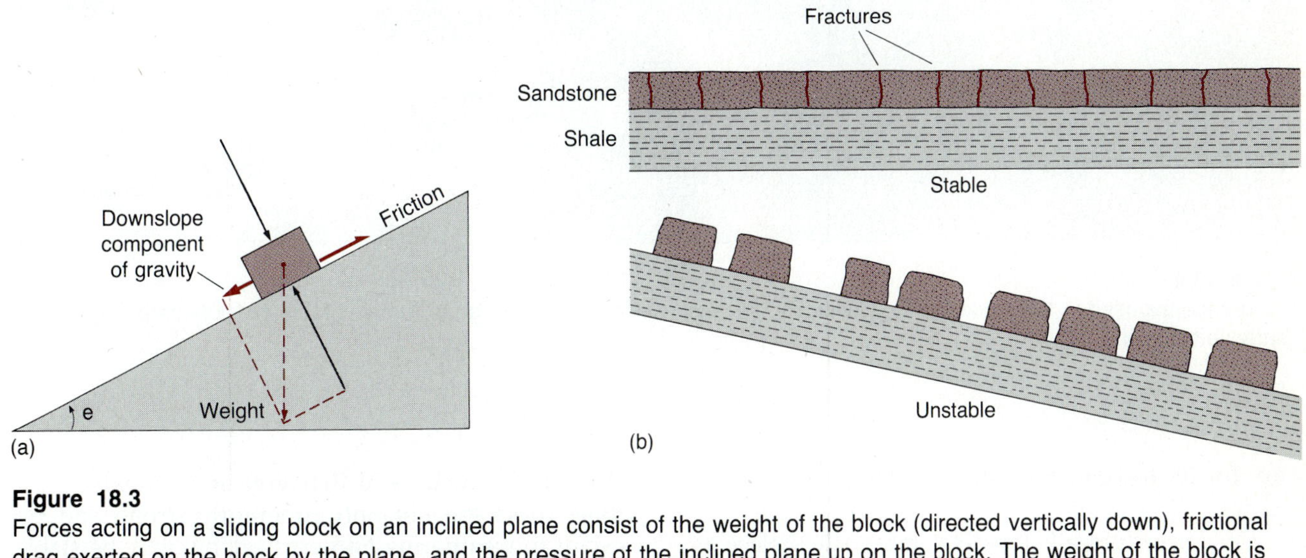

Figure 18.3
Forces acting on a sliding block on an inclined plane consist of the weight of the block (directed vertically down), frictional drag exerted on the block by the plane, and the pressure of the inclined plane up on the block. The weight of the block is shown resolved into two components—one perpendicular to the inclined plane and the second acting down the plane. This principle applies to any scale from centimeters to kilometers. (b) The block on an inclined plane situation described in (a) often occurs in nature. In the model shown here, a fractured sandstone or quartzite layer slides on a wet shale. This condition might be triggered as a result of a stream or a glacier, or a contractor cutting through the sandstone at the lower edge of the slope.

National Archives

Figure 18.4
The granite exposed in Yosemite Valley, California, is massive and strong enough to remain stable in 1000-meter-high, nearly vertical cliffs.

clay, for its resistance to shear drops sharply when it is wet.

Of course, not all slope instability takes place as rock slides down inclined planes, but the inclined plane situation makes it clear that gravity acts more effectively on steep slopes than it does on low slopes. In general, the steepness of the slope of the ground is a good indication of how stable it will be. But the character of the materials is another critical factor.

Physical Character and Behavior of Materials

Some rocks are not only inherently strong and resistant to weathering and erosion, but may be stable even in almost vertical cliffs. The deep valley at Yosemite Park, California, which glaciers cut into a massive granite, has relatively stable valley walls that are nearly vertical for hundreds of meters (Fig. 18.4). The mineral components of most such masses either form an interlocking network of crystals or are fragments tightly cemented together. The bonds between the minerals of such rocks are strong in all directions through the rock mass. This is not the case in most other rocks. Weakness in massive rocks like granite is usually the result of fractures that split it into blocks of various sizes (Fig. 18.5). Among metamorphic rocks, planes of weakness may also arise as a result of compositional layering or other types of mineral alignment.

The physical properties of sedimentary rocks are often uniform within layers, but vary greatly from one layer to another. These bedding planes are usually planes of inherent weakness. Some layers such as limestone or gypsum may be soluble; others (such as sandstone or conglomerate) may have open interconnected pore spaces in which water can accumulate; some, like shale, may expand or become plastic on wetting. All of these factors make the rock masses prone to downslope movement.

Unconsolidated sediments also differ widely in their physical properties and stability. Sediments composed of angular fragments and a mixture of sizes tend to be more stable on a steep slope than those that are rounded and uniform. Cemented materials are more stable than uncemented materials; dry sediment is more stable than wet sediment; and fragments with rough surfaces are more stable than those that are smooth.

In the sections that follow, it will become clear that very different processes are involved when slope failure occurs in cohesive masses of rock; in dry fragmental materials; in clays, which may absorb water and become plastic and slick; in water-saturated sediment; or in mixtures that are frozen together.

Canadian Pacific Railroad

Figure 18.5
The large blocks of rock shown here are bounded by fractures along which the rock separated. These blocks are part of a huge rock slide at Frank Mountain in Alberta, Canada.

Movement of dry fragmental materials. The behavior of dry fragmental materials is similar whether the fragments are sand or blocks of rock. Most of us learned a lot about this behavior playing in sand as children. We discovered that sand dribbled through the cupped hand forms a cone-shaped pile, and we might even have noticed that the slope of these piles is nearly constant for a given type of sand. That slope is known as the **angle of repose** of the sand. All sediment has a natural angle of repose, usually about 30 to 35°. In unconsolidated sediment, steeper slopes are usually unstable. Attempts to increase the slope of a material by adding more to the top of the pile or by removing the edge of the base of the pile almost invariably fail. Fragments roll and slide down the slope until the angle of repose is reached once again. Dry, angular blocks and mixtures of different-sized fragments have higher angles of repose than rounded, wet, uniform-sized particles. Thus, piles of rectangular blocks can sometimes be very steep.

When dry blocks are in contact with one another, the character of their surfaces influences the friction between them and thus also how difficult they are to move. For example, a wooden block moves easily across a polished table top but resists pushing across a piece of coarse sandpaper. The resistance that unconsolidated materials offer to shear-type displacements is referred to as the **internal frictional resistance.** Materials with high frictional resistance have high angles of repose and are generally more resistant to slope failure.

Rock fragments are produced by the physical disintegration of most rock types. Fragments are loosened and wedged off bare rock outcrops most effectively in high mountains, where temperature changes are often extreme, and where freezing, thawing, and frost-heaving are dominant weathering processes. In high mountains, large outcrops of rock exposed on steep slopes are also formed as streams and glaciers cut into uplifted rock masses. Erosion is so rapid in such areas that soil does not accumulate readily; and vegetation is sparse, allowing rock outcrops to persist.

When fractures are widely spaced, blocks of rocks several meters across may be formed, but usually the fragments are much smaller. Once a fragment is dislodged, it falls, rolls, or slides downslope, dislodging others as it goes. Finally, it comes to rest on a pile of similar fragments that accumulate on the slope. This type of fragmental material, **talus,** may accumulate in sheets (Fig. 18.6) or cone-shaped masses (Fig. 18.7) on steep mountainsides where the source of loose fragments is localized or where the fragments are channeled into a narrow zone. When talus is derived evenly along a cliff, it may accumulate as a steep, sloping sheet of loose rock banked up against the side of the valleys.

Converging piles of talus may coalesce and form a tongue-shaped projection down the valley if the slope of the valley floor is sufficient. These lobes of talus do move, but the movement is so slow that their motion can be detected only by checking the position of the end of the lobe over a period of years. They may extend considerable distances down valley, and because they resemble valley glaciers they are called **rock glaciers,** but they contain little or no ice.

National Archives

Figure 18.6
Fragments of rock, mostly 5 to 20 cm across, form a sheet of talus on this mountain slide. Here the rock fragments are breaking off of the irregular outcrops seen in the upper part of this photograph.

Figure 18.7
The cone-shaped pile of rock fragments shown here in the Alps is composed of rocks that are funneled through the narrow valley at the top of the pile. Some are washed into the pile; all move downslope as a result of gravity. The pile is more than a hundred meters high.

Most rock glaciers occur in high mountains where valley slopes are steep, as in Fig. 18.8, where freezing and thawing occur repeatedly, and where both ice and water may become incorporated with the masses of freshly broken rock that make up both talus and rock glacier accumulations. Rock glaciers may resemble ice glaciers, but they differ in that ice glaciers are composed mainly of ice and are formed from accumulations of snow. Rock fragments may fall onto the surface of an ice glacier as it moves down through a valley in a mountainous region, giving the glacier a superficial cover of rock. In contrast, rock glaciers are slowly moving masses composed mainly of rock fragments throughout.

Movement of wet materials. Streams, glaciers, and dry mass movements are all driven by gravity. One might reasonably expect to have little difficulty in distinguishing a stream from a glacier, and either of them from a dry mass movement, but the distinction may not be so clear-cut. Glaciers carry heavy loads of rock, soil, and debris, as do streams; and mass movements usually contain some proportion of water or ice. Thus a gradual transition from the flow of ice to stream flow and to the flowage of completely dry masses of rock and debris exists in nature.

We are accustomed to thinking of flow in terms of behavior of water, but water is not an adequate

U.S.G.S.

Figure 18.8
Talus from the cliffs around the sides of this valley in Colorado have combined with deposits on the floor of the valley to form a rock glacier.

model for all types of such movement. Water, like all true liquids, offers little resistance to shear. We can get an idea of this resistance to deformation by shear by considering the resistance offered to the movement of one piece of glass over another separated by a thin layer of liquid. Water has a very low **viscosity,** or resistance to shear deformation; one sheet of glass will slide freely over the other. Molasses has a relatively greater viscosity; with it between the sheets of glass, considerable effort is needed to shift them. When a liquid flows, it typically moves in one of two ways—either as **laminar flow,** which can be described as the slippage of one layer over another, or as **turbulent flow,** in which the water moves in irregular patterns, often loops or swirls as in most fast moving streams. The problem in applying these simple models to the flow of most materials is that rock, mud, and soil all contain more solids and are much more heterogeneous than water or molasses.

Some water is usually present in the ground, and an excess of water is frequently important in causing the ground to become unstable. A number of different mechanisms are involved, but the lubricating effect of water is perhaps most important. When water accumulates in surficial materials, it fills the pore spaces, adding weight and at the same time reducing the friction between the particles of the materials and increasing the tendency of the whole mass to flow.

As the soil or debris on a slope becomes wetter, the clay fraction absorbs water; the contacts between solid fragments are lubricated; and, eventually, if water content becomes high, the fragments are buoyed by the water. Under these conditions, the mass flows much like a viscous liquid. Its internal frictional resistance decreases. **Mudflows** of the types that caused extensive property damage and the loss of thirty lives near San Francisco in 1982 (Fig. 18.1) are examples of this type of mass movement.

Mudflows are most frequently found in dry areas, especially deserts, where the fine particles formed by weathering are dry most of the year. When rain does fall, it often comes in large quantities, perhaps with much or all of the total annual rainfall falling in a single storm. Water seeps into the weathering products and saturates them; and the mixture begins to move, gathering momentum as it moves down steep slopes along existing channels and gullies. Velocities may be very rapid, as much as one or more meters per second. Although a mudflow is certainly not dry, the mass does not necessarily contain much free water, and it may be thick enough to carry large boulders in suspension. Mudflows are favored by intermittent water supply, lack of vegetation, and abundant unconsolidated rock debris.

Soil flowage, called **solifluction,** and the downslope movement of soils, rock debris, and other water-saturated fragments over a frozen base frequently occur in climates where the ground is solidly frozen in the winter and thaws only partially during the summer months. When thawing occurs, upper layers of the soil, which have been forced up by frost-heaving, are bathed in meltwater, but the lower layers of the soil remain frozen. Water lubricates the interface between frozen and thawed layers and facilitates slow movement of the upper layers over the lower frozen ones, even on low slopes. The moving mass of soil and debris takes sheet, lobe-shaped, or tonguelike forms as it moves, following the slope of the ground to lower elevations.

Plastic behavior. Clays and mixtures of sediment containing a high percentage of clay exhibit plastic behavior when they are wet; beds of salt are plastic even when they are dry. A **plastic** material differs from most others in that it appears solid, but if it is subjected to pressure—if it is loaded—above a certain limit, it will gradually deform, as diagrammed in Fig. 18.9. This yielding takes the form of a solid flow—the kind of flow a block of asphalt shows in summer.

If clay minerals are present, the plasticity of soil or sediment can be increased by adding water to it. Water seeping down along fractures or through pore spaces into a claystone or mudstone may induce plastic flow. The failure of one of the largest earthen dams ever built, the Fort Peck Dam in Montana, was caused by swelling and flow of clays under the dam when water accumulated in the reservoir and seeped down into the clay layers, causing them to swell and flow and the dam to give way. Shale can be altered to clay by water, and limestone can be dissolved by ground water, leaving clay residue. Fault breccias are subject to similar weathering effects.

Plastic flow can be initiated when construction adds weight at the surface, as has happened in road-building over salt beds in the salt basins of Utah and over clay beds in the Coastal Plain. During construction of the Panama Canal, the clays under the canal became plastic and began to flow when dredged material was piled along the sides of the canal.

Movement of frozen materials. In middle and high latitudes and at high altitudes, the upper part of the soil or sediment contains ice during part or all of the year. Freezing may extend to depths of several meters in regions of permanently frozen ground—**permafrost**—and alternate freezing and thawing of the soil may occur daily where the ground is exposed to sunlight.

The expansion that takes place when water freezes dislodges rock, breaks up and loosens soil, and heaves the soil upward where layers of ice form, as described in Section 17.2. On a slope, the upward heaving moves the soil up perpendicular to the slope, but when the ice melts, gravity pulls the soil straight down (Fig. 18.10). Thus, each cycle of freezing and thawing moves the soil slightly downslope. Because

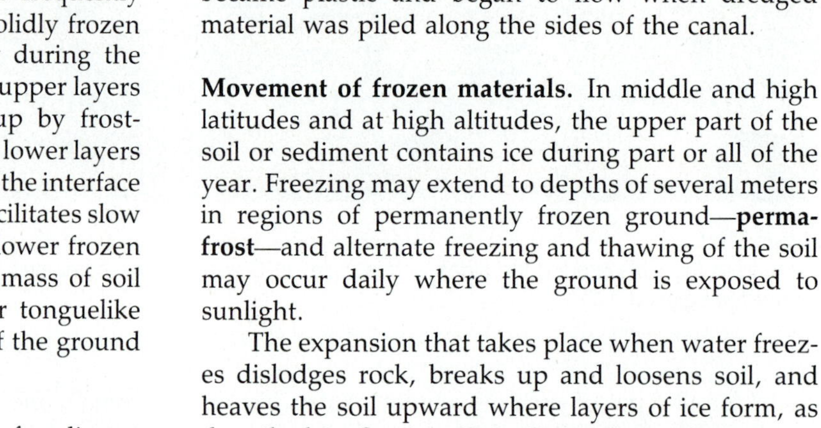

Figure 18.9
Plastic materials, like these shown here, behave elastically (that is, they return to their original shape when the weight is removed) if the pressure applied is below the yield point, but deform continuously when the pressure is at or above the yield point.

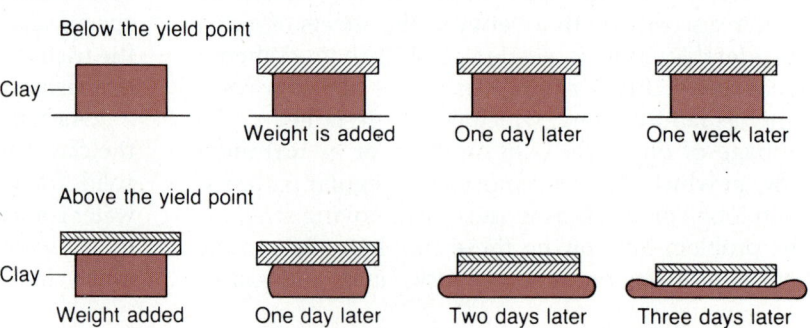

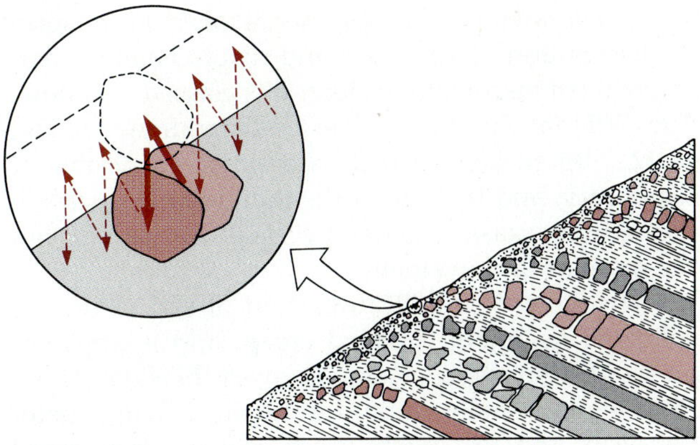

Figure 18.10
Freezing and thawing of ice in the upper layers of soil are responsible for slow downslope movement of the soil. Particles in the soil follow the path shown in the enlarged section as a result of repeated freezing and thawing. The enlarged section represents the upper four centimeters of soil.

the soil is bathed in meltwater as it sinks, any clay in it may become plastic and induce flowage.

Glaciers, whose mass remains below freezing year round, also move by a kind of solid flowage that is described in Chapter 24. Any mass of material that contains substantial amounts of ice can yield in a similar manner.

18.3 TYPES OF MASS MOVEMENT

We have already made reference to several important types of mass movement in the course of the discussion of principles. In this section other important and common types of movement are introduced, and distinctions between types of mass movement are given.

If the mass in motion travels most of the distance through the air, the movement is called a **fall.** Included are free falls, bounces, and rolling of rock and debris fragments without much interaction among fragments. Talus accumulations commonly result when falls continue to occur at one site for a long time.

Slides, movements caused by shear failure along one or several surfaces, may occur suddenly, or they may entail slow or intermittent movements continuing for years. The movements may be controlled by surfaces of weakness such as fractures, faults, or bedding planes. Movement may be a **block-glide movement** (Fig. 18.11*a*), in which the units remain more or less intact and move along a more or less

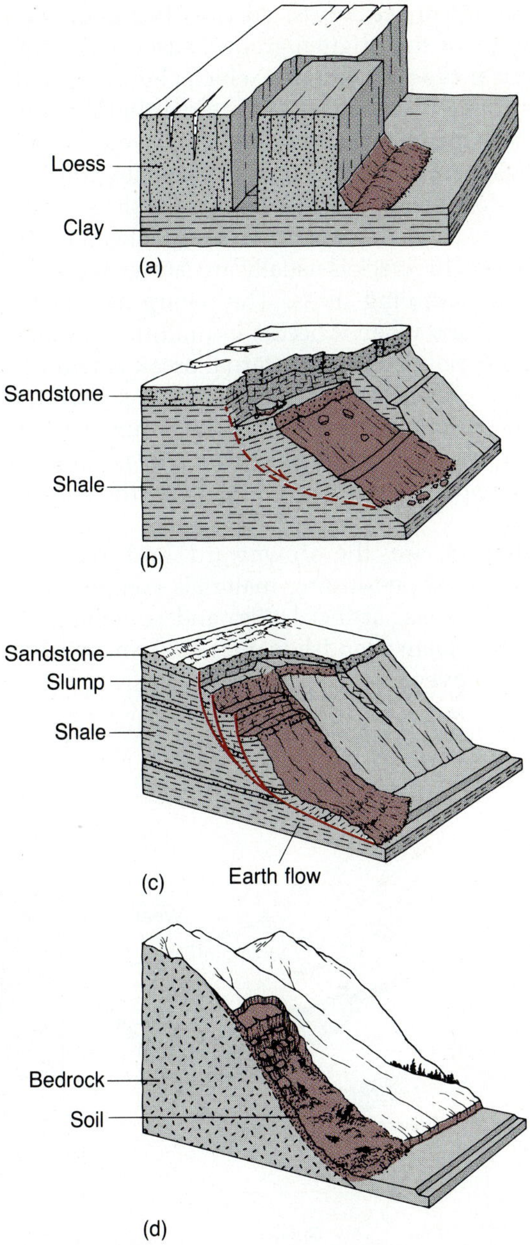

Figure 18.11
(a) Block-glide involves the slow movement of blocks over a base with a low coefficient of friction—loess over clay in this case. (b) A rotational movement takes place when a slump occurs. Movement is at slow to moderate rates. (c) Slump and earth flow frequently are combined as shown here. (d) Sliding of debris—rock, soil, and clay mixtures—over solid bedrock may occur slowly or very rapidly. All of these features may occur in a wide range of scales commonly from tens to hundreds of meters.

planar surface of weakness, such as bedding, fractures, faults, or the original ground surface. **Slumps** are a special type of slide, in which the movement involves rotation of a mass of material that remains more or less intact as it slides along a curved internal break (Figs. 18.11*b* and 18.11*c*) that is usually concave upward. Movement occurs on an internal slip surface, appearing as a crack in the ground at the head of the slump mass. This crack is usually arcuate and concave toward the slumping mass. The slump assumes a spoonlike shape when it occurs in uniform, granular material. Friction along the internal break is reduced when water flows into cracks at the head of the slump. As the slump mass rotates, the ground surface on the slump block is tilted, and water is directed toward the open cracks, funneling it into the zone of movement.

Landslides are the downward and outward movement of slope-forming materials composed of natural rock, soils, artificial fills, and combinations thereof. Material in a landslide does not remain intact during the movement.

If the moving mass of material takes on the form or distribution of velocities and displacements characteristic of viscous fluid flow, it is classed as a **flow.** The different varieties of flow are diagrammed in Fig. 18.12. Slip surfaces usually are either not visible or temporary; and the size of the individual particles in the moving mass is generally small compared with the amount of movement.

The slow downhill movement of soil or unconsolidated sediment is called **creep,** and is apparent on close inspection of almost every hillside. These movements usually amount to no more than a meter per year and are often much slower. Weathered remnants of rocks and boulders are drawn out into long, lens-shaped masses. Even fences and telephone poles set on slopes give indication of these surface movements as they slowly tilt downhill. Creep takes place in part as a result of the kind of freezing and thawing cycle described in the preceding section, the rotation of particles in the soil, the drawing out of plastic materials, and other types of internal readjustments in surficial materials. The process is often

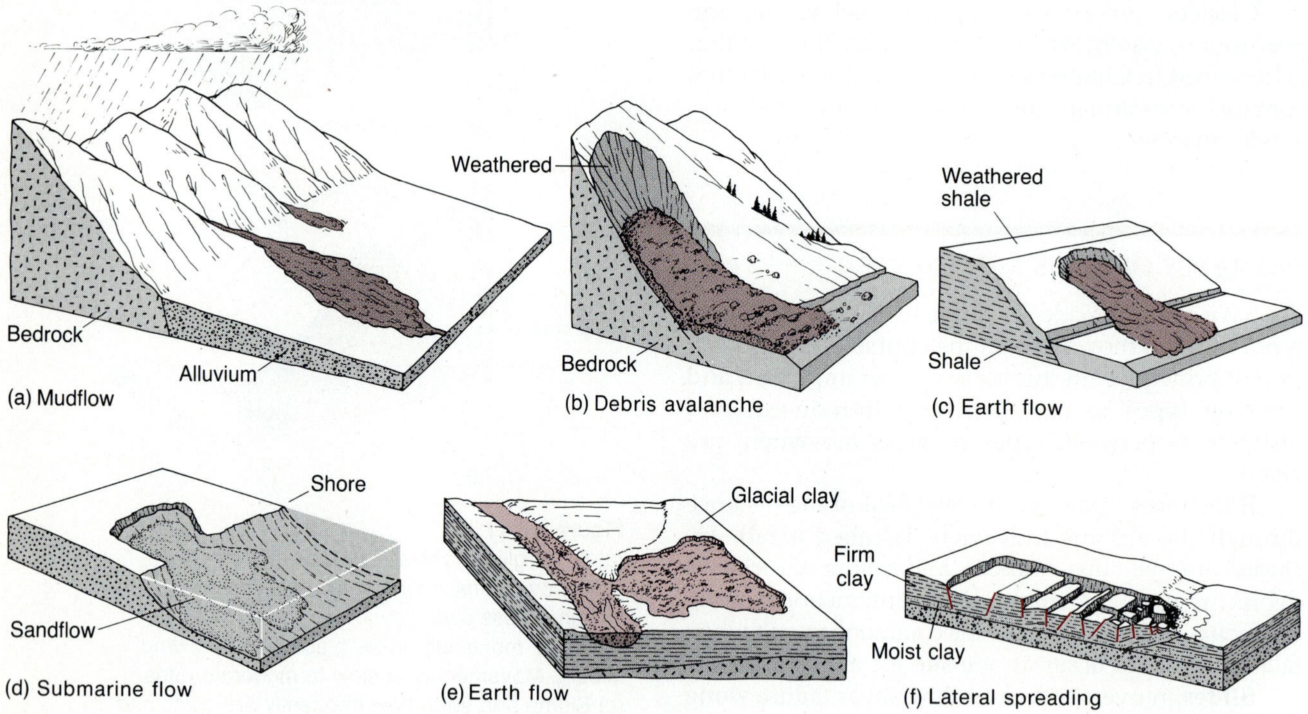

Figure 18.12
(a) A mudflow involves a very rapid movement in which there is a high water content in materials, usually consisting of a mixture of decayed rock and soil. (b) In the flow of a debris avalanche, dry to moderately wet debris moves rapidly over bedrock. (c) Earthflow in this case involves flow of wet weathered shale over solid unweathered shale. (d) When disturbed, water-saturated sand or silt may flow rapidly underwater. (e) Earthflow may occur in wet soil, such as the glacial clay and silt shown here. (f) Lateral spreading occurs most commonly and most rapidly in soft wet clay as depicted here.

Figure 18.13
The soil on this hillside is slowly creeping downslope. What started as cow paths around the hill are now rolls of soil. The fence posts provide a scale.

accelerated by animals. Small, parallel rows, originally paths used by grazing animals, circle many hillsides and show this creeping downslope movement (Fig. 18.13).

18.4 SUBSIDENCE AND COLLAPSE

The gradual lowering of surface materials into space opened as a result of removal of underlying rock is known as **subsidence.** If the displacement is more sudden it is referred to as **collapse.** Both subsidence and collapse are important where removal of rock or fluids such as water or oil takes place on a large scale. Underground mining is perhaps the most obvious potential cause of subsidence, and while it has caused much property damage it is neither as widespread nor as significant a source of damage to the ground surface as is the withdrawal of water or the solution of rock by groundwater. Although most mines involve small areas, mining has caused serious problems in some places. A few mines, however, are now using this collapse potential for profit. The famous copper mines at Butte, Montana, include a network of tunnels estimated at thousands of kilometers long. In recent years, the mining company has taken over the surface above these tunnels and is mining by a method involving controlled subsidence and collapse. The broken-up rock is allowed and encouraged to collapse into an area from which it can be removed through a tunnel and shaft.

Where surface rights and mineral rights are owned separately, mining has often produced more serious consequences. Removal of coal seams under some communities has caused surface property, including houses and other buildings, to subside and in a few cases to collapse (Fig. 18.14). Unless large blocks of coal are left underground, the roof of a shallow mine may subside, lowering the overlying rock and the ground surface. Obviously, it is to the advantage of the mining company to remove as much coal as possible, but it may be necessary to leave as much as a third of it underground as large pillars to support the roof. In a few instances, as in Scranton and Centralia, Pennsylvania, uncontrolled fires in coal seams underground have almost completely removed the coal, resulting in subsidence of the surface and much physical damage to structures located at the surface.

In areas underlain by gypsum ($CaSO_4 \cdot 2H_2O$), salt ($NaCl$), and anhydrite ($CaSO_4$), subsurface water can actively dissolve and remove rock, effectively undermining the surficial materials. Solution may be sufficient to induce large-scale subsidence or collapse. If the rock being removed is strong or overlain by rock that is strong, a large cavity may be developed before failure takes place. When failure does occur, it is likely to be sudden and consist of the collapse of the

U.S.G.S.

Figure 18.14
Subsidence pits resulting from collapse of the surface into voids left by underground coal mining in Wyoming. The mine was abandoned in 1914.

roof into the cavity. The broken rock forms a coarse breccia.

If the underlying rock is unconsolidated or semiconsolidated, withdrawal of water, oil, or gas may cause the surface to subside. Sometimes small localized collapse or subsidence results from the blowout of a high-pressure gas pocket or collapse of a cave, but subsidence of a large area occurs more frequently. In the Los Angeles area, where careful level surveys reveal the amount and direction of movement by markers, subsidence is associated with active oil fields. In the Wilmington field, subsidence of nearly 16 feet took place between 1928 and 1951. This type of subsidence appears to be caused by the withdrawal of fluids and accompanying compaction of the sediment from which they were drawn.

Regional subsidence has also taken place in the San Joaquin valley of California, where over three thousand square miles have been affected by the rapid drawdown of groundwater levels to support agricultural activity. Drawdowns amounting to several hundred feet have been accompanied by surface subsidence of 13 to 26 feet or more. Many problems go along with this type of subsidence, especially in coastal areas. Buildings may tilt and crack, and areas near sea level may become wet and subject to flooding. All forms of fluid transport, such as canals, irrigation systems, and pipelines, are affected. The solution to this problem is relatively simple in concept—stop pumping and recharge the aquifers. But this is not easy to do, because the rapid drawdown was caused in the first place by water shortages, and alternate sources of supply are still not readily available.

Subsidence also occurs in certain soils that collapse when they become very wet. The wet soil particles become more closely packed, reducing the volume of the soil and making it more compact. This problem is also acute in some parts of the San Joaquin valley.

18.5 CASE HISTORIES

Madison Canyon Landslide

About midnight on August 17, 1959, a severe earthquake described earlier (Section 8.4) shook the region west of Yellowstone Park. Surface waves rocked the water in nearby lakes, threw the soil up in waves, and cracked the highways and ground surface over a large area. This movement was caused by displacement along a fault that can be traced near the narrow and steep canyon of the Madison River. The earthquake triggered a movement in decayed soils and rock debris on the side of the canyon, which is 400 meters deep where it passes through this high mountainous region. In the chaos that followed, about 27 million cubic meters of rock, soil, and trees slid off into the canyon, leaving a scar from the top of the mountains almost to the river (Figs. 18.15 and 18.16).

U.S.G.S.

Figure 18.15
This gigantic landslide, triggered by an earthquake in Montana, dammed the Madison River.

U.S.G.S.

Figure 18.16
Aerial view of the top of the Madison slide shows cracks and displacements typical of many slumps and slides.

Debris engulfed the valley, the highway, a campground, and 28 campers spending the night there. It buried an area of about 135 acres under a layer of rock estimated to be 45 meters thick. The movement was extremely fast—perhaps as great as 96 kilometers per hour. The mass forced air in the valley aside, setting up a gale force wind. When the slide hit the river, it splashed water out, leaving evidence of mud in trees high up on the sides of the valley and sending a wave more than 1.5 kilometers up the river. Material broken loose moved with such impetus that it continued up the opposite side of the valley wall for 122 meters. Material in the slide ranged in size from blocks nine meters across to fine soil particles.

The potential for mass movement in the Madison Canyon had existed for many years. Slopes had been oversteepened to angles of 40 to 60 degrees as a result of erosion by the Madison River. Deeply weathered Precambrian gneisses and schists were exposed on the slopes (Fig. 18.17). These deep soils were held on the slope by a massive layer of dolomite (the calcium-magnesium-carbonate rock), which is resistant to weathering in semiarid climates and which outcropped at the bottom of the valley. All of the rocks—the gneiss, schist, and dolomite—sloped toward the river on the south side of the valley and were fractured. When the vibrations caused by the earthquake movements began, blocks of the dolomite broke loose, releasing the main mass movement.

The Vaiont Reservoir Landslide

The Vaiont Reservoir (Fig. 18.18) lies in the Alps, where intense deformation has folded, fractured, and faulted the sedimentary rock units. Mountain-building was followed by glaciation, which carved deep U-shaped valleys in the rock masses (Fig. 18.19). Previously formed fractures near the ground level opened, and new fractures probably developed as the ice melted, releasing the pressure due to its weight. The floors of the U-shaped valleys were subsequently cut down to form a narrow and deep inner V-shaped valley, and new release fractures formed along the sides of this valley. These events led to the development of the structural and topographic conditions depicted in cross section in Fig. 18.20.

The combination of a particular rock structure and surface processes created the conditions for the catastrophic landslide that occurred on Mount Toc. On October 9, 1963, 240 million cubic meters of rock and soil slipped from the side of Mount Toc near Longarone, Italy into the reservoir of the Vaiont Dam, the world's second highest dam. The reservoir filled for a distance of almost 1.6 kilometers, and waves estimated to be 244 meters high were set up. The mass of debris moved at rates of 25 to 30 meters per second, so fast that a blast of highly compressed air was forced into the valley; the water from the reservoir poured over the dam, destroying the town of Lon-

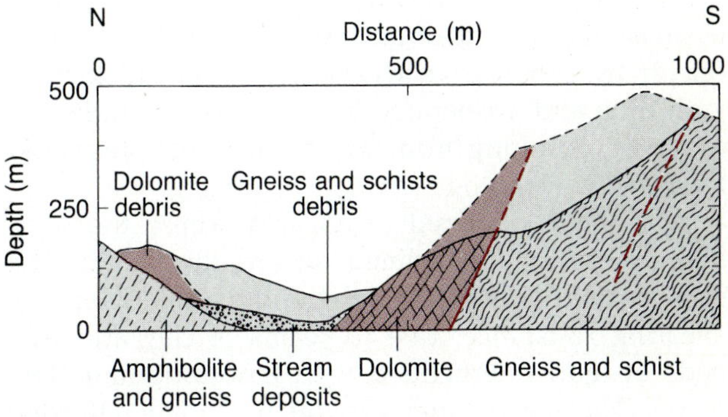

Figure 18.17
Profiles of the Madison Canyon before and after the slide show the massive changes that occurred. Weathered gneisses and schist were held on the mountainside by a steeply inclined bed of dolomite.

Figure 18.18
Sketch map of Vaiont Reservoir in Italy, showing the 1963 landslide, which created waves that overtopped the dam and caused flooding and destruction over large areas downstream.

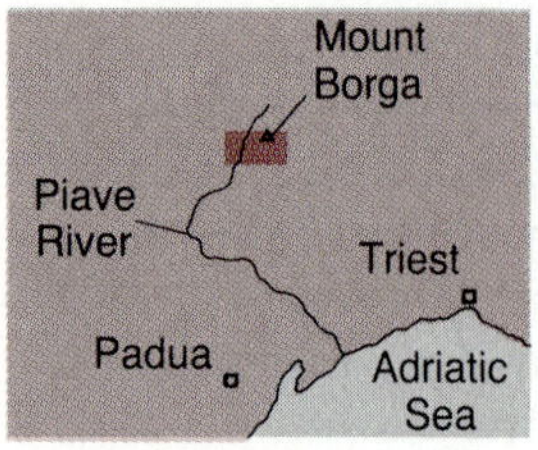

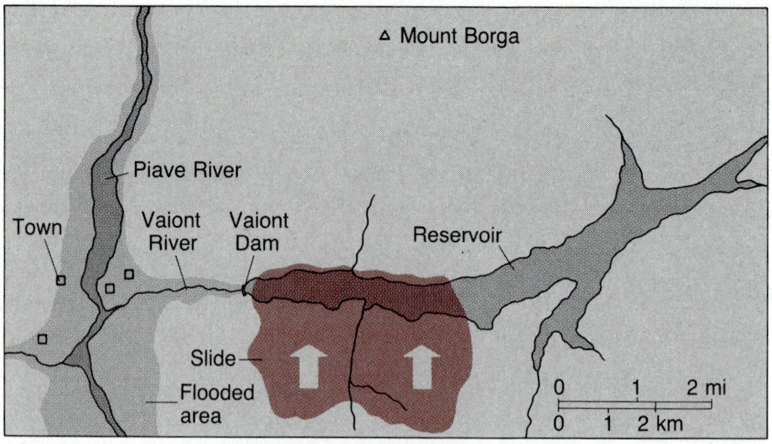

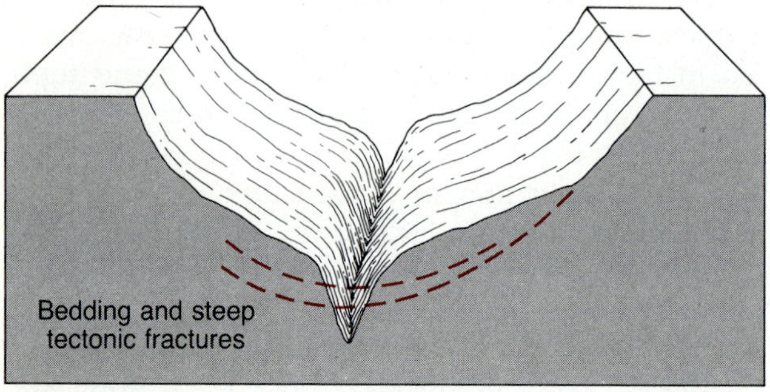

Figure 18.19
On this sketch of the Vaiont Canyon are shown remnants of the outer glacial valley and two sets of fractures formed by stress relief within the walls of the valley to depths of 100 to 150 meters.

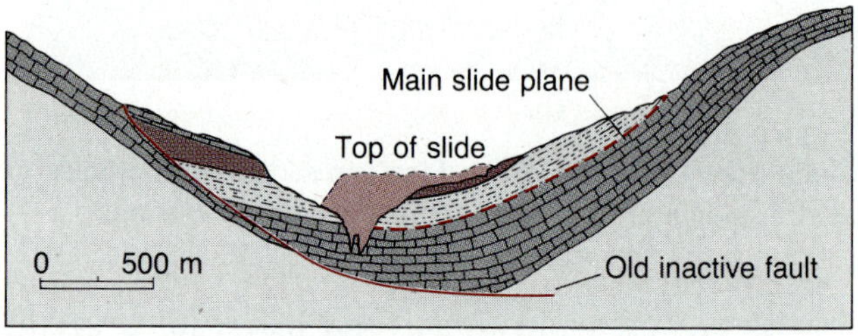

Figure 18.20
Geologic cross section of the Vaiont slide and reservoir canyon, extending from north to south, showing principal features of the slide plane and rock units.

garone, where 2600 lives were lost within minutes of the slide.

This disaster, the worst of its type, was caused by mass movement of an unstable mountain slope, set off by progressive weakening of the rock mass with time and accelerated by excessive groundwater recharge, resulting from heavy rains and the water level in the reservoir.

The slide occurred in an area where the sedimentary rocks are inclined toward the valley. The rock units involved are inherently weak with low shearing resistance, due to seams of clay and thin beds of marl interbedded with limestone and claystone. The steep inner canyon no longer afforded support for the units, which were completely eroded through by the combined effects of ice and streams. Thus, the rock masses above the zones of low shearing resistance became free to move downslope. Furthermore, the whole mass had been substantially weakened by the open fractures. The open fractures and the presence of limestone led to the development of extensive solution cavities, caves, and tubes, and a system of surface depressions—sink holes—formed on the upper slopes of the valley. These acted to catch surface water and direct it underground, where it lubricated the clay zones, weakened bonds, and increased hydrostatic uplift.

All of the above conditions existed before the construction of the dam. But as the reservoir filled, water moved into the open fractures along the sides of the inner canyon, artificially raising the groundwater level. The level of the reservoir rose about 20 meters in the months immediately preceding the slide, partly due to heavy rains in August, September, and October, suggesting that water levels were important in upsetting the equilibrium.

Landslides have been common in the Vaiont Valley. A large slide of about a million cubic meters had taken place three years earlier, in 1960; this movement was accompanied by the development of creep over a large area and by the opening of ground cracks upslope from the slide. These cracks ultimately outlined the October 9 slide (Fig. 18.18). The slide area moved very slowly throughout the spring and summer of 1963; a rate of 1 centimeter per week was measured. About September 18, some of the survey stations constructed after the 1960 slide started moving 1 centimeter per day. Heavy rains started September 28 and continued until after the landslide. About October 1, animals grazing on the north slopes of Mount Toc moved away. At first it was thought that only isolated blocks were moving, but as the rate of movement over the large area accelerated from 1 centimeter per day in September to 80 centimeters per day just before failure, it became apparent that several stations were moving together and that the motion involved an area five times larger than initially estimated. On October 8 the outlet tunnels of the Vaiont Dam were opened in an attempt to lower the level of the reservoir, but the amount of rainfall was too great and overbalanced this outflow.

A witness to the slide, whose house stood 255 meters above the reservoir on the opposite side from the slide, reported that he was awakened at 10:15 P.M. by the sound of rolling rocks, not uncommon in this area; but the sound grew louder, and at 10:40 P.M. his house was struck by a strong wind that broke the windowpanes. The house shook violently as the roof of the house was lifted; rocks and rain fell in; then the roof collapsed, and all was quiet. A massive wall of water 69 meters high at the mouth of the Vaiont Canyon buried Longarone at 10:43 P.M., destroying everything in its path. Astonishingly, the Vaiont Dam, a thin-arch concrete structure, withstood the force of the water and did not fail. By 10:55 P.M. the flood had receded.

Panama Canal: Slides, Slumps, and Plastic Flowage

Construction of the Panama Canal (Fig. 18.21) was plagued by many costly earth movements. At times they were so extensive and costly to remove as to raise doubts about the feasibility of completing the canal. They eventually bankrupted the company that originally contracted to build the canal. Slumping of the sides of the canal was particularly bad where the canal was cut through the Cucaracha formation, a sedimentary layer partially composed of a highly plastic clay. Some slides were caused by cutting the sides of the canal too steeply; material slid when the slope exceeded its angle of repose. Conditions worsened when the sides were cut down to lower slopes, because the material removed from the cut was dumped just above the top of the cut. The additional weight of this debris at the top of the unstable materials only accelerated the movements. When the canal was dredged to deepen the channel, the mixture of water and clay was pumped up on the sides, often into small channels cut to drain the mixture away. The weight of this mass compressed the clay in the underlying formations, squeezing the plastic clay and causing it to move until it came to a point where the pressure was less—this point, of course, was where the canal had been cut. Water in the dredged material facilitated this plastic behavior. So lumps of clay continued to rise in the canal. This process continued until most of the clay layer was removed.

Movements Triggered by the 1964 Alaskan Earthquake

Anchorage. The Good Friday earthquake of 1964 (Section 8.4) initiated extensive mass movements in south central Alaska that cost over a hundred lives and hundreds of millions of dollars in property damage. Most of this damage was caused by movements in the unconsolidated sediment on which the city of Anchorage is built. These effects were described by Gordon Oakeshott (*1*):

> The city [Anchorage] lies on a glacial outwash plain which slopes from the base of the high and rugged Chugach Mountains down to the sea, where it drops off abruptly along a series of bluffs of seacliffs about 30 to 50 feet above sea level. Under the entire city is the flat-lying formation called by geologists the Bootlegger Cove clay. The Bootlegger Cove clay is gray clay, silt, and mud deposited from melting glaciers in lagoons during the Pleistocene Epoch [the last 2 million years of earth history]; it ranges in thickness from about 100 to 250 feet and lies only a few feet below the land surface under most of Anchorage. When water-saturated, as it is in the Anchorage area, this material completely lacks structural strength. When the tremendous shaking of the earthquake occurred, the Bootlegger Cove clay failed partly by sliding out toward the waters of Cook Inlet on slip-surfaces of especial weakness, and probably also partly by "liquefying" to flow at low

(a) National Archives

(b) National Archives

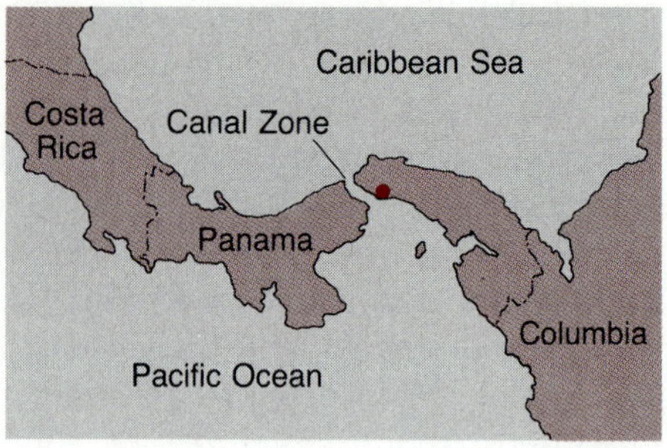

(c)

Figure 18.21
The steep walls of this cut, the Culebra cut, were the site of some of the largest slides and slumps encountered during the construction of the Panama Canal.

angles toward the waterfront . . . Great landslide cracks opened back of the bluff [Fig. 18.22]; some of the houses were carried toward the sea as far as 500 feet. About 200 acres of the terrace surface broke into a series of blocks each a few feet across, tilted at various weird angles to form a chaos of blocks of the bluish-gray Bootlegger Cove silt and clay, brownish peat and muskeg, black soil, and road pavement [Fig. 18.23].

. . . In front of some of the advancing blocks compressional "pressure ridges" were formed. Most of the damage in Anchorage was of this type; that is, due to direct landsliding and to landslide fractures, cracks, and in some cases, to the pressure ridges built up along the fronts of the slides.

The Sherman landslide. The Sherman landslide was another of many movements triggered in surficial materials by the Alaska earthquake of March 27, 1964. This one is of special interest because the landslide debris moved across and covered a large area of the

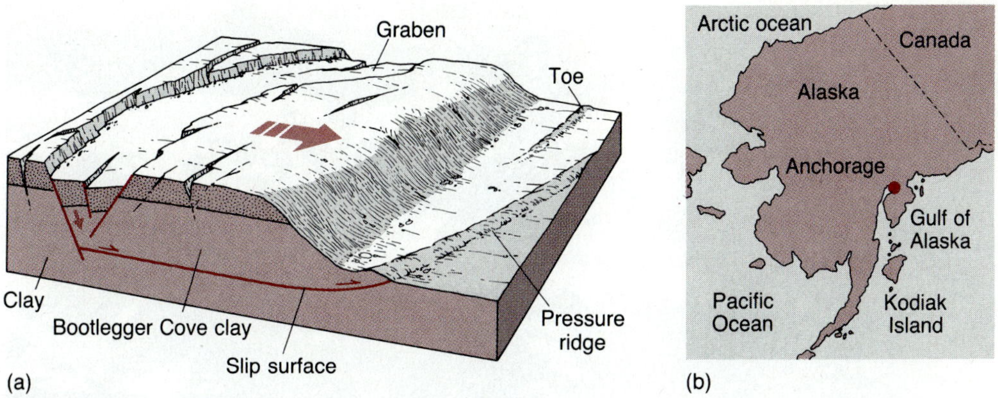

Figure 18.22
(a) Failure of the ground in Anchorage, Alaska, during the March 1964 earthquake was caused by lateral spreading of a wet clay layer beneath the city. As the clay spread, pressure ridges formed where a slip surface came to the ground surface, and grabens formed where the moving clay pulled apart. (b) Locator map.

Sherman glacier, and because the debris may have traveled on a layer of trapped and compressed air, a variety of slide that is known as the **Blackhawk-slide type.**

When the earthquake occurred, a massive slab of sandstone and shale that stood as a steeply dipping, pervasively jointed mass on the side of a steep peak was shaken free along a bedding surface. The mass of loose rock 450 meters long, 300 meters wide, and 150 meters deep plunged down the 40° slopes of the mountainside and slid up to 5 kilometers across the snow-covered surface of the Sherman glacier (Fig. 18.24). The debris spread out in an irregular lobe 1 to 3 kilometers wide with a remarkable uniform thickness of 3 to 6 meters. Despite the thinness of the layer, it was traveling with sufficient momentum to climb 25 meters up the opposite valley wall, 5 kilometers from its source.

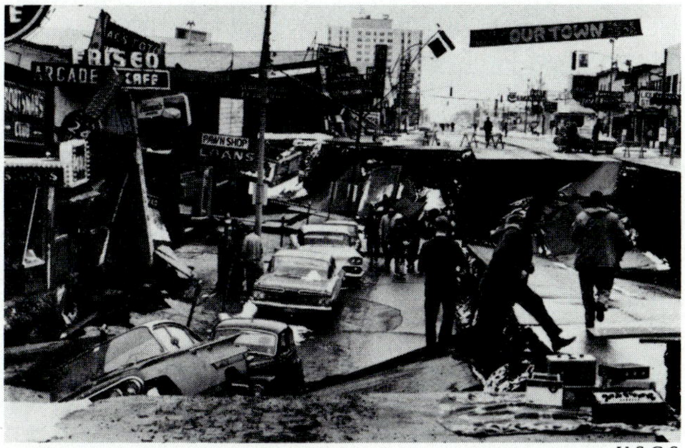

U.S.G.S.

Figure 18.23
Ground disruption and damage at the upper part of one of five major landslides that developed in Anchorage, Alaska, during the 1964 earthquake.

A number of the features of this slide throw light on its mode of origin. For example, the topography of the suface underlying the slide is faithfully revealed by the surface of the slide. The surface of the slide contains distinct compositional bands. The outer edge of the slide forms a rim where the slide debris apparently plowed into the glacier surface; ice and snow make up much of this rim, which is up to 15 meters high, 15 to 150 meters wide, and characterized by a chaotically hummocky, lumpy topography. Where crevasses have opened through the debris, its lower boundary can be seen to be sharp, with the angular boulders of the slide material in contact with ice or evenly stratified snow.

The surface of the slide is marked by longitudinal grooves up to 8 meters wide, 2 meters deep, and 600 meters long. Some of these end at boulders, indicating that the movement of the boulders may have formed them; but more commonly there are no such boulders. The origin of such grooves is uncertain, but they might be splits in the slide sheet caused by the lateral spreading of the sheet as it advanced. Transverse fissures, many concave toward the outer edge of the slide, mark the surface of the slide mass. Among the most unusual features of the slide are small cones of

U.S.G.S.

Figure 18.24
The Sherman landslide in Alaska formed in March 1964, as 30 million cubic meters (a volume equivalent to a 1000-foot cube) of rock dropped 600 meters and moved laterally 5 kilometers, partly over the Sherman glacier. The longitudinal grooves are approximately 8 meters wide and average 600 meters in length.

debris on top of boulders, as though the debris were dropped onto the slide and accumulated only above the larger boulders.

The low temperatures in the area rule out the possibility of lubrication of the slide by water and mud; however, dry snow or a mixture of snow and air are possibilities in the Sherman slide. In a report on the slide, R. L. Shreve (2) proposed that compressed air, trapped below rapidly falling rock as it reaches a break in slope, formed a lubricating layer on which the slide rode. If air were a primary lubricant, then the loss of air at the leading edge would account for the pronounced increase in friction there and for the snow consequently plowing up into ridges. Lubrication of some sort is required to explain the sheet's high speeds (80 and 180 kilometers per hour) and thinness.

SUMMARY

The downslope movement of masses of rock and soil occurs under the ever-present influence of gravity. Whether it occurs sooner or later, and slowly or calamitously fast, is determined by the slope of the surface, the degree of consolidation of surface materials, the presence of lubricating water or ice, and the presence or absence of stabilizing ground cover.

Slides and flows most often occur where a slope has been steepened beyond its stable angle of repose, either by natural erosion or by human construction projects. The breaks along which slides occur are often between strata in sedimentary rock, but they may also be fractures in igneous rocks.

Talus accumulations on high mountain slopes and in valleys are typical of dry sediment falls. They may flow down mountain valleys like exceedingly slow, dry glaciers. Mass movements of wet material may take the form of glaciers, mudflows, or solifluction, depending upon the climate and materials involved. All these are characterized to some extent by a motion like viscous fluid flow.

Some most disastrous slides and flows are caused by the yielding of plastic materials, such as salt deposits and, especially, wet clay under continued pressure.

In regions where the surface freezes, the repeated freezing and thawing may loosen rock and soil, contributing to the slow downslope creep of soil.

Slides are classified according to the break surfaces along which they move. Block-glide movement takes place along a more or less planar surface. Slumps involve a rotational movement along a curved surface, usually concave upward. They are often lubricated by the flow of water into cracks at the slip surface's end.

Where underlying rock has been removed by solution or by human mining or pumping of ores, petroleum, or water, the level of the ground surface either subsides gradually or collapses precipitously into the space below. Some of the broadest regional subsidences are caused by the rapid drawing of water from underground deposits.

KEY TERMS

angle of repose
Blackhawk-slide type
block-glide movement
collapse
creep
debris
earth
fall
flow
internal frictional resistance
laminar flow
landslide
mass movement
mass wasting
mud
mudflows
permafrost
plastic
rock
rock glacier
slide
slump
soil
solifluction
subsidence
talus
turbulent flow
viscosity

STUDY QUESTIONS

1. What information would you need in order to determine the stability of a claystone to be used as a foundation for a heavy pier?
2. Prepare an outline of the characteristics of each major rock type that may have a pronounced effect on slope stability (for example, slaty cleavage in slate).
3. How might you recognize an area that is in danger of slumping?
4. If you had a topographic map and a map showing aerial distribution for rock and sediment type, how might you recognize an area of potential landslide? What other information would be most helpful in identifying such an area?
5. What methods would be effective in controlling mudflows in an area, such as the San Gabriel Mountains in California, that is subject to them? In controlling solifluction?
6. Why do most walls built to stop downslope movements often prove ineffective in stopping slumps? Refer to Fig. 8.11(*b*).
7. What factors contributed to the mass movements that caused the failure of the slopes above the Vaiont Reservoir?

REFERENCES

1. Gordon B. Oakeshott, "The Alaskan earthquake," *California Division of Mines and Geology Mineral Information Service* 17:7 (1964), *119–125*.
2. R. L. Shreve, "Sherman landslide, Alaska, " *Science* 154:12 (1966), *1639–1643*.

SUGGESTED READINGS

Eckel, E. B., ed., *Landslides and Engineering Practice.* Washington, D.C.: National Academy Sciences–Natural Research Council Publication 544, 1958.

Fletcher, F., "A terrifying equality: The story of the Vaiont Dam disaster," *Susquehanna University Studies* 8:4 (1970), *271–300*.

McDowell, B., and J. E. Fletcher, "Avalanche! 3,500 Peruvians Perish in Seven Minutes," *National Geographic* 121 (1962), *855–880*.

Report of the Committee of the National Academy of Sciences of Panama Canal Slides, v. XVIII, 1924.

Varnes, D. J., *Slope Movement Types and Processes in Landslides: Analysis and Control.* Washington, D.C.: Transportation Research Board, National Academy Science–Natl. Res. Council Spec. Report 176.

STREAM ACTION

Chapter 19

Earth is sometimes referred to as the "water planet" because it is the only planet of the solar system on which water presently occurs in all three physical states—solid, liquid, and gas. Moreover, 60 percent of the earth's surface is covered by water. Although there are no streams on Mars now, some of the landforms there were probably formed by running water; but nowhere is the influence of stream action as pervasive or important in shaping the landscape as on earth.

Water is a central theme in this and the three following chapters. Here our concern is with the principles governing the flow of streams on the surface. In the following chapter, the effects of stream action in shaping the landscape are considered. The movement of underground water and its effects, both below and on the surface, are topics for Chapter 21; and finally, in Chapter 22, water is viewed as a crucial natural resource and as a cause of incalculable damage due to flooding. For both of these reasons, control and management of water on the surface and underground have become necessary.

19.1 THE HYDROLOGIC CYCLE

Streams are but one element in a complex and continuous system of movement and interchange of water on earth. The total of all these processes of exchange among the atmosphere, oceans, and land constitute the **hydrologic cycle** (Fig. 19.1). The sun and wind cause evaporation of large quantities of water from the surfaces of the oceans, streams, lakes, and marshes, and significant amounts from the land surface itself. Water vapor is also transpired directly into the atmosphere from plants. At any time the atmosphere holds an amount of water equivalent to that in the top two or three centimeters of the surface of the ocean. The moisture is caught up in circulation of air masses and eventually condenses or freezes and falls back to earth. This precipitation promotes rock weathering and erosion through glaciation or stream action. It may dissolve soluble minerals as it percolates through the soil, or it may be absorbed by and become involved in chemical reactions with minerals of rock or soil. Some water is absorbed by plant roots and returned to the air through the plant's leaves. It may infiltrate through the soil and move along cracks and through pore spaces in rocks as part of the groundwater flow. Or it may collect as runoff on the land surface and move in streams. Eventually, much of the water, along with sediment and dissolved substances, finds its way into the ocean; and the cycle is complete.

Another way of viewing the hydrologic cycle is as a system of interchanges of water between natural reservoirs. Water is stored in the oceans, in the atmosphere, in lakes and river systems, underground in pore spaces and cavities, and as ice in glaciers and pack ice at sea. The processes of evaporation, transportation, precipitation, condensation, freezing, melting, percolation, and gravity drainage are continuously transferring water back and forth among these reservoirs.

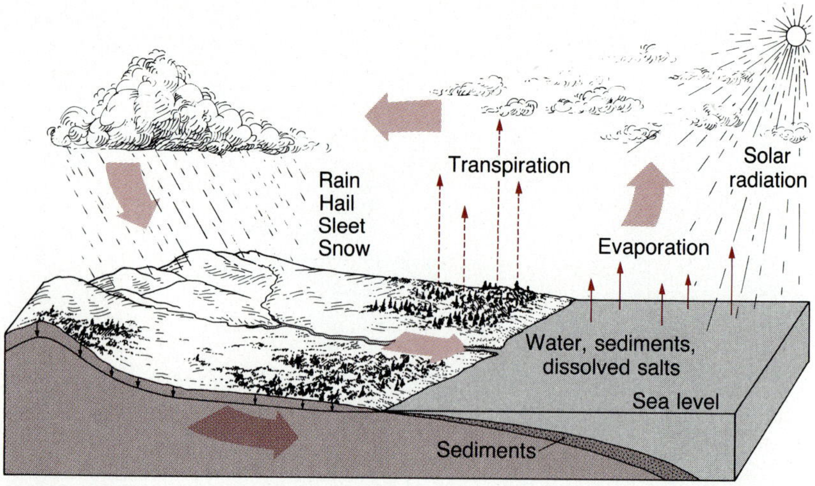

Figure 19.1
The hydrologic cycle. Water is an important agent of change near the surface of the earth, because it is recycled through the oceans, atmosphere, and land. Solar radiation causes evaporation of water from the oceans. Plants release water vapor directly into the air by transpiration. This water vapor then moves through the atmosphere until conditions favor its precipitation as rain, hail, sleet, or snow. Some of the water that falls to earth becomes runoff, which carries sediment in streams and erodes the land. The water that seeps into the earth moves much more slowly through the soil and bedrock, promoting solution and decomposition of rock. The broken sediments and dissolved solids are carried by the runoff and ground water into the ocean, where they are deposited as sediments on the sea floor.

Water Budget

As the need for increased water resources in the heavily populated parts of the world has grown, much more careful analysis has been made of the factors that affect the availability of water. It is useful to analyze the water supply of an area in terms of what is called its **water budget.** This is an attempt to inventory the water in the area plus water that is moving into and out of the area. At any given time, some water is held in storage lakes, in the ground, in storage compounds such as reservoirs, and in streams within the area. Water is continually coming into the area and going out via streams and underground flow. It is moved by pipelines and canals as well. Water is naturally supplied to the area through precipitation (rain, snow, and the like) and naturally lost from the ground by evaporation and transpiration. Some of it infiltrates through the soil to become groundwater, and the remainder runs off the land surface in streams.

In making a full-scale survey of water resources, all the above-mentioned sources and losses must be considered. Consumption of water and artificial movements of water, such as pumping of groundwater into streams or artificial recharge of groundwater, may be important in some areas. But for the moment we will consider the water budget in terms of factors that are important in natural streams. Those five factors are: precipitation, evaporation, transpiration, infiltration, and runoff. The amount of water available for surface runoff in an area is approximately equal to the amount of precipitation less the amounts lost by infiltration, evaporation, and transpiration.

The values for each of these variables can be determined, some with much greater ease and accuracy than others. Precipitation is measured by rain and snow gauges. The potential evaporation can be calculated by observing the amount of evaporation that takes place from open pans. This value may be used to calculate the losses from lakes and stream surfaces. For other land areas, it is usual to combine the losses through evaporation and transpiration. This is a difficult value to obtain because it involves water losses from the various types of soil and plants present. The level of the water table is the best and most direct indication of the amount of water that is infiltrating into the ground. And runoff can be easily calculated from stream gauge data. If the average velocity through a section of known area is measured, the discharge can be calculated. Channel profile, water depth, and velocity are periodically measured at many gauging stations along streams in the United States. From these data, the discharge can be estimated on the basis of the gauged water depth.

The U.S. Geological Survey has made estimates of the amount of water involved in the water budget for the country as a whole. According to these estimates, the average rainfall over the country is about 75 cm each year. (One centimeter is the amount of water required to cover an area to a depth of one centimeter.) Of that, about 30 percent runs off in streams to the ocean, and about 60 percent goes back into the atmosphere through evaporation and transpiration. An unknown amount of water is stored in the ground. A very small amount of this seeps into the ocean.

For a natural drainage basin, isolated from adjacent basins and unaffected by human development, an equilibrium condition exists when, over a period of time, the amount of precipitation is equal to the sum of the runoff, the evaporation, the transpiration and the amount of infiltration. This type of water balance is an oversimplification of what exists in most places, because it overlooks two factors that are especially important in many places: the effects of human changes in the natural system and the movement of water underground. Both of these must be considered carefully in the development of groundwater resources.

19.2 THE FORMATION OF STREAMS

How water gets into streams is not as obvious as it might at first seem. **Runoff,** water that does not sink into the ground but moves across the surface, usually accounts for a large part of the water in a stream, but even small streams continue to flow long after the last rain and after all the water on the surface and in near-surface soil has drained off. This flow is sustained by water reaching the stream from beneath the ground surface. Some of it essentially flows through the soil, but part rises in springs and even in the stream channel from the underlying bedrock. This movement will be considered later.

The most immediate source of stream water is runoff. A host of factors determine the amount of surface runoff and how much of the water evaporates, is transpired, or passes underground. A high proportion of the rainfall becomes runoff in areas that: have steep slopes; are underlain by ice, sediment (such as clay), or rock through which water moves slowly, if at all; and areas that have low rates of evaporation. Several of these factors are directly or indirectly related to climate; slope and the nature of the soil are a function of bedrock composition; and slopes are determined at least in part by the regional base level.

Runoff starts as a thin sheet of water on the ground surface. This **sheet wash** flows downslope and is directed into small channels that coalesce to form larger and larger streams (Fig. 19.2). The surface of the ground in most areas is a network of small, barely noticeable drainage channels that are dry most of the time. Only the larger ones appear as distinct valleys with recognizable stream channels; and of these, only a few contain water all of the time. In deserts, the stream channels are usually dry. Even in humid climates most valleys at higher elevations contain streams that flow only intermittently.

Thus, **permanent streams**—those that flow all year—are but a part of a much more extensive network of channels that drain the surface of the earth. The area drained by a stream is called its **drainage basin** (Fig. 19.3). The perimeter of a basin is defined by an imaginary line, the **drainage divide,** separating areas that drain into one stream from adjacent areas draining into other streams. This divide, while imaginery,

NASA

Figure 19.2
Most drainage systems consist of a network of small streams that feed into increasingly larger streams. Although this drainage system is dry at the time this picture was taken, it is well developed. This photograph, taken from the Gemini 4 spacecraft, shows part of the Arabian peninsula. The distance across it is several kilometers.

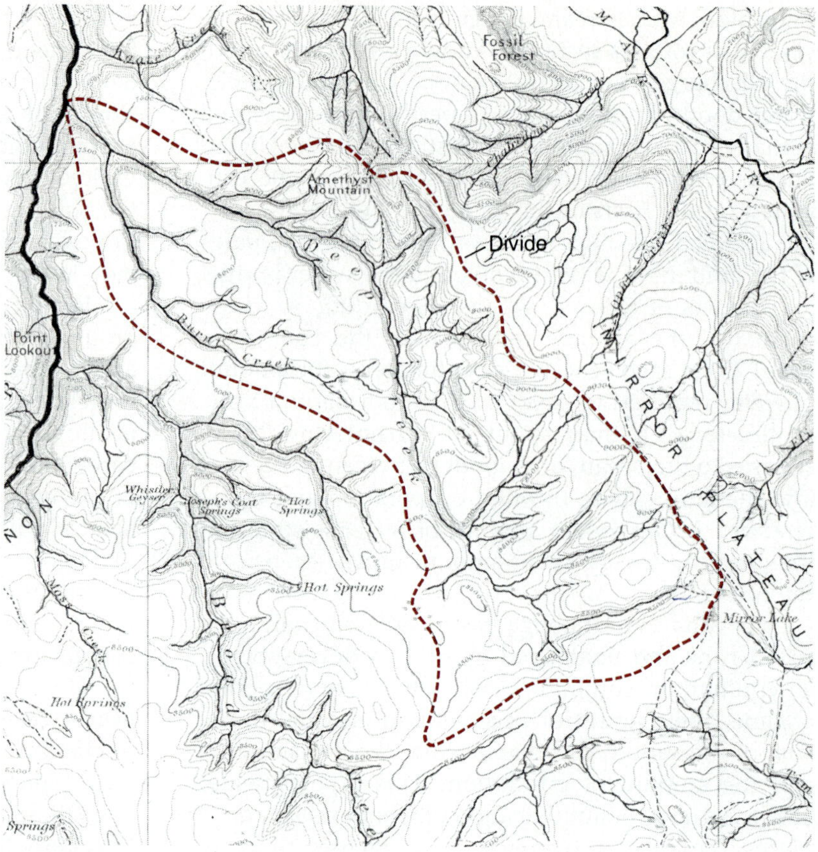

Figure 19.3
A small drainage basin is outlined on this map, and the location of the divide is indicated. The distance across the map is 14 kilometers.

has a very real meaning in that runoff on either side of it flows in opposite directions. The position of this divide can be plotted on maps, and as erosion changes the land surface, the position of the divide changes. Drainage basins are not necessarily topographically low nor are drainage divides always located on ridges. In fact, the divide often lies on a surface of low relief. In the following section, we will examine the processes in stream channels by which streams shape their basins.

19.3 PRINCIPLES GOVERNING STREAM ACTION

Even the casual observer is impressed by the diversity of streams and the landforms encountered along them. These differences include not only the quantity of water; the stream's width, depth, velocity; and the shape of the channel; but the pattern of the stream, the slope of the channel, and the relationship of the stream to the valley in which it flows. We will examine the interplay between the water in the stream, the channel, and its contents first, before turning to the relationship between the stream and the surrounding landscape.

The Stream and Its Channel

Channel form. Most permanent streams flow in well-defined channels that can be described by measuring their width and by sounding the depth at enough points to draw a cross section of the channel. From this it is possible to determine the cross-sectional area. If the average velocity is also determined, it is possible to calculate the stream's **discharge,** the amount of water passing through that section, expressed in volume per unit of time.

Channel shapes and discharge differ greatly from one stream to another and, as might be expected, even at different places along the same stream. Figure 19.4 shows that channels also differ greatly at the same cross section from one time to another, especially when the discharge changes due to heavy or sparse precipitation or to seasonal changes. In part, the alteration in cross section is due to the fact that the

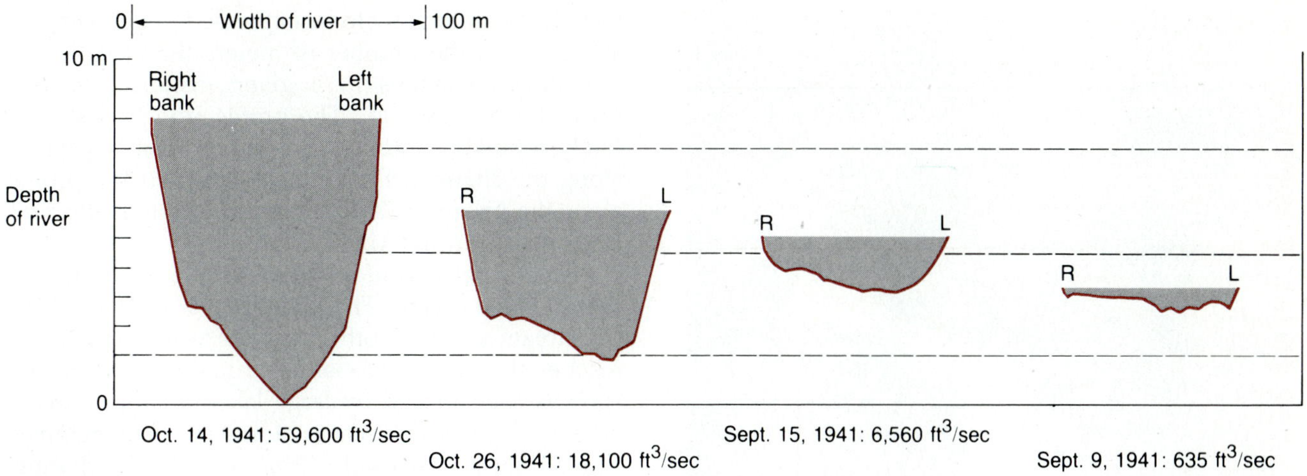

Figure 19.4
Changing channel cross sections during progress of a flood September through October 1941 at San Juan River near Bluff, Utah.

water level rises when discharge increases, but in most channels the shape and depth of the channel also slowly change. These changes must be the result of erosion and deposition of material in the stream channel. This loose, unconsolidated sediment, often consisting of silt, sand, and mixtures of gravel, is called **alluvium** (Fig. 19.5).

Most major streams flow in channels cut into alluvial deposits that fill the bottom of the stream valley. The channel's shape may be a near-perfect half-circle in section, but it is more likely to be assymetric, and, under some conditions, the channel consists not of one stream bed but of a large number of shallow, interwoven, and rapidly changing channels like those of Fig. 19.6, referred to as a **braided** pattern. Still other streams flow over bedrock in channels essentially devoid of alluvium (Fig. 19.7).

Because discharge changes, alluvial deposits along streams are not confined to the stream channel. During floods, the discharge is exceptionally high. At such times, the stream both erodes its valley beyond its usual channel and, as discharge decreases, deposits alluvium across the valley floor. In this fashion, a nearly flat surface, called the **flood plain,** is built up along most streams. Although all streams experience floods, not all have flood plains. This is

Figure 19.5
(a) A cross section through the Hudson River at the George Washington Bridge in New York City. The channel is largely filled with sediment that has been moved in the past by the river. During floods, much of the sediment may be picked up from the channel bottom and the depth of the river increased. (b) Locator map.

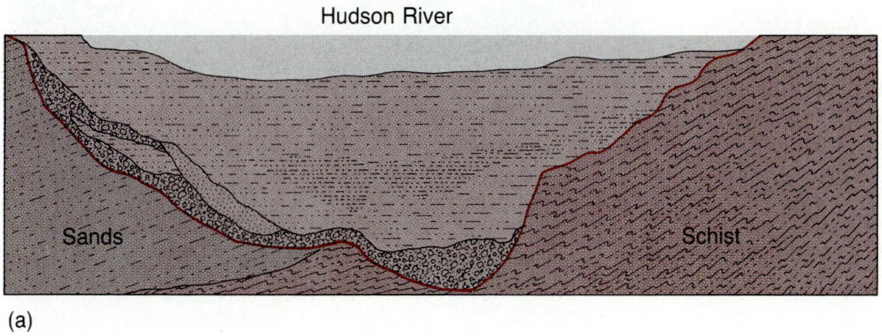

(a)

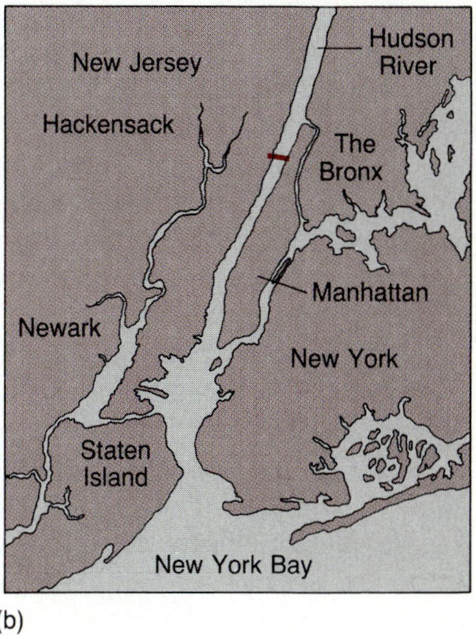

(b)

Royal Canadian Air Force

Figure 19.6
This river in the Canadian Rockies flows in a braided channel. Water depth in such channels is rarely more than a few meters, and the channel form changes frequently. The edge of the alluvial plain is clearly shown here. This stream is about a hundred meters wide.

especially evident where streams flow in canyons or on bedrock surfaces (Fig. 19.8).

Channels clearly vary along the stream from the **head** of the stream, where it originates at the higher elevations in the drainage basin, to its **mouth,** where the stream flows into another stream, a lake, or the ocean. Except in arid climates, channels tend to become bigger downstream as the area being drained increases and, with it, the discharge. Usually, the channel becomes deeper and shaped for more efficient movement of the water as well.

Streams flow because water moves downslope. Furthermore, the velocity of the water increases with slope. Consequently, the slope of a stream channel is an important feature of any stream. The slope of the stream, the **stream gradient,** is determined by measuring the difference in elevation and the distance between points along the stream. The slopes are usually much too low to be expressed in degrees and are given as the number of meters the elevation of the stream changes in a given distance (meters/kilometer or feet/mile). The profile along the stream can be found by plotting the elevation against distance along the channel. Figure 19.9 shows such a profile along the Arkansas River, from its head in the Rockies to its mouth in the Mississippi River.

The upper end of a stream profile is defined by the drainage divide. The lower end is located where the stream enters another body of water or when it reaches the base level—either a regional base level, such as that formed by large lakes, or the ultimate base level, the sea. Streams cannot cut their channels beneath these base levels. Consequently, the channel ends and the gradient becomes zero when the stream reaches base level.

Stream flow. The nature of the flow of water in streams undergoes many changes in the course of its journey to the sea. In mountains, where the water is confined to narrow channels filled with boulders and

Figure 19.7
This stream, with many others in the central Appalachians, runs in a bedrock channel and is able to remove all of the load supplied to it. The channel is about 3 meters wide.

Figure 19.8
The Gunnison River flows in a deep V-shaped valley in Colorado.

rock debris, it literally jumps, boils, and foams as it rushes down steep slopes. At the foot of the mountains, the stream is more likely to follow a sinuous course through rolling landscapes. The water is less turbulent, except for occasional floods or where it passes through rapids. As the stream approaches sea level, it may follow a broad meandering path across a flat countryside. The water flows gently with only mild disturbances of the surface.

Water sometimes moves as though it is made up of fine layers free to slip over one another. Objects suspended in this type of **laminar flow** will follow smooth, streamlined paths. Laminar flow is most closely approached when water is moving slowly through a smooth-sided channel or when velocity increases rapidly. In this situation, a lip forms on the water surface where the water, passing over a suddenly steepened slope, runs faster than that upstream (Figs. 19.10 and 19.11). The surface of the water drops and becomes smooth.

Where obstructions interfere with the smooth flow of water, the stream becomes turbulent—showing eddy currents, whirlpools, and boiling movements. Extreme tubulence occurs where water cascades down a rough stream channel. The flow in most streams on medium to low slopes is slightly turbulent and is called **streaming flow.** The surface of the water is smooth in some places, slightly undulating in others. A mild, slow eddying and boiling may be visible.

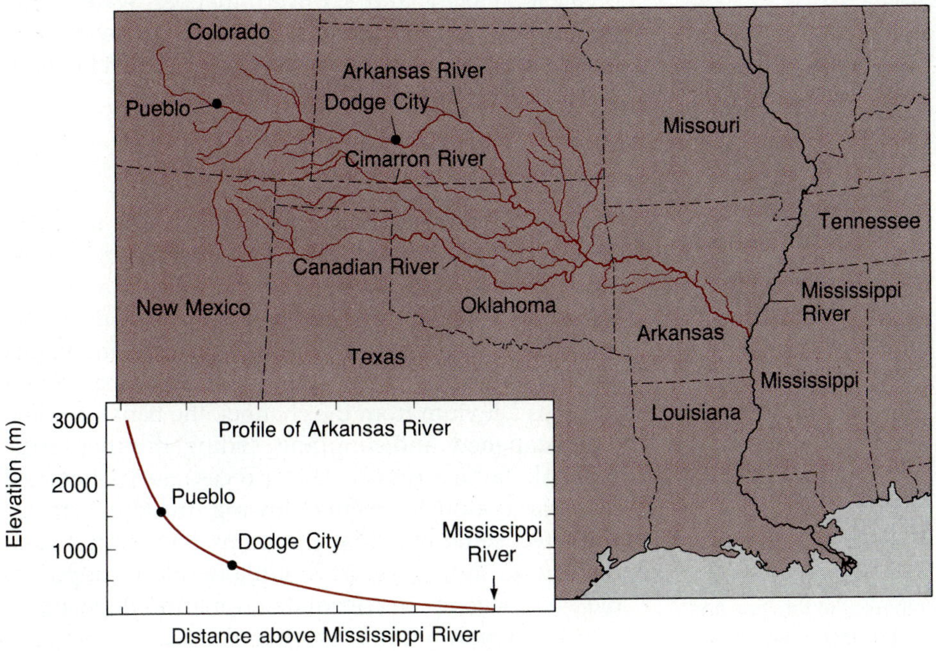

Figure 19.9
This longitudinal profile of the Arkansas River is representative of profiles of many major streams.

Figure 19.10
Shooting flow is often seen at the lip of a waterfall, such as this one at Niagara Falls.

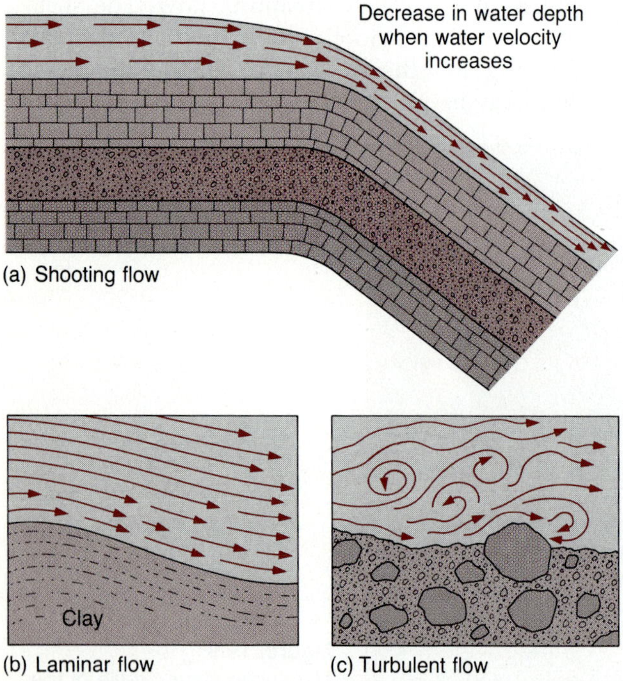

Figure 19.11
The path followed by water is quite different in laminar and turbulent flow. Smooth laminar flow often occurs where the slope increases.

If the stream is straight and the channel symmetrical, the greatest velocity is usually toward the middle and about one third of the way down from the surface (Fig. 19.12). But if the channel is asymmetric, as it usually is on a curve, the region of greatest velocity shifts toward deeper water. The deepest part of the channel is generally on the outside of a curve, and maximum turbulance lies just beneath and to either side of the maximum velocity.

Stream Load

Sources of stream loads. Except in high latitudes where glaciers remove material, streams are almost solely responsible for the removal of soil and rock from the continents. The **load of a stream** is the material transported by the stream. Stream banks usually provide a ready source of load. As a stream removes alluvium from the channel, the banks become oversteepened, and slumping, sliding, or other forms of bank failure occur. This process is particularly important along streams flowing through unconsolidated sediment. Many streams flow across such deposits as they cross old floodplains, glacial deposits, old lake beds, or unconsolidated marine deposits.

As a result of sheet wash and mass movements, the stream channel itself usually comes to hold an

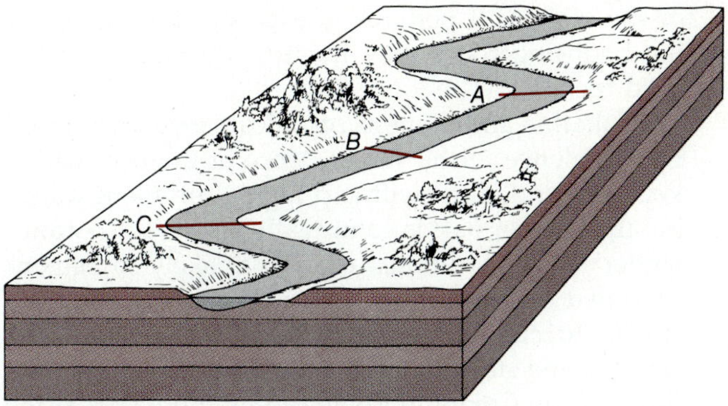

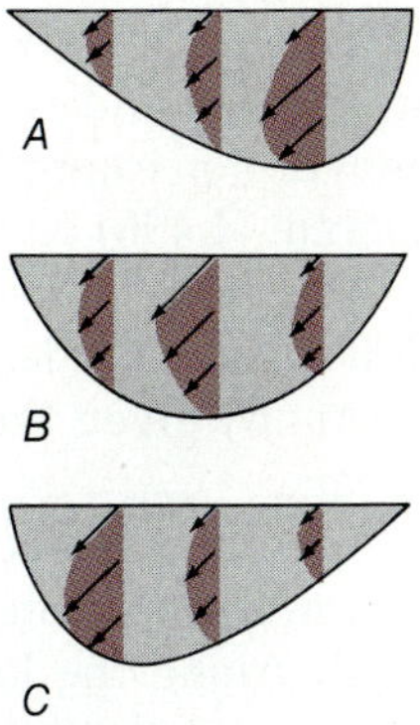

Figure 19.12
The velocity in most streams is greatest in the deepest part of the channel and in the upper third of the water, as indicated by the arrows (the length of the arrows is proportional to the velocity). The maximum velocity shifts toward the outside of the curve in streams following curved paths.

abundant supply of sediment. Ordinarily the channel is partially filled with alluvium left from the last flood or from former periods of high water when greater quantities of load could be moved (Fig. 19.13). This potential load may again be picked up and moved when high water comes again. At such times even the bedrock or sediment below the channel may become a source of load as the channel is cut deeper.

Load may also be derived from outside the immediate area of the drainage basin. Near modern glaciers the meltwaters from the glaciers carry suspended fragments into streams. Wind-blown sand and silt may account for large parts of the load of streams located in deserts, near lakes, or along coasts, where strong prevailing winds and sources of loose sediment are available. Volcanic ash and dust are locally important sources of stream load during eruptions.

Figure 19.13
This mountain stream channel is filled with a few boulders and much smaller fragmented material. At the time this picture was taken, discharge in the stream was too small to move most of its load. The coarser part of the load is moved in the spring when snow melts.

Movement of load by running water. When heavy surface runoff sweeps large quantities of soil and weathering products into streams, water appears muddy because part of its load is being transported in **suspension.** The lightest particles may float, and everything in the stream is buoyed up by a force equal to the weight of water displaced. Suspended load usually consists of silt- and clay-sized materials, but larger particles may also be suspended if the water is turbulent.

Some weathering products are carried in **solution,** especially in streams flowing through regions underlain by bedrock composed of limestone, dolomite, or salt beds. In streams in humid climates, most of this dissolved load comes from groundwater that finds its way into surface streams. But streams flowing out of deserts, where large quantities of soluble minerals may exist at the surface of the ground, also have large loads in solution. The load in solution is ordinarily a small part of the total load, but slow, sluggish streams, capable of carrying little in suspension, may be an exception.

The remainder of the stream load is moved along the bottom of the stream channel in what is called **traction load.** When the velocity of the water is great enough, sand, pebbles, and small cobbles may bounce along the bottom, hitting irregularities and traveling as suspended material for a short distance until the pull of gravity brings them back to the bottom again. Larger cobbles and boulders may roll along the bottom or slip and slide downstream. (Since most rocks have roughly the same density, size and weight are closely correlated.)

Experimental evidence indicates that the largest size that will roll along a gently sloping stream bottom varies as the sixth power of the velocity of the water (approximately). Thus, a slight increase in velocity brings about a great increase in the size and weight of rocks that the stream can move. For example, a stream that can move a two-kilogram (4.4 lb) boulder at a velocity of 100 cm/sec will be able to roll a boulder weighing 64 kg at a velocity of 200 cm/sec. The largest sized particle a stream can move at any given time is called the **competence** of the stream.

Most streams carry some of their load by each of these means, but one mechanism may dominate the others in the amount transported. Along the lower Mississippi River, for example, most of the load is carried in suspension and solution, and only a small part in traction. The load of most mountain streams, on the other hand, is carried almost entirely in traction. The amount of load in transport depends in part on the availability of debris. Streams do not always have enough debris coming into them to fill them to **capacity**—the total amount that they could transport under the existing conditions. Similarly, a stream may not be filled to capacity because it cannot pick up the available size of debris. This applies both to boulders that are not moved because of their mass and to clay that is held together so tightly by cohesive bonds that it is not readily dislodged. In both cases, the stream may have unused capacity in one of the three methods of transport.

A clear relationship exists between a stream's velocity and its ability to transport and erode. Figure 19.14 shows the velocities of water at which various sediment sizes can be transported. Fine sand and coarse silt are the most easily moved sizes. Clays, the finest of solid particles, remain in suspension even when water is not moving; as the figure shows, it is eroded only at very high stream velocity. Because wet clay is cohesive, a much higher velocity is necessary to bring clay into suspension than to pick up sand. Also, channels formed in clay are much smoother; the flow is more nearly laminar; consequently, less turbulence is present, and less erosion results.

Stream Erosion

A drainage basin is eroded by sheet wash, by mass movements, and by processes that take place in the stream itself. The pressure of the flowing water against rocks may cause them to roll, and if their velocity is high, they may bounce as they hit obstructions. Once this bounding, or **saltation,** begins, the bouncing rocks dislodge other sediment in the bed of the stream. If bedrock is exposed, the impact of the bouncing rocks may fracture and break up the solid rock.

Saltation and suspension of load are both affected by turbulence in the water. More of the bed load is kept suspended in swift and highly turbulent water, but the heavier part is continually hitting the bottom. Under these conditions, erosion is likely to be concentrated on the bottom of the channel, causing the stream to cut downward toward base level. As a result, some streams flow in narrow deep gorges like those of the Colorado River and its tributaries at the Grand Canyon, Arizona (Fig. 19.15). In such places erosion of the sides of the channel is negligible. The rivers are cutting high above regional base level, and the streams flow rapidly with great turbulence down steep gradients, cutting valleys in resistant rock.

Lateral cutting tends to be prominent in streams flowing in winding courses on low slopes. At each turn, water is shifted toward the outside of the turn,

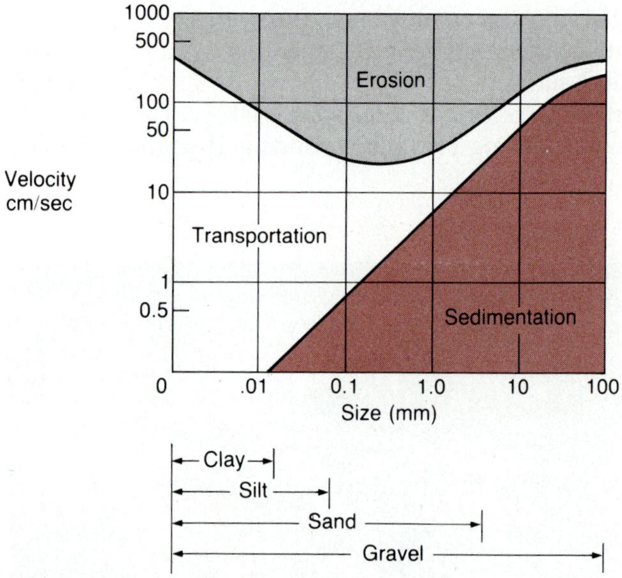

Figure 19.14
Relationship between stream velocity, sediment size, erosion, transportation, and deposition. The area marked "Sedimentation" shows combinations of stream velocity and sediment size for which the stream is unable to move the sediment. The field marked "Erosion" indicates combinations for which the stream may pick up and start new materials in transport. Note that very high velocities are required to erode both large sizes and clay. Clay is difficult to erode because of its cohesiveness.

Figure 19.15
One of the deep gorges in the Colorado Plateau. The stream has cut a narrow deep channel because it is so high above the base level.

where turbulence and erosion are concentrated. The force of the rapid turbulent stream is analogous to the erosion of soil by a stream of water from a hose. The result is to deepen the channel and steepen the bank on the outside of the curve. This makes the outside bank unstable, and it slumps into the stream (Figs. 19.16 and 19.17).

Where streams flow across bedrock, the stream channel is often marked by rounded depressions of the kind shown in Fig. 19.18, called *potholes*. These almost invariably contain a few rocks or coarse sediment. When the velocity of the stream is high, the rocks in the hole are moved around in the hole and undoubtedly abrade it. For this reason, the holes are thought to result from long-continued circular motion of eddy currents carrying sand, silt, and gravel. The surface of the hole is abraded by the grinding action of sediment in the hole. Dissolved and suspended products are washed up and out of the hole. Most potholes are small, but occasionally they become six meters or more deep and are an important process, causing erosion of bedrock channels.

Where water becomes charged with air bubbles, collapse of the bubbles may be an important cause of erosion. It almost certainly plays a role in erosion along rapids and at the base of waterfalls. Bubbles collapse as the water pressure on the submerged bubble increases. Collapse of this sort, called an implosion, creates a strong shock wave that may be capable of causing rapid erosion even in hard materials. Dam spillways must be constructed to prevent or minimize this effect.

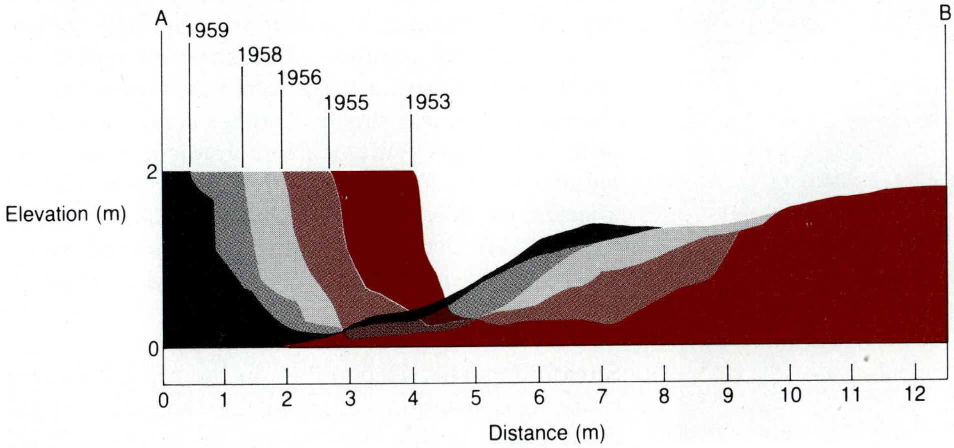

Figure 19.16
Successive profiles across a meandering stream show the gradual movement of the channel to the left. Bars are built up on the right at the same time the bank is eroded on the left.

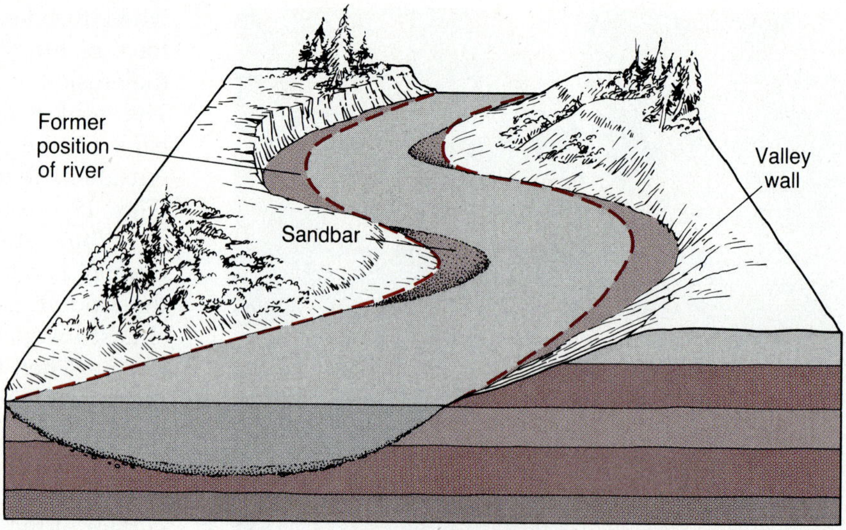

Figure 19.17
Over a period of time, the stream shifts its channel by eroding the bank on the outside of the curve. Compare this with Fig. 19.16.

Stream Deposition

Exactly how much sand, silt, and rock is transported by a stream depends on how much load is potentially available and the capacity of the stream. Stream capacity is determined mainly by the amount of water flowing through it, discharge, and the velocity of the water, which is very sensitive to the stream gradient. Deposition occurs whenever stream capacity is less than the load available to the stream. Thus, deposition is likely to take place when either the velocity or discharge of a stream is reduced or when the load supplied by sheet wash, wind, or mass wasting exceeds the ability of the stream to move it. In humid regions, discharge always increases downstream, except where surface water is lost to underground drainage, and that is not a widespread condition. Streams in arid regions are often loaded beyond their capacity by fine sediment washed or blown into the channel. The channel silts up, as shown in Fig. 19.16, and the stream, unable to move its load, usually takes on a braided form. The discharge of these streams also commonly decreases downstream as a result of infiltration of water from streams into stream alluvium and bedrock and because evaporation rates are high.

Figure 19.18
Potholes are cut into a stream bed by the abrasion and impact of rocks caught in eddy currents. This picture was taken when the stream was dry, shows rocks under water in the pothole, which is over a meter across.

Deposition caused by decrease in velocity is found in both wet and dry climates. Velocity decreases are caused by a number of conditions, especially reductions of stream gradient. The sharpest reductions occur where a stream flows into a lake, reservoir, or the ocean, but less drastic changes commonly occur when a stream with a steep gradient flows into another with a lower gradient. Velocity may also be affected by decrease in discharge, because smaller channels are usually less efficient in moving water. Thus decrease in discharge usually causes a reduction in velocity as well.

Stream deposits. Depending on the conditions that cause it, deposition occurs in the channels, on the flood plains beside the channel, or in the body of water into which the stream flows. Channel deposits in streams that are actively eroding the underlying bedrock tend to consist of small, thin, scattered bars of sediment. These are usually transient features that are moved during the next period of high discharge.

The sediment fill in most channels fluctuates in thickness as the discharge and velocity in the stream vary. The lower part of such fill may be left undisturbed for long periods between major floods. And if the valley in which the stream is flowing is lowered as a result of tectonic movements, the valley may begin to silt in. A similar buildup of sediment can occur if the load being supplied increases substantially relative to the discharge. Such conditions can be brought about by a change in climate, volcanic activity, or wind action. If this happens, the older part of the stream channel gradually fills with sediment and the part of the channel containing water rises higher and higher in the valley. This condition is markedly different from that occurring during floods.

Flood waters are heavily charged with sediment because the increased velocity and discharge both increase stream capacity. The waters flood over the banks of the stream, spreading out across the flood plain. The waters on the flood plain move more slowly than those in the channel. More obstructions, especially trees and plants, are normally present, and the broad thin layer of water experiences more drag. As the water velocity drops and as the floods subside, capacity drops and the suspended sediment settles out, leaving a thin layer of new sediment across the flood plain. Sometimes a thicker and coarser wedge-shaped deposit, called a **natural levee,** forms along and close to the channel (Fig. 19.19). The renewal of sediment is one reason flood plains and deltas provide such rich agricultural land.

The deposits built by streams where they flow into still bodies of water or into streams with much less capacity are called **deltas.** Some, like the Nile delta (Fig. 19.20), are triangular in shape like the Greek letter delta (Δ), but many are rounded like the delta of the Yukon River (Fig. 19.21). A dusting of fresh snow brings the veinlike pattern of **distributary** streams on the Yukon delta into clear view. The main river breaks up into many smaller streams, which carry the water and its suspended load out onto the delta surface, a broad, flat, marshy land. Some distributaries may have natural levees, and all contain a traction load; but when water floods over the delta, the suspended load settles out as it does on a flood plain.

NASA

Figure 19.20
Gemini 4 spacecraft view of the Nile delta, Egypt, with the Mediterranean Sea to left, and the Suez Canal and Red Sea in background. The dark area, which is cultivated land, corresponds closely to the deltaic deposits.

The beds of sediment deposited on top of the delta, both by the streams and their flood deposits, form layers called **top-set beds.** At the front of the delta, sediment carried in the channel is deposited on the steep slopes built out into the still water. Sedimentary layers formed here are called **fore-set beds.** Very fine sediment that stays in suspension long enough to drift into deeper water in front of the delta forms **bottom-set beds.** As the delta grows, inclined fore-set beds are deposited on top of the bottom-set beds. The top of the delta cannot be raised above the level of the ocean or lake into which the delta is built. Consequently, the top-set beds are thin

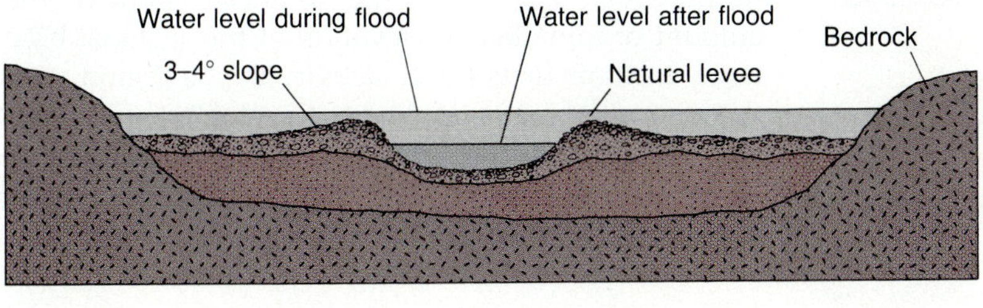

Figure 19.19
Cross-sectional view of a stream showing the formation of a natural levee. Ordinarily the stream is confined to the deeper part of the channel. The levees are built as floods recede.

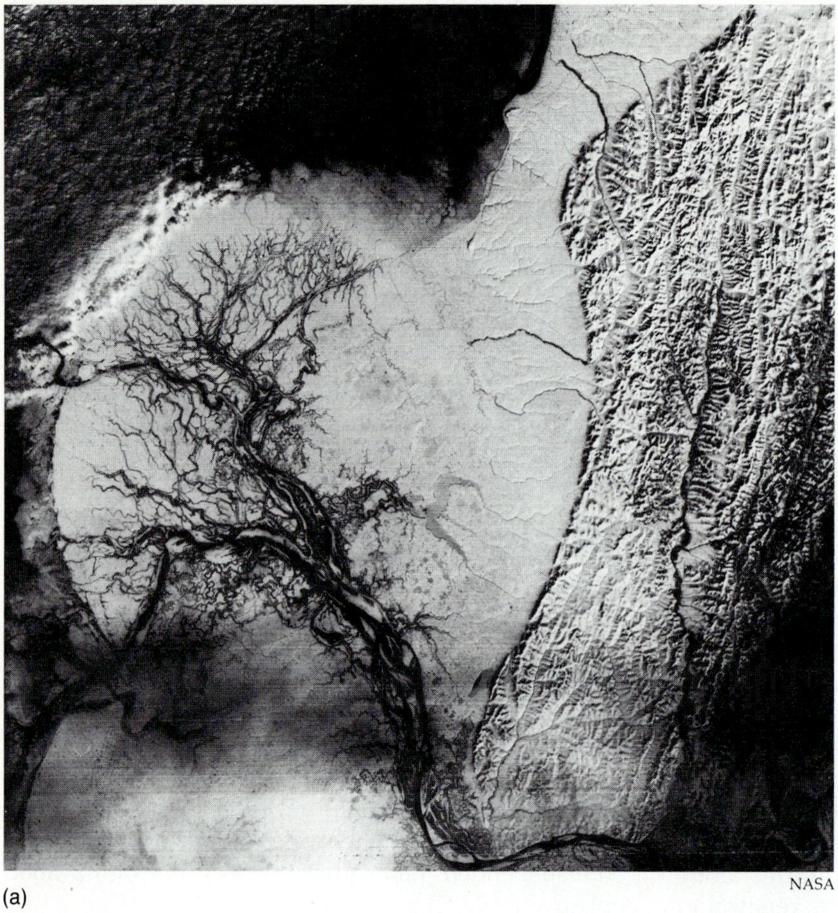

(a)

(b)

Figure 19.21
(a) The distribution of the Yukon River in Alaska stands out unusually well on this satellite image, because the land had received a snow cover. The water is dark. Unlike the Nile and Mississippi deltas, this one is very rounded. This image covers an area nearly 160 km across. (b) Locator map.

and often cut off as the distributaries shift channels. The result of this process, if continued long enough, is the formation of cross-bedding of the type shown in Fig. 19.22. Because this type of cross-bedding is so frequently found in deltas it is called **deltaic cross-bedding.**

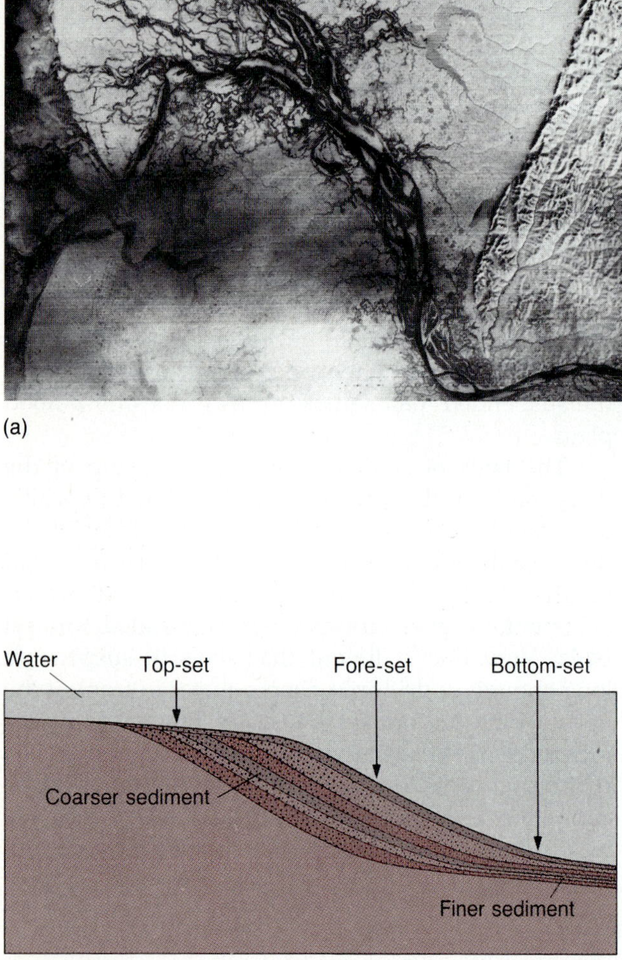

Figure 19.22
Ideal deltaic cross-bedding consists of thin beds of fine sediment—the bottom-set beds; a thicker and coarser set of inclined beds; the fore-set beds; and a flat-lying top set. The top-set beds are thin or may be missing. This type of cross-bedding may occur on scales ranging from centimeters to tens of meters.

19.4 GRADED STREAMS

From our understanding of erosion, transportation, and deposition in streams, it is clear that stream action involves a rather complicated interplay of a number of variables. Some of these—notably the amount of rainfall, the elevation of the regional base level, tectonic or volcanic activity in the region, and the amount of load supplied to the stream—influence stream action, but are independent of the processes of erosion, deposition, and transportation acting in the stream. Other variables—such as the gradient, elevation of the water surface and the streambed, velocity, and capacity—are interdependent. For example, when gradient (slope of the stream) increases, velocity and capacity also increase. These variables are also dependent, especially on discharge, as seen in Fig. 19.23.

All of these variables are continually changing. Some change rapidly over the time of a thunderstorm. Others, related to climatic changes or tectonic activity, bring about slow, long-term changes in the stream and its valley.

Many of the short-term changes can be analyzed in terms of what Hoover Mackin (1948) described as a **graded stream.** According to Mackin (*1*), a graded stream is one that has, over a period of years, achieved an equilibrium involving all the factors that affect it. The slope of the stream is delicately adjusted to provide just the velocity required to transport the load supplied from the drainage basin at the prevailing discharge and channel conditions. The equilibrium can be recognized by the fact that any change in one of the controlling factors will displace the equilibrium in a way that tends to absorb the effect of the change. This model of stream behavior allows us to predict the series of changes in other variables that will follow a change of any one of the variables. For example, if load is increased in the upper reaches of a stream as a result of construction, the gradient in that part of the stream is steepened, because erosion from the

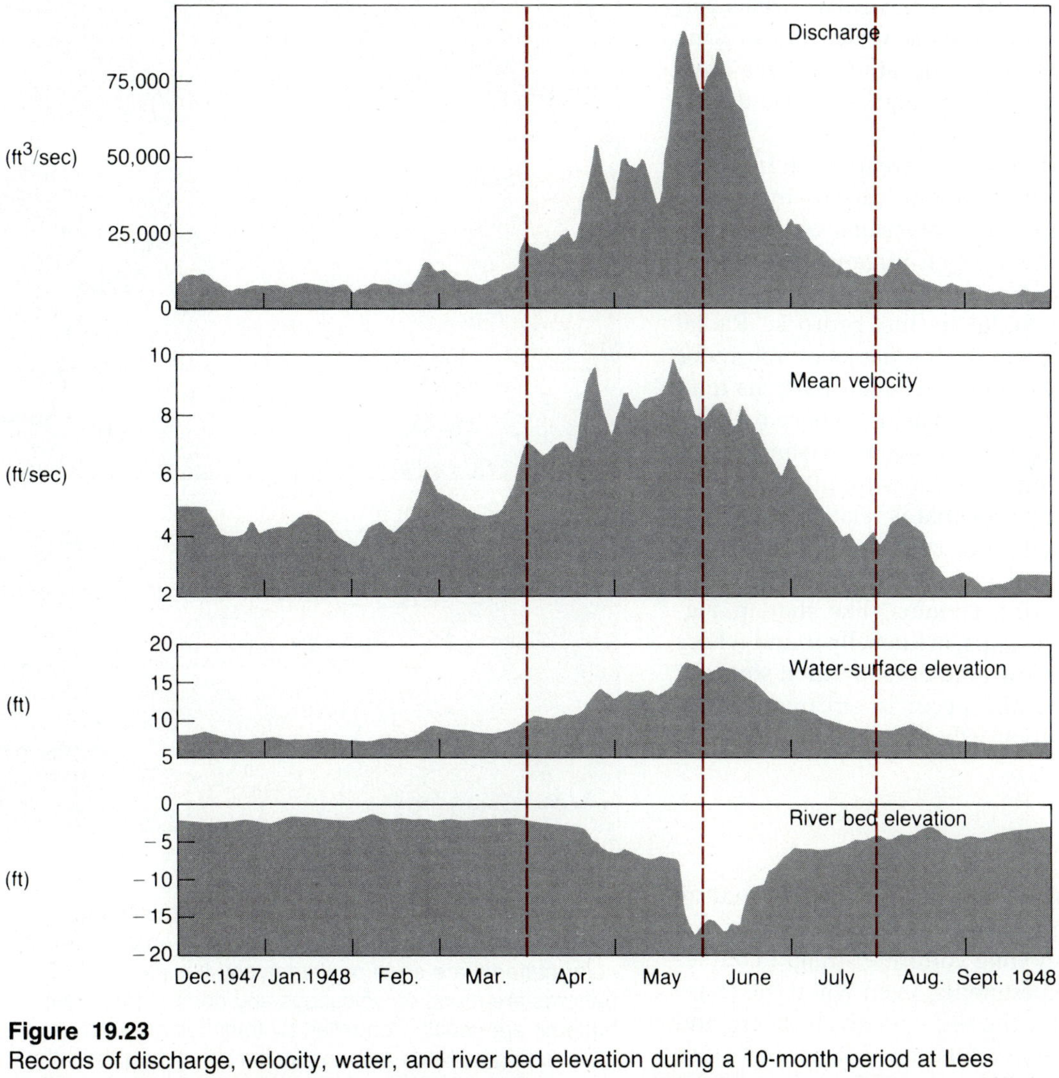

Figure 19.23
Records of discharge, velocity, water, and river bed elevation during a 10-month period at Lees Ferry on the Colorado River. These records show the close relationship between discharge, which is affected mainly by rainfall and snow melting, and the other variables.

site fills the upper end of the stream channel with sediment. The steeper slope causes the velocity of the stream to increase, and this increases the capacity of the stream. As capacity increases, the load is removed, lowering the gradient and, in turn, the velocity.

All streams exhibit these types of adjustments, and many approach the model closely.

19.5 STREAM PATTERNS

Although no two basins are identical, certain patterns of drainage recur over and over. The most common of these are the branching or treelike pattern, referred to as **dendritic** form, and rectangular, trellislike, parallel, or criss-cross patterns in which the streams follow nearly straight lines. Still other streams have braided, radial, concentric circular, or meandering patterns (Fig. 19.24).

Careful comparison of the stream patterns with the surface and bedrock geology reveals that these patterns are characteristic expressions of particular ground conditions. Most often streams that follow straight-line geometric patterns (Fig. 19.25) are following fractures or faults in the bedrock. Radial patterns (Fig. 19.24) form on the flanks of volcanoes and domal uplifts. Concentric, circular patterns tend to develop around domes or basins, where rocks of different resistance to erosion occur in similar patterns. Dendritic patterns are most frequently found in terrain underlain by rock that is relatively uniform in resistance to erosion. They commonly occur where flat-lying rock layers are uniform in composition over large areas. Meandering streams, like that in Fig. 19.26, defined by large loops, are usually found where streams flow on very low gradients. Braided streams, like that in Fig. 19.6, also occur in streams flowing on low slopes, but they usually contain more load than they can carry.

Origin of Meanders

Streams seldom follow straight courses in nature unless they are controlled by joints or faults in the bedrock. Nor do they long continue straight courses during laboratory experiments, even when the channel is initially straight, the slope relatively steep, and the bank material uniform. Natural obstructions cause the flow pattern to be broken and deflected, and once deflected, the water tends to follow a sinuous path

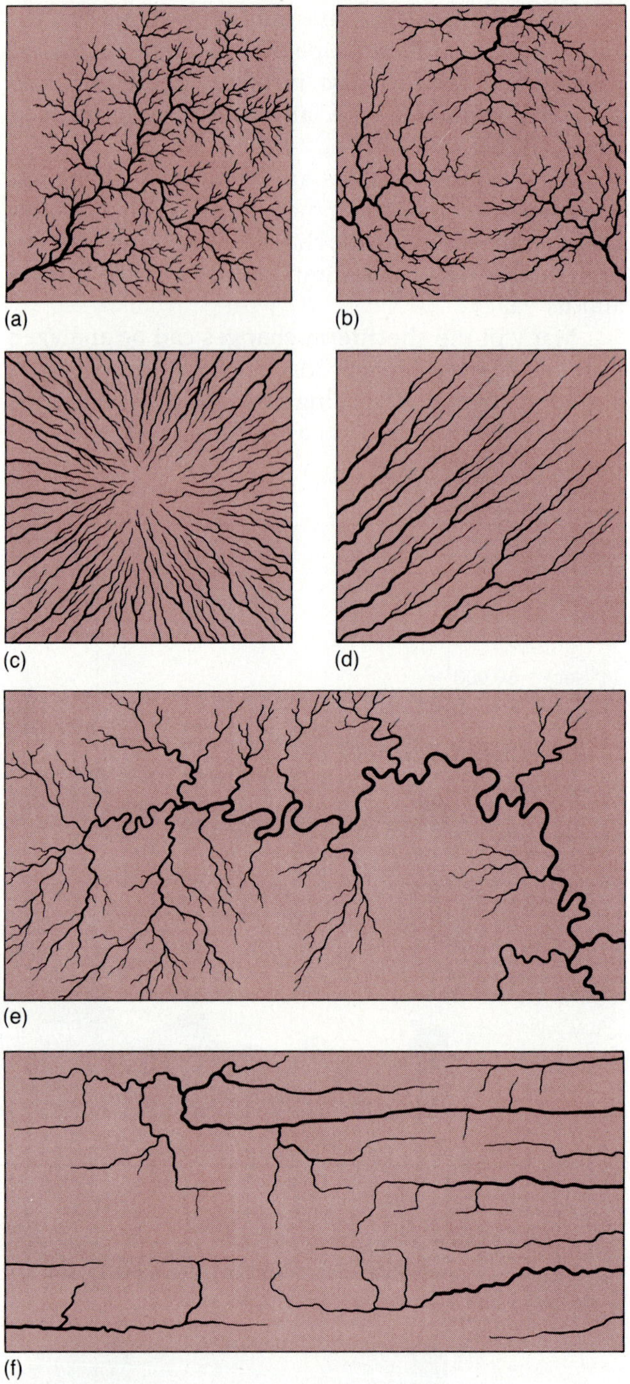

Figure 19.24
Common drainage patterns. (a) Dentritic drainage is common on rocks of uniform resistance to erosion. (b) Annular patterns are seen on domes and basins. (c) Radial patterns are found on volcanoes and domes. (d) Linear patterns are usually controlled by fractures or faults. (e) Meandering patterns are most common where gradients are low and resistance uniform. (f) Trellis or rectangular patterns are found in fold belts and along fractures.

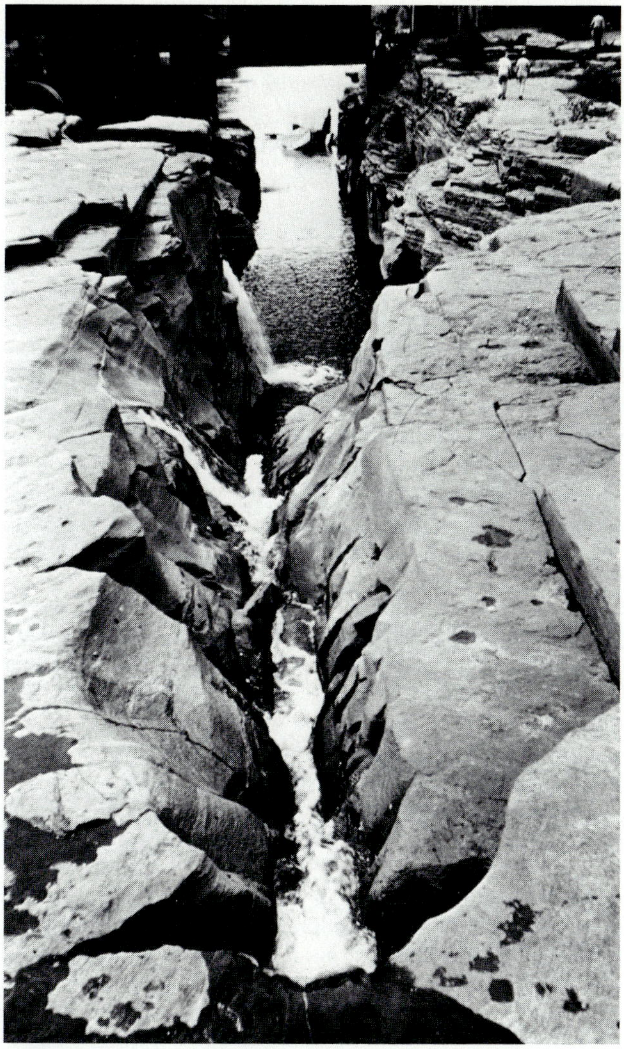

Figure 19.25
The course of this stream is controlled by fractures in the bedrock, some of which can be seen on the right. The men in the upper right give an indication of the scale.

(more accurately, a helical path). This helical flow is formed in experiments even in the absence of obstructions. As soon as a curve is established, the velocity maximum is shifted toward the outside of the curve, and the water surface takes on a slight slope toward the inside of the curve, which directs the flow back across the channel and against the opposite bank. Soon deposits form on the inside of the curve, and the initial conditions leading to broad, looping stream curves called meanders are started.

In nature, meandering streams are usually found where the gradient is low, as in Fig. 19.26. One effect of the meandering pattern is to lengthen the stream. Thus, the development of meanders may be thought of as an internal adjustment along a stream that effectively lowers the stream gradient and increases the resistance to flow on the curves. Meandering paths appear to be normal patterns for streams flowing on moderate or low gradients in valleys containing unconsolidated sediment, such as alluvium. Experiments conducted by the U.S. Army Corps of Engineers confirm this conclusion. In one experiment, shown in Fig. 19.27, an initially straight channel cut in unconsolidated sand is quickly modified to a winding form. The bends result from local bank erosion on the outside of curves and deposition on the inside of curves. Clearly the resistance of the bank determines how rapidly it can be eroded. This type of experiment shows that channel depth of a meandering stream also depends on the resistance of the banks to erosion. Resistant banks usually result in deeper channels; easily eroded banks form shallower channels. When the bank materials are unconsolidated sediment, discharge is low, and load is great, streams in experiments assume a braided pattern. When some or all of these factors are changed, the stream channels revert to a single, central, sinuous channel. Thus, the behavior of these streams is determined by the interaction of the rate of bank erosion, the amount of load, the channel form, dis-

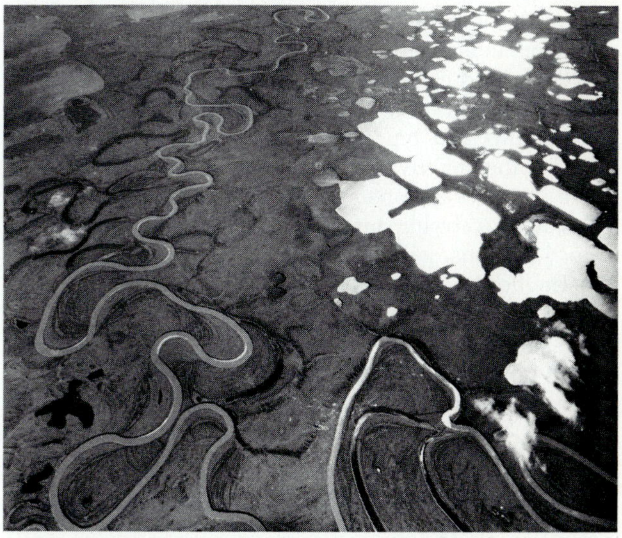

Royal Canadian Air Force

Figure 19.26
Although this area in Canada is far removed from the Mississippi, it shows many of the same landforms—meanders, sand bars on the inside of the curves, and cutoffs, as seen in the foreground.

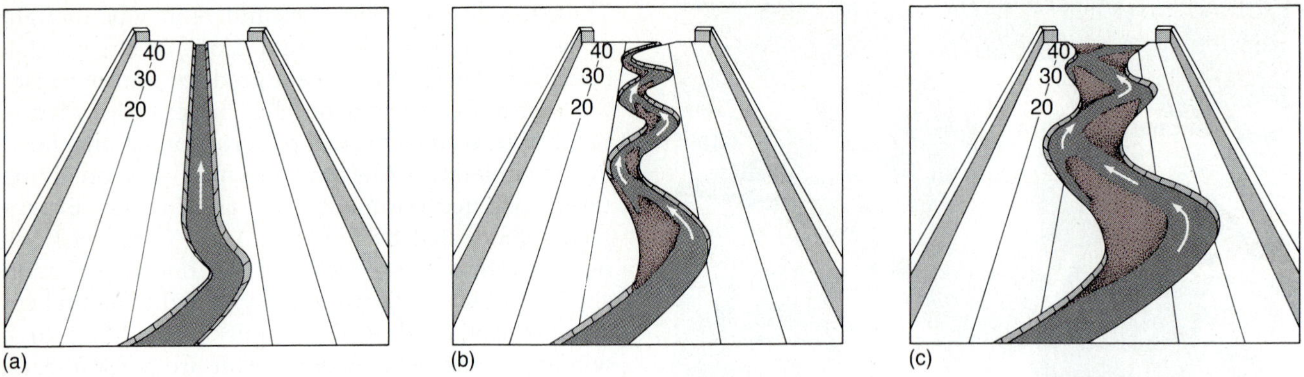

Figure 19.27
The development of a sinuous stream pattern for a straight channel took only six hours in the experiment illustrated here. Erosion of the outside of curves and deposition inside curves is clearly the most important process in this change.

charge, and gradient. The results of this type of experimental work with streams is applied to the Mississippi River in the following chapter.

SUMMARY

The action of streams represents a balance between the many factors that affect stream flow and deposition. Climate, slope, and the presence of ground cover all help determine the amount of water that enters a stream from runoff or underground sources. Permanent streams flow in well-defined channels, whose form is also the result of the interaction of several factors including stream discharge, the amounts and types of alluvium in the channel, the nature of the underlying bedrock, and the stream gradient. These variables give rise to channels ranging from broad low valleys carrying braided streams to steep-sided V-shaped canyons.

The flow of water within the channel shows a similar range from smooth laminar flow in unobstructed paths to mildly turbulent streaming flow and the highly turbulent flow of rapids and falls.

As a primary agent of erosion, streams carry loads of silt, sand, rock, and dissolved material, largely provided to them by sheet wash and mass movement. The load is carried in solution, in suspension or in traction; the amount carried by each mechanism is determined by the climate of the stream's drainage area and by the velocity and turbulence of the flow.

The combination of moving water and its load act to abrade the stream channel, cutting it downward toward base level or laterally on low slopes.

Where the velocity or volume of flow drops, so does the stream's capacity to carry load. Deposition therefore is likely to occur where the stream gradient decreases suddenly or where a stream overflows its banks forming flood plains, or enters a large body of water, depositing its load in a series of graded delta beds.

The drainage patterns formed by the set of streams in an area reflect the topography and bedrock of the region. Most common are treelike systems of streams, forming a dendritic pattern. Domes or volcanic cones show radial drainage patterns; and straight-line, rectangular patterns form where fractures and faults in bedrock define stream channels. Wide looping stream channels characterize the pattern of meanders formed on areas of low slope.

KEY TERMS

alluvium
bottom-set beds
braided
capacity
competance
delta
deltaic cross-bedding
dendritic
discharge
distributary
drainage basin
drainage divide
flood plain
fore-set beds
graded stream
head
hydrologic cycle
laminar flow
load of a stream
mouth
natural levee
permanent streams
pothole
runoff
saltation
sheet wash
solution
stream gradient
streaming flow
suspension
top-set beds
traction load
water budget

STUDY QUESTIONS

1. Explain the reason for the shape of the curve between the fields of erosion and transportation shown in Fig. 19.14.
2. Under what conditions might a stream profile measured along the stream be convex upward?
3. What factors affect the shape of a stream channel?
4. Describe the effects both upstream and downstream of building a dam on a graded stream.
5. What factors influence the map pattern of a drainage system?
6. In what ways does the presence of turbulent motion in the water in a stream affect the shape of the channel?
7. In what ways does the load in most streams get into the channel?
8. Under what conditions do streams flow across bedrock rather than on alluvium?

REFERENCE

1. Joseph Hoover Mackin, "Concept of the graded river," *Bulletin of the Geological Society of America* 59:5 (1948), 463–511.

SUGGESTED READINGS

Coleman, James M., *Deltas: Processes of Deposition and Models for Exploration*. Champaign, Ill.: Continuing Education Publication Company, Inc., 1976.

Gregory, K. J., and D. E. Walling, *Drainage Basin Form and Process, A Geomorphological Approach*. New York: John Wiley & Sons, 1973.

Morisawa, M., *Streams: Their Dynamics and Morphology*. New York: McGraw-Hill, 1968.

Schumm, Stanley A., *Drainage Basin Morphology*. Stroudsburg, Pa.: Dowden, Hutchinson & Ross, Inc., 1977.

Schumm, S. A., *The Fluvial System*. New York: Wiley, 1977.

SHAPING OF THE LAND BY STREAMS

Chapter 20

For most land areas, streams are the most important factor in sculpturing surface forms. Running water, combined with downslope movements and weathering, brings the bedrock structure into relief in some areas and buries it in others. Where water is abundant, surface drainage is well developed, streams are numerous, and large quantities of sediment are transported off the continents and into the sea. Where water is scarce, surface drainage is poorly developed, but streams and surface wash are important in shaping the land nonetheless.

In this chapter we shall follow the development of landforms by streams through case history studies of the drainage in three quite different parts of the United States: the arid region of the southwest, drained mainly by the Colorado River; the meander belt of the lower Mississippi River; and the humid region of the middle Atlantic states. Because these areas differ sharply in climate, in bedrock structure, in the relief of the regions above base level, and in the evolution of the drainage, they will give us some understanding of the range of effects involved. First we will consider some general questions about the relationship of streams to the landscape through which they flow.

20.1 THE STREAM AND ITS VALLEY

Most valleys owe their origins to the streams flowing in them. The form of the land in drainage basins also is in large part the result of processes of running water, weathering, and downslope movement of surficial materials imposed over a long period of time. In fact, as we saw in Chapter 14, the earth is old enough for its surface features to have been eroded away many times over. Streams play an important role in this continued shaping of the landscape. But the recent history of the earth has been marked by glaciation, changes in sea level, volcanic activity, and crustal deformation, which have affected and altered drainage systems over much of the earth in the last few thousand years. As a result, in certain areas where streams are today the primary geomorphic agent, the landforms were shaped in large part by other geomorphic agents. For this reason, many streams are not as closely related to their valleys as might be expected. Striking examples are found in high mountains, where valley glaciation has been active. Glaciers still occupy the valleys in these mountains at the higher latitudes, and small glaciers still exist even in middle and low latitudes. In these areas valleys, like that in Fig. 20.1, have forms that are relatively unaltered by any stream action.

The ice ages have had many influences on streams and their valleys. When the last large-scale advance of continental ice sheets ended about 15,000 years ago, glaciers extended as far south as the Missouri and Ohio Rivers. In fact, those rivers flow in channels that were established along the edge of the ice sheet. Of course, all drainage north of the ice margin was disrupted by the advance of the ice. The ground surface there is largely covered by glacial deposits, and the landforms were shaped primarily by glacial deposition, not by stream erosion. South of the former

IGS Photo

Figure 20.1
A small stream flows through a large valley shaped mainly by glaciers in Portshire, Scotland.

edge of the ice sheet, deposits from streams formed by melt water are abundant, and large areas are covered by layers of silt blown from these deposits. Streams are now cutting into these deposits, but the process is in an early stage of development.

The climate changed during the glacial advances and retreats. The changing patterns of atmospheric circulation, temperature, and precipitation had marked influence on stream behavior. In areas where streams had much greater discharge than at present, valleys were shaped by streams of a different character from those that now flow. The balance between load, discharge, and other variables in streams has undergone many changes during and following the ice advances and retreats. Near the coasts, this has been complicated by changes in the water–ice balance and consequent long-term fluctuations in sea level, the ultimate base level.

Volcanism has also played a role in disrupting modern drainage systems, although to a much smaller extent than glaciation. Volcanic activity, especially the filling of valley floors by lava flows, ash falls, and mud flows, has drastically changed the surface drainage in many parts of the United States. And both the Hawaiian Islands and the Aleutians owe their existence to Cenozoic volcanic rocks. The drainage in vast areas of the Cascades and Columbia River Plateau and in smaller areas throughout the region west of the Rockies has been altered by Cenozoic volcanism.

Regional tilting and warping, caused by movements of the crust related to glaciation, are still affecting the areas once covered by ice sheets. In addition, slow regional warping and abrupt movements on faults due to other causes affect large parts of the western United States. The history of these movements in the West extends far back in geologic time, and it has had an important influence on stream action and landform development. Movement on a fault may change a stream's gradient, create a waterfall, block drainage, or even divert streams from one drainage basin into another.

Thus, landforms are a product not only of the action of geomorphic agents now at work but also of those active in the recent past. Forms are related to the stability or instability of the area, to changing patterns of climate, and to other factors. In short, landforms are the product of an area's geologic history as well as of modern processes.

20.2 THE EFFECTS OF TIME

The shape of the land surface changes gradually as erosion takes place. Do these changes lead to orderly progressive changes if continued for millions of years? If so, is there an ultimate situation in which changes essentially stop? Can we predict the types of landforms that might be expected to evolve over a long period of time? William Morris Davis, a famous theoretical geomorphologist, believed that such a pattern does exist and he elaborated one of the grand concepts of geomorphology—the ideal fluvial cycle (Fig. 20.2), described below.

Davis's Cycle

Davis's fluvial cycle of landscape evolution commences with rather rapid uplift of sedimentary rocks from the sea floor. The landscape evolves in a temperate, humid climate in which erosion is accomplished mainly by running water and downslope movement of materials.

A number of widely separated, deeply incised, and narrow valleys containing streams with many rapids, waterfalls, and no flood plains are formed at first, in the stage Davis termed **youth.** By the time the equilibrium situation of a graded stream is achieved, most irregularities, such as waterfalls and cascades, will have been removed from the river channel, and lateral erosion will have become more important. Valleys are broader and contain flood plains; the drainage pattern is better developed; broad upland

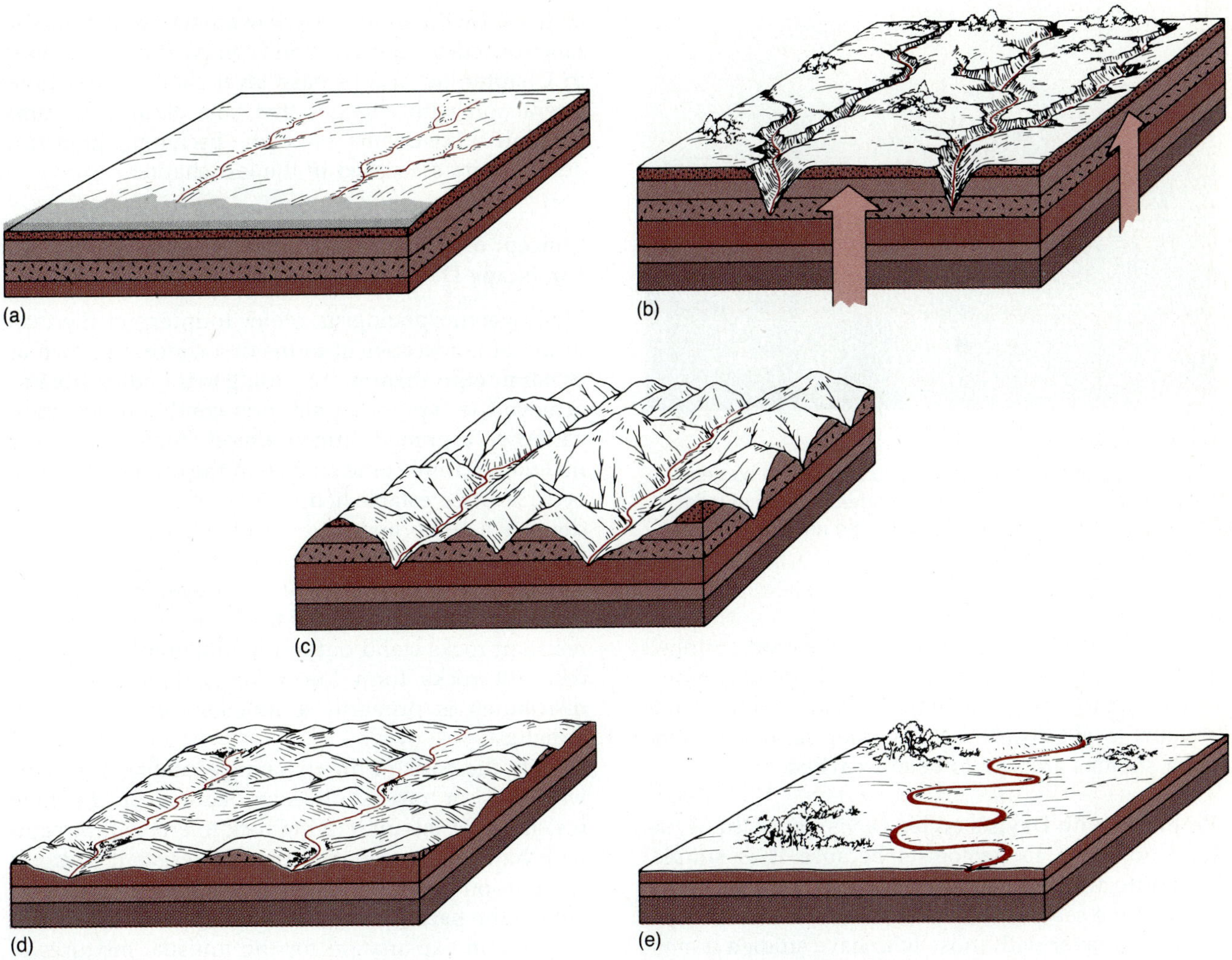

Figure 20.2
William M. Davis thought that the landscape undergoes a systematic evolution. He postulated that following uplift of a flat land near sea level (a) streams cut down forming deep canyons (b). As these become larger the topography looks like (c). Continued erosion reduces relief and slopes to form mature topography (d). Finally, the land surface is again mainly flat (e).

areas near divides disappear; and streams follow sinuous courses. At this time, the landscape is said to be **mature,** and streams cut their channels deeper at a slower rate. Valley slides continue to be reduced by weathering, mass movement—particularly surface creep and wash—and lateral erosion. Gradually, valley sides are reduced to lower and lower slopes. Divides between drainage basins are likewise lowered until they are almost completely flattened. According to Davis, a lowland of faint relief, called a **peneplain,** is the ultimate phase in the erosion cycle. At this stage, called **old age,** running water ceases to be an effective agent of erosion.

Davis believed that compound or multicyclic evolution is more common than single cycles. Topography characterized by features of old age or maturity, but containing elements of youthful landscape, is found in many places. According to Davis, the landscape can revert to youthful characteristics through a process called **rejuvenation,** which can occur in a number of ways. For example, an old or mature topography could be uplifted as was suggested for the Appalachains (Fig. 20.3), or sea level could change, effectively lowering the regional base level. Either of these would steepen the gradient and cause streams to cut down toward the new level. Some degree of

Figure 20.3
The surface of the valley of Virginia appears to be nearly flat, but actually has nearly two hundred meters of relief. This surface was long considered to be an uplifted peneplain by geologists.

rejuvenation should also result if the load supplied to a stream decreases, or if discharge increases. According to Davis, any of these changes could cause streams to cut down, forming deep narrow valleys. Such streams are said to be **entrenched.**

Problems with Davis's cycle. Davis's hypothesis has long appealed to geologists because it is simple, straightforward, and can be used to interpret much of what we see in the landscape, but in recent years it has lost favor with those who have studied it most closely. The concept of the peneplain is perhaps the most frequently challenged idea in this evolutionary scheme. It is difficult to identify good examples of modern peneplains. Many of the erosion surfaces called peneplains are plains formed by the removal of surficial materials down to a resistant flat-lying layer. Some are ancient erosion surfaces, once buried by sediment but subsequently uplifted and stripped by erosion down to the resistant underlying rocks. This type of surface often occurs where the erosion surface developed on resistant igneous and metamorphic rocks, as it did on the Beartooth Mountains north of Yellowstone Park (Fig. 20.4). Other plains, like the Great Plains of North America, are plains of deposition. It is even harder to prove that the high elevation erosion surfaces called peneplains were ever reduced by stream erosion to an elevation near sea level, as postulated by Davis. One by one, the nearly flat surfaces once considered to be peneplains have been reinterpreted. The demise of Davis's cycle has coincided with an increased awareness of the importance of bedrock structure and composition, discussed in Chapter 16, and of base level changes and their interaction with changes that take place in streams through adjustments of load, discharge, gradient, and velocity discussed in the last chapter.

Concept of Dynamic Equilibrium in Landscape Development

Many geomorphologists prefer to interpret the evolution of landscapes in terms of a concept known as **dynamic equilibrium.** According to this idea, streams achieve the type of equilibrium condition described earlier as a graded stream almost immediately and maintain it until some change in the drainage system occurs. Once established, the landforms are eroded down at a rate primarily determined by the resistance of the underlying bedrock to weathering and erosion. The landforms found in a drainage system are related primarily to the geology of the basin. Thus the more resistant rocks stand out in the topography. The less resistant rocks form lower lands. How these are distributed is primarily a function of the regional structure.

On the basis of the dynamic equilibrium theory, waterfalls or rapids develop because of the high resistance of the underlying rock to erosion. Streams with high slopes usually occur on resistant bedrock; streams on weak rocks generally have low slopes. One of the particular advantages of this theory is that it offers an explanation for the unusual mixtures of topographic features that would belong to different stages of Davis's erosional cycle. And it is unnecessary to postulate periodic uplifts to explain the various flat erosion surfaces found in mountain systems.

The dynamic equilibrium theory does not abandon the idea that the landscape undergoes evolutionary development. Rapid evolution would follow certain types of changes, such as tectonic uplift, variation in the type or quantity of load available, or discharge. Load or discharge might be changed by variation in climate or by exposure of a weak rock unit as an overlying resistant unit is removed. Where tectonic activity is going on, an equilibrium condition may not be found, but in long-stable areas such as the Appalachian Mountains, the necessary adjustments have been completed and equilibrium achieved. We will touch on this again as the streams in the mid-Atlantic region are examined.

The photographs of Figs. 20.5, 20.6, and 20.7 give some sense of the evolutionary changes predicted by

(a)

(b)

Figure 20.4
(a) The plateau surface shown here is nearly 3500 meters high in the Beartooth Mountains of Montana. This surface is underlain by ancient metamorphic rocks from which the sedimentary cover has been stripped by streams and glaciers.
(b) This flat surface in southwestern Colorado, at the Mesa Verde National Park, coincides with the structure of the underlying rock.

U.S.A.F.

Figure 20.5
A broad flat plain seen here is in early stages of erosion. Note the wide divides between drainage. Relatively few streams cross the area, and those that do flow in deep gulleylike valleys. The light and dark rectangular areas are cultivated fields. A small village is seen in the foreground.

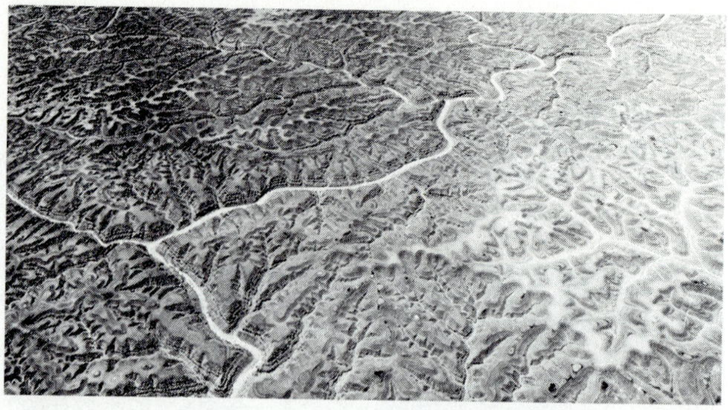

U.S.A.F.

Figure 20.7
The area shown here is underlain by horizontal sediments, as is that in Figs. 20.5 and 20.6, but this area in Saudi Arabia is more completely eroded. Broad flat areas are no longer present along the divides. Most of the ground surface is sloping; very little is flat. The area shown is several kilometers across.

U.S.A.F.

Figure 20.6
This photograph, taken in the same region as that in Fig. 20.5, shows part of the area that is more completely dissected by streams. The heads of the streams are cutting into the plain at left. At left divides are becoming narrow. Rectangular areas are cultivated, as are the terraced slopes seen in the foreground. The distance across the area shown is several kilometers.

the dynamic equilibrium hypothesis. All three of these areas are underlain by flat-lying sedimentary rocks that are relatively uniform in composition and structurally simple. All are about the same distance above base level, and all have been shaped mainly by running water. Each represents a different stage in the progressive dissection of the strata. Broad flat uplands remain in Fig. 20.5, but only a few vestiges are seen in Fig. 20.7. The streams are expanding their drainage system by extension of the heads of the streams into the flat uplands, a process called **headward erosion.**

According to the dynamic equilibrium hypothesis, the streams shown here are graded and responding to changes in load and discharge. This will continue as the region is eroded, until changes in the variables affecting the basin cease. However, erosion will proceed at a slower pace as the higher relief disappears. Streams are less efficient at cutting downward as the gradients are reduced and the level of the land approaches base level.

20.3 FORM AND PROCESS IN ARID CLIMATES

Although arid regions throughout the world are characterized by very sparse rainfall, streams play an important role in shaping the land. As in humid regions, stream action in deserts operates in conjunction with weathering and downslope movements. The amount of vegetative cover and the type of weathering products available are important differences between these climatic regions. In humid regions vegetation promotes infiltration of water and holds surficial materials in place. Absence of this cover means that most precipitation in deserts runs off on the surface giving rise to flash floods. Coupled with the abundant supply of fine weathering products in arid regions, the lack of vegetation frequently leads to the creation of mudflows following the occasional heavy rainfalls found in deserts. These same conditions promote the development of **arroyos** or dry gulleys in deserts.

Because the amount of readily available load in deserts is so great and the amount of water is small, streams are usually choked with sediment. As we saw in the last chapter, this condition causes streams to become braided. Most streams are **intermittent**—they flow only part of the time. For a few hours, they may be rushing torrents with tremendous transporting power; then the water level falls almost as quickly as it rose, and the stream is reduced to a shallow braided channel. Eventually, even this water may infiltrate the ground or be evaporated in the heat of the day.

Often we think of deserts as great expanses of sand, and some deserts do contain large areas of windblown sand. We will examine the character of these special landforms when we study wind action in Chapter 23. In this chapter, our main concern is with the effects of water. First we will take a look at some of the forms found in the arid regions of the American west, where an arid climate and particular bedrock structure combine to produce some of the most spectacular scenery on earth. Then we will compare this region with two other distinctive regions: the lower Mississippi River valley and a part of the Appalachians. As we will see, some of the ideas developed about the origin of the landforms in the arid west can be applied to explain what we see in the humid east.

Arid Regions of the American West

Much of the arid region in western North America is located in the two physiographic provinces, the Basin and Range and the Colorado Plateau, outlined in Fig. 20.8. These provinces are bordered on the west by the Sierra Nevadas and the mountains along the west coast of southern California. To the north lie the young lava flows of the Columbia River Plateau and the Rocky Mountains, which also form the eastern border of the Colorado Plateau.

The Basin and Range Province is characterized by highly irregular topography consisting of broad, flat to gently sloping basins, separated by blocklike mountains, many of which rise more than a thousand meters above the adjacent basin floors. Some of these mountains are high enough (3000 to 4000 meters) to hold snow on their peaks well into the summer months. Both the basins and mountain ranges tend to be elongate in a north–south direction.

The bedrock of the mountains in the Basin and Range is composed largely of consolidated sedimentary rocks, although some is also composed of younger lavas and igneous materials. The internal structure of these mountains is highly complex, but, because most of the rocks are resistant to weathering and erosion, the internal structure is not expressed very clearly in the topography. Extensive faulting during the Cenozoic era has played a major role in defining the forms of the region. Many of the ranges (Fig.

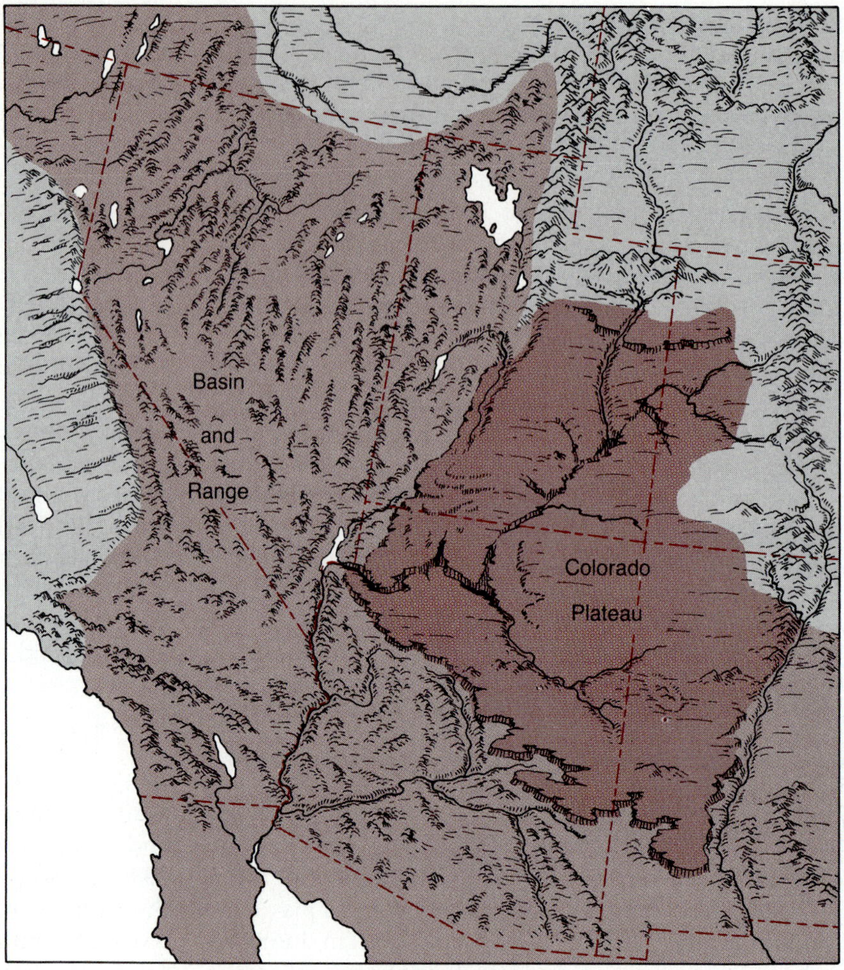

Figure 20.8
The Basin and Range and the Colorado Plateau physiographic provinces. Of the few streams that drain the vast areas included in these provinces, the Colorado River is by far the largest. Much of the Basin and Range is not drained by streams that flow out of the basins. One of these basins, Death Valley, is below sea level.

20.9) are bounded by high-angle faults, some of which are still active. Some of these ranges are horsts or tilted fault-block mountains (see Section 11.1); others are anticlinal uplifts (Section 11.1). The basins are located where the crust is bent down or where blocks of the crust are low relative to adjacent mountains, as a result of movements on steeply inclined faults. Many are grabens (see Section 10.2) or down folds. The bottoms of these basins constitute the regional base level. The floor of Death Valley, one of the regional base levels, is about 90 meters below sea level. Obviously streams do not flow out of such basins toward the sea. In fact, in this arid region few permanent streams exist, and those that do flow from the mountains into adjacent basins, where the waters evaporate and disappear into the ground.

In contrast to the Basin and Range, the Colorado Plateau has a more unified bedrock structure. The plateau is largely underlain by sedimentary rocks that are flat-lying or gently dipping. These are offset in places by high-angle faults or broad flexures, such as those shown in Fig. 20.10, but for the most part the plateau is much less deformed than either the Basin and Range to the west or the Rockies to the north and east.

Despite the slight internal deformation of rock under the Colorado Plateau, the region as a whole has been uplifted during the deformation of western North America. This uplift has been sufficient to elevate rocks originally deposited below or near sea level to heights between one and two thousand meters above sea level. Thus the plateau surface is nearly two kilometers high.

The landscape of both the Basin and Range and the Colorado Plateau is characterized by extensive flat to gently sloping surfaces or plains. But as we shall see, not all of these plains formed in the same way.

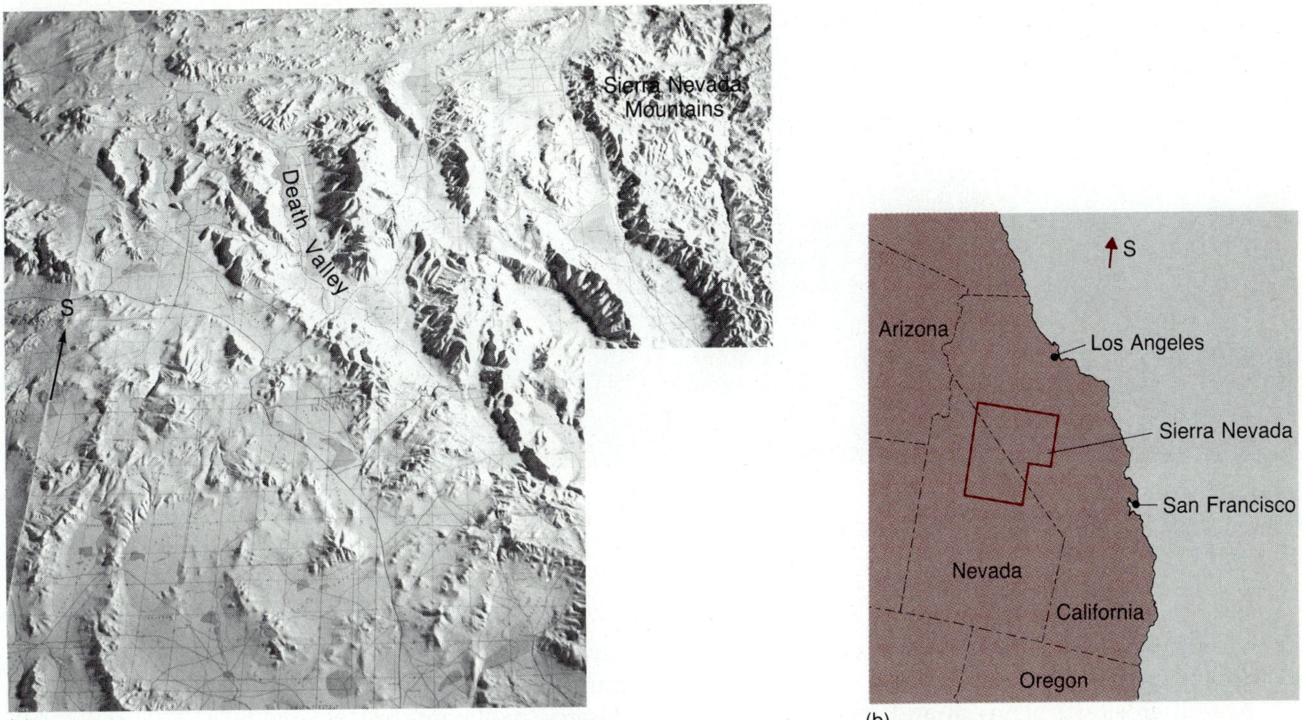

Figure 20.9
(a) Characteristic topography of the Basin and Range physiographic province includes elongate mountain ranges separated by flat floored basins. The Sierra Nevadas lie on the western edge of the Basin and Range province. Like many of the mountains, they are bounded by a steep fault. The entire province is arid. (b) Locator map.

Plains produced by erosion. Prominent in the landscape of the Colorado Plateau and the Basin and Range Province are plains formed by erosion. These plain surfaces are typically covered with loose materials, such as sand, silt, and gravel (Fig. 20.11), moved across the ground surface by sheet wash following flash floods. Much of the surface is cut by shallow braided channels of dried-up streams that flow for only a few hours after a heavy rain.

In many parts of the Colorado Plateau, the bedding is flat or gently dipping, and the plain is formed by the gradual retreat of a cliff at the edge of the plain. The underlying sedimentary strata is gradually exposed by surface wash and erosion. Plains formed in this way are flat primarily because underlying rock layers are nearly flat, and the layers are being stripped by the gradual but progressive removal of the face of the cliff. The flat surface of the plain above the Rio

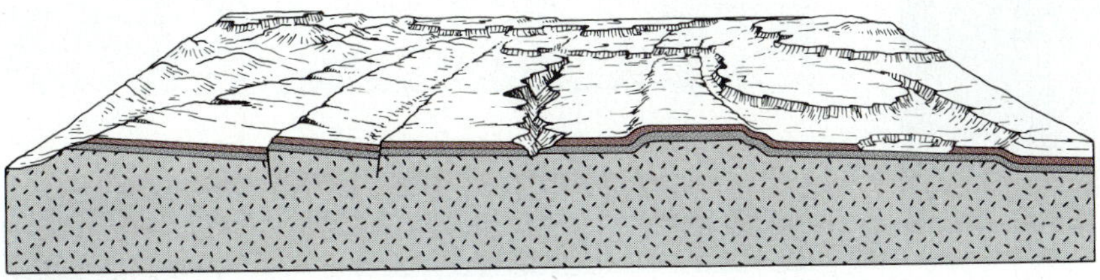

Figure 20.10
The Colorado Plateau is a large block of continental crust that has been uplifted along steep faults. The ancient igneous and metamorphic rocks are exposed in the bottom of the Grand Canyon, but most of the plateau is covered by slightly deformed sedimentary rocks.

Figure 20.11
Typical Basin and Range topography consists of gently sloping surfaces, often terminated abruptly at the edge of mountains by a cliff.

Grande (shown in Fig. 20.12) is an example of such a stripped surface.

Not all erosional plains in arid regions coincide with structural surfaces. Few plains in the Basin and Range province originate as structural surfaces. Here the gently sloping surfaces cut across outcrops of different rock types, often of different resistance to erosion. These plains, called **pediments,** (Fig. 20.13), slope gently up toward mountains or cliffs and are covered by a thin veneer of sediment that is in transport by surface wash or in shallow channels. Because pediments are covered by so little sediment, it seems likely that the slope of the pediment is approximately the slope needed by runoff to move the products of weathering and erosion across the pediment. The most active erosion takes place at the upper edge of the pediment, where the cliff or mountain front is subject to mass wasting as well as erosion by water. Debris eroded from the cliff accumulates at its base and on the upper edge of the

Figure 20.12
Flat surfaces, like these in the Colorado Plateau near Taos, New Mexico, are often the result of stripping of flat-lying strata. Note also the gently sloping erosion surface rising to the base of the cliffs. The canyon of the Rio Grande is over a hundred meters deep at this locality.

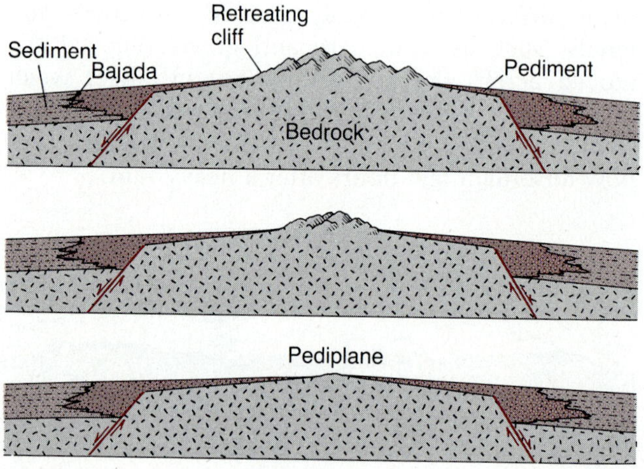

Figure 20.13
Evolution of topography in the Basin and Range is thought to involve gradual slope retreat and growth of the pediment until, in the final stage, a gently sloping plane, a pediplane, is all that remains.

pediment, producing a steeper slope there. Over time this accumulation is removed by the surface wash and gradually the cliff is eroded back. Retreat of cliffs and the ensuing growth of the pediment is a common feature of arid climates. If the process of slope retreat continues long enough, the high ground forming the cliff will disappear, as shown in Fig. 20.13. The resulting gently sloping plane is called a **pediplane.**

Planes of deposition. Large fan-shaped deposits of stream alluvium form where mountain streams flow out into basins, larger valleys, or onto pediments. The streams flowing on steep gradients in the mountains are abundantly supplied with load from the steep valley walls. The sudden change in gradient that occurs at the mountain front drastically reduces the capacity of the stream. Consequently, much of its load is deposited. As deposits of alluvium build up, the stream shifts from one position to another, breaking up into a system of distributaries reminiscent of those found on the front of a delta. Channels are usually shallow, braided, and temporary features. Eventually, the alluvium builds out into the basin, creating a large, gently sloping, fan-shaped deposit, such as the ones shown in Figs. 20.14 and 20.15, called an **alluvial fan.** Adjacent fans may coalesce at the foot of the mountains to form a broad surface. If the supply of sediment is great enough, fans may cover and bury large areas of previously formed pediments.

Depositional plains also form farther away from the mountain front, typically at the foot of the pediment and in the central part of the basin below the pediment. This depositional surface, called a **bajada,** forms where the sediment carried across the pediment is dropped out as the water evaporates or infiltrates into the ground. These sediments form the thick deposits in the valley floor between mountains. The central part of the basins may also be the sites of intermittent lakes, called **playa lakes,** which form for a time after a heavy rainfall but gradually evaporate, leaving a flat area of fine sediment and often salt deposits (Fig. 20.14).

Lessons from study of the Colorado River basin. The Colorado River and its tributaries flow across one of the most diverse and spectacular regions in the world. With headwaters in the high mountain peaks of western Colorado and as far north as Wyoming, the river flows to the Gulf of California. As it crosses the Colorado Plateau, it cuts one of the deepest gorges on earth, the Grand Canyon, and brings water to the parched deserts of Arizona and southern California (Fig. 20.8). Along all but its headwaters, the Colorado flows through arid and semiarid lands.

Scientific exploration of the Colorado began with the famous expeditions of John Wesley Powell, which began in 1867. Powell, a largely self-taught scientist with a keen interest in nature and especially in rivers, spent his early years exploring, teaching, and collecting. He lost an arm at the Battle of Shiloh during the Civil War. Following the war, Powell returned to teaching. Powell's first trip was spent largely in the upper reaches of the Colorado in Wyoming and Utah, but he soon resolved to explore farther south. In 1869, with support from several Illinois science groups and the U.S. Army, he set out to travel the Colorado River by boat through the canyons. Despite the loss of one boat smashed in the rapids and the refusal of several members of the expedition to go through some of the rapids in the canyon (two of them were killed by Indians as they climbed out of the canyon), Powell continued. In following years, he returned with greater financial support; and, finally, by 1879, he convinced Congress of the need to establish a national geological survey. Powell's explorations were very important in opening the west to settlement. He brought back the first reliable and comprehensive reports concerning the western lands, their natural resources, and the Indians who inhabited those lands.

Powell proved to be an outstanding scientist as well as a keen observer and hardy explorer. As a result of his efforts to encourage the government to take a more active role in the exploration and development of its western lands, he was appointed as the first director of the newly created U.S. Geological Survey. Some of his observations on the Colorado led him to formulate theories about the relation of streams to the land they drain that are still in use. He was the first to recognize and explain the concept of **base level** (see Section 14.4):

> We may consider the level of the sea to be a ground base level, below which the dry lands cannot be eroded; but we may also have, for local and temporary purposes, other base levels of erosion which are the levels of the beds of the principal streams which carry away the products of erosion . . .

This important concept provides a basis for understanding many aspects of the relationship between streams and the form of the land through which they flow. It is critical to understanding the behavior over the last million years of streams that flow near sea level. We will examine this behavior in Section 20.4

◀ **Figure 20.14**
White salt deposits in the center of Death Valley basin are clearly visible. Alluvial fans in the foreground cover the pediment. No clear break is visible between these fans and the deeper basin deposits. (Photo by John Shelton.)

when we consider the history of the Mississippi River. It is important to note here, however, that unlike the Mississippi, the Colorado River flows well above its base level throughout much of its length. This is clearly one reason the Colorado and its tributaries flow through so many and such long canyons (Fig. 20.16).

Powell saw that the streams in the Colorado River basin were not all alike in their relationship to the structure of the bedrock on which they flow. He proposed a classification of these relationships that is still in use. It contains four categories; consequent, superimposed, antecedent, and subsequent streams.

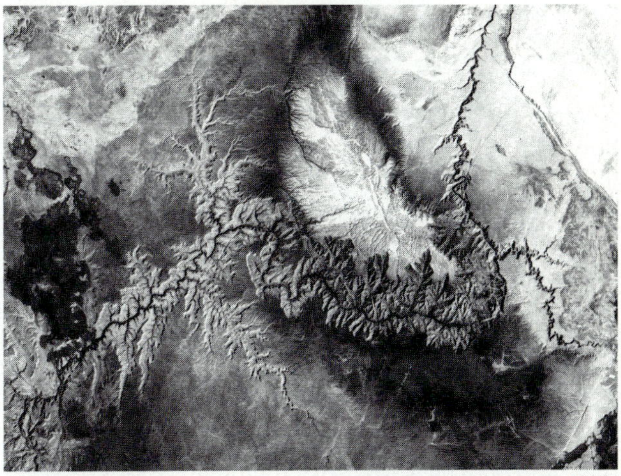
NASA

Figure 20.16
Part of the Grand Canyon of the Colorado River as seen by satellite. The different shades of gray indicate areas where different strata are exposed. Most of the surface area is formed by flat-lying strata from which the cover has been stripped by erosion. The black areas are outcrops of volcanic rocks, basaltic lava flows. The distance across the area shown is about 120 kilometers.

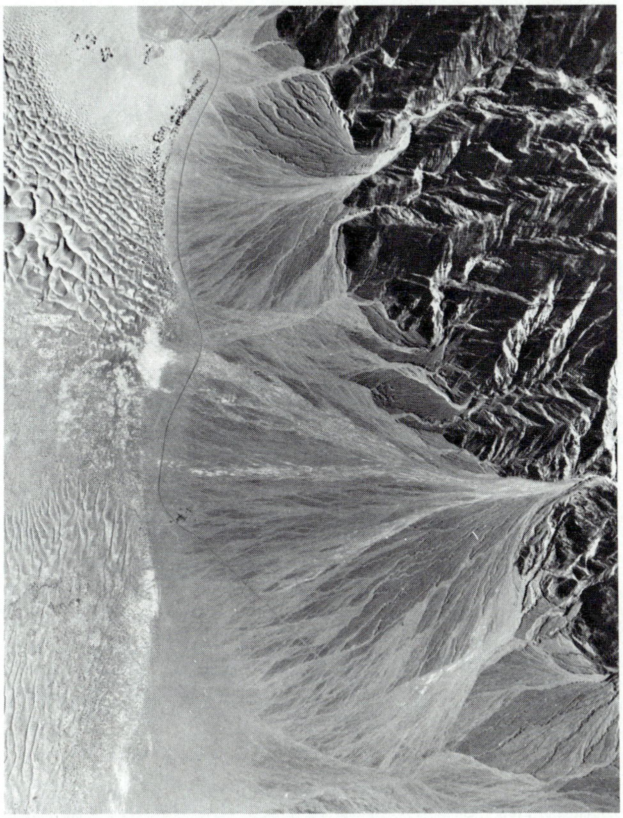
U.S.G.S.

Figure 20.15
This vertical photograph shows clearly the fan shape and braided distributory channels found on alluvial fans. Such fans may cover pediment surfaces eroded before the fans formed. The distance across this area is about 8 kilometers.

Powell proposed the name **consequent** for streams that flow down the initial slope of a newly formed surface. Streams flowing off volcanoes or down recently emerged portions of the continental shelf are examples. In the Colorado plateau, streams flowing down the bedding surface of newly tilted or folded strata would be examples.

Superimposed streams are those that have cut down to their present courses through an overlying cover. For example, a stream flowing across flat strata might, as it cuts down, encounter and cut through an angular unconformity. Its course in the rocks beneath the unconformity might then be determined by its earlier course, established when the stream was flowing on the overlying cover.

The third class of streams, called **antecedent,** are streams that persist on their course on the land surface even as the bedrock beneath them is deformed by folding, uplift, or faulting. Such streams cut their channels down rapidly enough to maintain a slope across a rising structure.

A fourth class, **subsequent streams,** are those that follow courses dictated by the resistance of the underlying bedrock. Such streams follow the paths of least resistance. They generally flow in shales or other weak rocks or in fault zones rather than across resistant rocks.

Antecedent drainage. Of course all streams flow downslope, but often the stream bed does not slope in the same direction as the underlying bedrock—the usual condition of consequent streams. Powell was very impressed by this discordance. In many places throughout the Colorado Plateau, he found streams that followed courses through canyons and across the structural features of the underlying bedrock.

The course of the Green River, a tributary of the Colorado, especially intrigued Powell where the river cuts across the Uinta Mountains in northern Utah (Fig. 20.17). The Uinta Mountains are formed by an east–west trending anticlinal uplift in which ancient (Precambrian) gneisses have been raised and subsequently exposed along the crest of the fold over 3000 meters above sea level. Younger sedimentary rocks dip off the uplift on both sides. Despite this imposing uplift, the Green River flows south across Wyoming to the edge of the Uinta uplift, where its course is slightly deflected to the east; but it then turns and flows across the Uinta uplift in a spectacular gorge, the Ladore Canyon, shown in Fig. 20.18. Powell knew that the uplift of the Uintas was relatively young because many of the youngest sedimentary layers in the region show effects of the folding. He concluded that the Green River was flowing south into the Colorado River before the Uinta uplift took place, and that the main stream was able to cut its channel downward fast enough to maintain its course across the uplift even as it formed. In contrast, many of the smaller streams flowing off the Uintas are consequent—they flow off the uplift both north and south.

Figure 20.17
The Green River flows across the eastern end of the Uinta Mountains. Powell thought the river maintained its course while the mountains were uplifted. Most smaller streams in the area are consequent streams.

National Park Service

Figure 20.18
This Green River flows through this spectacular gorge, where it cuts across the Uinta Mountains. Here it flows directly across an anticlinal structure

Recent studies of this drainage by U.S. Geological Survey scientists indicate that the evolution of this drainage is much more complex than Powell suspected, but his idea of antecedent drainage is still considered to be a valid concept of geology.

Stream piracy. The origin of valleys apparently shaped by stream action but now dry was one of the major insights of a geologist named G.K. Gilbert, who worked with Powell in his explorations of the west. Gilbert saw many canyons similar to the Unaweep Canyon (Fig. 20.19). This canyon, about one and a half kilometers wide, is shaped like a river valley and contains river gravels, but only small streams now flow out of it at both ends. Gilbert concluded that the stream that cut this valley has been diverted by a process he called **stream piracy.**

Piracy occurs when a stream with a channel below the level of another stream intersects the drainage of that second stream. The intersection occurs because the head of the pirate stream is eroding more actively. It grows by headward erosion into the drainage basin of the adjacent stream and may eventually intersect the main channel of the captured stream. Probably, the Colorado River once flowed through the Unaweep Canyon and the river flowed across a rising anticline.

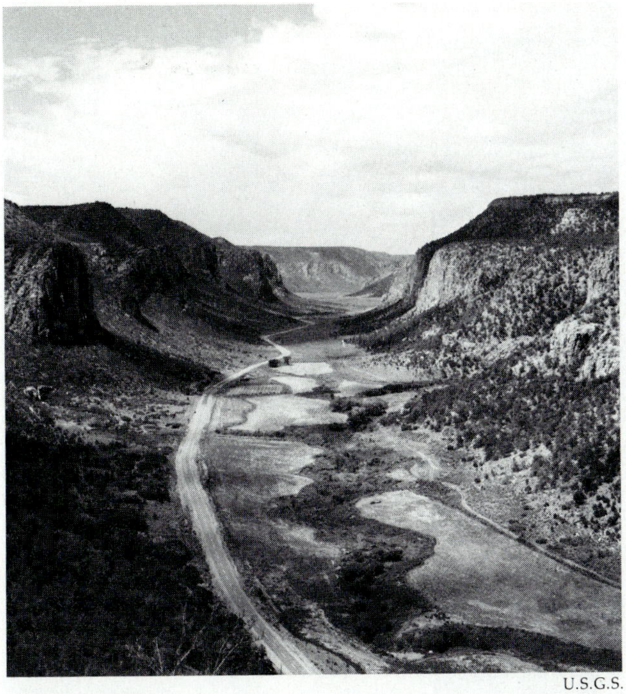

Figure 20.19
Unaweep Canyon, the abandoned canyon of the Gunnison River and probably of the Colorado River, across the uplift at the Uncompahgre Plateau. The mile-wide canyon is in Precambrian gneissic rocks. The uplift that elevated this canyon and diverted its drainage occurred about 2 million years ago, in the Pliocene epoch.

Eventually, the headwaters of the Colorado were diverted into a stream that flowed around the rising arch, as shown in Fig. 20.20. As the drainage changed, the old course through the Unaweep Canyon was abandoned, leaving it dry.

Headward erosion and stream capture is often suggested as one way streams cut across resistant ridges. Canyons through ridges of resistant rock are called **water gaps** if streams flow through them. In some places, water gaps are located along faults, fracture zones, or other lines of structural weakness that would be easily cut by the headward erosion of an active stream. Water gaps may be formed in other ways, as we shall see in the discussion of Appalachian drainage.

Entrenched meanders. Along most of its path to the sea, the Colorado River and its tributaries flow through a succession of steep-walled canyons. One of the most spectacular sections of the Colorado and Green Rivers is that section of the Colorado Plateau south of the towns of Green River and Moab where both rivers follow meandering paths through steep-walled canyons hundreds of meters deep, as shown in Fig. 20.21. Only one explanation seems likely—that the rivers were flowing in meandering courses on a nearly flat surface, and that they continued this course as the region was slowly uplifted. The flat surface on which they flowed is still flat, but it is now high above them on the plateau. Even today, the rivers appear to be concentrating their channel erosion downward rather than laterally. Such river courses are said to be entrenched. Entrenched meanders, alluvial fans, and abandoned stream channels are

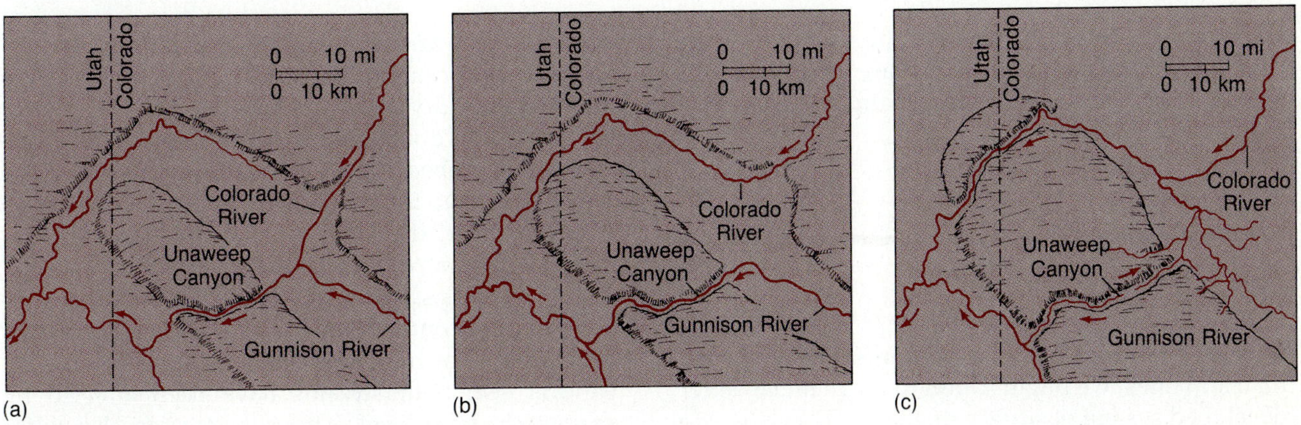

Figure 20.20
(a) The Unaweep Canyon was formed when the Colorado River followed a course through it. (b) Later, the main stream was diverted. (c) Finally, the Gunnison River was also diverted, leaving the Unaweep dry.

(a)

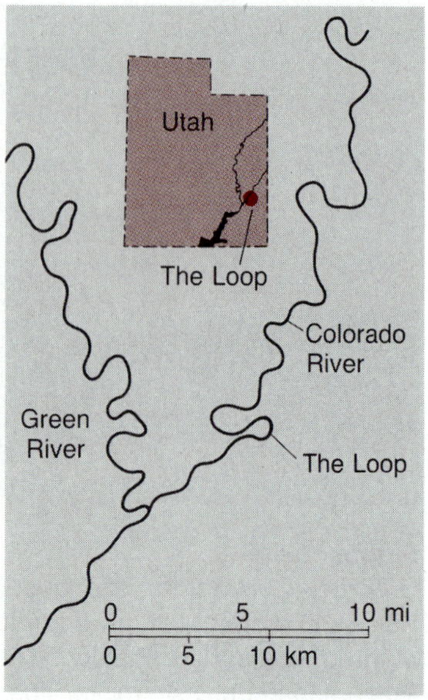

(b)

Figure 20.21
(a) Crossing the Canyonlands section of the Colorado Plateau, the Colorado and Green Rivers flow through deep canyons in Jurassic, Triassic, Permian, and Pennsylvanian rocks. This view of the Loop is about 6 miles above the junction with the Green River. The gorge here is about 500 feet deep. (b) Locator map.

displayed in the dry climates of the Colorado River Basin, but these same features are also found in humid climates, as we will see in the following sections.

20.4 FORM AND PROCESS IN HUMID CLIMATES

Because humid regions receive so much more rainfall than arid ones, we might expect to find a much better developed system of surface drainage. Usually we do see many more streams, deeper and better developed valleys, better-formed and larger tributaries, and of course more permanent streams. While many of the large valleys in arid areas contain intermittent streams only the upper parts, the heads, of small valleys are intermittently dry in humid areas.

Although drainages in humid areas share the features described above, they often differ from one area to another for other reasons. In the examples described here, the lower Mississippi and the Potomac, the streams differ in several important respects. The lower Mississippi flows near base level on a very low slope. In contrast, only in the last few kilometers is the course of the Potomac close to base level. As we will see, both streams have been affected by changes in base level. The two streams also differ remarkably in that the lower Mississippi flows in a broad and deep alluvial plain as it crosses the Gulf Coastal Plain, while most of the course of the Potomac lies in the folded and faulted rocks of the Appalachian Mountains, described earlier in Section 11.1.

The Mississippi River

The Mississippi River drains more than 3.3 million square kilometers with a network of rivers more than 100,000 km long, the largest drainage basin in North America. It discharges an average 15,360 cubic meters of water per second across its delta, and in flood stage it may discharge almost four times as much. The Mississippi has three major tributaries: the Ohio drains eastern interior lowlands and the Appalachian Plateau region; the Missouri headwaters reach into central Montana and drain the central Rocky Mountains; and the Arkansas River heads in the high mountains of southern Colorado. The upper Mississippi drains the north central states (Fig. 20.22). This vast system of rivers and tributaries flows across and through many different geological settings. We will look at some of the features along only one section of the river, its lower part.

Figure 20.22
The Mississippi River carries runoff from a vast area of the central United States. The river flows over a wide alluvial surface (darker tone) or windblown (lighter tone) deposits in the areas shown by patterns.

The lower section of the Mississippi flows across the Coastal Plain from near its junction with the Ohio River to the Gulf of Mexico (Fig. 20.23). Topography throughout the Coastal Plain is subdued; relief is rarely over a hundred meters, which is approximately

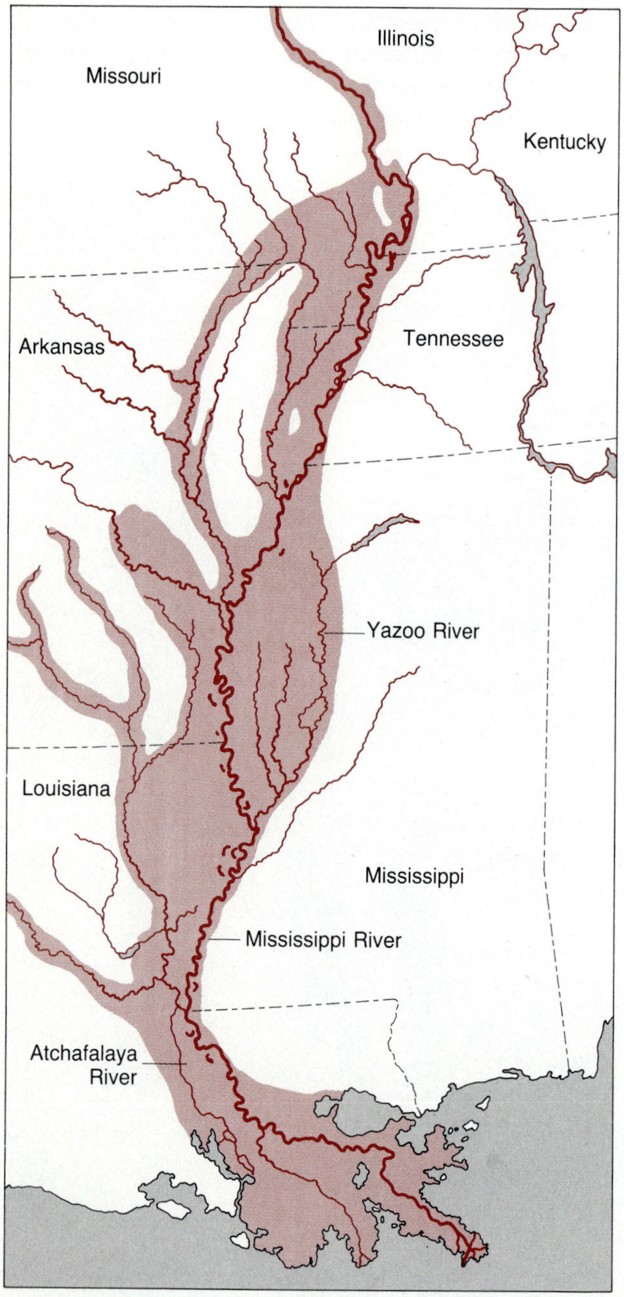

Figure 20.23
The lower Mississippi is characterized by meanders, which flow in a broad alluvial plain. This part of its course is in the Gulf Coastal Plain.

Figure 20.24
Deposits of silt carried into the Mississippi River valley by strong winds during the Ice Ages. These deposits form cliffs where cut by streams or roads.

the elevation of the junction of the rivers at Cairo, Illinois.

For all its long, meandering course to the delta, the Mississippi flows on a broad alluvial plain (Fig. 20.23), composed of stream channel, flood plain, and marsh deposits. At a few places, the river is at the edge of the flood plain and is cutting into unconsolidated and semiconsolidated sediment, composed mainly of silt blown into the region from the edge of ice sheets that covered the northern states during the last glacial advances. These deposits form a blanket-like sheet over large areas of the Mississippi Valley and form cliffs several tens of meters high where the river approaches them as it does at Vicksburg, Mississippi (Fig. 20.24).

The channel forms and stream patterns of the lower Mississippi are typical of those found along many meandering streams. The present belt of meanders is about 10 to 20 kilometers wide, but it is obvious from aerial photographs and maps that the meander belt is not fixed. The stream meanders travel by two types of movement—the meanders tend to migrate downstream, and the meander belt as a whole shifts slowly across the alluvial plain.

These changes occur as a result of erosion on the outside of curves and deposition on the inside of curves. The outside of curves are gradually undercut because velocity and turbulence are greatest there (see Section 19.3), and sediment builds up as **point bars** on the inside of curves. The successive buildup of point bars produces a characteristic pattern, seen in Fig. 20.25. Occasionally, during floods, a channel, called a **chute,** will reopen between point bars.

Over a long period of time, the enlargement of meanders by concentration of erosion on the outside of curves causes the river to intersect itself; a **cutoff** is formed as the course changes. The abandoned channel remains as a loop of slow-moving or stagnant water, which is eventually silted in by the influx of sediment from small streams and by sediment from the main channel that settles in the quiet water. Lakes formed in this way are called **oxbow lakes.** The pattern of former oxbow lakes can often be seen long after they are filled by silt. Excellent examples are seen in Fig. 20.25.

Some shifts in the position of the meander belt take place gradually as a result of lateral erosion. Others occur rapidly during floods, when the main flow of water can be diverted into other channels on the flood plain. In the lower reaches of the Mississippi the main flow has, at times, been directed into another river, the Atchafalaya, which flows south parallel and close to the Mississippi. Abrupt channel changes have also taken place as a result of movements associated with earthquakes. Such changes occurred during the New Madrid earthquake in southeast Missouri (see Section 8.4).

Natural levees are well developed along the lower Mississippi; these contain moderate increases in the river level, but they also keep small streams on the flood plain from draining into the main channel. The result is a region of sluggish drainage on the flood plain. The meandering streams with little or no movement are called **bayous.** Some large streams flow parallel to the main channel and drain flood plains, as does the **Yazoo River** (Fig. 20.23), for which this type of river is named.

Natural levees are beautifully developed on the Mississippi delta, where the river flows in a channel slightly higher than the surrounding country. This relief, never more than a few meters, causes drainage changes to occur rapidly during floods and aids in the development of the distributary pattern (Fig. 20.26). Mats of organic material, called **peat,** mixed with mud, cover most of the top of the delta (Fig. 20.27). Within this are long thin deposits of sand formed in the channel and as natural levees. Over

U.S.A.F.

Figure 20.25
Meanders, a partially filled-in oxbow lake (foreground), and curved point bar deposits that leave scars in the alluvium are typical features of the lower Mississippi Valley. The distance across this area is several kilometers.

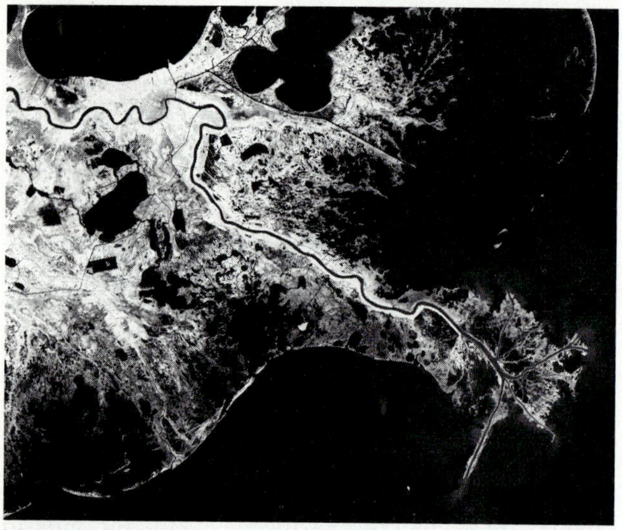

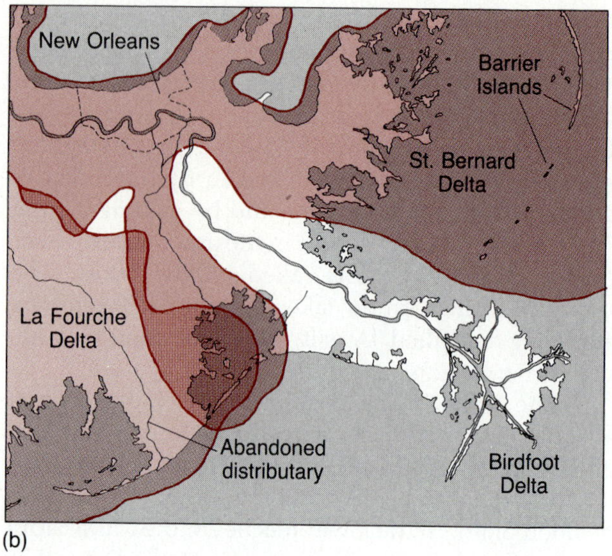

Figure 20.26
(a) This satellite image of the Mississippi shows the distributaries at the end of the modern Birdfoot delta, at right. Older deltas, like Lafourche, are located farther west, and the St. Bernard is to the east (right). Beach deposits now outline these older deltas. (b) Schematic drawing of the area.

long periods of time the delta grows, as the sediment load on the order of 2.4×10^{11} kilograms is carried onto the delta each year. The deposits laid down on top of the delta gradually fan out and cover those laid in front of it.

Evolution of the lower Mississippi River. The Colorado, its tributaries, and the landscape of the Colorado Plateau have clearly been greatly affected by changes in base level related mainly to tectonic uplift of the region. Base level changes have also affected the Mississippi drainage, but there the changes have been of a different kind. The base level of the Mississippi was drastically affected by a drop in sea level that took place during the last glacial advance. During the maximum advance of ice the level of the oceans was nearly 100 meters lower than it is today, as water from the oceans was stored in glaciers on land. Of course, the Colorado was affected by this too, but the elevation of the Colorado along most of its course was so great that the drop in base level was not significant except along the lower part of its

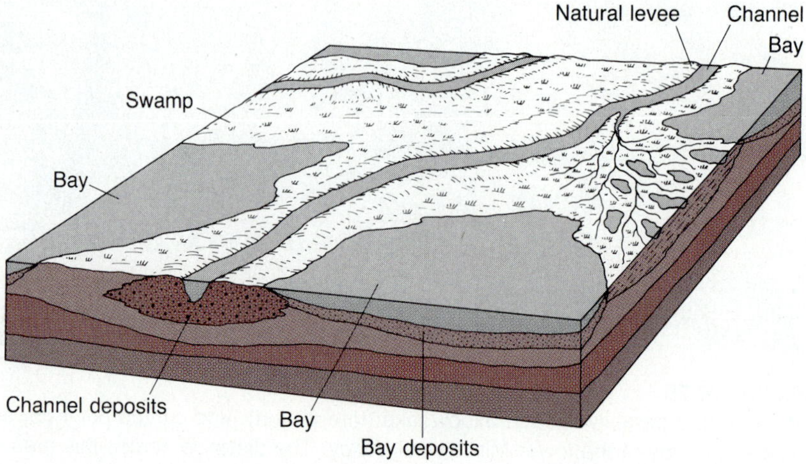

Figure 20.27
Deposits on modern deltas include the types shown here. Sand and silts make up the stream channel and its natural levees. Swamp deposits rich in organic matter and clay flank the main distributaries and open bays. A break through the natural levee is shown at right.

course. The effects there were similar to those we will consider for the Mississippi.

The lowered base level increased the gradient of the Mississippi. The consequent increase in velocity caused the river to cut its channel down rapidly as sea level dropped (Fig. 20.28). At the glacial maximum, the stream most likely flowed in a sinuous channel in a valley several times deeper than the present valley. As the ice began to melt, large volumes of meltwater were added to the normal runoff, and discharge must have increased dramatically. At the same time, a huge influx of load from the melting glaciers came into the drainage. The load exceeded the capacity of the stream, and the channel became choked with debris. The stream became braided. Sands and fine gravels from the glaciers were carried in the bed load as far south as the present Gulf Coast.

The deep valleys cut during the glacial maximum gradually filled as more and more sediment came into the valley. The rise in sea level that was taking place at the same time speeded this filling process, as it decreased the grade of the river toward its mouth, causing more rapid deposition of deltaic deposits there. As the gradient became still lower and the load supplied decreased, the river gradually assumed a meandering pattern similar to the one we see today.

During these changes the course of the lower Mississippi has not remained static. The main channel has shifted position, and its point of entry into the Gulf has changed. For this reason, large deposits of deltaic sediments, similar to those on the modern delta, are found along the southern Louisiana coast (Fig. 20.26). These deltas merge, the younger ones superimposed on the older.

Rivers of the Middle Atlantic States

Most of the streams that flow into the ocean along the middle Atlantic states originate in the folded

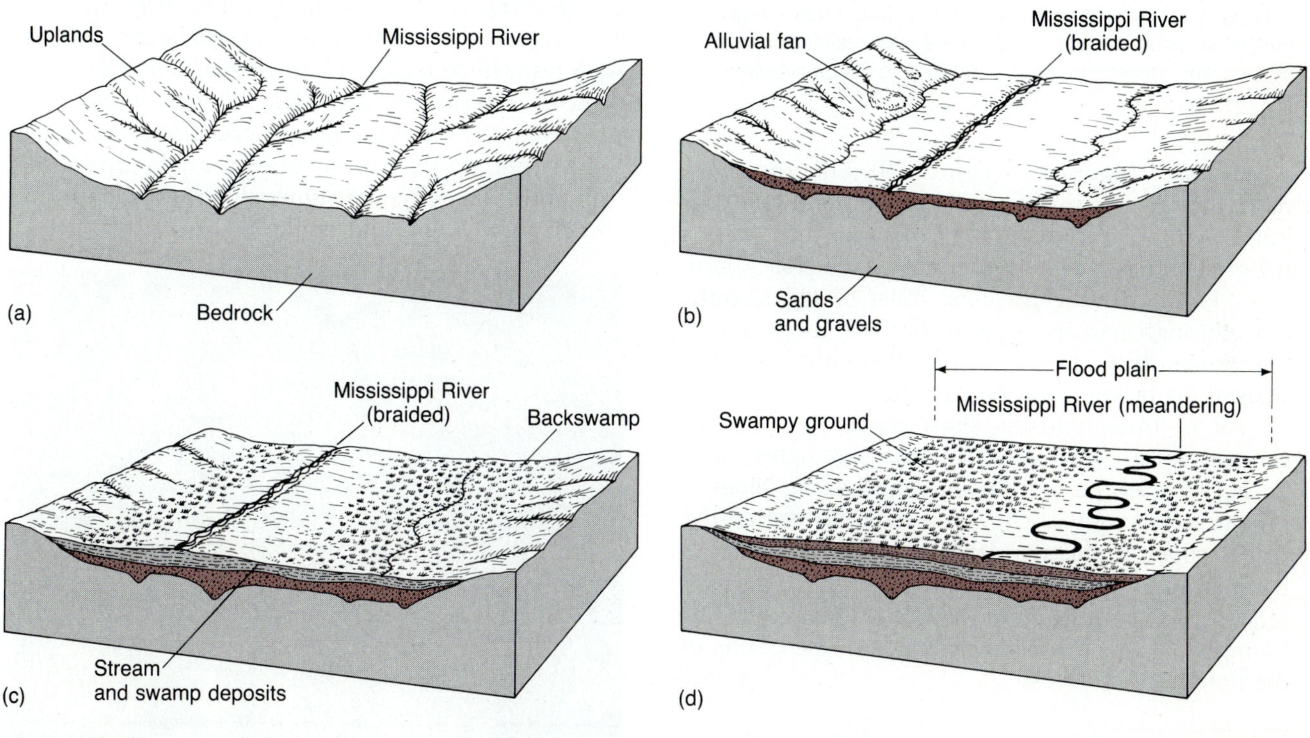

Figure 20.28
Four stages in the evolution of the lower Mississippi during the last part of the Ice Age. (a) When glaciers reached their maximum extent about 18,000 years ago, sea level was more than 100 meters lower than it is at present. Because of this lower base level the Mississippi cut its channel deep into the Coastal Plain sediments. (b) As base level rose, the valley was filled in by sediments washed in from the glacial deposits. The river probably had a braided pattern as shown. (c) When sea level rose close to its present level, swamp deposits began to fill in the valley. (d) Finally, when sea level reached its present stage the river took on a meandering pattern, and the valley fill is close in elevation to the land outside the valley in many places.

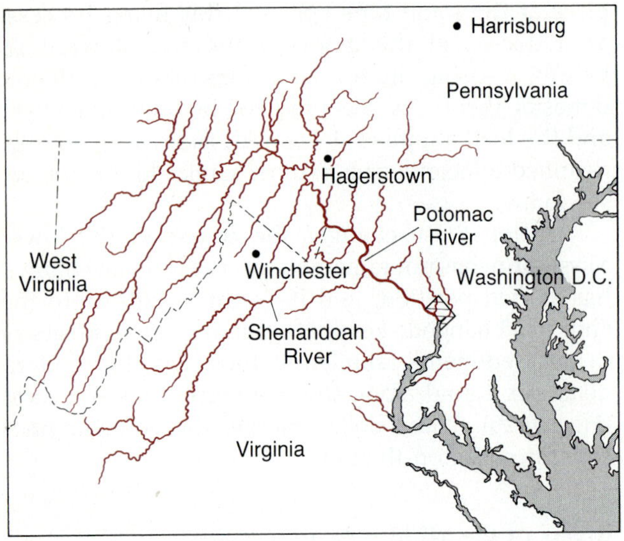

Figure 20.29
The major streams in the central part of the Appalachians, like the Potomac river shown here, flow into the Atlantic Ocean. The major streams flow southeast across rocks of varying resistance to erosion, but their tributaries follow northeast–southwest courses on the less resistant rocks, shales, and limestones in the Valley and Ridge province.

Appalachian Mountains (Fig. 20.29). Like the Colorado, they flow across several quite different geological provinces, but unlike the Colorado, they cross a region that has been tectonically stable for a long time. From their headwaters, most of the streams flow through narrow gaps in the Blue Ridge across the metamorphic terrain of the Piedmont, over the Coastal Plain, and into the Atlantic.

All of these streams show markedly different channel forms, drainage patterns, gradients, and loads, as they pass from one of these geological provinces to another. They all drain eastward off the Appalachian highlands, and they all flow across bedrock provinces that differ tremendously in resistance to weathering and erosion. We will look at some examples of this drainage drawn from the basins of the Potomac River basin.

Subsequent drainage of the folded Appalachians. The fold belt in the Appalachians known as the Valley and Ridge province consists of ridges held up by quartzites and sandstones and valleys floored mainly by limestones and shales. A close relationship exists between the structure of these beds and the topography, but many valleys are formed in shales exposed along the eroded crest of long anticlinal folds, and some ridges are the troughs of synclines held high by folded quartzites (Fig. 20.30). This condition, in which upfolds or anticlines are valleys and synclines are ridges, is referred to as topographic inversion.

Drainage in the Valley and Ridge is closely adjusted to rock structure in general. Small streams flow down steep slopes, on the flanks of the ridges, into larger streams on much lower gradients. The larger streams generally flow in valleys flooded by less resistant rocks, such as shale or limestone (Fig. 20.31). Where these streams cross more resistant rock belts, they flow through water gaps.

Origin of the transverse drainage of the Appalachians. Many of the major streams in the Appalachians flow transversely across the structures of the region. This transverse drainage is especially dramatic where rivers cut across the metamorphosed and igneous rocks of the Blue Ridge, much as the Green River cuts through the Uinta Mountains. The origin of these **transverse streams** has long been debated. The overall drainage pattern—subsequent drainage with transverse major streams—has long been interpreted in terms of the ideas Powell and Gilbert proposed in the west. The main trunk streams appear to be superimposed on the underlying bedrock; their tributaries have become adjusted to the structure and composition of the underlying bedrock.

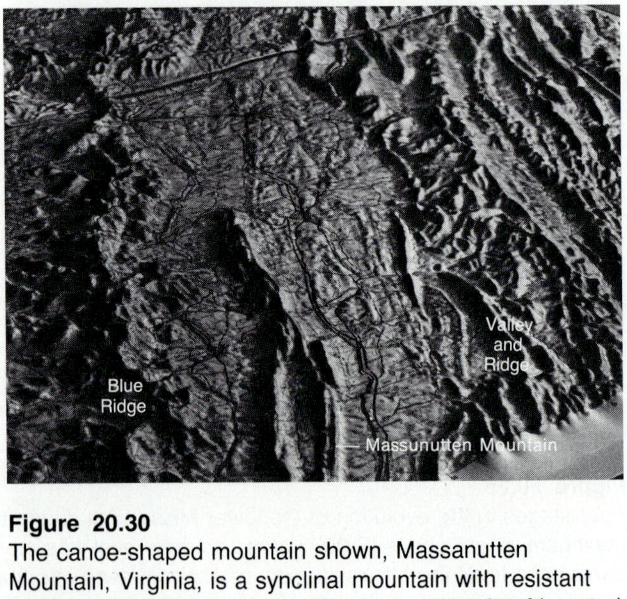

Figure 20.30
The canoe-shaped mountain shown, Massanutten Mountain, Virginia, is a synclinal mountain with resistant quartzite forming the ridges. This is an example of inverted topography, which is common in the Appalachian Mountains.

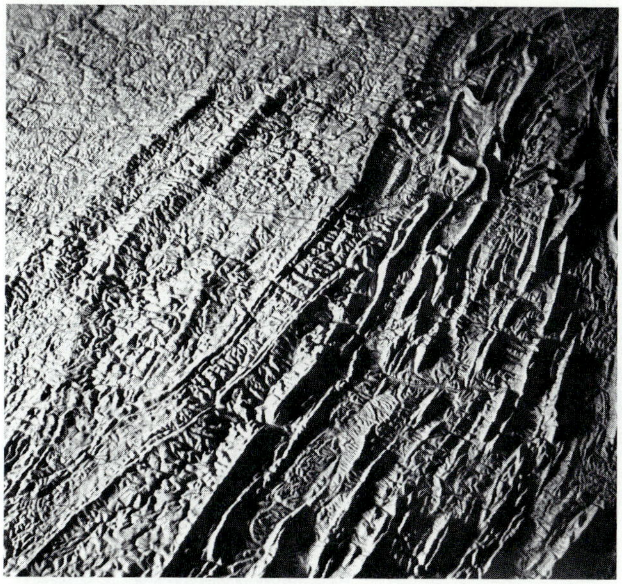

Figure 20.31
This map covers a large part of the drainage of the Potomac River, shown in Fig. 20.29. Comparison of these figures will clearly demonstrate the relationship between the subsequent drainage in the Appalachians and the landscape. Most of the ridges are held up by quartzite.

For many years it was argued that the main streams became superimposed on the underlying rock structure as they cut down through a cover of sediments, like those now found in the coastal plain. This downcutting was presumably initiated as the region, which was thought to be flat and near sea level, was arched upward. Unfortunately, no remnant of this cover remains anywhere in the Valley and Ridge or on the Blue Ridge. So, the only real evidence for this theory of superposition lies in the drainage system, especially the transverse streams, and in the many entrenched meandering streams. The transverse drainage can, however, be easily interpreted in at least one other way.

That alternative is that the drainage in the Valley and Ridge was captured as streams advanced into the high region by headward erosion. Numerous examples of undisputed stream capture can be found in the Appalachians, but all of these are of streams much smaller than the major transverse streams like the Potomac. These diversions have left high-level gaps in many ridges, now dry and called **wind gaps.**

The character of the streams changes dramatically as we follow them from one geologic province to another across the Appalachians to the ocean. Many streams in the Valley and Ridge flow on bedrock (Fig. 19.7). Both the steep slopes and the high discharge and velocity during floods make it possible for these streams to transport their loads easily. The bedrock channels are eroded slowly, because the rocks are hard and consolidated, and many are highly resistant to erosion. Even so, the streams in the valley floors flow in meandering courses and often in channels entrenched 30 to 100 meters in the bedrock.

The major rivers cross the resistant rocks of the Blue Ridge in narrow and deep gaps. Rapids and low waterfalls mark the steep gradients found in these gaps where the streams pass from the Valley and Ridge to the Piedmont.

As the streams pass from the high mountains onto the metamorphic rocks of the Piedmont region (Fig. 20.32), a number of prominent changes occur. The topographic relief decreases abruptly, from approximately 1000 meters in the highlands to 100 meters in the Piedmont. The bedrock, despite its metamorphic and igneous origin, is relatively uniform in resistance to weathering and erosion, and it is usually deeply weathered. The main streams have lower gradients, and contain more load and more channel fill; flood plains are more prominent, and the tributary system has a dendritic pattern. Most streams show little or no adjustment to the bedrock structure or composition of the underlying crystalline rocks.

Passage of the main streams from the Piedmont to the Coastal Plain is also marked by pronounced changes. The contact between the Piedmont and the semiconsolidated sediments of the Coastal Plain is so often characterized by waterfalls and rapids that it is called the **fall line;** it is the upper limit of navigation on the main streams. Because of this, many cities—including Columbia, Richmond, Washington, Baltimore, Philadelphia, and Trenton—developed along the fall line.

The falls represent a sharp increase in gradient clearly related to differences between the bedrock composition of the Piedmont and that of the Coastal Plain. Below the falls the streams flow across uniform and more easily eroded sediment. The Atlantic Coastal Plain is similar to the Gulf Coastal Plain in structure and in landforms, but, unlike the Mississippi, none of the Atlantic streams have built deltas. Instead, their lower reaches have been flooded by the sea, and their sediment loads are effectively dispersed by currents along the coast.

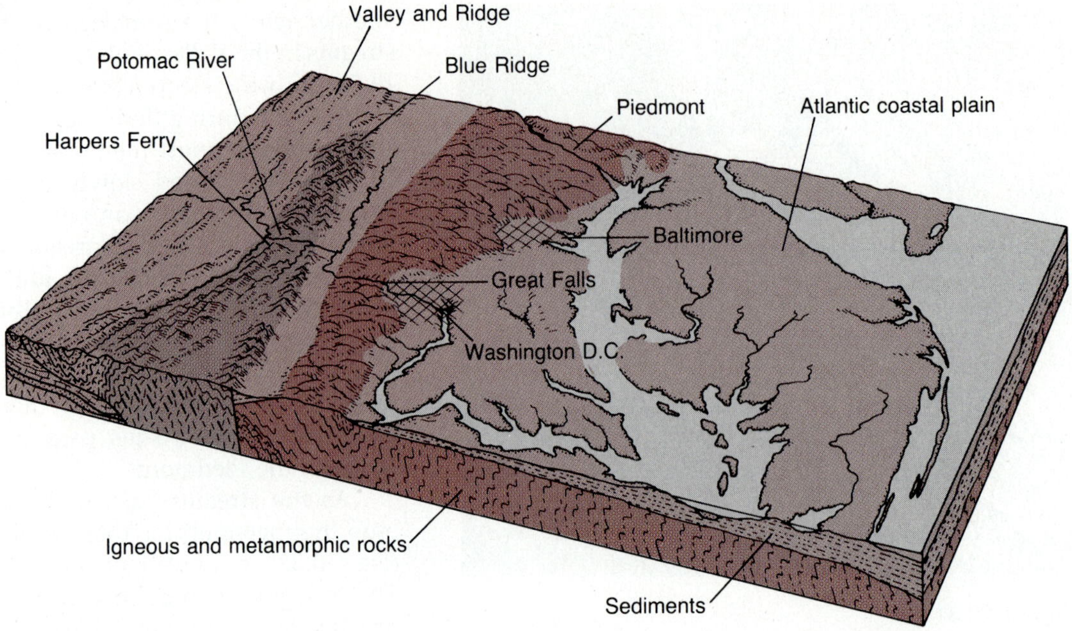

Figure 20.32
The Potomac River, which originates in the folded Appalachians, flows across the Blue Ridge at Harpers Ferry and across the metamorphic complex of the Piedmont. The Great Falls of the Potomac are formed where the river crosses onto the Coastal Plain just outside Washington.

SUMMARY

Streams are a primary agent in sculpting the earth's surface. The forms that result from stream action depend on the volume of water available, the nature of the underlying bedrock, the relief above base level, and the time span over which drainage has developed. Most streams flow through valleys that they have carved, but some flow through topography determined by other geomorphic agents. The most important of these is glaciation, which cut deep valleys and left extensive deposits through which fairly young streams now flow. The course of stream action can also be affected by volcanic activity, isostatic adjustments, and folding or faulting of the crust.

Davis described the cycle of stream development with time as youth, maturity, and old age. In youth, the stream is characterized by deep, narrow valleys with falls and rapids but no flood plain. By maturity, flood plains have developed and valleys are broader, cutting into the uplands between streams. Also, the stream's course has become more winding. In old age the stream flows slowly through a broad, flat, eroded peneplain.

Today many geomorphologists prefer to approach the development of streams as a dynamic equilibrium among all the variables that describe a stream. According to this theory the gradient of the stream is related to the resistance of the underlying bedrock and load and discharge change in response to climate and tectonic activity. The state of a stream represents a delicate balance capable of readjusting quickly to changes along its course.

The Basin and Range province of the western United States is a dry region of high mountains separated by deep basins. The area features plains produced by erosion and others (pediments) produced by deposition. Alluvial fans, deposited as mountain streams flow out onto valley or basin floors, may eventually grow together to form broad flat areas with temporary, shallow, braided channels in times of rainfall. No streams flow out from many of the basins; instead water evaporates or sinks into the ground, leaving deposits of silt and salt. Study of the Colorado basin in the Basin and Range province led Powell to propose a classification scheme for rivers: consequent streams are those that develop on newly formed slopes; superimposed streams are those that have cut down to the present level from higher levels; and antecedent streams are those that cross folds or other uplifts because they eroded their channels fast enough to maintain their course through the rising structure. Subsequent streams are those whose path is determined mainly by bedrock structure. Streams flowing through resistant ridges may also form by headward erosion through the ridge and capture of streams on the other side.

Slow uplift of a broad flat area bearing a meandering river may produce the deep winding canyons of entrenched meanders, like those along part of the Colorado River.

The drainage basin of the Mississippi differs sharply in climate and topography from that of the Colorado. The Mississippi flows through a broad alluvial plain of stream channels, flood plains, and marsh deposits. This plain lies within an earlier deposit of silt blown in from the edge of the North American glaciers. Its channel is a series of meanders formed by erosion on the outside and deposition on the inside of curves in the channel. Eventually, these loops are cut off by erosion and the river's course changes, leaving behind isolated oxbow lakes. Natural levees along the lower Mississippi help contain the river in times of high water, but they also block small streams in the flood plain, creating a region of poor drainage.

KEY TERMS

alluvial fan	**oxbow lakes**
antecedent	**peat**
arroyos	**pediments**
bajada	**pediplane**
base level	**peneplain**
bayous	**playa lake**
chute	**point bar**
consequent	**rejuvenation**
cutoff	**stream piracy**
Davis's fluvial cycle	**subsequent streams**
dynamic equilibrium	**superimposed**
entrenched	**transverse stream**
fall line	**water gaps**
headward erosin	**wind gaps**
intermittent	**Yazoo River**
mature stream	**youthful stream**
old stream	

STUDY QUESTIONS

1. In what ways do the landforms of arid regions differ from those of humid areas?
2. What types of evidence might you find that would support the concept of peneplanation, as described by W. M. Davis?
3. What evidence would indicate that a wind gap was once a water gap?
4. How does antecedent drainage differ from consequent and subsequent types of drainage?
5. How can streams grow longer?
6. What would be the likely effects on the drainage of the Colorado, Mississippi, and Potomac Rivers if sea level dropped a hundred meters.? (Assume the discharge remained constant.)
7. Based on your knowledge of the formation of meanders, how can an entrenched meander form?
8. The braided patterns of many streams in the Great Plains suggests a process of formation for these plains. What is it?

SUGGESTED READINGS

Bloom, A., *Geomorphology: A Systematic Analysis of Cenozoic Landforms.* Englewood Cliffs, N.J.: Prentice-Hall, 1978.

Davis, W. M., "The geographical cycle," *Geographic Journal* 14 (1902), *481–504.* Reprinted in *Geographical Essays.* New York: Dover, 1954.

Douglas, Ian, *Humid Landforms.* Cambridge, Mass.: The M.I.T. Press, 1977.

Hack, J. R., "Interpretation of erosional topography in humid temperate regions," *American Journal of Science* 258-A (1960), *80–97.*

Lohman, S. W., "The geologic story of Colorado National Monument," *Geological Survey Bulletin 1508* (1980).

Mabbutt, J. A., *Desert Landforms.* Cambridge, Mass.: The M.I.T. Press, 1977.

Schumm, S. A., ed., *River Morphology.* Stroudsburg, Pa.: Dowden, Hutchinson & Ross, 1972.

GROUNDWATER: FORM AND PROCESS IN AREAS OF SOLUBLE BEDROCK

Chapter 21

Many communities around the world—not only those in arid or semiarid regions where surface water supply is uncertain—depend on water drawn from underground for their water supply. Without the conditions that make it possible for water to be stored and replenished, it would be impossible for these regions to develop economically. In other areas, especially where rainfall is abundant and the bedrock is soluble, underground water is the single most important agent shaping the land surface. Some of the most extraordinary natural shapes found anywhere are formed by solution of bedrock and its redeposition in caves.

21.1 UNDERGROUND WATER

Origin of Water in the Ground

Water in the ground originates from three distinctly different sources: most of the water with which we are concerned infiltrates into the ground from the surface, where it falls as precipitation or is part of the surface runoff; at greater depths, water trapped during sedimentation may still be present; and some water derived from deep within the earth has come up during volcanic and igneous activity.

Water that falls as precipitation, including rain, sleet, snow, and hail, is called **meteoric water.** On the average, precipitation over continents amounts to nearly 74 centimeters per year. This is taken up by plants, evaporation, and infiltration into pore spaces in the soil and rocks, and it comprises most surface runoff. Meteoric water is usually encountered in the relatively shallow wells drilled for water supply.

Connate water is water trapped in sediments as they are deposited. Since most sediments originate in marine water, connate water is usually salty, and therefore a source of problems if it mixes with water used for drinking or irrigation. Connate water is often encountered in deeply buried sedimentary layers, and it often occurs with oil, which floats on it and rises upward through the water-saturated sediment until it is trapped by some natural barrier.

Water produced from rock melts, or **magmatic water,** has a distinctive range of compositions because it contains components that are rarely, if ever, found in connate or meteoric water. Nevertheless, it is hard to determine how much water from this source is coming to the surface of the earth. Because many volcanoes are located underwater, and many more are found around the margins of the ocean, the magmatic water is rapidly mixed and diluted. More than 90 percent of all the gases coming out of volcanoes is steam. Some of this is clearly new, called juvenile water, but part of it is simply being recirculated from the ocean.

Water Storage Underground

Water is stored in soil and rocks in many ways. Some underground streams do exist, but most are small in size and restricted to limestone and other soluble bedrock and volcanic terrains. Most water is stored in pore spaces in rock and soil, in open fractures, and in solution cavities.

Fractures are present in all types of rock, and they may be an important source of groundwater. Sometimes they are open, especially if they formed

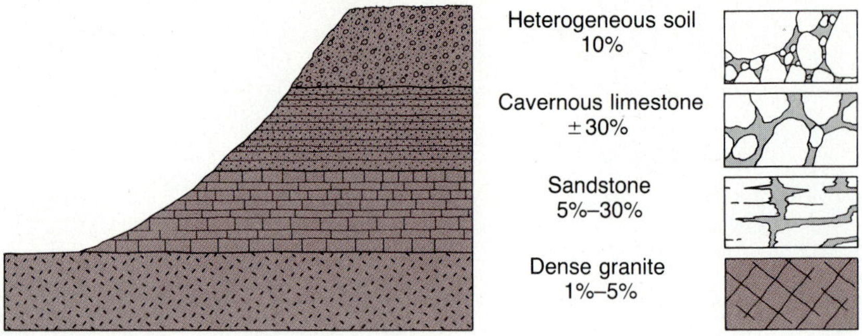

Figure 21.1
Various types of porosity and representative values for pore spaces as a percentage of the rock volume.

as a result of extension; and usually water can seep down along the fracture surface even if it is a small crack. Fractures and the breccias formed along faults provide storage space for water; they are the main reservoir in most igneous and metamorphic rocks.

The volume of pore space in a sediment or rock is called **porosity** and is expressed as a percentage of the total volume (see Fig. 21.1). Some pore spaces are formed at the time of deposition, particularly in sedimentary rocks. Others are created when the sediments are compacted and cemented into rock or even later by weathering or deformation. The amount of pore space depends on the shape, packing, size, and degree of sorting of the rock's components. It might seem likely that a boulder conglomerate with its wide range of particle sizes would be more porous than a sandstone, but the reverse is usually true. Large particles are likely to be mixed in with smaller sediments that fill up spaces between them. Thus, a great assortment of sediment sizes is unfavorable for the development of maximum porosity.

Great differences in porosity may result from different packing arrangements even when all of the particles are of the same size and shape. If every grain is directly above the center of another, maximum porosity results; minimum porosity occurs when the grains are offset. If the particles in a rock are elongate, the porosity will depend on the degree of alignment of the particles. The percentage of pore space in some common clastic sediments given in Table 21.1 will give some idea of the range of porosities found.

Pore spaces may be closed or filled in by natural processes. Groundwater usually dissolves substances that may be deposited, cementing cavities, pores, or fractures, and preventing future storage or passage of water. The most common cementing materials are calcium carbonate and silica. Human activity also may destroy the porosity of some water-bearing sediments by withdrawing water out of them too quickly or too completely. If water is removed, sediments, particularly those composed of clay, become more consolidated and compacted. Usually this process is not reversible.

Although clay has a high porosity, the water held in clay behaves quite differently from that in silt or sand. Clay colloids absorb and hold water so that it cannot readily drain out. Thus, porosity is not always a reliable guide to the amount of water that can be produced: because water does not move through it readily, clay does not yield much of its water content.

The pattern of water storage in cavities formed by the solution of limestone and dolostone is much more irregular than that in intergranular pore spaces. In some caves, subterranean lakes and channels are filled with water, but most solution pores and passageways are small. Usually solution cavities are found along fractures or where soluble minerals have been etched out.

Infiltration of Water into the Ground

If the pore spaces of a rock are not interconnected, rock can be porous and still not allow fluids to move through it. This happens when the pore spaces are closed. In a few extreme cases, such as the volcanic

Table 21.1 Percentage of Pore Space in Common Clastic Sediments.

Sediment	Pore space (%)	Permeability
Soils	50–60	good
Clay	45–55	poor
Silt	40–50	excellent
Sand	30–40	excellent
Gravel	20–40	good
Sandstone	10–20	good to poor
Shale	1–10	poor
Limestone	1–10	good to poor

rock pumice, the rock is so porous and the pore spaces are so completely sealed that the rock will float in water. The ability of a material to transmit fluids is referred to as its **permeability.** Flow of water through a rock is usually easiest when the pore spaces are interconnected and large. Most well-sorted granular rocks (those in which particle size is uniform) have both high porosity and high permeability. Partially cemented rocks, those composed of clay particles, and massive igneous and metamorphic rocks tend to be less permeable unless they are fractured.

Virtually all near-surface water comes from precipitation, finding its way from the surface of the ground down into the permeable rocks below (Fig. 21.2). The water is pulled downward by the force of gravity and pushed down by the weight of water above it. Its downward movement is retarded by surface tension of the water, as it clings to the particles of soil through which it much pass, and by buildup of water ahead of it. Gradually, it seeps into and through the soil. In this zone, water is mixed with air; together they bring about decay of the soil and its rock fragments. After passing through the soil, the water reaches the contact between soil and unaltered rock below. If the rock is porous and permeable, water continues into the rock. Such a water-bearing rock is called an **aquifer.** If the rock is impermeable, water either accumulates above it or moves laterally downslope along the contact of this rock with a more permeable one.

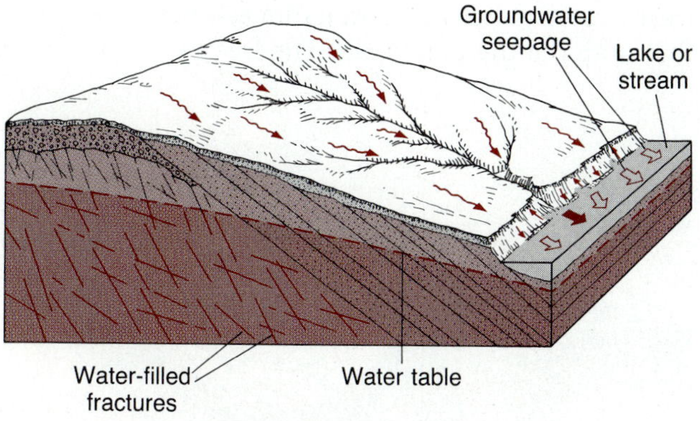

Figure 21.2
Water percolates into the soil and through the soil into the pore spaces and fractures in the bedrock. In massive igneous rock, the water moves mainly in fractures, but in sediments such as sandstone it can move through the pore spaces as well. This water flows out of the ground into streams and lakes. Below bodies of water and at depth in the rock, pore spaces are saturated with water.

Water moving through a homogeneous, porous, and granular material eventually reaches a depth below which the rock is saturated with water. That part of the soil and rock containing both air and percolating water is called the **zone of aeration.** The term **water table** is used to define the surface that separates the zone of aeration from the zone below it, in which pore spaces are filled with water. Despite its name, the water table is not flat, nor does it stay the same level or maintain the same shape all the time. The water that saturates rock and soil below the water table is known as **groundwater.**

Several factors influence the amount of water infiltrating the ground during and after a rain. The total amount of precipitation is a rough guide to the maximum possible amount of infiltration, but usually only a fraction of that total gets into the ground. Infiltration and potential storage of ground water is greatest under the following three conditions: when the precipitation falls at a rate slow enough to enable the water to infiltrate, rather than escape as surplus runoff water on the ground surface; when the soil and the underlying bedrock are porous, permeable, and unsaturated; and when the rates of evaporation and transpiration are low.

21.2 CONFIGURATION OF SATURATED ZONES

Unconfined Water

The distribution of saturated zones in the ground is determined by the porosity and permeability of the rock body and by its shape. When the structure and lithology form a more or less continuously connected hydraulic system in which water can move freely, the water is said to be **unconfined** (Fig. 21.3). When the distribution of permeability is such that water moves into layers or zones partially surrounded by barriers to free movement, the water is said to be **confined.**

The best example of the unconfined condition is a region underlain by a thick granular sedimentary sequence that is porous and permeable. The bottom of the unconfined water is formed by some deep or underlying impervious rock. The lower boundary may be bedrock composed of crystalline rock in which little or no water is expected, or it may be an impervious sedimentary bed below which other water-bearing layers containing confined water may occur. The upper surface of the saturated zone in unconfined

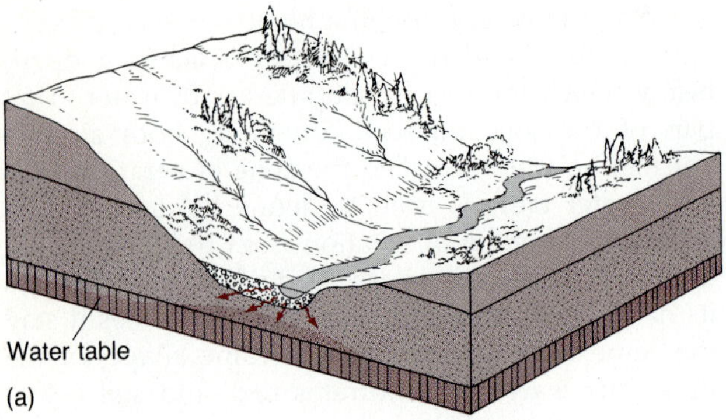

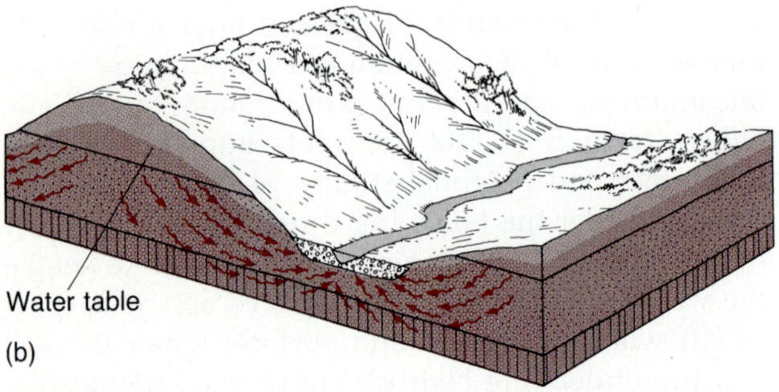

Figure 21.3
Unconfined groundwater. Movement of water below ground level depends to a great extent on the level of the water table. (a) When the water table is deep, water seeps out of bodies of water and flows down to the water table, a condition called influent. (b) But when water is plentiful, it flows as shown by the arrows toward bodies of water; this is an effluent condition. Note that the water table often resembles the topography in shape.

conditions is the water table. The shape of the water table is usually a somewhat suppressed version of the topography. It is depressed most under hills, and it comes closest to the surface of the ground at the edges of lakes and along streams. The shape of the water table may be represented by contour lines drawn using the depth at which water is encountered in wells. The level of the water table moves up and down with the rate of supply of water as water moves within the zone of saturation. Usually the water table reaches a high stage in the spring and a low stage in the fall or winter.

Normally in regions of abundant rainfall water moves from beneath the water table into lakes and streams. When this happens, an **effluent** condition is said to exist. However, during periods of little rainfall or when excessive pumping lowers the water level, the water table may become depressed below the level of the stream and lake bottoms. Water will then infiltrate from the lake or stream toward the water table in what is called an **influent** condition (Fig. 21.3). Influent conditions may last long enough for lakes and streams to dry up entirely.

When some impervious barrier prevents water from moving down to the general level of the regional water table, it creates a localized near-surface zone of saturation and a **perched water table** (Fig. 21.4). Water is usually lost from a perched water table through springs located where the impervious layer reaches the ground surface. Local barriers such as dikes or veins may prevent the free lateral movement of water underground, and these may give rise to the local differences in the level of the water table in an otherwise unconfined groundwater situation.

Water Tables in Contact with Salt Water

Seawater moves through porous and permeable sediment or rock just as readily as fresh water does. In coastal areas, fresh water percolates down into the ground on the landward side of the shore, while salt water moves into pore spaces under the ocean. Because of its lower density, fresh water floats on salt water where the two come in contact. Consequently, a salt-water wedge extends from the shoreline under the edge of the fresh water beneath the land, as shown in Fig. 21.5. Along this contact, the fresh water is contaminated by salt, making it unsuitable for drinking purposes. This condition poses a problem on many islands and in Florida, where fresh water is held in porous limestones and sands that underlie the peninsula. The fresh water forms a

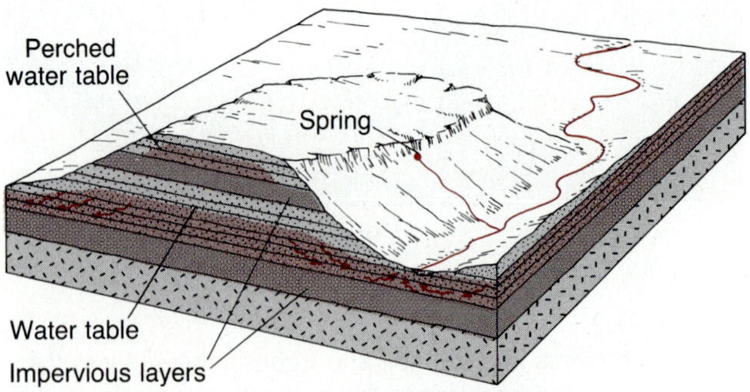

Figure 21.4
One of the conditions that can give rise to a perched water table. Water falling on the mountain cannot move down because of the impervious layer; so water builds up over that layer.

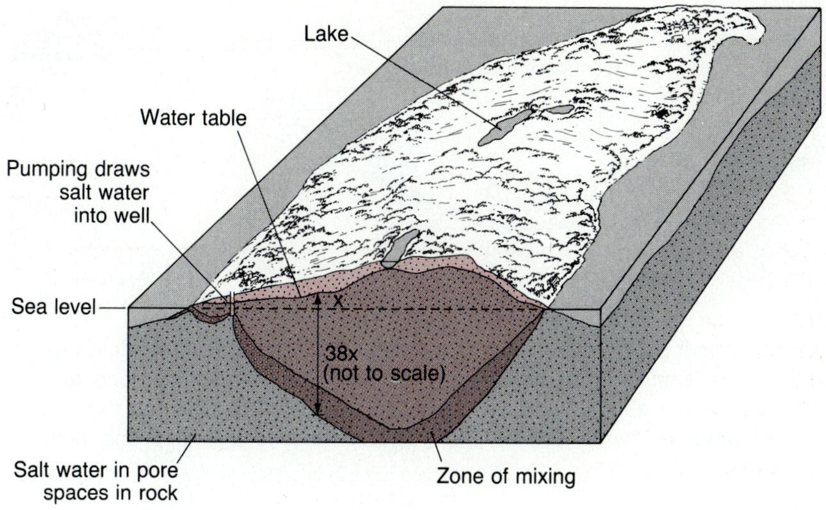

Figure 21.5
Storage of water in sediments in contact with the ocean. The fresh water tends to float on the more dense salt water. For every meter of elevation of the water table above sea level, 38 meters of fresh water lie beneath sea level.

large, dome-shaped mass underlain by similar rocks saturated with salt water. When the rate at which water is pumped from the ground exceeds the rate at which fresh water is being replenished by infiltration or by lateral movement of water toward the well, the well water may become salty. Pumping of water causes the water table around the well to take on the form of a cone, as shown in Fig. 21.6. The size of this **cone of depression** depends on how rapidly water is withdrawn and on how rapidly it is replaced. Excessive pumping near salt water may either bring the salt water closer to the surface or cause the wedge to move inland. Either endangers the supply of fresh water. This problem has already reached serious proportions along portions of the east and west coasts of the United States. Water supplies from Miami to Atlantic City and even Martha's Vineyard are now endangered.

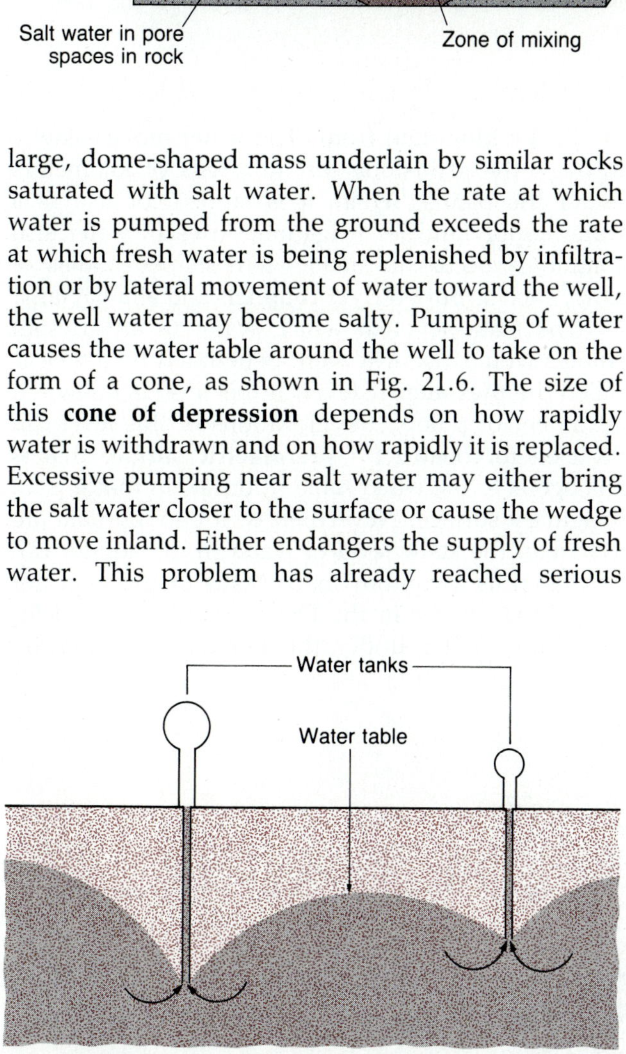

Figure 21.6
The water table is drawn down around a pumped well. This cone of depression grows until the amount of water removed is equal to the amount replenished naturally. When wells are close together, these cones intersect.

Hydrostatic Pressure

Because groundwater usually moves so slowly, static conditions are approximated. The pressure exerted at any point in a body of still water or in a water-saturated sediment by the weight of the overlying column of water is called the **hydrostatic pressure** at that point. This pressure is equal at all points at the same level regardless of the shape of the container. We often express this by saying that water seeks its own level, a fact that can be demonstrated by connecting tubes or pipes in irregular configurations and filling them with water. One such system is illustrated in Fig. 21.7. The pressure at any point in this system can be expressed in terms of what is called the **hydrostatic head**—the height of the column of water that is supported by the pressure at that point. Thus the pressure in the water system of the house in the valley is high, while that in the house at the top of the hill is almost zero.

Confined Water: Artesian Conditions

Artesian conditions occur when groundwater is confined in such a way that hydrostatic pressure builds up enough in parts of the confined water to cause it to rise above the zone of saturation. When water is trapped in a porous and permeable layer between impervious layers, artesian conditions can develop.

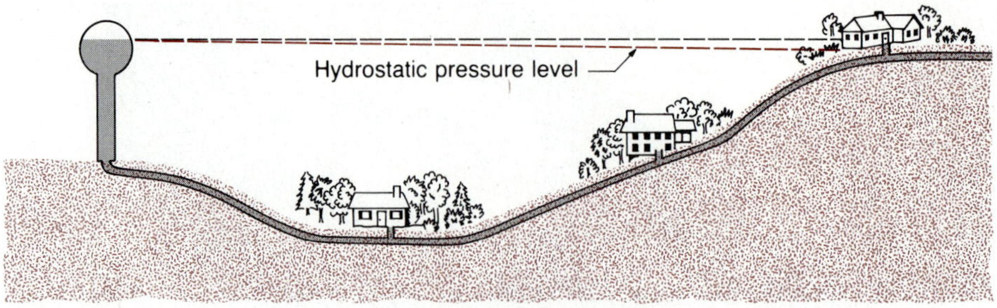

Figure 21.7
Water rises throughout pipes connected to a reservoir to a level almost equal to the level of water in the reservoir. The amount of pressure in the line is proportional to the elevation of the reservoir above any outlet. Thus the house in the valley has high water in its lines while water will barely rise into the lines in the house on the hill and then only when the reservoir is full.

Water enters such a layer where it is exposed at the ground surface or where it is connected with other water-bearing units and free movement of water is possible (Fig. 21.8). This is called the **recharge** area. As water moves down into the aquifer, the available pore space is filled and hydrostatic pressure gradually builds up. If the aquifer is tapped by a well at an elevation below the level to which it is saturated, water will flow from the well under hydrostatic pressure. In Fig. 21.8, for example, water could rise in the well or in pipes connected to it almost to the elevation of the level of saturation in the recharge area.

Artesian conditions exist in extensive strata such as the Dakota sandstone (Fig. 21.9), a porous layer of rock that lies under the Great Plains. Water infiltrates into it where it is tilted up and exposed along the Rocky Mountain front. The water moves slowly through the sandstone for hundreds of kilometers under the plains, where it is tapped for use as a water supply and for irrigation. Artesian water also commonly occurs in smaller lens-shaped bodies of sand, such as buried river channels and alluvial fans.

The static level to which water in an artesian aquifer would rise if it were penetrated by a well is called the **piezometric level.** If this level is known at a number of points, the elevations of this level can be used to construct a **piezometric surface** that is analogous to the water table. It differs in that it is an imaginary surface. Nevertheless, if enough data are available, the static level of water in the aquifer can be drawn as a contour map. Maps of this type for the artesian water in the Dakota sandstone and for the main aquifer under the Florida peninsula are

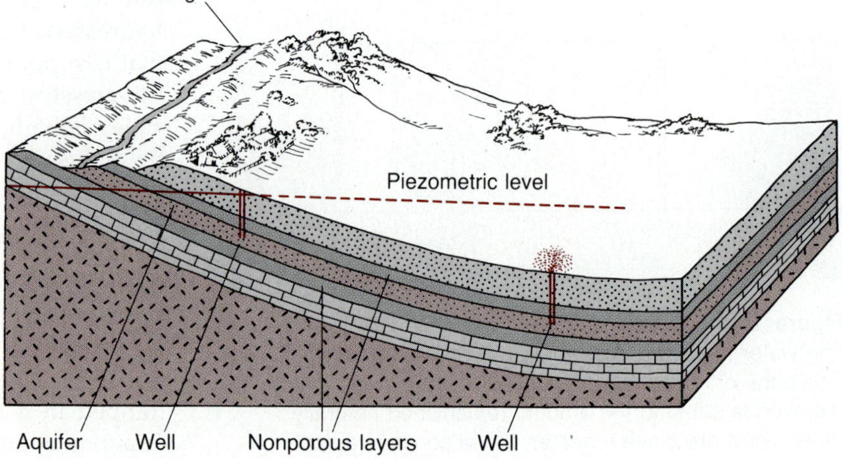

Figure 21.8
Artesian conditions frequently result as shown. An aquifer exposed in the mountains takes in water from the soil and streams. Water remains in and fills the aquifer if it is sealed above by a nonporous or impervious layer. When wells are drilled into the aquifer, water may rise as high as the elevation of the top of the zone of saturation in the aquifer. These conditions exist in the Great Plains.

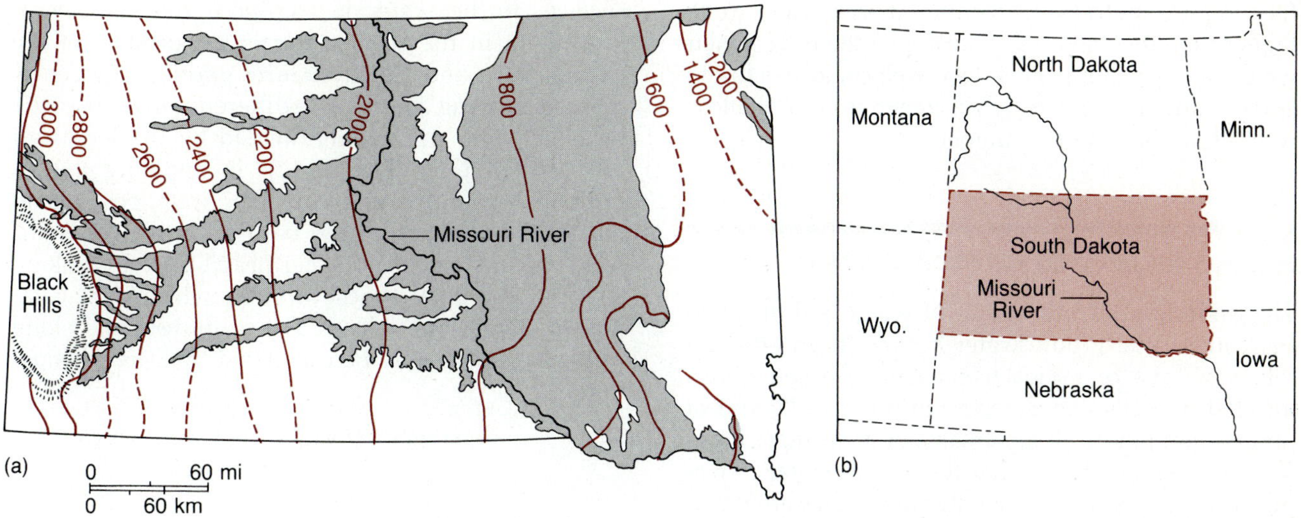

Figure 21.9
(a) The areas shown in gray in North Dakota are underlain by water under artesian conditions. Recharge takes place around the Black Hills, and water will rise to the elevations shown by the contour lines. (b) Locator map.

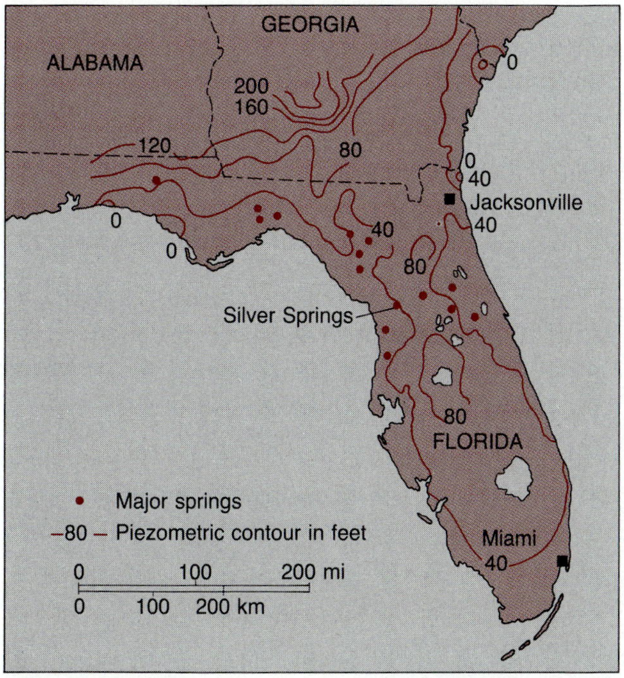

Figure 21.10
Florida is underlain by aquifers containing water under pressure. The contours show the levels to which this water can rise, and major springs are indicated. Some parts of the main aquifers are contaminated by salt water along the coast and in south Florida.

shown in Figs. 21.9 and 21.10. These maps are useful in the exploration and development of water supplies.

It is important to remember that confined water in aquifers is usually overlain by unconfined water and a water table developed in near-surface materials. This configuration is represented in Fig. 21.11 for conditions found in the Atlantic Coastal Plain where older tilted strata, including aquifers, are covered unconformably by unconsolidated Quaternary sands and gravels. If the area where the artesian aquifer is recharged is higher than the valley floors, it is possible

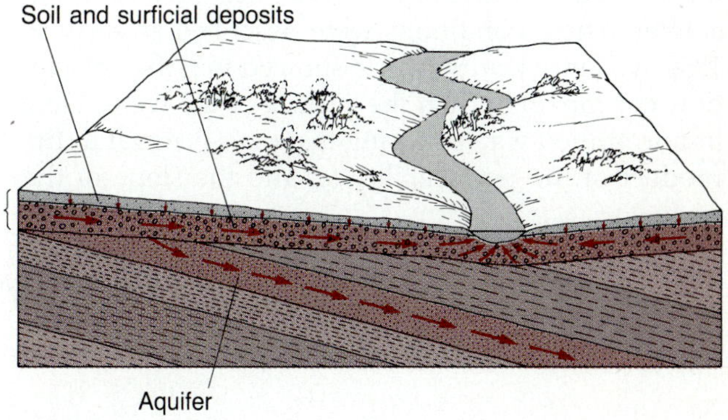

Figure 21.11
Throughout most of the Coastal Plain, water infiltrates through soil or gravel deposits into slightly inclined layers, some of which are aquifers. A single inclined aquifer is shown, but a number may occur close together.

for the piezometric level for the artesian water to be higher than the water table surface of the unconfined water. Thus, water from a deep well could rise to the surface under pressure while water from a shallow well would have to be pumped.

21.3 MOVEMENT OF GROUNDWATER

The underground movement of water is difficult to track in natural groundwater basins. In an effort to follow it, dye or radioactive tracers are sometimes injected in wells and detected in other wells and at observation points along streams or at springs. From such studies, we know that the rate of water movement in the ground is generally slow. Rates are much higher where groundwater flows through underground streams, as it may in limestone terrain, than where flow is through pore spaces. Rates as low as 1 to 10 meters per year are found in some rocks with low porosity and permeability; rates as high as several hundred meters per day may occur in coarse alluvial gravels. Where water is trapped in **confined aquifers** with no outlet, where the water table is nearly flat, and where porosity and permeability are unusually low, the water may be almost static.

Much of what we know about groundwater movement is derived from theoretical analyses and laboratory experiments modeled after existing natural conditions. The experimental model is then used to predict the response of the system to changes that can be expected to occur in the natural system. These predictions are then compared with field observations. Simple examples are the flow of water in unconfined sand and flow of water in a tilted confined aquifer. These conditions were described as early as 1856 by Henry Darcy (*1*). He showed that the velocity of water moving down the surface of the water table in unconfined water-bearing materials is equal to the product of the permeability (k) and the slope of the surface. If the water is confined, the velocity of movement in the aquifer equals the product of the permeability and the **hydraulic gradient,** which is defined as being equal to the difference in hydrostatic head between two points divided by the length of the path of flow (Fig. 21.12). In applying this to ordinary groundwater problems, it is possible to measure the length of the path (L) and hydrostatic head (h), and to determine permeability (K) by making tests on the rock types through which the water passes. Thus, the velocity (V) of movements along particular paths may be calculated by Darcy's equation:

$$V = \frac{Kh}{L}.$$

21.4 RISE OF GROUNDWATER TO THE SURFACE

Springs

Water flows from the ground where the water table intersects the land surface, where water under artesian conditions finds its way to the surface, and sometimes where the water table or artesian water rises underwater along coasts. The hydrostatic pressure on groundwater causes it to issue from **springs.** For this reason, they are usually located at low places in the topography, but the structure of the rock and the way porosity and permeability vary in the bedrock are also important factors controlling spring location.

A list of all the conditions that can cause springs to come into existence would be very long indeed, but a few examples will illustrate common occurrences. Springs often issue on valley sides where some especially permeable zone in the bedrock comes to the ground surface, as shown in Fig. 21.13. The most common situation is that found where bedding planes, fracture zones, or faults crop out on a hillside.

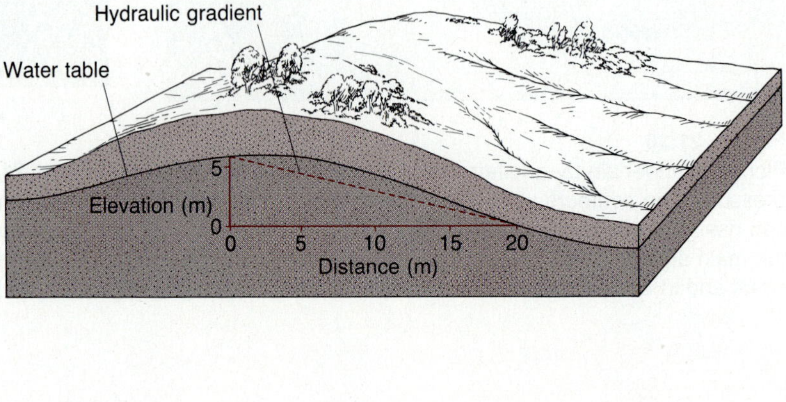

Figure 21.12
The hydraulic gradient between two places on the surface of the water table is the ratio of the difference in elevation between those points divided by the horizontal distance between them.

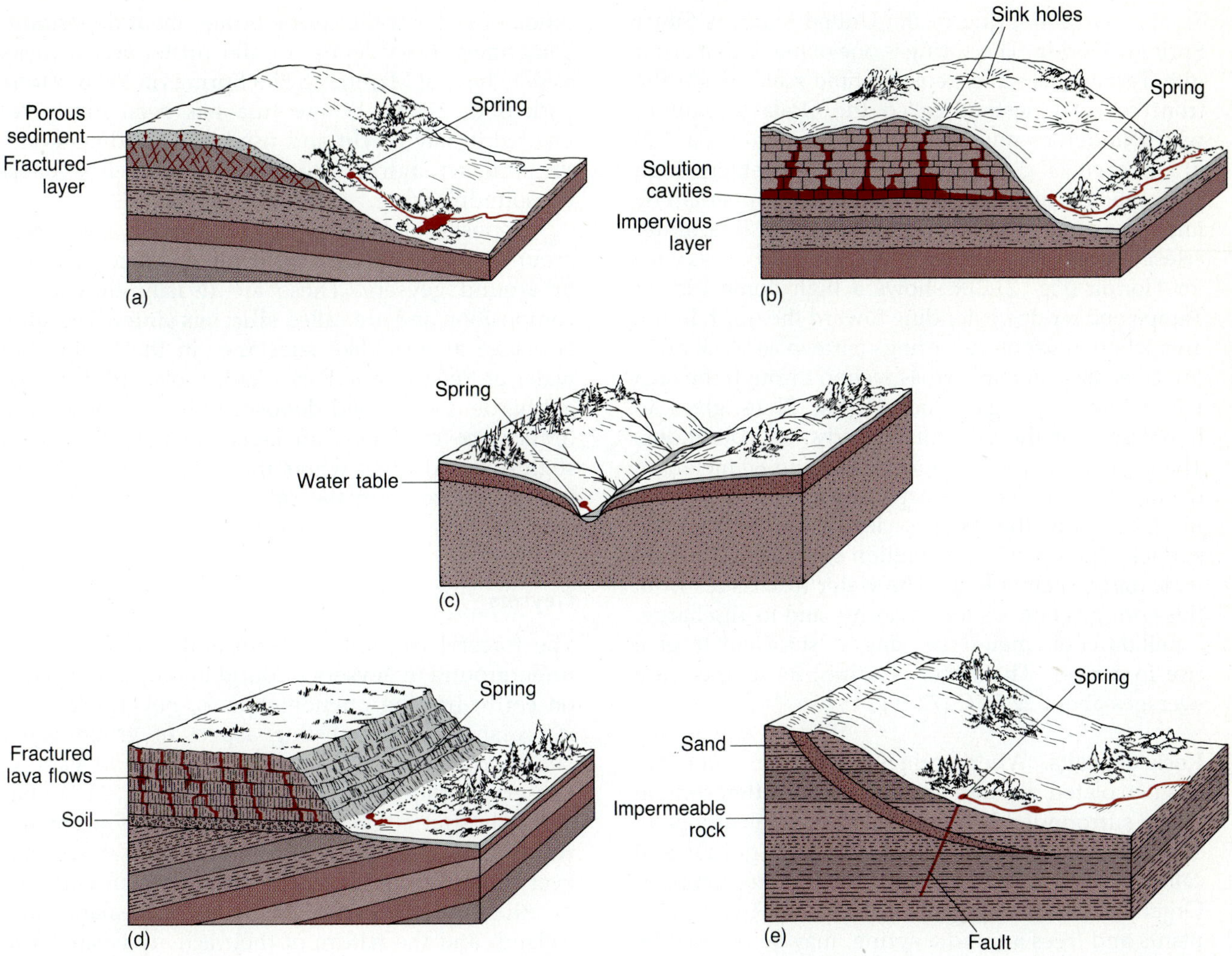

Figure 21.13
Springs occur where the water table intersects the surface of the ground. The geological conditions vary greatly as shown here. (a) Fracture spring; water in highly fractured rock flows out above a watertight layer. (b) Solution spring; a solution cavity drains water from porous rock. (c) Depression spring, characterized by unconfined conditions with a very high water table. (d) Contact spring, where water flows out at the base of a lava flow. (e) Artesian spring; a fault provides a way for water to flow out.

In the case of bedding planes, especially, impervious layers may block the downward movement of water and cause it to move laterally along the bottom of a porous and permeable layer (Fig. 21.13*a* and *b*).

In limestone terrains, springs often occur where cavities formed by solution along fractures, beds, or faults come to the ground surface. The rocks themselves are often nearly impervious, and water is readily funneled through the solution openings. Somewhat similar conditions are found in lava flows, which are composed of nearly impervious rock. More often, however, lava contains highly porous and permeable zones within blocky flows, lava tubes, or cooling cracks. Thus, springs often issue from the base of lava flow outcrops, especially where they are underlain by less permeable rock. Of the largest springs in the United States (those with flows exceeding 100 cubic feet per second), thirty-eight are in volcanic rocks, twenty-four are in limestone, and three issue along faults. Artesian springs are not common, but they do occur where a natural break occurs in a confined aquifer. Such breaks can result from faults and fractures or from erosion of the overlying impervious cover.

The largest spring in the United States is Silver Springs, Florida. The spring is one of many that occur in a Tertiary (see the geologic time scale inside the front cover) limestone called the Ocala formation, which underlies much of the Florida peninsula. Solution cavities give this limestone extremely high permeability. In most places, the Ocala is overlain by impervious sediments, and the water in it is under artesian pressure. A map of the piezometric surface for Florida (Fig. 21.10) shows a high dome east of Tampa and a ridge extending toward the north in the area where most of the springs (shown as black dots) occur. A diagrammatic cross section through the area where Silver Springs is located (Fig. 21.14) gives an impression of the general groundwater conditions. The main pool at Silver Springs is about 80 meters in diameter and 12 meters deep; the main opening from the Ocala into this pool is clearly visible from the surface. The opening, a solution cavity, is 20 meters wide and 4 meters high. The water that issues from this spring is famous for its clarity, and its discharge, 2 million cubic meters per day, is sufficient to give rise to a river. The water is warm; its temperature averages about 22.5 °C (72 °F).

Rock deposits. When springs are fed by water that has percolated through limestone, they often contain or are surrounded by deposits of calcium carbonate. These deposits are called **tufa** if they are porous and cellular in structure and **travertine** if they are compact. Organic matter, especially leaves and twigs from plants and trees around a spring, may be covered by the calcium carbonate deposits and become incorporated in the tufa, making it a porous and poorly consolidated material. The deposition of calcium carbonate takes place because the water is saturated with respect to calcium and carbonate ions. Evaporation of water at the spring brings about deposition. Sometimes these deposits build up terrace deposits as they have at Mammoth Hot Springs in Yellowstone Park (Fig. 21.15). Where tufa has been firmly cemented it may be cut and used as a building stone, the porous, rough texture making it uniquely valuable for soundproofing.

Deposits of a different composition may occur around springs issuing from hot volcanic materials or around geysers. These are usually siliceous in composition and are called **siliceous sinter**. The silica is visible as a gel-like substance in the boiling hot water at Yellowstone Park. As it cools and drys, the gel hardens to a solid deposit. Other deposits form terraces covered by thin layers of hot water with slightly raised edges where the water cools; and silica is precipitated from the saturated water as it flows over the rim.

Geysers

The forceful ejection of steam and hot water from underground in **geysers** is found in only a few places on earth, but the phenomenon is, nevertheless, an impressive and interesting aspect of groundwater behavior. Since its discovery in 1870, Old Faithful geyser in Yellowstone National Park (Fig. 21.16) has regularly spurted 38 to 45 cubic meters of steam and water to an average height of 40 meters about once every hour. Few geysers are this regular, but a number of others exist in Yellowstone, New Zealand, and Iceland, and the nature of their activity is similar to that of Old Faithful.

Following an eruption, most of the water returns to the ground at the vent or through nearby holes. A small part runs off in small streams flowing away from the cone of siliceous sinter built up around the

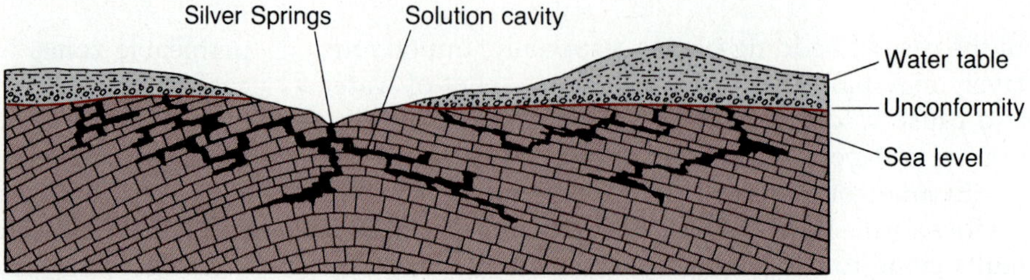

Figure 21.14
A number of unusually large springs occur in Florida. These springs are fed by water that infiltrates through porous sediments at the ground surface. The water then moves into solution cavities in the Ocala limestone, an arched layer separated from the surface sediments by an angular unconformity.

(a) National Archives

(b) National Archives

Figure 21.15
(a) These terraces were built up by the deposition of tufa and travertine at hot springs in Yellowstone Park. (b) Geyserite deposits built up around a geyser in Yellowstone Park.

(a) National Archives

Figure 21.16
Old Faithful in eruption. Boiling hot water sprays to a height of nearly 100 meters. Deposits of siliceous material, called geyserite, form at the base where the people are standing. (b) Locator map.

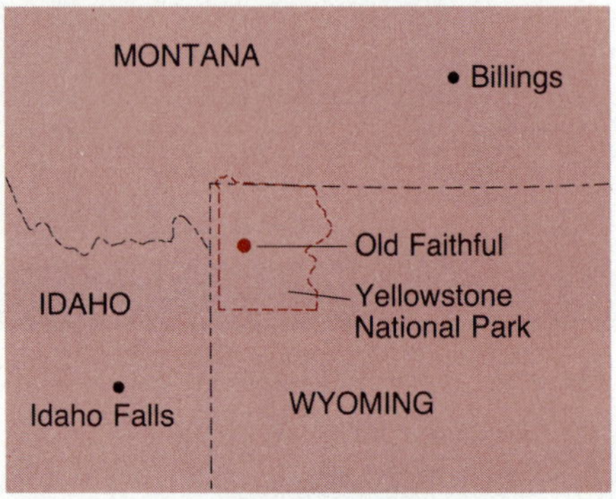

(b)

vent. During the period of quiet that follows, only a few whiffs of steam rise from the vent. After about 30 or 40 minutes, steam starts to rise once again and a little water appears at the surface. Water and steam rise and flow out of the vent in a series of small outpourings. Some of this runs off, but most goes back into the hole. Several small eruptions occur, throwing bubbling water out of the vent to a height of several meters above the ground. After several minutes of these minor eruptions, a column of water and steam spurts forth, occasionally as high as 76 meters. The eruption lasts several minutes, dies down, and the cycle is repeated.

Most of the water coming from Old Faithful and other geysers is meteoric water, but it contains small quantities of common volcanic gases and solutions. As it infiltrates the ground, it is warmed by heat from a pluton deep under the park area. Surface waters flow into the fractured and porous volcanic rocks that underlie Yellowstone Park. Beneath the surface, the vent is connected to an irregularly shaped network of fractures and cavities dissolved by the hot waters, from which the ejected water comes. After eruption, time is required for water to percolate back into the channels and refill them. The water, which has just been cooled by its ascent into the air, is rewarmed quickly. Since the heat in the rocks increases with depth, the water in the lower parts of the network becomes hot more rapidly than that closer to the surface. Water at the surface boils when it reaches 100 °C (actually slightly less at Yellowstone because of its high elevation). But water at the bottom of the system is under the pressure exerted by the weight of the column of water above it. Thus, it must be heated to a much higher temperature before it will begin to boil. Eventually, somewhere in the underground network, the boiling point is reached and steam forms. The rapid expansion of the water as it changes into steam exerts sufficient pressure to force the overlying water out at the top of the vent. Just a little water is forced out at first, then more, until finally enough is removed to lower the pressure on the network, permitting large quantities of water to convert suddenly into steam. The main part of the eruption starts when this explosive generation of steam occurs at depth.

21.5 LANDFORMS SHAPED BY GROUNDWATER SOLUTION

Solution of bedrock by groundwater and by streams diverted underground is an important process shaping the land in many areas around the world where soluble bedrock is at or near the surface. Most rocks are not highly soluble, but salts, gypsum, and the carbonate rocks, limestone and dolostone, are. Carbonate rocks, originally deposited as carbonate banks in the shallow continental platforms, are widely distributed across the continents. The topography that results from solution effects is quite varied and often strikingly unusual in form. It is characterized by caves, enclosed depressions, stream valleys that end suddenly, natural bridges, and tunnels. The name **karst topography,** taken from a region in Yugoslavia, is applied to areas that exhibit these types of landforms.

Karst topography is widely distributed across the United States (Fig. 21.17). The largest areas are located: in the Florida peninsula, the southeastern Coastal Plain, and Texas where young (Tertiary) semiconsolidated limestones, such as the Ocala, are present; in the folded Appalachians, the Appalachian Plateau, and the interior lowlands where thick layers of older (Paleozoic) consolidated carbonates are found; around the Ozark dome in Arkansas; and in the Carlsbad Plateau region of New Mexico. Other small areas of karst are scattered in many other areas as well.

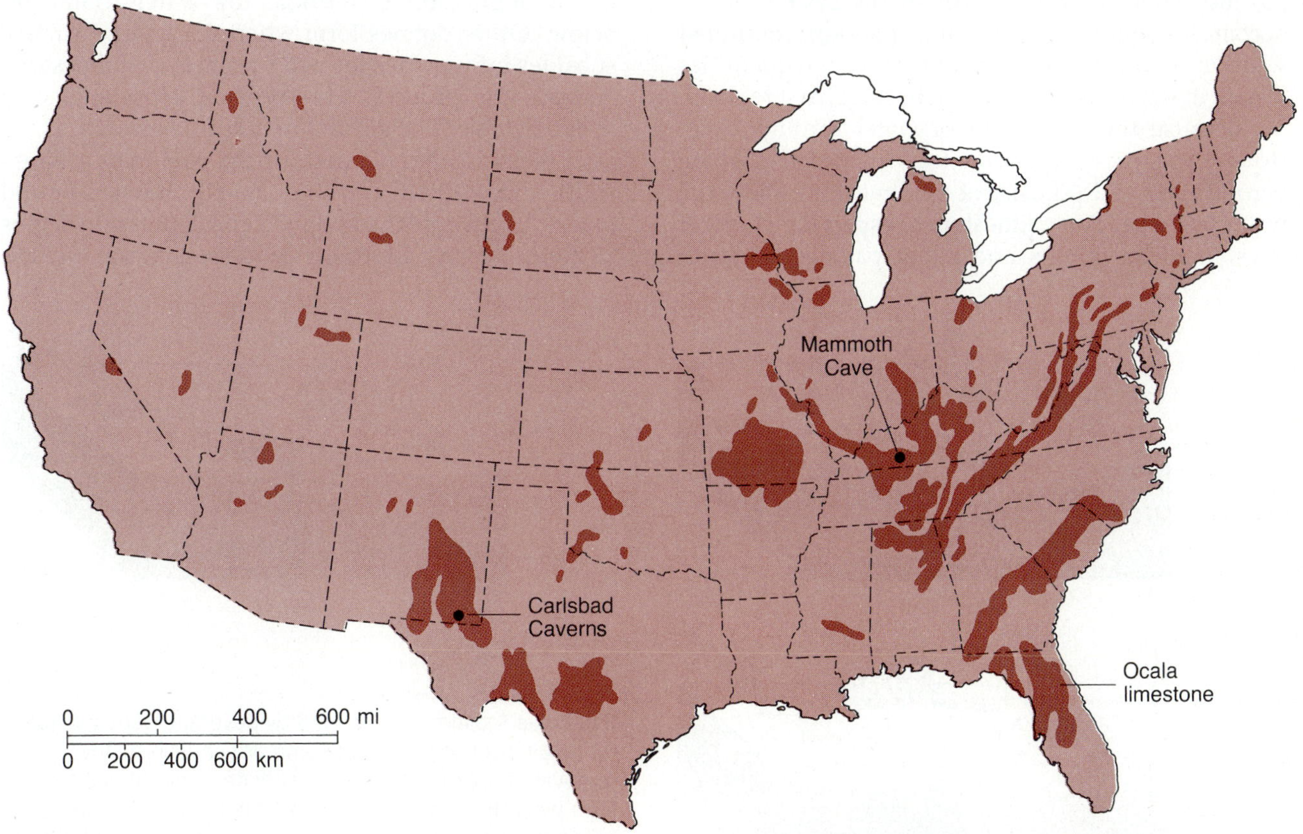

Figure 21.17
Large areas of karst topography, shaped by the solution of carbonate across the United States.

Solution

We have seen in Section 17.2 that limestone, largely calcium carbonate, is dissolved by rainwater containing carbon dioxide from the atmosphere. Some slight solution will occur even in neutral distilled water, but the process is accelerated by the presence of acid. Estimates of the rate of solution have been based on everything from the observed dissolving of tombstones to water hardness (the dissolved mineral content in water). As might be expected, limestones vary considerably in their solubility; so no single value can be applied universally. Estimates for the rates of corrosion of limestone terrain range from 5 mm per thousand years to about 0.5 mm a year. Undoubtedly, acid rainfall is accelerating the process even more today.

The dissolution of limestone in rainwater is not sufficient by itself to cause the effects we observe in karst areas. Limestone will cease to dissolve as soon as the water is saturated with calcium or carbonate ions. For the process to continue, the saturated water must move out and be replaced by unsaturated water. So, movement of water through the system is important for solution to continue; and long, continued solution is clearly necessary to remove the quantities of rock that have been dissolved in most karst regions.

Because the movement of water through limestone is so critical, solution effects are greatest where permeability and porosity of the limestone allow the water to move. Some limestones, especially younger ones, may be porous and allow passage of water throughout. This is true of limestones composed of broken shell fragments and other carbonates of biological origin, but many older limestones are fine-grained and compact; they have been recrystallized. They have little or no intergranular porosity. Water moves through these limestones primarily along fractures and bedding planes.

Landforms

The land surface in karst regions is never shaped by solution alone. Weathering, downslope movements caused by gravity, and stream action almost invariably work together with ground water.

The most characteristic features of karst regions are enclosed depressions on the ground surface called **dolines** or **sink holes** (Fig. 21.18). These occur in a great range of sizes and may be formed in several quite different ways. If the limestone has intergranular porosity, or if the rock is broken along closely spaced fractures, dolines may form by solution distributed through the rock. The effect is similar to that of dripping water on a sugar cube. Where the drops hit, a depression is formed; the water drains off below. Other dolines form when the ground surface subsides into cavities formed by removal of limestone beneath the ground. This type of process is very unpredictable. The ground may subside slowly as the cave is gradually enlarged, or the roof may collapse suddenly without warning as big fracture-bound blocks drop from the ceiling. Once a depression begins to form by any of these processes, water on the

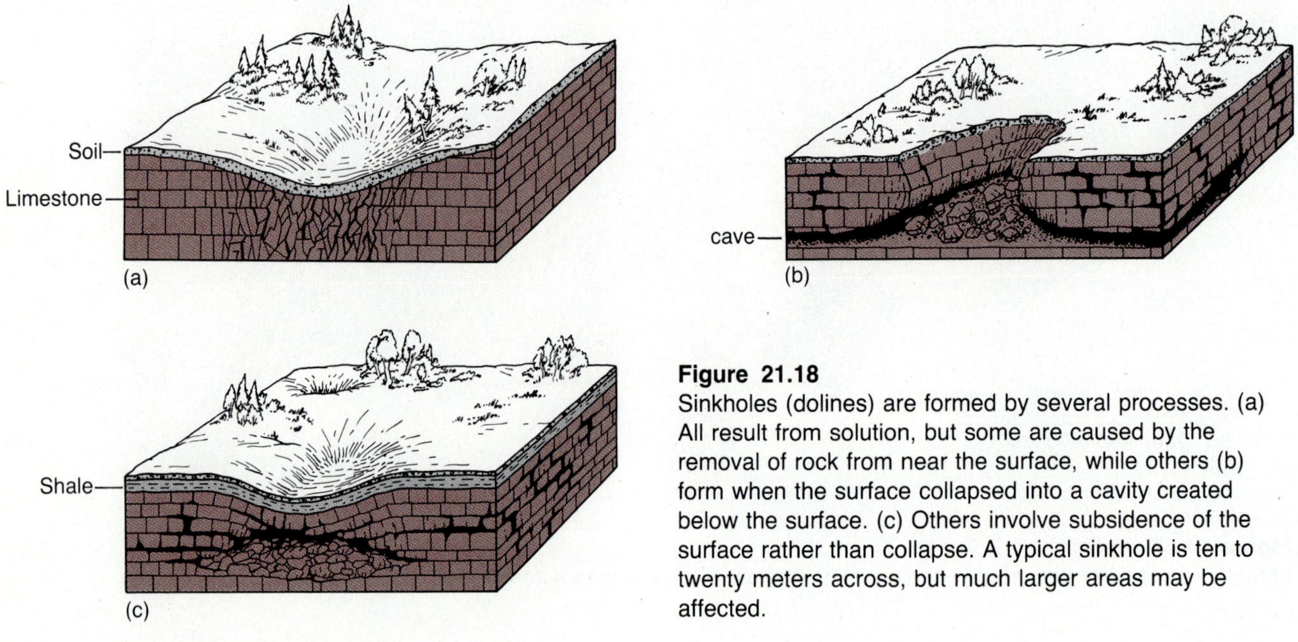

Figure 21.18
Sinkholes (dolines) are formed by several processes. (a) All result from solution, but some are caused by the removal of rock from near the surface, while others (b) form when the surface collapsed into a cavity created below the surface. (c) Others involve subsidence of the surface rather than collapse. A typical sinkhole is ten to twenty meters across, but much larger areas may be affected.

Photo by Richard Deuerling

Figure 21.19
The sudden and rapid growth of sinkholes is demonstrated by this one which formed in Winter Park, Florida, in 1981.

ground surface is diverted into the depression and the solution of underground limestone accelerates as more and more water is diverted into the sink. The formation of sinkholes may be initiated by too-rapid withdrawal of groundwater. This is probably a major factor in the sudden appearance and growth in 1981 of the sinkhole shown in Fig. 21.19 at Winter Park, Florida.

Most limestones contain clay and other insoluble impurities, which are left behind as the limestones dissolve and form residual red, clay-rich soil called **terra rosa.** The depressions formed by sink-hole development become natural traps for surface-washed soil or even alluvium. When subsidence or collapse occurs, the hole contains an irregularly shaped and uncompacted mixture of soil, partially weathered rock, and blocks of fresh limestone. The mixture is very unstable and may be saturated with water.

Movement of groundwater in karst regions differs from that described earlier, mainly in the progressive concentration of flow along solution passageways. Groundwater that flows through limestone or into sinks moves down toward the water table where it joins in the normal circulation of the saturated zone. The caves we map and visit are above the water table, so the streams through them are directed toward the zone of saturation; but caves certainly exist in many places below the water table also. It seems clear that caves form both above and below the water table.

Streams that are diverted underground usually re-emerge elsewhere and join surface streams. In some cases, the water comes to the surface as springs with large discharges; and in others, waters collected underground emerge as a river. Such arrangements amount to shortcut paths for the drainage of surface waters toward the sea. Occasionally, a cave system is located at approximately the same level as the stream it feeds. Then the cave may be flooded when the river is high and drained as discharge drops. Under these conditions, solution in the cave may be accelerated by the periodic filling and discharge of water from the cave.

Solution cavities occur in artesian systems, especially if the water from the aquifer has a low-level outlet, an artesian spring, that makes it possible for the water to move more or less continuously through the aquifer. The big springs in the Ocala limestone in Florida are outlets for water moving under artesian conditions. Most of the caves formed in artesian aquifers are formed in the zone of saturation by water flowing through it. Cavities also form in soluble rock as a result of the flow patterns described for unconfined conditions.

Where the soil is washed off limestone being dissolved, a highly irregular, fluted surface like that shown in Fig. 21.20 is seen. This is especially well formed where solution is controlled by fractures.

Diversion of Surface Drainage

As dolines develop, more and more of the surface drainage is diverted underground. Where they grow in stream channels, dolines can divert the entire stream underground if there are passageways to carry

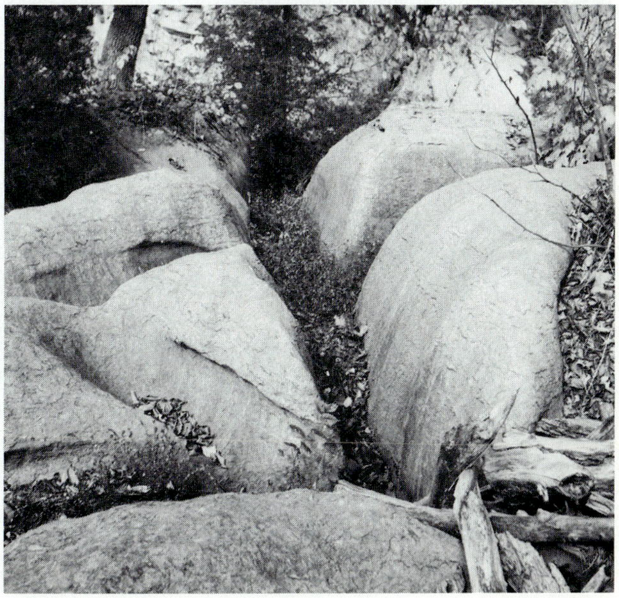

Figure 21.20
The smooth surfaces seen here were developed just below the soil as a result of solution by groundwater. The soil that filled these crevasses has been washed out.

the discharge. When this occurs, the process of stream erosion as well as solution effects shape the underground passages. The surface in these areas takes on some strange forms. The valley abandoned when a stream is diverted underground stops being eroded by stream action. Consequently, products of mass wasting and surface wash accumulate. The valley is no longer cut down and may appear to end as the stream is swallowed. Such a valley is called a **blind valley.** Some such streams flow through what are called **natural tunnels** or, if only a remnant of the tunnel remains, a natural bridge, like that in Fig. 21.21.

Karst that has developed in tropical regions where the soluble rock is thick or flat-lying often exhibits such extensive solution that the land consists of moundlike or towerlike hills of limestone rising above large sinks or hollows from which the limestone has

Natural Bridge of Virginia, Inc.

Figure 21.21
Natural Bridge, Virginia. The natural bridge may once have been a long tunnel. The stream, which once flowed as an underground river, is again flowing on the ground surface as a result of the collapse of the roof.

National Geographic

Figure 21.22
Tower karst, characterized by the steep-sided and isolated mountains rising above a plain, represents an advanced stage of erosion by groundwater solution. The area depicted here is in southeastern China.

been removed (Fig. 21.22). The abundant rainfall in these areas may account in part for large-scale solution effects. This type of topography represents a late stage in the evolution of a karst region.

The valley floors in some of these terrains are composed of flat-lying less soluble rock. The steep-sided hills are, thus, erosional remnants of the limestone. Their form is significantly controlled by rock structure and the base level of streams.

Caves

A cave is any naturally formed underground cavity or void. Most are formed as a result of solution of carbonate or other soluble rocks, but they also exist where fluid lava has drained out of a lava flow, leaving an underground passageway, called a lava tube, as in Fig. 21.23. Caves also occur in salt, and some of the most beautiful formations are found in caves formed in gypsum beds.

The conditions favorable for formation of caves have been discussed in the preceding section concerning the development of karst topography. Although it has been argued that caves form in certain positions relative to the water table—above it, just below it, or well below it—the process of solution that leads to cave formation almost certainly is not confined to any one of these positions; caves probably form in all three.

The present location of any particular cave relative to the water table is not necessarily an indication of where it formed. Movements of the ground caused by tectonic activity, changes in sea level, and the downcutting of stream channels in response to changing base levels have all affected the position of caves relative to the water table. Some caves formed below the water table have undoubtedly been drained and others have been submerged and flooded.

Proponents of the idea that caves form above or near the water table base their opinion on the following points. First, solution along fractures is usually evident. The pattern of passageways in some caves follows very closely the alignment of fractures in the rock. These fractures and bedding surfaces are important as the prime avenues for movement of water through rock that often has almost no porosity or permeability except along fractures. Blocks of rock bounded by fractures and bedding planes are found collapsed into caves.

Second, subterranean streams are present in many caves. Often these can be traced back to the ground surface. Some are known to be streams diverted from the surface into caves only to emerge again at a lower elevation. Streams in caves often carry sediment introduced from the ground surface in sinks or **swallow holes,** openings in a stream channel that lead underground; and sediment-laden streams are effective in accomplishing erosion. Potholes, grooves, and even meanders are found in caves.

Other observations favor the idea that caves form below the water table and are subsequently drained. Many caves now beneath the water table and filled

U.S.G.S.

Figure 21.23
Caves in lava flows usually do not involve solution effects. Instead they are created by the drainage of hot lava from beneath a cooled roof. The roof of the lava tube seen here has collapsed.

with water are found in the Ocala limestone of central Florida. Water-filled cavities have also been encountered beneath the water table in the Appalachians. The walls of many caves are lined with crystals that apparently grew under water. The internal network of branching tubes, pockets, and irregularly shaped cavities of many caves does not possess a pattern readily explained in terms of stream action. The cavities have a three-dimensional, spongelike pattern with blind pockets and dead-end tubes rather than a streamlike pattern.

Cave deposits. The calcite deposited within caves sometimes takes spectacular and characteristic forms. This calcium carbonate is dissolved from the limestone through which the groundwater passes as it enters the cave. This water was originally rainwater, acidic enough to dissolve the limestone, as described in Chapter 17. Precipitation of calcite as a cave deposit comes about when the water either evaporates or loses carbon dioxide. Because the groundwater was saturated or nearly saturated with respect to calcium and carbonate ions, loss of water or acidity causes the calcium carbonate to precipitate. The rate at which cave formations build up varies considerably. In active caves supplied with large quantities of water, the rate is high—as much as a centimeter a year—but where the supply of water is limited, the rate is slow or deposition may cease altogether.

Moisture seeps into caverns and drips from the ceiling and sides of the passageways to form pools and streams on the floor. Places where water drips from the ceiling become sites of calcite deposits, often having the shape of thin straws. These hanging deposits, called **stalactites,** are seen in various stages of development. As a drop of water hangs from the ceiling, some of the water within the drop evaporates, leaving a thin ring of calcite. This ring is gradually lengthened by more deposits until it becomes filled with calcite. Water then moves as a thin film over the surface of the straw. Calcite slowly builds up layer on layer until a large tapering cone, like those in Fig. 2.27, is formed.

Water that drops from the end of a stalactite falls to the floor, where it evaporates further and precipitates more calcite. Slowly, a stump-shaped deposit, called a **stalagmite** (Fig. 21.24), rises from the floor. Usually it has a small saucer-shaped top into which drops of water fall from the ceiling. Overflow from this saucer runs down the sides of the stalagmite, building it out. Eventually, stalactites and stalagmites may grow together to form columns.

Photo by New Mexico Tourist Bureau

Figure 21.24
Strawlike stalactites hang from the ceiling here in Carlsbad Cavern located in southeastern New Mexico (see 21.17). The huge columns of calcium carbonate that have been built up from the floor are surrounded by flowstone on the floor of the cave.

Other formations grow on the sides and floor of most caverns. Droplets of water on the walls evaporate, leaving small, rounded, knoblike projections of calcite on the walls. Calcium carbonate deposited from flowing water in caves is called **flowstone.** Such deposits form beautiful solid cascades on the side or floor of a cavern. Rims of calcite **(rimstone)** are deposited around the edge of the pools, and formations composed of flow- and rimstone give rise to large terraced deposits.

In the past few chapters we have explored the geological effects of water—its ability to carry material and to carve the surface and near-surface regions of the earth. As human beings we have a further interest in the hydrologic cycle: our very lives and everything we consume depend on the presence of adequate supplies of water. And, on the other hand, our lives and constructions are periodically affected by an overabundance of water. In the next chapter we shall see what factors influence our water resources and susceptibility to floods.

SUMMARY

Underground water is of three types: meteoric (derived from precipitation), connate (trapped in sedimentary rock), and juvenile or magmatic (from rock melts). It may be stored in rock and soil within pore spaces, solution cavities, and open fractures.

Most near-surface water is meteoric water that has infiltrated soil and rock. Below some depth, known as the water table, the rock and soil are saturated with groundwater. The water table tends to follow the surface topography, but in a subdued form, as it is generally deeper on hills and closer to the surface in low areas.

Where groundwater can move freely through rock and soil, it is said to be unconfined; if impermeable layers bound the porous saturated zone, the groundwater is confined. Impervious layers may also create perched water tables or give rise to artesian conditions if the area of recharge is above most of the aquifer.

Where the water table intersects the land surfaces, the hydrostatic pressure of the groundwater causes springs to issue. If the groundwater has percolated through limestone, calcium carbonate deposits of tufa and travertine may form around the springs.

Underground deposits of limestone or other soluble rocks may form the solution caves, stream valleys, natural bridges, and tunnels that are typical of karst topography. Sink holes or dolines form when the surface subsides or collapses into these solution cavities. Within caves, calcite is often redeposited as stalactites, stalagmites, and columns, as groundwater saturated in calcium carbonate drips from the ceiling of the cave; or as flowstone and rimstone by underground streams and pools.

KEY TERMS

aquifer
artesian conditions
blind valley
cone of depression
confined water
connate water
doline
effluent
flowstone
geyser
groundwater
hydraulic gradient
hydrostatic head
hydrostatic pressure
influent
karst topography
magmatic water
meteoric water
natural tunnel
perched water table
permeability
piezometric level
piezometric surface
porosity
recharge
rimstone
siliceous sinter
sink hole
spring
stalactite
stalagmite
swallow holes
terra rosa
travertine
tufa
unconfined water
water table
zone of aeration

STUDY QUESTIONS

1. Although clay has a porosity range three to four times as great as sandstone, sandstone usually yields more water. Why?
2. Distinguish between permeability and porosity.
3. Why is salt-water encroachment an important problem in central and western Long Island?
4. After a cave begins to form, a sink hole may occur over it. Why does this speed up the expansion of the cave into what might eventually be a large underground system?
5. What steps would you recommend to help conserve water and protect groundwater supplies in the area where you live?
6. The porosity of sediments containing mixtures of clay, sand, and gravel may vary greatly. What can cause such variations?
7. Explain the geologic conditions in Florida that cause artesian wells and springs. How do these differ from the geologic conditions causing artesian flow in the Great Plains?
8. Describe the occurrence of groundwater in the area where you live.

REFERENCE

1. Henri Darcy, *Les fontaines publiques de la ville de Dijon.* Paris: Victor Dalmont, 1856.

SUGGESTED READINGS

Anderson, J., *Cave Exploring.* New York: Association Press, 1974.

Bouwer, H., *Groundwater Hydrology.* New York: McGraw-Hill, 1978.

Freeze, R. A., and J. A. Cherry, *Groundwater.* Englewood Cliffs, N.J.: Prentice-Hall, 1979.

Hack, J. T., and L. H. Durloo, "Geology of Luray Caverns, Va." *Virginia Division of Mineral Resources Report of Investigations 3* (1962).

Jennings, J. N., *Karst.* Cambridge, Mass.: The M.I.T. Press, 1971.

Leopold, Luna B., "Surface water and ground water," in *Water: A Primer.* San Francisco: W. H. Freeman, 1974.

Sweeting, Marjorie M., *Karst Landforms.* New York: Columbia University Press, 1973.

WATER RESOURCES AND FLOODING

Chapter 22

Water, soil, and air are our most important natural resources. In this chapter we shall examine the occurrence of water resources and some of the effects of our use of water. We rely on both ground and surface water, and with few exceptions, groundwater and surface water supplies are closely linked. Case histories are used to illustrate the interrelationship of these two types of reservoirs.

The basic problem in water management is to provide adequate supplies of clean water for human use while avoiding the devastation of floods and a variety of other problems that may arise when large quantities of water are removed from the ground.

22.1 WATER: A RESOURCE AND A HAZARD

Those of us who have spent most of our lives in areas of abundant rainfall rarely think of fresh water as a valuable natural resource. At a time when industrial and agricultural demands for water are so great, when populations are increasing, and when pollution of fresh water has reached a critical stage in many streams and lakes, we should be more aware of the value of this resource, take steps to determine the extent of our holdings, and make careful plans for its future use and conservation.

That water is stored in the soil and rocks of the earth's crust is apparent from springs that issue from valley sides, from geysers, and from wells that produce much of our drinking water. So common is the experience of obtaining water from the ground that we may be lulled into a belief that groundwater supplies are inexhaustible. However, many cities in the United States have already experienced periods of short water supply. Each year over a thousand communities are forced to ration water, and more widespread shortages almost certainly lie in the future.

It is also easy to overlook the great value of streams. Water for drinking, irrigation, and transportation led to the early concentration of cities and farms along rivers and stream valleys, particularly on the flood plains. All streams are subject to changes in discharge as a result of seasonal fluctuations in precipitation. Therefore, the size of the channel required by a stream to carry its discharge varies from day to day and from one season to the next. Unusually high discharge leads to high water and to flooding of flat land adjacent to the river (Fig. 22.1).

These **flood plains** along rivers have for centuries been particularly attractive for agricultural and trade activities. Unfortunately, most human activity has, as we shall see, increased the danger of flood by increasing runoff; and streams that in their natural state were only rarely subject to flooding have become sites of floods with increasing regularity. Not only do more floods occur, but their cost in damage to property steadily increases, for the development itself causes higher flood levels.

Our investment in property on flood plains is already so great we cannot allow streams to revert to their natural state. Like it or not, we have no acceptable alternative to flood control and water mange-

Virginia Department of Highways and Transportation

Figure 22.1
The James River at flood stage in Richmond, Virginia, during the flood in 1969 when Hurricane Camille caused record rainfall in the James River basin. The effects of this flood are described later in this chapter.

ment. This is not an easy or simple alternative, because the behavior of water underground and in streams is complex. Failure to understand the principles at work has doomed many attempted solutions in the past and provided convincing evidence that a better understanding is needed for the future.

22.2 GROUNDWATER RESOURCES

Many communities and individuals depend on water drawn from underground for their supplies, and groundwater is also important for irrigation. For these reasons, increasing attention is being given to the most effective ways of developing groundwater supplies, especially in arid and semiarid regions and near population centers where the annual recharge of groundwater does not equal the rate at which it is being withdrawn. Studies of water supplies in these areas have been most successfully approached on the basis of comprehensive analysis of natural basins. A **groundwater basin** may be thought of as an area bounded by natural geologic boundaries that more or less separate water within it from adjacent basins. Such conditions are met particularly in the western states, where many structural basins are separated from one another so that little water moves underground from one basin to another.

Even when an area is not an enclosed basin, the groundwater supply can be analyzed if the **water budget** for the area can be accurately determined. The construction of reservoirs, dams, and artificial lakes; the importation and exportation of water; irrigation; water consumption; construction of pipe lines, canals, paving, and buildings: all influence the water budget of an area. These influences must be evaluated for intelligent analysis of the water supply. An equation for this more comprehensive water budget is given in Table 22.1.

The purpose of most water supply analyses is to determine how much water is available for use. The amount of water that can be withdrawn without producing undesired results is called the **safe yield.** Undesired results clearly include depletion or pollu-

Table 22.1 Equation of Hydrologic Equilibrium.

surface inflow through streams	increase in surface storage
+ subsurface inflow	+ increase in subsurface storage
+ precipitation	− surface outflow
+ importation by pipes & canals	− subsurface outflow
− decrease in surface storage (reservoirs)	− consumptive use
− decrease in ground-water storage	− exported outflow
balance =	balance

tion of the water supply. They may also include damage to the aquifer by destroying its porosity and permeability by promoting salt-water intrusion, or by causing subsidence or cracking of the ground surface, such as that shown in Fig. 22.2. Undesirable effects result from a variety of actions, but by far the most common cause is long-term overdraft that lowers the water table.

Water quality is as critically important as the quantity of water. Water quality has raised questions that are just as difficult to resolve as who gets how much water in areas where it is available in limited quantity. Along coasts, overdrafts cause salt-water intrusion. In some deep aquifers far inland, similar intrusion of seawater, trapped in the ground millions of years ago when the sediment was deposited, may also be induced by excessive pumping. Use of water for irrigation may produce undesirable effects where the water filters through soils containing salts in arid areas. The direct injection of toxic substances into the ground and the burial of toxic chemicals in leaky containers are unfortunately far more common problems than most of us realized only a few years ago.

Obviously, very difficult decisions are being faced in many communities across the country today. Some must decide whether to allow the lowering of water quality through irrigation or industrial contamination or to suffer the economic effects of causing industries or agriculture to fail because of inadequate water supply. In some communities, local and regional planning commissions are allowing development and issuing building permits in areas where there is no long-range water supply. It becomes clear that water supply must be studied in terms of the economics of the industries, housing, and agriculture dependent on water, as well as on the quality of water, the legal rights to its use, and the quantity available for use.

In more and more areas of the world, people must face the fact that the supply of water underground is not unlimited. For any given area a certain amount of water is held in the ground, and a certain amount is annually recharged into the ground by natural processses. To remove more water from the ground for drinking, agriculture, or industrial use than is normally recharged is to "mine" the water. Thus, the amount of water in the ground, the rate at which it is naturally recharged, and the ability of people to supplement or recharge the water supply by bringing water into the area create a limit to the potential groundwater development of that area.

Long-range planning should be based on the proposition that the water budget must be balanced, that an equilibrium must be maintained. One remedy for water shortage is importing water. It is now economically feasible to recover fresh water from sea water along coasts in arid areas, and this will undoubtedly be part of the solution to future water needs in the coastal areas of the world. For some coastal areas, water may be obtained by towing in icebergs. The idea has been discussed; plans are

U.S.G.S.

Figure 22.2
Tension cracks formed in the ground surface as a result of withdrawal of water from alluvium. An area of about 8000 km² in south central Arizona is affected by this type of cracking. Compaction of the aquifer over irregularities in the underlying bedrock causes the cracks.

underway; and this may be a water source for a few places in the near future. Water can also be transported from one groundwater basin to another. Water for southwestern states may eventually be imported through pipes, canals, and rivers from the Canadian Rockies. But the cost of transporting water inland over mountains and long distances makes it imperative for us to develop existing groundwater supplies first.

The Quality of Groundwater

High-quality water is needed especially for drinking and for many industrial and agricultural purposes as well. Water quality is determined by means of analyses that reveal contamination from pesticides and other chemicals as well as bacteria. The kinds and quantities of ions present are used as a guide to the purity of the water and the uses for which it is suitable. The total amount of dissolved solids can be estimated quickly by a test measuring the electrical conductivity of the water sample. Conductivity is directly related to the total salt content of the water. This is not the same as water **hardness,** which refers to the amount of calcium and magnesium ions present. These ions are generally derived from the solution of carbonate rocks. For this reason, groundwater in limestone terrain is unusually hard, and rainwater, by contrast, is very soft. Bacteriological analyses are made of drinking water to insure that there has been no contamination from sewage. Other tests may be made to determine temperature, color, odor, and taste.

The U.S. Public Health Service sets standards for most drinking water supplies. Permissable concentrations of ions are expressed as parts per million (ppm). One ppm is one part by weight of the ion to a million parts by weight of water. Mandatory maximum limits are set for ppm of lead (0.05), fluorides (1.5), arsenic (0.05), selenium (0.01), chromium (0.05), and mercury (0.002); and a maximum total solids concentration is set at 500 ppm. In recent years dangerously high concentrations of mercury and other highly toxic elements have been discovered in some water supplies. Desirable limits for many less dangerous elements are also provided, along with standards for physical characteristics and bacterial quality.

Pollution of groundwater by various chemicals has become a widespread and serious problem. Some of these chemicals are not biodegradable—they will not break down in the natural environment. The solvent trichloroethylene is an example. This substance has been recognized in groundwater around many eastern cities; and, especially in a number of places in Pennsylvania, wells have been permanently poisoned and for that reason have been closed by Federal order.

Many industries require extremely pure water or water containing low concentrations of particular ions. Water standards vary greatly from one industry to another, but they are a major consideration when the industries locate new plants. Obviously, serious problems arise if undesirable elements enter the water supply.

Even water used for agricultural purposes must meet certain standards. Some plants are sensitive to concentrations of particular salts. For example, apples, pears, oranges, plums, peaches, and strawberries all have low salt tolerances, as do green beans, celery, and radishes. Date palms, spinach, asparagus, sugar beets, and barley can stand concentrations of three or four times the tolerable limits for the plants previously listed. This problem becomes particularly important in arid regions where irrigation is necessary, because water found in such regions is likely to have high salt content. Furthermore, pumping water to the surface for irrigation may cause an increase in the salt content of all the water in the vicinity. Irrigation water dissolves salts from the soil and carries the added salt load back to the zone of saturation as it infiltrates, contaminating other water supplies, encrusting the soil, and decreasing fertility. Continued recycling of water in this way will eventually make the water unsuitable for irrigation. Excessive pumping may also cause deeper water with a high salt content to come to the surface. Usually deeper water has been in the ground longer and has a higher content of dissolved mineral matter.

Many other conditions also lower groundwater quality. Inadequate treatment of sewage before water is returned to the ground; dumping of sewage into streams; disposal of waste food products, oil, chemicals, and tailings from mines and mineral processing plants; all are sources of pollution. Water can also be contaminated by natural causes. Magmatic waters, for example, naturally contain high concentrations of salts and other more toxic chemicals.

The Geology of Groundwater Occurrence

The occurrence of groundwater in the United States varies greatly from one part of the country to another, as shown in Fig. 22.3. In all of these areas, a close relationship exists between the local geology, as represented by rock types, structural configuration,

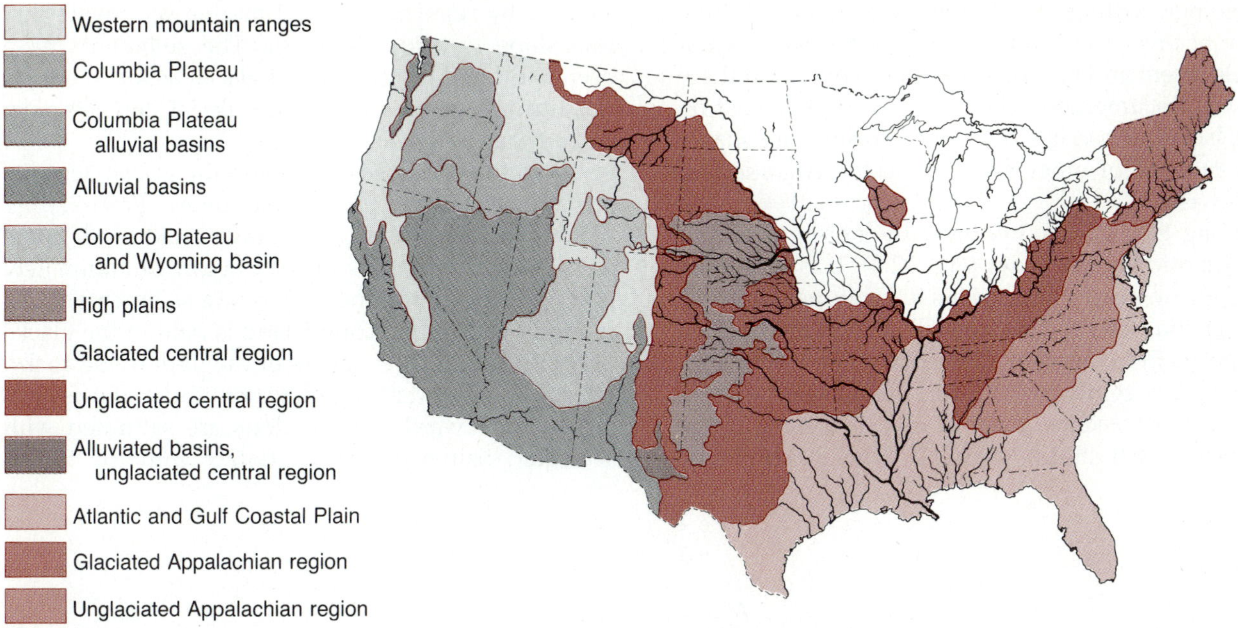

Figure 22.3
Groundwater provinces of the United States. The heavy black lines mark important watercourses where groundwater is replenished by perennial streams. Other important provinces are alluvial basins and glacial deposits. In the Atlantic and Gulf Coastal Plains, water is produced from sedimentary units, sands and silts, of Cenozoic age. In the unglaciated Appalachian region, water is held primarily in fractures. In the central region, bedrock is composed of late Paleozoic and Cenozoic sedimentary units. The Columbia Plateau is largely underlain by lava flows. In the western mountains, most water supplies come from surface streams. The Colorado Plateau and Wyoming basin are underlain by Cenozoic and Mesozoic sedimentary units.

and surficial deposits, and the manner of water occurrence. Groundwater conditions are uniform over large regions in some parts of the country, but more generally water occurs in several different geological situations within any one area.

Large water supplies exist in the alluvial deposits along the courses of the major streams such as the Mississippi, Ohio, and Missouri Rivers. Other important supplies occur in unconsolidated sediments of glacial origin in the northern part of the country, in the alluvium of the basins in the Basin and Range region, in the surficial alluvial deposits in the Great Plains, and in the Coastal Plain. Water in these deposits is usually found in sands, gravels, and other clastic sediments. In some places, especially in the Coastal Plain, the aquifers are extensive layers of relatively uniform thickness. Other places, such as those with glacial deposits, have porosity and permeability that vary over short distances.

Large supplies of groundwater are less frequently found where water occurs in consolidated rocks. Some sandstones have relatively high porosity and permeability due to intergranular pore spaces. Limestones may contain solution cavities, and shales may be broken up along cleavages. All of these contain some water in fractures. These types of occurrence are found in the Colorado Plateau, in the bedrock of the continental interior, and in the folded Appalachians.

Water may be abundant in some lava flows where it is found in the gas pockets and lava tubes, between flows, and in blocky lava. Much of the groundwater in the Pacific northwest and in Hawaii occurs in this way. Much more limited supplies of water occur in fracture and fault zones in the igneous and metamorphic rocks of the Sierras, the high Rockies, and the Appalachians.

A much better understanding of the nature of groundwater occurrence is obtained through the detailed examination of particular areas. The following case histories illustrate some of the different types of occurrence and the problems that exist in these areas.

Case Histories

Groundwater supply on Long Island. The water supply on Long Island is of special interest because its geologic setting includes aspects of several quite different provinces, and because the history of water development and the consequences of overdraft there are well documented. The problems experienced on Long Island, especially salt-water intrusion, are shared by a number of other areas along the Atlantic and Gulf coasts.

Long Island is located near the northern end of the Atlantic Coastal Plain (Fig. 22.4). It is underlain at depth by nearly impervious ancient metamorphic and igneous rocks. This crystalline complex, which is exposed in Manhattan, Connecticut, and New York state, slopes gently toward the Atlantic, as shown in Fig. 22.4. It reaches a depth of 600 meters at the southern coast of Long Island. This basement is covered by semiconsolidated marine sediments that also slope toward the ocean. The sediments have been truncated by stream and glacial action at a level close to present sea level, and glacial deposits now cover them.

Both the older sediments and the glacial deposits contain porous and permeable zones. Precipitation directly infiltrates the glacial gravels and sand. Some of this water seeps back into the ocean, but ultimately part of it passes into the underlying sediments. Most groundwater under Long Island is held in the glacial deposits and in three layers of the underlying rocks. These strata contain fresh water under the island, but their seaward continuations are saturated with seawater. Salt water is in contact with the lower-

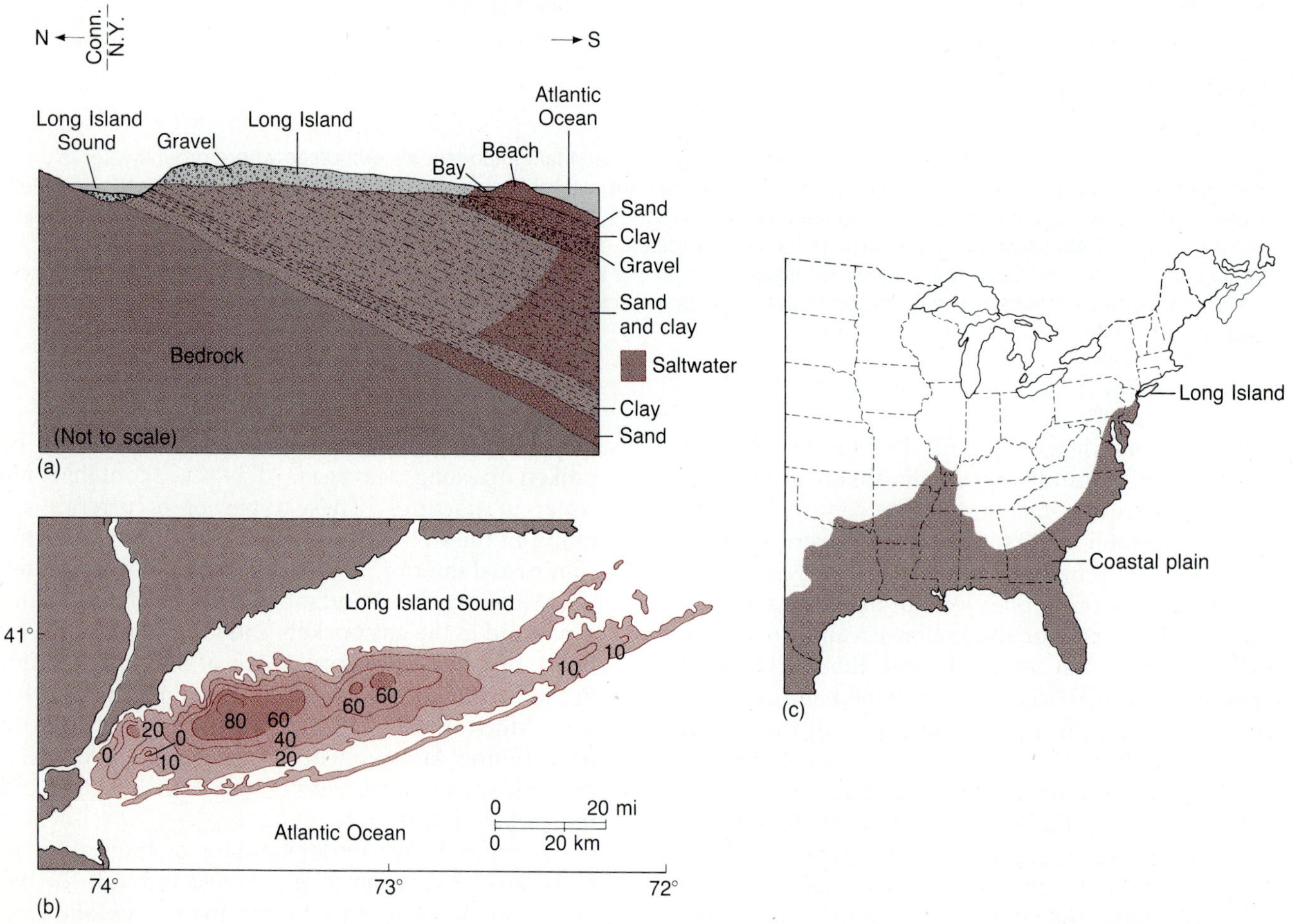

Figure 22.4
(a) The bedrock geology of Long Island resembles that of many other parts of the Atlantic Coastal Plain, but it differs in that the surface of Long Island is covered by glacial deposits. It is also one of the most densely populated parts of the Coastal Plain. The dip of the strata below Long Island is greatly exaggerated in this section. The strata actually dip at less than 10 degrees. (b) The water table on the island stands nearly 20 meters above sea level in the center of the island. Fresh water moves into underlying sediments, but this fresh water reservoir is in contact with salt water both on the oceanward and the Long Island Sound sides of the island. (c) Locator map.

density fresh water under both Long Island Sound and the Atlantic Ocean. The water table for the glacial deposits follows roughly the configuration of the land surface.

Before Long Island became so heavily populated, the long-term average groundwater recharge and discharge were about equal. The subsurface interfaces between the fresh and salty water in each of the aquifers were stable, reflecting an overall hydrologic balance. The average infiltration of rainfall, about 1.5 million liters per square kilometer, on Long Island moved into the glacial deposits and was discharged from them into the sea or onto the ground surface. An estimated 90 percent of the total recharge was discharged from the glacial deposits into the sea. Some of that eventually infiltrated down into the underlying Tertiary and Cretaceous rocks.

As the population of Long Island grew and the demand for water increased, more wells were put down into each of the water-bearing units. At first many individual wells and sewage disposal units were used. These wells took water at shallow depths from the glacial deposits. Soon the disposal of sewage into these same deposits created a problem of cesspool pollution in the shallow water supply. Wells had to be drilled deeper into the Tertiary sediments. At this stage, private water supplies gave way to municipal water development. The highest population and greatest use of water for industrial purposes occur at the western end of the island. It was in this area that large-scale groundwater withdrawal from the deeper aquifers first disturbed the position of the fresh-water/salt-water interface. As steadily increasing amounts of fresh water were withdrawn, the salt-water/fresh-water interface moved farther inland. The intrusion of the salt-water wedge was accelerated by the gradual expansion of sewage systems designed to discharge waste into the ocean, and by the reduction of rainwater infiltration by construction of buildings, streets, and so on. In the late 1930s groundwater levels of fresh water under Brooklyn had been depressed locally to as much as 10.5 meters below sea level. This resulted in large-scale contamination of the aquifers by salt water.

The deep wells had to be discontinued, and now much of the western part of Long Island receives its water from the New York City supply, which is obtained from surface reservoirs in upstate New York.

Today the eastern end of the island is operating as a balanced water system in which recharge is roughly equal to or greater than withdrawal. Most water is produced from shallow wells in that area, but salt-water encroachment is a continuing problem in the central and western portions of the island. A number of steps have been taken to slow or even reverse the salt-water intrusion. Laws have been passed requiring that water removed from the ground for air conditioning be pumped back into the ground. Other efforts have been made to increase recharge both by pumping water into the ground and by designing leaky ponds that trap surface runoff and direct it back into the ground.

Las Vegas groundwater basin. The Las Vegas, Nevada, groundwater basin is located in the Basin and Range province, an arid region with average annual rainfall of 12.5 centimeters. The mountains rise to elevations of 3000 meters and stand about 2000 meters higher than the intervening basins. Many of these basins, including the Las Vegas basin, are almost completely enclosed by mountains composed of crystalline rocks and tightly cemented, nearly impervious sedimentary rocks (Fig. 22.5). The basins have internal drainage; no surface streams drain surface water into or out of the basin from adjacent basins. Thus each basin forms a closed system in which little water moves in or out either on the surface or underground.

Water occurs in the sediments that have been deposited in the basin. The oldest sediments deposited in the Las Vegas basin are Tertiary (2 to 70 million years old). They consist of thin sand lenses and fine gravels interbedded with thick layers of clay. The sand and gravel lenses contain silt and clay and are partially cemented. These sediments are overlain by younger alluvial fans (see Section 20.3) composed of poorly sorted heterogeneous mixtures of boulders, gravel, sand, silt, and clay. Long stringlike deposits of well-sorted gravel that is highly porous and permeable tend to accumulate in the stream valleys, and the interstream areas are covered by mudflows and sheet-wash deposits that are relatively impervious. At the outer edge of the fans, the coarser materials give way to fine-grained sands, silts, and clay, which are interbedded with lake-bed deposits formed in intermittent lakes (playas). These deposits have been built up over the last few million years (possibly 10 to 20 million) and now are over 300 meters thick.

During much of the last two million years, the mountains in this region were covered by ice. The basin floors were sites of lakes that formed as the deposits in the basins became saturated by greatly increased rainfall and meltwater. Deposits from these former lakes, impervious materials composed of clay, silt, and fine sand, are up to 15 meters thick. Recent gravels, sands, silts, and playa deposits lie over the lake bed.

(a) NASA

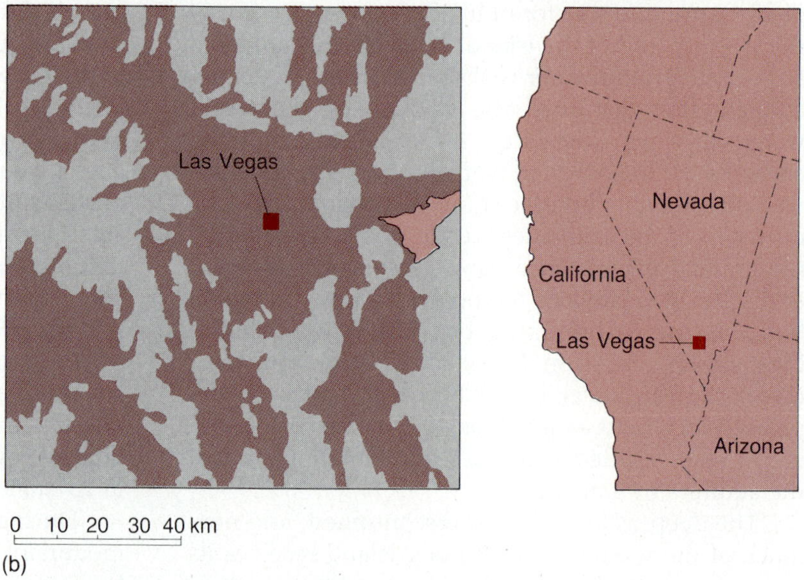

(b)

Figure 22.5
(a) Las Vegas, Nevada, derives its water supply from an enclosed basin containing alluvial fill. Many other areas in the west have a similar geologic setting (see Fig. 22.3). (b) The mountains shaded gray in the sketch surrounding the basin contain only moderate water supply. They do collect snow and rain that infiltrates the unconsolidated sediment in the basin.

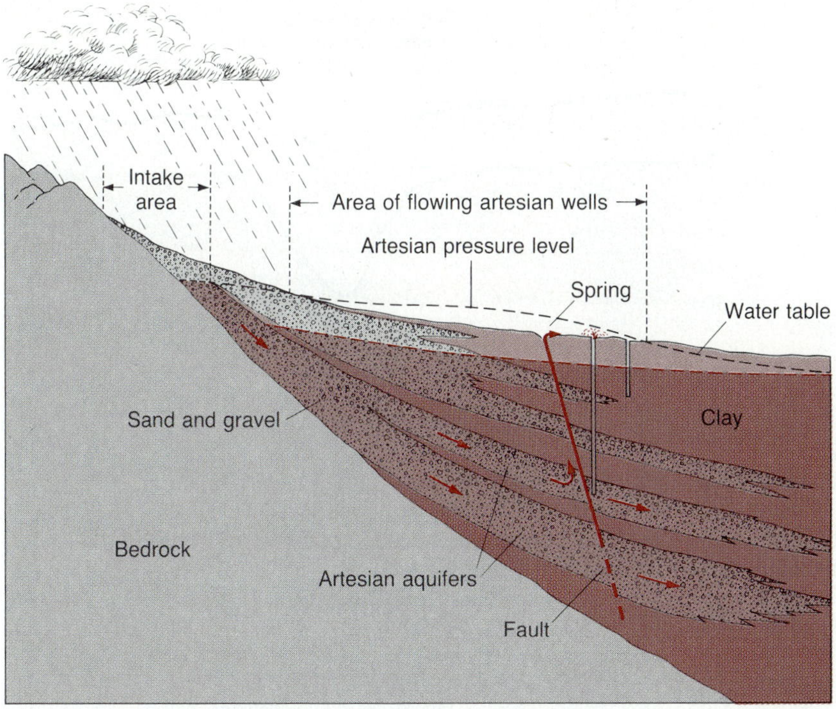

Figure 22.6
The water supply in the Las Vegas area consists of both unconfined and artesian water. Water enters the deeper parts of the alluvium on the sides of the surrounding mountains and saturates the coarse sediment in the alluvial fans.

Porous and permeable beds in unconsolidated alluvial deposits (Fig. 22.6) form the principal groundwater reservoir in this basin. The aquifers are not continuous because the water-bearing layers are interbedded with irregularly shaped impervious beds. Several zones contain water under artesian pressure, but most of the aquifers leak water into other beds, producing a more or less continuous hydraulic system. Water on or near the ground surface moves toward the lower parts of the basins, but the quantity of water lost by evaporation and transpiration there is almost equal to the total annual rainfall. Only in the mountains does precipitation occur in quantities sufficient to allow infiltration. There, around the edge of the basins, water moves down into the aquifers and becomes confined under layers of clay or other impervious beds.

High-angle faults cut many of the alluvial deposits. Some artesian layers have been severed as a result of movements on these faults; and in places the porous layers are offset so that they terminate against impervious units, trapping the water and preventing its migration across the fault.

The largest withdrawals of groundwater are concentrated near Las Vegas. This excessive pumping has caused a reduction in the artesian pressure in the aquifers and threatens a serious overdraft.

In addition to the reduction in availability of water, careful releveling shows that the ground surface in the region is dropping slowly (Fig. 22.7). This subsidence is caused in part by the weight of the water stored in the nearby Lake Mead reservoir, but apparently also in part by compaction of fine-grained sediments as water is withdrawn from them. This porosity and the storage capacity it represents are lost once compaction occurs and cannot be opened again without use of expensive techniques such as injection of water at high pressure. Similar effects have been observed in the San Joaquin valley of California, and may also be destroying the capacity of aquifers to store water in the Great Plains and even in the Atlantic and Gulf Coastal Plain, where water is comparatively abundant.

Water crisis in the High Plains. Overdraft of water from a rock layer known as the Ogallala formation has now reached the stage where it is discussed in newspapers and popular magazines all across the country, as well as in technical water-supply papers and farm journals. The Ogallala is the primary aquifer from which irrigation water has been pumped in the high plains states from Texas to Nebraska and South Dakota (Fig. 22.8). Water from the Ogallala is used to irrigate and raise nearly 12 percent of the nation's cotton, corn, and wheat. This water is drawn from an estimated 150,000 wells drilled into the layer over an eight-state area. The extensive pumping is lowering water levels in the aquifer so rapidly that by

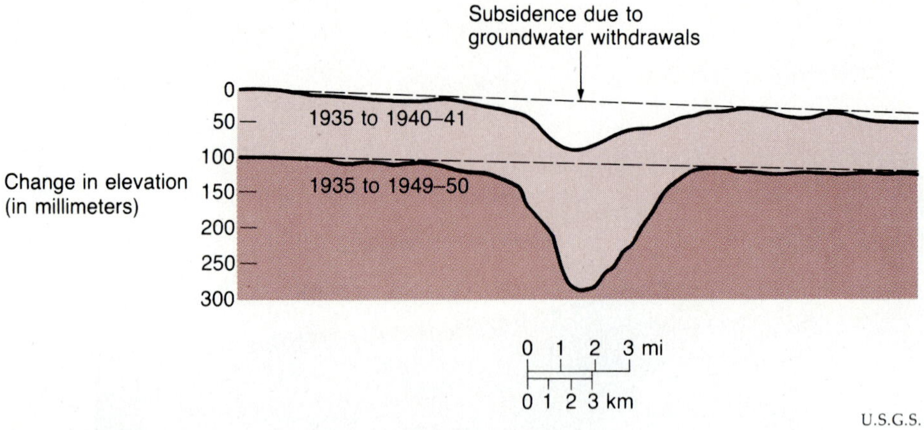

Figure 22.7
The surface of the land in the area near Lake Mead (near Las Vegas in Fig. 22.5*c*) changed level as a result of the weight of water in the reservoir. This accounts for the general tilt seen in these curves. The more pronounced depression is attributed to subsidence caused by pumping water out of the ground.

the year 2000 it will contain enough water to irrigate less than half the number of acres now watered from it.

From the area covered, it is obvious that the Ogallala formation is an extensive, sheetlike deposit located in the region of the plains close to the Rocky Mountains. In fact, the sands, silts, and lenses of gravel that make up most of the water-bearing part of the Ogallala were derived by erosion of the high Rockies during the later part of the Cenozoic era. The sheet of sediment is not uniform in either composition or thickness over this region. It varies from zero to slightly more than 200 m thick.

In some places the Ogallala formation is exposed at the ground surface, but generally, it is covered by thin deposits of more recent alluvium, sand dunes, glacial deposits, or wind-deposited silt. Where water infiltrates directly into the Ogallala through one of those covers, its permeability greatly affects local recharge. Unfortunately, rainfall in the plains and recharge from streams are not nearly adequate to replace the water pumped out. Nor does all the water pumped out for irrigation purposes infiltrate back into the ground. Much of it is lost to evaporation and transpiration.

Water supply in the Ogallala is far from uniform. Most variations are directly related to the thickness of the water-bearing beds. But porosity, permeability, and local recharge also vary within the unit. Because of this, the thickness of the saturated zone varies by nearly 200 meters. Most wells hit the water table at depths of 20 to 150 meters, but as overdrafts continue,

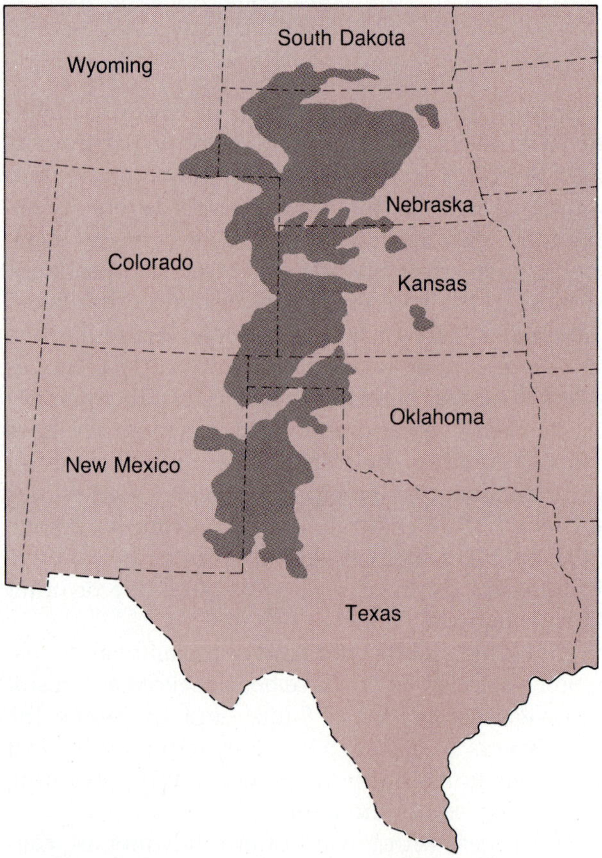

Figure 22.8
The Ogallala formation is an important aquifer throughout the shaded area. Serious problems face the farmers in this area because water is being pumped out of the Ogallala faster than it is being replenished naturally.

this depth and the cost of pumping water to the surface continually increases. The water may become economically unavailable in some areas long before it is exhausted.

The outlook for the future is bleak. Water in the aquifer is being "mined"—drawn down 10 to 15 times faster than it is being replenished by recharge. Cost of pumping is increasing, and the cost of importing water by canals or pipelines would be exceedingly high. Some hydrologists predict that we may one day see vast areas in the region given over to capturing water and recharging the Ogallala at a rate close to that at which water is withdrawn.

Hazards of Groundwater Development

A number of potential natural hazards are involved in groundwater development. Many of these have been mentioned in earlier discussions, but because of their importance, they are restated briefly here:

Long-term overdrafts (when pumping exceeds recharge) lead to exhaustion of the supply. When exhaustion is reached, water can be drawn only at the same rate as the supply is recharged.

When the rate of pumping of water from an aquifer exceeds the rate of recharge, the aquifer can be damaged by compaction and destruction of porosity.

Where salt water is present in an aquifer or in adjacent rocks, as it is along many coasts, salt water can be drawn into and contaminate the supply if fresh water is withdrawn faster than it is recharged.

If wells are located too close together, the cone of depression of the water table around the wells will intersect and reduce production of water.

Excessive pumping of water from karst areas may induce subsidence or collapse of the ground surface.

Sewage or chemical or radioactive waste disposed of underground in water-bearing rock or sediment will contaminate groundwater and be carried with it as it moves through the aquifer. Safety of water in such an aquifer can be endangered for many years, and in the case of radioactive waste, damage may be essentially permanent.

The salt content of groundwater is increased where water infiltrates salt-encrusted soil in arid regions. For this reason, recycling of groundwater by irrigation tends to increase its salinity.

Replenishing Aquifers

The areas where people are most dependent on groundwater are usually also areas of low rainfall. It is now clear that if we are to avoid massive economic disaster in many parts of the country, we will have to give greater attention to groundwater supplies in the next decade than we have in the past. At some point, it will be necessary to restrict withdrawal of groundwater in order to keep the water budget in balance.

Partial solutions to this water supply problem will come from techniques designed to increase recharge of groundwater from permanent streams and in the natural areas of recharge. In Fig. 22.9, recharge of an aquifer is increased by drawing water levels

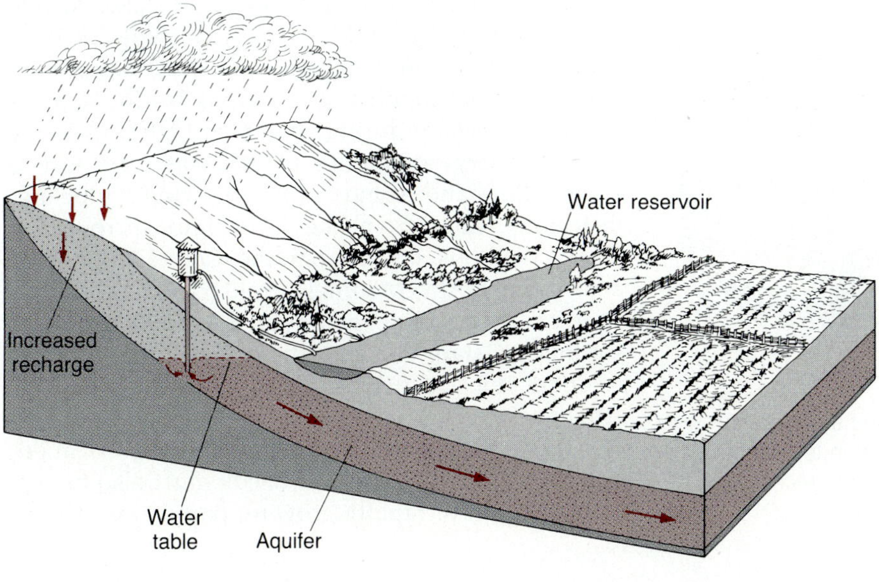

Figure 22.9
One method of inducing recharge in an aquifer is to lower the water table in the recharge area during wet seasons. The water withdrawn could be put in storage in reservoirs or reinjected elsewhere.

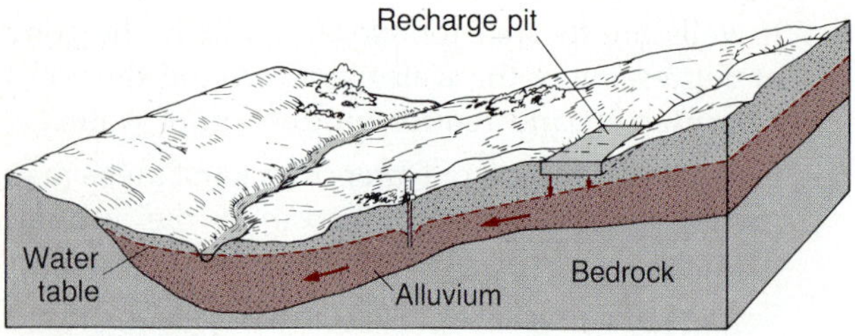

Figure 22.10
Artificial recharge of aquifers can be accomplished by constructing pits filled with porous and permeable rock in the valley of stream tributaries.

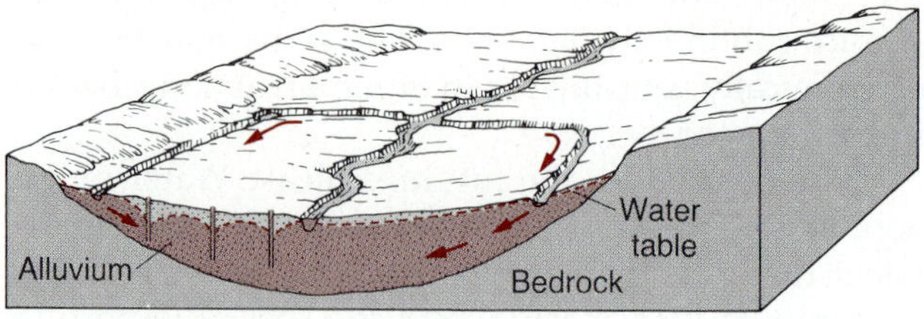

Figure 22.12
Canals can be used to recirculate streamflow across porous and permeable alluvium, increasing infiltration.

down in the natural area of recharge. By keeping the water table lower, infiltration is increased and less water runs off on the surface. In the second technique (Fig. 22.10), recharge pits or basins are constructed in tributary valleys of major drainage. Small dams would serve this purpose. Unlike most reservoirs, however, they are located and constructed in such a way that their contents leak into the soil. In the third example, (Fig. 22.11), wells located near major streams are pumped to draw the water table down below the stream and promote infiltration from the stream. This system would be most effective near cities where flood runoff is often excessive. Finally, Fig. 22.12 shows a system designed to reuse water by diverting water from a stream channel into leaky canals that increase recharge of the valley fill. These and other methods are already in limited use. Some communities are even using a technique of spraying sewage onto the ground surface rather than treating it and releasing the water in streams. Some of these methods will become common practices by the year 2000.

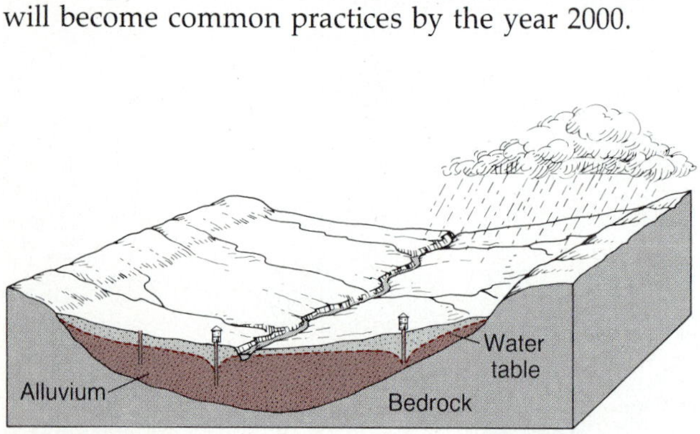

Figure 22.11
Infiltration along a stream can be increased artificially by pumping water from wells near the stream. In this way the heavy runoff from urban areas can be used to recharge groundwater supplies.

22.3 SURFACE WATER RESOURCES

Most domestic and industrial water supplies in the eastern half of the United States come from surface waters. Because rainfall in this region is generally adequate to support agriculture, little of the supply is diverted to agricultural use. Water is stored in reservoirs held by dams. The primary problems in the East, pollution and finding new dam sites, are very different from those in the West where rainfall is insufficient. The areas are similar in that surface water is heavily utilized. In the West, all of the available supply is still not enough.

The Colorado River basin (Fig. 20.8) is a good example of the difficult situation found in semiarid regions. Rainfall over most of the basin is slight—far below the potential evaporation. Most surface water comes from the high mountains in the Rockies. Even this is reduced by local removal of water for irrigation. Within the main channel, a number of exceptionally large dams have been constructed at Lake Powell, Lake Mead, and Lake Mohave. As recently as 1980, some of the stream beds between these dams were dry in summer months. The purpose of the dams is to impound water during periods of high flow in the river, but large quantities of this water are lost to evaporation and some infiltrates. Huge pipelines distribute water from the lakes into surrounding areas. Water is taken from the Colorado by pipe across the Rocky Mountain front ranges to supply Denver. An aqueduct and a canal remove water into southern California, and the largest project of all is under construction to carry water into southern Arizona.

These waters are used extensively for irrigation, but domestic and industrial needs must also be met in this region of rapidly growing population. Unfortunately, little or no water is left in the Colorado by the time it reaches the Gulf of California. Rights to the available water have been a source of litigation for years. Although legal action may continue for years, the chance of increasing the amount of available water is not good.

22.4 FLOODING

When discharge exceeds the amount the stream can carry in its channel, water overflows the banks and streams flood. The causes of flooding are more varied and subtle than they might at first appear. Excessive rainfall over a short period of time is certainly the main cause of flooding. But a number of other variables determine how much of that rain runs off in the streams and how much is consumed in other ways.

The part of rainfall that becomes runoff is much greater on steep slopes than it is on low slopes; it is greater on impervious rock than on permeable soil; and it is greater where plants are sparse than it is where a thick plant cover exists. Of these factors, the slope of the land does not change rapidly. There are certainly no seasonal variations. Most mountainous areas are subject to flash floods when unusually heavy rains fall. These may be triggered by thunderstorms, hurricanes, or stationary weather fronts accompanied by heavy moisture supplies. The ease with which water infiltrates is important here, as it is in other places, but steep slopes predispose mountains to rapid runoff. Low porosity and permeability are universally important causes of flooding. In northern areas many floods are caused by heavy spring rains falling on frozen and occasionally snow-covered ground. The impermeability of the frozen ground prevents infiltration, and plants are still dormant so transpiration is negligible.

Flat areas in general, and especially flat areas underlain by impervious soils or shallow bedrock, are frequently flooded by heavy rain. Such areas simply lack sufficiently well-developed drainage to move surface runoff out of the area.

Floods are much more common now than they were even a few decades ago, and property damage due to flooding increases steadily. Most of the increase is caused by housing and industrial development. Some basins have been almost completely paved over by roads, parking lots, and buildings. This pavement converts most of the water, which would normally infiltrate, into runoff, insuring both maximum runoff and damage to the structures that themselves literally cause the flooding.

The removal of plant cover that occurs in the clear-cutting of forests and in clearing land for highway, residential, and urban development also increases runoff. Because the ground surface is disturbed in the processes that remove plant cover, erosion is accelerated and soil washes into streams. In the following case histories, we will see these causes in operation.

Case Histories

1973 flood on the Mississippi basin. Efforts to control flooding and improve navigation on the Mississippi River began before the Civil War. Billions of dollars have been spent to build levees and dams, dredge channels, drain the flood plain, and divert the river. These efforts have undoubtedly saved large areas from flood damage. Yet in 1973, one of the worst floods on record covered 12 million acres of land, damaged 30,000 homes, and produced a record of 77 days of flooding at St. Louis, 63 days at Memphis, and 88 days at Vicksburg, Mississippi. Parts of ten states were flooded. Fifty thousand people had to be evacuated, and property damage was estimated at more than 400 million dollars.

The spring floods in 1973 were preceeded by an unusually wet fall in 1972. Reservoir levels were high, and most streams in the central and southern part of the drainage basins were already well above the normal stage (Fig. 22.13). Unusually high precipitation continued in February and March. Much of the earlier snow accumulation in the northern part of the basin began to melt, while the ground remained frozen and unable to absorb the meltwater. Flooding in the upper Mississippi River started in early March; and as heavy rains continued, flooding extended south. By April, the entire mainstream of the Mississippi south of Cairo, Illinois, was above flood stage (Fig. 22.14).

Precipitation in the continental interior greatly exceeded normal levels during March and April. Large areas received one-and-a-half to two times their normal rainfall. These rains were associated with cold fronts and cyclonic disturbances. Although these conditions are in themselves normal, unusual amounts of warm moist air from the south and cold arctic air from Canada intruded into the region over the Mississippi basin, triggering exceptionally persistent and large amounts of rainfall.

As streams exceeded their flood stages, water spread out onto the flood plain (Fig. 22.15). To some

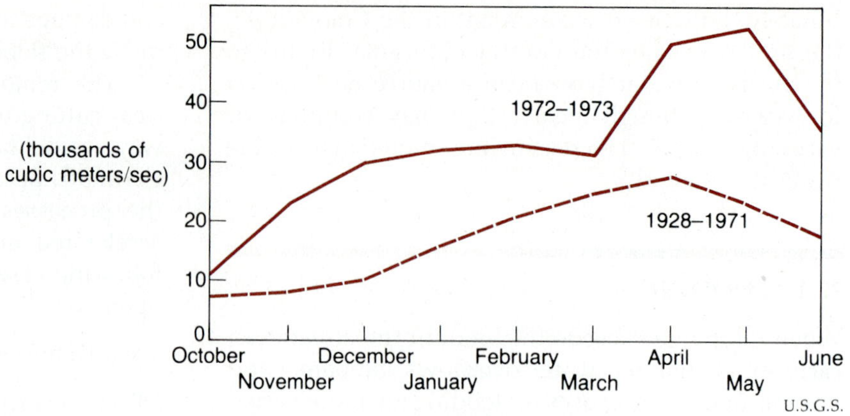

Figure 22.13
The long-term averages for discharge in the Mississippi River, measured at Vicksburg, Mississippi (bottom curve), show a seasonal increase in the late spring. During 1972–1973 (top curve) discharge was well above average all year, but reached a peak in April.

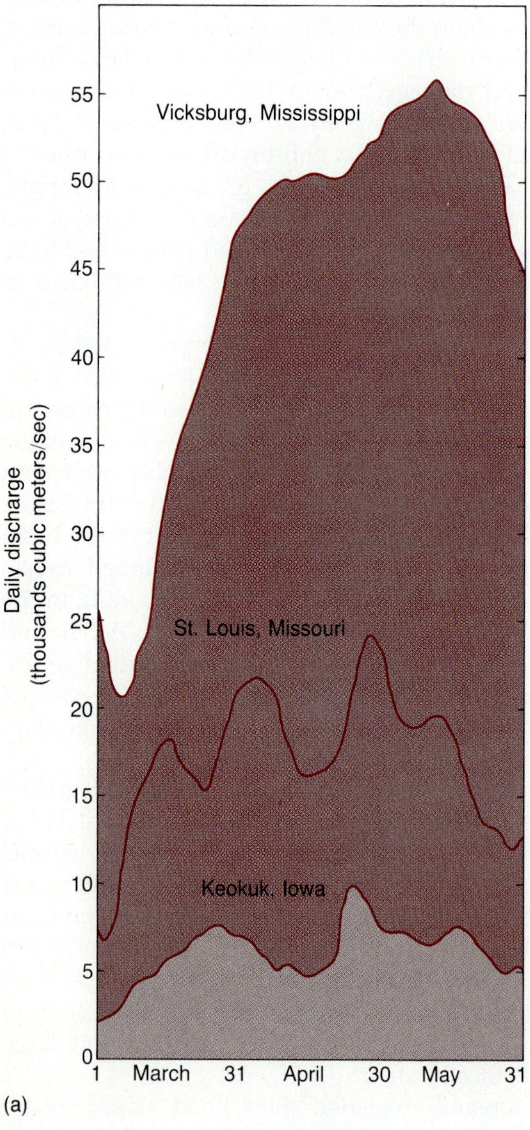

Figure 22.14
(a) These records of discharge show clearly the downstream increase in discharge along the Mississippi during the 1973 flood. Note also that the peak discharge arrives progressively later, and that flooding lasts longer downstream. (b) Locator map.

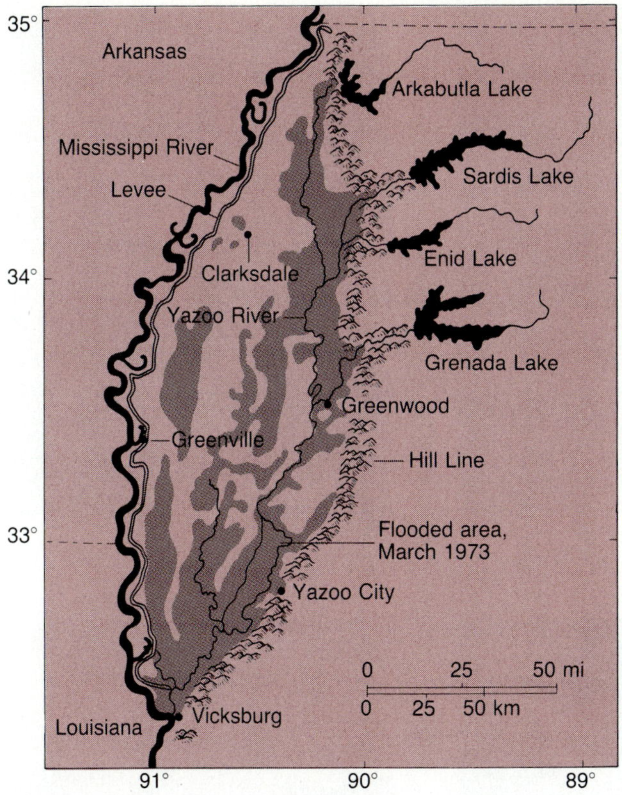

Figure 22.15
During the spring of 1973, the Mississippi overflowed its banks all the way from Cairo, Illinois, to the Gulf of Mexico. Flooding was especially severe in the section of Mississippi shown here.

extent, water was contained by networks of levees constructed along the river, but the rains were so general that floodwaters built up on both sides of some levees, covering vast areas. Satellite photographs such as those in Fig. 22.16 clearly document the rise and fall of the waters.

Although the immediate cause of the flooding was unusually heavy precipitation over a period of months, some hydrologists conclude that this flood, was caused in large part by levees constructed to keep the stream in its channel. They also prevented water outside the levees from returning to the main channel. While levees help prevent flooding when the floodwaters originate farther upstream, they do not solve the problem of locally heavy runoff.

Hurricanes Agnes and Camille. On August 20, 1969, rains caused by Hurricane Camille set records for the James River basin in Virginia, causing a flood of the severity that generally occurs less than once a century—what hydrologists call a **hundred-year flood.** Three years later, in June 1972, a second **hundred-year flood,** caused by Hurricane Agnes, struck the James. While the designation "hundred-year flood" means that, statistically, the flow should be expected only once in a hundred years, it is clear that hundred-year floods don't necessarily come one hundred years apart.

Hurricanes frequently cause floods in the eastern United States because the storm track of hurricanes originating in the Caribbean usually follows a northeast path up the coast. Many storms are over water, but others, like hurricanes Camille and Agnes, move over land. Generally, hurricanes lose intensity over land because the water supply from the sea is cut off, but these two combined with other weather conditions to produce unusual precipitation.

Camille came ashore in the Gulf near New Orleans (Fig. 22.17) and followed a path up the Mississippi Valley where heavy rains fell. Then it turned east and crossed Kentucky and Virginia to the coast. Over the mountains of Virginia, Camille encountered a region already overlain by heavy moist air. This apparently caused the storm to intensify, and the resulting low pulled moist air from the Atlantic inland. Rainfall exceeded 28 inches in an eight-hour period in one county—an amount of rain that might be expected about once in a thousand years.

The pattern of rainfall along the track of the storm shows highs that die out rapidly to either side, but the path crossed the James River drainage basin in a way that allowed most of the heaviest rain to occur there. The results were disastrous throughout most of the basin. Some of the heaviest rains fell on the steep slopes of the Blue Ridge. The mountain soils became saturated and slid off of the slopes, leaving thousands of long scars up the sides of the mountains and exposing bedrock where the soil had been (Fig. 22.18). These slides buried some homes and added great quantities of debris to the rising streams at the foot of the mountains. Many people who lived on the flood plains were caught as their houses and trailers were washed away. The death toll exceeded 150 persons.

The flood waters swept downstream, as shown by the discharge hydrographs in Fig. 22.19. These are typical of the types of records obtained during many floods. Streams in the upstream portion of the drainage basin show very sharp increases in discharge in the hours during and immediately after the main

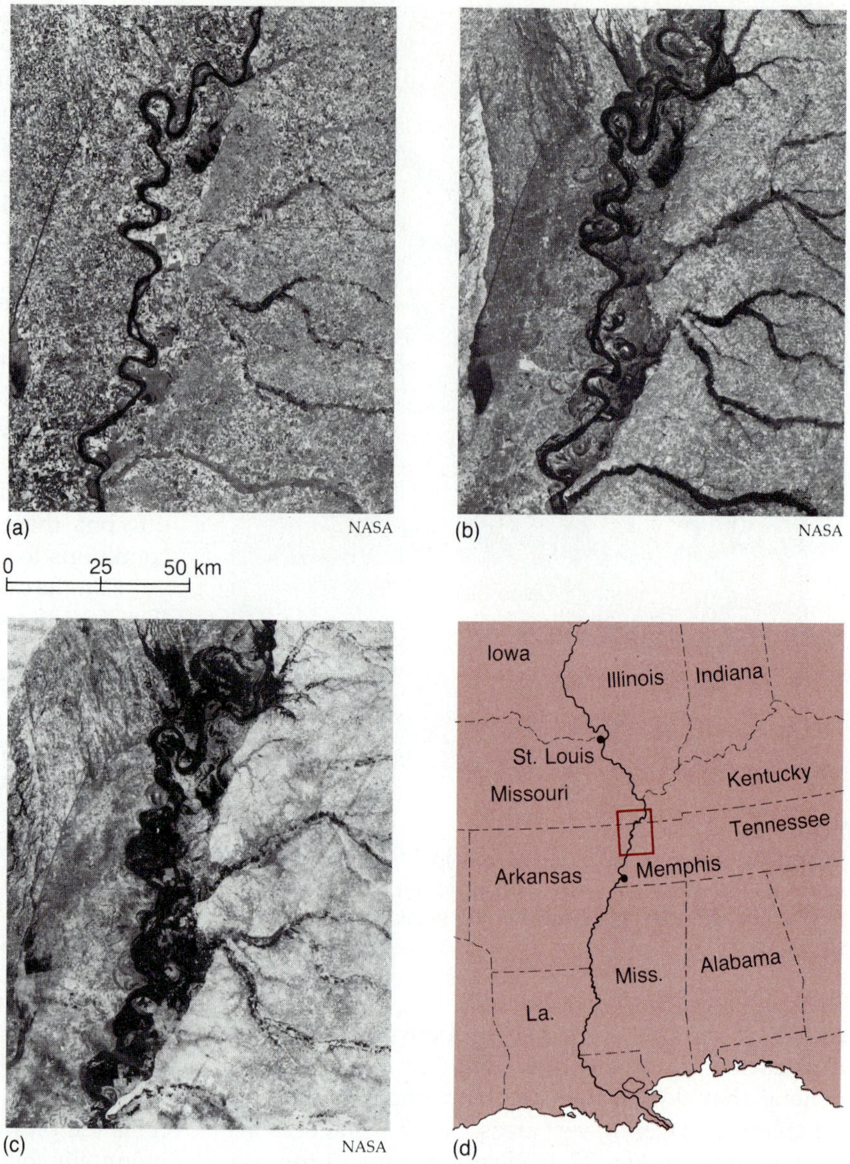

Figure 22.16
Satellite photographs provide an exceptionally useful tool for studying flooding. These three images of the area shown in the locator map show the progressive flooding of the area. (a) Preflood image; (b) flood waters rising on both tributaries and the main stream; (c) waters are beginning to recede in the tributaries, but the main stream valley is flooded; (d) locator map.

storm. Downstream, the discharge increased later because it took time for the flow to reach downstream stations. The discharge was also higher downstream because more tributaries had contributed their discharge to the streams, and the peaks became broader as flood stages were maintained for a longer period of time than upstream. This is simply an effect of the backup of water in the stream system.

The effects of this storm included not only the high water, deaths, and destruction of property, but landslides on the steep slopes, inundation of towns and fields by mudflows, removal of topsoil from large areas of the flood plains in the upper part of the basin, scouring of the channels, and finally inundation of the flood plains throughout the basin; all added to Camille's toll.

Like Camille, Hurricane Agnes in June 1972 followed an unusually long overland path and caused especially intense floods because it acted in conjunction with other weather systems. A large low-pressure system located west of Agnes followed a parallel path as Agnes moved up the coast (Fig. 22.17). The counterclockwise flow of air around Agnes pumped moisture-laden air into the east coastal region, and the effects of this rainfall were compounded by rain from the other low-pressure center.

The path of Agnes followed the east coast, and the effects were felt over a much larger area than

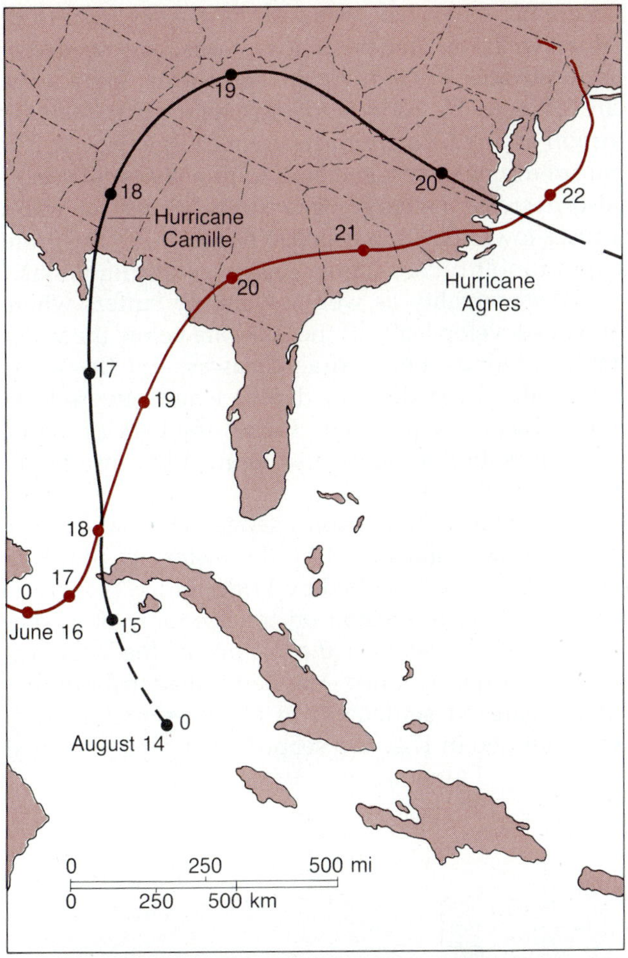

Figure 22.17
The paths followed by hurricanes Agnes and Camille. The circles and numbers indicate the position of the storm center on successive days.

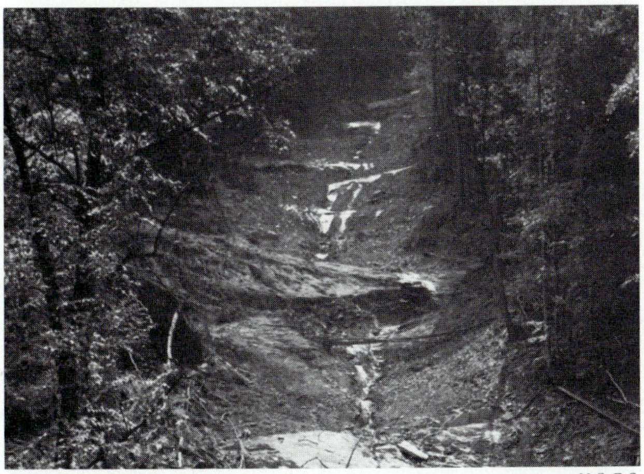

U.S.G.S.

Figure 22.18
Mountain soil became saturated with water during the record rains accompanying Hurricane Agnes. The soil cover slid and was washed out of many steep stream valleys, like this one, leaving thousands of scars on the Blue Ridge.

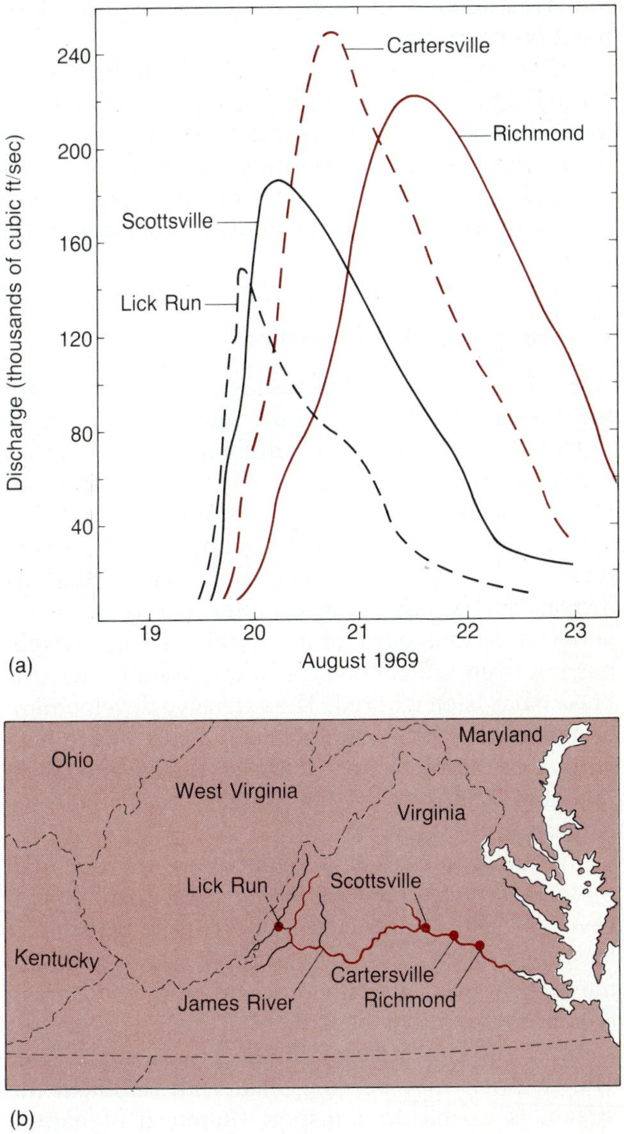

Figure 22.19
(a) These records of discharge allow us to trace the movement of the flood crest for Hurricane Agnes down the James River in Virginia. Note that discharge increases downstream, and that the flood crest is wider downstream. (b) Locator map.

Camille; but the James River basin was again one of the sites of heavy precipitation. The total rain on land from Agnes is thought to have set a new record. This rain followed an unusually wet spring; the ground was saturated at high levels, and runoff was excessive even where the slope of the ground was fairly low.

Over a twelve-state area, hundred-year-plus flows occurred. Resulting property damage was estimated at 3.1 billion dollars, making this the most destructive flood in the history of the country—mainly because the areas flooded are sites of high-density development on flood plains.

Discharge from the James River at Richmond exceeded that observed during Camille. Although the maximum observed rainfall from Camille was more than twice that observed during Agnes, heavy rains covered much larger areas. Record discharges were also recorded on rivers in Pennsylvania and New York.

Development and Urbanization

When people first took advantage of the rich soil of the flood plains for agriculture and the easy access to water by building towns and cities there, a direct conflict between natural processes and artificial modification of the environment was started. This conflict and its ramifications have burgeoned over the centuries as our ability to modify the environment has increased. Too often the complex interrelationship between various parts of a natural drainage system has not been understood; or, more recently, what is known has been ignored. The extensive development of agriculture and some forestry practices have hazardous implications for drainage basins and flood potential, and even greater dangers exist where urbanization is taking place.

When land is put into cultivation, the potential for increased erosion exists, because the natural plant cover is destroyed and the land is laid bare for at least part of the year. At that time, heavy rains can move the loose soil easily, and gullying where wash is concentrated can be greatly accelerated. As this soil gets into local streams, their behavior is modified. The channels may become filled with sediment the stream is unable to transport. Removal of natural vegetation may increase surface runoff, increasing discharge and leading to floods downstream. The same problems of erosion and excessive runoff exist where forests are destroyed by fire or are cut down.

The impacts of urbanization are even more complex and often produce severe consequences. When an area is urbanized, water use is sharply increased. More wells, or larger-capacity wells, are needed if groundwater is used. At the same time, large areas are covered by roofs and pavements that increase runoff and reduce groundwater recharge (Fig. 22.20). Storm-drainage systems built to provide rapid runoff also decrease recharge. The increased use of water causes lower flows in drought periods, and the more rapid runoff causes higher discharge during floods.

Water quality as well as quantity suffers where sewers develop leaks. If the leaks are below the water table, groundwater is drained away and the water table falls. Regardless of depth, when sewers leak, groundwater is polluted. Large numbers of septic tanks in suburban areas also frequently cause pollution of shallow aquifers.

The way urbanization changed the nature of a stream is well illustrated by the history of the Anacostia River, as described by Withington (*1*).

The effect of erosion on the economy of a community can be seen in the history of Bladensburg, Prince Georges County, Maryland. Bladensburg, the first organized settlement in the Washington area, was founded in 1742 as a seaport on a deep protected

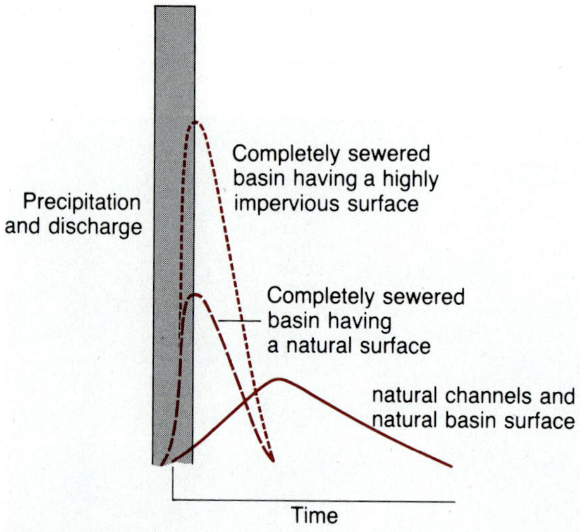

Figure 22.20
The curves show schematically the effects of urbanization on the discharge in a drainage basin. The amount of rainfall is represented by the area in the rectangle at left. In a normal channel the discharge takes place over a long time and has a low peak. As the basin is paved, the peak is higher and runoff takes place in a shorter time.

harbor at the head of navigation on the Anacostia River. At first, ocean-going vessels reached Bladensburg with little difficulty, and the town became a major center for tobacco export. As the land around Bladensburg was cleared for crops, unconsolidated coastal plain sediments were exposed to erosion, and the channel of the Anacostia began filling with sediment. By 1800, the maximum depth at low tide had been reduced from the original 24 feet to 14 feet (from 7.3 to 4.3 meters), and the width of the channel was so narrowed that navigation was difficult. By 1830, the river became unnavigable to all except the shallowest draft ships, and Bladensburg was abandoned as a port. Few floods had occurred at Bladensburg before the 1930s; but by then, urbanization upriver from the town had increased the silt load in the river. The silt clogged the channel so that floods occurred with increasing frequency. By the 1950s, extensive flood-control measures were constructed.

Three severe drainage problems seem invariably to accompany urbanization: streams flood more frequently; flood stages are higher; and increased sediment load occurs during construction phases. The first two of these problems are difficult to solve. They occur because runoff is increased due to the construction of impervious cover on the ground. The most effective ways to resolve this problem are to increase infiltration artificially and to design channels or storm sewers that are adequate to carry the increased discharge. In general, neither of these solutions is carried out until floods demonstrate the magnitude of the problem.

Silting of streams can be effectively reduced by building codes and zoning ordinances that require the maintenance or quick replacement of plant cover and use of various "screens" to reduce erosion and surface wash on construction sites. The U.S. Geological Survey made a study of an area just north of Washington from 1962 to 1974. Rainfall during this period remained relatively constant from year to year, but as urbanization took place, starting especially in 1969, both water and sediment discharge increased dramatically, reaching a peak in 1972 at a time of extensive construction, as shown in Fig. 22.21. In the following years the amount of sediment in the stream decreased dramatically.

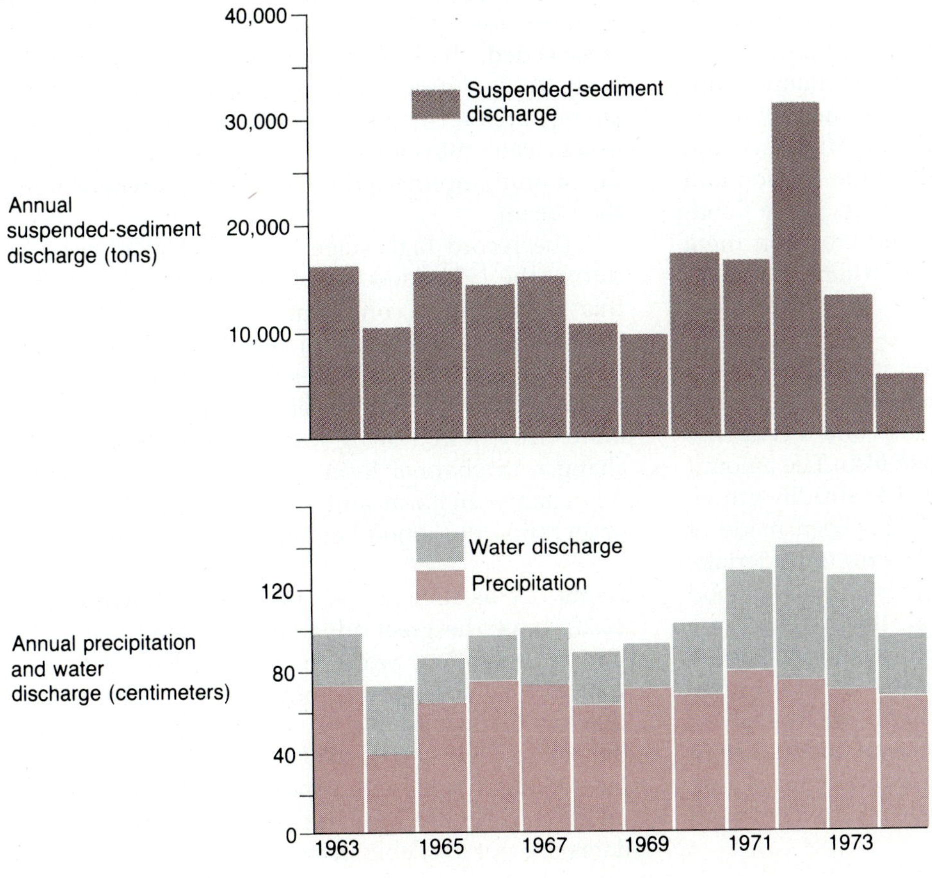

Figure 22.21
The annual rainfall in the basin of the Anacostia River remained relatively constant during the years 1963–1974, but both the runoff and the suspended sediment load show peaks in 1972 at the height of a construction boom.

This remarkable turnaround is directly related to a mandatory program for sediment control in Montgomery County. A system of control initiated in 1971 requires minimum disturbance of the plant cover, use of temporary vegetation and mulch, construction of sediment basins on sites, planting of permanent cover at the end, and selection of sites that have the least erosion potential. In 1970, 41 percent of the construction sites had approved sediment control; in 1972, 56 percent were approved; and in 1974, 74 percent were in compliance. The results show up clearly in the graph.

Modification of river systems through construction of towns, docks, levees, and so on has proceeded too far for us to consider allowing most streams to revert to their natural state. The question now is what can be done to insure the proper future development of drainage systems. It is for these reasons that much of the quantitative study of streams has been done.

Flood Control

Flooding is characteristic of most streams. With few exceptions, channels have adequate cross-sectional area to contain the waters that normally flow along the stream. But when excessive rains fall or when the ground is frozen and rainwater cannot infiltrate into the ground, the excessive water fills the channel, tops the banks of the stream, and floods some part of the adjacent valley floor. Indeed, flood plains are built over a period of time as a result of the erosion and deposition that occurs during flooding. These flood-plain deposits are often so rich and crops on them so bountiful that early settlers were willing to endure the hardships or the occasional floods in order to use the land.

Until the 1930s, many houses in the flood plain of the Mississippi were built on stilts or with a ground-level basement. Of course towns, and eventually cities, sprang up on the flood plain also. The amount of damage to buildings and roads has steadily grown as more and more intensive use has been made of the flood plains. The political pressure to undertake projects designed to protect the property on the flood plains has grown as well. At first, the projects were designed to provide local protection, such as a levee around a town, but as engineering capabilities increased, and with the advent of hydroelectric power, major flood-control projects have been undertaken to control large parts of whole river systems. These efforts have been subjects of great controversy from their inception.

The primary objective of flood control is to hold discharge along as much of the stream channel as possible at a level that can be contained in the channel. This amounts to reducing the level of the peak flow, such as that we have on the flood hydrographs for the James River (Fig. 22.19). In theory, at least, a number of simple techniques can be used to accomplish this. Flow can be regulated by construction of dams behind which excess water is stored in times of flood and released during dry spells. The edge of the channel can be raised by constructing levees. The channel characteristics can be changed (straightened or deepened, for example) to allow water to move more rapidly along a section of the stream, and steps can be taken to reduce or slow runoff by increasing infiltration. Most of these techniques involve major projects that are expensive, require continual maintenance, and drastically change the character of the streams on which they are used. Each has its own set of advantages, limitations, and serious problems.

Levees. When a levee or flood wall is built, it prevents drainage into the main stream as well as from the stream onto the flood plain. Construction cost increases greatly with height, and the levee must be designed for some particular flood level. If that level is exceeded, the levee is topped and may be at least partially destroyed. A levee also cuts off a natural storage area for flood water on the flood plain. That excess water must go somewhere else—into the channel or onto unprotected flood plains elsewhere along the stream.

The record high stage levels on the Mississippi during the 1973 flood have been attributed in part to this type of flood control measure. Flood-stage discharges on the Mississippi have steadily increased since the late 1800s. This indicates that more rapid runoff is taking place, possibly because channels are more efficient in moving water as a result of artificial changes in channel form. Levees help bring about this change in form and also may account for the observation that flood heights are increasing.

Dams. Dams such as the Oahe Dam, shown in Fig. 22.22, have the great advantage of being useful for power generation, water storage, and recreation as well as for flood control, but they too have serious drawbacks. Because a dam stops the flow of water, the stream deposits its load behind the dam, where the sediment will eventually either fill the reservoir or have to be dredged out. Also, suitable sites for dams are not available in many places. A good dam

U.S. Army Corps of Engineers

Figure 22.22
Dams are one of the most effective flood control devices, provided the reservoir is not too full when the flooding occurs. This photo shows the Oahe Dam on the Missouri River in South Dakota.

site must meet special foundation requirements, the reservoir must be water-tight and of large capacity, and building a dam should not flood out or endanger a town. Most of the best sites in the United States have already been used or, like the Grand Canyon, are protected. In addition, dams are exceedingly expensive to build, and in dry periods when water is drawn down, large mudflats often form a border around them. Despite these drawbacks, dams are an important and effective means of flood control. If the reservoir is large enough, a dam can be used to regulate the flow downstream. If the flood waters originate above the dam, it can be used to reduce flood damage downstream by holding some of the water as the peak discharge enters the reservoir. That water is then released slowly later. If the flood originates downstream, the dam is used to reduce normal flow in the stream by stopping discharge from the reservoir. This too decreases peaks downstream.

An important question, still largely unresolved, is the comparison of the cost-effectiveness of a few large dams versus many smaller dams on the small tributaries of major streams. The smaller dams are cheaper to build; they help reduce the buildup of sediment in large reservoirs; and they would improve infiltration. But they are less easily controlled. At this time, few streams in this country remain in their primitive state. Most of the best dam sites have been used, extensive levees line the Mississippi, and many thousands of miles of stream have been run into channels or pipes. Still the problem of flooding remains in many areas.

Some modern floods result from rainfalls that produce water levels that exceed the design specifications of the engineering works: most dams and levees are designed to handle hundred-year floods. But many floods are a direct result of human activities. Flooding in urban areas can often be traced directly to the vast areas being covered with asphalt and other impervious surfaces. These increase both the amount and speed of runoff, as both infiltration and transpiration are drastically reduced. In other areas, floods on one portion of a stream can be traced to flood control of some other part. Gradually, more and more students of this problem conclude that the cost of adequate flood control along much of the remaining unprotected flood plains exceeds the benefits to be derived by development on that land. This has given rise to flood-plain zoning that restricts the uses of flood plains to those compatible with occasional flooding.

The provision of flood insurance by the federal government for buildings on flood plains is perhaps the most significant development in recent years. Before the program went into effect in 1968, insurance protection against flood damage was difficult to obtain and very expensive. Under provisions of the program, insurance is now available in participating communities. In order to qualify, a community must agree to follow certain procedures in the management of flood-prone areas. In general, communities identify flood-prone areas by mapping the levels of the hundred-year flood and manage these areas through a flood-plain zoning ordinance. The effect has been to make people and communities much more aware of flood hazards.

Flood Prediction

The cost of floods and the importance of knowing how much water is available in every part of a stream system have led to demands for a precise understanding of stream behavior. The factors that affect streams have long been recognized. Now they are being evaluated quantitatively and experimentally. The most important hydraulic factors are discharge, velocity, channel characteristics, gradient, and load.

Discharge and velocity are easily measured at any place on a stream. Usually they have been correlated with the stage of the river and can be determined simply by recording the height of the water at a gaging station. The stream load is more difficult to determine, especially the amount of bed load. Consequently, most studies concentrate on the suspended load. Channel characteristics are difficult to evaluate quantitatively. The shape and roughness of channels and their drag on the flow of water are hard to measure, but they can be approximated. The stream gradients can be determined by measuring the elevation of the channel bottoms at various places along the stream. Even the slope of the water can be approximated from gage heights.

We discussed the interplay of these factors in a qualitative way in terms of the concept of the graded stream. But each of these factors can be measured, and the relationships among them can be expressed mathematically. Thus, for any particular cross section across a stream, the width, height of water, average velocity, discharge, and suspended load can be related to one another mathematically. These relationships change systematically down the stream.

In real streams, the problem is complex because discharge usually increases downstream due to the addition of flow from tributaries. Also, rainfall is continually changing throughout the drainage basin, so that flows are not constant. Nevertheless, when enough data are available for the system, it becomes possible to construct both mathematical (computer) and physical models for predicting the behavior downstream that may be expected to result from any specified change upstream. Figure 22.23 shows a physical model of the Mississippi River near Vicksburg, Mississippi. This and computer models are used to predict how long it will take for a flood crest to move downriver and what the flood stages at various places will be. These models are now sufficiently accurate to provide adequate warning of most floods on major streams. The result has been a dramatic decrease in the number of deaths during floods. In some instances it has even been possible to raise levees downstream to avert flooding.

During the last two decades we have become alert to at least some of the threats to our water supply. Happily, in this time, it has also become apparent that most of the problems of water pollution can be solved—though often at a very high cost.

The relationship between urbanization and flooding is now clearly understood by most geologists and engineers. We even have a good grasp of the more difficult problem of maintaining a water balance in arid regions. The next step is to devise sound and economically feasible policies that can guide us through what promises to be exceedingly difficult years ahead. In order for this to proceed, a better understanding of these problems and their potential solutions is needed by lawyers, the courts, legislators, the commissions that make decisions about water supply, and the public.

U.S. Army Corps of Engineers

Figure 22.23
This scale model of the Mississippi River is used to predict the course and effects of flooding. The small pegs simulate drag on water in the flood plain, and the dark rows of wire, which resemble stacks of cards, simulate trees.

SUMMARY

In arid and semiarid areas much of the water for human consumption and for industrial and agricultural uses is drawn from groundwater. If the amount removed exceeds the aquifer's safe yield, the water supply may be depleted or polluted by the incursion of salt; or the aquifer itself may by physically damaged by compaction and loss of porosity. The problem is aggravated by the fact that increased pumping from the aquifer is generally accompanied by increased surface development; the paving over of land for roads, parking lots, and housing simultaneously decrease recharge of the aquifer by limiting infiltration. Techniques for increasing the rate of recharge are being developed to meet the accelerating use of groundwater supplies.

Across the United States, groundwater is stored in various rock formations in different regions. Along the major river systems, water is stored in alluvial deposits; and in glaciated areas, unconsolidated glacial deposits hold major water reserves. In consolidated rock such as that under the Appalachians, more limited supplies of water are in fractures, in pores of rock such as sandstone, and in solution cavities within limestone. Porous lava deposits are drawn upon in the Pacific Northwest and Hawaii.

When discharge exceeds the amount a stream can carry within its banks, flooding occurs. The abnormally high discharge generally results from an increase in runoff that is in turn caused by heavy rains combined with factors that limit infiltration. Steep slopes, impervious surfaces of rock or pavement, frozen ground, and absence of plant cover can all contribute to stream flooding. Since human activity has tended to increase the amount of impervious surface and to strip away the protective plant cover, it has increased the severity of flooding along many rivers. Improved understanding of the factors in flooding has caused many communities to require construction practices aimed at increasing infiltration in order to avert flooding. Elsewhere artificial levees, dams, and holding reservoirs have been built in an effort to contain floodwaters or regulate river discharge.

KEY TERMS

flood plains
groundwater basin
hardness
hundred-year flood
safe yield
water budget

STUDY QUESTIONS

1. How can infiltration of rainwater be increased in an area like the Great Plains?
2. Why is groundwater pollution a serious problem in areas with karst topography?
3. What steps could be taken to stop salt-water encroachment in a coastal plain?
4. How can artificial levees contribute to flooding in the area where they are built?
5. In what ways may the construction of a large development cause locally higher flood levels?
6. What precautions should be taken to avoid groundwater pollution at waste disposal sites?
7. Why do so many floods occur in the late spring?
8. What is meant by the expression "safe yield"? Explain what conditions might affect the amount of the safe yield in the area where you live.

REFERENCE

1. C. F. Withington, "Geology—Its role in the development and planning of metropolitan Washington," *The Journal of the Washington Academy of Science,* 57 (1967), *189–199.*

SUGGESTED READINGS

Boslough, John, "Rationing a river," *Science 81* (1981), *26–38.*

Cohen, P. O.; O. L. Franke; and B. L. Foxworthy, "An atlas of Long Island's water resources," *N. Y. Water Resources Commission Bulletin 62* (1968).

Ford, R. S., "Ground water—California's priceless resource," *California Geology* 31:2 (1978), *27–32.*

Leopold, L. B., "Hydrology for urban land planning," *U. S. Geological Survey Circular 559* (1968).

Lindorf, D. E., and K. Cartwright, "Ground-water contamination—problems and remedial actions: Environment Geology Notes," *Illinois State Geological Survey No. 81* (1977).

Piper, A. M., "Has the United States enough water?" *U. S. Geological Survey Water Supply Paper 1797* (1965).

LANDFORMS SHAPED BY THE WIND

Chapter 23

In the vast sand-covered areas of the world's largest deserts, aptly called the sand seas (Fig. 23.1), wind is the most important geomorphic agent. The deserts of the earth occur in belts located at latitudes 20° to 40° both north and south of the equator (Fig. 23.2). In the northern hemisphere, they are most extensive in a great swath that crosses North Africa, Arabia, Iran, and into China. In the southern hemisphere, the largest deserts are in South Africa and Australia. The effects of wind action are not confined to the earth. Photographs of Mars show that wind is an important geomorphic agent there also.

In this chapter, we will first briefly consider some of the types of movements that take place in the atmosphere, and how these movements can, in turn, pick up and move the materials on the ground surface. Obviously, the availability of movable materials is an important determinant of where wind action is effective. For this reason the wind is an important geomorphic agent along coasts, at the edge of glaciers, and over volcanoes, as well as in the desert. We shall investigate both the mechanism by which wind accomplishes erosion and the features that result from this erosion. Finally, the nature of wind deposits and their forms and characteristics will be examined.

23.1 MOVEMENTS OF THE AIR

Air moves mainly in response to pressure variations in the atmosphere and to the drag effects between the rapidly rotating solid earth and the envelope of

U.S.A.F.

Figure 23.1
Large areas in northern Africa and south central Asia are covered by sand. Complex dunes such as the ones seen here from the air in Algeria are common. These dunes are tens of meters high.

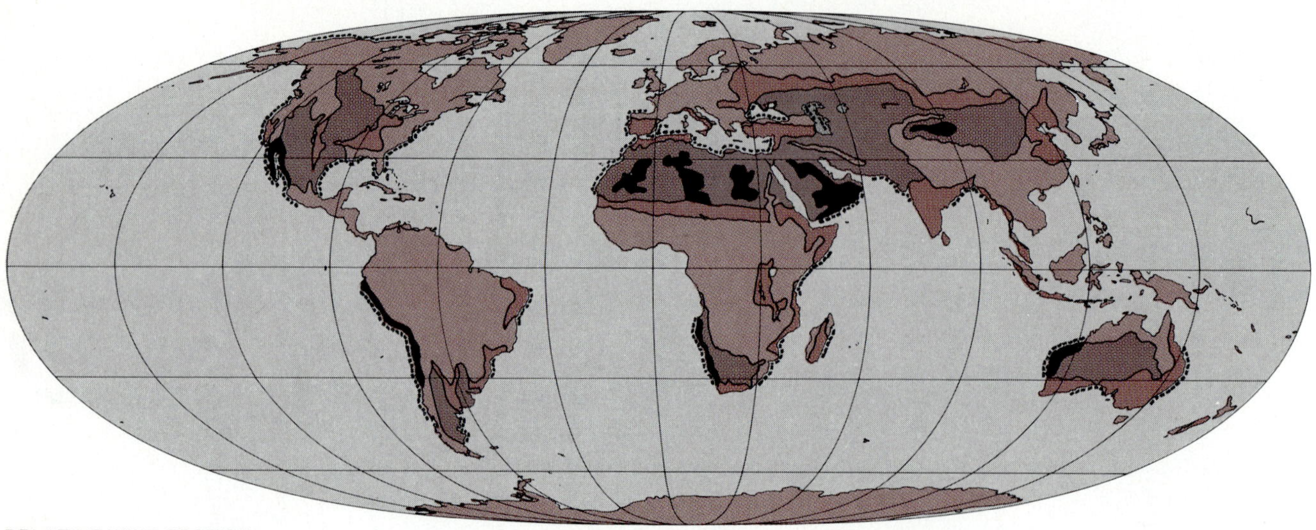

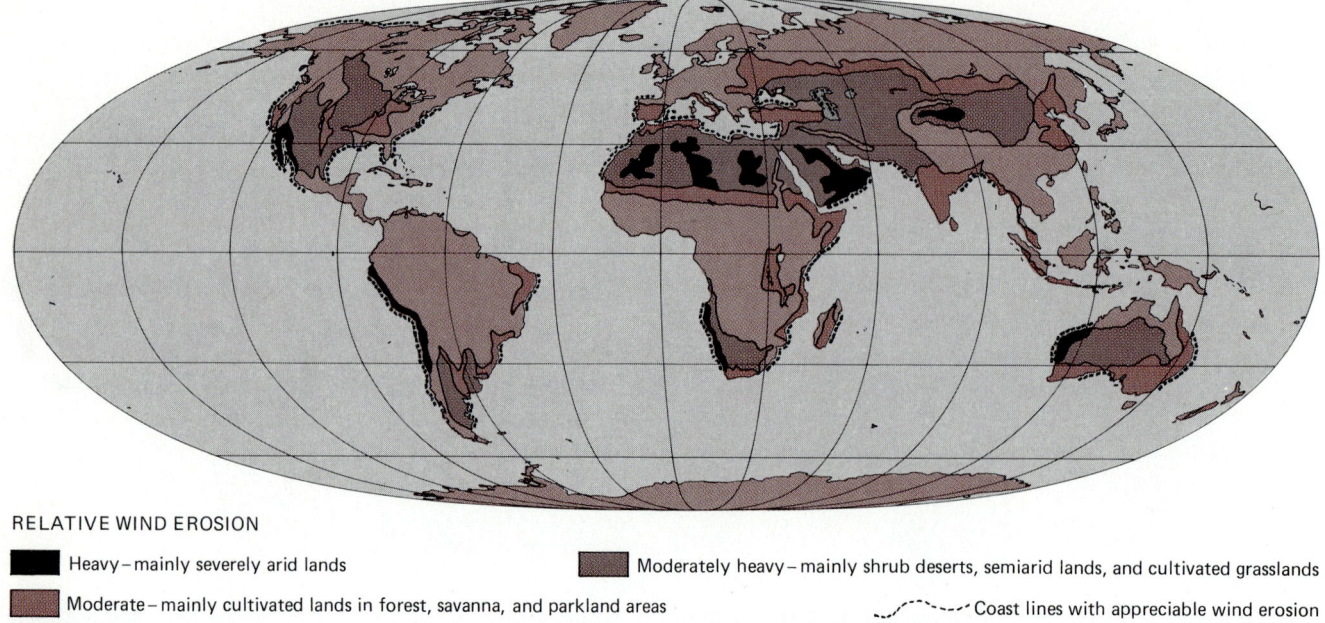

RELATIVE WIND EROSION

Heavy – mainly severely arid lands

Moderately heavy – mainly shrub deserts, semiarid lands, and cultivated grasslands

Moderate – mainly cultivated lands in forest, savanna, and parkland areas

Coast lines with appreciable wind erosion

Figure 23.2
Areas of the earth's surface where wind erosion is moderate to heavy.

air around it. Because the earth rotates from west to east, the main air currents at high altitudes flow in this direction. Strong westerly flows dominate the upper atmosphere in the middle latitudes and give rise to the persistent high-speed jet streams located there.

At lower altitudes, the prevailing wind directions are greatly influenced by the development of high- and low-pressure cells over the continents and oceans. Lows form over warm land or water; as heat is transferred to the overlying air, it becomes less dense and rises. As it goes up, air is drawn in from surrounding areas due to the decreased pressure. Air moving toward a low is deflected as a result of the earth's rotation and begins to move in a counterclockwise spiral in the northern hemisphere or clockwise spiral in the southern hemisphere. Flow directions are exactly opposite where high pressure circulation cells develop causing air to flow outward.

Low-pressure conditions are likely to arise over continents in the summer, at the same time highs form over the cooler oceanic areas. In the winter, highs tend to form over the cold land areas, which experience greater temperature extremes than water-covered areas. At the ground surface, the prevailing winds are largely governed by the location relative to these major high- and low-pressure cells. Few places experience winds that are always from the same direction. Wind velocity also fluctuates and generally changes with the season. For this reason it is difficult to describe the wind at any particular place in a simple way. Often the winds blow from one direction, but they also blow out of other directions frequently.

Local winds are affected by the surrounding topography and along coasts by the movement of air to and from the sea. Consequently, wind action at any particular place must be considered in terms of the local conditions, and the general patterns of atmospheric circulation may bear little relation to the erosion or deposition of sediment by the wind there.

Movements in clear air are usually not apparent, but it is possible to see the flow pattern where smoke is introduced or where dust is in the air. Smoke is used in laboratory experiments to trace air movements. These experiments prove the existence of laminar and turbulent patterns similar to those observed in water. Turbulence exists where air moves around obstructions that break up and deflect the streamline pattern of laminar flow. High-velocity movements are also unstable and may break up spontaneously to produce turbulence.

Wind velocity and turbulence strongly influence **sand drift,** the movement of granular materials across the ground surface. The amount of sand moved is affected by the wind velocity, the size of the sand, the roughness of the surface, the amount and kinds of vegetative cover, and the amount of moisture in the sand. As might be expected, high wind velocity, small sand size, smooth surfaces, and little cover or moisture all favor high drift rates. Later in this chapter, we shall see how this concept applies to formation of the most familiar wind-shaped landforms, sand dunes.

23.2 SOURCES OF WIND-BLOWN MATERIAL

Potential sources of wind-blown material exist wherever small particles of soil or sediment are exposed and not held down. Most modern beaches and many streams present an abundant supply of fine sediment. Even ancient beach deposits may become sources of sand where such strata are exposed to weathering and erosion. Both stream deposits and ancient sandstones are most likely to become sources of wind-blown material where they are exposed in arid or semiarid regions. Plant and vegetative cover tend to prevent removal of material by wind, as does soil moisture, which acts as an adhesive; but in dry areas the soil moisture is low and few plants hold the sediment in place. As we saw in Chapter 20, deserts are also sites of playa lakes, bajadas, and pediments, where weathering products are abundantly exposed and subject to wind erosion.

Many of the modern sources of wind-blown materials are related to glaciation. Glaciers grind great quantities of rock into sizes that can be moved by the wind. These fine materials are then carried by meltwater and deposited beyond the margins of the glaciers to form huge plains of sediments. When the glaciers disappear, vast lakes may be left both in the regions beyond the former edge of the ice and also behind deposits created at the ice margin. As the ice melts and the water evaporates, these sediments dry out and are exposed to the wind. The Great Salt Lake of Utah is a remnant of one such large inland lake.

Volcanoes are also locally important sources of material. The dust and ash blasted into the air by volcanic eruptions can be carried at a much higher altitude than is usual for most other wind-blown deposits.

Human activity has often played an important role in exposing otherwise unavailable material. The problem is especially serious where lands receiving barely enough precipitation to support crops are brought into cultivation. Such lands are usually prone to dry periods, sometimes even for a number of years in succession. The southern Great Plains were devastated by such droughts in the early 1930s. The region had been extensively developed in the preceeding two decades, and vast areas were being cultivated each year. During the early part of the dry years, plowing continued, accelerating the evaporation of the soil moisture and loosening the soil. As the drought continued, the plant cover was destroyed; new plants failed to grow; and the fine, dry unprotected soil was removed by the persistent winds.

Similar problems arise as a result of over-grazing land that is too dry. Remarkable aerial photographs taken in northern Africa show places where the desert is separated from dry but scrub-covered land by perfectly straight lines—fence lines that separate ungrazed lands from regions grazed by herds of cattle until the entire plant cover is stripped away.

23.3 MOVEMENT OF MATERIALS BY THE WIND

Under the extreme conditions that exist during tornados and hurricanes, when wind speeds rise above 100 kilometers per hour, large objects can be moved by the wind. The center of a tornado is an intense low-pressure column of rapidly rising air. High-velocity winds rotate around this center and the air spirals into the rising column under tremendous pressure. Winds associated with these relatively small centers have derailed trains, picked up houses and their furnishings and carried them for tens of kilometers, and flattened forests.

A pattern of flow similar to that of a tornado but covering a much larger area occurs in hurricanes. Both of these storms have winds capable of moving large objects, and both are usually accompanied by severe damage due to the movements of material by the wind. Most tornados affect narrow surface paths and occur where the ground surface is covered by vegetation. Consequently, they are not especially important as geomorphic agents. Hurricanes originate over water-covered areas, but they follow long paths and, as we saw in Chapter 22, frequently cross over coastal zones. Thus they are important geomorphic agents along the shorelines.

High winds are capable of moving pebbles although they rarely are lifted off the ground. Usually they roll, possibly even bounce, along the ground. Boulders are occasionally rolled as the supporting sediment is dislodged and moved away from the back side of the rock by eddy currents. Smaller particles—granules (2 to 4 mm), sand (1/16 to 2 mm), silt (1/256 to 1/16 mm), and dust (less than 1/256 mm)—make up most of the load normally moved by the wind. What sizes are actually moved depends on what is available and on the velocity of the wind.

The movement of granules, sand, and silt takes place by much the same mechanism. Fragments of particular sizes begin to move when the wind velocity reaches a critical level for that size. At first, the particles roll along the surface and bounce into the air when they hit one another. Once bouncing begins, the process accelerates, because grains that bounce into the air are propelled forward by the wind and given more energy. When they hit the surface again, the impact dislodges and causes other grains to bounce into the air, setting up a chain reaction. Soon a zone above the ground surface measuring from a few centimeters to as much as a meter thick is filled with bouncing grains involved in **saltation** (Fig. 23.3). Both the size of the particles moved and the height of the zone of saltation increase with increasing wind speed.

The surface of most sand deposits is covered with a ripple pattern (Fig. 23.3*b*), caused by the rolling and saltation of the sand. The instability of this surface is shown by how quickly the ripples change in response to changes in the wind. Ripple crests form perpendicular to the wind direction, and this alignment shifts quickly as the wind changes direction.

The succession of impacts the grains experience

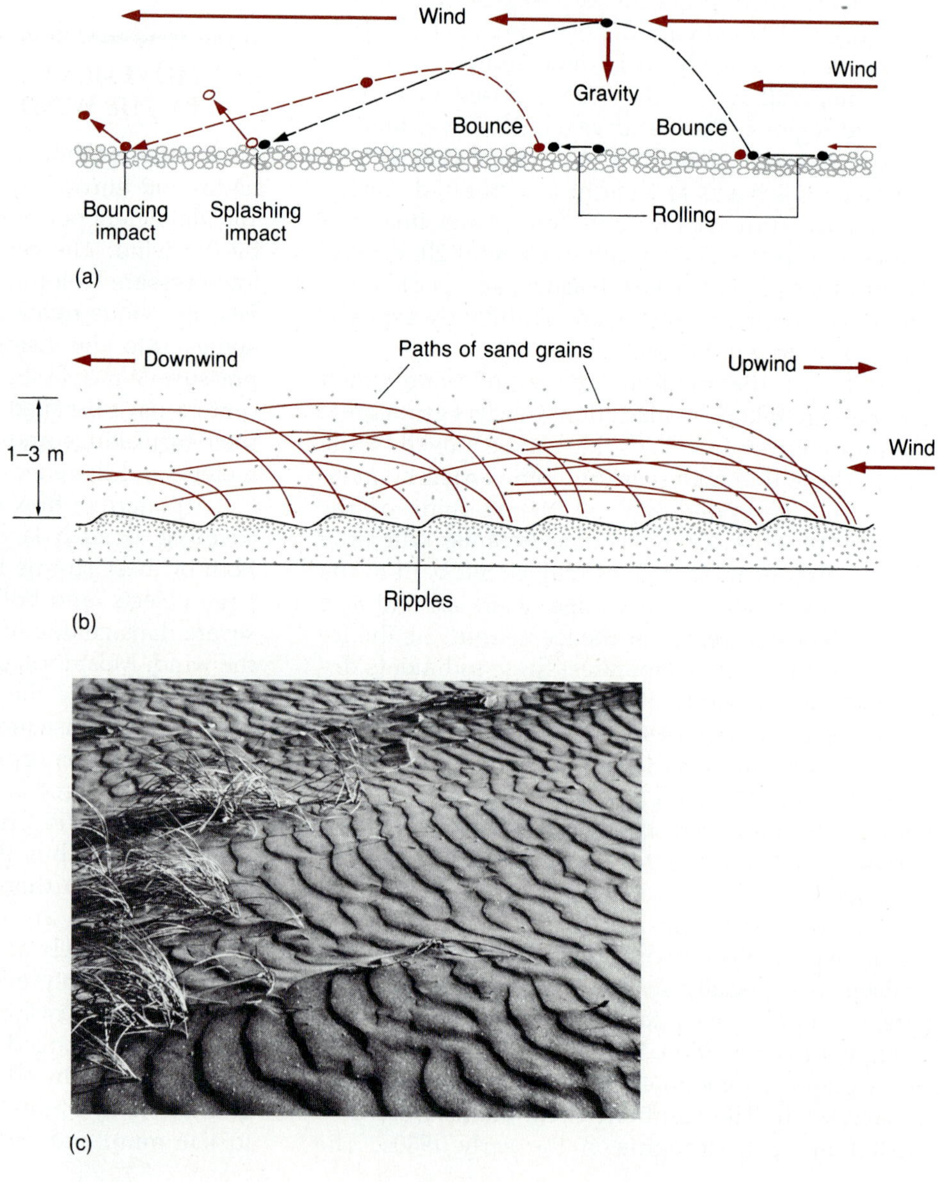

Figure 23.3
(a) A strong wind causes many grains to follow this pattern. (b) The result is a layer of sand-filled air close to the ground. (c) Usually the sand surface takes on a rippled form. These ripples are asymetric; the side facing downwind is steeper than the upwind side. The distance from crest to crest of ripples is commonly several centimeters.

Figure 23.4
These smooth, polished rocks owe their shape and surface texture to the abrasion and impact of wind-blown sand. Note that the rocks have flat sides.

tends to round the grains and to give them pitted surfaces. Pitting is also caused by chemical alteration of the surface of the grains.

The finest particles in the air—dust, cloud particles, and smoke—are moved at higher levels and in a different way. Dust is often composed of thin, flat particles with surface areas that are large relative to the particles' weight. This makes it possible for the upward force of wind currents to more than counterbalance the downward pull of gravity on the particles. Consequently, dust is easily wafted higher in the air by slight updrafts (Fig. 23.4). Large quantities of volcanic dust carried aloft by explosions can remain in the upper levels of the atmosphere for several years and be carried around the earth repeatedly. Small particles, like dust and ash, may be blown directly into the air, or they can be lifted up by differences in pressure in the air. Such pressure variations commonly occur where wind velocity changes abruptly between adjacent layers of air. Lower pressure exists in the high-velocity layers. Thus, if wind flows over a thin, still layer of air at ground level, pressure in the moving layer is lower than that below, creating an updraft that can carry dust with it. Pressure drops of this type are also found where the wind blows over or around obstacles.

23.4 EROSION BY WIND

Wind accomplishes erosion by **deflation** (blowing away of materials), corrosion, abrasion, and impact. Sand, silt, and dust carried by the wind are effective tools of wind erosion. They scrape and abrade as they move over surfaces, but even more important is the long-continued effect of impact—the sand-blasting effect of wind-driven sand. When a sand grain is blown into or against a rock surface or into the soil, the momentum of the impact dislodges other granular material; or it may break the cement of a rock, freeing other sand grains. Sand-blasting is used to frost glass, to clean stone buildings, and even to cut letters in polished stones as hard as granite.

Where sand and silt have blown over rocks for a long time, the rocks become worn from the abrasion, achieving smooth, polished surfaces. Stones like those in Fig. 23.5, called **ventifacts,** are often found in deserts. They are recognized by their polished surfaces and the flat sides cut by wind action.

U.S.G.S.

Figure 23.5
A dark cloud of dust rises over the surface of the Texas Panhandle.

Landforms Produced by Wind Erosion

Wind is so effective in removing small particles that many of the landforms caused by wind erosion are the result of the removal of fine, loose sediment. Where mixtures of different sizes of sediment are present, the finer ones are selectively removed, leaving the coarse fraction behind. This coarse material often forms a thin cover of gravel, called **lag gravel,** across pediments and bajadas. Sometimes the lag forms a tight mosaic cover, called **desert pavement** (Fig. 23.6), which protects the underlying surface from erosion.

A similar effect is found in deserts where stream gravel has been deposited in channels that have been cut across finer sediment by intermittent streams. If the fine particles are removed over a long period of time, the channel gravels may eventually be left as a sinuous ridge when the surrounding areas have been lowered by deflation.

U.S.G.S.

Figure 23.6
Desert pavement composed of angular fragments of quartzite.

New Mexico State Tourism & Travel Dept.

Figure 23.7
The shallow depressions in this hillside and the arch were formed mainly by wind erosion. Note the trees for scale.

Where the ground is covered by relatively thick sediment and overlain by vegetation that is not uniform, pockets of the sediment may be removed by wind, leaving **blowouts,** most of which are broad, shallow depressions. These commonly occur along coasts where sand deposits inland from the shoreline are partially covered by grasses or trees. The uncovered areas become sites of blowouts.

The solid rocks exposed in deserts sometimes show forms due, at least in part, to the effect of wind action. Where persistent wind loaded with a supply of sand blows against a solid rock, the rock will eventually be smoothed by abrasion. Because erosion due to this type of sand-blasting effect is concentrated close to the ground, rocks may become undercut. Broad shallow notches (Fig. 23.7) often form, and sometimes unusual forms referred to as table rocks or pedestal rocks are created where an isolated mass of rock is left supported by a thin column. The small column is often composed of a more easily eroded rock than that which it supports.

23.5 DEPOSITION FROM THE WIND

As wind velocity drops below some **critical velocity**—the minimum wind velocity required to move a particle of a given size—grains of that size stop moving or settle out. Because the critical velocity depends so strongly on size, wind does an excellent job of separating particles according to size, shape, or weight, a process known as **sorting.** Velocity determines the maximum size of material the wind can transport by rolling and saltation and in suspension. Thus, a distinct separation can be accomplished. Smaller sizes are suspended and carried away; heavier particles lag behind on the ground; sand and silt bounce along. Suspended and saltation loads are usually further separated, because some sizes in saltation are moved more rapidly than others. Thus, the wind's load becomes sorted while it is in transport. Because the degree of sorting is usually better in wind-blown deposits than it is in marine or stream deposits, the two can usually be distinguished.

Most wind-blown deposits are sheetlike and cover large areas with relatively thin layers of sediment or dunes. However, even the dunes can become so close together and cover such large areas that they too are sheetlike when considered as a whole. Within sheets, thickness varies. Often a sheet thins toward its margins and away from the source area. Two excellent examples of sheetlike deposits are the layers of ash and dust that settle out downwind from volcanic eruptions and the layers of silt that are removed from the plains of sediment often deposited by meltwater beyond the edges of glaciers.

Loess

Many of the silts that derive from glacial action form deposits known as **loess** (lo'es). Loess is unconsolidated, unstratified, and homogeneous sediment composed of small, angular particles. It is porous, easily crushed, and buff to yellowish in color. Evidence that loess is deposited from the wind and not by water is based on the fact that it often contains small shells of a type of snail that lives only on land. Loess may also show small tubes of calcite that apparently have filled the long thin holes left after the decay of tall grasses. These holes are thought to be upright because the silt and dust sifted down over the grass so slowly that the blades were not bent and forced down. Because they are unconsolidated, loess deposits are easily eroded by streams, but the angularity of the particles makes them less prone to slump than most unconsolidated sediments. Nearly vertical cliffs are common where loess is dissected by streams, and vertical road cuts in it have remained stable for many years (refer back to Fig. 20.24).

Most of the loess in the United States is found in the Great Plains and Mississippi Valley (see Fig. 20.22). During the Ice Ages, glaciers ground rocks under them into a powdery material that was then carried out to areas in front of the ice sheet by meltwater and deposited. Winds blowing off the ice sheet dried the sediment, picked it up, and carried it southward where it settled out, forming loess.

In Europe, glacially derived loess is also exposed in a long belt extending from France through Russia. A thick, extensive cover of loess occurs over vast areas in and around the Shansi province of China, derived from the Gobi Desert rather than from a glacial source.

Sand Deposits

Accumulations of sand freed from their source rock or blown off the beach or out of other unconsolidated deposits are typical features along coasts and in deserts. Usually the sand is heaped in a form, called a **sand dune,** that is constantly moving and changing shape. Isolated dunes rarely occur; they are usually part of a much larger mass of sand that may exhibit simple or extremely complex forms. In no case are they permanently fixed in position or tied in any way

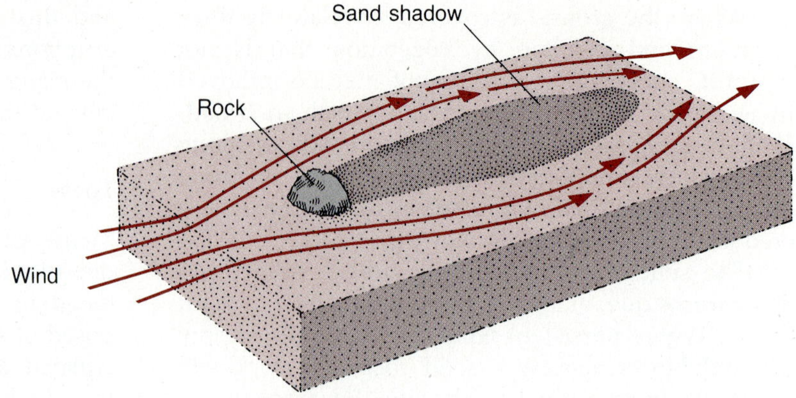

Figure 23.8
Sand is deposited in the "wind shadow" behind a rock. This is often the initial stage in the growth of a sand dune.

to a particular location. However, if the climate should change or if vegetation becomes established on dunes, their movement may be slowed or even stopped. This, of course, is the aim of most sand-stabilization programs.

Another type of deposit, a **sand shadow,** accumulates downwind from and sheltered by some type of obstruction such as a boulder, a grove of bushes, or a cliff (Fig. 23.8). Unlike dunes, shadows do remain fixed as long as the supply of sand holds out and the obstruction remains. In a gap between two obstructions yet another type of deposit forms. Such a break acts as a funnel through which the sand-laden wind blows, leaving a trail of sand downwind (Fig. 23.9).

Most sand is composed of the mineral quartz because it is both insoluble and hard. It is one of the few common minerals that can survive both the weathering process and the mechanical breakdown that goes with stream or wind transportation over long periods of time. Nevertheless, sands of other compositions are found in many places. Sand composed of the mineral olivine, derived from volcanic rocks, is found on the beaches of Italy. Sand-sized pieces of volcanic ash make up the sand on many Hawaiian beaches. The White Sands National Monument in New Mexico contains sand composed of calcium sulfate, the mineral gypsum. Even ice makes up sand dunes in parts of the Antarctic. No matter what their compositions, these sands all form dunes of the same general shapes. Clearly, then, the size of the particles and nature of the movement process shape dunes.

Sand-dune formation. Sand dunes may be formed wherever strong winds, either constant or intermittent, and a source of sand occur together. Dunes can originate as sand is removed from a source such as a decaying sandstone outcrop or from a sandy beach. Often sand begins to accumulate where an obstruction interferes with the movement of sand by the wind. Such objects create a sand shadow of the type we saw in Fig. 23.8. This pile of sand may grow until the top of the pile reaches the height of the obstruction. The pile may begin to move away from the obstruction as a result of a slight change in wind direction.

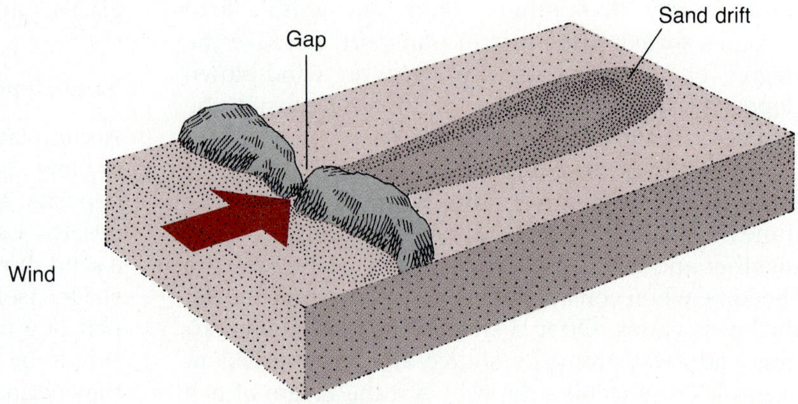

Figure 23.9
Sand dunes may also originate where sand blows through a depression between obstructions, such as a gap in a rock outcrop or a break in vegetation cover. In profile, the sand drift shown here has steeper sides downwind. Dunes occur on scales from centimeters to hundreds of meters in length.

The initial form of a sand dune may be a streamlined pile with nearly equal slope on both up- and downwind sides, over which sand moves almost uniformly. But as the dune grows, the streamlined pattern of flow over the dune breaks up at its crest. Eddies are set up on the downwind face, and the dune begins to assume an assymmetrical form (Fig. 23.10). After assymmetry develops, sand moves up the low slope of the windward side by saltation. Just beyond the crest, wind velocity drops and saltation on the lee side of the dune stops. A back slope forms as a result of slumping and sliding of sand as it assumes the characteristic angle of repose of loose sand, about 35°. This surface, called the **slip face** (Fig. 23.11), is barely stable; any addition of sand at the top causes small slides on the back slope. Sand on the windward side is continually blown over to the back slope, causing it to become oversteepened and triggering slumps and slides that correct its angle of repose. Once the dune is formed, it advances as

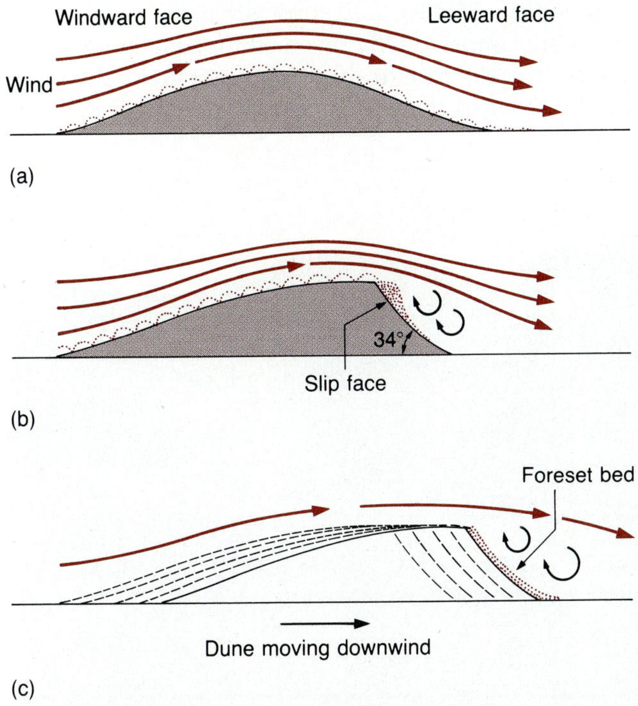

Figure 23.10
Stages in the evolution of a sand dune. (a) The dune is smooth and rounded as the wind flows over its entire surface. (b) The sand has built up until the main stream passes over the slip face where eddy currents are shown. (c) The sand is carried up the low-sloping surface of the dune to the top, where it is dropped over the side. The slip face has an angle equal to the angle of repose of the sand, usually about 35°.

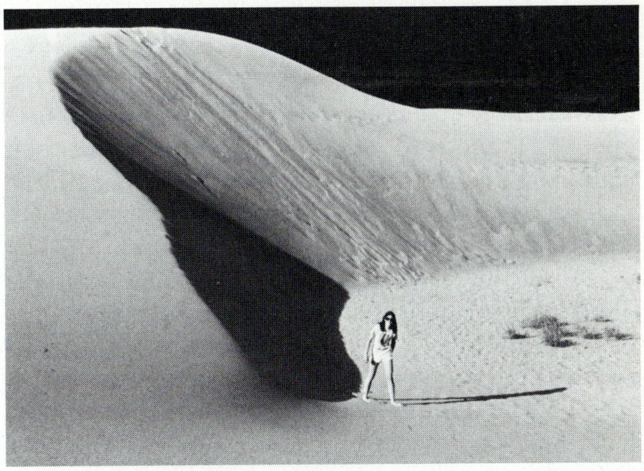

(a)

Oregon Department of Transportation

(b)

Figure 23.11
(a) The steep face of this dune indicates its direction of movement. (b) Sand slips down the steep face of most dunes, as seen here, leaving long scars down the face until they are covered or slip again. This dune is about 12 meters high.

sand is blown up the windward side to the crest, slides down the leeward side, and eventually is exposed on the windward side again where it starts the process over again. Individual dunes move at rates of as much as 30 meters a year.

Most sand dunes exhibit low slopes and slip faces of the type just described. But the shape the sand dune takes is ultimately determined by the supply of sand, the wind velocity, the constancy of the wind direction, and the amount and distribution of vegetative cover. Individual dune shapes are much easier to see when only a limited amount of sand is present. Isolated and perfectly shaped dunes are rare when sand is abundant, because the dunes coalesce, losing their individual characteristics (Fig. 23.12).

Classification of dunes. Wind-laid deposits, including sand dunes, are classified on the basis of their plan view (their shape as seen in an aerial view), the number of slip faces, and overall form of the deposit. The main types are illustrated in Fig. 23.13.

All of the dunes shown in Fig. 23.13 are simple dune types, but compound dunes formed where dunes of the same or even different types are superimposed are common. The combinations lend a feeling of almost infinite variety to the seas of sand; yet, most sand bodies can be placed in one or the other of these categories.

Three types of wind deposits show no slip face: a **sheet,** a broad, flat wind deposited surface; a **stringer,** a long, narrow strip deposit; and a **dome dune,** a circular or elliptical mound with no identifiable slip face.

Barchans, barchanoid ridges, and transverse dunes (Fig. 23.13) are the most common dune types and what we most often think of as dunes. All three are characterized by a single slip face, and all are oriented at right angles to the prevailing winds, which come from a single dominant direction. Isolated **barchan**

Figure 23.13 ▶
Block diagrams depicting the forms of the most common sand dunes. In each figure, arrows indicate the prevailing wind conditions that give rise to the dunes. Dunes range greatly in size from centimeters to tens of meter in height.

dunes (Fig. 23.13*b*) exhibit a beautiful and symmetrical crescent shape with the points or wings of the dunes pointing downwind. When a number of barchans coalesce, they create the **barchanoid ridge** (Fig. 23.13*c*), which often changes into the long assymmetrical form of the transverse dunes (Figs. 23.13*d*). These three form gradational sequences that show progressive change related to the supply of sand (Fig. 23.14). **Transverse dunes** occur where sand is abundant and where a balance exists between the supply of sand and the wind velocity. As the supply of sand diminishes, the transverse form breaks up and eventually isolated barchans, products of a sparse sand supply, are all that remain.

One type of transverse dunes, called **reversing dunes** (Fig. 23.13*e*), develops when the wind directions are almost exactly reversed for equal times. This causes the slip face to shift from one side of the ridge to the other as the winds change, and gives rise to a

Figure 23.12
Sand dunes in the Sand Dunes National Monument, Colorado. Note the two people on the right for scale. Wind blows from left (west) to right.

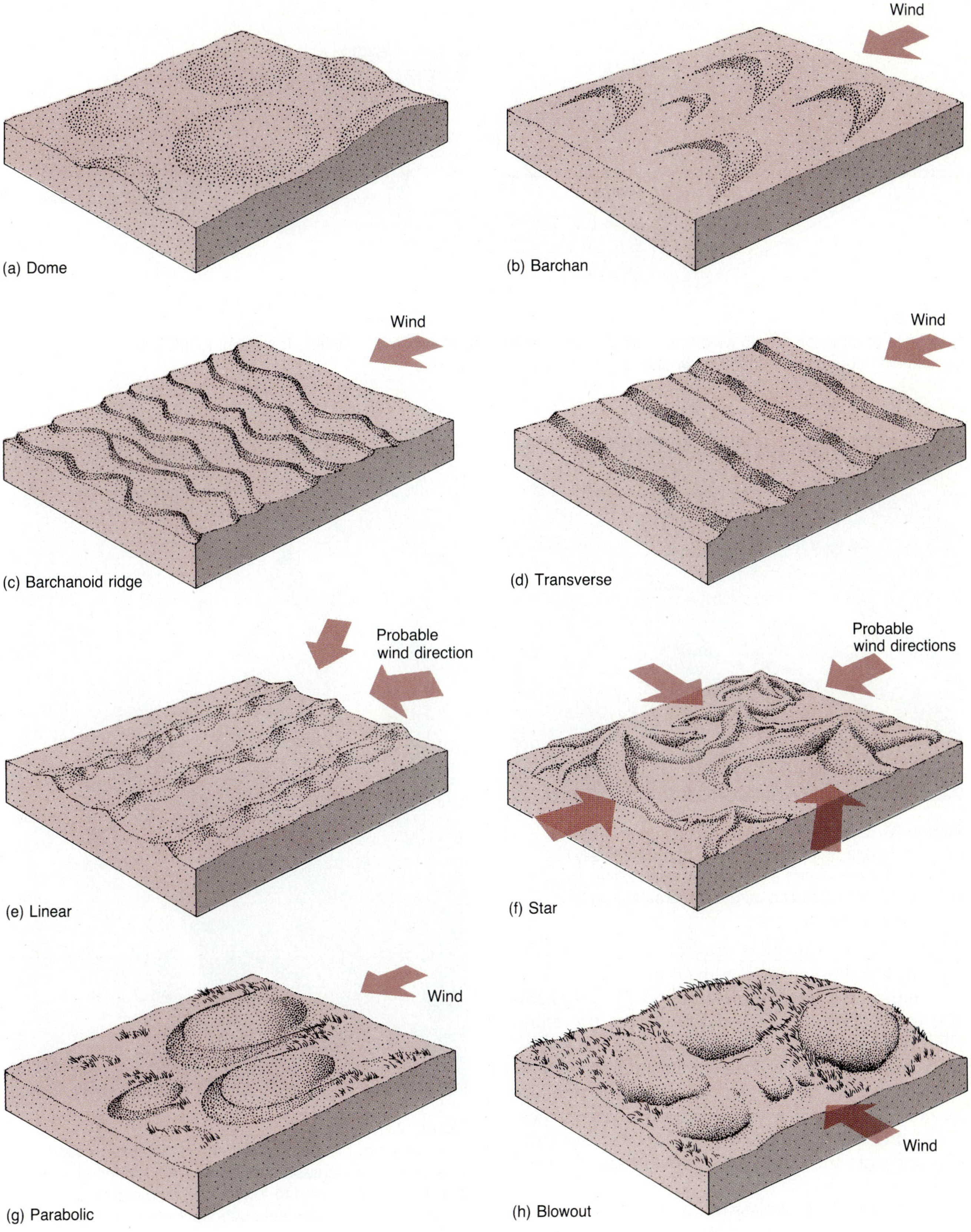
Wind
Wind
Wind
Probable wind direction
Probable wind directions
Wind
Wind
(a) Dome
(b) Barchan
(c) Barchanoid ridge
(d) Transverse
(e) Linear
(f) Star
(g) Parabolic
(h) Blowout

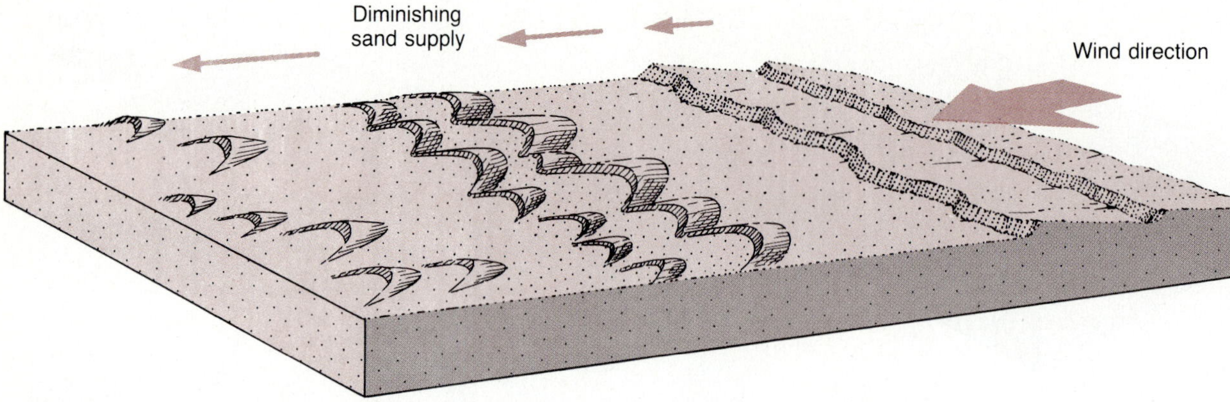

Figure 23.14
Sequence of dune types with wind from a single direction, with sand supply diminishing from right to left: transverse, barchanoid ridge, and barchan. Arrow shows prevailing wind direction.

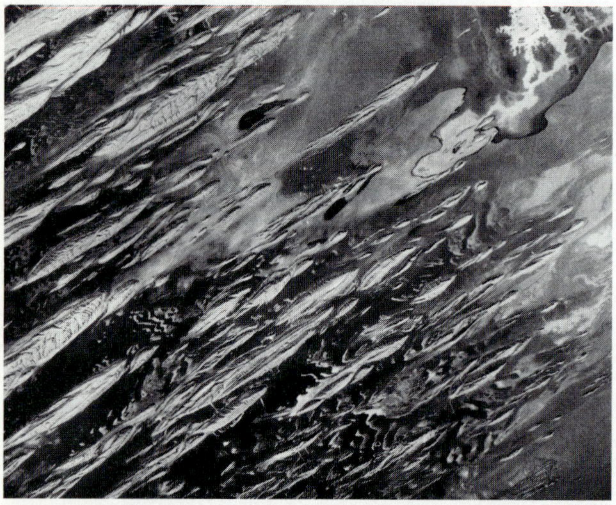

NASA

Figure 23.15
Northeast edge of the Shahr-e-Lut, Iran. Beautifully developed linear dunes are shown in this vertical photograph. The distance across the area shown is about ten kilometers.

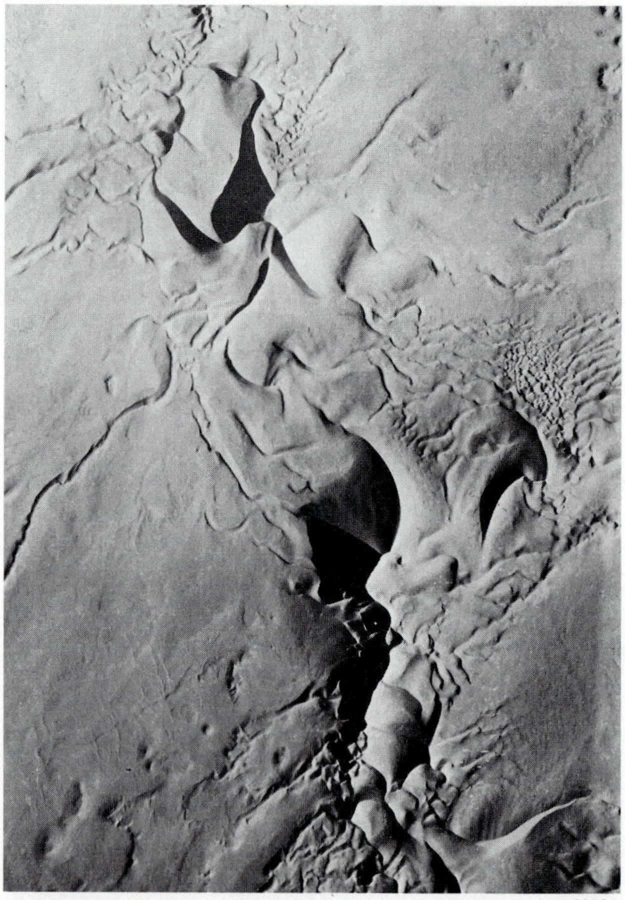

NASA

Figure 23.16
This portion of the plains of Algeria is covered by sand. This vertical photograph shows two well-developed star dunes, which tend to form when wind changes direction often. The distance across the area shown is about five kilometers.

ridge with two slip faces, one of which is usually much better developed than the other.

The origin of **linear dunes** (Figs. 23.13*e* and 23.15), also called **seifs,** remains uncertain. The prevailing wind associated with seifs changes direction during the year. Nevertheless, long straight ridges more or less parallel to the prevailing wind direction are formed. Slip faces develop on both sides of the ridges, suggesting that the ridge results from wind sweeping the sand onto the ridge first from one side, then the other.

A third important class of dunes is characterized by starlike shapes with a central high peak from

Figure 23.17
Ancient sand deposits formed in dunes are preserved on this mountainside in Zion National Park, Utah.

which ridges extend in several directions. **Star dunes** are often high, and they form where the wind direction is unusually variable (Fig. 23.13*f* and 23.16), creating several slip faces.

In addition to the types of dunes described above, which exhibit relatively simple forms, many dunes are complex. Often two or more of the simple types are superimposed. Sometimes these can be identified, but in other cases, especially in sand sheets, simple forms cannot be distinguished.

Internal structure of sand dunes. Slight variations in grain size, color, and composition give rise to laminated structures in dune deposits. The layers usually show **cross-bedding** (Fig. 23.17). The different orientation of the laminae are due to the differences in the slope of the slip face, typically 35°, and the windward face, typically 5° to 10°. But these orientations change as the wind shifts direction and where the sand from one dune encounters and advances onto either topographic features or other sand bodies.

SUMMARY

The movement of air that is caused by differences in atmospheric pressure is an especially important geomorphic agent in regions of little water and poor ground cover. Wherever small particles of soil or sediment are exposed and not held down by moisture or vegetation, there is potential for erosion and redeposition of material by the wind. In addition to deserts, glacial margins, volcanoes, and land under cultivation all offer sources of wind-blown particles.

Part of the air-borne material, the finest particles, is carried suspended in the air. Much more is carried near the surface, rolling and bouncing in a zone of saltation whose thickness depends on the wind speed. Both types of particle make the wind an effective agent of erosion. This sand-blasting by the wind may form table, mushroom, or pedestal rocks. Removal of smaller particles often leaves behind a rocky surface of desert pavement.

Where the wind speed diminishes, the air-borne load is dropped in well-sorted deposits. Silts picked up from glacial outwashes have been found redeposited across wide areas as loess. Accumulations of wind-blown sand form sand shadows downwind of obstructions or a wide variety of sand dune forms—among them barchans, transverse, linear (or seif), and star dunes—depending on prevailing winds and the available supply of sand.

KEY TERMS

barchan dune
barchanoid ridge
blowouts
critical velocity
cross-bedding
deflation
desert pavement
dome dune
lag gravel
linear dune
loess
reversing dunes
saltation
sand dome
sand drift
sand dune
sand shadow
sand sheet
sand stringer
seif
slip face
sorting
star dune
transverse dune
ventifacts

STUDY QUESTIONS

1. What are the main sources of sand?
2. Why is wind such an excellent sorter of sand and silt?
3. What determines the slope of the slip face of a sand dune?
4. What evidence would you look for in trying to determine whether an ancient sandstone was deposited by wind action?
5. What does each dune type tell you about prevailing wind direction and abundance of sand?

SUGGESTED READINGS

Bagnold, R. A., *The Physics of Blown Sand and Desert Dunes.* London: Methuen, 1941.

Cutts, J. A., and R. S. U. Smith, "Eolian deposits and dunes on Mars," *Journal of Geophysical Research* 78:20 (1973), *4139–4154.*

Doehring, D. O., ed., *Geomorphology in Arid Regions: Publications in Geomorphology.* Binghamton, N.Y.: SUNY, 1977.

McKee, E. D., ed., "A study of global sand seas," *U. S. Geological Survey Professional Paper 1952* (1979).

GLACIERS AND GLACIATION

Chapter 24

As early as the first part of the eighteenth century, naturalists were studying and speculating about the origin of alpine glaciers. In 1723, J.J. Scheuchzer, a Swiss geologist, recognized that ice in high mountain valleys moves as a result of the expansion of water as it freezes in cracks in the glaciers. Most of his contemporaries thought the only ice movement was due to sliding downslope. Many who lived near them were familiar with the ridges of debris at the front edge of glaciers. Some also recognized that blocks of rock with polished and scratched surfaces had been moved far from their sites of origin by glaciers. But this evidence was not formally presented until 1821 when Ignaz Venetz, an engineer, read a paper at the Swiss Society of Natural History in which he detailed evidence that the glaciers in the valleys of the Swiss Alps had formerly occupied positions much farther downvalley than their present margins. A few years later he suggested that all of northern Europe had been covered by ice.

During the next thirty years other scientists became convinced of the validity of Venetz's idea. Notable among these was Louis Agassiz, who realized the implications of the widespread glacial deposits and later advanced the idea of "ice ages." Agassiz correctly identified the extent of glacial deposits in Europe and later in North America. Glacial theory developed rapidly in the middle of the nineteenth century, but it was much later before the extent of modern-day ice caps was known. The cause of ice ages is still a subject of debate.

During most of the earth's history the entire surface has been ice-free. Nearly 10 percent of the land area of the earth is covered by ice today. Almost 1.5 percent of the earth's water is now in the form of ice. Glaciers occupy valleys of many mountains in high latitudes; and some glaciers, like those on Kilimanjaro (elevation 5968 meters), are located in equatorial latitudes. A thick sheet of ice covers most of one entire continent, Antarctica (Fig. 24.1), and other sheets cover a large part of Greenland, Iceland, and islands in the Arctic Ocean (Fig. 24.2). We clearly live in an ice age.

For many years the glacial deposits that are so prominent in northern Europe and the bedrock containing marine fossils were thought to be debris left by the biblical Flood in the time of Noah. If we could not actually walk across these glaciers, watch them in action, and see deposits forming at their margins, it is doubtful that we would ever believe that the glacial deposits that cover the northern United States and northern Europe had been left there by an advancing sheet of ice. We probably would not consider accumulation of such large bodies of ice a reasonable possibility.

Although glaciers have been present during only a small part of the 4- to 5-billion-year history of the earth, they have been extremely important in producing the present configuration of our landscape. Most of the present topography of the world has formed within the past few million years, and for the past 2 to 3 million years, ice has existed in large quantities at the poles and in mountains all over the world. The last major advance of ice sheets reached a maximum between 17,000 and 21,000 years ago. At that time, 30 percent of the land surface of earth was

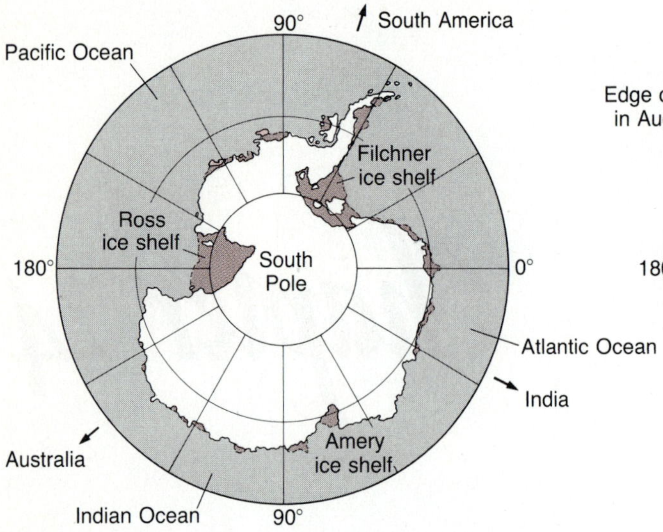

Figure 24.1
The largest accumulation of ice on earth at the present time is in Antarctica. Ice is nearly four kilometers thick near the South Pole. Ice shelves, remnants of once vast sheets of ice, or sea water remain in areas shown.

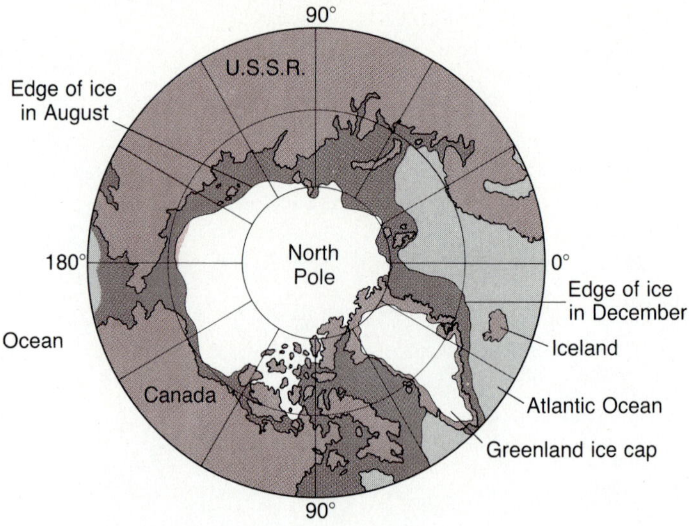

Figure 24.2
An ice sheet covers Greenland, but most land areas in the northern hemisphere are free of ice and snow in the summer months. The central part of the Arctic Ocean is covered by ice in midsummer, and this sheet covers all the areas shown in color in midwinter.

covered by ice. Deposits and erosional effects of this glacial advance have left a clear and well-documented record of the immense importance of glaciation as a process of gradation in this last period of geological time.

24.1 TYPES OF GLACIERS

The large perennial masses of ice formed on land by compaction and recrystallization of snow that constitute **glaciers** are conveniently classified according to their form, geography, and temperature. The two main groups, **valley glaciers** (Fig. 24.3) and **ice sheets** (Fig. 24.4), are distinguished on the basis of their shape and location. The main difference between them is that valley glaciers are confined to valleys, generally in mountains; ice sheets may reach continental dimensions and bury whole mountain ranges. Glaciers from several valleys often coalesce along a mountain front, forming an intermediate type of glacier called a **piedmont glacier.**

Temperature is an important determinant of the rate at which glacial ice melts or accumulates, the character of the end of the glacier, and its ability to erode bedrock. Glaciers are now sometimes identified as either temperate (warm) glaciers or polar (cold) glaciers. The wisdom of this distinction lies in the effects temperature has on bedrock erosion by a glacier; the character of the lower end, or terminus, of the glacier; and the movement of the ice. Warm ice flows, yields to stress, and recrystallizes more readily than cold ice. Even the internal structure and texture of the two types differ.

The temperature of glacier ice is primarily determined by climate. Temperature is affected by the amount of solar radiation reaching the ice and by the reflection and radiation from it, by air temperature, wind, evaporation, and percolation of water. The upper temperature limit of a glacier's existence is the freezing point of water, 0 °C, but glaciers in Greenland and Antarctica may have internal temperatures 10 to 20 °C lower than that. Surface temperatures of most glaciers vary widely with the seasons, but conduction is negligible once the upper few inches are at 0 °C. So the interior of the glacier tends to have a more stable temperature. Warming the deeper portions of glaciers depends primarily on the percolation and refreezing of meltwater. Heat is liberated when water freezes (80 calories per gram), and this heat raises the temperature of the surrounding ice. Thus, a warm glacier may be at 0 °C at the surface and only slightly colder at the bottom. Cold glaciers are −10 °C or less at the surface, and slightly warmer at depth.

National Park Service

Figure 24.3
Valley glaciers originate from snow fields high in mountains. The ice flows along pre-existing drainage lines, enlarging the valleys. The ice from numerous valleys coalesces as it moves, creating larger glaciers at lower elevations. Mt. Russel and Yentna Glacier, Alaska, are shown here.

National Park Service

Figure 24.4
Ice sheets, such as this one in Harding ice field, Alaska, can become so thick that they flow across even low slopes or flat land.

The Polar Ice Masses

At present, most of the land and water area covered by ice is located in polar latitudes. These regions have great influence on both the climate and weather in the rest of the world and on the circulation in the ocean basins. About 88 percent of the world's ice is contained in the Antarctic ice sheet. This mass of ice, about 13.5 million square kilometers, covers all but a small percentage of the area of Antarctica.

The ice in Antarctica is more than 4000 meters thick in the thickest part (Fig. 24.5). Mountain ranges, one 1500 meters high and several hundred kilometers long, are buried beneath ice as much as 1500 to 3000 meters thick. Clearly, the continent would have a very different geography if the ice melted. The shape of Antarctica would resemble that shown in Fig. 24.6 if the ice were instantly removed. The bedrock surface under a large area in the western part of Antarctica is below sea level. If the ice were not present, the lowest point would be 1800 meters below sea level, and the eastern part of the continent would be a high plateau standing about 2000 meters above sea level.

Glacier ice from the Antarctic ice sheet flows into the sea, producing ice shelves unsupported by land, some reaching 1000 meters in thickness. The Ross and Filchner Seas are covered in this manner. In

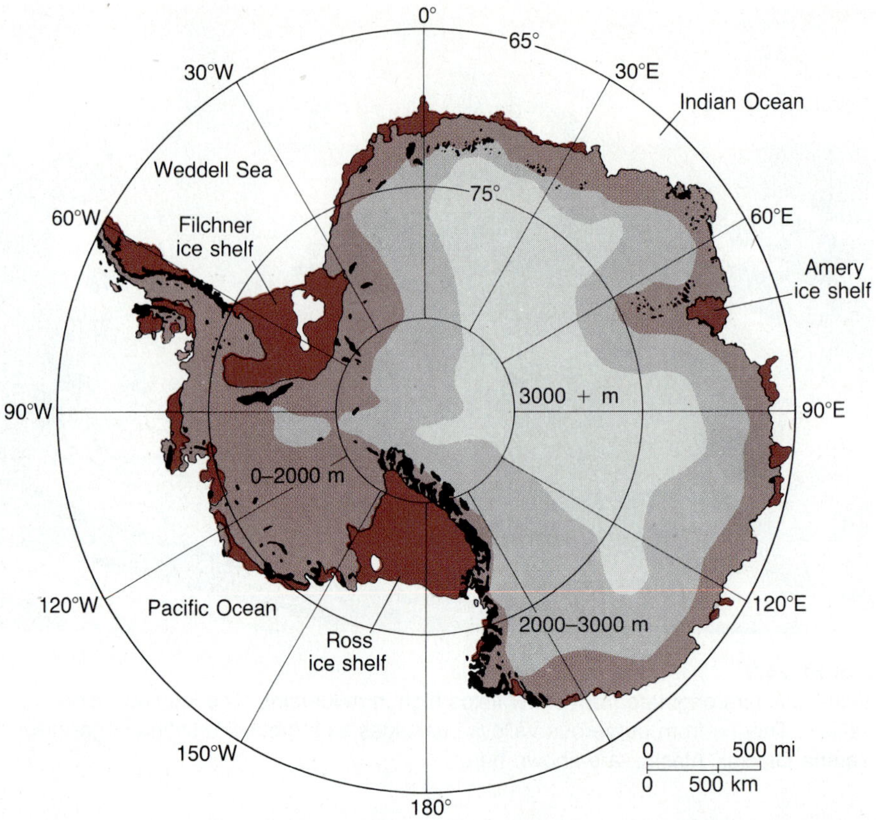

Figure 24.5
Generalized map of ice thickness in Antarctica.

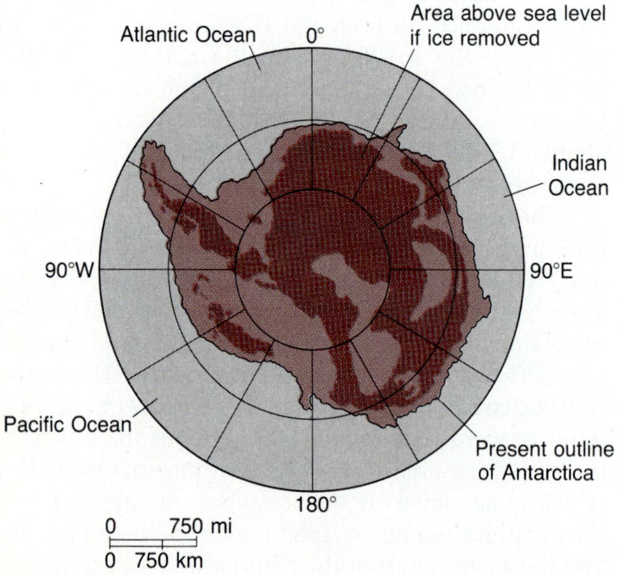

Figure 24.6
Physiographic diagram of Antarctica with ice cover removed. This map is slightly modified from a U.S. Geological Survey map of the bedrock surface. Hachures show continental slope and major mountain areas, stippled regions are above sea level.

U.S.A.F.

Figure 24.7
Pack ice like that shown here covers vast areas in the Arctic Ocean. This picture, taken looking straight down, shows ice-free areas between large sheets of ice. Note the wrinkles in the ice, caused by pressure that builds up during winter. The distance across the area shown is estimated to be between one and two kilometers.

addition, **pack ice,** floating ice consisting of pieces of ice driven together (Fig. 24.7), covers most of the surface of the Antarctic Ocean south of 70°S latitude. At its greatest extent, at the end of winter in September, pack ice covers most of the ocean south of 65°S latitude.

24.2 THE ICE BUDGET

A budget concept, analogous to the water budget, is applied to glaciers to account for the balance between moisture supplied to a glacier or ice cap and that lost from it. A loss signals melting (wastage) and decay. Estimates of the supply and loss of ice and snow over such a vast area as Antarctica are extremely difficult. However, recent estimates indicate that the Greenland ice sheet has a nearly balanced budget with a net loss of only 100 cubic kilometers per year of water equivalent, compared with a growth of about 1100 cubic kilometers per year on Antarctica. This increase in the size of the ice cap in Antarctica is something of a surprise since sea level has been rising during most of the last 15,000 years, suggesting a net shrinkage in the amount of ice on land.

The highest accumulation rates in Antarctica (the equivalent of about one meter of water per year) occur along the coastal areas. Losses are due only in minor part to melting, as we might expect in an area with a mean annual temperature of −30 °C. Much more ice is lost through the process of **sublimation,** by which ice at the surface evaporates directly from the solid to the gaseous state; but the greatest loss is from the breaking away of ice from the ice sheet where it protrudes into the sea. **Icebergs** are formed in this way (Fig. 24.8). Some ice is also lost as a result of melting at the bottom of the ice shelves.

Figure 24.8
Icebergs form by large chunks of ice breaking off from the edge of glaciers where they flow into the sea. The tall icebergs shown here are several tens of meters high. Far more ice lies beneath the water.

24.3 FORMATION OF GLACIERS

During the last glacial advance, 18,000 years ago, a sheet of ice covered most of Canada and over a third of the United States. It extended southward into the United States, where its terminus is now marked by ridges of debris, called **moraines,** deposited near the front of the advancing ice (Fig. 24.9). Moraines form low ridges that can be traced more or less continuously across Long Island and Staten Island into New Jersey. Features formed at the margin of the ice can be further traced across Pennsylvania to the Ohio River. The southern margin of the ice cap is marked approximately by the position of the Ohio and Missouri Rivers.

Because no ice sheet occupies central Canada today, we are prompted to wonder how climatic conditions today differ from those when ice did cover much of North America, and if there is any chance that glaciers might return. It appears that the only condition necessary for the growth and expansion of glaciers is that the amount of snow and ice accumulation must exceed the amount of ice loss through melting and sublimation for an extended period of time.

Growth of glaciers is favored by increased precipitation in the form of snow and sleet and by temperatures low enough for the ice to be preserved throughout the year. The extremely low temperatures (−30 °C average) at Antarctica are certainly not necessary for the formation and growth of continental glaciers. In fact, glaciers are most likely to grow when temperatures are only slightly below freezing. Air at that temperature can hold much more moisture than can air at extremely low temperatures. Hence, even a drop of just a few degrees in the average annual temperatures over the face of the earth would almost certainly bring about formation of large masses of ice on the continents. These would form first in high latitudes and in mountains, where average temperatures in many places are already nearly low enough to promote large-scale mountain glaciation. The lowest altitude at which snow persists year round, the

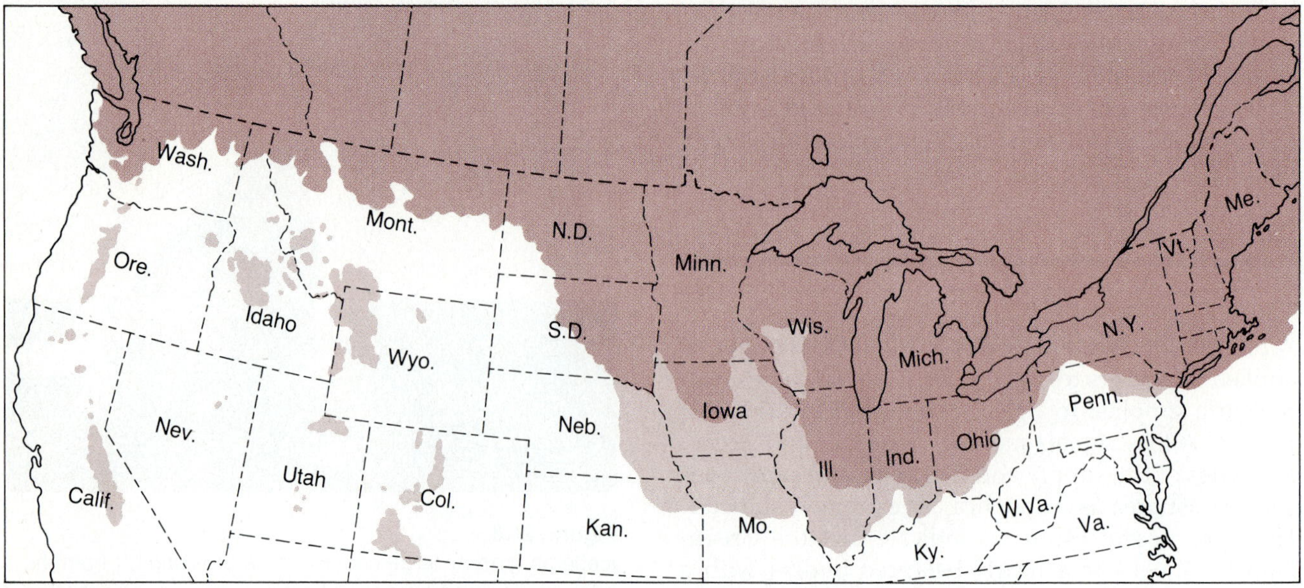

Figure 24.9
Ice sheets advanced into the United States repeatedly over the past one and a half million years. The most recent advance (which reached its fullest extent 17,000 to 21,000 years ago) is known as the Wisconsin period. Ice margins of earlier ice advances (Illinoian and Kansan) are also indicated.

snow line, fluctuates over a period of years, reflecting changes in temperature and amount of snowfall. The area above the snow line, where snow remains from year to year, is called a **snow field** (Fig. 24.10). Snow fields of ice sheets are vast, with accumulations of snow and ice completely burying all topographic features. This must have been the condition during the recent glacial advances in central Canada, where relief is low. In mountainous regions, however, the snow fields may lie below the summit levels of peaks. This is true of the present snow fields in the northern Rockies and in some of the mountains of Antarctica. Peaks or isolated hills that protrude through the snow fields are called **nunataks.**

New Zealand Publicity Service

Figure 24.10
Snow fields high in the New Zealand Alps. Valley glaciers originate here in these deep, snow-filled valleys. Mount Cook stands to the right.

Formation and Movement of Glacier Ice

Snow is a crystalline form of water and is, consequently, a mineral. Snowflakes display an almost infinite variety of delicate and beautiful hexagonal patterns. If temperatures after snowfall rise above freezing, the snow at the surface melts and water percolates deeper into the accumulation where it may freeze again in a more compact form. If the temperatures remain below freezing after a snowfall, individual flakes recrystallize. The extremities of the delicate patterns melt under the pressure of the overlying snow and the water moves toward the centers of the flakes until all the ice has recrystallized as small grains. This partially consolidated snow and granular ice is called névé or **firn** (Fig. 24.11). As snow accumulates, the pressure compacts the ice grains together. The rounded grains of firn tend to

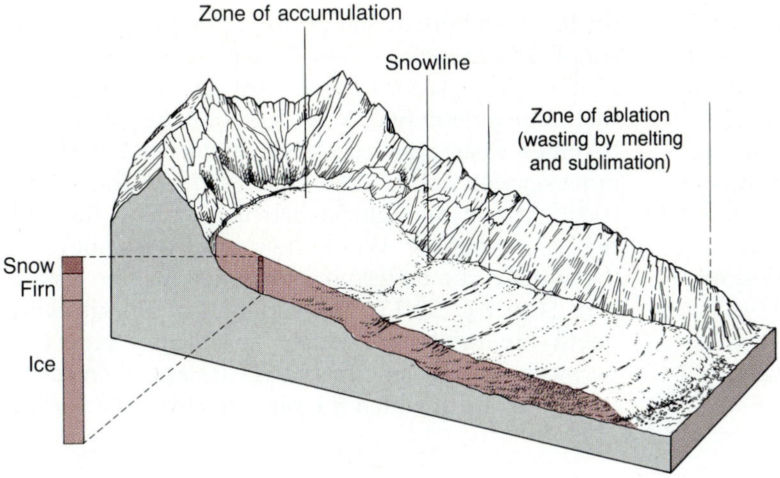

Figure 24.11
Snow is transformed by melting, recrystallization, and compaction to form what is called firn and, at greater pressure, glacier ice. The glacier ice flows under the pressure of the overlying snow and ice. Cracks are often seen in the upper part of the glacier where the ice is brittle. Most modern glaciers can be divided as shown here into zones of snow accumulation and a zone where the ice is wasting.

rotate and move downslope under the weight of the overlying snow. Large crystals of ice grow as one grain fuses into another until the firn compacts into a solid mass of ice.

Ice behaves as a brittle, elastic solid until the pressure on it is equal to the weight equivalent of 60 meters of ice. At that load it will flow continuously. Movement of the ice layers is accomplished in part by slippage between layers of molecules within the crystals, as shown in Fig. 24.12. This gliding mechanism is accompanied by recrystallization. Movement is also accomplished by shearing along planes through the ice. Thus, deformation deep in a glacier is thought to involve both external and internal movements of the individual ice crystals and recrystallization. The net effect is flow much like the flowage of plastics and of viscous liquid. The ice sheets that covered Canada accumulated to a great thickness, probably comparable to that in the Antarctic today. Once the thickness became great enough, pressure on lower layers from the weight of the overlying mass caused pseudo-plastic flow near the bottom of the ice sheet, and the sheets slowly spread over irregularities driven by the pressure of their own weight.

Because the behavior of ice changes as pressure is increased, a typical glacier has a brittle upper crust 45 to 60 meters thick and a lower plastic zone (Fig. 24.11). Fractures and crevasses form in the brittle portion of the glacier as it moves. These crevasses open and close in response to shearing, compression, and extension of the ice as the lower part flows over irregularities and around obstructions in its path.

If snow fields form in mountains, existing stream drainages become the course of least resistance for the ice as it begins to flow. Then tongues of ice begin

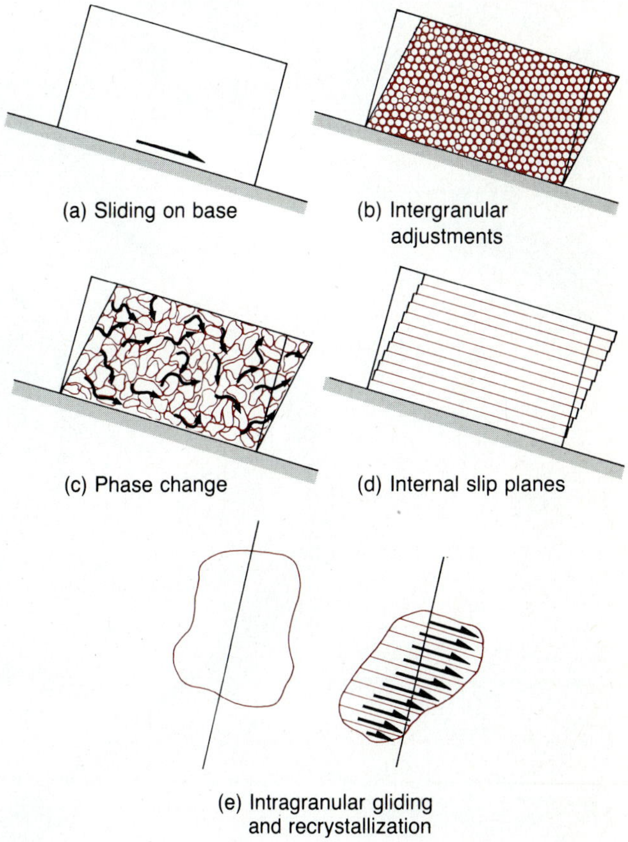

Figure 24.12
The mechanisms by which glaciers move are shown diagrammatically. Intergranular adjustments and phase changes are most important near the top of the glacier. Movement along slip planes and by gliding occurs deep in the ice.

their slow movement down the valleys as valley glaciers. How fast and how far these glaciers move depends on the rate of accumulation of snow in the snow fields, the temperature, the shape of the surface over which the ice moves, the thickness of the ice, and, of course, the rate of ice loss at the terminus of the glacier. The rate of advance increases as more snow accumulates in the snow field and as the ice thickness increases. It also increases on steeper slopes and across streamlined topography.

Movement in valley glaciers. Many existing valley glaciers move at slow but measurable rates. One of the first techniques used to study this motion consists simply of driving a series of stakes across the surface of the valley glacier in a straight line and observing their movement over a period of time. It quickly becomes clear that ice at the center of the glacier moves much faster than that at the sides, where friction along the contact between ice and the rock retards movement. Where new snow does not cover the surface, the pattern of movement may be visible, as it is in Fig. 24.13. The variation of velocity with depth can be determined by inserting a pipe in a perfectly vertical hole in the glacier and then measuring the rate at which the pipe tips from the vertical

U.S.G.S.

Figure 24.13
A flow pattern is clearly visible in the glacier ice of the Gilkey glacier in southeast Alaska. Two valley glaciers have joined near the top of this scene, but each continues to move as a separate glacier. Note the velocity distribution in the glacier at right.

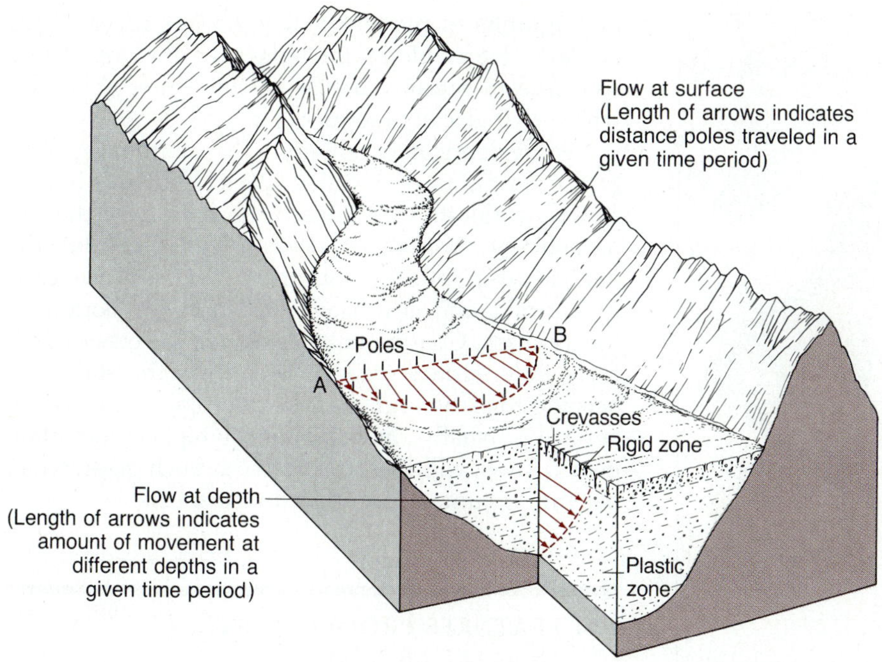

Figure 24.14
The velocity of ice movement is generally highest in the center of a glacier and retarded by frictional drag around the ice perimeter.

(Fig. 24.14). For most glaciers, the maximum velocity is well above the middle of the glacier.

Some glaciers move so slowly that trees and other plants grow in the debris on top of them; others move at rates as high as ten meters per day and, in exceptional cases, over thirty meters per day. Hassanabad Glacier in the Karakoram Mountains in Kashmir once advanced ten kilometers in 2.5 months—nearly 130 meters per day. Such extremely rapid movements of glaciers, called **surges,** usually do not continue for long periods.

Other indications of movement in glaciers are provided by the arrangement of crevasses, the internal structure of the glacier ice, and the form taken by debris on the glacier surface.

The prime mover in all glaciation is the force of gravity. Like valley glaciers, ice sheets move downslope; but if their volume is sufficient, they also flow over extensive flat regions and even up over irregularities. In part, this is accomplished through the push exerted on the front of the lobe of ice by ice moving downslope behind it.

How glaciers transport load. Evidence of the great erosive capacity of glaciers is seen in the large quantities of debris that they deposit, in the changes they cause in glaciated terrains, and in the abrasion and polishing of bedrock surfaces over which they move. The front edges of both valley glaciers and ice sheets push and scrape material in front of them—a sort of bulldozer effect. This is particularly effective in removing soil and semiconsolidated sediments, but most erosion cannot be attributed to this process.

Most glaciers carry much of their load near the base of the ice. Big blocks of rock may be dragged along under the ice, but most of the debris under an advancing glacier becomes frozen into the ice as it is overridden. Ice under great pressure is squeezed into cracks and dislodges even blocks of the underlying bedrock as they are pulled away by the moving glacier. The processes by which blocks of bedrock are lifted out of place is referred to as glacial **plucking** or **quarrying.** Fractures and bedding planes in the bedrock play an important role in these processes. Such planes of weakness allow water to enter. There it freezes, wedges the block up, and makes it possible for the blocks to become incorporated in the glacier. The concentration of this rocky load in the lower part of the ice is extremely important in facilitating erosion.

Because of this bed load in glacier ice, bedrock surfaces in glaciated regions are usually abraded and polished. The surfaces of bedrock exposed to the ice commonly bear striations (lines) and grooves carved by the scraping of rocks in the ice across the bedrock. Some of these, such as the ones on Kelley's Island in Lake Erie, are deep enough (2 meters) to use as a giant, long bathtub, but most are much smaller. Because these abrasion marks are aligned in the direction of glacier flow, movements of ice sheets can be determined. When the orientation of glacial stria-

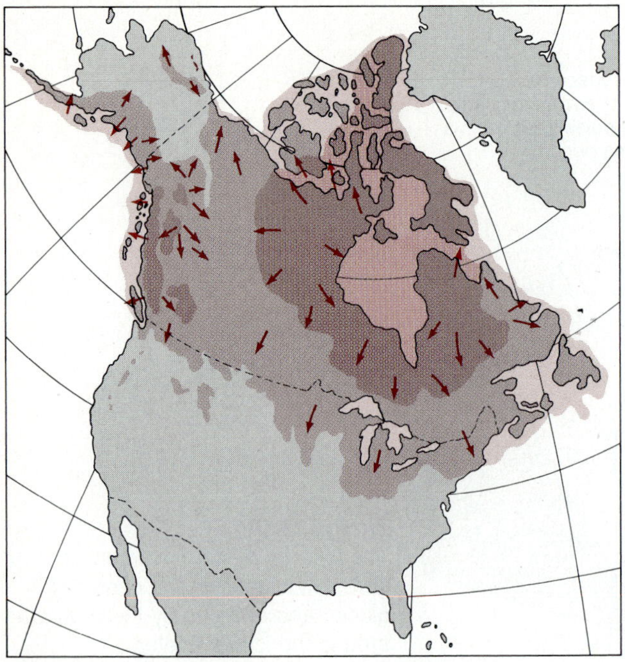

Figure 24.15
This sketch map of North America shows the areas covered by ice at the height of the Pleistocene epoch. Arrows indicate the general direction of ice movement.

tions are plotted for a large region, as is done for the parts of North America in Fig. 24.15, a good impression of the overall ice movement pattern is obtained. As we will see later, many areas have been subjected to multiple glacier advances. New striations and grooves are produced each time. Thus, some care must be taken to distinguish marks from different glaciations.

Valley glaciers also leave striations, but little question exists there about the direction of ice movement. However, erosion by valley glaciers has some characteristics not shared by ice sheets. Usually the tops of the mountains are not overridden by ice. Mountain peaks and ridges commonly rise above the ice. For this reason, the sides of the valley are scaped, loose debris in the valley and on its slopes is removed, and blocks are plucked and removed from the sides of the valley. This leads to undercutting of the valley walls and produces unstable slopes. Slumping, sliding, and debris avalanches bring great quantities of debris onto the top surface of the glacier.

Increased frost action accompanying glaciation promotes mass movements down the slope, causing the production of large quantities of talus, which fall onto the ice on both sides of valley glaciers. Much of this load is carried on the surface of the ice. Air photographs of modern valley glaciers reveal strips of dark debris along the edges of the ice. These converge to produce a pile on the center of the ice where two glaciers merge.

Some debris may remain in the ice from the head to the terminus of a valley glacier, but much of it slowly works its way toward the bottom of the ice, particularly through crevasses. This load plus the debris frozen into the bottom of the glacier ice leads to accumulation of a heavy load near the bottom of the ice mass. Continental glaciers, on the other hand, carry little surface load because once the sheet has become thick enough to flow over flat areas, its top surface is usually above all but the highest mountain peaks. Thus, valley glaciers carry a much greater load per unit volume of ice than do ice sheets.

24.4 FEATURES PRODUCED BY GLACIAL EROSION

Many small-scale erosional features are common to ice sheets and valley glaciers. Striations, grooves, and chipped and polished surfaces are found both on the bare bedrock over which the ice moved and on the rocks carried in the bottom of the ice.

Large-scale erosional features produced by ice sheets are not as distinctive as those produced by valley glaciers. Ice sheets tend to streamline preexisting topography, removing irregularities like peaks and sharp ridges. Mountainous topography streamlined by valley glaciers is illustrated in Fig. 24.16. In Canada, large areas such as the one in Fig. 14.5a have been flattened and smoothed by the erosion of ice sheets. One result of this is the production of streamlined hills called **roches moutonnées,** most commonly

Figure 24.16
The distinctive streamlining of this mountain in New Zealand was caused by erosion of a large ice sheet that formed in the New Zealand Alps and moved east.

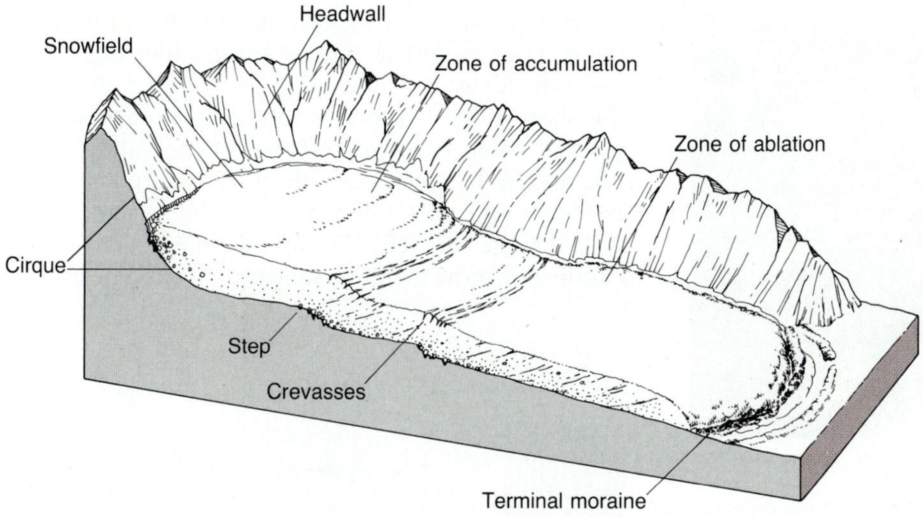

Figure 24.17
Valley glaciers accomplish erosion by removing rock from the head, sides, and floor of the valley. Rock plucked from the valley head forms the bowl-shaped feature called a cirque. Plucking is also shown here where the ice moves over irregularities in the valley floor, creating steps. The surface of the glacier usually has crevasses in it over these steps.

formed by ice sheets. These hills, often largely solid rock, are asymmetrical. A smoothed and streamlined surface forms on the side from which the ice came. A steeper and more angular side, produced by glacial quarrying, will face away from the incoming ice.

Erosion by Valley Glaciers

A number of distinctive and characteristic landforms are produced by valley glacier erosion. Figure 24.17 summarizes many of these features. The heads of most valley glaciers occupy hollowed-out amphitheater-shaped features called **cirques,** like those shown in Fig. 24.18. These are sites of glacial quarrying where the glacier starts its movement down the valley. Walls produced by this quarrying action become precipitous. They may drop almost vertically thousands of meters before they curve out into the floor of the valley. Cirques usually form at the heads of the glaciers near former drainage divides of streams that predated the glaciers. The already-steep slopes here are accentuated by the further removal of soil,

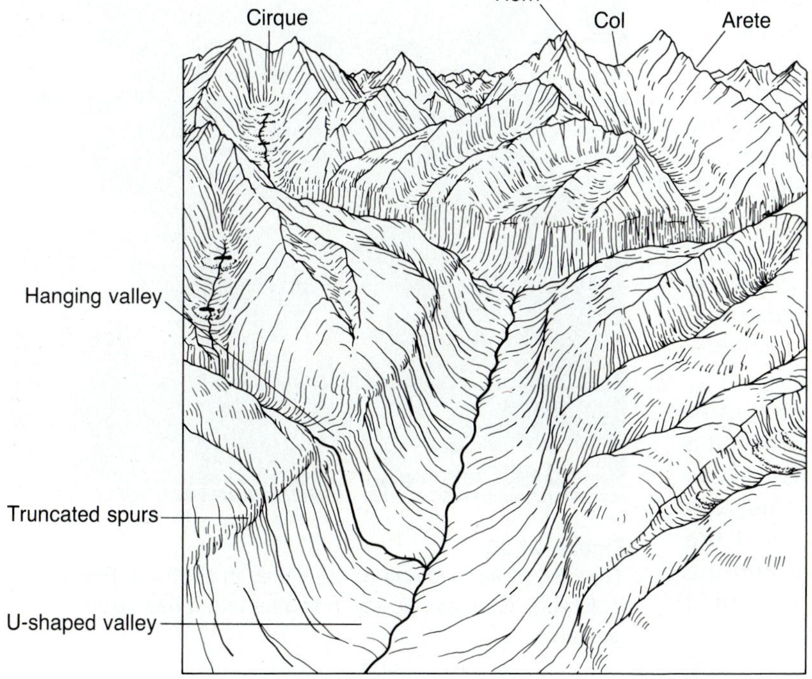

Figure 24.18
After the ice is gone, the features produced by glacial erosion in high mountains are clearly visible. Note the deep U-shaped valley, the hanging valleys, the cirques, horns, cols, and aretes. Glacial lakes, tarns, are left in the valleys where glaciers deepened the valley.

U.S.A.F.

Figure 24.19
These mountains in Alaska show distinctive forms caused by glaciers. Sharp ridges, aretes, separate a series of parallel valleys. These ridges are truncated where they enter the large valley at right.

talus, and eventually bedrock. The bottom of the cirque is usually gouged out, leaving a depression in the valley floor. After the glacier melts, a small lake, a **tarn,** usually occupies this depression.

The natural course for a glacier to follow is the old drainage pattern through the mountains. Thus, the initial movement of glaciers follows the line of least resistance, but the behavior of streams and glaciers is very different, and stream valleys are not suited for optimum movement of glacier ice. The ice cannot readily move around sharp bends. Erosion is concentrated at these places and the ice eventually cuts the valley sides at bends, streamlining the valley and leaving ridges, called **truncated spurs,** that are cut off at the edge of the glaciated valley, Fig. 24.19. The valley is streamlined by cutting off or smoothing out curves and removing obstacles, resulting in a U-shaped valley, which is more efficient for ice movement. These are distinctive features of glaciated valleys (Fig. 24.20) and quite different from the V-shaped valleys characteristic of mountain streams.

Valleys of tributary streams also become sites of glaciers, but these glaciers are smaller and erode at a slower rate, and consequently do not cut their valleys as deeply as that of the glacier to which they are tributaries. When the glaciers recede, valleys of the tributary glaciers are exposed. If they open into the larger valley above the valley floor, they are called **hanging valleys** (Fig. 24.21).

Sharp and rugged ridges, called **arêtes,** form along the crest of mountain ranges and on the ridges that separate adjacent valleys (Fig. 24.19). In both places the sharp ridge is produced by glacial plucking and gradual enlargement by erosion of the valley sides. The divides between the valleys are reduced in width until nothing more than narrow, sharp, and steep-walled ridges are left. Continued enlargement of the cirque and movement of the cirque walls toward the

Royal Canadian Air Force

Figure 24.20
The distinctive U-shape of the valley in southern British Columbia, Canada, clearly indicates that it has been glaciated.

U.S.G.S.

Figure 24.21
Yosemite Valley, in the Sierra Nevada Mountains of California, owes its distinctive forms to valley glaciation. The massive granitic rock stands in near vertical cliffs.

divide between adjacent valleys eventually causes the cirque walls to join. A pass, called a **col,** is formed where two cirques converge cutting into the same wall and lowering it below the level of the remaining ridge.

Sharp peaks that project above the surrounding area and are shaped by glacial erosion are called **horns.** The Matterhorn in Switzerland (Fig. 24.22) is a classic example. Horns are usually produced where several valley glaciers have heads in the same area. As the cirques enlarge and eventually converge, an isolated peak more or less surrounded by cirque walls is produced. Horns are formed late in glaciation, after the summit of the pre-glacial mountains has been reduced to a few isolated peaks.

Fjords (Fig. 24.23) are glaciated valleys, the floors of which are below sea level; some are as much as several thousand meters deep. They form where a valley glacier flows from the mountains into the sea. This is due partly to rises in sea level since the last major glacial advance and partly to the glacier's cutting its valley below sea level. If the ice moves fast enough, the glacier is able to carve its valley below sea level before the ice melts in the sea water or breaks away to form icebergs. Because the ice is buoyed up by the sea water, it is usually unable to cut a valley far off shore. Fjords develop where glaciated mountains are located on the coast. The coasts of Norway, Greenland, British Columbia, Patagonia, and southwest New Zealand are all famous for their spectacular fjords.

Figure 24.23
A hanging tributary valley and the distinctive U-shape of this Norwegian valley indicate its glacial origin. In this case, the floor of the valley is filled with water from the ocean making it a fjord.

TWA

Figure 24.22
The Swiss Matterhorn is a classic example of a peak formed where cirques join. The surrounding mountains have been eroded to lower levels, leaving the peak isolated.

24.5 FORMATION OF GLACIAL DEPOSITS

The load carried by a glacier is likely to include samples of every rock type that crops out in the terrain over which the glacier moves. Soil, alluvium, and any other unconsolidated materials encountered by the moving ice will be incorporated into it, along with solid rock plucked from bedrock and materials produced by grinding and abrasion of bedrock.

When the ice melts, the load is deposited. Deposition occurs most rapidly at the terminus of the glacier, where there is a balance between melting and forward movement of the ice. The position of the terminal edge of the glacier is determined by this balance between loss and supply of ice. The terminus may move slowly forward even as melting occurs; it may remain in one position with the amount of **ablation** (melting and sublimation) just equal to the amount of ice moving to the end of the glacier; or melting may exceed the rate of movement, causing the glacier to recede. Much of the glacial load is deposited at the end of the glacier, but meltwaters carry finer materials far beyond this point; and still other deposits are formed under glaciers or are developed as the ice finally melts, lowering its load to the ground after forward movement stops.

All glacial deposits are generally referred to as **drift,** but two quite different categories of deposits are formed. One is composed primarily of water-laid sediment called **glaciofluvial deposits.** The other consists of unstratified sediment called **till.** Each type occurs in a number of forms and at different locations relative to the ice. Some develop beyond the ice margin; others take shape under the ice, in it, or on it; and still others are deposited only when the ice melts.

Most till is a heterogeneous mixture of different sizes and shapes. It may contain a mixture of soil, stream deposits, and blocks of "quarried" bedrock. Usually, stratification is poorly developed or entirely absent, and the deposits are poorly sorted.

Unstratified deposits. All till is of glacial origin, but two distinctly different processes of deposition are involved. The glacial debris transported near the base of the ice or under the ice is subject to considerable pressure and may become pressed or plastered in the deposits under the glacier. Such deposits show no sorting, but oblong rocks are usually aligned with their long axis in the direction of ice movement. Some parts of the till may be crushed, and the deposits are often shaley in appearance.

A second type of till is formed where debris transported within or on the ice is gradually lowered to the ground as the glacier wastes away. The resulting deposit is loose; it shows no signs of pressure effects, and is called **ablation till.**

Stratified deposits. Glaciofluvial deposits laid down under and immediately beyond the glacier by meltwaters differ from most other sediments in that the minerals of which they are composed are largely fresh. They may consist of undecomposed fragments of minerals such as hornblende or feldspar, which ordinarily decompose relatively rapidly in water. Except for their composition, glaciofluvial deposits have the same characteristics as other water-deposited sediments. They are stratified, sorted, and may have cross-bedding or other primary sedimentary structures. A milky water usually issues from streams flowing out from under both valley and continental glaciers. The powdery silt-sized material in this water is called **rock flour.** The streams carry rock flour beyond the end of the ice where it gradually settles out to become part of what is called **outwash** sediment.

Landforms Produced by Glacial Deposition

The landforms commonly produced by deposition from valley glaciers and ice sheets are shown in Fig. 24.24 and 24.25. Isolated blocks of rock carried by both valley glaciers and ice sheets and dropped as the ice melts are among the most distinctive features formed by glacial deposition. Such blocks are called **erratics,** because they often are far removed from the site where they crop out, and they are usually different rock types from the bedrock beneath them. They are out of place—erratic. Such rocks, originating in Africa but found in South America, were one of the early lines of geological evidence that these two continents were once connected.

A number of glacial landforms result from a combination of glacial erosion and deposition.

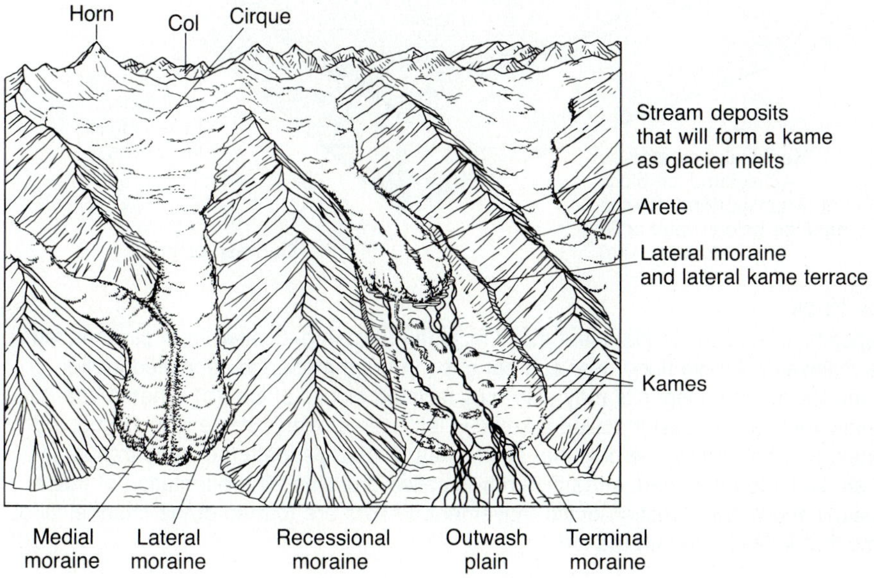

Figure 24.24
Deposits formed by valley glaciers include the terminal and recessional moraines found at the end of the glacier. Outwash deposits are laid beyond the ice by streams. Lateral moraines are composed of debris that accumulated on the top of the ice along the sides of the valley. This is often mixed with kame terraces, formed by stream deposits along the margin of the ice. Isolated hills of debris that collected on top of the ice or were funneled through the ice are called kames. Deposits in streams flowing beneath the ice sometimes leave sinuous ridges called eskers.

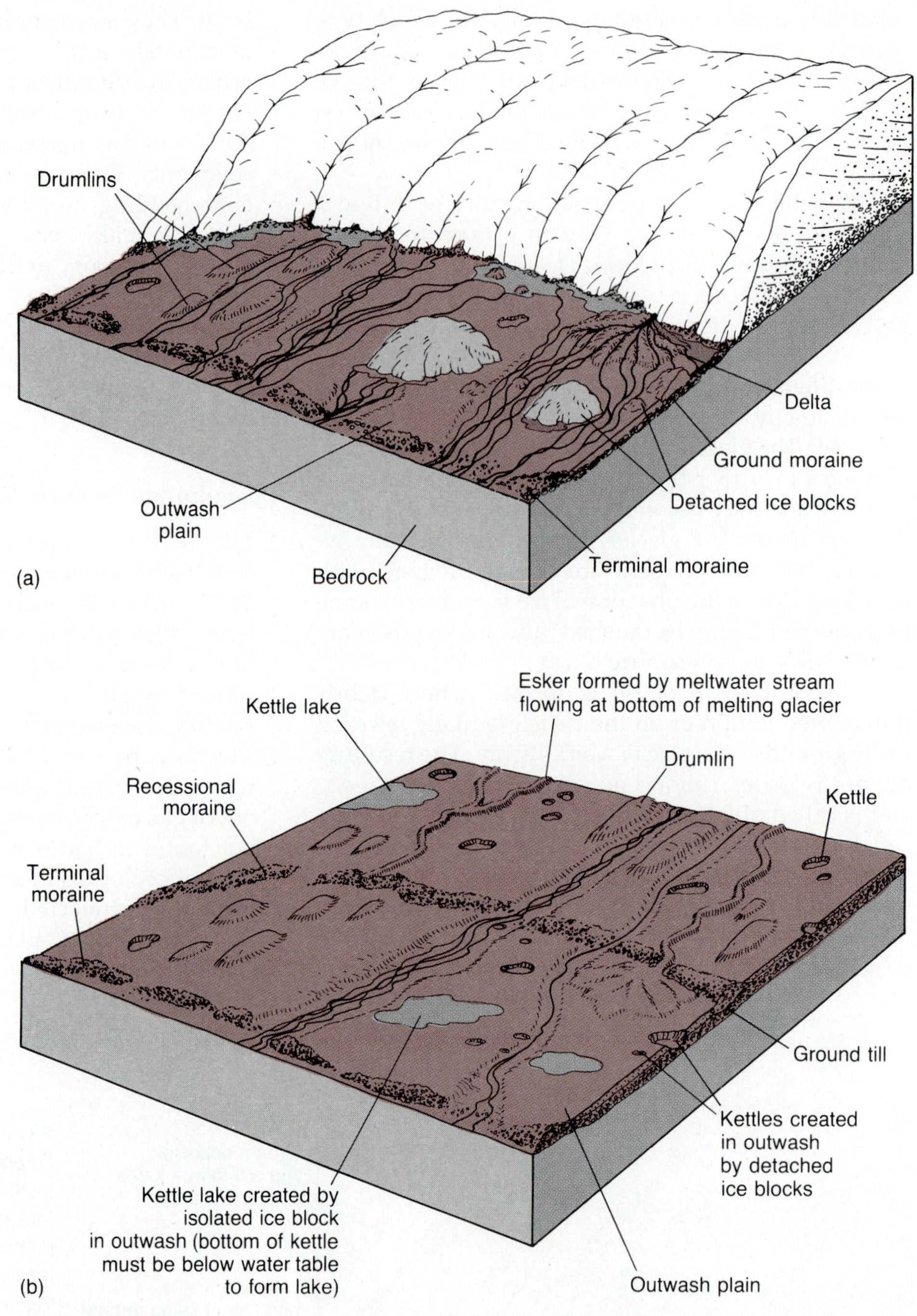

Figure 24.25
(a) Deposits formed as an ice sheet melts include outwash deposited from streams fed by the meltwater. These flood over and cut gaps through moraines deposited earlier along the ice margin. Deposits may also bury isolated blocks of ice. (b) When these eventually melt, a depression called a kettle is left. Lakes often form, especially behind moraines, and deltas may be built into such lakes. Debris beneath the ice is frequently found shoved into elongated, streamlined hills called drumlins. The orientation of these hills clearly shows the direction of ice movement. Eskers are formed of the channel fill of streams that flowed beneath the ice.

Streamlined hills shaped like whale backs and called **drumlins** are among these. These piles of till are formed of the material overridden by ice sheets. They are elongated in the direction of glacier movement and often occur in great numbers.

Terminal moraines (Fig. 24.26), is the name given to the ridges of drift (largely till) deposited along the edge of both valley glaciers and ice sheets at the position of the glacier's furthest advance. They form as debris continually melts out of the glacier, while the balance between ice supply and loss causes the edge of the ice to remain in one location. Meltwaters usually flow down the edge of the ice and over the terminal moraine while the glacier is close to it. When the glacier begins to retreat, the large quantity of meltwater usually forms a lake. Eventually, the lake tops the dam made by the moraine, cutting narrow valleys through the moraine, and water drains out, carrying finer debris beyond the terminus. Broad, low alluvial fans may grow from the notches cut into moraines. Beyond the terminal moraine, sediment-choked streams typically flow in braided patterns during the summer. But because they are subject to seasonal freezing, they may dry up or be buried by snow in the winter months.

Recessional moraines (Fig. 24.24) are formed if, during retreat of a glacier, the margin of the ice remains in one position long enough for a ridgelike mass of drift to accumulate from the debris dropped by the glacier. Older moraines are usually destroyed if the glacier readvances, and the till from them is redeposited in a new position.

Lateral moraines are formed along the sides of valley glaciers where debris accumulates on the glacier

Figure 24.27
The sharp ridge in the foreground is composed of debris that came down the mountainside and banked up against glacier ice when the top of this glacier near Mount Cook, New Zealand, was at this level. Glacier ice can be seen in a few places at left, but most of the ice is covered with debris carried in and on the glacier. The large boulders in the foreground are the size of a car.

surface as a result of the normal downslope movement of talus and surface wash from the sides of the valley. This debris is present on top of many glaciers. As the ice melts, a ridge composed of till is left along the foot of the steep valley sides (Fig. 24.27).

Ridges of till are formed when two or more valley glaciers flow together. As two valley glaciers join, the lateral debris on the adjacent sides of the two glaciers merges to form a single mass that will continue down the valley on top of the ice as a single medial ridge (Fig. 24.28). When the glaciers melt, these medial ridges are lowered and may be recognized as ridges of till called **medial moraines.** This merging of glaciers occurs wherever tributaries of a glacier join, so a number of medial ridges are often formed, as shown in Fig. 24.28.

When melting of either valley glaciers or ice sheets has progressed far enough or when the supply of new ice decreases, the glacier ice ceases to move forward. As melting continues, the debris on the surface, within the ice, and caught near the bottom of the ice is slowly lowered to the ground. The resulting deposits, called **ground moraines,** have low relief and consist of till or mixtures of till and water-laid deposits, (Fig. 24.29).

Both ice sheets and valley glaciers partly melt during summer even while glacial advance is in progress, and meltwaters are abundant as the climatic conditions giving rise to the glacier begin to wane.

U.S.G.S.

Figure 24.26
The end moraine of Columbia glacier in Alaska. Sediment-laden streams flow into the sea.

Figure 24.28
Debris destined to become medial moraines when the ice melts is prominent on this view of Kaskawulsh glacier in the St. Elias Mountains of Alaska. (U.S.G.S. photo.)

Water flows on the surface of the ice, seeps down along the margins, and even moves through holes in the ice. Water accumulates on the surface and beneath the glacier and forms streams. These streams are different from ordinary streams in that their banks are composed of ice. All sorts of material being transported by the glacier may get into the stream. The fine sizes like rock flour are carried easily downstream, but the coarser boulders and blocks are not.

Figure 24.29
Hilly ground is composed of moraine left as the glacier that once filled this valley near Mount Cook, New Zealand, retreated. Note the large erratic in the center and the horizontal markings on the mountainside that indicate how deep the ice once was. The present glacier is not seen, but outwash from it makes up the flat plain in the middle of this picture.

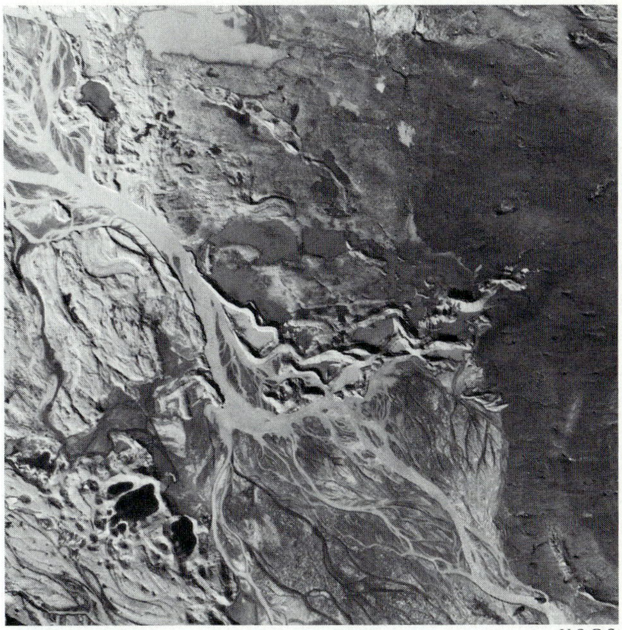

U.S.G.S.

Figure 24.30
Eskers are ridges of stream deposits left after glaciers melt. The channel was probably formed in the ice at the bottom of the glacier. The eskers shown here are exposed a short distance from the end of Casement glacier at Glacier Bay, Alaska.

The channels in and under glaciers often have walls of ice containing the gravel and boulders. After the glacier melts, a sinuous ridge of this channel fill may remain, marking the former stream channel; such ridges are called **eskers** (Fig. 24.30). They also form where a stream beneath the ice cuts into sediment or weak rock beneath the glacier, leaving deposits of the coarser part of its load. Later, if the sediment on either side of the stream deposits is more easily eroded than the material in the channel, the stream deposits may be etched out to stand as a ridge.

The term **kame** is widely used to describe glaciofluvial deposits laid down in contact with ice. Such deposits may form in a variety of positions, commonly as alluvial fanlike deposits along the ice margin. The term is also applied to glaciofluvial deposits formed by sediments washed into depressions on the top surface of a glacier by surficial meltwaters. As the ice melts, the material that formerly filled depressions on top of the glacier is dropped, leaving small hills. Similar kames form beneath the ice where sediment washed down through crevasses and holes in the ice accumulates in small hills. Glaciofluvial deposits also form flat benchlike forms, called **kame terraces,** along the sides of the valley where they may coalesce with lateral moraines.

In the late stages of the melting, debris in the lower part of an ice sheet and on top of a valley glacier completely covers the ice. As the ice disappears, blocks of ice may be trapped and protected from melting by this cover. When these last remnants of ice melt away, an enclosed depression called a **kettle** is left. Most such features are formed near the edge of the ice (Fig. 24.31).

When a glacier first begins to melt, the meltwater is incapable of transporting all the debris present. Part of it is deposited as broad alluvial fans or, in the case of ice sheets, as a series of coalescing alluvial fans called the **outwash plain,** located beyond the terminal or recessional moraines (Fig. 24.32). Here rock flour, along with most of the sand and clay carried from the glacier, is deposited. Because the steams are heavily loaded, and because they are shifted by often-changing conditions, braided-stream patterns characterize the outwash-plain streams. Later, if the rate of melting begins to increase, the streams may establish more permanent channels, through which large amounts of the debris are moved beyond the immediate region of the glacier.

U.S.A.F.

Figure 24.31
The irregular surface of this outwash plain is caused by blocks of ice that were buried by the outwash before they completely melted. Note the road and houses at the right for scale.

Figure 24.32
Outwash plains are often composed of fine sediment ground up by glacial action. The finest part can be picked up by wind and blown far from the glacier. This outwash plain is in New Zealand.

24.6 THE ICE AGES

Judging from the scarcity of glacial deposits in the rock record of past times, periods of glaciation are very unusual. The remains of corals and other animals that ordinarily favor warm waters are found in rocks of Paleozoic age in the Arctic and Antarctic. What has caused glaciers to form on earth is an intriguing question that has given rise to a great deal of speculation. Scientific literature abounds in hypotheses that try to answer the question. Some of them are based on assumptions that have now been proved fallacious, and the validity of others remains uncertain.

For any hypothesis to be acceptable it must explain, or at least be compatible with, the observed nature of glaciers and the major features of glaciation. For example, an acceptable hypothesis must take into account the known history of glaciation and the record of climatic changes. The geological record of climate consists of a variety of indicators. The temperature of the near-surface waters of the oceans may be indicated by the oxygen-isotope and paleontological studies described in following sections. Fluctuations in sea level serve as a guide to the amount of water tied up in the form of ice on land. The size of a continental ice sheet may be estimated from the extent of the glacial deposits laid down at the margin of the ice. Moraines and lake-bed deposits formed behind them are especially useful. The distribution of fossils of temperature- and moisture-sensitive plants may also be useful. Plant pollen preserved in sediments is often used in this way.

The occurrence of till containing striated and grooved pebbles assures us that several periods of glaciation predate the one in which we now live. The oldest of these deposits are found in Precambrian rocks, and it is possible that more than one period of glaciation occurred during Precambrian time. These oldest ice sheets occupied an area in the southern part of the Canadian shield. Younger glacial sediments are also found in widely separated areas in rocks that lie just beneath the Cambrian system. These tills are more than 600 million years old. They are scattered over the Canadian shield in Utah, Norway, Australia, eastern Greenland, India, and in the Appalachian Mountains. The third major period of glaciation occurred during the late part of the Paleozoic era in the Carboniferous and Permian periods. At this time, about 230 million years ago, glaciers covered parts of South Africa, South America, Australia, Antarctica, and India. In Australia, five separate sheets of till of late Paleozoic age are interbedded with coal. This relation indicates cold periods of glacial advance separated by warm, moist interglacial periods during which plant life flourished. The late Paleozoic glaciation took place before the breakup and separation of the continents in the southern hemisphere. The areas affected by this glaciation were close enough together to be covered by a single continental glacier, as suggested in Fig. 12.17. The orientation of glacial grooves and striations on these now widely separated areas is compatible with the hypothesis of a common ice center when the position of the continents is restored as shown.

The rock record of these earlier glaciations is scattered and poorly preserved. In contrast, we have a relatively complete understanding of the events that have taken place during the Pleistocene epoch, the most recent glacial period. This knowledge imposes important limitations on the theories of glaciation. Pleistocene glaciation started between 2 and 3 million years ago. Since its beginning, at least four major intervals of glacial advance and three interglacial ages have occurred. We now live in what will likely prove to be an interglacial period.

The subdivisions of the Pleistocene epoch in North America are based on the stratigraphic analysis and dating of the deposits formed during advances of ice and during interglacial periods. Each is named for a region or locality where these deposits are well exposed. The American divisions are as follows:

Wisconsin glaciation
  Sangamon interglacial—named for deposits in Sangamon County, Illinois
Illinoian glaciation
  Yarmouth interglacial—named for deposits in Yarmouth, Iowa
Kansan glaciation
  Aftonian interglacial—named for deposits in Afton Junction, Iowa
Nebraskan glaciation

Although the deposits from each of these ages are of the same general type, they can still be distinguished. The standard methods used in stratigraphy are applied to these unconsolidated sediments. For example, older tills are found buried by younger outwash and ground moraines. Deposits formed in lake beds that existed during an interglacial period are found covered by moraines of a renewed advance. Older moraines are found partially destroyed and incorporated in the tills of younger advances.

In addition, soil profiles developed in glacial drifts have been used as an effective means of identifying tills of different ages. The depth of oxidation and leaching, the amount and kind of chemical weathering, and the extent of decay are used as means of estimating relative ages of tills. The most recent glacial tills are fresh in comparison with older deposits. For the younger deposits, radiocarbon dating based on the percentage of radioactive carbon-14 in organic material has been especially valuable (see box on "Carbon-14").

The last great resurgence of ice reached a maximum about 18,000 years ago. During periods of advance, the ice sheets have covered most of Canada, the northern United States (Fig. 24.9), Greenland, northern Europe, northern Siberia, Antarctica, and high mountains throughout the world. Each of the major Pleistocene advances took place essentially simultaneously throughout the Northern Hemisphere. Whatever caused the glaciation must have affected this huge region at about the same time. Even within each of the major glacial advances, fluctuations in the position of the ice sheet and minor advances and retreats of relatively short duration took place. These advances and retreats were clearly neither regular nor uniform. Ice sheets grow and decay over periods of a few thousand years—a short interval in geological time. Most glaciologists now believe that ice sheets decay nearly five times as fast as they grow.

CARBON-14 Well known for its application to archeological findings, the carbon-14 method of dating depends on the presence of radioactive carbon in organic matter. Carbon-14 is formed in the upper atmosphere when nitrogen is bombarded by cosmic rays. Nitrogen, when hit by a cosmic ray, emits a proton and becomes carbon-14. Carbon-14, along with other carbon, combines with oxygen to form carbon dioxide, which is absorbed by all living matter. When a plant or animal dies, it stops absorbing carbon-14, and the carbon-14 present in the organism begins to decay, forming nitrogen again. The half-life of carbon-14 is about 5710 years. In order to date organic remains, the amount of carbon-14 in the organic matter is measured and compared with the amount found in living matter. The ratio of the carbon-14 content in the dead and living matter is a measure of the length of time carbon-14 in the dead matter has been decaying, and therefore of its age. For example, if exactly half as much carbon-14 is found in a piece of wood that has been buried in an old river deposit as in a living tree, then the age of the old wood can be estimated. Assuming that the old wood had the same concentration of carbon-14 in it when it was living as the present-day tree has, then one-half has decayed. Thus the age is equal to the half life of carbon-14, or 5710 years. Because of its relative short half-life, carbon-14 is useful in dating relatively young material. After about 50,000 years, so little of the isotope remains that the age of an object cannot be estimated very accurately.

Formation of the Great Lakes

Each of the major glacial advances during the Pleistocene epoch affected the north central states. It is clear from the map of Fig. 24.9, showing the location of deposits associated with each advance, that the area now occupied by the Great Lakes was repeatedly covered by ice. In fact, these lakes owe their existence to the erosion and deposition of the great lobes of ice that pushed across the region. The region and the deposits from the glaciers have been studied carefully, and a relatively clear description of the evolution of the Great Lakes region is now possible. Details of the Wisconsin retreat, shown in Fig. 24.33, are best documented.

Before the first glacial advance, no lakes of unusual size were present. The region would have been characterized as an upland of low relief located on the edge of the Canadian shield. The area now occupied by Lake Superior was probably low because it contains rocks that are slightly less resistant to erosion than surrounding areas. As the ice sheet formed and pushed south, soils were stripped and the less resistant rocks gouged out. The tremendous thickness of ice in central Canada caused the lithosphere to subside slightly and the area of the Great Lakes to tilt slightly toward the advancing ice to the north. Lakes formed behind dams of earlier moraines that existed in the region during the interglacials. These older lakes may have been in the same general vicinity as the modern lakes but, undoubtedly, had somewhat different shapes.

Like earlier advances, the Wisconsin glaciation consisted of a number of relatively minor advances and retreats. During each of these advances, the older moraines and other deposits were partially destroyed by the overriding ice. By the time of the Cary advance, 16,000 years ago, depressions had already been formed where the lakes were destined to reside. Ice lobes at the edge of the ice sheet ended at the margins of both Lake Michigan and Lake Erie.

Three more advances took place after the Cary event, and the region finally became ice-free about 5000 years ago. The lakes occupy depressions gouged out by glacial advance and dammed up by recessional moraines.

Since the last glaciation of the region, crustal (isostatic) rebound has caused the area once covered by ice to rise. This has resulted in the tilting of moraines and lake deposits as described in Chapter 9.

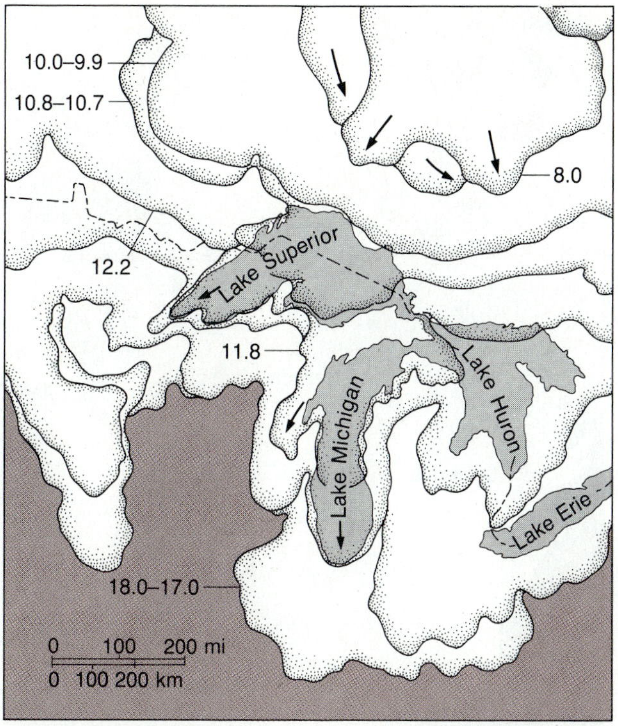

Figure 24.33
The Great Lakes were formed primarily by glacial action. The position of ice margins varied from time to time. During the last retreat, moraines were left as shown here. Note that wood in these moraines has been dated, and the ages (in thousands of years) of a few of these and moraines are shown.

Both the ice and the glacial rebound have drastically altered the drainage patterns of the region. The present drainage is shown on Fig. 24.34. Clearly some of the modern streams must now flow across morainal deposits. Most of these are relatively small tributaries of the Ohio and Missouri Rivers, both of which flow along the southern margin of the glacial deposits. Before glaciation, the main stream south of the Great Lakes was a now-buried river called the Teays. This stream had its headwaters in the Appalachian highlands of North Carolina. It followed the course of the New River in Virginia and West Virginia (Fig. 24.34), but its course departed from the modern drainage in southern Ohio, and it flowed north across Ohio, Indiana, and Illinois. There it flowed into an arm of the Gulf of Mexico that extended up the Mississippi Valley into southern Illinois. This is but one example of the changes that took place all along the ice margin.

Effects of Continental Glaciation: Beyond the Ice Margin

Continental glaciation has had many important effects far removed from the actual extent of the ice. One of the most prominent of these is change in sea level. As we shall learn in the next chapter, the worldwide sea level has apparently been rising for the last 15,000 years (as will be illustrated in Fig. 25.3). That rise began from a level about 130 meters below modern sea level and has been proceeding at slower rates for the last 5000 years than for the previous 10,000 years. If all the present ice melted, sea level would rise by an amount conservatively estimated as being between 25 and 60 meters. These changes in sea level have had pronounced effects on the evolution of coastal landforms, as we will see in the next chapter, and they have dominated the recent history of the continental shelves.

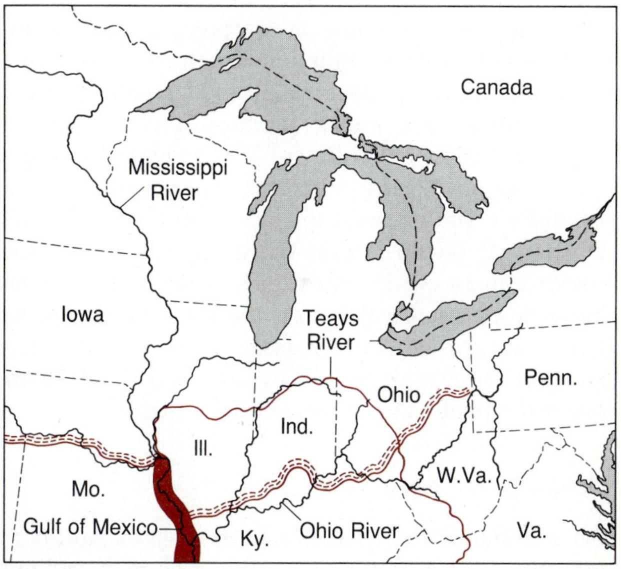

Figure 24.34
Modern drainage south of the Great Lakes has been determined by events in the Pleistocene. During the last interglacial epoch, waters of the Gulf of Mexico filled the Mississippi River valley as far north as the Missouri River. At that time, the region just south of the Great Lakes was drained by a river known as the Teays. The western side of the central Appalachian Mountains drained into the Teays. During the last glacial advances, the Teays was filled with sediment as the ice moved south. Following this last advance, the main drainage, the Ohio River, remained over the terminal moraine.

Changes in climate, patterns of atmosphere circulation, precipitation, and water supply have had dramatic effects over vast areas beyond the limits of the ice sheet. Lakes such as glacial Lake Agassiz in Canada, which covered more area than the present Great Lakes, were formed as the retreating ice margins left natural dams in the form of recessional moraines (Fig. 24.33). The Great Lakes owe their presence to moraines that prevent drainage toward the south, to deepening due to scouring and removal of rock and debris by the glaciers, and to depression of the crust by the weight of the ice sheet. The readjustment to the removal of this weight is still going on. Glacial processes also deepened lakes in the western United States by as much as 300 meters. Other lakes formed beyond the margins of the glaciers, partly due to meltwater, but primarily as a result of increased precipitation and lower evaporation rates caused by the nearness of the ice mass. The Great Salt Lake is a remnant of one such lake, Lake Bonneville, located far from the edge of the ice and covering approximately 80,000 kilometers (Fig. 24.35).

Lakes were common in the present arid regions of the southwestern United States and in the Sahara, Arabian, Kalahari, Atacama, and Patagonian deserts. These areas were moist, perhaps fertile, and well watered during these times. These climate changes were brought on by both the general atmospheric conditions that caused the glaciers and by secondary effects that resulted from the presence of large ice-covered areas. As the air over ice is cooled, it becomes more dense, and it flows off the ice. Changes of air temperature over large masses of ice clearly affect atmospheric circulation. They almost certainly compressed climatic zones toward the equator.

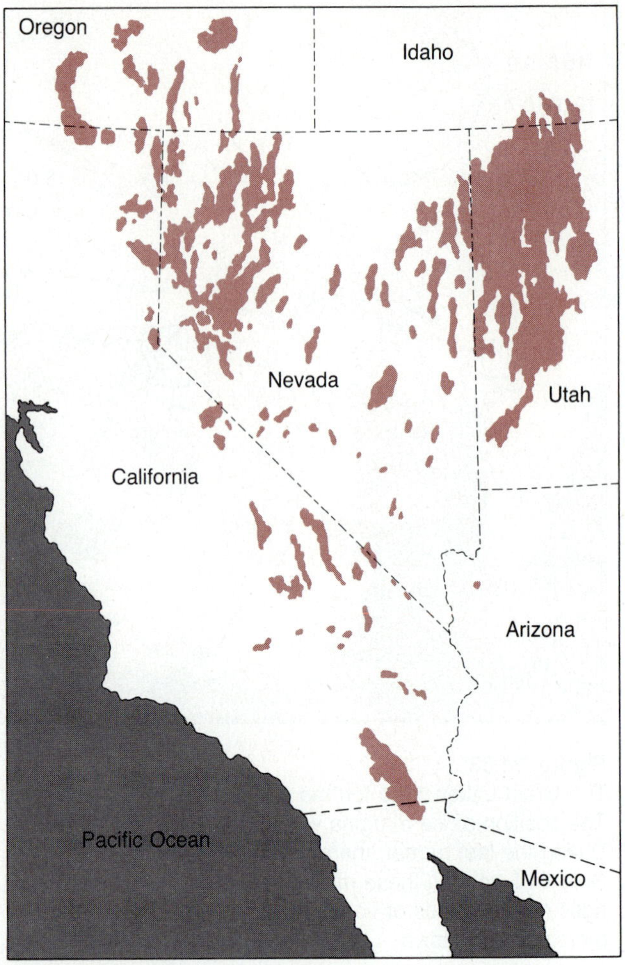

Figure 24.35
Many large lakes, including the Great Salt Lake of Utah, formed south of the ice margin, filling as the ice sheets melted farther north and in nearby mountains. Most of these lakes have long since dried out.

Causes of Ice Ages

An acceptable theory of glaciation must explain both why the earth was free of ice for periods of millions of years between glaciations and what caused climatic fluctuations occurring over periods of a few thousand or tens of thousands of years during periods of glaciation. It seems clear that a slight decrease of mean and annual temperatures, for example 5 °C, at the present time would cause substantial growth of existing ice sheets on land and pack ice at sea. Such a change might even trigger the growth of continental ice sheets once again.

Thus, the first stage in explaining causes of glaciation lies in determining what processes could cause a decrease in mean annual temperatures of this magnitude.

Decreasing temperature on earth. A decrease in the amount of radiation produced by the sun would certainly reduce the temperature on earth. Although the idea may seem reasonable, the observed fluctuations (about 3 percent) in the solar radiation are much too small to produce the observed effects, and we have no other evidence to suggest larger-scale changes in the energy produced by the sun.

The amount of solar radiation intercepted by the earth and the way it is distributed over the surface does change over long periods of time as a result of variations in the path and tilt of the earth as it moves around the sun. Three distinctly different effects are known. First, the shape of the elliptical path of earth around the sun has changed—it has varied from being nearly circular to being more eccentric (more distinctly elliptical) over a period of 93,000 years. When the path is most elliptical, the winter seasons are more extreme than they are when the orbit is more circular, because the earth is farther away from the sun during winter. Second, the inclination of the earth's axis to the plane in which earth moves around the sun varies. A maximum variation in tilt of 3° (from about 22.1° to 24.5°) occurs about once in every 41,000 years. Third, the axis of the earth wobbles and this causes a change in tilt that recurs with a 21,000 year period.

Both the change in inclination and the wobble of the axis have the effect of tilting the polar regions farther away from the incoming solar radiation. When tilt is greatest, the days are shorter, the area of total winter darkness is greater, and less radiation hits the polar regions during winter.

Conditions favorable for growth of ice sheets in polar latitudes are greatest when these effects are additive—when the maximum tilt of the axis coincides with the most eccentric orbit.

The effects of each of these three variations on surface temperatures can be predicted with mathematical precision. In 1938, Milankovitch prepared curves showing the combined effects of these variables on temperatures at latitude 65°N for the past 150,000 years (Fig. 24.36). When these three effects combine to lower temperatures, the result is a reduction of average summer temperatures of about 3 °C, almost certainly enough to cause glaciation.

Decreases in temperature in the lower part of the earth's atmosphere would also be brought about if incoming solar radiation were reflected or absorbed by something coming between the sun and the earth's surface. Both comets and clouds composed of interstellar dust have been suggested, although little evidence of this can be found. Volcanic emanations such as carbon dioxide, dust, and ash seem a more likely possibility. Carbon dioxide absorbs radiation of a longer wave length, and explosive eruptions such as that at Mount St. Helens project dust high into the atmosphere. The small particles that remain suspended for years intercept solar radiation, absorbing some and reflecting more back into space. The importance of volcanism is still being evaluated. Cores taken in the ocean reveal a number of widespread layers of ash that must have been laid down during times of massive eruptions. Some of these occurred about the time of the beginning of the Pleistocene era. But geologists are divided in their opinion of the significance of this timing. Some believe volcanic dust contributed to the start of the ice ages. Others conclude that ash is found on plates approaching the volcanic centers along subduction zones, but that the frequency of ash layers on the more stable plates is no greater in the Pleistocene epoch than in earlier Cenozoic records. A large amount of volcanism could certainly cause enough cooling of the atmosphere to reduce temperatures several degrees.

To this point we have considered conditions that affect the amount of solar radiation reaching the earth. Climate is also greatly affected by the amount of heat actually absorbed by the earth. Snow-covered land and ice-covered seas reflect almost all radiation striking them. Thus, snow and ice cover tend to cause further cooling. Once glaciation begins, it is probably increased by this effect.

The position of continents. Continents and oceans have pronounced effects on climate, and ocean currents play an important role in distributing warm

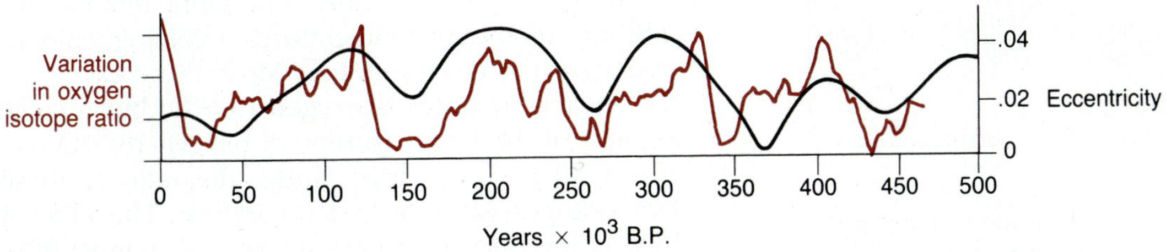

Figure 24.36
Predicted variations in global temperature averages based on changes in the orbit of the earth and its inclination were first compiled by the Yugoslavian meteorologist Milankovitch. His theory is now widely accepted as the most probable primary cause of global cooling and warming.

equatorial waters to high latitudes. If the lithosphere were static, these factors would perhaps deserve less consideration, but we know now that continents and ocean basins are continually changing position on the globe. Thus, both the patterns of oceanic circulation and the position of the continents relative to the poles are important considerations. A dip in mean annual temperature of even 10 °C would not be sufficient to create an ice sheet in equatorial latitudes. If continents were not located in high latitudes, we almost certainly would not have had glaciation of continents in the Pleistocene epoch.

Continental drift shifted the continents of North America and Eurasia toward the polar regions during the Cenozoic era. This is certainly a major contributing cause of glaciation, if not, indeed, the main cause. The effects of this drifting have been twofold: first, it constricted and reduced the circulation of waters in the Arctic, making it easier to cool them; and second, it shifted large land areas into latitudes where less solar radiation hits them in winter and where temperatures are more extreme.

The importance of having continents in polar positions in order to achieve glaciation is confirmed by the late Paleozoic glaciation. The continents affected at that time were all adjacent and located near the south pole, Fig. 24.37.

Continental drift is clearly important, along with variation in insolation as explained by Milankovitch, as a first-level cause of glaciation. Neither, however, is sufficient to explain the glacial–interglacial variation within ice ages.

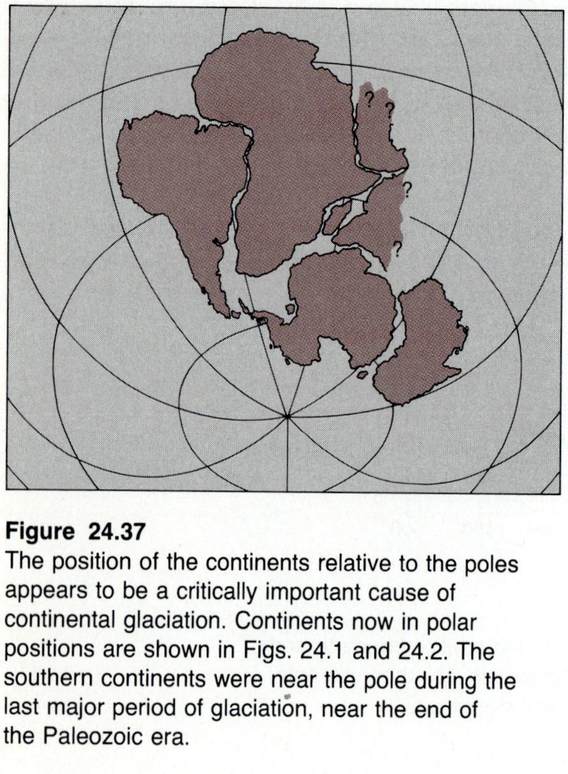

Figure 24.37
The position of the continents relative to the poles appears to be a critically important cause of continental glaciation. Continents now in polar positions are shown in Figs. 24.1 and 24.2. The southern continents were near the pole during the last major period of glaciation, near the end of the Paleozoic era.

Methods used to determine past temperatures. Most of the methods used to determine past climates provide only a general indication of what conditions were like. For example, the presence of salts in deposits of a certain age indicate that conditions were favorable for evaporation (generally high temperature, low humidity, and often high winds).

Other climate indicators, such as tree rings, suggest the severity and length of the winters. The spacing of tree rings indicates the amount of growth. Rings are thin and close together during cold and dry years. Although tree-ring measurement is most useful for the last few hundred years, analysis of tree rings provides a record for more than a thousand years. The cold period from about 1350 to 1800, called the "Little Ice Age," shows up clearly in the tree-ring measurements, as shown in Fig. 24.38.

In order to probe older climates, totally different methods must be used, many of them based on the fossil record. Especially valuable are analyses of the abundance of certain species of microscopic marine animals, the foraminifera. Some species are temperature sensitive and live only in cold waters; others inhabit warm waters. Thus the presence of large numbers of cold-water foraminifera in sediments deposited at high latitudes is an indication of warmer ocean waters. Because the temperature of the oceans is slow to change, variations in water temperature are thought to be a good index of temperature changes on the earth as a whole.

On land areas, pollen provides a reliable source of climatic data. Because many plants live in restricted environments, the presence of their pollen is a good climate indicator.

In recent years a new and more quantitative indicator of ancient temperatures involving isotopes of oxygen has come into use. Most of the oxygen in the air and in water is oxygen-16, ^{16}O, but a small amount of the heavy isotope of oxygen-18, ^{18}O, occurs with it. Most importantly, the ratio of these two isotopes varies with temperature. The ratio of $^{18}O/^{16}O$ increases 0.02 percent per 1 °C temperature drop.

The reliability of this ratio as a temperature indicator can be checked by seeing how accurately it reveals the temperature in the modern environment. Thus, the ratio of oxygen isotopes found in the air,

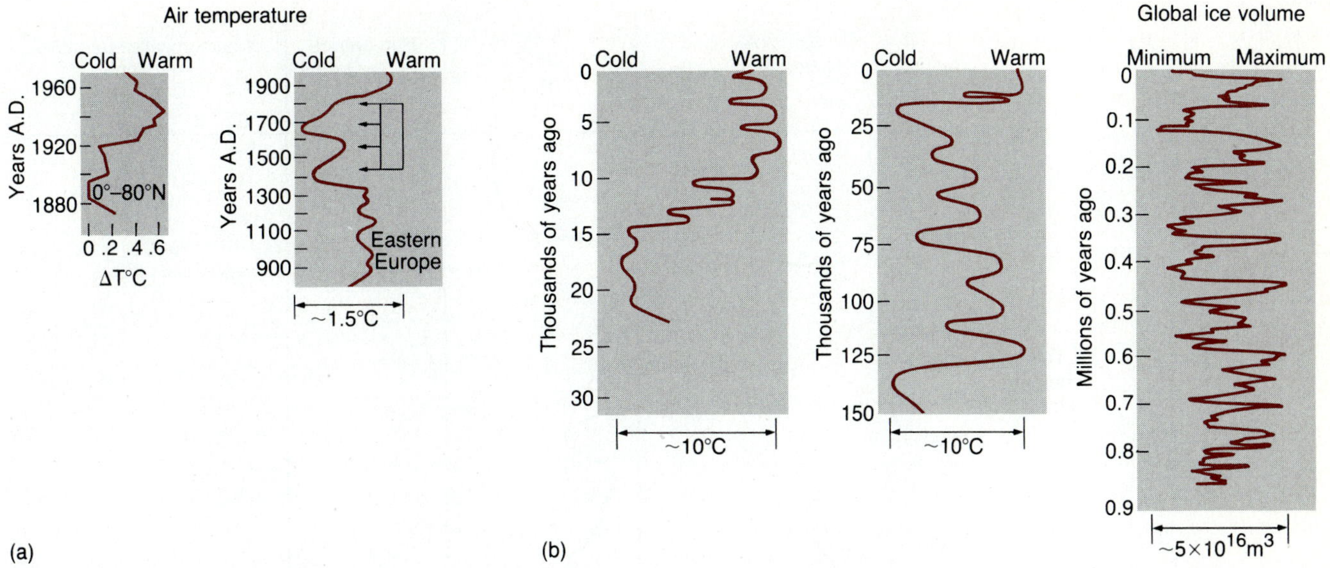

Figure 24.38
The documented history of climatic changes. (a) The earth has been getting cooler since the 1940s, but it is still much warmer than it was in the late 1800s or in the late Middle Ages, when it was so cold the period was named the "Little Ice Age." (b) When the record is extended farther back it is clear that we are enjoying one of the warmest periods and one of the times of minimum ice cover in the last million years.

in water, in snow and ice, in wood, and in shells ($CaCO_3$) can be directly compared with the temperatures of their environments. The ratio is found to be in excellent accord with modern temperature measurements, and can be used to measure the rate of accumulation of layers of new snow and ice on modern glaciers. The application of the method to wood and shells is even more significant for the purposes of tracing glacial history.

Marine organisms take oxygen from sea water to construct their shells. Thus, the shells of such microorganisms as the planktonic foraminifers have shells made of $CaCO_3$ in which the ratio of $^{18}O/^{16}O$ is a function of the temperature of the near-surface water masses in which these organisms live. As these organisms die, their shells sink and are preserved as part of the sediment on the sea floor. The more recent parts of this sediment can be dated by radiocarbon dating techniques, and the older parts can be dated by their fossil content. Using these two types of observations—temperature and age—it is possible to construct a graph, such as the one in Fig. 24.39, showing the manner in which water temperatures have varied through time.

The results of this type of study can be compared with other types of observations that give an independent but qualitative estimate of temperature. For example, the temperature curve (Fig. 24.39) based on oxygen isotope data is in good agreement with a qualitative curve based on the type of foraminifera.

Comparing the evidence. We now have a number of very different types of data that can be compared as the search for possible causes of glaciation continues. These data include: the Milankovitch curves showing variations in the earth's orbit; the position of the continents through time; layers of volcanic ash; oxygen isotope data; data from temperature-sensitive organisms; changes in sea level due to ice–water balance, as indicated by deposits and erosional features; and pollen analysis for temperature sensitive plants.

Good correlations exist among much of these data. An especially good correlation exists between oxygen isotope data, analysis of ancient environments, cold/warm water foraminifera, and pollen analysis for the period for which carbon-14 dating can be used (Fig. 24.39). Good correlations also exist for the amount of solar radiation reaching earth, evidence of cold/warm water fauna, high stands of sea level, and time of heavy precipitation in deserts for the last 150,000 years, if a lag of 8000 years

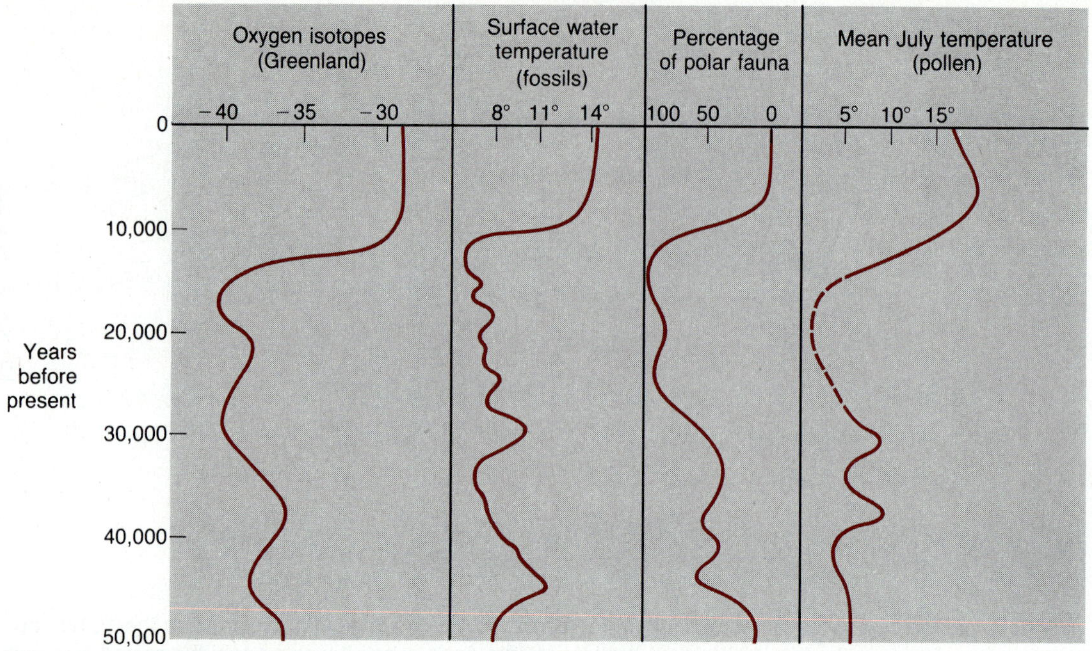

Figure 24.39
The results of four very different methods for estimating climate are compared in these graphs. Although the curves differ in detail, they are in good agreement about the general climatic trends.

between cooling and glacial effects is allowed (Fig. 24.39). We shall see in the next section one model that might explain such a long lag.

These curves leave little doubt that the Wisconsin glaciation took place during a cold spell, not a warm one. And that the cold temperatures followed a period during which sunlight was reduced as a result of the earth's tilt and orbital position. Most glaciologists are now convinced that the prime first-level cause of glaciation is the coincidence of having continents in polar regions at the time insolation in high latitudes is reduced because of the earth's orbit. The cause of the glacial/interglacial episodes during ice ages is less certain, but it is probably related to axial tilt, axial precession, and orbital eccentricity.

Causes of the Interglacial Episodes

Several important problems arise in trying to study the earlier part of the Pleistocene epoch. The record is not as complete as that for the Wisconsin stage. Age determination is perhaps the most serious, because carbon-14 is only reliable for the last 50,000 years; in older materials the remaining amount of C-14 is so small that dating of earlier events becomes uncertain. Ages are based on estimates of rates of sedimentation and other methods that are even less reliable as ways of getting absolute ages. Thus, the relative ages of much of the data are known, but their absolute age (in years) is not.

A good correlation is suggested by some workers between the solar radiation curves based on the orbit and many events in the geological record of the pre-Wisconsin ice ages. Other workers have suggested that somewhat different processes may be involved. One of these is described in the following section.

Effect of an ice-free Arctic Ocean. The amount of ice in the Arctic Ocean influences the growth and decay of continental ice sheets in the Northern Hemisphere, according to a hypothesis advanced by Maurice Ewing and William Donn (*1*). They suggest that the numerous advances and retreats of ice sheets over the past million years have been caused in part by the availability of moisture in the Arctic. Most of the moisture that falls on the land areas around the Arctic is derived from the evaporation of seawater from the Arctic Ocean. When that ocean is largely frozen over, as it is now, very little moisture is available to fall as snow on adjacent lands. At the present time, northern

Canada and Siberia, areas that were covered by ice during the last glacial advance, are desertlike in their low precipitation, averaging only 10 to 25 cm (4 to 10 inches) per year. Yet, the presence of ice sheets in Greenland indicates that temperatures in these high latitudes are low enough to support extensive glaciers. They may now be absent in Siberia and North America only because there is not enough snow falling to replace the amount of loss during the summer and due to sublimation.

Looking farther back in geological time, the Arctic Ocean was ice-free before the Pleistocene epoch. At that time, sea level was higher and there was a much better interchange of water between the Arctic and the Pacific through the Bering Straits. It is probable that the opening between North America and Eurasia was much larger and that it became progressively narrower as a result of continental drift. This closing off of the Arctic Ocean restricted the interchange of waters with warmer water from farther south. This resulted in the cooling of the water in the Arctic Ocean.

Although the Arctic waters were cold, they continued to provide a source of moisture for snowfall around the Arctic as the Pleistocene epoch began. This snowfall led to the formation and growth of continental ice sheets. As snow accumulated and ice sheets grew, temperature dropped and so did sea level. This drop in sea level caused the Arctic waters to become increasingly isolated from the other oceans, cooling the Arctic even more and eventually leading to freezing of the surface of the Arctic. This cut off the most immediate source of moisture in the region, resulting in decreased snowfall and the end of growth of the ice sheets. With the snowfall cut drastically, the ice sheets began to waste away, leading to a retreat of glaciers.

As the huge glaciers melted, sea level rose and once again an interchange of water between the Pacific Ocean and the Arctic took place. Gradually, the Arctic thawed out and an interglacial period ensued. However, the presence of open waters in the Arctic set the stage for increased evaporation, more precipitation in the form of snow, and the start of a new glacial advance.

Although we have found reasonable answers to many of the questions about glaciers and glaciation, some of the most intriguing questions remain. Of these, what will happen in the future is foremost. This becomes especially interesting because we do not know how rapidly the climatic changes associated with glaciation occur. Much evidence from a variety of types of data suggests that climatic changes are often rapid. It is especially intriguing to note that we are living in one of the warmest periods in many thousands of years (Fig. 24.39), and that much colder periods have repeatedly followed such warm periods in the past. We know that the results of climatic changes—the changes in sea level as the ice–water balance varies, the changes in temperature, and the advance of ice sheets—are drastic. For this reason, we should take a special interest in the possible effects of human activities that have the potential for changing climate. In the following chapter, we will examine some of the direct effects of Pleistocene climates on sea level and the indirect changes associated with variations in sea level.

SUMMARY

The large masses of ice formed by recrystallization of snow that accumulates year after year are called either valley glaciers or ice sheets, according to their locations and sizes. Valley glaciers are found today in high mountains at all latitudes; ice sheets are confined to the polar regions. Most of the world's ice is in the Antarctic ice sheet, which is about 13 million square kilometers in area and as much as 4000 meters thick. Where it meets the sea, ice shelves and pack ice extend out into the water.

Glaciers grow and advance when the climate causes the amount of precipitation they receive to exceed the losses by melting, sublimation, or the breaking off of icebergs. Today the Arctic ice sheet seems to be very nearly at equilibrium, while that in the Antarctic is growing at the rate of more than one thousand cubic kilometers per year. The most recent major glacial advance reached its maximum extent about 17,000 to 21,000 years ago, when an ice sheet covered most of Canada and the northern third of the United States. Its terminus is marked by deposits of moraine extending from Cape Cod to the Missouri River Valley.

Snow that persists from one year to the next is gradually compacted and recrystallized under the weight of the later snows to form first firn and eventually the solid ice of the glacier's base. Slippage and shear within crystals and recrystallization of the ice produce a viscous plastic flow in the glacier. Valley glaciers flow downhill through existing stream beds. Ice sheets flow outward under the pressure of the overlying mass on the lowest layers. Because the upper layers of a glacier are under less pressure and often are also colder, they are more brittle then the base and may crack to form crevasses as the ice mass flows.

Because of friction with the valley floor and walls, a valley glacier flows most rapidly near the center of its surface. Along the sides and bottom it picks up and carries large and small blocks of bedrock, abrading both the valley and its rock load as the glacier moves forward. The striations formed on bedrock point the direction of flow both for valley glaciers and ice sheets.

Glaciers leave other evidence both in the erosion features they produce and in deposits left as they recede. Valley glaciers produce steep U-shaped valleys with cirques at their head. After the glacier melts, a small lake known as a tarn may form in the cirque. Truncated spurs and hanging valleys may be left where tributary glaciers enter the main valley. Sharp ridges called arêtes form between tributaries; and where several glaciers head around a single peak, they may leave a steep horn.

Glacial deposits, drift, are of two types: glaciofluvial deposits are water laid, and consequently show sorting; till material carried by the ice itself is a poorly sorted mixture of soil, stream deposits, and bedrock left under the glacier or dropped by melting at its terminus as terminal or recessional moraine. Lateral moraines are left along the sides of valley glaciers and consist of debris that falls from steepened valley walls; where two valley glaciers flow together, they may produce a medial moraine.

Streams of meltwater within the glacier form eskers, deposits of the coarser till. Where streams leave the glacier they may form alluvial fans. Small hills dropped when a glacier melts, leaving behind a pile of sediments that accumulated in a surface depression, are called kames.

Although ice ages are relatively rare and short-lived events, in the history of the earth, there have been several such periods of glaciation, and they played an important role in forming the topography of North America. Three major periods of glaciation have affected the Northern Hemisphere. The most recent, which began to recede only about 18,000 years ago, left behind the Great Lakes and drastically altered the drainage pattern for the continent.

The onset of an ice age would require average annual temperatures lower than those today. Possible causes for such cooling included reduced solar radiation, because of increased distance from the sun or increase in tilt of earth's axis, both known to occur periodically. The plate movement that shifted land masses to higher latitudes is believed also to be important in indicating a glacial age. The shorter interglacial periods within ice ages are believed by some to result from changes in sea level and their climatic effects in the Arctic, and from the increase in reflection (and decrease in absorption of solar radiation) that results from the growth of snow fields and ice shelves like that around Antarctica today.

KEY TERMS

ablation
ablation till
arête
cirque
cols
drift
drumlin
erratics
esker
firn
glacier
glaciofluvial deposit
ground moraine
hanging valley
horns
iceberg
ice sheet
kame
kame terrace
kettle
lateral moraine
medial moraine
moraine
nunatak
outwash
outwash sediment
pack ice
piedmont glacier
plucking
quarrying
recessional moraine
roche moutonnee
rock flour
snow field
snow line
sublimation
surges
tarn
terminal moraine
till
truncated spur
valley glacier

STUDY QUESTIONS

1. Summarize the data that must be known about an ice sheet to make an accurate determination of the ice budget of the sheet.
2. What erosional features are similar in valley glaciers and ice sheets? What distinguishes these two types of glaciers?
3. How might you distinguish two moraines of different ages that occur in the same area?
4. What is the effect beyond the ice terminus on streams during a glacial advance?
5. In what way is continental drift related to the cause of continental glaciation?
6. How might you expect to discover that the earth had ice ages before the Pleistocene epoch?
7. What types of effects are produced beyond the ice margin by the growth of ice sheets?
8. In what ways would today's high mountains differ in shape if there had been no glaciation?
9. In what ways are climatic conditions recorded in rocks?
10. Briefly outline the origin of the Great Lakes. How does the origin of the Great Salt Lake differ? Why is water in the Great Lakes fresh while that in the Great Salt Lake is salty?

REFERENCE

1. Maurice Ewing and W. L. Donn, "A theory of ice ages II," *Science* 127 (1958), *1159.*

SUGGESTED READINGS

Beaty, C. B., "The causes of glaciation," *American Scientist*, 66 (1978), *452–459.*

Calder, Nigel, "Head south with all deliberate speed—ice may return in a few thousand years," *Smithsonian,* 8:10 (1978), *32–40.*

Dyson, J. L., *The World of Ice.* New York: A. Knopf, 1979.

Embleton, C., and C. A. M. King, *Glacial Geomorphology.* London: Edward Arnold, 1975.

Flint, R. F., *Glacial and Quarternary Geology.* New York: Wiley, 1971.

Imbrie, J. and K. Imbrie, *Ice Ages: Solving the Mystery.* Short Hills, N.J.: Enslow Publishers, 1979.

Matsch, C. L., *North America and the Great Ice Age.* New York: McGraw-Hill, 1976.

FORM AND PROCESS IN COASTAL ZONES

Chapter 25

Continental margins mark a transition between the deep seas and the continents. The line of contact between the sea and the land, the **shoreline,** is a precise boundary (Fig. 25.1) separating the **coast,** a broad zone landward from the shoreline, from the submerged continental shelf. The forms and processes acting to shape this transition zone are as varied as any we have considered earlier. In fact, all of the processes described in earlier chapters affect the coast in one place or another. But, in addition, the coast is shaped by waves, currents, and marine organisms. The effects of these agents are closely tied to the line of contact between the land and the sea. This line of contact is not a stable one; its position is constantly changing. In the short term, it moves as waves and tides come and go. In the longer term, which is of special concern in geology, it has varied greatly as the level of the sea has changed.

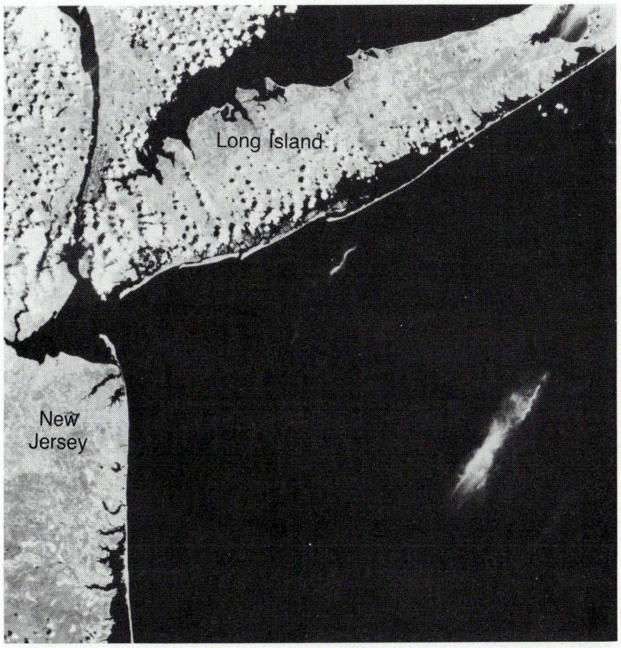

NASA

Figure 25.1
Sand beaches form an almost continuous strip along the coast from Long Island, shown here, to Texas. Water is black on this satellite image, and clouds cover part of the land. The origin of offshore beaches like the ones seen here is described in Section 25.3.

When sea level rises, waves, currents, and marine organisms—the agents that shape coasts at the shore—come in contact with land previously shaped by terrestrial agents. When sea level drops, marine agents begin to alter sea floor that was formerly too deep to be affected by their actions. The configuration of the land against which the sea rests at any time is clearly one of the most important factors determining the shape and character of the coastline. Thus, some coasts have shapes that clearly reflect the influence of the geomorphic agents working on the land—streams, glaciers, wind, and the like. Other coasts are dominated by the forces that build up the crust—volcanic activity and tectonic activity, such as folding or faulting. Still other coasts are characterized by features produced mainly in the marine environment—reefs, marine deltas, and beach deposits.

25.1 SEA LEVEL

The position of the contact between the sea and the land is almost continually changing as waves come ashore, as the tides come and go, and as the atmospheric pressure varies. These types of variations can be measured and averaged out over a long period of time to give a position for **mean sea level.** We often speak of mean sea level as if the surface of the ocean had the characteristics of a plane surface, with the same elevation everywhere. As we saw in Chapter 9, however, this is not the case. Ocean water is held to the earth by gravitational attraction. The force of this attraction varies from place to place as a result of differences in density of the crustal rocks and elevation of the crust. Water in the ocean basin is actually pulled toward the continents.

Sea level is important as a reference level because most topography, both submarine and continental, is surveyed and its elevation described in terms of elevations to mean sea level at certain selected points along the coast, and these points are related to one another and to other points across the surface of the earth.

Glacial Influences on Sea Level

The periodic level changes caused by the tides occur against a background of much longer-term movements of mean sea level. Abundant evidence indicates that during Pleistocene time the surface of the sea has varied as much as 100 meters above and below its present level. This is mainly the effect of the changing balance between the earth's ice and water as the climate shifts between glacial and interglacial periods. If all of the ice now on earth melted, sea level would rise about 100 meters.

It is hard to envision the impact such a change in sea level would have, not only on the overall shape of the continents but also on the character of the coasts. In North America, for example, the entire Florida peninsula would be submerged, as would most of the Atlantic and Gulf Coastal Plains; huge bays would form along the west coast. Coastal cities would include St. Louis, Missouri; Macon, Georgia; Austin, Texas; Harrisburg, Pennsylvania; Albany, New York; and Fresno, California. Seas did occupy these areas during the Pleistocene era at the height of some interglacial periods. But generally sea level was not so high.

Tectonic Effects

Sea level changes attributable to glaciation, which take place simultaneously throughout the world, have had the greatest influence on the shape of modern coasts. But the relative level of the sea and the land varies in response to other processes as well. The movement of sediment from land areas into the ocean basins must also cause some slight changes in the level of the seas. And in the history of the earth, sea level rose as new waters were added to the surface from within the interior. It has been difficult to evaluate the importance of such additions in the more recent parts of earth history, but most geologists do not think much new water is being added in this way.

Far more significant than either of these are movements of the lithosphere. According to plate tectonic theory, both the size and shape of the ocean basins are slowly but continually changing. The Atlantic appears to be increasing in size at the expense of the Pacific. The sea floor is affected by the upward movements in the vicinity of spreading ridges, by the downward movements at subduction zones, by the deformation associated with plate collision, and by the growth of new volcanic centers at hot spots. We do not yet have a precise estimate of the net effect of these changes on sea level. Most of them would be slow relative to fluctuations in the ice–water balance during the Pleistocene era but other tectonic movements are much more rapid. Many relatively localized sections of the coast and continental shelf are being uplifted or downwarped by crustal movements. Some reefs in Indonesia, formed during the Pleistocene era, now stand hundreds of meters out of water—despite the recent rise in sea level. Terraces formed at sea level are found elevated along much of the coast of the western United States (Fig. 25.2). At other localities, the movements are downward.

Changes in Sea Level

Evidence indicating the position of earlier stands of the sea takes a number of forms, some much more precise and easier to date than others. Some evidence is based on the form of the land; some is based on marine fossils. Beaches, notches cut in cliffs at sea level, and terraces cut by wave action at or just below mean sea level are among the forms most frequently used to indicate former sea levels. These are found both above and below modern sea level. Fossil evi-

Figure 25.2
Sea cliffs like this one in northern California are common features of the west coast of the United States. Note the rocks exposed in the water at the foot of the cliff. Wave action slowly wears the cliff back leaving a shelflike terrace at sea level. There is another wave-cut terrace, high above present sea level, at the top of this cliff.

dence includes the remains of trees that grow only in shallow water, shells that live attached to the sea floor only in water of a certain depth, and other types of remains of animals that live in restricted depths or only in salt water.

Several of these types of evidence usually occur together. A clear record of a former stand of sea level is likely if sea level held that position for a long time and then rose or dropped rapidly to another level. These forms are widely recognized both on the inner part of the modern continental shelves and on the emerged part of the continental margins.

Recent Sea Level Trends

Figure 25.3 plots the changes in sea level over the last 35,000 years of the earth's history. This curve is based on the types of evidence described in the last section. Many of the dates are based on carbon-14

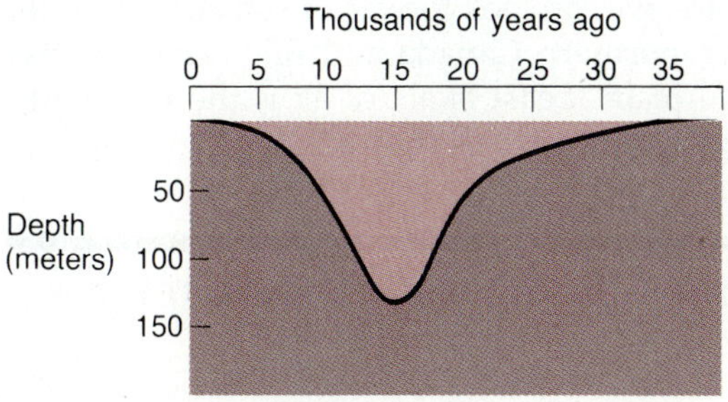

Figure 25.3
The sea along the Atlantic coast is now at the highest level it has reached in the past 35,000 years. Sea level was nearly 120 meters lower during the last glacial advance, which ended about 15,000 years ago. Level changes are not the same on all coasts, because the land itself is not stable everywhere. Because the Atlantic coast is relatively stable, this curve is probably a good representation of worldwide sea level changes.

dating of plant and animal remains found in shallow water. The changes indicated by this plot are related to the ice–water balance. Sea level started to drop about 35,000 years ago and reached a low point during the last maximum glacial advance, between 17,000 and 21,000 years ago. Thereafter, sea level rose rapidly at first and then slowly during the last 5000 years. Recent measurements indicate that sea level is again rising rapidly. Since 1940 the rate of rise has been nearly 0.1 inch per year, that is, three times the rate of rise between 1890 and 1940. The rise continues today, causing many problems along the heavily populated portion of the east coast where beaches have been intensively developed.

Future Sea Level Trends

The direction of future climatic trends is not yet clear. On these trends hinge the important question of whether sea level will continue to rise. This rise, in combination with the threat of hurricanes, puts the beaches along the eastern coast of North America and the developments on them in serious jeopardy. In the short term, this may be the most important coastal problem.

In the long term, in a geological time frame of hundreds or thousands of years, we can look forward to cultivating the broad expanse of emerging continental shelf off the southeastern states and accommodating the shift of population from the frigid north if we pass into a new glacial period. Or, if the earth continues to become warmer, as it has for the past 15,000 years, we can look forward to an agriculturally productive northern Canada and fine sailing across the coastal plain. Least likely of all is the possibility that things will remain as they are.

25.2 PROCESSES SHAPING THE SHORE

Tides

Tide is the name applied to the periodic rise and fall of sea level. Even in ancient times a relation was recognized between tides and the phases of the moon. As humans learned to measure the **period** (time between successive high water), **range** (variation in level of the water surface), and variability of tides, they came to recognize the complexity of the earth's tides. The two dominant factors are gravitational attraction between the water in the ocean and the moon and sun, and the effects of the shape of the ocean basins on the moving water. Yet even today tidal observations cannot be completely explained by tidal theory. An appreciation of the variability of tides will be gained by study of the charts (Fig. 25.4) that show fluctuations of sea level at various coastal points over a period of time. Each locality has distinctive tidal effects characterized by a certain combination of topography of the shore and the period, range, number of tides per day, and pattern of variation with time.

Astronomical theory of tides. The sun and moon are the only bodies that exert significant gravitational force on the surface water of the earth. Although it is 149 million kilometers away, the sun is such a massive body that it exerts nearly 167 times as much attractive force on earth as does the moon. Therefore, it seems unlikely that the moon should be more closely linked to tidal phenomena. Yet this is the case, and the reason is that the attraction of the moon for water on opposite sides of the earth differs to a much greater extent than does that of the sun. Water on the side toward the sun or moon is 2440 kilometers closer to it than water on the opposite side. The 2440 kilometers makes a significant difference in the distance to the moon (382,200 kilometers) but only a very little difference in the much greater distance to the sun.

To understand how the earth's tides are affected by the gravitational attraction of the moon and sun, assume that the earth is a rotating, revolving sphere covered to a uniform depth with water. Because the moon and the earth revolve about a common center of gravity, the forces of attraction between the two bodies are counterbalanced by an opposed force, called **centrifugal force,** directed away from the center of rotation of the earth–moon system, as shown in Fig. 25.5. Two tidal bulges occur on our model earth: one on the side toward the moon, where the moon's gravitational force of attraction is reinforced by centrifugal forces; and a second on the opposite side, where the centrifugal force exceeds the force of attraction. These two forces are close but not equal in strength, giving rise to a slightly higher bulge on the side toward the moon.

The effect of the sun is similar in character, but slightly less than half as strong as that of the moon. The sun and moon seldom lie on a straight line with the earth. Thus, the two effects are usually out of phase. It may be useful to envision two bulges of water on earth, the size and shape of which gradually change as the water responds to the slow changes in the position of the sun and moon relative to the earth.

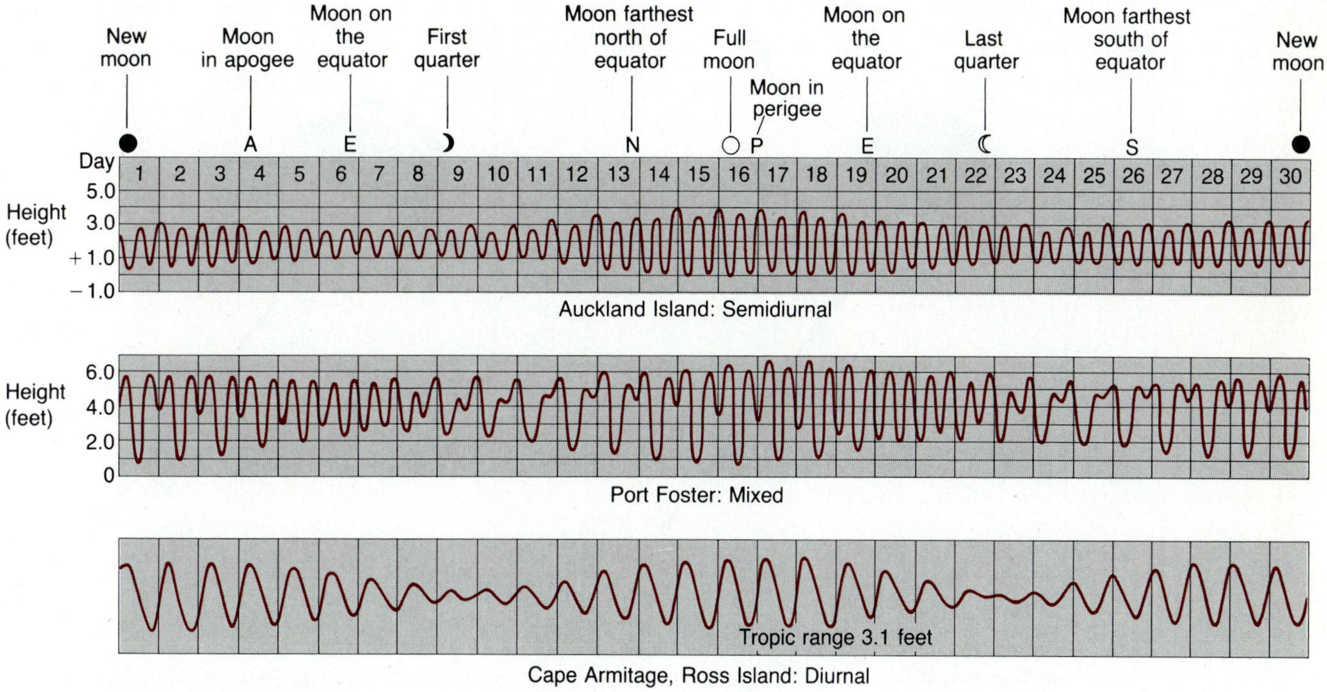

Figure 25.4
Tide records from these localities illustrate three very different types of tidal conditions. Note the relationships between the phases of the moon and the tidal ranges.

Two high tides and two low tides should occur each day as a result of the earth's rotation about its axis. But on most coasts, the interval between high tides is 12 hours and 25 minutes, the slight difference being due to the revolution of the moon around the earth. Because of this, the first high tide occurs about 50 minutes later each day. These tides are generated by the moon, but they do not normally coincide exactly with the passage of the moon overhead. Frictional drag usually causes the tide to arrive after the passage of the moon. The interval varies from place to place, but for a given locality, it is relatively constant.

When the sun and moon lie in line with the earth, as at the times of new and full moons (Fig. 25.6), the attraction of the two bodies is cumulative, and unusually high and low tides, called **spring tides** but having no relation to the season, result. When the sun and moon are located at right angles to the earth, the tide-producing forces of the sun and moon act at

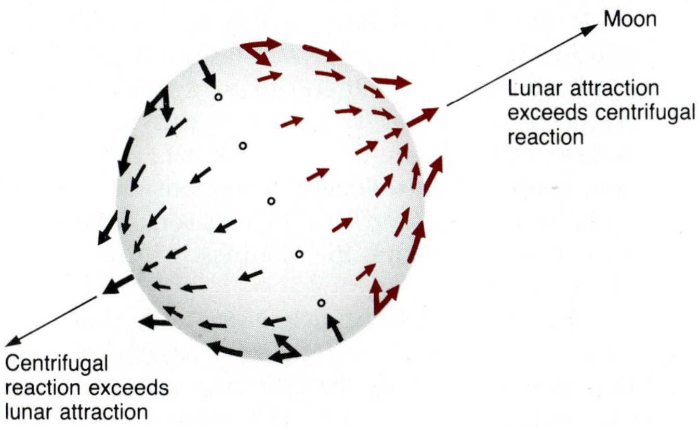

Figure 25.5
Two tidal bulges are formed in response to the forces acting on the earth. One bulge, on the side toward the moon, is pulled out by a combination of gravitational attraction between the water and the moon plus centrifugal force. The bulge on the far side of the earth is caused by centrifugal force.

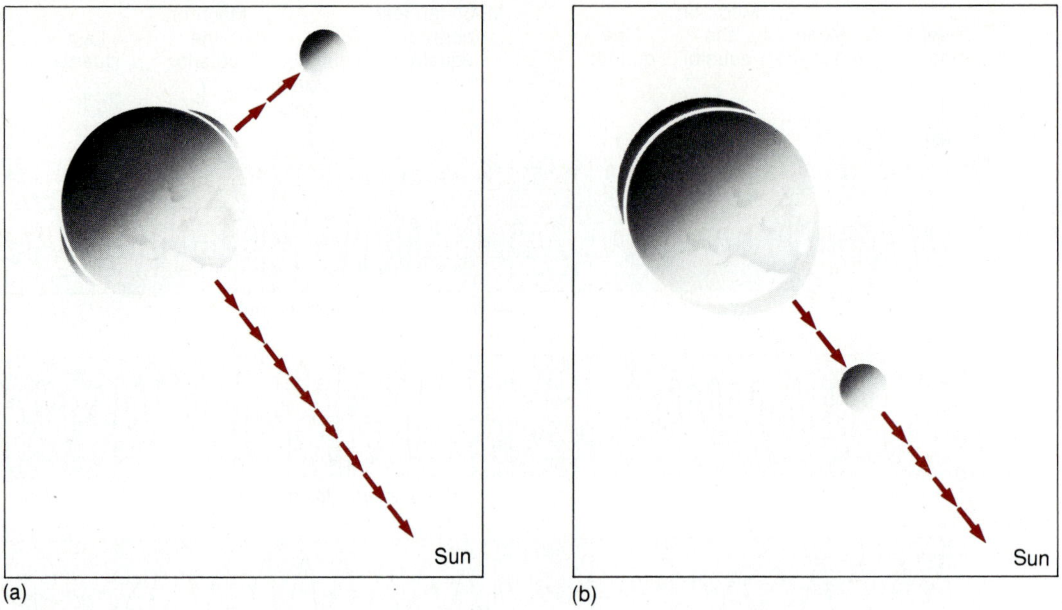

Figure 25.6
(a) The lowest tidal ranges (neap tide) in most places occur when the conditions shown, called quadrature, exist. (b) The highest tides, spring tides, occur when the earth, moon, and sun are aligned.

right angles, tending to cancel each other out and giving rise to unusually low tidal ranges, called **neap tides.** Both these conditions occur twice each month, confirming the close tie between tides and gravity effects.

Many tidal observations cannot be reconciled with the gravitational theory. On the Pacific coast of North America, for example, where the direction of the coast line is essentially north–south, high tides sweep from south to north. High tides normally come to Cape Flattery in northern Washington three hours later than they do to San Diego in southern California. Similar wavelike motions are found in many places and cannot be directly related to the movements of the earth, sun, and moon. These effects can be explained in terms of wave motion related to the size and shape of the ocean basins.

As the tidal bulges move across the ocean they encounter the coasts. A **standing wave,** not unlike the wave in a rocked cup of coffee, is created. Because the earth is rotating, these standing waves move and their crests sweep around a central point where tides, if observed at all, have small ranges. Thus, although tides are generated as described by the astronomical theory, many features of tides cannot be explained without taking into consideration this standing-wave motion.

Geological effects of tides. Most points on the world's sea coasts are affected by tidal movements. Typical tidal ranges along the coast of the United States are two to seven meters. At certain locations, as in the Bay of Fundy in Nova Scotia, even higher ranges are known. As the tide approaches the coast, it creates a horizontal current, called the **flood tide,** which moves into bays and rivers and, in some instances, continues upstream great distances. The Hudson River, for example, is affected by tides as far as 208 kilometers upstream at Troy, New York, where the tidal range of the river is 1.5 meters. In a number of rivers where tidal ranges are particularly high, the movement of the water takes the form of a turbulent wave, called a **tidal bore,** which moves with high velocity. The tidal bore formed on the Amazon River is 4.8 meters high and moves at a velocity of 22 kilometers per hour. The currents that result from these movements are effective agents of marine erosion, keeping fine sediment in suspension and preventing its deposition. For this reason, tidal bores help scour and shape the channels of the streams where they occur.

The force of tidal attraction on the hydrosphere is not confined to the surface of the oceans. Water at all depths experiences the pull, and it is likely that deep tidal currents result. The principal effects of these currents are thought to occur where the deep water off the continental slopes is pulled landward onto the continental shelf. These tidal currents may be responsible for the shifting and grading of the sea-floor sediments near margins of the shelves. The incoming tidal currents would bring with them little or no sediment at all; but as they reversed direction, they would carry away the finer sediments. This may explain why sediments at the outer edge of the continental shelf are often coarser than those closer to the coast.

Waves

The tides may be viewed as a special class of very long **waves.** However, most of the waves we see in the open ocean are small elongated humps or pointed peaks separated by troughs. Often waves of several different sizes are seen together, and in this respect natural waves in open water differ from the long, regular, simple waves produced in simple laboratory experiments. Figure 25.7 shows a profile model of a theoretically perfect wave. This model is approximated in nature by some waves, especially by wind-generated waves that have traveled out of the area where they formed—waves called **swell.**

Most waves in the ocean are of a type called **progressive waves.** The wave form moves across the surface of the sea, but the water does not travel with the wave form. If it did, seas would be impassable to most surface vessels. Instead, the water moves in a nearly circular path, like a bobbing cork, as waves pass. The size of the circular orbit decreases with depth until the motion is negligible at a depth equal to half the distance between consecutive crests, the wavelength (Fig. 25.8). Below this depth, wave movements have no effect either in the water itself or on the ocean bottom. Usually, a submarine submerged to this depth experiences no movement even if violent storm waves occur on the surface. However, wavelengths of some storm waves are great—several thousand meters—and these can be felt on the bottom at depths of hundreds of meters.

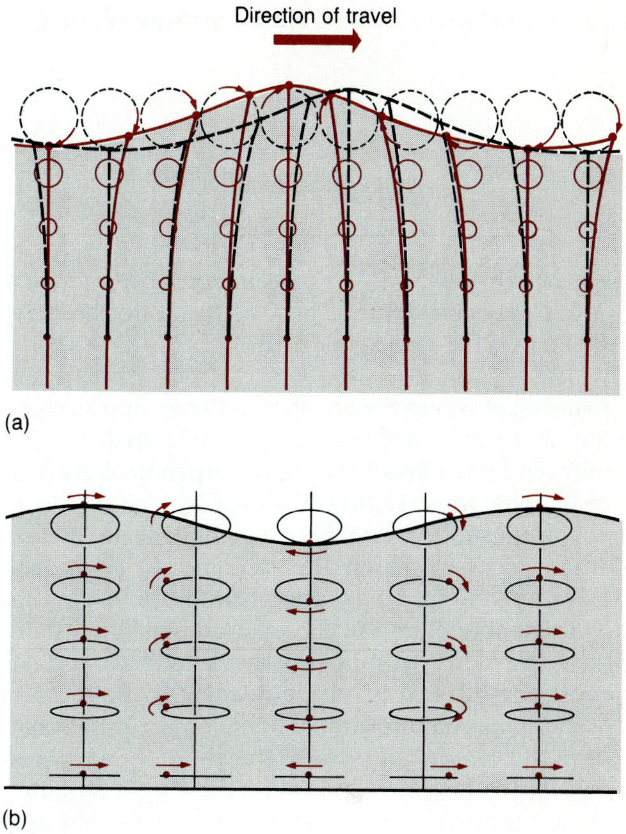

Figure 25.8
Water movement when water is shallow relative to the wave length: as progressive waves move into shallow water, (a) the circular motion of the water changes, and (b) the water moves in elliptical paths.

When progressive waves travel across water that is so deep that the wave motion does not cause a disturbance of the bottom, the velocity of movement of the wave form is given by the following relation between wave length and the time required for the

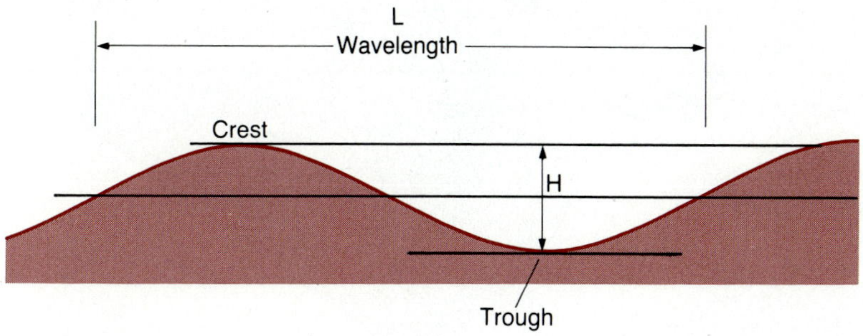

Figure 25.7
Theoretical wave form, seen in cross section. The movement of water beneath the wave surface dies out at a depth equal to about one-half the wave length. Particles of water in the wave, shown here for a progressive wave in deep water, move in slightly elliptical paths.

passage of two successive crests, the **wave period,** T:

$$v = \frac{gT}{2\pi}$$

But when the wave moves into water that is shallow relative to its wavelength, the velocity is

$$v = \sqrt{g \cdot d}$$

where v = velocity, g = acceleration due to gravity, and d = water depth. Thus, as the water becomes shallower, the velocity decreases.

Bending of waves toward shore. Waves approaching the shore obliquely are bent, or **refracted,** so that they tend to approach the shore perpendicularly (Fig. 25.9). The wave is slowed down in shallow water where subsurface water involved in the water motion is slowed by drag along the bottom. The part of the wave moving in deeper water continues to travel at its original rate until it, too, reaches shallow water.

Refraction is also of importance in that it focuses energy of the waves on the **headlands,** those parts of the shore that protrude into the ocean, rather than on bays or recessed parts of the shore, as shown in Fig. 25.10. Energy of a wave is evenly distributed along the wave front while it is out at sea, but as it comes into shore and is refracted, the part of the wave's energy entering a bay is spread out and distributed over a long shoreline. The part of its energy directed against the headlands is concentrated along a short portion of the shore. For this reason, headlands become eroded back by wave action much more rapidly than does the shoreline as a whole.

Despite the refraction, waves do not arrive exactly perpendicular to the shore. A component of motion in the water is directed along the shoreline and sets up a movement close to shore called a **longshore current** (Fig. 25.9), which is important in transporting sand and silt. Because of it, some sand on the beaches of the outer banks in North Carolina originated along the coast of Maine.

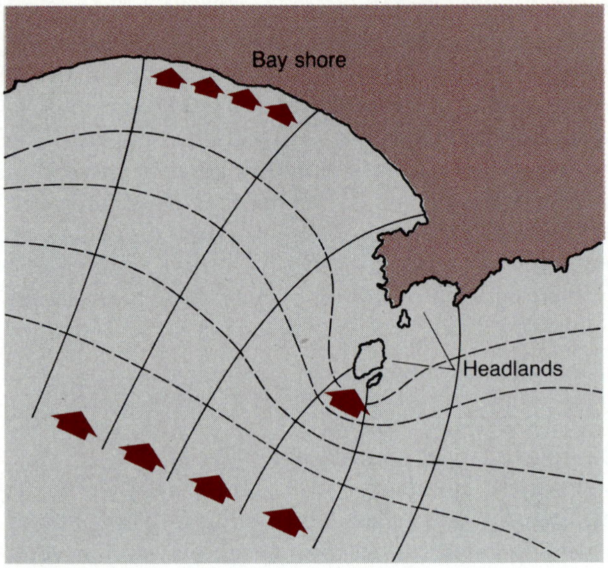

Figure 25.10
As waves approach the shore they slow down when they begin to reach shallow water. This slowing becomes noticeable where the water depth is about half the wavelength. The crest of the wave bends in response to water depth. As a result of this waves spread into bays and focus on headlands as shown here. The dashed lines are wave crests; the solid lines indicate the direction of movement of different parts of the wave.

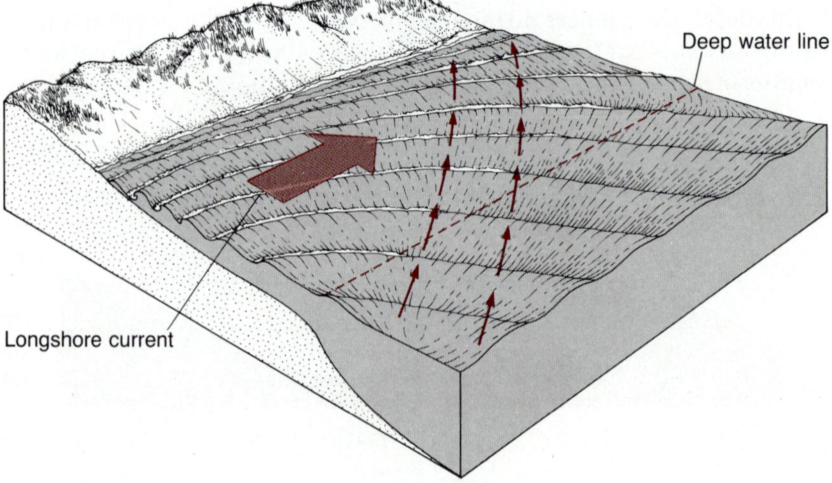

Figure 25.9
Because wave velocity decreases as waves enter shallow water, waves bend when they approach shore. The longshore current is set up in the zone of breakers as shown here.

Breakers. As a wave enters increasingly shallow water, water movement accompanying the wave motion is restricted. The circular paths of water movement are forced to flatten into flat ellipses, and movement is reduced by drag on the bottom. The restriction of freedom of water movement causes the wave height to increase and wave velocity to decrease. This change usually takes place rapidly when the water depth is between one-tenth and one-twentieth of the wavelength. The velocity of the top part of the wave begins to exceed the forward velocity of the wave as a whole; the top outruns the rest of the wave and spills over in front of it, forming a **breaker** (Fig. 25.11). The wave form is lost; water is thrown onto shore, often with great force, as a rushing, turbulent mass. Sediment on the sea bottom is disturbed where waves break. Out beyond the zone of breakers little more than a slight shifting of sediment back and forth by wave movement takes place; but within the zone of breakers, sediment is churned up and brought into suspension in the water. A slight depression, or trough, may form within this zone as sediment from the depression is piled up on the landward side to build a **bar** and eventually, in some instances, a **beach.**

Figure 25.11
The top of this wave near Cape Hatteras, is plunging forward onto a sandy beach covered by a thin layer of water that is flowing back toward the ocean and under the breaking wave.

Mechanics of wave erosion. Where the water is deep enough for the waves to reach shore before they break, rock and sediment exposed along the coast undergo an almost constant attack by waves. The effectiveness of wave action as a process of erosion is particularly evident when we can follow the destruction of the land. A new volcanic cinder cone was formed at the edge of the volcano Fayal in the Azores in 1957. This cone grew to be several hundred meters high but was largely destroyed by wave action within a few weeks after the volcanic eruption subsided. Wave erosion is especially rapid in such cases because the volcanic debris is loose, granular, and readily eroded and removed.

Much wave erosion is accomplished in the zone of breakers. Where waves break on a sandy beach, the water digs several centimeters into the bottom sediment as the water plunges forward in the breaker. Sand is churned up and can be easily moved in the currents. Water in the top part of the breaker is thrown forward toward the shore. Where the waves break against a cliff, erosion results from hydraulic pressure built up in cracks in the rock and by the impact of sediment picked up and hurled against the cliff.

The force of the water alone can dislodge fractured blocks. Water is forced into the fractures where it becomes highly compressed by the thrust of the water behind it. The amount of this pressure is greater than is generally recognized. A moderate-sized wave, 1.5 to 3 meters high, can exert pressures from 24,000 to 48,000 kilograms per square meter (one kilogram = 2.2 pounds) against the rocks exposed where it breaks. This is enough to extend or enlarge pre-existing fractures and to dislodge loose blocks.

Other processes of erosion act along with impact phenomena. Weathering effects, solution of rock and cementing materials, chemical and biological weathering, and the alternate wetting and drying caused by sea spray are all at least indirectly due to wave action, and all serve to loosen the rock and make it more open to erosion.

Carrying the sand, pebbles, or even blocks that form rocky shores, the waves rapidly break down loosely consolidated or weathered rocks, sometimes undermining more resistant units, but the impact of large boulders that may be moved in storm waves, such as those in Fig. 14.7*a*, is capable of breaking up even the most resistant granites and massive sedimentary rock units.

The effectiveness of wave erosion is determined by the size of the waves, where they break, and the type of rock or sediment present along the coast. Larger waves carry more energy, and are capable of picking up larger "tools" and causing more erosion. Because where a wave breaks depends on the water

(a)

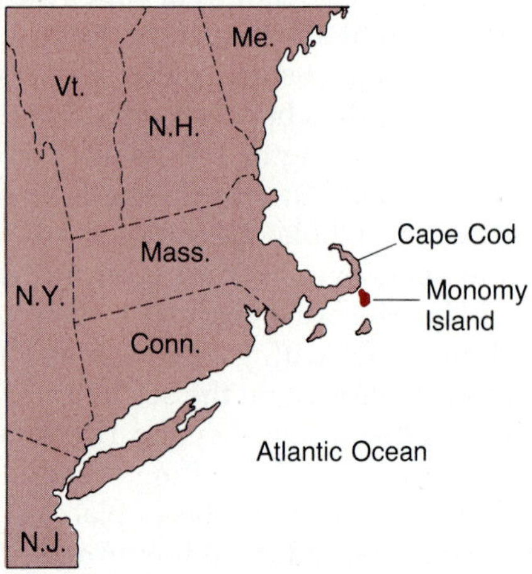

(b)

Figure 25.12
(a) This barrier beach at Monomy Island was cut by storm waves during a hurricane in 1978. All the beaches in the eastern United States are subject to this type of damage. (b) Locator map.

depth and the wavelength, erosion is usually most pronounced where deep water comes in close enough to the shore so that the largest waves actually break against or near exposed rock.

Features Formed by Wave Erosion

Erosion is the dominant process along many stretches of the North American coast, and the rate is not always slow. For example, the south shore of Nantucket Island retreats as much as 1.5 or 2 meters per year. A 1500-meter-wide breach (Fig. 25.12) opened in the beach on nearby Cape Cod during a single storm in February 1978. And wave action continues to undercut the sea cliffs along the west coast, causing landslides and slumps and damaging homes and highways, as shown in Fig. 25.13. These cliffs, and others throughout the world, were initially formed by wave action.

Cliffs and terraces. Cliffs and the flat surfaces at their bases formed by wave erosion are referred to as wave-cut cliffs or wave-cut terraces. Most sea cliffs are steep, and many are hundreds of meters high. Typically, a sudden break in slope occurs at the foot of the cliff, and a nearly flat bedrock surface, often

Figure 25.13
Storm waves have battered this section of the coast of the state of Washington, eroding and cutting the highway. Similar undercutting of heavily populated areas in southern California has caused extensive property damage where homes as well as highways have collapsed into the sea.

IGS Photo

Figure 25.14
A gently sloping wave-cut terrace at the base of the chalk cliffs of Dover, England, is exposed at low tide. Only a thin veneer of sand covers the bedrock.

covered with a sediment veneer, slopes gently seaward (Fig. 25.14). This platform or terrace may be partially covered by the rocks that have been dislodged from the cliff or it may be the site of a sand deposit part of the time. The sea uses material eroded from the cliff for tools to further undercut the cliff.

The effect of the breaking waves is negligible a few meters below the surface of the water. For this reason, the cliff stops abruptly just below water level, giving way to a flat wave-cut terrace. The width of the terrace is limited by the fact that the water is shallow. As the terrace becomes wider, more and more of the wave energy dissipates before the wave reaches the base of the cliff, and its ability to erode the cliff is reduced. Normally, the terrace width is narrow, but if sea level rises over a long period of time, the terrace can become much wider. The part of the terrace formed first gradually submerges and waves can continue to move across the terrace and break near the shoreline.

Stacks, arches, sea caves, notches. Irregularities of several types develop along the coast as a sea cliff is eroded back. A number of these and some types of beaches found on rugged coasts are depicted in Fig. 25.15. The development of such features may be due to differences in where the wave energy is concentrated, or to differences in the hardness of the rock, or to the rock structure. Often retreat of a wave-cut cliff leaves columns of rock standing isolated as islands just off the shore. These features frequently resemble smoke stacks sticking out of the water, and are called **stacks** (Fig. 25.16), although they are not necessarily either small or round.

Sea caves and **arches,** like that shown in Fig. 25.15, are likely to develop when the rocks of a cliff are stratified sediments of varying hardness. The weaker rock units are undercut rapidly, sometimes leaving the resistant capping layers as a natural bridge or arch over them. Such features, like the stacks, are transient. They will eventually disappear and new ones often develop farther inland at a later date.

Even if the cliff is composed of some massive, resistant rock type, wave action may produce a **notch**

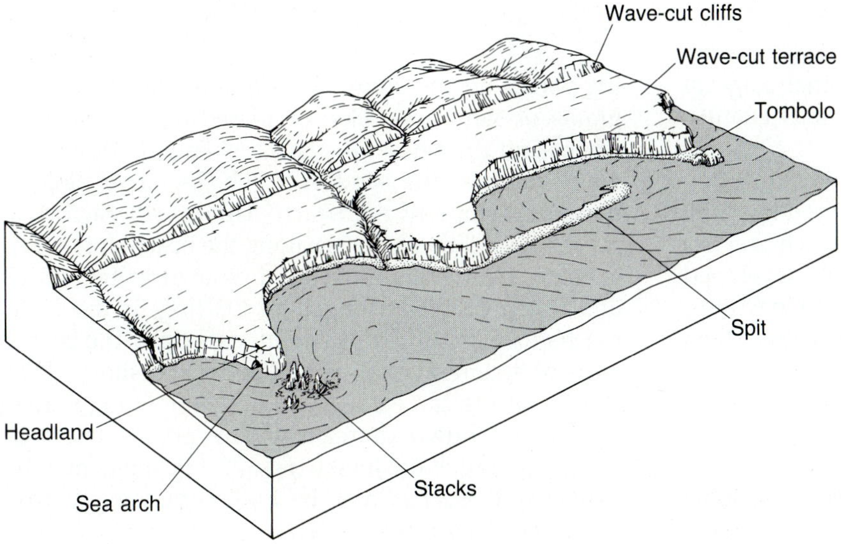

Figure 25.15
Typical features on rugged coasts, such as these on the west coast of North America, include wave-cut cliffs and terraces, sea arches, stacks, spits, bay head beaches, raised cliffs and terraces, and islands tied to land by beaches (tombolos).

(a) IGS Photo (c)

(b) (d) Oregon Department of Transportation

Figure 25.16
Features of shorelines with cliffs. (a) Steep cliff, a slight notch at high tide, and a beach of coarse sediment mark this view of the cliffs at Dover, England. (b) A notch and wave-cut terrace are prominent features of this New Zealand shore. (c) A wave-cut terrace elevated above present sea level and an arch at the shoreline mark this part of the California coast. (d) A sea stack rises off the beach at Cannon, Oregon.

near the base. Such a feature is the initial stage of undercutting; the rock above the notch may eventually break away and fall into the sea.

Beaches of sand, pebbles, or cobbles are often found along portions of coasts characterized by cliffs, stacks, and other wave erosion features. Such beaches are often temporary features, being well developed during summer but often completely absent in winter, when waves generated by winter storms destroy them and erode the cliffs.

Movement of Sediment in Shallow Water

Breaking waves and longshore currents cause most movement of sediment. Even after refraction, incoming waves hit the shore at a slight angle. Pebbles or grains of sand are rolled or tossed in the breaking waves and thrown on the beach obliquely. When the water returns, the sand and pebbles move directly down the slope of the beach. Thus, with each new wave, the sediment moves along the beach, following a zigzag path. This movement, shown in Fig. 25.17 and called **beach** or **longshore drifting,** is the main mechanism by which sediment, especially the coarse sizes of sediment, are moved along the shore. That beach drifting can extend beaches over a long period of time is shown dramatically by photos taken at Little Egg Bay, New Jersey (Fig. 25.18). Inlets may be closed and beaches may be built completely across bay mouths by this process.

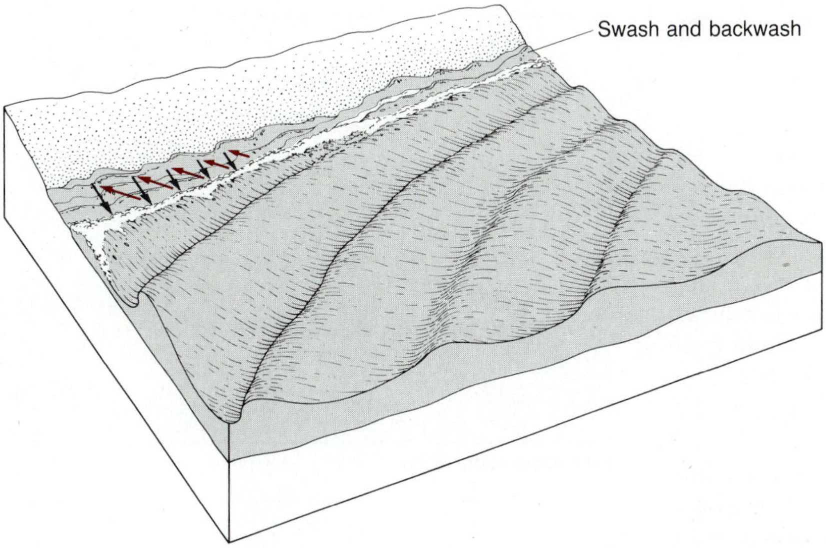

Figure 25.17
Sand moves along the shore in a zigzag path, thrown obliquely up on the beach as waves break. Then it moves back downslope in the backwash until it is caught in the next breaking wave.

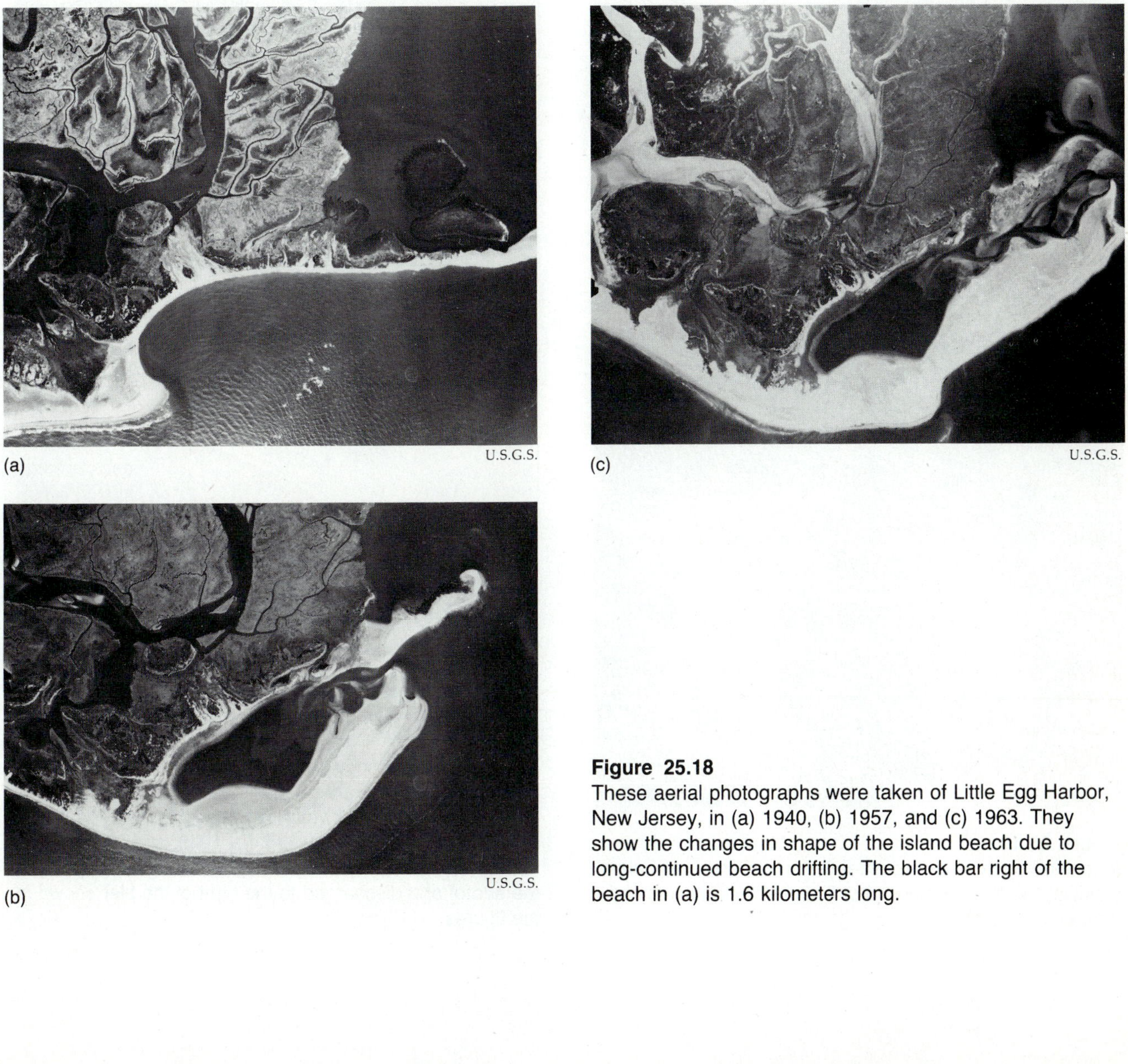

Figure 25.18
These aerial photographs were taken of Little Egg Harbor, New Jersey, in (a) 1940, (b) 1957, and (c) 1963. They show the changes in shape of the island beach due to long-continued beach drifting. The black bar right of the beach in (a) is 1.6 kilometers long.

25.3 MARINE DEPOSITS

Beaches

Beaches are transient features. The sandy beach that appears permanent in the summertime may be reduced to a narrow strip covered by coarse pebbles in the winter. In some places all sediment is removed, exposing the underlying bedrock. Most beaches are made up of sand-sized materials that are easily moved by waves and can be transported in moderate currents. Even pebble or cobble beaches, such as the one shown in Fig. 25.19, can be washed away by storm waves.

But storms are not the only mechanism responsible for the movement and destruction of beaches. Any upset in the supply of sand to the beach or in the currents along the beach influences beach growth or depletion. Attempts are often made to control erosion and deposition along the shore by building jetties, piers, or breakwaters. But care must be taken in planning such obstructions, because jetties occasionally introduce side effects that cause more damage than they prevent. Building such obstructions may protect and build up one segment of the shoreline while cutting off the normal supply of sand and subjecting another to increased erosion (Fig. 25.20).

Figure 25.19
A cobble beach on the coast of Brittany, France. At Calais, a similar beach is composed of flint pebbles and is mined for ball mills (grinding).

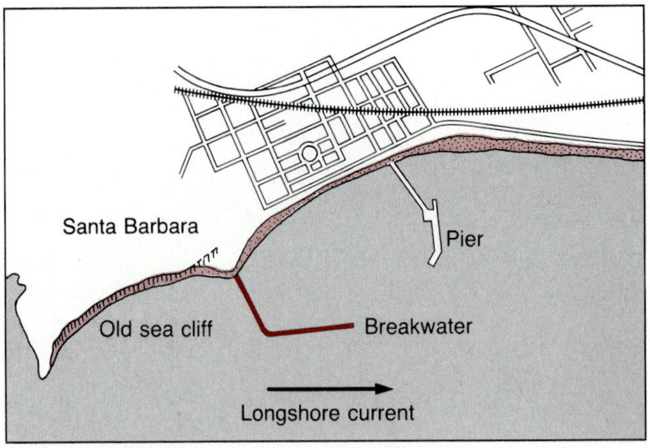

(a) Construction 1928

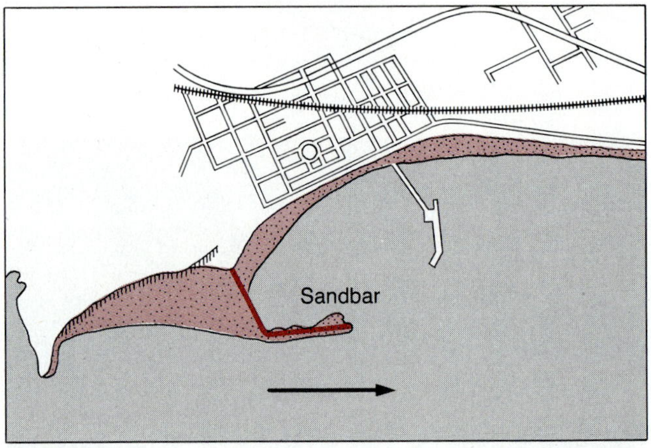

(b) 1948

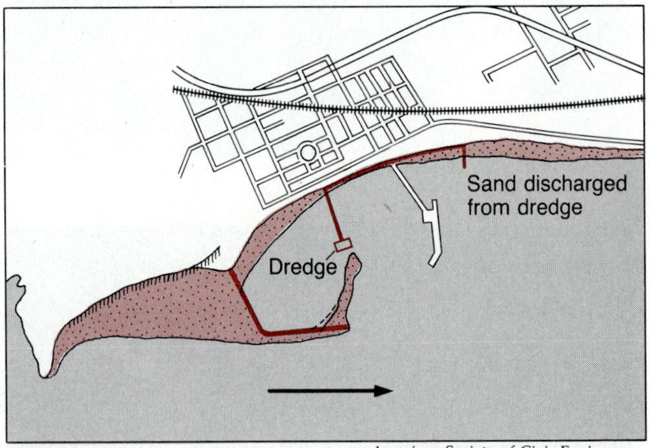

(c) 1965

American Society of Civic Engineers

Figure 25.20
The sketch maps show the changes that took place along the coast at Santa Barbara, California, following construction of a breakwater in 1928. The shoreline is shown as it looked in (a) 1928, (b) 1948, and (c) 1965. The harbor is now kept open by dredging; sand is dredged at the end of and discharged to the right of the pier shown in the figures.

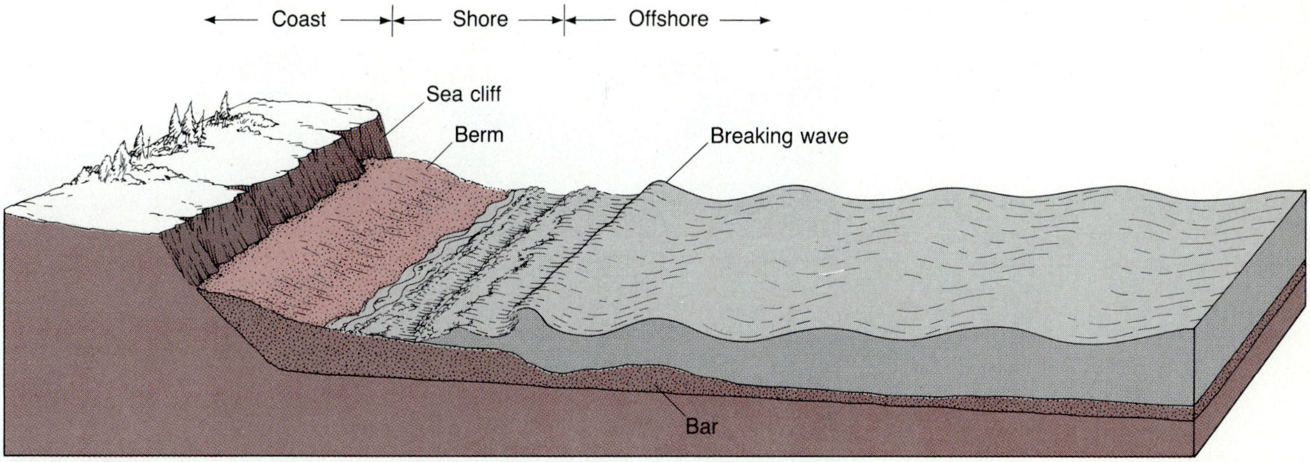

Figure 25.21
The contact between water and land is called the shoreline. This line shifts back and forth across the beach as tides and waves come and go. A ridge of sand, the berm, is built when the sea is high. Often a trough forms in the zone of breakers and a bar may be found outside this trough.

Where the water depth increases rapidly off the beach, an obstruction may force the sand involved in beach drift to move out into deep water as it moves along the obstruction. In some places the sand then slips down into deeper water and is not returned to the beach. Off the California coast, sand is lost in this way when it slips or flows into deep basins close to shore. This is less common on the east coast, where water depth increases slowly and beaches are more or less continuous.

The near-shore region of most beaches can be divided into three parts, as shown in Fig. 25.21. **Offshore** is the portion from the position of low tide seaward. The **foreshore** is the zone between low tide and the point where the beach becomes either horizontal or slopes landward. The **backshore** is that part of the shore that is level or sloping landward.

The profile of a beach changes from hour to hour and particularly from one season to another. The short-term variations are caused by changes in winds and tides that are rapidly reflected in wave size and in the amount of beach drifting. Seasonal variations arise from differences in the amount of storm activity and the balance between sediment supplied to the shore and rates of beach drift.

Seasonal variations in the profile of a beach at Carmel, California, are shown in Fig. 25.22. During the summer, sand is deposited on the beach, making it wider. Offshore sand fills in irregularities, producing a smooth bottom. The much larger waves that strike the beach in the winter months churn up the sand, which then moves into deeper water where bars are built.

Formation of beaches. Beaches are characteristic of shores dominated by deposition, but they occur as patches along even the most rugged shores. Of prime importance is a source of sediment supply. Most of the sediment comes either from streams entering the ocean from the continent or from wave erosion and breaking down of rock units exposed along the shore. In addition, during storms, sediment from the sea floor may be brought in from deeper waters offshore, but ordinarily the movement of this material is minor. Locally, the sediment may come from volcanoes or melting glaciers, or be blown in by the wind.

If beaches are to persist, the rate of supply of sediment to the shore must equal or exceed the rate at which it is removed either by storm waves or by longshore currents. If sediment is abundant, the beach may extend for hundreds or even thousands of kilometers along the shore. If the amount of sediment is limited, or if longshore drift is rapid, beaches develop only along protected stretches of the shore.

Along irregular coast lines, cliffs and stacks are often present where the land protrudes out into the sea, at headlands, but beaches commonly occur at the head of bays or inlets. There the **bay-head** beaches, as they are called, are relatively protected from wave action and from the longshore current. Headlands projecting out on either side of the bay prevent the rapid removal of beach materials by beach drifting and the longshore current (Fig. 25.23). And protection from wave action in the bays comes from the fact that the wave energy is spread out over a long stretch of beach due to wave refraction along the sides of the bay (Fig. 25.10).

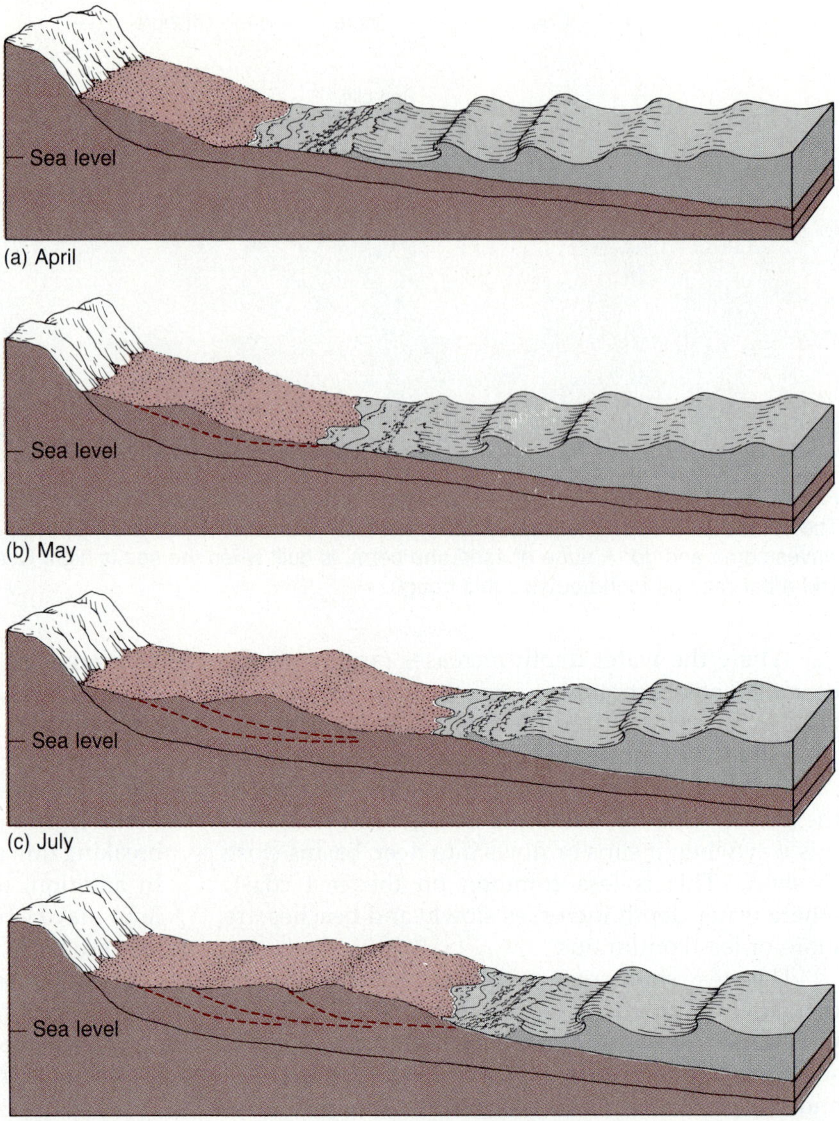

Figure 25.22
During summer months, breaking waves gradually shift sand from offshore onto the beach. By the end of the summer the beach is much wider than it is in spring. During winter months the sand is shifted into deeper water by heavy winter storm waves.

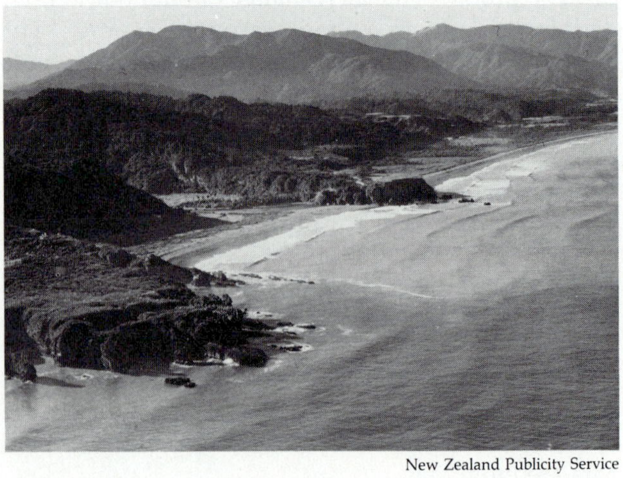

New Zealand Publicity Service

Figure 25.23
Headlands of bedrock projecting into the ocean separate protected bayhead beaches here on the coast of New Zealand.

Figure 25.24
A beach has been built almost completely across the mouth of this bay on the west coast of North America. Breaking waves can be seen on the far side of the beach.

The headlands are the sites of most intense wave erosion, but when the supply of sediment is great, beaches may develop even at the foot of sea cliffs at the headlands. Such beaches are destroyed during storms, and they come and go as the supply of sand changes. Beaches may even be built across the mouth of the bays. At first, a submerged bar is built out from the tip of the headland beach in the direction of beach drifting. With the addition of more sand, this may emerge and form a projection from the land, as we saw in the case of Little Egg Harbor. The ends of these projections, called **spits,** generally curve back into the bay (Fig. 25.18). If the supply of sediment is adequate, the spit may grow until it extends completely across the bay, forming a **bay-mouth** bar (Fig. 25.24). Such beaches are unlikely to cross large bodies of water, because rivers empty into most of them, and this water must find an outlet into the sea even though the outlet may be no more than a narrow channel.

Spits also form along coasts with low relief, like the one illustrated in Fig. 25.25. Here a spit has developed in the foreground. That beaches on this part of the coast have been building out into the sea is indicated by the pattern formed by older deposits.

Sometimes a beach will be built between islands and the mainland. This is called a **tombolo** (Fig. 25.15). The islands are generally stacks. When the rate of wave erosion is reduced or the supply of sand increased, the small strait between the island and the land becomes silted in. The island shields the area immediately behind it from wave action, and as more sand enters the zone, the submerged bar rises above sea level.

Miles Hayes

Figure 25.25
The buildup of deposits along the coast is dramatically shown in this aerial view of a spit at low tide near the entrance to the Strait of Magellan, Chile. The beach here is composed of gravel. This part of the spit is about 0.5 km wide.

Case history: The barrier beaches of the Atlantic and Gulf coasts. The sand beaches that mark the shore of North America from Cape Cod to Mexico comprise one of the longest essentially continuous beaches in the world (Fig. 25.26). The continuity is broken only where streams, such as the Hudson and Delaware, enter the ocean; where the great mangrove swamp of the Everglades lies along the southwestern tip of Florida; and where the deltas of the Mississippi are built into the Gulf. Everywhere else the beach is present. Along some sections, the beach lies along the shore of the mainland. In other sections, the beach stretches as a long, thin ribbon of sand separated from the mainland by a lagoon. These are called **barrier beaches.**

These beaches lie along the shore of the geological province known as the Coastal Plain (described in Section 16.3). Except off the southeast coast of Florida where the Gulf Stream has removed sediment close to the coast, the continental shelf is nearly flat and gently sloping seaward. The supply of sediment that nourishes these beaches is derived in part from the rivers flowing into the Atlantic and in part from the sediment on the continental shelf itself. Part of the sand in the northern section is derived from glacial outwash deposits, such as those exposed on Cape Cod and Long Island, which lie in shallow water on the shelf. Farther south, water from the Hudson, the Delaware, Chesapeake Bay, and small rivers along the coast supply sand produced by weathering and erosion of the Appalachians and the coastal plain sediments.

A marked change in the beach sand occurs toward the south—the sand is composed increasingly of shell fragments and less and less of quartz. This, of course, signifies the change in the source of the sand. Both quartz and shells are important in the warm, highly productive waters of the Gulf.

The shoreline beaches are easily explained in terms of the accumulation of sand carried by beach drifting and in the longshore currents, but the origin of the barrier beaches is not so simple. Some barriers may originate by a mechanism similar to that by

NASA

Figure 25.26
Beaches along the east coast of the United States are supplied with sand by streams flowing off the continent. These sediments are derived from glacial deposits north of New Jersey. The sands come from older sediments in the Coastal Plain and in the Appalachians, in the middle Atlantic states. In Florida the sand is composed almost entirely of organic materials.

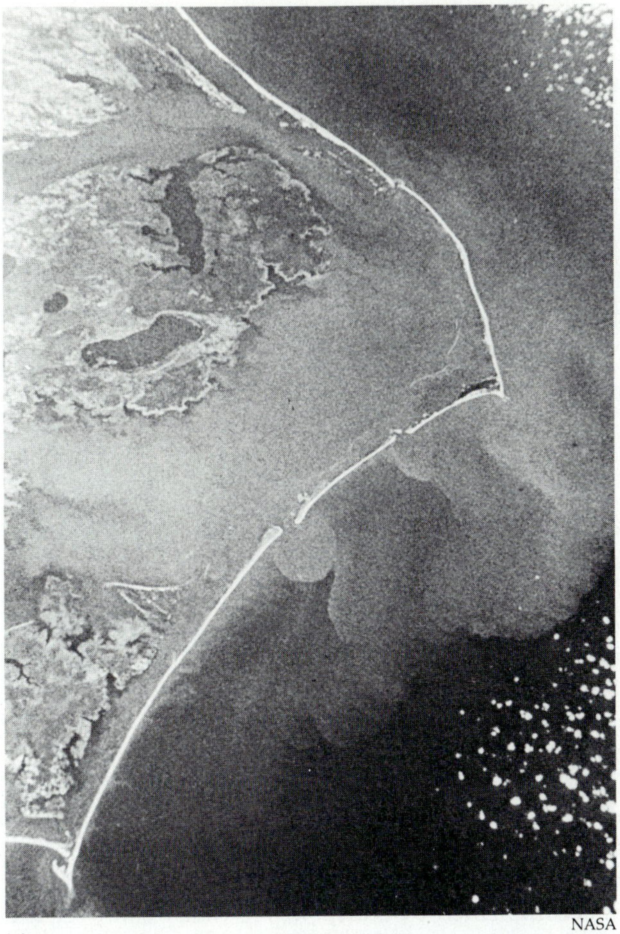

(a)

which bay mouth beaches and bars are formed. But the long barriers, some of which stand several kilometers offshore, such as the Outer Banks of North Carolina (Fig. 25.27), are better explained by other mechanisms.

Most of the barrier beaches along the Gulf and Atlantic Coasts have a similar setting. They are composed almost entirely of sand, and are capped by sand dunes, the highest of which was used by the Wright Brothers for their initial flight at Kitty Hawk, North Carolina. The beach area, rarely more than a kilometer wide, slopes down on the landward side into a shallow lagoon or bay. These bays vary greatly in width from one place to another, but because they are all situated on the edge of the Coastal Plain, the landward side is marked by low slopes and relief. The sedimentary rock layers exposed in the Coastal Plain extend under the bays, the barrier beaches, and the continental shelf.

The beach is subjected to the constant attack of waves. Sand is often added to the beach during the summer and fall and removed into deeper water during the winter and spring, when storm waves are more common. The more protected landward side of the beach is covered by grasses and some trees. The bay is a totally different type of environment. The water there is generally quiet although it is subjected to the daily rise and fall of the tides. Strong tidal currents carry sand in and out through inlets that cross the barrier beaches. The bay is also the site of

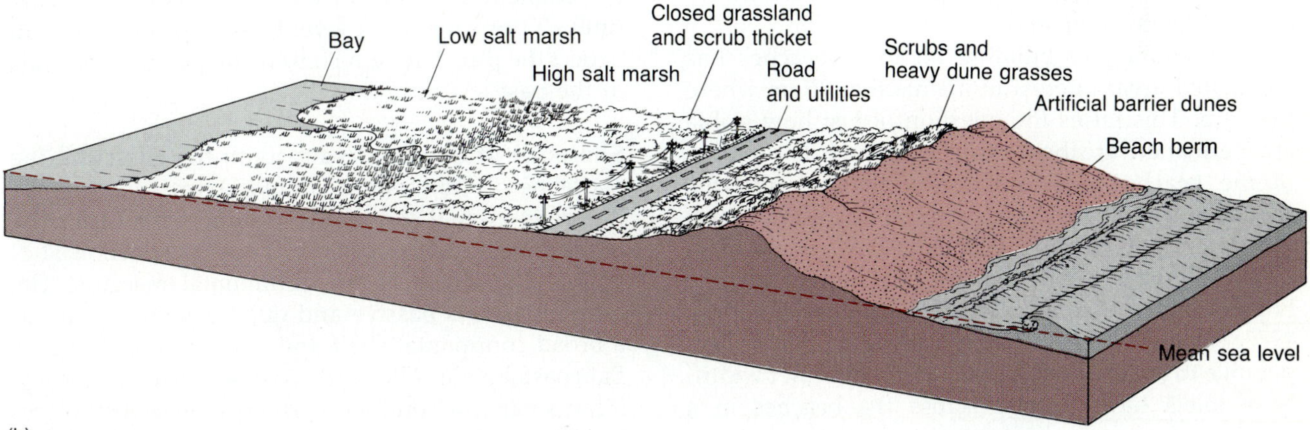

(b)

Figure 25.27
(a) This satellite image shows the Outer Banks barrier beach at Cape Hatteras, North Carolina. The beaches shown here separate the Atlantic, at right, from the shallow bay area at left. The distance across the area shown is about 100 km.
(b) Cross section of a typical barrier beach of the type found on the east coast of the United States. The sand dunes are high in many places as a result of efforts to stabilize them by use of fences and plantings.

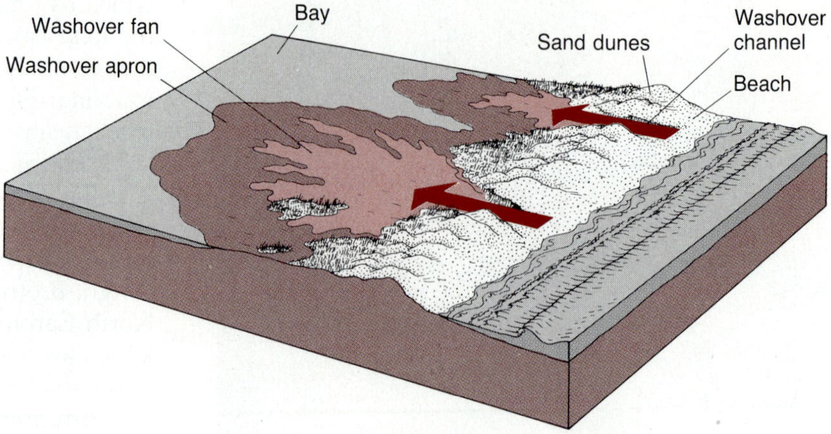

Figure 25.28
Sand from barrier beaches is often washed over into the bays that separate them from land during storms. Sometimes these branches become inlets that remain for many years. Photographs of such inlets are shown in Figs. 25.12 and 25.29.

deposition of rich organic muds that come into the bay from the mainland. The environment is one of the most productive of all marine environments, and a rich fauna is present. Parts of the bay are tidal flats; parts are covered by open water; grasses cover large areas; and tree-covered swamps are present in other places.

Most students of the barrier beaches recognize the significant influence of the recent changes of sea level in their formation. They are viewed as having formed primarily as a result of the gradual buildup of sand from the sea floor, first as a bar, and later as an island. This type of construction is initially due to the effects of wave action, and later due to the additional effects of beach drifting. The process begins with breakers churning up sediment that settles out as a bar. This deposit is built up by storm waves until a beach begins to emerge.

The amount of building on these beaches has stimulated great interest in maintaining them where they are. This raises the question of just how stable and permanent are these stringers of sand. The history of the beaches does not leave much doubt. Six thousand people drowned when a hurricane struck the beach at Galveston Island in 1900. A storm in 1962 did half a billion dollars of damage between Long Island and Cape Lookout, North Carolina. Since 1977, Miami has spent more than six million dollars per mile to restore their beaches to their former width. New inlets have formed across the beaches in a number of places where hurricanes crossed them. And a constant struggle goes on at many places to maintain the beach and save the homes, hotels, and towns built at the water's edge.

Not only are the barrier beaches subject to devastation during hurricanes and as a result of inadequate supply of sediment, but most if not all of them are slowly migrating toward the bay. This became clear when it was recognized that many of the shells eroding out of the sand on the beach side were of invertebrate animals that lived in the bays. In addition, layers of peat and even forests that once stood on the bay side are being exposed on the beach side of the islands.

Landward migration of the islands is the result of several processes. Perhaps the most important is the removal of sand from the beach and deposition of it in the bay during storms. Much of the transfer is accomplished by storm waves that wash over the beach as shown in Fig. 25.28 and by tidal currents acting through inlets (Fig. 25.29).

Unless there is a sudden increase in the amount of sediment available to the coast, which is highly unlikely, or unless sea level begins to drop, we can expect the processes which have shaped these islands in the past to continue.

Shoreline of the west coast. Both the coast and the shoreline of the western United States are strikingly different from those of the Atlantic and Gulf. The two differ in their tectonic settings (see Section 11.1) and in the shape of the continental margins. The eastern coast is passive and quiet tectonically; it has a broad continental shelf and is flanked inland by a flat coastal plain. The west coast is active tectonically; it has a narrrow continental shelf cut in several places by deep canyons that head near shore, and recently uplifted mountains and active volcanoes lie inland. Tectonic setting and activity are probably the most important factors causing the differences in the coast along the two sides of North America. Most of the east coast has been progressively submerged during

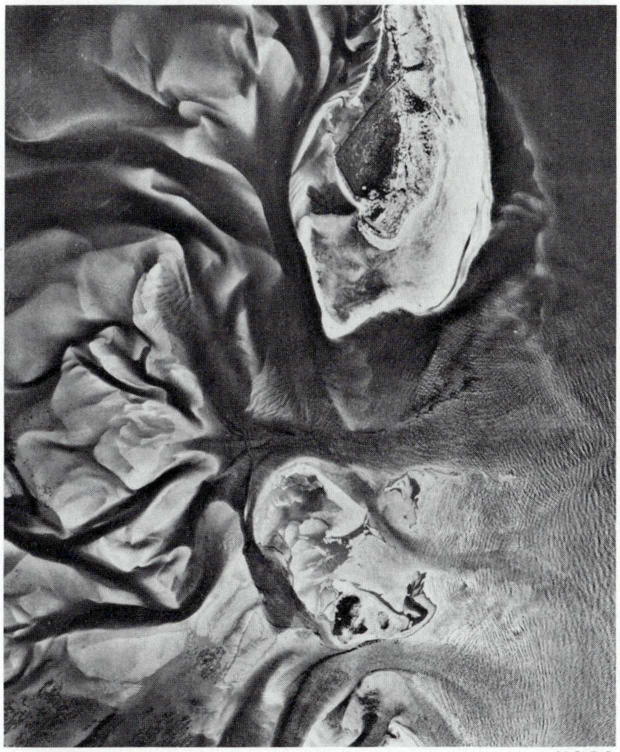

U.S.G.S.

Figure 25.29
This vertical photograph shows the western edge of Hatteras inlet, North Carolina. This is one of the breaks in the long series of islands and beaches known as the Outer Banks. Open ocean is at the bottom of the photograph. Bars in the inlet cause the unusual pattern in the lower part of the photograph. The scale of the photograph is about 1 : 20,000.

the rise in sea level that has accompanied the melting of glaciers during the last 15,000 years. But during this time, crustal movements on the west coast have continued.

Relief is the most obvious difference in the landforms along these two coasts. Sea cliffs, stacks, and wave-cut terraces are common along the west coast; even the beaches are juxtaposed against high cliffs. On the east coast the land inland from the beach is a lagoon, a marsh, or at least nearly flat from Cape Cod south. Thus, the beaches in the east can transgress the land as sea level rises. Most of those in the west cannot.

Sand for west coast beaches is carried to the shore by rivers (Fig. 25.30), and it is spread along the coast by longshore drift, forming beaches, spits, and bars of the types described earlier. But in a number of places (see Fig. 25.31), the beaches disappear, and rocky headlands line the coast. Mapping of the sea floor and exploration by divers has shown that the sand moving in the longshore current is diverted from the shore into submarine canyons. Sand cascades down the steep submarine slopes, as shown in Fig. 25.32, and disappears into deep submarine basins where it contributes to the construction of large fan-shaped deposits—submarine counterparts of the alluvial fans described in Chapter 20.

Tidal flats. Low-lying land that is near sea level, yet protected from wave erosion and strong currents, may develop tidal flats like those in Fig. 25.33. They are generally located near an abundant source of sediment, such as the mouth of a large river. Because many invertebrate animals such as gastropods, worms, and pelecypods inhabit the tidal-flat environment, their remains are an important part of the sediment. Fine silt, clay, and some sand mix with varying amounts of shell fragments, sea urchin spines, and fine plant matter to produce the soft, water-soaked mud of the tidal flats. In some areas, excrement of worms and other mud-ingesting invertebrates makes up a significant part of the sediment.

These deposits are usually stratified and cross-bedded, reflecting reworking of the sediment by shifting currents. Sources of the muds that make up the flats vary from place to place. Some tidal flats originate almost entirely from deposition of suspended matter in streams, but others appear to be swamp or glacial deposits reworked by the forces acting in the tidal zone. The amount of sediment and the rates

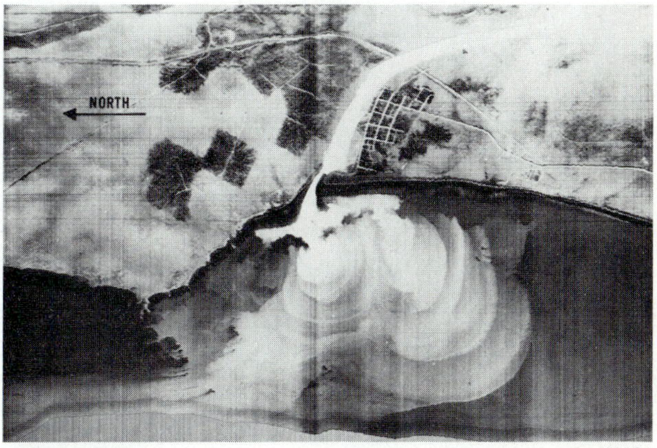

NASA

Figure 25.30
Plumes of sediment and warm water are clearly visible in the ocean off the mouth of the Quinault River in Washington. The film used to obtain this image was sensitive to infrared radiation.

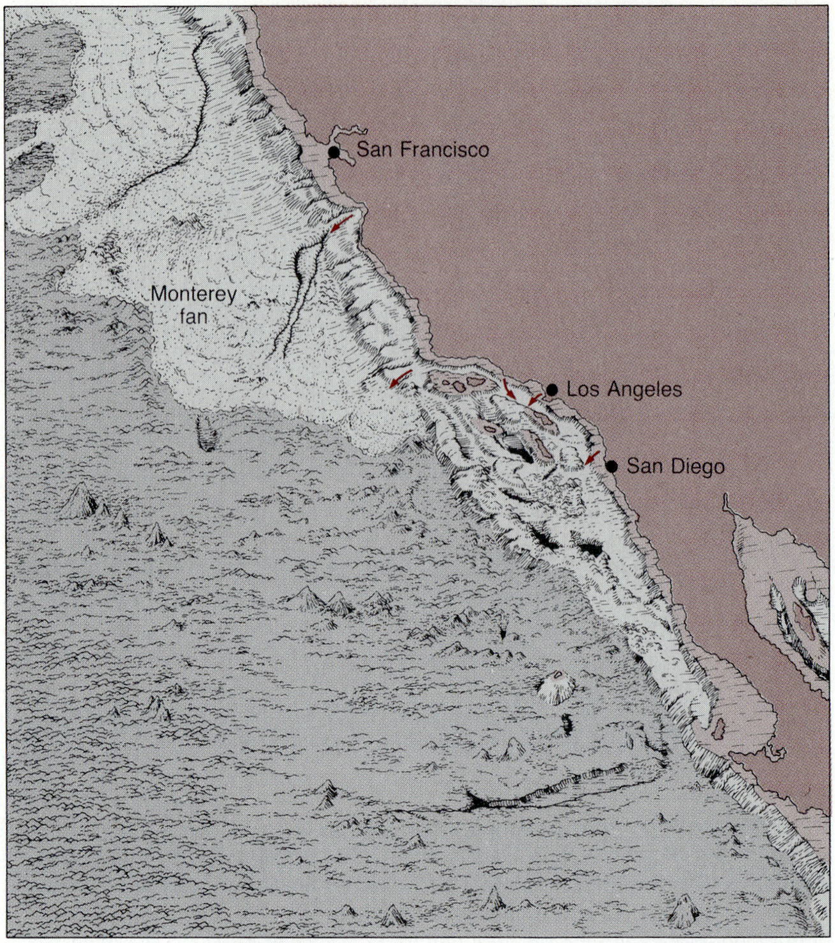

Figure 25.31
The continental shelf along the west coast of North America is narrow. Sand supplied to the beaches from streams draining the Coast Ranges moves southward along the coast. In several places submarine canyons intercept the longshore drift of sand and divert it into deeper water.

Figure 25.32
Sand diverted into a submarine canyon from the beach forms a sandfall here at Cape San Lucas Canyon in Baja California. The fall is about ten meters high.

Scripps

Embassy of France

Figure 25.33
This view is across the tidal flats around Mont St. Michel on the coast of Brittany, France. At high tide, the ancient abbey is surrounded by water except for a causeway. At low tide, it is possible to wade across the mud flats from the mainland.

of sedimentation on modern tidal flats are surprising. As much as 7.3 meters of sediments were deposited within three years in the harbor entrance to Wilhelmshaven, Germany. The rate is sufficient to make continuous dredging necessary to keep the channel open for shipping.

Where storm waves have built up tidal flats close to sea level or slightly above it, the deposits may be drained and used for farm land. Extensive dikes and elaborate drainage facilities are used in Holland for this purpose.

Marine Swamp Environment

The names **swamp, bog,** and **morass** are applied to low, spongy land generally saturated with moisture. Abundant rainfall or some other water supply and in impermeable substratum that prevents drainage are necessary for the continuance of a swamp. This type of land is not confined to areas near sea level or even to flat land, although it is much more likely to form and less likely to be drained in one of these locations. Some estimates put the amount of land covered by swamps at more than 2.6 million square kilometers. Not only are they important environments today, but there is a rich and extensive record of this type of environment in the geologic past. At various times swamps have covered much larger areas than they do today. Late in the Paleozoic era, conditions favorable for development of swamps became so extensive throughout the world that the period of time from about 350 to 270 million years ago is named for the carbonaceous sediments formed from the swamp deposits. This time is called the Carboniferous period.

Marine swamps are located along coasts where both brackish and fresh water are present. This particular environment is commonly preserved in the rock record because relatively slight elevations of sea level submerge the swamps. Where this happens, the swamp remains may be preserved under marine deposits that protect them from rapid decay.

Marine swamps are common along coasts of the eastern and southern United States. Many have formed behind offshore bars where lagoons have filled with sediment from the continent and with plant remains. Plant life is most abundant where the water is quiet. Thus, the most favorable areas for swamp conditions are regions without strong tidal currents and where tidal range is low.

Quite a few swamps are at least partially, and some completely, covered by sea water during high tide. Within a marine swamp, different plants from those found in fresh-water swamps prosper. In fact, most fresh-water plants are killed by salt water. In tropical climates, for example, marine swamps contain mangrove or cypress forests, such as those in southern Florida (Fig. 25.34). However, many marine swamps are partially covered by moss, and most contain grasses and reeds in abundance. As plants die, they fall into the water and begin to decay. If they are buried rapidly enough by other plant remains or by marine sediments, there is a good chance that they will be preserved and may eventually be transformed into coal. If burial is not rapid, oxidation and bacterial action soon break them down. For this reason, only thin accumulations of swamp deposits are found if the level of the swamp remains stable. Probably no more than a few meters of plant matter can accumulate, since water depth is seldom great. The thick sequences of swamp sediments so common in stratigraphic sequences of the Carboniferous period indicate changes in the position of sea level in relation to the swamp. A slowly subsiding area is the most favorable site for preservation of great thicknesses of such sediment. It is estimated that about 38 meters of plant matter are needed to form one meter of coal. Thus, the coal, often several meters thick, found in Pennsylvania, West Virginia, and other coal-producing areas indicates sites of long-continued subsidence and swamp formation. One coal seam in Russia is nine meters thick.

Fringing and Barrier Reefs

Most modern reefs are built of corals, although many other invertebrate animals contribute to their mass. They are located in the parts of the oceans where waters are warm and corals flourish. Most modern coral reefs are situated right along the shore and are called **fringing reefs,** or they are located out on the continental shelf, separated from the mainland by a body of water, and are called **barrier reefs.**

The reef usually consists of a narrow zone of living coral, the top surface of which is flat and located close to mean sea level. This surface may be exposed above water at low tide, but the corals must remain wet most of the time. Water depth usually increases rapidly off the seaward side of the reef. During storms the top edge of the living reef may be broken, and the loose materials slip down the steep seaward slope to accumulate at its base. The landward side of the reef is usually a lagoon where more delicate forms of coral and other invertebrates are protected from the effects of breaking waves. The lagoons are only a few meters to a few tens of meters deep, but they are not as densely populated with animal life as the reef flats. Algae and fine sands derived from the reef proper are usually a main part of the sediment in the lagoon.

The animals and plants that make up reefs are types that flourish in near-surface waters. Many, like corals, favor warm water that is relatively free of suspended dirt. Corals are attached to the sea floor; they are not free to move about to gather food. Thus, they must live where food is brought to them. Clean, agitated waters near the surface are most favorable. Where muddy waters enter the ocean from streams flowing off the land, breaks in the reefs are found.

Case history: The Great Barrier Reef of Australia. The Great Barrier Reef of Australia, the largest in the world, covers nearly 200,000 square kilometers and extends 1900 kilometers along the east coast of Australia (Fig. 25.35). The continental shelf is a complex of scattered reefs separated by channels. This complex is not a single, continuous reef mass, although one reef in it is 960 kilometers long. Long reefs aligned parallel to the coast form an outer barrier near the edge of the continental shelf. A relatively open body of water called the Barrier Reef Channel, 240 kilometers wide near its southern end, lies be-

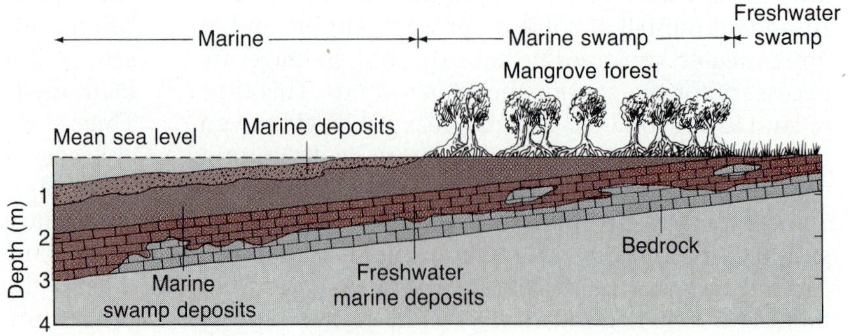

Figure 25.34
A geologic cross section of the marine swamp environment off the coast of southern Florida. Older deposits formed in the swamp are now buried beneath younger marine deposits. As sea level rises, the swamp is submerged and the deposits are preserved.

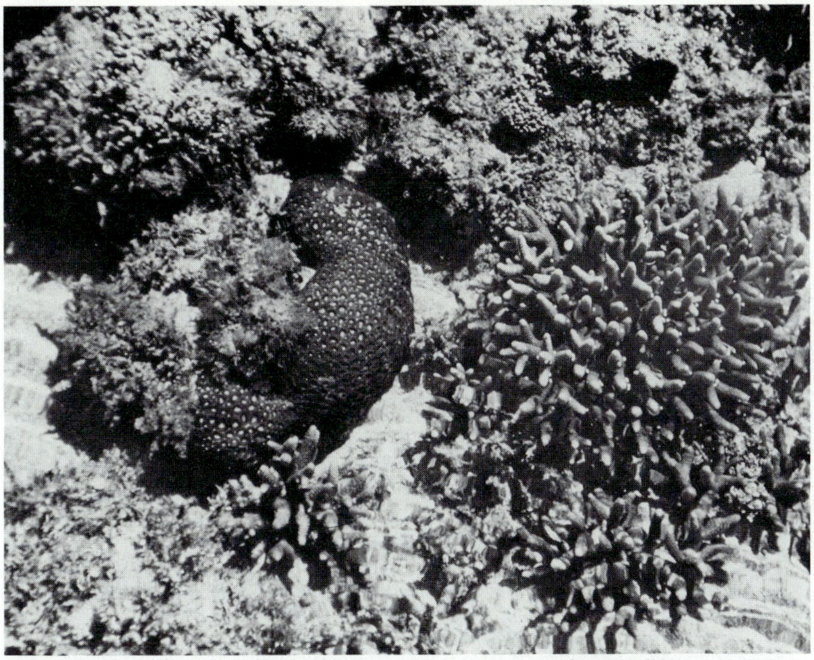

(a) Queensland Government Railways

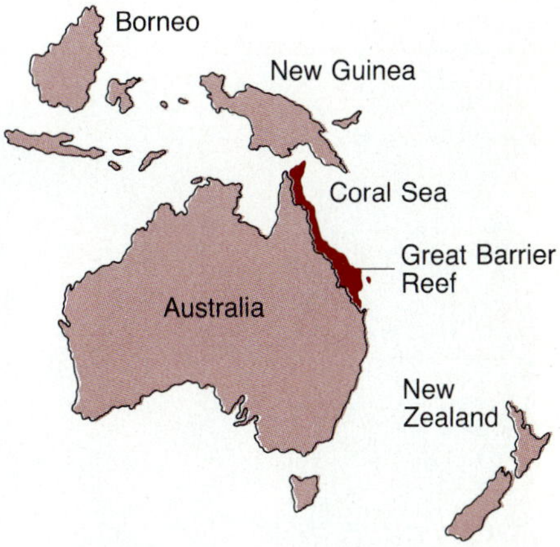

(b)

Figure 25.35
(a) At low tide, the reef flat of this part of the Great Barrier Reef of Australia is exposed. Some species of coral can survive short periods of exposure. (b) Locator map.

tween the outer barrier reefs and the mainland. Some six hundred islands are found in this channel. A number of these are small islands composed of coral sand and fragments; others are composed of rock masses like those on the mainland. Most of these islands are surrounded by fringing reefs.

During the 1960s, students of the reef noticed an unusually large population of a giant starfish, Acanthroaster (Crown of Thorns), which eats coral. Over a period of a few years, this population exploded and large areas of the reef were destroyed by the starfish. Studies were quickly started in an effort to determine the reason for this population growth. These focused on the natural predators of the starfish. Among these are the coral shrimp, which eat larva of the starfish. The coral shrimp are highly sensitive to the pesticide DDT, which has been used extensively in Australia. Runoff from land areas sprayed by DDT went into the breeding grounds of the shrimp, killing them off and, in turn, setting off the population explosion among the starfish.

Water wells drilled on some islands on the reef complex penetrate hundreds of feet of coral debris. This supports the idea that the reef complex has been built up as sea level rose following the last glacial advance and accompanying drop of sea level.

Atolls. Ring-shaped coral reefs that enclose lagoons are called **atolls** (Fig. 25.36). Darwin suggested that

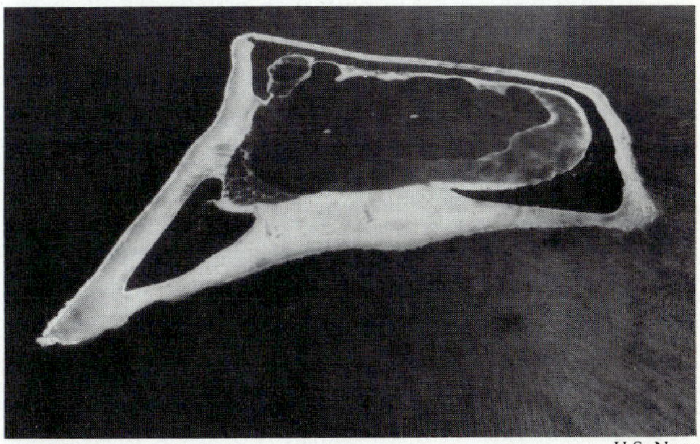

U.S. Navy

Figure 25.36
This small island, Namorik Atoll, is located in the Marshall Islands. The light area is the reef flat, which is covered by living coral. The dark areas are low islands composed of coral sand covered by shrubs and woods.

U.S. Navy

Figure 25.37
A coral reef surrounds the island of Bora Bora in the Society Islands. If sea level rose, the volcanic island would be submerged, but the barrier reef would continue to grow upward, forming an atoll.

these form when an island with a fringing barrier reef, such as that in Fig. 25.37, is totally submerged. The reef continues to grow upward to maintain the living organisms at the necessary water depth. Eventually, the island is covered by water, leaving the circular atoll around a water-covered depression.

An alternate hypothesis was advanced by R.A. Daly in 1910 (*1*). He reasoned that during the Pleistocene era, when sea level was more than 100 meters lower than it is now, the ocean was cooler than at present and thus less favorable for the growth of corals. As sea level dropped, islands and their fringing reefs were exposed above sea level. Corals could not grow around the margins of the island fast enough to protect the islands from wave erosion. Thus, the islands were leveled off by wave erosion to form platforms. When sea level began to rise from the melting of glaciers and oceans began to warm up, new reefs formed near the margins of these platforms. These reefs maintained their positions during further rises in sea level, creating the atolls, most of which surround flat-bottomed lagoons.

At the present time, hundreds of volcano-shaped mountains, called seamounts, are submerged in the Pacific Ocean. All of these are thought to be of volcanic origin, and many are known to be volcanic. Some of them have flat tops and are called **guyots.** These are interpreted as volcanoes that became dormant, had their peaks leveled by wave action, and were later submerged. Some became sites of atolls. Others may have submerged too rapidly for coral growth to maintain the reefs, and still others in high latitudes are in waters too cold for coral to live and grow.

Case history: Bikini Atoll. Bikini Atoll is an oval ring of islands about 42 kilometers long and 24 kilometers wide (Fig. 25.38), probably best known as the site of early American tests of atomic bombs. Bikini is the most intensively studied coral atoll. Marginal reefs up to about a mile wide are almost continuous around the atoll. The central lagoon, which has an average depth of 48 meters, covers approximately 648 square kilometers. The islands that surround the lagoon stand at a maximum height of eight meters above sea level. A number of passes one to two kilometers wide cut through the reef.

Reefs on the seaward side of the islands have an abundant and varied fauna. The open sea brings in a constant supply of water rich in food and nutrient salts. The seaward margin of the reef and the assem-

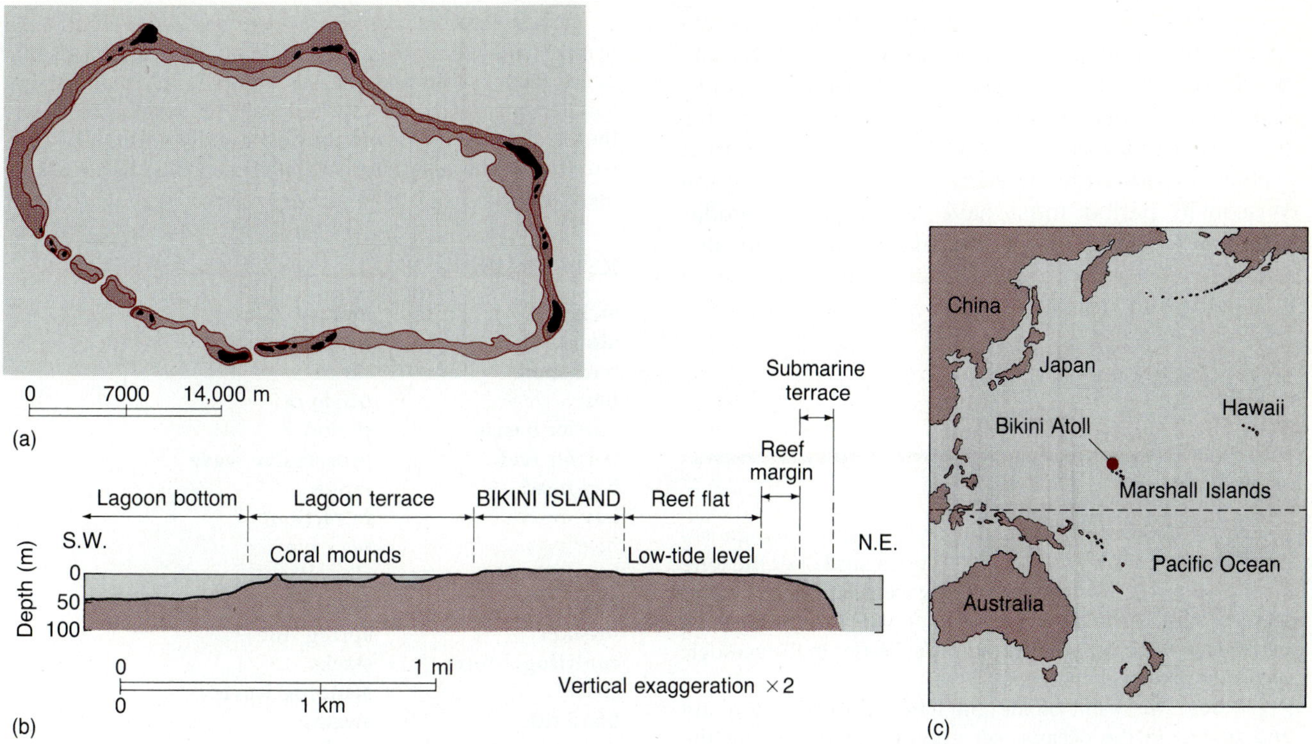

Figure 25.38
(a) Bikini Atoll, famous as the site of the early U.S. nuclear bomb tests, is one of the most intensively studied atolls in the world. Like many others, it is made up of a few islands connected by coral reefs surrounding a shallow lagoon. (b) Cross section. (c) Locator map.

blages of organisms that live there are determined by the slope on which the reef is located, the prevailing waves, the currents, and other ecological conditions.

The reef is strongly grooved on the seaward side. These grooves form where the water returns to the sea after waves break on the edge of the reef. Ridges between grooves are flattened and covered by red algae, which thrive in the strong, steady surf. Growth of corals is profuse on the reef flat where marginal reefs are low and water circulation is good.

The topographic setting of Bikini (Fig. 25.39) shows that the atoll is located on a flat-topped mountain that is isolated except for an adjacent guyot. The shapes suggest the base of a volcano, and drilling on Bikini confirms that it is underlain at depth by volcanic materials, but the thickness of the coral came

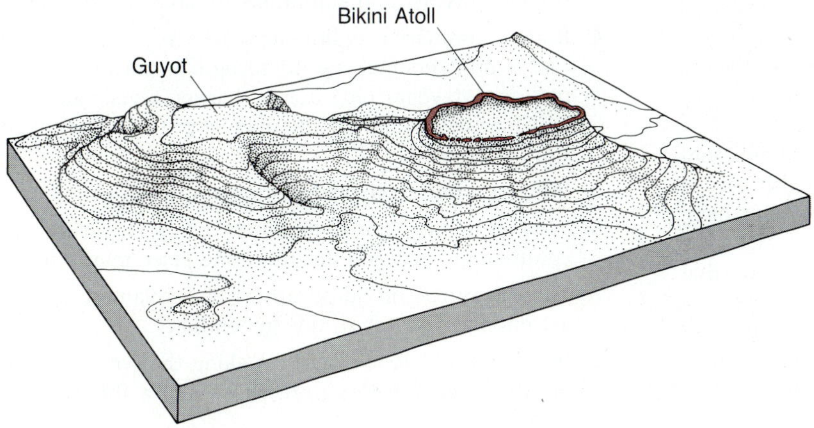

Figure 25.39
Bikini Atoll is located on a flat-topped seamount. The shape of the base of the atoll and that of the nearby guyot suggest it is a volcano. This has been confirmed by drilling through the coral deposits near the surface.

as a surprise. A well penetrated more than 700 meters of coralline limestone. This thickness is obviously much too great to be explained by sea level changes due to water–ice balance. In order to explain this much reef limestone, subsidence of the sea floor as well as a worldwide lowering of sea level during the Wisconsin period must have taken place. Similar regional warpings of the sea floor have been also called on to explain why many guyots are now submerged 100 to 1000 meters.

Where waters are warm, another type of offshore feature, coral reefs, may grow. The plants and animals that build these carbonate masses flourish in clear, warm, agitated, near-surface waters. When sea level rises slowly the coral may build reefs upward rapidly enough to keep pace with the rise, forming reef deposits hundreds of meters thick.

SUMMARY

The level of waters in the oceans is constantly changing—over the short term as the result of tides, and over the long term as the result of changes in the ice–water balance and tectonic effects. Because the level of the sea affects many geological processes, there is ample evidence in the geologic record of sea level changes.

Tides, the result of the gravitational force of the sun and moon on the oceans, set up currents on and in the oceans. The deep-sea currents are believed responsible for sorting sediments at the margins of the continental shelves.

The changing ice–water balance that accompanies the advance and retreat of major glaciations has produced sea-level changes of as much as 100 meters about the present sea level. Even greater differences in the elevation of continents relative to sea level are the result of tectonic activity. The evidence of these changes is seen in beaches and wave-cut cliffs or terraces now far above or below sea level and in fossil remains of marine animals now found at elevations of hundreds or thousands of meters.

One of the most effective agents in determining the form of the shore is waves. They act to erode headlands at a faster rate than they do bays, and thus help even the coastline. Oblique waves set up longshore currents that transport sediment along the coast, often for hundreds of kilometers. The turbulent water of a breaker and the sand and pebbles it carries are an effective agent of shoreline erosion forming sea cliffs and wave-cut terraces. As the sea cliff is eroded back, stacks, sea caves, and arches may be left offshore as remnants of earlier stands of rock.

The products of shoreline erosion and stream sediment are deposited, picked up again, and redeposited by ocean waves and currents in transient beaches, spits, and bars. Sometimes these deposits form barrier beaches, leaving the protected lagoons between the beach and shoreline that are one of the most productive marine environments.

Land that is near sea level but protected from wave erosion may become tidal flats. Marine swamps, which also form near sea level, are the deposition site of plant remains from which coal deposits form.

KEY TERMS

arches	**morass**
atoll	**neap tide**
backshore	**notch**
bar	**offshore**
barrier beach	**period**
barrier reef	**progressive wave**
bay-head	**range**
bay-mouth	**refraction**
beach	**sea cave**
beach drifting	**shoreline**
bog	**spit**
breaker	**spring tide**
contrifugal force	**stacks**
coast	**standing wave**
flood tide	**swamp**
foreshore	**swell**
fringing reef	**tidal bore**
guyots	**tidal flat**
headland	**tide**
longshore current	**tombolo**
mean sea level	**wave**

STUDY QUESTIONS

1. What evidence proves that sea level has had much higher stands than it has now?
2. Describe the causes of the changes shown on the three photographs of Little Egg Harbor.
3. Why don't all volcanic islands in the Pacific have barrier reefs? What differences cause some volcanoes to become atolls, some to become seamounts, some to become guyots, and still others to have barrier rims?
4. Briefly summarize sedimentary rock types and forms that would enable you to recognize ancient (many millions of years old) deltas, barrier reefs, beaches, and marine swamps if the evidence is in cores taken from deep wells.
5. What conditions favor development of sea cliffs?
6. Why does the type of sediment being deposited on continental shelves differ from one place to another?
7. What features of tides cannot be explained by the astronomical theory of tide generation?
8. In what ways do the beaches along the west coast of North America differ from those along the Atlantic coast?

9. How could the construction of dams on major rivers cause beaches to disappear along the coast?
10. Explain how efforts to restore or preserve beaches at one place on the coast may cause damage to beaches at other places on the coast.

REFERENCE

1. R. A. Daly, "Pleistocene glaciation and the coral reef problem," *American Journal of Science,* 4th series, 30 (1910), 297–308.

SUGGESTED READINGS

The Conservation Foundation, *Barrier Islands and Beaches.* Washington, D.C.: 1976.

Fisher, J. S., and R. Dolan, eds., *Beach Processes and Coastal Hydrodynamics.* Stroudsburg, Penn.: Dowden, Hutchinson & Ross, 1977.

Leatherman, S. P., ed., *Barrier Islands.* New York: Academic Press, 1979.

Shepard, F. P., *Submarine Geology,* 3rd ed. New York: Harper & Row, 1973.

OUR FINITE EARTH AND RESOURCE-DEPENDENT SOCIETY

Part 5

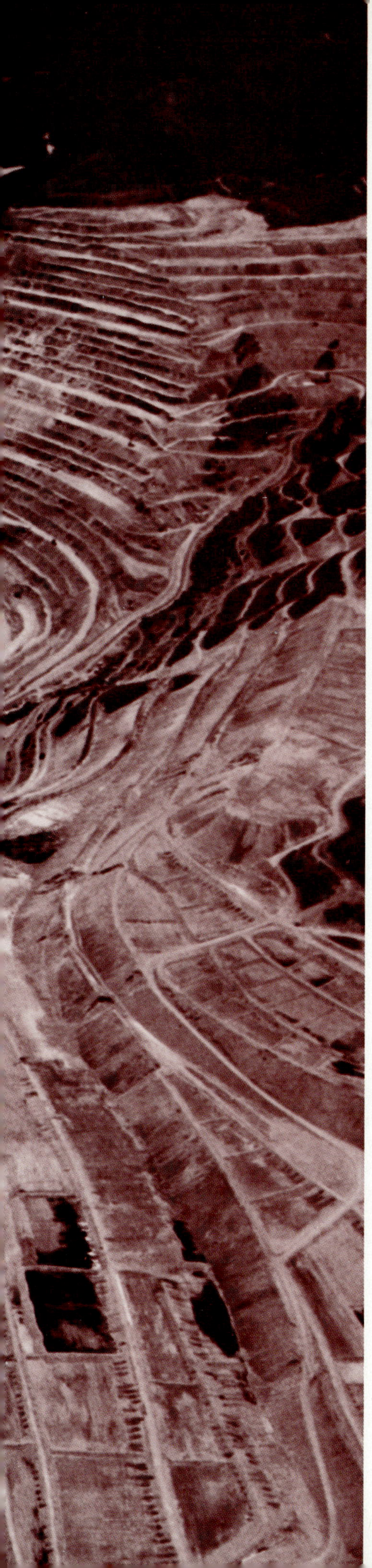

The quantity of natural resources consumed per person is one of the most obvious distinctions between developed and undeveloped nations. The United States, with about 5 percent of the world's population, consumes about a third of the total energy produced each year. If the standard of living in the developing nations is to be raised, either we must share the resources we now consume or great increases in production of natural resources must be achieved. The projected growth of the world's population, especially in the less developed countries, will make attainment of this goal most difficult, perhaps even impossible. The economic, political, and human implications of this situation are certainly among the most important problems facing this generation.

From geology we can obtain a better understanding of the origin and occurrences of rock, mineral, fuel, and water resources. This understanding provides a basis on which we can better assess the resource problems facing our society.

THE ORIGIN AND OCCURRENCE OF IMPORTANT RESOURCES

Chapter 26

All of us share an interest in natural **resources,** because they are so closely related to the economic well-being of society. Geologists are especially concerned because they are directly involved in the exploration for and the development of natural resources, many of which occur as rocks and minerals or, like water and oil, reside in the crust of the earth. Water and petroleum are quite different from the others in the nature of their occurrence. These fluids are not exactly part of the rock. Although they occur in rock, they are free to move, and it is unnecessary to remove the rock in order to obtain them. Mining and destruction of the rock are involved in the production of virtually all other resources obtained from the crust. In some uses, almost all of the rock removed is consumed, as it is with building stones (Fig. 26.1) or coal. In other instances, tremendous volumes of rock must be mined and processed to obtain a small amount of the material

National Archives

Figure 26.1
Building stones of various types are quarried in most countries of the world. Recently, nearly fourteen million tons of stone have been produced each year. The methods being used here to split stones are still in use in many places.

sought. Disposal of the wastes generated in concentrating the desired element has become a serious problem to the public, because of the potential for pollution of land, air, and ground and surface water, and to mining companies because of the cost of avoiding pollution. Only in the last two decades has the tradeoff between need for resources and damage to the environment come into clear focus. Increasing demands and dwindling supplies will make the decision on what price to pay for resources more and more difficult in future years.

26.1 COST AND AVAILABILITY OF RESOURCES

Usually the costs ultimately determine whether a particular rock or mineral resource will be developed. If the cost of removing the material from the earth, processing it, and moving the desired part to the place where it will be used becomes greater than the cost of the same material or a suitable substitute from some other locality, the **mineral deposit** at the first locality will probably not be developed. It is not "economically available." For this reason, some natural resources are not moved far from their point of origin. Concrete aggregate and road construction materials are good examples. Crushed stone (Fig. 26.2) is generally the best material for these uses, but most localities possess something that can be substituted if the cost of crushed rock rises high enough. For this reason, concrete aggregate may consist of crushed limestone in one place, stream gravel in another, and even sea shells in a third.

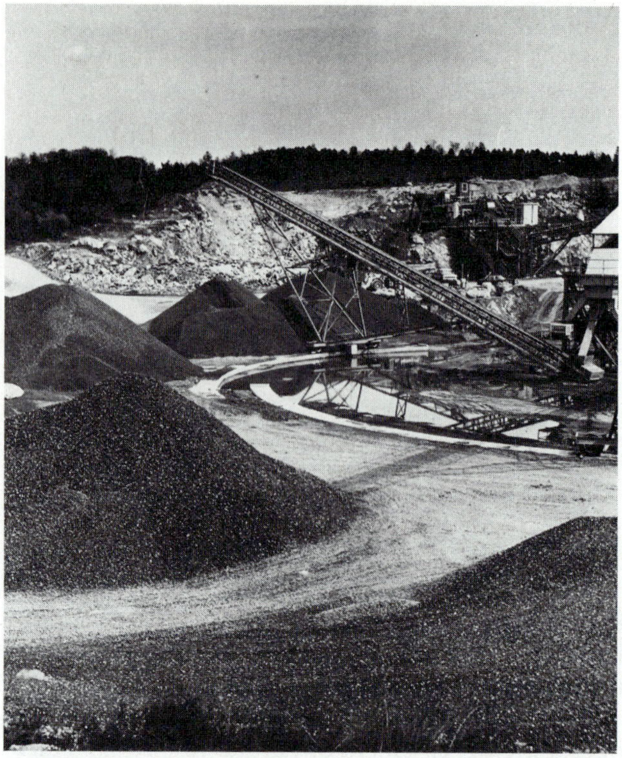

Figure 26.2
Over three billion tons of crushed stone are produced each year. A third of that is produced in the United States. This stone is used for road bases, as concrete aggregate, and for fertilizer, iron, and steel production.

On the other hand, materials that have a high value, are not common in their occurrence, or are strategically important may be transported around the world. Thus, diamonds, emeralds, or mica may be mined and moved anywhere economically because their volume is not great and their value is high. Although it is bulky and heavy, coal mined in the eastern United States is shipped to Japan, which needs coal for heating and production of steel. We, in turn, import all our chromite, manganese, and tin because we do not possess deposits of **ore,** rocks or minerals that can be mined and refined to yield the metal at a profit.

Unfortunately, many important resources have very limited occurrence. And even among those that occur widely, most are not economically available because the desired material is not concentrated enough to make it possible to mine, process, and deliver it at a profit. This principle applies to all types of resources. You may recall from the discussion of water resources in Section 22.2 that deep water from the aquifer in the Great Plains is not pumped in places where the cost of pumping is too great. Huge quantities of coal are also left in underground mines because the cost of removing the coal that supports the mine roof is too great. Many underground mines close not because the ore is depleted but because the cost of mining it is too high relative to the current price of delivery. For example, many of the gold and silver mines in the western United States closed in the 1890s because these metals could not be produced at a profit. At today's prices, many of these mines have become profitable again and are now producing ore. Potential offshore oil and gas deposits have not been drilled because, even at today's oil prices, the cost of erecting and maintaining the drilling platform makes them uneconomic.

A few natural resources are both widespread and readily available, but most occur in relatively restricted geological settings. In the following sections we will consider the geological circumstances in which important resources occur and the way some of these are naturally concentrated.

Resources and Reserves

In discussing the availability of natural resources it is important to bear in mind the degree of uncertainty that exists about how much of the resource is present on earth. Our knowledge is limited by the simple fact that many parts of the earth's surface have not been explored even superficially and very few areas have been explored at depth. For these reasons, it is useful to distinguish resources that have been identified or at least inferred to exist and that can be produced economically, which we will call **reserves,** from the total amount of a resource that may be present.

If we use this definition of reserves, it should be clear that the amount of reserves can change drastically as new discoveries are made or as a particular deposit becomes economically available. A deposit may become economically available either as a result of lower cost of production or as a result of a higher price.

It is also useful to make a distinction between reserves that are so well known that they have been measured, called **proven reserves,** and those that are unmeasured or only inferred to exist.

26.2 THE OCCURRENCE OF ROCK AND MINERAL RESOURCES

In some mineral deposits, the whole rock or most of it is used, and in other instances an economically important mineral is disseminated throughout the host rock. More often, the deposit consists of a body of mineral matter that has been concentrated by some geological process, and in the cases of water, oil, and gas, the host rock is nothing more than a reservoir in which the resource is stored.

Some common rocks that are mined and used in only slightly modified form are listed in Table 26.1. The value of whole rock materials used in large volume greatly exceeds that of any of the more concentrated mineral deposits. Huge volumes of crushed rock are used throughout the developed nations for road building and concrete aggregate. Granite, limestone, basalt, conglomerate, sand, and quartzite are all used in this way.

A greater variety of rocks are used as building stones than for crushed stone, but the suitability of rocks for these uses is not always the same. For example, travertine (spring deposits) is cut and used because of its soundproofing qualities, and serpentine is used for building facings. But neither is suitable for road-building or for concrete. Travertine breaks down physically, and serpentine is slick and unstable.

The geological conditions under which rocks such as basalt, slate, and marble originate were discussed in Part I of this text. The special conditions involved in the origin and concentration of some economically important minerals involve processes of igneous activity, metamorphism, and sedimentation not unlike those discussed in Chapters 5, 6, and 7. Usually economically available ore deposits (sources of metals) have been concentrated by some natural process. Some of the most important of these occur in magma.

Magmatic Differentiation

Magmatic differentiation is a general expression for all the processes that lead to the separation and usually to the concentration of some minerals in a crystallizing magma. Chromite, the main source of

Table 26.1 Uses of Some Common Rocks.

Rocks	Uses
Igneous Rocks	
Granite, Syenite, Diorite	Crushed stone, building stone
Rhyoilte	Crushed stone
Basalt	Crushed stone
Ash	Insulation, fill
Metamorphic Rocks	
Gneiss	Crushed stone, building stone
Marble	Crushed stone, building stone
Slate	Crushed stone, building stone
Serpentine	Building stone
Sedimentary Rocks	
Limestone	Crushed stone, building stone
Sandstone	Crushed stone, building stone
Clay	Bricks, drilling muds, ceramics
Salts	Chemicals
Gypsum	Construction materials
Travertine	Building stone

chromium, which is used to strengthen steel in alloys, is concentrated in this way. Although chromite is a minor constituent of most magmas, it has been sufficiently concentrated in a few places to be mined. Concentration occurs because when chromite, a heavy mineral, crystallizes early in the cooling history of magmas, it settles through the mafic magma in which it is most abundant and accumulates in a layer near the bottom of the intrusion (Fig. 26.3).

Pegmatite Deposits

The liquids that remain as a magma crystallizes often contain relatively rare and sometimes valuable elements. This last fraction of the magma fills irregular openings in and around the intrusion, forming pegmatites, Fig. 26.4. The minerals found in pegmatites usually include quartz, feldspar, and mica. Some pegmatites are mined for feldspar, which is used in ceramics, and for mica, which is used for insulators; more exotic minerals containing rare elements such as boron, fluorine, beryllium, and lithium may also be present. Uranium-bearing minerals and gemstones are also among the host of unusual minerals that occur in this way. Through the processes of crystallization these constituents have become concentrated in the magma.

Veins of quartz containing native metals such as gold and silver are of special interest among the products of igneous activity. Quartz veins often found as dikelike intrusions in the country rock around granitic intrusions sometimes bear gold or silver. They too are late-stage products of the crystallization of magma.

Hydrothermal Deposits

The hot aqueous solutions that rise from magmas deep in the earth often carry the elements of important ore minerals. Some of these solutions carry metals from within the magma itself, eventually concentrating it as some metal-bearing mineral. Often the minerals form as a result of a reaction between the hot solution and the rock through which it passes. These are referred to as **hydrothermal deposits.** In this type of metamorphic reaction, the economic mineral is deposited in pore spaces or fractures; or it reacts with the rock to form a new mineral; or mineral matter in the solutions replaces the host rock. Such deposits are usually classed according to the depth (and temperature) at which the reactions occur. The solutions generally move along fracture and fault zones. Consequently, the mineral deposits formed this way are often located with such zones. A notable example of this situation occurs at the Cor d'Alene district in Montana, where copper, manganese, lead, silver, and a number of other metals are found.

Hot waters rising from fractures near some oceanic spreading centers carry metals and demonstrate clearly that a number of metals may be transported

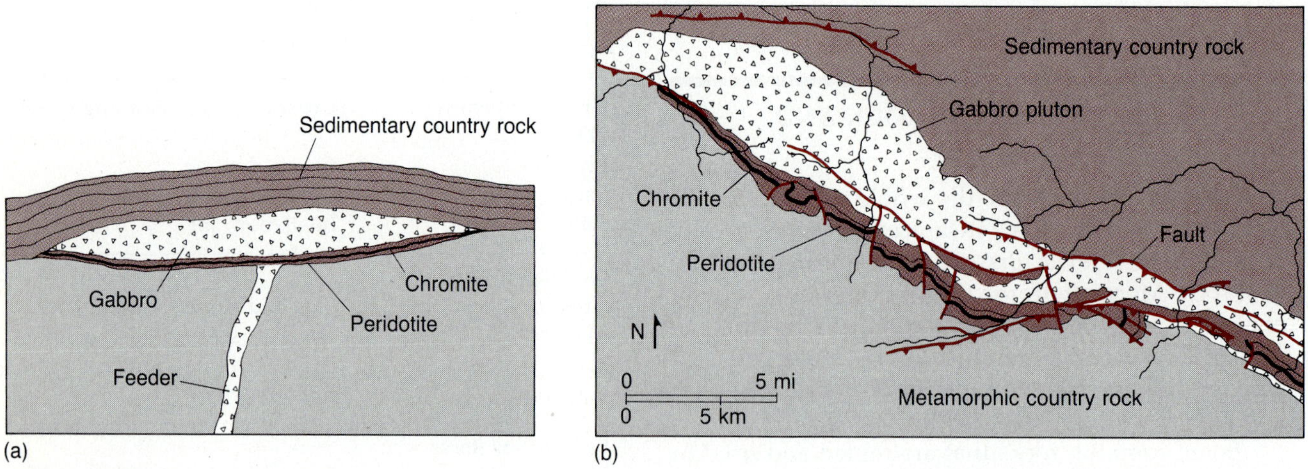

Figure 26.3
Some metals are concentrated in cooling magmas when early-crystallized heavy minerals sink through the melt, as in this large intrusion in Montana, known as the Stillwater complex. Chromite occurs as a layer near the bottom of this lens-shaped pluton. The pluton has been uplifted, rotated so the layers that formed horizontally are now vertical, and exposed by erosion. (a) At the time of intrusion, the pluton probably resembled the form shown in cross section. (b) The pluton in map view, as it appears today.

Figure 26.4
Pegmatites are composed mainly of the common minerals, quartz, feldspar, and mica, but many rare and valuable minerals also occur there. Most pegmatites, like the one shown here, cut across the country rock.

from deep sources toward the surface in this way. Metals are carried from depth toward the surface in subduction zones as well. This upward migration of sulfides from the mantle is thought to account for many mineral deposits in mountain systems such as the Cordilleran Mountains. Similarly, gases emitted from fumaroles in some volcanic centers deposit metals where the fumaroles issue from the ground. Magnetite, an iron ore; galena, a lead ore; and sphalerite, an ore of zinc, are all found in deposits around fumaroles at the Katmai volcano in the Aleutians.

Residual Concentrates

Other mineral deposits become concentrated by processes acting at or near the ground surface. Weathering, for example, is important in the formation of economically important deposits of some ores and nonmetals through processes of residual and mechanical concentration. **Residual concentration** is a process by which some valuable mineral constituents accumulate while undesirable constituents are removed by weathering and erosion. Among the economically important deposits formed in this way are those of iron in the Great Lakes region; manganese in India, the Gold Coast, Brazil, and Egypt; bauxite in Arkansas; residual clay for use in ceramics; nickel in New Caledonia; phosphates in Florida; and zinc in Virginia and Tennessee. Many of these deposits are of minerals or rocks that owe their origin to chemical reactions that occurred during weathering. In these cases, weathering is responsible not only for concentration but also for the origin of the ore or nonmetallic minerals themselves. This is clearly the case with regard to bauxite and clay.

Laterite is a special type of clay-rich soil that has a tendency to harden into a bricklike solid. It is found where the rainfall is highly variable, causing large fluctuations in the level of the water table from within half a meter of the ground surface to depths of many meters. Silica is extensively leached from such soil, and hydrous aluminum oxides are concentrated, as shown in Fig. 26.5. These soils contain so much aluminum that they are potentially valuable sources of the metal. Further alteration of some laterites, leading to the formation of the principal aluminum ore, **bauxite,** has occurred in many tropical areas. The deposits in Jamaica and in parts of the Gulf Coastal Plain are especially important as sources of American bauxite.

The nickel deposits of New Caledonia, the world's largest source of nickel, provide another example. The nickel occurs in silicate minerals in large bodies of peridotite altered to serpentine by reaction with

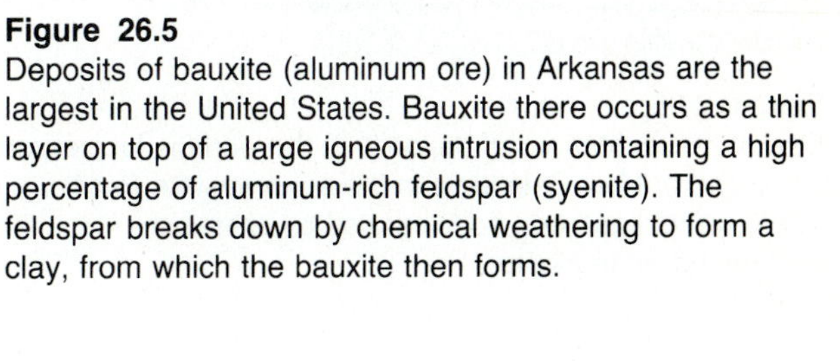

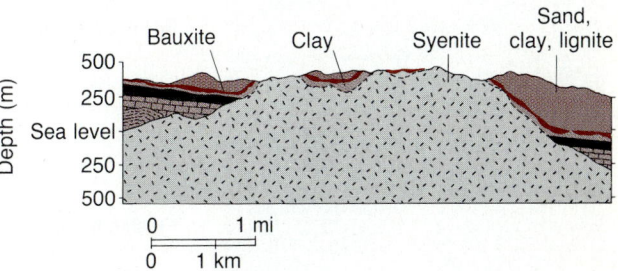

Figure 26.5
Deposits of bauxite (aluminum ore) in Arkansas are the largest in the United States. Bauxite there occurs as a thin layer on top of a large igneous intrusion containing a high percentage of aluminum-rich feldspar (syenite). The feldspar breaks down by chemical weathering to form a clay, from which the bauxite then forms.

hot water. These bodies have been weathered under tropical conditions to form a lateritic soil. The peridotite has decomposed, silicate minerals have broken down, and hydrous silicates of magnesium and nickel have formed. These are concentrated in ore bodies in the lateritic soils, as represented in Fig. 26.6.

Secondary Enrichment

Chemical weathering of ore minerals by near-surface waters may oxidize some of the minerals to yield acids that dissolve or react with other minerals. Thus, the host rock containing an ore mineral may be leached of valuable elements that move downward in the ground. If these metal-bearing solutions encounter water or rocks, such as limestone, with which they can react, the metals may be deposited. As this process continues over a long period of time, the deep deposit becomes enriched in the metals leached from above. The ores pyrite (FeS) and chalcopyrite ($CuFeS_2$) are important because they oxidize to limonite and sulfuric acid, which is highly reactive. Most deposits of this nature are characterized by a limonite-rich (an iron oxide) leached cap, called **gossan,** and an enriched zone in which carbonates of the metal are concentrated at depth (copper and silver deposits may be enriched in this manner). Gossans are easily recognized by their yellow color in regions of sparse soil and thin plant cover, and they have been successfully used by prospectors in exploration for copper and other deposits, particularly in the southwestern United States.

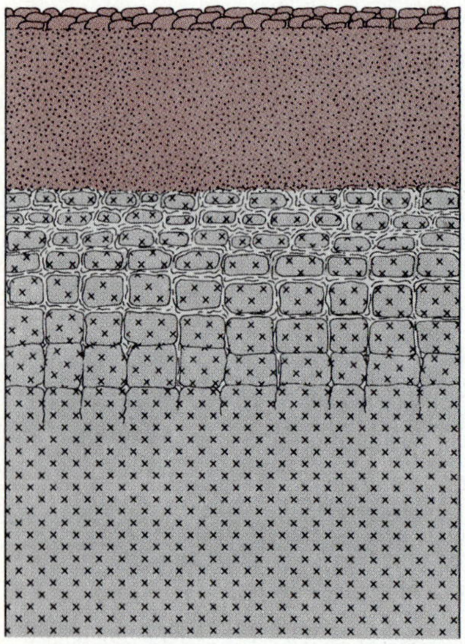

Figure 26.6
One of the largest nickel deposits in the world is located on the island of New Caledonia in the southwestern Pacific. There nickel has been concentrated by chemical weathering of peridotite.

Deposits from Near-Surface Water

Near-surface mineral deposits (those less than a kilometer deep) may also be deposited from shallow circulating solutions. Such solutions and their metal content may originally come from deeper sources, including magmatic sources, but often the deposits are far removed from sources of magmatic solutions. In these cases, circulating groundwater is probably responsible for picking up the valuable elements that may have been released from the host rock by weathering or even by the groundwater itself. This type of deposit takes the form of vein-filling; the ore may occupy fractures, fault-zone breccia, or even pore spaces in solution cavities filled with collapsed roof rock. Hot springs may be surface manifestations of underlying networks of such fractures and cavities. Gold, silver, antimony (used in alloys), and mercury are among the resources that occur in such deposits.

Mechanical Concentration

Native gold and cassiterite (a tin mineral) are two minerals that may be freed from the rocks in which they originally occur and concentrated by mechanical processes. The chief mechanism for concentration is gravity separation of the heavy particles by streams, coastal currents, or wind. Only minerals that are heavy, resistant to chemical weathering, and hard, malleable, or durable are likely to survive the process of **mechanical concentration.** Deposits formed in this way are called **placer deposits**; among the other minerals found in them are platinum, diamond, magnetite (an iron ore), chromite, ilmenite (titanium ore), copper, and monazite (a phosphate-bearing mineral). The stream placers of gold were particularly important in the development of the American west. Gold, which is malleable (can be beaten into shape), weathered from gold-bearing quartz veins and accumulated in alluvial gravels along streams flowing out of the mountains where the gold veins cropped out. Gold accumulated behind boulders and other natural barriers across the streams and in gravels in stream channels (Fig. 26.7). Even today, prospectors in Alaska and elsewhere work to trace such placer deposits back to their origin upstream.

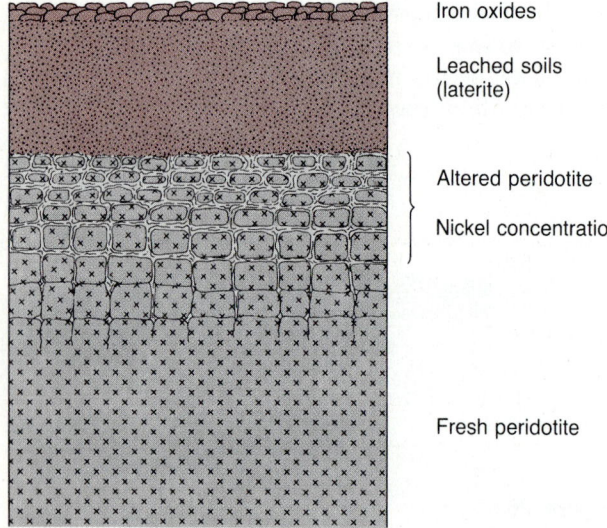

U.S.G.S.

Figure 26.7
Gold, which originally formed in quartz veins, is sometimes found in stream deposits like the ones being mined here. Because gold does not break down in water, once freed from the quartz it remains in the sediment. Here the gold has been dredged up and separated from other sediment.

26.3 ENERGY SOURCES FROM WITHIN THE EARTH

Most of the energy used in the developed countries of the world is produced from fossil fuels. The sources of energy used in the United States are shown in Fig. 26.8. Thirty years ago, about 90 percent of our energy came from fossil fuels and 10 percent from nuclear, hydropower, and geothermal. The growth of our energy demand, also depicted in Fig. 26.8, has increased steadily and quite rapidly until recently; but since the early 1970s, most of this increase has been supplied by imported fuel. United States oil and gas production reached a peak in the early 1970s and has shown a decline since then.

Where energy supplies for the future will be found is one of the crucial questions for the next decade. If our demand for energy continues to grow as it has in the past, this problem is likely to become increasingly difficult to solve. Foreign reserves of oil and gas are sufficient to provide our needs for a few decades at least, but the price of foreign oil and its availability are uncertain. And it is clearly undesireable for America to become ever more dependent on other countries for so vital a resource. To insure energy independence, we must look for domestic supplies or to alternate sources.

Among the alternate sources: tar sands and oil shales are both fossil fuels derived from rocks; nuclear energy is produced from rocks and minerals containing uranium or other radioactive isotopes; and geothermal energy is derived mainly from naturally hot waters near the earth's surface. Other sources include solar energy, wind, tide, wave-generated energy, and plant fuels. The sources from within the earth are examined in the following sections.

Coal

Coal is vegetable matter that has been altered both physically and chemically through natural processes to a black, rocklike substance. The chemical changes involve a loss of moisture and volatile constituents such as oxygen and hydrogen, some rearrangements of molecules in the remaining matter, and an increase in the proportions of carbon and ash. Physical changes include a darkening of color with increased hardness and density. Not all coal is alike; in fact, organic materials can be found in varying stages of alteration

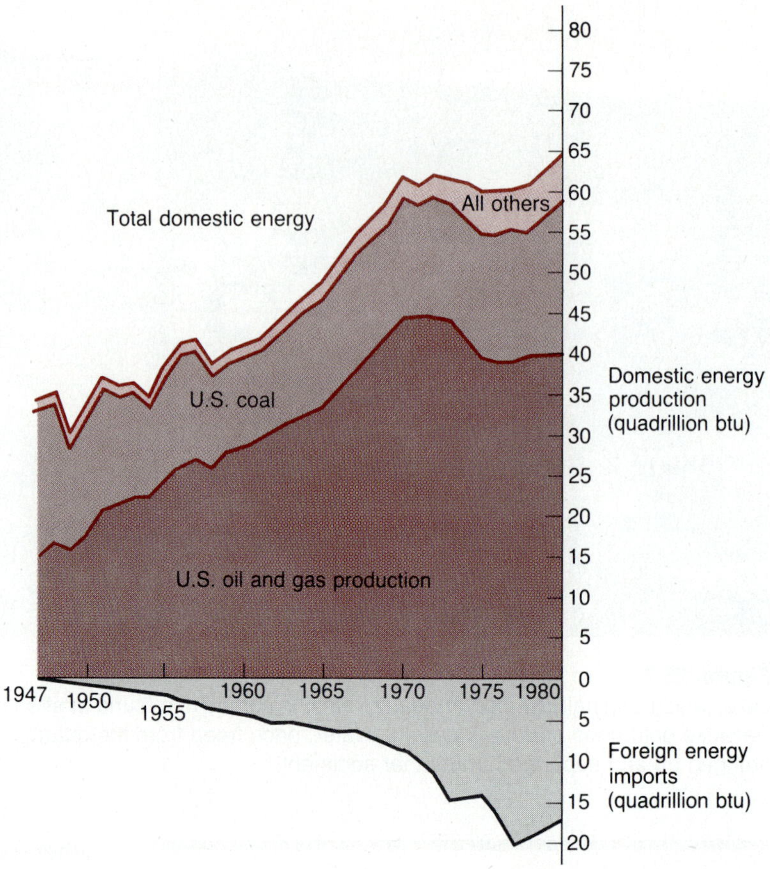

Figure 26.8
The total consumption of energy in the United States has continued to grow dramatically in the last thirty years. Oil, gas, and coal continue to predominate as our main energy sources.

ranging from decaying mats of organic matter in swamps to the hardest, most brittle variety of coal, **anthracite.** If we take the progression one step further, the material formed is almost pure carbon, graphite. There is no single place where all the varieties can be found to grade into one another, but the field evidence of progressive alteration is so convincing that it has been widely accepted. In recent years anthracite coal has been produced experimentally from cypress wood at Pennsylvania State University. Varieties or ranks of coal, as they are called, are listed from low to high rank:

Peat is an accumulation of vegetable matter that is in the initial stages of disintegration and decomposition. The water and oxygen content of peat is high. The material ranges from a distinctly fibrous and woody light-brown material to a dark-brown and black jellylike substance. Peat grades into mixtures of peat and sediment in swamps or bogs where it is most commonly found. It may be cut out in blocks, allowed to dry, and used for fuel, but such use is infrequent today.

Lignite and brown coals are more compact than peat, and although they may be woody or amorphous in texture, twigs and branches can usually be found and identified with them. When dried, they tend to split into slabs.

Subbituminous coal ranges from a glossy black coal to lignite. The coal is usually black with a pitchy luster, and it parts along a surface nearly parallel to the bedding, forming slabs.

Bituminous coal is pitch black to dark gray. It burns with a long, yellowish flame and gives off a strong odor. It is more or less laminated, and it breaks into small blocky fragments. It may be resinous, silky, pitchy, or dull and earthy in luster, and it soils the hands.

Subanthracite coal is intermediate in properties between bituminous coal and anthracite. It is formed by alteration of lower ranks of coal and is often classified as a metamorphic material.

Anthracite coal is dark black in color and dull to brilliant, even submetallic in luster; it will not soil the fingers because it is too hard; it burns with a short, blue flame and emits little odor. It is the hardest coal, having a hardness of about three on Moh's scale. In places, anthracite has some graphite mixed in with it. It always has a high carbon content and a low percentage of volatile matter.

Analyses of coals are made to determine their suitability for use as fuels as well as for their special uses in the chemical and steel industries. Figure 26.9 illustrates the variation in composition among the different coal ranks, as well as their energy content.

Origin of coal. Coal tree trunks rooted firmly in the underlying sediment prove that coal is an alteration product of plant matter. The old soils on which the plant matter grew are sometimes found under coal seams. Such coal must have been formed where it is. Other coal beds have almost certainly formed from plant matter that was transported before it accumulated and was buried. For coal to be derived from peat, not only must the peat be buried and preserved, but it must also accumulate in a thick layer. Modern peat bogs have been found in which there are tens of meters of peat. Thick accumulations might form in regions where regional subsidence or perhaps an advance of the sea allows the peat to become buried under new peat. Unless the peat is buried in some way, it is oxidized, and bacteria tend to break down the woody matter, ultimately leaving only ash. But if it is buried, the weight of the overburden will begin to compact and squeeze water out of the partially oxidized and decomposed plant matter. Biochemical processes also play an important role in the initial breakdown of plant matter to coal. Bacteria and fungi are most important, though their action is confined to the upper portions of any peat accumulation because acids form in the lower portions, where they cannot survive.

Compaction, depth of burial, heat, and escape of volatile constituents are probably important in forming coal, and anthracite occurs in sedimentary sequences that have been subjected to high pressure (usually the strata are tightly folded and faulted). This suggests that deformation may play a role in its development.

Coal is of economic importance not only as a source of heat energy but because it is necessary in the manufacture of steel. Special types of coal are used to produce a porous material, coke, that is used in the furnace with the iron ore to allow circulation of heat through the ore and to provide the carbon needed in the steel. It is also used in the manufacture of a wide range of synthetic materials.

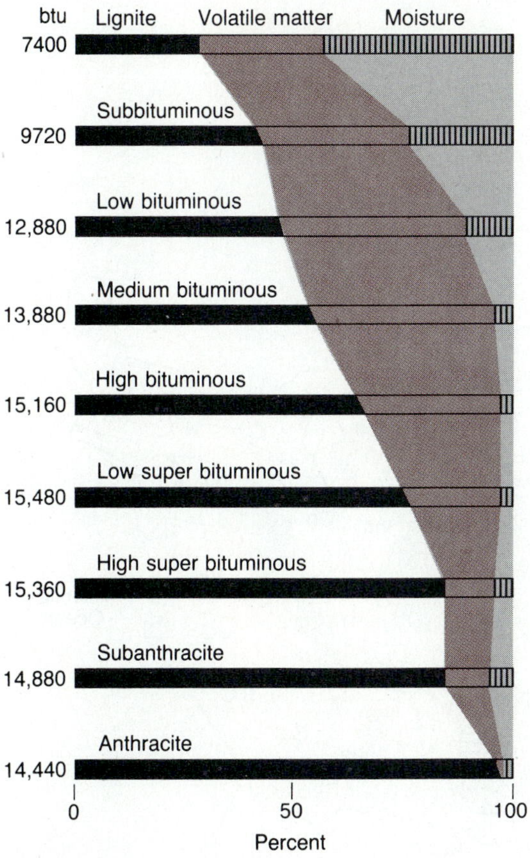

Figure 26.9
The amount of moisture and volatile matter progressively decrease as lignite changes first to bituminous and then to anthracite coal. The heat value of the various ranks of coal is expressed in Btu units (one British thermal unit is the quantity of heat needed to raise the temperature of a pound of water one degree Fahrenheit).

Coal resources. For many years coal was the primary source of fuel energy in the world. In the present century, petroleum products have gradually replaced it in that role, but it remains essential for the manufacture of steel. As supplies of cheap oil and gas dwindle and as imports become less certain, coal is increasingly being looked to, once again, as a source of energy. One reason for this is that the western world possesses huge reserves of coal, as indicated in Table 26.2. It is estimated that the earth contains about six times as much energy stored in coal as in oil and gas.

Table 26.2 World Coal Reserves in Billions of Metric Tons. (Data from the U.S. Department of the Interior.)

	Measured Reserves
United States	396
Canada	13
West Germany	100
United Kingdom	99
Remainder of Western Europe	26
Japan	3
Other noncommunist countries	140
U.S.S.R. and Eastern Europe	349
China	201
total	1327

Most of these deposits reside in the rocks of just three countries—the United States, the Soviet Union, and China. Additional coal may be found as exploration for all fuels increases, but it is unlikely that additional deposits of the size found in these three countries will be found. Of these, nearly half the coal in the western world is located in the United States.

The large coal fields of the United States are located in the continental interior (Fig. 26.10). Vast areas of the Appalachian Plateau and the central states from Texas to Michigan are underlain by coal formed in the late Paleozoic era. Most of these coal beds are little deformed. These flat-lying or slightly folded layers of coal crop out at the ground surface in places, but more generally they are buried by younger Paleozoic sedimentary rocks. The economic

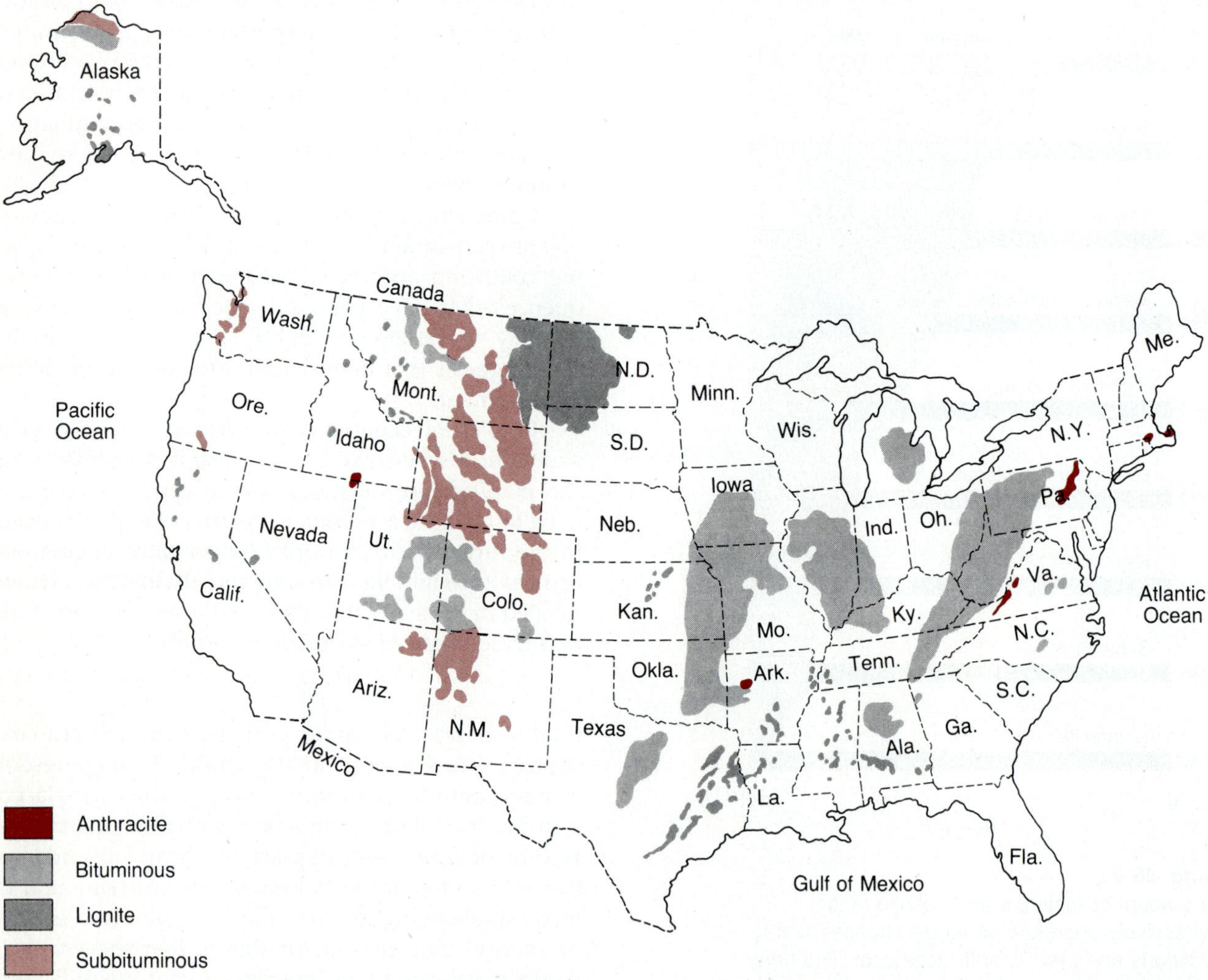

Figure 26.10
The main coal fields of the United States. Anthracite occurs mainly in Pennsylvania, but small deposits are also found in Arkansas, Colorado, New Mexico, and Virginia. Metamorphosed anthracite is present in Massachusetts and Rhode Island.

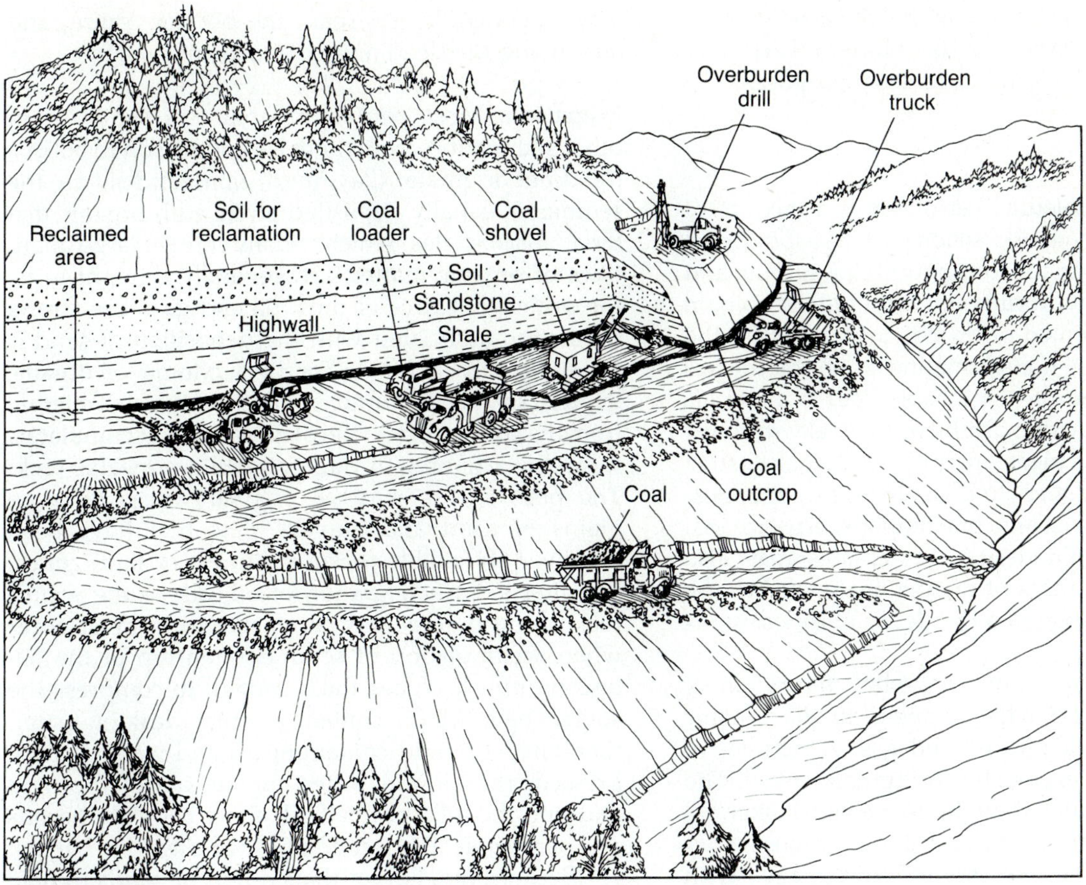

Figure 26.11
A variety of strip mining known as highwalling is illustrated here. Mines of this type are common in the mountains of the Appalachians. When reclaimed, the mountain side is terraced with flat, soil-covered areas. The cliff remains.

availability of these coals is determined by their thickness and quality, and the depth of the overlying rock. High-quality, thick coal seams can be mined economically by underground mining methods, which are too expensive for thin beds or low-quality coals. In general, it is less expensive to produce coal by surface or strip-mining methods. In these methods the cost of production is closely related to the thickness of the overlying rock and soil that must be removed before coal is reached, and, in recent years, to the cost of restoring the land to meet environmental standards.

Coal beds of younger age, mostly Tertiary subbituminous coal and lignite, are found in the Gulf Coastal Plain and in many of the basins of the Rocky Mountain states. Many of these western coal beds are close enough to the ground surface to allow strip mining. As prices rise, more of the deeper coals will also be economically available for strip mining.

Increased use of coal raises a number of concerns. Among them are the increasing levels of carbon dioxide in the atmosphere and the effects of this on our climate; acid rain; sulfur dioxide pollution; and the loss of lives in mines. Strip mining involves the removal of the topsoil and takes large areas of land out of use for other productive purposes, at least during the mining operation. It is possible to restore and reclaim the land for agricultural and other uses; but, depending on the topography, restoration may be prohibitively expensive. For example, in areas of high relief where mining is done on mountainsides, stripping often creates a vertical "high wall" (Fig. 26.11), which is cut back into the mountainside until the cost of removing the overburden becomes too great. At this point, mining ceases and a mine consisting of a terrace and steep wall remain. It is impractically expensive to restore such land to "its original contour," as is called for in some reclamation legislation. However, soil can be replaced on the terraces, and the land can be brought back into production for agricultural or forestry purposes.

Petroleum

The origin of petroleum. Petroleum is a mixture of natural gas, crude oil, and solids such as asphalt and waxes, all of which are compounds of carbon and hydrogen, called hydrocarbons. Although petroleum is known to be formed of once-living materials, the precise steps in the transformation to petroleum are still uncertain. The most probable source materials for petroleum are marine plankton, especially diatoms, fungi, bacteria, and algae; marine invertebrates, such as the protozoans; and, to a lesser extent, sponges, corals, worms, bryozoans, brachiopods, crustaceans, mollusks, and echinoderms. Vertebrates may also provide lesser amounts of organic material, but the notion of oil forming from the decay of dinosaurs is pure myth.

The principal questions regarding the origin of petroleum deal with what happens to the organic matter to convert it into oil and gas. A number of processes may be responsible for the conversion. The mechanism generally thought to be most likely is bacterial action in conjunction with one or more other processes. Bacteria are found universally with petroleum, and it has been definitely established that they can produce hydrocarbons from organic matter.

Knowing the source materials, we can easily envision the general environments in which oil might begin to form. Areas similar to the shallow waters of parts of the continental shelves and the Gulf of Mexico are probably localities. Droplets of oil have been found near the top of the sediments deposited in the Gulf of Mexico. Although these droplets have had neither the time nor the necessary conditions to accumulate in quantity, their presence indicates that relatively short periods of time are necessary for oil to form.

Bodies of marine water with shallow connections to the ocean should be particularly suitable for the accumulation of hydrocarbons. Free-floating and swimming plants and animals would be abundant at the surface; and as they died, their remains would sink to the bottom. Because the body of water is enclosed, circulation would be poor, and an oxygen deficiency would exist at the bottom. This would hinder oxidation, and poisonous gases produced by the decaying organic matter would form toxic conditions. Thus, scavengers could not roam the bottom, eating and destroying the dead animal and plant remains.

Migration. Most petroleum occurs in sands or in other sediments where there is good reason to believe it did not originate. Clay, mud, sand, and silt are the sediments usually deposited along with organic material. Black shales, which usually contain large quantities of organic matter, are generally thought to be the principal source of most oil and gas. The strong circulation associated with the deposition of sand is generally not favorable for preservation of organic matter, so the oil found in sandstone must migrate from the source rock to the place where it accumulates. It can migrate only through porous, permeable rock. The pore spaces may be original spaces between grains of sand and gravel, spots that have been dissolved in soluble rocks, fractures, or fault zones. The first step in migration of oil, out of the source rock into a porous and permeable rock, must occur when accumulation of sediments on top of the oil-bearing unit becomes thick enough to compress the source bed. Muds contain a great deal of water; therefore, if a mud containing oil and water is compressed, the oil and water migrate out of the mud and into the adjacent beds. If these beds are porous and permeable, the second phase of migration begins.

Because it is lighter than water, oil tends to float on it. Oil moves up slowly around the sand grains or through other pore spaces until it reaches a barrier that prevents further movement. This barrier is usually a layer of less porosity or permeability. Many such barriers are composed of impervious, clay-rich rock. Salt, cemented sedimentary rock of all types, compact limestones, and igneous or metamorphic rocks all occasionally form such barriers. In some places, petroleum acutally migrates to the ground surface. When it does, the gases held in under pressure at depth are released, and the petroleum is altered naturally to asphalt. Of course, asphalt is impervious, and as it forms in the pore spaces of the oil-bearing strata, the top of the layer is sealed close to the ground surface. Oil and gas below that will then accumulate as its upward movement is halted. We say that the oil migrates until it becomes trapped. Gases that have been produced are lighter than the oil and tend to move to the top of the oil accumulation.

Petroleum traps. The geological conditions that lead to the entrapment of petroleum can usually be broadly classed as structural or stratigraphic, or some com-

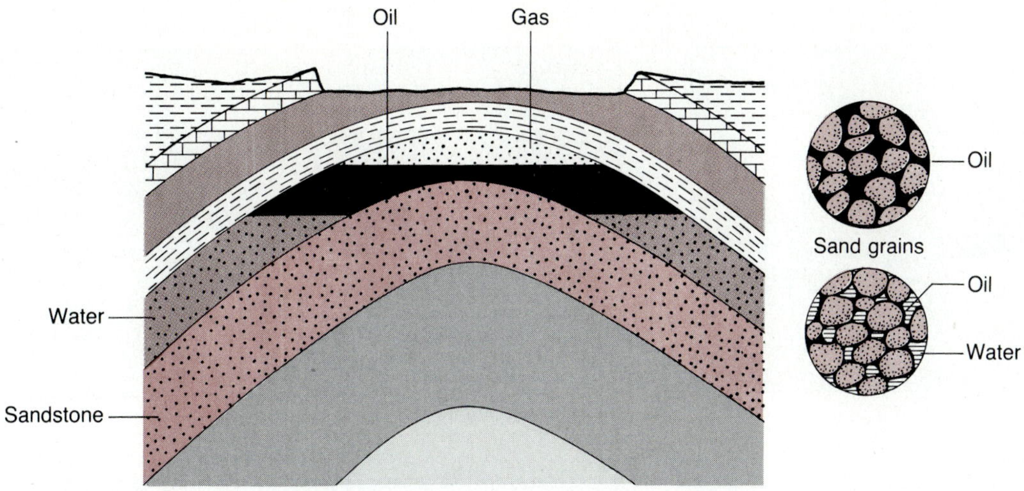

Figure 26.12
Oil is stored in pore spaces between sand grains, as shown in the enlarged sections at right. The oil probably gets into the sandstone from black muds or shales. Because oil tends to rise through water, it moves up in sandstone units that contain water. This migration continues until the oil becomes trapped, as shown in the anticlinal structure. Some anticlines are tens of kilometers or more across.

bination of these. The anticlinal or domal structures illustrated in Fig. 26.12 are among the simplest and most common structural traps. Petroleum migrates into an arched or folded "reservoir rock," such as sandstone, and migrates up through it until it reaches the crest of the fold. If the dome is capped by an impervious layer, upward migration is halted and the oil accumulates. Gases may rise through the oil and form a cap at the top of the fold, leaving the fluids in the sandstone stratified according to their density (gas, oil, water). Traps are also often formed where an impervious bed is faulted into contact with a layer through which petroleum migrates (Fig. 26.13).

Along the Gulf coast of the United States, many traps are found associated with large plugs of rock salt that have risen from great depth through the overlying sediments (Fig. 26.13). A thick layer of salt was formed in the Mesozoic throughout a large area along the Coastal Plain, under the present continental shelf. This less-dense salt is squeezed by the weight of accumulated sediments and flows upward. No petroleum originates in the salt, but the petroleum that is relatively abundant in some of the sediments penetrated by the salt may accumulate in traps formed in and around the salt domes. Oil is trapped in the domes over the salt, against the salt, at faults formed by the rising salt, and by various types of stratigraphic traps formed as the salt moves up.

Stratigraphic traps are caused by lateral or vertical changes in the porosity and permeability of layered sedimentary rocks. Many factors may cause such variation, most of them occuring as the sediments are laid down. For example, all sedimentary layers are finite in size, and most of them thin or "pinch out" toward their margins. If a permeable layer is overlain and underlain by impervious rocks, a trap forms where it pinches out. Reefs, which have unusually great porosity, are commonly associated with shales, evaporite rocks such as salt, or other impervious beds that can hold petroleum. Stream channel deposits are usually composed of sand and gravel, often encased in impervious beds, forming traps. Finally, lateral variation in the composition or grain size within layers often creates potential traps.

Frequently, structural and stratigraphic traps occur together as the result of deformation of sedimentary deposits. Many of the petroleum-bearing sediments originated on continental margins. As we have seen, these margins may become involved in plate interactions, including mountain-building. Other petroleum-bearing sediments form in sedimentary basins that form on the continental platforms. This is true of many of the important oil accumulations in the central United States. These strata are generally not intensively deformed, although they may be broadly warped or affected by high-angle faults.

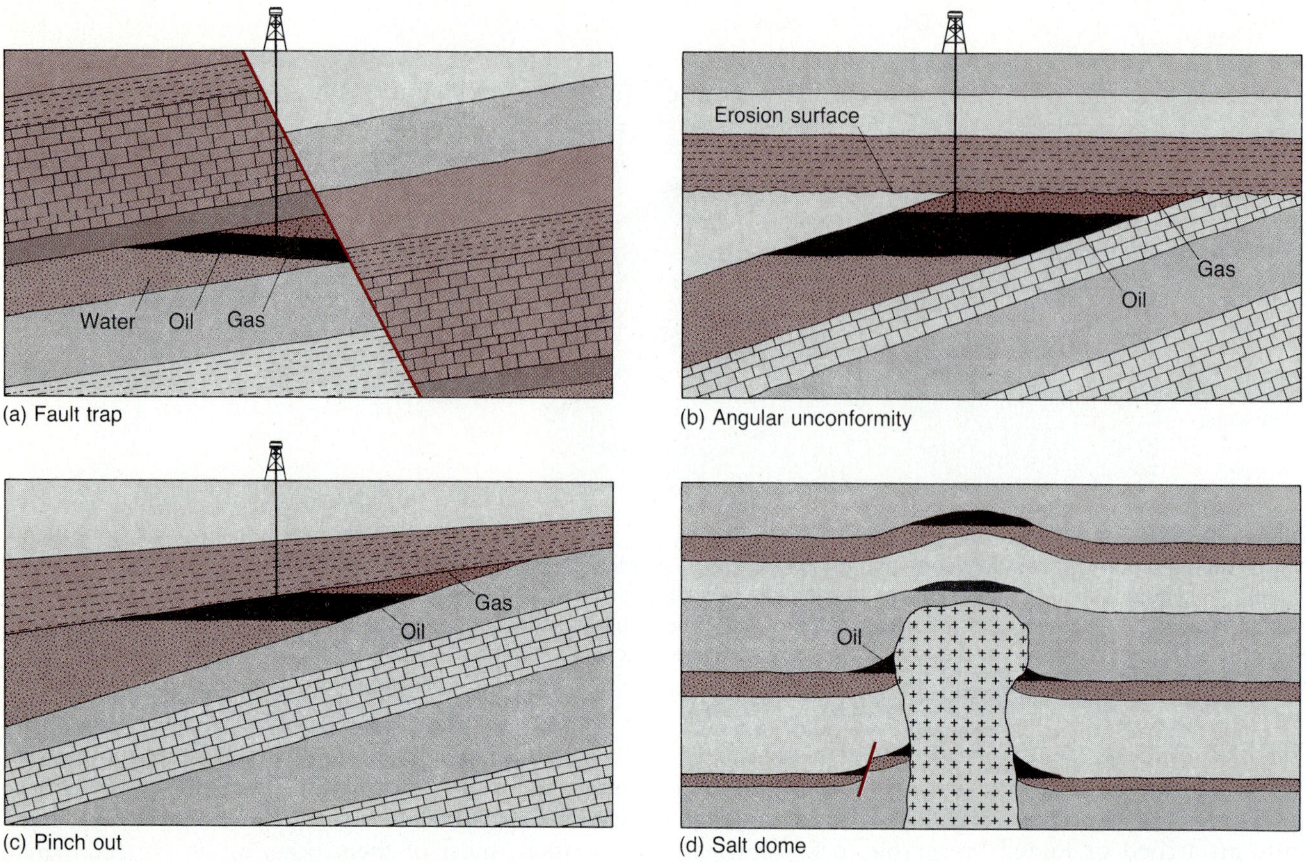

Figure 26.13
In addition to the anticlinal trap shown in Fig. 26.12, oil is trapped in a number of other ways. Among these are (a) fault traps, formed where oil and gas are sealed into a rock unit because they cannot pass through an impervious rock across a fault; (b) traps formed by unconformities; and (c) those formed where the oil-bearing layer pinches out. (d) Many important oil fields along the Gulf coast are associated with traps formed over and around plugs of salt that have penetrated the sediment over them.

Geology of oil and gas. Most of the world's proven reserves of oil and gas occur in a relatively small number of major sedimentary basins (Fig. 26.14). The major proven reserves are located in the countries listed in Table 26.3. These figures, compiled in 1977, indicate the quantities of oil that were then known and could be produced with the technology and at prevailing prices. As methods of producing oil improve, a larger proportion of the oil in the ground may be recovered from many of these fields. If our present understanding of the origin and occurrence of oil and gas is correct, almost all future discoveries of these resources will be made in environments similar to those of known reserves.

Within the United States, oil and gas occur in several quite different geological settings (Fig. 26.15). The Gulf Coastal Plain contains petroleum accumulations associated with salt domes, faults, and folds, as well as in stratigraphic traps in the Mesozoic and Cenozoic sediments of the region. Older, Paleozoic sedimentary rocks contain important reserves of oil and gas in the Great Plains, especially in Texas, Oklahoma, and Kansas; in large sedimentary basins in the interior lowlands of Illinois, Kentucky, and Ohio; and in the Appalachian Basin.

Additional significant reserves are found in the intermountain basins of the Rockies and in the much younger sedimentary basins in southern California. The most recent major discoveries have been made on the north slope of the mountains in northern Alaska.

Prospects for discovery of new giant oil fields in North America remain highly uncertain. The most prominent shallow prospects on the continent have

Table 26.3 Location of Major Sources of Proven Oil Reserves (in billions of barrels). (Data from the American Petroleum Institute.)

Saudi Arabia	150.0
Soviet Union	75.0
Kuwait	67.0
Iran	62.0
Iraq	34.5
United Arab Republic	32.4
United States	29.5
Libya	25.0
China	20.0
United Kingdom	19.0
Nigeria	18.7
Venezuela	18.2
Mexico	14.0
Indonesia	10.0

already been drilled. Likely sites of future major discoveries are at greater depth in existing producing areas, or more likely, beneath the continental shelf or rise. Exploration of these offshore areas is still in its early stages. The large new discovery east of Newfoundland holds out the hope that other major fields will eventually be found.

Tar Sands and Oil Shales

When the volatiles are lost from hydrocarbons, a thick, viscous asphaltic material called tar is left. Tar occurs where crude oil has migrated to the surface and as a pore-space filling of some sandstones. One notable formation, the Athabasca sands of Alberta, Canada, contains an estimated 300 billion barrels of tar. The tar can be removed from the sand by heating it; as the temperature rises, the tar flows. Once extracted from the sand, the tar can be refined to produce oils and gas. The cost of these processes has

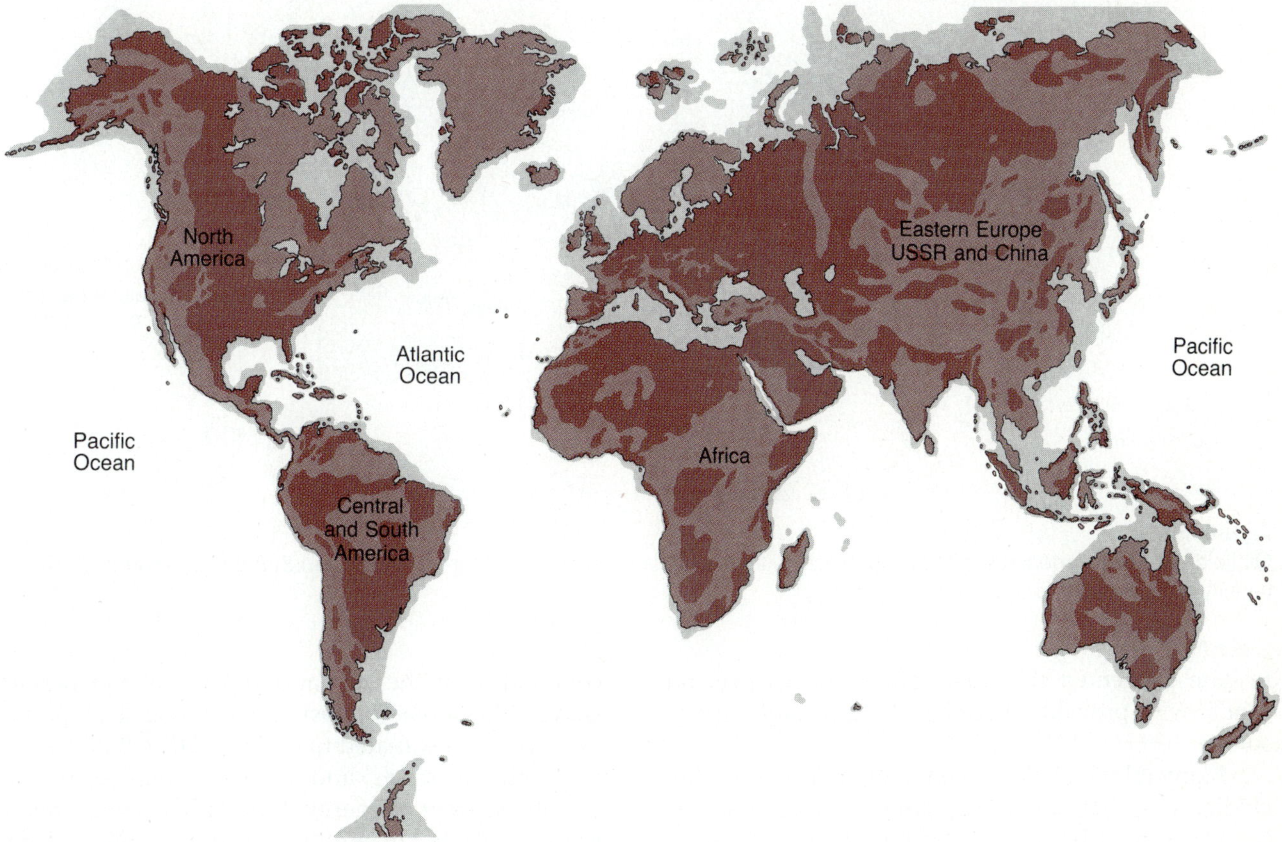

Figure 26.14
Almost all oil and gas discovered is found in sedimentary rocks. The sedimentary basins outlined here are the areas where future discoveries are most likely to occur. In addition, the continental shelves may be important reservoirs of petroleum.

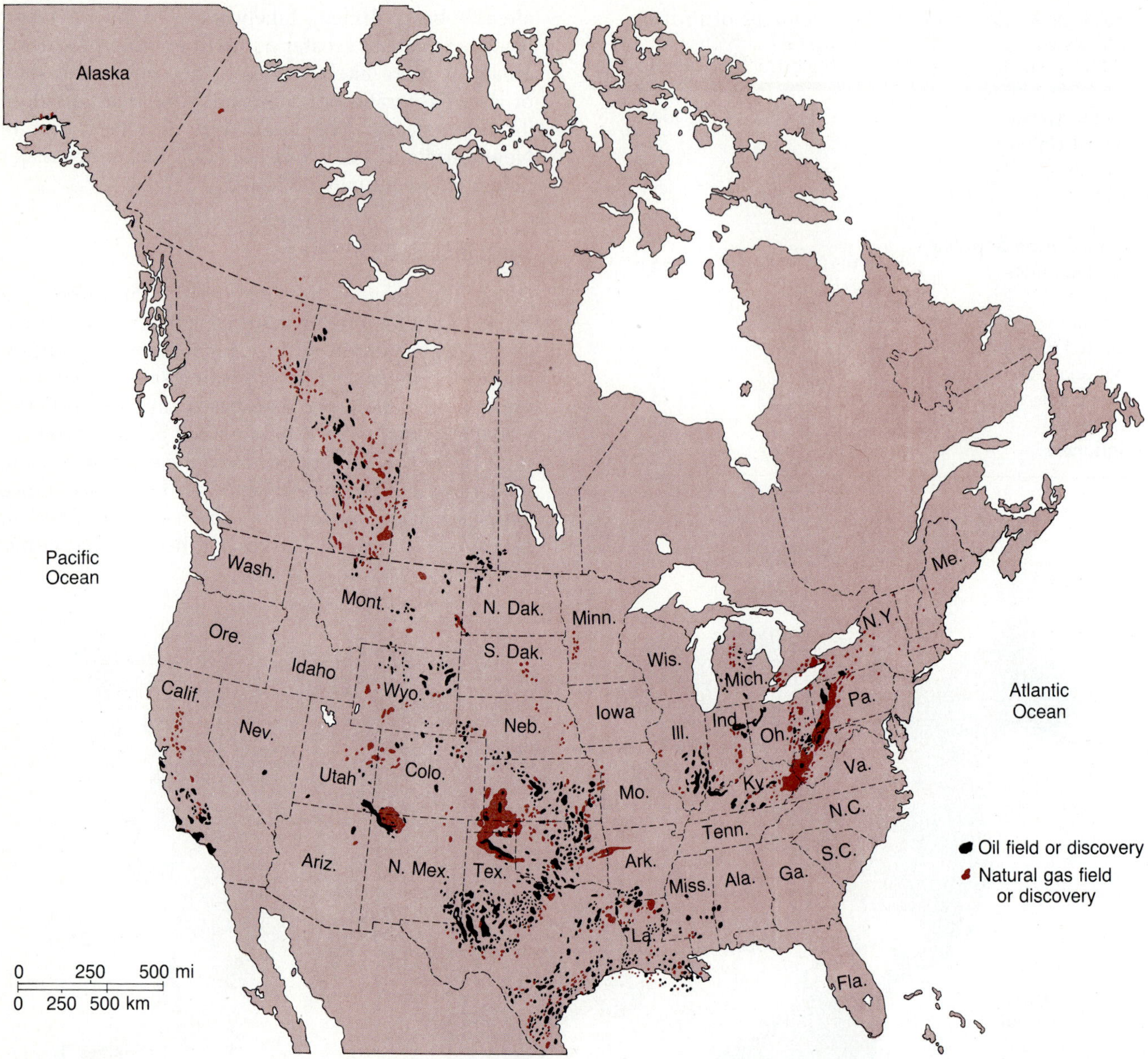

Figure 26.15
Oil fields in North America are shown in black, with gas fields in color. It is clear from this map that these resources are heavily concentrated in a few areas.

to date prevented their use, but as prices rise, tar sands will probably become a significant energy source.

Many black shales contain organic matter, including oils, that can be recovered. But like tar sands, the shale must be mined, heated, and processed to obtain the oil. The processes used to produce oil from shale also require large quantities of water, a scarce commodity in the semiarid regions of the western United States where most American oil shale occurs. Reserves are estimated to be about 200 billion barrels.

Both tar sands and oil shale may eventually become sources of energy, but neither is yet economically available. In addition, serious environmental problems are involved in the strip mining and disposal of the waste rock materials.

Uranium

Uranium, the principal nuclear fuel, is produced from igneous rocks; from pegmatite veins, which may contain the uranium ore pitchblende; and from sedimentary rocks. Most American deposits are of the last type. The uranium occurs as the minerals uraninite or carnotite in sedimentary rocks, where the uranium was apparently leached out of volcanic ash beds and then precipitated from groundwater. Precipitation occurs where uranium-bearing solutions encounter a reducing agent, such as organic matter. Thus, some of the important deposits in the Colorado plateau are found where buried logs have been replaced by uraninite. Precipitation has also occurred in organic-rich shales, such as a unit known as the Chattanooga shale in the southern Appalachians.

Extensive exploration for uranium deposits has gone on since the early 1940s. A large number of important deposits were located during the early stages of this exploration program, but additional large deposits have proved to be much harder to find. It seems likely that the large reserves anticipated in the 1940s may not exist.

The cost of constructing and maintaining nuclear reactors and of disposing of nuclear waste has made earlier estimates of our future use of nuclear energy highly uncertain. Problems at the reactor at Three Mile Island in Pennsylvania shook public confidence in the safety of reactors. This adds another element of uncertainty in the future growth of nuclear usage in America. However, other countries, notably France, are going ahead with major construction programs.

Geothermal Energy

The rocks beneath the surface of the earth constitute a huge reservoir of heat energy. Some of the heat found there is conducted from deep within the interior. Part of it is brought up by convection from within the mantle, but most of the heat in the granitic part of the crust is either produced by the decay of radioactive isotopes such as uranium, thorium, and potassium, or it is associated with igneous intrusions in the crust.

On the average, the temperature of rocks and of the water contained in pore spaces in them increases at the rate of about 1 °C for every 30 meters of depth. Thus, rock at a depth of 3 to 5 kilometers, depending on the local geothermal gradient, is hot enough to bring water to 100 °C. Unfortunately, the cost of pumping water up from that depth makes utilization of this heat uneconomic in most places. However, in many places the geothermal gradient is steeper than this average, around volcanic centers or buried intrusions. Boiling hot waters may actually come to the surface of the ground as geysers, fumaroles, or hot springs (Fig. 26.16), described in Chapters 15 and 21.

Heat resources in the ground occur in several different ways. Hot waters with temperatures in the range of 90 to 350 °C and water vapor with temperatures above 240 °C are easiest to use. Heat is also contained in partially molten rocks at temperatures above 650 °C and in hot, but dry, rocks with temperatures in the range of 90 to 650 °C. In addition, hot brines are found in many deeply buried sedimentary rocks.

Of these various heat sources, hot waters are the most suitable for development of geothermal energy. Heat energy around geyser basins and in a number of volcanic centers is already being used. In California, Japan, and especially New Zealand, naturally produced steam is used to generate electricity.

A survey of potential geothermal resources in the United States reveals three quite different types of sources. Fifty-eight volcanic systems in the contiguous states should contain hot waters that could be tapped at depths less than 3000 meters. An additional 85 volcanic systems are located in Hawaii and Alaska. A few of these have been developed; others are known to contain hot waters near the surface and

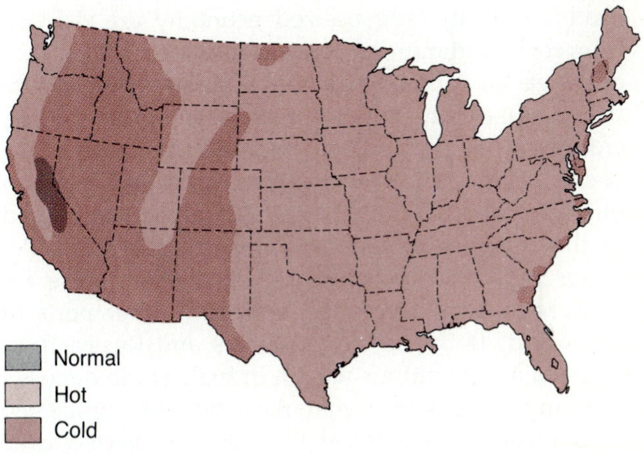

Figure 26.16
The areas of the continental United States where geothermal energy is most likely to be developed in the future are these areas where unusually high geothermal gradients are found.

can be easily developed. Many of them have not yet been drilled to determine the depth or volume of available hot waters.

A vast area of the western United States is underlain by hot, dry rock. It is estimated that some 95,000 square miles is underlain by rock at a temperature of 290 °C at depths of less than about 5000 meters. Here water might have to be introduced, circulated, and brought back to the surface in order to use the heat.

In the eastern United States, a number of areas along the Atlantic and Gulf Coastal Plains are probably sites of abnormally high geothermal gradients. These areas are places where thick blankets of water-saturated sediment lie over granitic plutons containing high levels of radioactive elements. Geothermal gradients in these places are expected to reach nearly 1.6 °C per 30 meters of depth.

It is still much too early to predict the extent to which these geothermal resources will contribute to our energy budget in the future. They are attractive because they are widely distributed and renewable, but they are not located close to large cities and are still in very early stages of development in the United States.

26.4 RESOURCES, POPULATIONS, AND ECONOMIC GROWTH

Both the raw materials and the energy sources used to produce, fabricate, and distribute most of the products of an industralized economy are derived from rocks and mineral deposits at or near the surface of the earth. A broad array of materials ranging from the metals—iron, copper, zinc, lead, gold, nickel, chromium, manganese—to minerals used as a source of chemicals (such as salt and potassium) and common building stones and road aggregate are encompassed in the category of rock and mineral resources. All major industrial countries either possess these resources or have access to them from other parts of the world. Both the United States and Russia have enjoyed a tremendous wealth in high-grade deposits of many of these raw materials, but the sufficiency of our resources to meet the needs projected over even the next few decades has been seriously questioned in recent years, and the debate about this vital and complex question continues. The seriousness of this problem for the United States is dramatically portrayed in Fig. 26.17, which shows the extent to which America has been dependent on imports for its supply of a number of mineral resources for more than a decade.

The realities of the problems of raw material supply have been impressed on the public most vividly as a result of the 1973 embargo placed on oil shipments by countries of the Middle East. This embargo proved dramatically that the United States is dependent on other countries of the world for a significant portion of its fuel supply—especially the fuel for automobiles—one of the most important products of our industry. We can be sure that a finite amount of petroleum exists in the rocks of the earth's crust, but we cannot be sure where it all is, or even approximately how much is left to be discovered within the United States. Estimates vary tremendously depending on the assumptions used to make the estimates.

In evaluating our reserves, it is important to distinguish between proven reserves, whose extent is known from drilling experience, and potential reserve estimates, based on assumptions about the probability of finding a resource in a given volume of sedimentary rock. Optimistic estimates of potential oil reserves in the United States are based on the possible existence of large quantities of oil in reservoirs off the east coast or at greater depth than most current drilling on the continent. Pessimists, on the other hand, point to the declining rate of discovery of new proven reserves. We have reached the point where more and more drilling is required for each new discovery. Thus, the outlook for our becoming self-sufficient in the long run in oil or many of the materials listed in Fig. 26.17 is very bleak indeed.

Pressures on Resources from Population Growth

In looking for causes of the resource supply crisis, two considerations are paramount—population growth and increased rates of consumption of resources. Estimating future population growth is uncertain at best, but the patterns that have emerged to the present give ample cause for alarm. A number of projections of future population growth for the world have been made, including three United Nations estimates (Fig. 26.18). These graphs illustrate a plausible range of population projections to the year 2000. The high estimate shows the population in 2000 to be nearly double the 1975 figure, and even the most optimistic estimate, based on a large reduction in fertility, shows

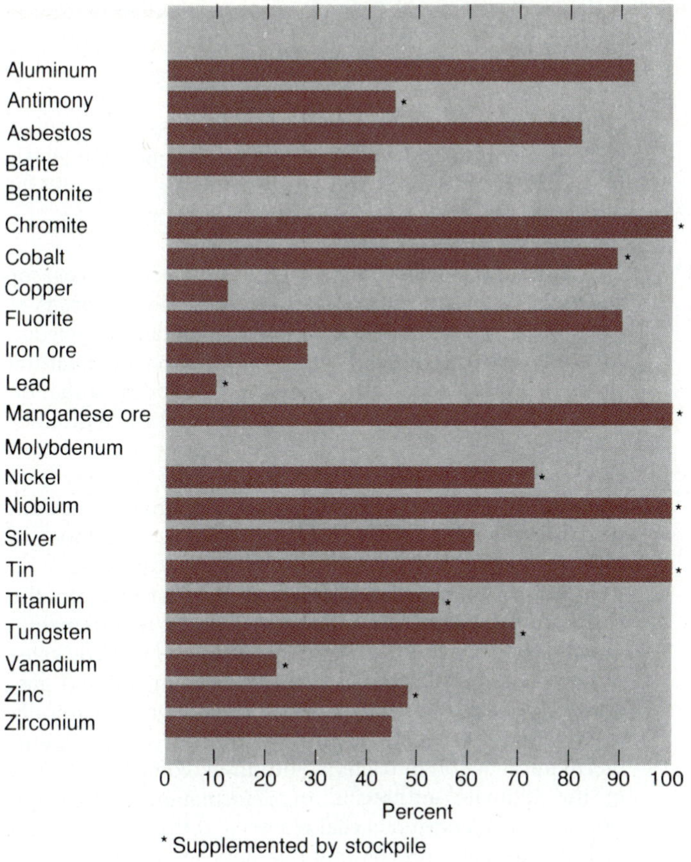

Figure 26.17
The extent of the dependence of the United States on foreign supplies of natural resources is clearly indicated by this graph. The figures shown here were determined for 1973. Since that time, the percent of imports has increased.

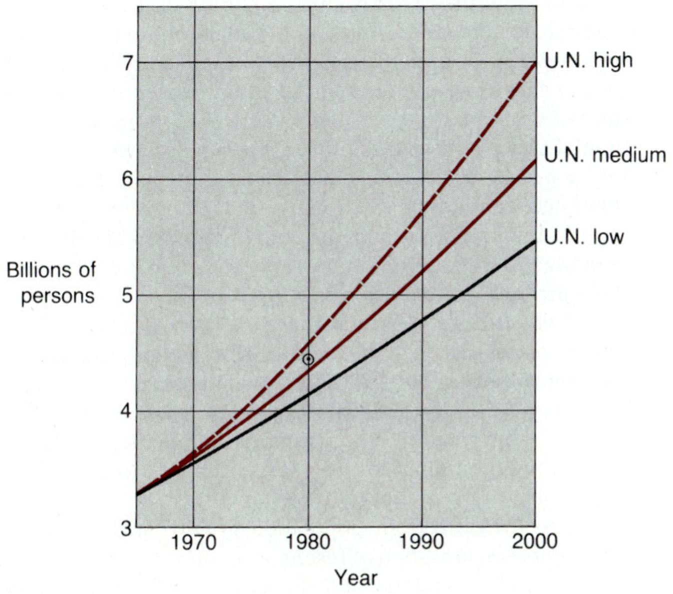

Figure 26.18
In 1965 the United Nations made a study of world population and its projected growth through 2000. Three growth rates were projected. In 1980 the world population was estimated to be about four and a half billion; thus the actual growth rate from 1965 to 1980 was between the U.N. high and medium rates.

almost a 20 percent increase. The demand that population increases of this magnitude will create for more natural resources is obvious. Even if population remained stable, vast quantities of all raw materials would be needed significantly to increase the standard of living of the present population.

In 1980 the United States government made a study entitled "Entering the Twenty-First Century." This study analyzes trends of population and natural resource usage. The following quotations are drawn from the conclusions of this study (*1*):

> If present trends continue, the world in 2000 will be more crowded, more polluted, less stable ecologically, and more vulnerable to disruption than the world we live in now. Serious stresses involving population, resources, and environment are clearly visible ahead . . . world population will grow from four billion in 1975 to 6.35 billion in 2000, an increase of more than 50 percent. . . . During the 1990s world oil production will approach geological estimates of maximum production capacity, even with rapidly increasing petroleum prices. The Study projects that the richer industrialized nations will be able to command enough oil and other commercial energy supplies to meet rising demands through 1990. With the expected price increases, many less developed countries will have increasing difficulties meeting energy needs. . . . While the world's finite fuel resources—coal, oil, gas, oil shale, tar sands, and uranium—are theoretically sufficient for centuries, they are not evenly distributed; they pose difficult economic and environmental problems; and they vary greatly in their amenability to exploitation and use. . . . Nonfuel mineral resources generally appear sufficient to meet projected demands through 2000, but further discoveries and investments will be needed to maintain reserves. . . . The quarter of the world's population that inhabits industrial countries will continue to absorb three-fourths of the world's mineral production. . . . Regional water shortages will become more severe. In the 1970–2000 period population growth alone will cause requirements for water to double in nearly half of the world. Still greater increases would be needed to improve standards of living.

From this and many other studies, we must conclude that our society faces serious resource and environmental problems. These are not problems far removed in the future, but ones this generation will have to face. It also seems clear that understanding the earth, the distribution of materials in the crust, and the processes that shape the surface are fundamental to the solution of the problems.

SUMMARY

Almost all the resources necessary to sustain an industrial society are drawn from the earth. Mineral deposits that can be exploited economically are said to be economically available. Rock or minerals from which metals can be derived at a profit are ores.

Although some mineral deposits, such as coal or building stone, are used in their entirety, most deposits of ore must be concentrated and refined to obtain the needed material. The ores themselves are often the result of natural processes of concentration, such as magmatic differentiation and pegmatite formation or deposition from heated groundwater solutions. Surface processes, such as residual concentration and secondary enrichment, may also serve to form economically available deposits. Weathering and stream deposition combine to form placer deposits, like those of gold that set off the gold rush in the American west.

Among the most important resources derived from the earth are those that fill our growing energy requirements. Today 90 percent of the energy used in the United States is from the so-called fossil fuels: petroleum, natural gas, and coal. Coal is vegetable matter that has altered chemically as the result of burial in an oxygen-poor environment, biochemical action of bacteria and fungi, compaction, and, in the ultimate anthracite, of deformation under high pressure. The enormous coal reserves of the United States include late Paleozoic bituminous deposits in the Appalachian Plateau and central states and deposits of younger, lower-rank coal on the Gulf Coast and in the Rocky Mountains.

Petroleum is believed to be of marine origin, formed largely from plants and invertebrates that grow in the oceans. How their remains are converted to the mixture of natural gas, crude oil, and tars that is petroleum is still unknown, but bacterial action is believed to be in part responsible. The environment most favorable to petroleum formation seems to be in the shallow waters of the continental shelves—especially those partially isolated from the open ocean. Most recoverable petroleum has, however, migrated from its site of formation into traps of porous and permeable rocks capped by some impermeable barrier. Among the more common trap structures are those formed by anticlines, faults, and salt plugs.

The earth's geothermal energy is now being tapped at many places around the world. Hot waters and steam around volcanic centers are being used to generate electricity in some areas. Future developments are planned for other areas of abnormally high geothermal gradients, but there are too few localities for geothermal energy to provide a major share of our energy needs.

The pressure from population growth and increased per capita consumption of resources point to a coming crisis in the supply of material and energy from the earth. Clearly our society will have to face different economic, environmental, and moral decisions on resource use in the coming decades.

KEY TERMS

anthracite coal
bauxite
bituminous coal
coal
gossan
hydrothermal deposits
laterite
magmatic differentiation
mechanical concentration
mineral deposit
ore
pegmatite
placer deposits
proven reserves
reserves
residual concentration
resources

STUDY QUESTIONS

1. What rocks and minerals are mined or quarried in the area where you live? What uses are made of them?
2. What processes of sedimentation cause the natural concentration of economically important minerals?
3. In what ways could groundwater bring about the concentration of economically important deposits?
4. How could knowing the way coal originates help in the exploration for coal?
5. Why should it be easier to calculate coal reserves than oil reserves?
6. What factors may cause a deposit to become economically unavailable?
7. Briefly summarize the advantages and disadvantages of each of the types of energy resources discussed in this chapter.
8. Identify some of the most important effects that might be expected if the demand for natural resources doubled. In what way would these effects differ if this doubling took place over a short time (a decade) rather than a long time (a hundred years)?

REFERENCE

1. Gerald O. Barney, study director, *The Global 2000 Report to the President: Entering the Twenty-first Century.* Washington, D.C.: The Council on Environmental Quality and the Department of State, April 1981.

SUGGESTED READINGS

Armstead, H. C. H., *Geothermal Energy.* London: Chapman & Hall, 1978.

Bowen, R., *Geothermal Resources.* New York: Wiley, 1979.

Brobst, Donald, and W. P. Pratt, eds., "United States mineral resources," *U. S. Geological Survey Prof. Paper 820* (1975).

Burk, Creighton A., and Charles L. Drake, "The impact of geosciences on critical energy resources," *American Association for the Advancement of Science, Symposium 21.* Boulder, Col.: Westview Press, 1978.

Cuff, D. J., and W. J. Young, *The United States Energy Atlas.* New York: The Free Press, 1980.

Landes, K. K., *Petroleum Geology of the United States.* New York: Wiley-Interscience, 1970.

Legget, R. F., *Cities and Geology.* New York: McGraw-Hill, 1973.

"Mineral Resource Perspective," *U.S. Geological Survey Professional Paper 940* (1975).

Ridge, J. D., *Ore Deposits of the United States, 1933–1967.* New York: AIME, 1968.

Skinner, Brian J., *Earth Resources,* 2nd ed. Englewood Cliffs, N.J.: Prentice-Hall, 1976.

APPENDIX A
MINERAL DESCRIPTION

I. Metallic Luster

Mineral and chemical formula	Hardness	Color	Description
Graphite, C	1	gray	Massive forms of graphite are most common: scaly, foliated, granular, or earthy. The color, dark gray, hardness of 1, one perfect cleavage, and greasy feel make it easy to identify.
Copper, (native), Cu	$2\frac{1}{2}$	copper red	Copper occurs in irregularly shaped masses in pore spaces in lavas and gravels. It is ductile.
Galena, PbS	$2\frac{1}{2}$	gray	Crystals with perfect cubic cleavage and lead-gray color combined with its high specific gravity make galena easy to identify, but it also occurs in granular aggregates.
Gold (native), Au	$2\frac{1}{2}$	gold	Gold usually is found as rolled scales, grains, or nuggets. Its high specific gravity and color distinguish if from most other metals. It is malleable and ductile and has no cleavage.
Bornite, Cu_5FeS_4	$2\frac{1}{2}$	bronze	This important copper-bearing mineral occurs in massive forms that are usually granular or compact. Bornite is opaque. The peacock-colored tarnish is most characteristic. It is found associated with chalcopyrite, malachite, and other copper-bearing minerals.
Chalcopyrite, $CuFeS_2$	3	brass yellow	Crystals are usually tetrahedra, but chalcopyrite also occurs in compact and granular forms. It may be distinguished from pyrite by its deeper color and greater softness. Common associates are pyrite, bornite, galena, sphalerite, and chalcocite.

I. Metallic Luster (*Cont.*)

Mineral and chemical formula	Hardness	Color	Description
Cuprite, Cu_2O	$3\frac{1}{2}$	red to black	Cuprite crystals are usually octahedral, but massive and earthy forms also occur. The red color, brownish-red streak, hardness, and translucence make cuprite distinctive. It is mined for copper.
Pyrrhotite, FeS	4	bronze	Pyrrhotite occurs in massive granular form, frequently mixed with chalcopyrite. It tarnishes readily and is slightly magnetic. It can be distinguished from pyrite and chalcopyrite by its color, tarnish, and magnetism. Nickel is often associated with it.
Chromite, $FeCr_2O_4$	$5\frac{1}{2}$	black	Chromite is usually found in granular masses with cubic grains. It is opaque, has a metallic luster, and is sometimes very slightly magnetic. It is an important ore mineral of chromium.
Hematite, Fe_2O_3	6	red or black	There are several important varieties of hematite. *Specularite* is characterized by its metallic luster, steel-gray or iron-black color, and reddish streak. It occurs as crystals, in a micaceous form, and as granular masses. *Oölitic* or *fossil hematite* is characterized by very small egg-shaped bodies. Hematite may occur as a cement in sandstone, as oölites, or as a replacement of fossils. Compact hematite occurs as kidney-shaped masses with fibrous radial internal structures. The luster is submetallic, and the color is iron-black or brownish-red. All hematite has a red streak.
Magnetite Fe_3O_4	6	black	Crystals of magnetite belong to the cubic system and are often octahedra. Magnetite also is found in compact, granular, and lamellar forms. Cleavage is indistinct. Unlike most other minerals, it is strongly magnetic.
Pyrite, FeS_2	6	brass yellow	Pyrite is commonly found in striated, cubic crystals, but it may also be found in massive granular forms. When weathered, pyrite may turn into limonite.
Cassiterite, SnO_2	$6\frac{1}{2}$	black	Cassiterite occurs as short, prismatic crystals or in granular forms, sometimes with radial fibrous structures. Cleavage is indistinct, and the color is highly variable, but usually is black.

II. Nonmetallic Luster

Mineral and chemical formula	Hardness	Color	Description
Talc, $Mg_3(OH)_2Si_4O_{10}$	1	variable	When in crystals talc is thin and tabular. More common are the foliated massive forms or fibrous, granular, or compact masses. Pearly luster and talc has a greasy feel.
Gypsum, $CaSO_4 \cdot 2H_2O$	2	white	Gypsum crystals are tabular or prismatic. Massive forms may be laminated, granular, fibrous, or earthy. It is usually colorless, white, or gray. Colorless transparent crystals or cleavage plates are called *Selenite.* Fibrous forms that have silky luster are called *satin spar.* Granular forms are called *alabaster.*
Clay Minerals			
Kaolinite, $Al_4Si_4O_{10}(OH)_8$; and montmorillonite, $Al_2Si_4O_{10}(OH)_2 \cdot H_2O$	2	white	Clay minerals cannot be distinguished from one another in hand specimens. X-ray diffraction is normally used to distinguish between them. Clay has a greasier feel than chalk, with which it may be confused. Clay will not effervesce in dilute acid; chalk will. *Kaolinite* is used in manufacture of paper. *Montmorillonite* is used in drilling muds. The mineral swells by absorbing water, producing a gel-like substance.
Sulfur, S	2	yellow	Granular, fibrous, crust, or compact forms of sulfur are most common. Straw-yellow color is characteristic. There is no cleavage.
Halite, NaCl	$2\frac{1}{2}$	colorless	Crystals of halite, common table salt, are cubic. Massive forms may be cleavable masses, granular, fibrous, stalactitic, or crusts. Colors are white, gray, or colorless. Because halite can absorb water it may feel damp and slick. The salty taste is characteristic, as is its perfect cubic cleavage.
Mica Group (complex potassium, aluminum silicates)			
Biotite (with Mg and Fe) Lepidolite (with Li) Muscovite (with K)	$2\frac{1}{2}$ 3 $2\frac{1}{2}$	black pink white	Members of the mica group are hexagonal in shape. Scalelike sheets formed as a result of the one perfect cleavage are the most characteristic occurrence in rocks. Each has a characteristic color: biotite is black, brownish, or blackish-green; muscovite is colorless, white, or yellowish; lepidolite is pink or lavender.
Chlorite, $Mg_5Al(OH)_8AlSi_3O_{10}$	$2\frac{1}{2}$	greenish	Crystals of chlorite are tabular, hexagonal, and often bent. Colors are shades of green. One perfect cleavage is responsible for the micaceous appearance in most occurrences. Chlorite occurs in metamorphic rocks and as a scaly coating on other minerals. The flakes are flexible but not elastic.

II. Nonmetallic Luster (*Cont.*)

Mineral and chemical formula	Hardness	Color	Description
Serpentine, $Mg_6(OH)_8Si_4O_{10}$	2–5	green or black	Serpentine is found only in massive forms that may be compact, columnar, fibrous, or granular. The luster is greasy or waxy. It is often spotted, clouded, or multicolored. Asbestos is a variety of serpentine.
Calcite, $CaCO_3$	3	variable	Crystals of calcite are found in many shapes, and although these crystals or granular aggregates are most common, it is also found in stalactitic form in caves. Color is highly variable. In clear specimens, double refraction of light may be observed—two images of objects below the calcite will be seen. There are three perfect cleavages, not at right angles. Calcite will effervesce in dilute hydrochloric acid.
Anhydrite, $CaSO_4$	$3\frac{1}{2}$	colorless to gray	Anhydrite usually occurs in massive, fine granular, or fibrous forms, rarely as crystals. There are three mutually perpendicular cleavages that distinguish it from gypsum, with which it is frequently associated. The mineral is translucent and has pearly to vitreous luster.
Azurite, $Cu_3(OH)_2(CO_3)_2$	$3\frac{1}{2}$	dark blue	The azure-blue color is characteristic of azurite, which will effervesce in hydrochloric acid. It is a minor ore of copper, most commonly found with malachite.
Sphalerite, ZnS	$3\frac{1}{2}$	brown yellow	Resinous luster is one of the important characteristics of sphalerite. Yellowish colors are also typical. Crystals, when found, are tetrahedral. Cleavage is prominent in six directions.
Dolomite, $(Ca,Mg)(CO_3)_2$	$3\frac{1}{2}$	variable	Dolomite is similar in appearance to calcite, but its crystals, of rhombohedral shape, have curved faces. It also occurs in massive forms. It is distinguished from calcite by its failure to effervesce in weak hydrochloric acid.
Bauxite, Al hydroxides	variable	variable	Bauxite is found in a massive, earthy, or claylike form. Rounded pea-shaped concretions are characteristic. It is the main ore of aluminum. Hardness is variable.
Malachite, $Cu_2(OH)_2CO_3$	4	green	Crystals of malachite are needlelike and form in groups. It is more commonly found in massive fibrous, stalactitic, kidney-shaped masses with internal banding, as velvety crusts or as earthy masses. The silky luster and bright green color are characteristic.
Fluorite, CaF_2	4	variable	Fluorite occurs in cubic crystals and in cleavage masses or in granular form. Color is variable, but purples and greens are most common. The cleavage is octahedral and is

II. **Nonmetallic Luster** (*Cont.*)

Mineral and chemical formula	Hardness	Color	Description
			perfect. Fluorite is most often confused with calcite, but differs in its hardness of 4.
Kyanite, Al_2SiO_5	4–7	blue	Long, bladed crystals that are curved and radially grouped are characteristic of kyanite, as are the bluish streaks or spots in it. The hardness varies with direction, being 4 in one direction and as much as 7 in the other. It occurs in metamorphic rocks.
Limonite, $FeO(OH) \cdot H_2O$	5	rust	Limonite is usually found in massive forms that are nodular, compact, earthy, or stalactitic. The rusty color is characteristic of the surface. The compact forms are fibrous, but earthy limonite is more common.
Apatite, $CaPO_4$ (with F, OH, Cl)	5	highly variable	Crystals of apatite are hexagonal prisms. It also occurs in modular or earthy massive form in crystalline limestone, ore deposits, and igneous rocks. Color is highly variable. The single cleavage is imperfect.
Hornblende (complex silicate with Ca, Mg, Fe, and Al)	$5\frac{1}{2}$	black	Hornblende occurs in long, prismatic crystals and in massive forms of small, bladed, fibrous, granular, or compact grains. There are two good cleavages at angles of 56° and 124°.
Augite (complex silicate with Ca, Mg, Fe, and Al)	$5\frac{1}{2}$	black	Augite is the most common member of the pyroxene group. It is most easily confused with hornblende, but the two can be distinguished by the angle between the cleavage faces. The acute angle between cleavages in augite is 87°, compared with an angle of 56° in hornblende. Most augite is altered to hornblende, and both cleavage angles may be observed on one specimen.
Feldspar Group			
Potash Feldspars:			
Microcline, $KAlSi_3O_8$	6	green	Both orthoclase and microcline feldspars contain potassium. They occur as thick tabular crystals and in massive forms that may be granular or cleavage masses. Common colors are white, gray, pink, and, for microcline, green. They may be distinguished from other feldspars by rectangular cleavage and by absence of fine striations on cleavage surfaces that are the result of twinning.
Orthoclase, $KAlSi_3O_8$	6	pink, yellow, or brown	
Plagioclase Feldspars:			
Albite, $NaAlSi_3O_8$	6	white or gray	Albite, labradorite, and anorthite are *plagioclase* feldspars. These contain various amounts of sodium and calcium. Albite is one end member of the group, containing only sodium. Anorthite, the other end member, contains only calcium. Other
Labradorite, $(Na,Ca)AlSi_3O_8$	6	gray or green gray	
Anorthite, $CaAl_2Si_2O_8$	6	colorless or white	

II. Nonmetallic Luster (*Cont.*)

Mineral and chemical formula	Hardness	Color	Description
			members of the group contain varying amounts of sodium and calcium, and the physical properties vary with the relative amount of each of these constituents. The plagioclase feldspars are all twinned. There are fine striations on crystal and cleavage faces. Labradorite is the most easily identifiable member of the group because it is usually gray and has a play of colors as it is turned in light.
Epidote (complex silicate with Ca, Al, Fe)	$6\frac{1}{2}$	green, brown, or yellow	Epidote occurs as elongated prismatic crystals, but is more commonly found in massive forms that may be columnar, fibrous, or granular. The greenish-brown, greenish-yellow, and yellow colors are characteristic. Common associates are feldspars, amphiboles, and pyroxenes.
Garnet, $(Ca,Mg,Fe,Al)(SiO_4)$	$6\frac{1}{2}$	variable	Cubic or dodecahedral garnet crystals are common, but it also occurs in massive granular forms. Color is highly variable, but deep-red colors are characteristic of some varieties. Cleavage is not distinct. Hardness is one of the best ways of distinguishing it from similar minerals.
Olivine, $(Mg,Fe)_2SiO_4$; fayalite, Fe_2SiO_4; forsterite, Mg_2SiO_4	$6\frac{1}{2}$	green brown and black white	Olivine is usually found in granular masses. The green color is typical, and striations of the crystal faces may be apparent in larger crystals, but are difficult to see in granular aggregates.
Sillimanite, Al_2SiO_5	$6\frac{1}{2}$	gray brown	Sillimanite occurs as long, thin needlelike forms or as radiating fibrous masses in metamorphic rocks. Silky luster is an important characteristic of most occurrences.
Quartz, SiO_2	7	variable	Crystalline varieties of quartz are hexagonal prisms, with horizontally striated crystal faces. This group includes rock crystal quartz and milky quartz. The conchoidal fracture, glassy luster, and hardness are characteristic. Cryptocrystalline varieties, in which the crystal structure is not apparent are also very common. They appear as massive forms that may be nodular, kidney-shaped, banded, concretionary, stalactitic, or compact masses. The luster of these forms is waxy or vitreous. *Chalcedony* has a waxy luster, hardness of 7, and conchoidal fracture. *Agate* is banded chalcedony. *Onyx* is agate in which the lines or bands are even and straight. *Jasper* is distinguished by its red color, which is derived from specks of

II. Nonmetallic Luster (*Cont.*)

Mineral and chemical formula	Hardness	Color	Description
			hematite in the quartz. Opal is an amorphous silica mineral that contains water.
Staurolite (a complex silicate with Fe and Al)	7	gray, brown, or black	Prismatic crystals of staurolite are common, usually in a crosslike shape. It is found in metamorphic rocks.
Tourmaline (a complex silicate of B and A)	$7\frac{1}{2}$	black, pink, or brown	Crystals of tourmaline are common as long or short hexagonal prisms characterized by a triangular outline in cross section. Color is sometimes distributed zonally along the crystals, which may vary from transparent to nearly opaque.
Corundum, Al_2O_3	9	variable	Corundum is frequently found in hexagonal and barrel-shaped prismatic crystals, but also occurs in massive granular forms. Color is variable, but the most common colors are gray, green, and blue. Hardness is the most distinguishing feature, but care must be exercised in determining the hardness because chlorite is often associated with corundum, which makes it appear softer than it is.
Diamond, C	10	colorless	Hardness of diamond is its most distinguishing feature. Diamond crystals are most commonly octahedra, cubes, and slight modifications of these shapes, but they are often rounded and distorted. Diamond has octahedral cleavages that are perfect. In addition to the colorless varieties there are yellowish, red, green, blue, and black diamonds. The brilliance of diamond is due to its high dispersion of light.
Carnotite (K,U,O,V,H_2O)	—	bright yellow	Carnotite occurs as a powdery incrustation in sand or sandstone. Canary-yellow color and radioactivity are the most distinctive characteristics.

APPENDIX B1
DESCRIPTIONS OF COMMON IGNEOUS ROCKS

COARSE-GRAINED INTRUSIVE ROCKS

Granite. A coarse- to medium-grained equigranular rock composed of orthoclase or microcline and quartz, with some biotite mica and hornblende, usually present in small amounts. Many other minerals may be present in small amounts. These include black specks of magnetite, honey-yellow crystals of sphene, red grains of garnet, and other less common minerals.

Migmatite. Rocks of mixed nature containing metamorphic- and igneous-appearing rock. Sometimes the term is restricted to injection of granite into schist along the layering or schistosity of the schist.

Granite pegmatite. A very coarse-grained rock with granitic composition. It generally contains microcline, quartz, and mica (biotite or muscovite). A great variety of rare minerals may be present.

Syenite. A coarse- to medium-grained equigranular rock composed primarily of orthoclase. Some hornblende, biotite, and pyroxene are present. Either a small amount of quartz or nepheline may be present.

Granodiorite. A medium- to coarse-grained plutonic rock that contains quartz, calcic plagioclase, and orthoclase as light-colored constituents, and biotite, hornblende, or pyroxene as mafic constituents. It contains at least twice as much sodic plagioclase as orthoclase.

Diorite. A coarse- to medium-grained rock composed of intermediate plagioclase and biotite, hornblende, or pyroxene. Small quantities of quartz may be present. The amount of dark minerals ranges from 12 to 36 percent of the total.

Gabbro. A rock composed of coarse- to medium-grained pyroxene, hornblende, and biotite. The amount of these may exceed the amount of feldspar, which is calcic plagioclase. The mineral olivine is usually present, and quartz is absent. Most gabbros are dark because dark minerals predominate. They are equigranular. It is not uncommon to find gabbro pegmatites.

Peridotite. A dark, coarse-grained, equigranular rock containing a large quantity of olivine and/or pyroxene or hornblende, but no quartz or feldspar.

Dunite. A rock composed almost completely of the mineral olivine; accessory pyroxene and chromite may be present.

Pyroxenite. A dark, coarsely crystalline rock composed mostly of pyroxenes.

Hornblendite. An igneous rock of coarse texture and nearly equigranular fabric composed mostly of hornblende.

FINE-GRAINED IGNEOUS ROCKS

Felsite (rhyolite). A general term applied to igneous rocks of fine-grained texture and light color, indicating a granitic composition. If large crystals are present in a felsitic groundmass, it is called a felsite porphyry. Colors of felsites range from white, gray, pink, yellow, or brown, to purple and light green. When phenocrysts of quartz are present the rock is termed a quartz felsite or, more commonly, a quartz porphyry.

Basalt. A dark-colored, fine-grained rock composed primarily of plagioclase and pyroxene, with or without olivine. Basalts are dark because they contain large percentages of hornblende, pyroxene, biotite, olivine, or other dark minerals. It is often an extrusive igneous rock and may have cavities formed by bubbles of gas.

Diabase (dolerite). A rock of basaltic composition, consisting mainly of plagioclase feldspar and pyroxene, and with a texture characterized by well formed crystals of plagioclase embedded in a fine-grained matrix of pyroxene.

GLASSES

Obsidian. A glassy rock compositionally equivalent to granite or felsite; a solid, natural glass containing no crystals. Most obsidian is black, but it also occurs in green and brown. It breaks with a conchoidal (curved) fracture.

Pitchstone. A variety of obsidian with a luster like that of a resin instead of glass. Pitchstone differs in composition from obsidian in that it contains 5 percent or more of water. Colors include black, gray, red, brown, and green.

APPENDIX B2
DESCRIPTIONS OF COMMON SEDIMENTARY ROCKS

MECHANICALLY DEPOSITED SEDIMENTARY ROCKS

Coarse-grained Fragmental or Clastic Rocks

Conglomerate. The name applied to a consolidated gravel. Fragments that make up conglomerates may be composed of any other rock. An appropriate prefix may be added to indicate the size of the major constituents of the rock; for example, pebble conglomerate or boulder conglomerate. Fragments in a conglomerate are rounded, usually as a result of rolling in a stream or in waves.

Breccia. A rock composed of angular fragments. Breccias occur in volcanic vents where sides of the vent are broken by explosions. They may form from fragments produced by the collapse of the roof of a cave, by fracture and deformation of brittle rocks that are folded, and by faulting.

Sand-sized Fragmental Rocks

Sandstone. Any rock composed of fragments of the size range 1/16 millimeter to 2 millimeters, regardless of composition. Most sandstone is composed of quartz grains, but many other minerals may be present or even predominate. Sand dunes composed of fragments of ice are found in the Antarctic; some beaches in Italy are made of olivine sand derived from volcanic rocks; and sand dunes are composed of gypsum at White Sands National Monument, New Mexico.

Arkose. Made largely of a mixture of quartz and feldspar fragments. It may contain small angular rock and mineral fragments. Frequently arkose is red or pink, colors derived from oxidation of iron-bearing minerals and pink orthoclase feldspar.

Graywacke. An impure sandstone consisting of quartz and feldspar fragments and small fragments of igneous, metamorphic, and sedimentary rocks. One common association is ash and volcanic dust with quartz and feldspar fragments. The color most commonly is gray, which comes from a matrix that is composed of a mixture of mica, chlorite, and quartz.

Silty Rocks

Siltstone. A consolidated rock composed of clastic fragments in the size range of 1/16 millimeter to 1/256 millimeter in diameter. Rocks composed of a large amount of silt are much less common than are sandstones or shales. Shales may contain up to 50 percent silt.

Loess. A light-buff, unconsolidated silt that is usually homogeneous and poorly stratified, or even unstratified. The particles of loess are very well sorted; they are almost all the same size, averaging from 0.01 millimeter to 0.05 millimeter. It is fine enough to be generally interpreted as a wind-blown deposit.

Clay-rich Rocks

Shale. Consolidated mud and clay. It often contains sand or lime in large amounts and usually has an earthy odor. Most shales are fissile; that is, they break or split along nearly parallel planes. Other important properties of shale are its low permeability (ease with which water moves through it), generally low porosity, particle size (less than 1/256 millimeter in diameter), and composition. Shales are composed of clay minerals, quartz, sericite, chlorite, feldspar, calcite, and small quantities of many other minerals. Unlike most other sedimentary rocks, shale owes its rocklike character primarily to compaction of the particles rather than cementation.

CHEMICALLY DEPOSITED SEDIMENTARY ROCKS

Siliceous Deposits

Some sediments containing silica are deposited directly from water; others are formed through processes of recrystallization of the siliceous remains of microorganisms in sediments during consolidation. The most common occurrences of chemically deposited siliceous sediments is as nodules in layers of limestone. Many of these nodules contain fossils at their centers, showing that they are of a secondary origin.

Chert (flint). A dense, hard, siliceous rock in which color ranges from white through gray to black. Chert is a form of quartz, SiO_2. It has the same hardness (7), a conchoidal fracture, and semivitreous luster.

Siliceous sinter. A chemical sediment formed at mineral springs. It consists of silica and is white or light colored and porous. When it is formed around the vents of geysers, it is known as *geyserite.*

Limestone and Other Calcareous Deposits

Many of the calcareous sedimentary rocks (largely composed of calcium carbonate, calcite) may be classed as organic rocks because they are composed of shells or are deposited by organisms. Others are chemical precipitates. Limestones tend to have many of the physical properties of their main constituent, calcite. They have a hardness of about 3, and they effervesce in hydrochloric acid.

Travertine and tufa. Limestones formed by evaporation of spring, stream, and groundwaters containing calcium carbonate in solution. The name *tufa* is applied to spongy, porous, fragile deposits with an earthy texture. The deposits frequently contain branches, twigs, and other debris that fall into the water. Dense, banded, and compact deposits are called *travertine.* Cave deposits are usually of this type.

Caliche (duricrust). Deposits formed in the soils of semiarid regions underlain by soluble rocks. Caliche is often composed of calcium carbonate, but the name is also applied to nitrates of soda and alumninous, ferruginous, and siliceous materials.

Dolomite (dolostone). A sugary-textured, dense rock that does not effervesce in dilute acid unless it is powdered. Dolomite is a calcium magnesium carbonate, $(Ca,Mg)CO_3$. Dolomites may be formed during recrystallization of calcareous sediments. Because calcium and magnesium ions are nearly the same size, it is possible for magnesium to replace calcium in the calcite structure. Dolomitized fossils have been found, but no animal is known to construct its shell of dolomite.

Iron-bearing Sediments

Bedded siderites. Rocks that are mined as iron ore in Michigan, but also commonly contain Mg and Mn as well as an association of siderite ($FeCO_3$) and chert. In this area both siderite and chert are thought to be direct chemical precipitates.

Iron-silicate sediments. Silicate minerals containing iron, occurring in mudstones and limestones. One of these, glauconite (iron-potassium silicate) is forming on the continental shelves today. It is commonly known as *greensand* and looks like other sands except for its color. Glauconite is closely related to the micas and is essentially an aggregate of colloids of hydrous potassium iron silicate formed through weathering of iron minerals.

Sedimentary hematites. Important iron ores. Some of these are composed of fossil fragments that have been replaced by hematite; in others, oolitic ores have cores of quartz grains around which layers of hematite were deposited.

Salts

Salts are deposits formed by precipitation from concentrated solutions of brines. Three salts are of particular importance: halite (common table salt, sodium chloride); and gypsum and anhydrite, which are calcium sulfates. These, along with less common salts, are precipitated in a sequence from waters that become saturated. Such saturation of sea water with salts may occur when a part of the sea becomes cut off, leaving an isolated body of water that evaporates. Similar conditions hold for inland bodies of water such as Great Salt Lake.

ORGANICALLY FORMED DEPOSITS

Siliceous Deposits

Siliceous deposits are formed from large amounts of silica.

Radiolarian oozes. A group of single-celled animals that construct skeletons of silica, which accumulate in great quantities and may form a large part of the sediment in areas where rates of sedimentation are slow, and particularly where water is deep (for example, at depths between 4,000 and 8,000 meters). Silica is stable at the cool temperatures and high hydrostatic pressures found in the deep seas, but calcareous shells tend to dissolve under these conditions.

Diatom ooze. Siliceous plants of microscopic size that may be rodlike, spherical, or look like a circular disc, forming rock with an earthy appearance and texture. It is a loose, fine, white, powdery rock, resembling chalk.

Calcareous Deposits

Calcareous deposits contain calcite or calcium carbonate.

Globigerina ooze. Calcareous deposits that are composed largely of the shells of protozoans, particularly one group characterized by globular-shaped shells and known as *globigerina.* These cover vast areas of the modern-day sea floor.

Fossiliferous limestone. Primarily those limestones formed from the shells of marine animals. These shells are composed of the mineral calcite, which the animals are able to take from seawater and build into structures that house their soft parts. Shells accumulate in quantity in shallow seas and eventually become cemented by calcite, silica.

Chalk. A fossiliferous limestone composed of the shells of protozoans and particularly fossil *globigerina.* Chalk is white and has the property of being so soft that it will easily mark most things.

Marl. Mixtures of shells and shell fragments with muds, clay, particles of calcite or dolomite, or sand. It is an impure limestone usually found in a semiconsolidated state, held together loosely.

Coquina. Sedimentary rocks composed of a loose aggregation of shell fragments cemented together as rock.

Phosphatic Deposits

Bones and bird excrement are two organic sources for phosphate. Neither is common in rocks. Economically important deposits of bird excrement, *guano,* are confined to a few islands where birds have lived in large numbers for long periods of time. Accumulations of bones are rarely found in large enough quantities to be called rocks.

Ferruginous Deposits

Certain bacteria and algae may cause deposition of ferric oxide and ferrous sulfide. The iron bacteria can extract iron from solution and deposit it around their cells. Others may perform the function of gathering the materials, which are then directly precipitated. An accumulation of bacteria casts and precipitated granules forms a rock called *bog iron.* As the name suggests, it is deposited in bogs, swamps, or marshes.

APPENDIX B3
DESCRIPTION OF THE COMMON METAMORPHIC ROCKS

Slates. The most perfectly foliated metamorphic rocks. The foliations or slaty cleavages are so closely spaced that a single piece of slate may be split into thin sheets. The mineral constituents are so small that they cannot be identified with the unaided eye.

Phyllites. Fine-grained schistose rocks identified by the lustrous, silky sheen that characterizes light reflected from chlorite and muscovite micas of which they are composed. Quartz and albite are often present. The grain size is larger than that of slates but finer than that of schists. They are usually greenish or red and may show the initial stages of segregation of some mineral constituents into layers.

Schists. Strongly foliated rocks of medium to coarse crystalline texture. Most of the mineral constituents are easily identified without the use of a microscope (this is not true of phyllites). Foliation is caused by parallel or nearly parallel alignment of micaceous minerals. The most common minerals in schists are quartz, feldspars, and micas. If one of the constituents makes up 50 percent or more of the rock, the name of that constituent is attached as a modifier (such as mica schist, quartz schist, or hornblende schist). If no constituent comprises 50 percent, the names of the two most abundant constituents are used (for example, garnetiferous-mica schist).

Gneisses. Characterized by compositional layering, which produces a banded appearance. Gneiss is medium- to coarse-grained; it usually contains quartz and feldspar interlayered with thin layers rich in hornblende or mica that induce planar weakness in the rock. The quartz, feldspar, and other constituents usually have interlocking boundaries. The layering is thought to develop as a result of segregation of mineral constituents during metamorphism, and it often corresponds to original bedding of metasedimentary rocks. When gneisses are known to be derived from igneous rocks they are called *orthogneisses.* When the parent material is sedimentary, the name *paragneisses* is applied.

Schists are distinguished from gneisses by the way schists break into thin platy slabs (1 to 10 millimeters thick) or into pencil-like columns, whereas gneisses break into much thicker slabs or even across the layering. Most gneisses also have a much higher feldspar content. Twenty percent feldspar is used to distinguish the two rock types in some classifications.

Marble. The metamorphic equivalent of calcite, limestone, or dolomite. Marbles often contain micaceous impurities that give them a foliation. The texture is usually that of an interlocking mosaic growth of calcite or dolomite crystals.

Quartzite. The metamorphic equivalent of quartz sandstones, composed of about 80 percent or more quartz. The grain boundaries of metaquartzites are interlocked and the rock often has some foliation. Tightly cemented sandstones, called *orthoquartzites,* may resemble *metaquartzites* in hand specimens. Orthoquartzites form as a result of cementation of sandstone by silica from groundwater in the absence of true metamorphism. The boundaries of the original grains of sand are usually visible. The cementing material is silica, and the bonds are so strong that, when quartzite is broken, fractures cut indiscriminately across sand grains and cement. Orthoquartzite is interbedded with unmetamorphosed sedimentary rocks.

Hornfelses. Fine-grained, nonfoliated, dense, usually dark rocks formed near the contacts of igneous intrusions. The minerals form a mosaic of interlocking grains. They commonly break into splintery fragments that have translucent edges, like horn. The term is most frequently applied to baked shales.

Amphibolites. Contain mostly plagioclase and hornblende, and often some biotite. The prismatic hornblende crystals lie in the plane of foliation and may induce a planar fissility or a linear schistosity. Plagioclase is often strongly aligned also. Quartz-free amphibolite is usually derived from basaltic or mafic tuffaceous rocks. When quartz is present it suggests the possibility of derivation from marls or other sedimentary rocks.

GLOSSARY*

Ablation. The combined processes, particularly melting and sublimation, by which a glacier wastes.

Ablation till. Glacial till deposits formed from the debris that is transported within or on the ice and gradually lowered as ablation causes the glacier to shrink.

Abyssal plain. Flat area in the deep ocean having low relief; irregular in shape and found in a great range of sizes up to widths of several hundred kilometers.

Accumulator plant. Plant that accumulates and stores certain elements or compounds.

Aggradation. The process of building up a depositional surface.

Alluvial. Referring to materials (sand, gravel, etc.) transported by and deposited from streams.

Alluvial fan. A fan-shaped deposit of stream alluvium laid down where a change in stream gradient occurs.

Alluvium. Material transported by streams and deposited in stream valleys.

Amorphous solid. A solid that does not possess a crystalline structure; for example, glass.

Andesite line. The line around the northern and western Pacific Ocean that separates volcanoes erupting basalt from those erupting andesite or other sialic volcanic products.

Angle of repose. The steepest slope on which a uniform, uncemented, granular material can maintain stability.

Angular unconformity. An erosion surface separating folded or tilted rocks below from less deformed rocks above.

Anomaly (gravity). A gravity measurement that is greater (positive) or less (negative) than would be observed at a locality that fits an idealized model of the earth.

Antecedent. A drainage that predates the origin of structural features, such as ridges or anticlines, across which the streams flow. The streams maintain their courses while the structure develops in the underlying rock.

Anthracite. Coal that is dark black in color, dull to brilliant and even submetallic in luster. It is hard, burns with a short, blue flame, and emits little odor.

Anticline. An upfold in which the limbs dip away from the fold axis. Beds exposed in the crest are older than those exposed on the limbs after erosion.

Anticlinorium. A fold system located on an uplifted area.

Aphanitic. A textural description in which individual crystals in a rock are too small to be seen or identified by the unaided eye.

Aquifer. A water-bearing strata or rock body.

Arc-trench gap. The horizontal distance in an island arc between the axis of the deep-sea trench and the center of volcanic activity.

Arête. Sharp ridge produced by glacial erosion.

Argillaceous. Containing clay. See also *lutaceous.*

Arkose. An arenaceous rock made largely of a mixture of quartz and feldspar fragments.

Arroyo. A dry stream channel; commonly found in the desert regions of the southwestern United States.

Artesian aquifer. A water-bearing rock unit that contains water under pressure.

Artesian conditions. Situation in which groundwater is under sufficient pressure to rise above the zone of saturation.

Ash. A designation in size ($\frac{1}{16}$ to 2 millimeters in diameter) given to fragmental eruptives.

Assimilation. The digestion or incorporation of materials originally in the wall rock of a magma chamber into the magma.

* The *Glossary of Geology and Related Sciences,* published by the American Geological Institute, has been followed as the authority for geological definitions in the text and in this glossary. The student should refer to the A.G.I. *Glossary* for words not defined here, for various meanings used for a word, and for original sources of the terms.

Asthenosphere. A zone in the earth some tens of kilometers deep in which the rock is plastic and where any stresses are removed by flowage.

Asymmetrical fold. A fold in which the limbs are not symmetrical about the axial surface.

Atlantic-type margin. Continental margins of the type found around most of the Atlantic Ocean; characterized by absence of volcanic, seismic, and tectonic activity.

Atoll. A ring-shaped island with coral reefs that enclose a lagoon.

Atomic number. The number of protons in the nucleus of an atom.

Axial plane. An imaginary surface that approximately bisects a fold into symmetrical halves.

Axis. An imaginary line formed by the intersection of the axial surface of a fold with a bedding surface.

Backshore. The part of the shore that is level or sloping landward, covered by water only during storms.

Bajada. The gentle, sloping ground surface located between a pediment and a playa or flats in arid regions; often restricted to surfaces formed by deposition.

Bar. A sand or gravel deposit submerged at shallow depth.

Barchan dune. A curved sand dune whose convex side faces the wind and whose wings point downwind.

Barrier beach. A beach deposit formed offshore, such as those along much of the eastern coast of North America between Long Island and southern Texas.

Barrier reef. The name given to a coral reef that runs parallel to the shore of an island or continent, separated from it by a lagoon.

Basalt. A dark, fine-grained rock composed primarily of plagioclase and pyroxene, with or without olivine.

Base level. A level or elevation below which land cannot be eroded by streams. Sea level is referred to as the ultimate base level.

Basin (structural). A circular or elliptical downwarp or structural depression, with younger beds in the center.

Batholith. A large pluton of intrusive rock, many square kilometers in areal extent.

Bauxite. The principal ore of aluminum, composed of hydrated alumina ($Al_2O_3 \cdot 2H_2O$).

Bay-mouth beach. A beach located across the mouth of a bay.

Bayou. Type of stream characterized by sluggish drainage, and located on the flood plain of a large river.

Beach. The gently sloping shore of a body of water, particularly the sea, which is washed over by waves or tides.

Beach drifting. Movement of sediment along the shore by combined effects of incoming waves and return flow from beach.

Beheaded stream. A stream that has lost part of its drainage system as a result of stream piracy.

Benioff zone. See *seismic shear-zone.*

Blind valley. Valley that leads into a hillside or gradually loses the characteristics of a valley as water from its streams is lost to subsurface channels.

Body wave. Seismic wave propagated throughout a three-dimensional continuum and not related to a boundary surface.

Bog. See *swamp.*

Bomb. Fragment (32 millimeters in diameter or larger) of material erupted from a volcano.

Bottom-set beds. The subhorizontal layers of sediment deposited in front of a delta.

Bouguer effect. The attraction that the slablike mass of material between the geoid or sea level and the elevation at which a gravity reading is taken has on the value of gravity at that elevation.

Bouguer gravity anomaly. Any residue when the theoretical value of gravity for a given latitude calculated by the international formula is subtracted from the observed gravity at a station corrected for free-air effect, Bouguer effect, and topography.

Braided stream. A type of stream pattern in which the stream consists of a number of small channels that cross back and forth over one another, producing a braided appearance.

Breaker. A wave breaking on the shore, formed when the velocity of the top part of the wave begins to exceed the forward velocity of the wave as a whole; the top outruns the rest of the wave and spills over in front of it.

Breccia. A fragmental rock containing angular, instead of rounded, pieces.

Brittle failure. Rupture or breakage characteristic of brittle substances, preceded by little flow or plastic deformation.

Calcareous. Containing calcium carbonate.

Caldera. A large enclosed or partially enclosed depression caused by collapse or explosion of a volcano.

Calving. A process in which masses of ice split off to form icebergs where an ice sheet extends into the sea.

Capacity. A measure of the ability of a stream to transport suspended and bed load.

Carbonaceous. Containing carbon.

Carbonation. Process by which carbon dioxide is added to oxides of calcium, magnesium, sodium, and potassium to form carbonates of these metals.

Carbonic acid. A weak acid formed by a combination of carbon dioxide and water.

Cave. A naturally formed underground cavity or void.

Cenozoic era. The most recent era into which geologic time is divided. It extends from the end of the Mesozoic era, about 70 million years ago, to the present.

Chalk. A fossiliferous limestone composed of the shells of protozoans and particularly fossil globigerina. It is white and soft.

Chemical weathering. The breakdown of rocks and minerals at the earth's surface as a result of chemical processes.

Chert (flint). A dense, hard rock composed of SiO_2, with color ranging from white through gray to black.

Chondrite. See *stony meteorite.*

Chute. A narrow passage of water between an island and the main bank of the stream; commonly found on the inside of curves along meandering streams.

Cinder cone. A cone-shaped hill formed from accumulation of cinders and ash around a volcanic vent.

Cirque. An amphitheater-shaped depression formed by glacial quarrying or plucking where the glacier starts its movement down the valley.

Clastic particle. Fragmented material such as sand and gravel that is produced by the mechanical disintegration of all types of pre-existing rocks.

Cleavage. (1) Mineral: The property exhibited by some minerals of breaking along definite smooth planes; a manifestation of the internal orderly arrangement of atoms in a mineral. (2) Rock: The property exhibited by some rocks of breaking along definite, smooth, subparallel, planes; caused by closely spaced fractures or by alignment of platy minerals.

Climate. The long-range averages of weather in a region.

Coal. Vegetable matter that has been altered both physically and chemically through geological processes to a black, rocklike substance.

Coast. A broad zone directly landward from the shore.

Coccolith. Minute calcareous plates formed on some marine flagellate organisms.

Col. Pass formed where two cirques converge, cutting into the same wall and thus lowering it below the level of the remainder of the summit area.

Colloid. A particular size of matter in the range of 10^{-5} to 10^{-7} centimeters in diameter.

Colluvium. A mixture of soil, stream deposits, and talus.

Compaction. Process by which material is pressed together (for example, as a result of the weight of sediment deposited on a given layer).

Competence. The largest size particle (measured by diameter) a stream can move.

Composite cone. Volcanic cone composed of both cinders and lava flows; such cones are circular in plan, and the sides are concave upward in profile.

Compressional waves. See *dilational waves.*

Compressive stress. A stress applied (as in a vise) in such a way that material is forced together.

Conchoidal fracture. A broken surface that is curved or shell-shaped, as exhibited by glass.

Concordant injections. Intrusions emplaced along planes of stratification, layering, or schistosity so that the borders of the intrusion are parallel to preexisting layers.

Confined water. Groundwater conditions in which the water in an aquifer is confined to that aquifer by an overlying impervious cover, which may prevent its upward movement.

Confining pressure. Term applied to nondirected pressure due usually to depth of burial in rocks.

Connate water. Marine or fresh water trapped in sediments when they are deposited on lake or sea bottoms.

Consequent stream. A stream that flows in the same direction as the inclination of the slope on which it formed originally.

Contact metamorphic aureoles. Concentric zones of alteration around plutons due to the effects of heat and chemically active fluids in the magma on the country rock.

Contact metamorphism. The alteration of country rock around an intrusion near its contact.

Continental accretion. The theory that continents have grown by incorporation of mountain belts around their margins.

Continental drift. The theory that continents have moved relative to one another.

Continental margin. The zone of transition from crustal structure of the continents to that of the ocean basins.

Continental rise. The gently sloping surface located at the base of the continental slope.

Continental shelf. A shallow, gently sloping surface of low local relief that extends from the shore line to the shelf break, where the seaward gradient sharply increases to greater than 1:40.

Continental slope. The relatively steep portion of the sea floor that occurs at the seaward border of the continental shelf.

Convergent boundary. In plate tectonics theory, the boundary between the plates that are moving toward one another.

Coordination number. Number of anions in contact with a cation in any given packing arrangement.

Coquina. Sedimentary rocks composed of a loose aggregation of shell fragments cemented together.

Core. The central portion of the earth's interior, bounded above by the mantle and thought to be composed of metals in a liquid state.

Coriolis effect. The deflection of any object in motion on a rotating sphere.

Country rock. The rock in which an igneous intrusion is emplaced.

Crater. A depression around a volcanic vent shaped like an inverted cone.

Craton. The stable portion of each continent; the shields and the surrounding areas.

Creep. (1) Rock: extremely slow deformation of materials subjected to stress below the elastic limit. (2) Soil: the imperceptibly slow movement of soil downslope.

Crest. The highest part of a fold form; also, the top of a wave form.

Cross-bedding. Successive, systematic, internal bedding that is inclined to the principal surface of accumulation.

Crust. That part of the earth above the *M*-discontinuity, approximately the outer 10 kilometers in ocean basins and the outer 20 to 40 kilometers under continents.

Crystal. A homogeneous body, the surfaces of which

have grown into smooth planes as a result of the internal orderly arrangement of atoms.

Curie point. The temperature at which a material loses its spontaneous magnetization; about 570 °C for ferromagnetic minerals.

Cutoff. The short channel formed where a stream cuts through the neck of a meander loop.

Debris. A general term applied to mixtures of rock, soil, plant matter, or mud.

Declination (magnetic). The horizontal angle at a locality between the direction to magnetic north and true (geographic) north.

Décollement. "Ungluing" of a bed or zone, during which beds above slip over those below. Usually the layers are composed of salt, gypsum, clay, or other weak rock.

Deep-sea trench. Long, narrow troughlike depressions in the floor of the ocean. The deepest of these, the Mariana trench, is over 38,000 feet deep.

Deflation. Blowing away of materials.

Degradation. The general lowering of the land surface by erosion.

Delta. Deposit of alluvial material, sometimes triangular, formed at the mouth of a stream.

Dendritic. Having a branchlike outline.

Denudation. Laying bare of the surface of the earth; now applied to reduction of the level of the land by processes of erosion.

Desert pavement. A tight mosaic cover formed by lag gravels that protects underlying material from wind erosion.

Diastrophism. Deformation of the earth's crust; the processes by which the crust is deformed and the results of this deformation.

Diatom ooze. A deposit consisting of siliceous remains of diatoms.

Diatom. Simple siliceous plant of microscopic size with one of many different shapes, may resemble a rod, sphere, or circular disc.

Dike. An injected body with parallel or subparallel walls that is narrow relative to its lateral extent and discordant with respect to pre-existing layers.

Dike swarm. Term used when many dikes, often of similar trend or orientation, occur together.

Dilational wave (*or* primary, P, compressional, *or* longitudinal wave). Seismic wave propagated like sound waves with movements in the medium through which it passes parallel to the direction of propagation.

Diorite. A coarse- to medium-grained rock composed of intermediate plagioclase and biotite, hornblende, or pyroxene.

Dip. The angle between a plane and the horizon, measured in a verticle plane at right angles to the strike.

Dip-slip fault. A fault in which the direction of displacement is up or down the dip of the fault.

Directed pressure. Application of a pressure, stress, or force in such a way that it has direction.

Discharge. A measure of the quantity of water passing through a cross-sectional area of a stream per unit of time.

Disconformity. A break in a succession of sedimentary layers caused by a period of erosion or nondeposition. Layers above the break are parallel to those below.

Discordant intrusion. Intrusion injected across planes of stratification or schistosity.

Disharmonic fold. Fold form whose geometry varies in an irregular way with depth.

Displacement. The straight-line distance after movement between two points that were originally adjacent on opposite sides of the fault.

Distributary. Branch of a stream, usually at the mouth of a river on a delta, formed by the breaking up of a single stream into smaller streams.

Divergent boundary. In plate tectonics, a boundary between plates that are moving away fron one another.

Doline. A synonym for sink hole.

Dolomite. A sugary-textured, dense rock that does not effervesce in dilute acid unless it is powdered. Dolomite is a calcium magnesium carbonate.

Dome. A circular or elliptical uplift; older beds are exposed in center after erosion.

Drainage basin. The area drained by a stream.

Drainage divide. An imaginary line separating the area drained by one stream from the areas drained by adjacent streams.

Drift. Glacial deposit.

Drumlin. A streamlined hill shaped somewhat like a whaleback and generally composed of glacial till.

Dune. Accumulation of sand that is mobile and independent of obstructions.

Dunite. A rock composed almost completely of the mineral olivine.

Dynamic metamorphism. Metamorphism produced when deformation is significant in the transformation of rock. The term is applied when effects of directed stress are important in metamorphic processes.

Earth. A general term describing disintegrated rocks and loosely consolidated sediments.

Eclogite. Metamorphic rock composed primarily of a red garnet and a green pyroxene, having a bulk chemical composition that could only have formed from rocks of basaltic or gabbroic composition.

Effluent. Water that moves from beneath the water table into lakes and streams.

Elastic limit. The limit of stress beyond which a rock cannot fully recover its original shape after a deforming force is removed.

Elasticoviscous. Material behavior characterized by a combination of elastic and viscous properties.

Elastic rebound theory. All theories of earthquake wave

generation that rely on the concept of waves caused by sudden release of stored strain through elastic recovery of strain.

Electron. One of the fundamental particles of which atoms are composed. The mass is 9.107×10^{-28} grams and is assigned a relative negative electrical charge of one unit, which equals 4.803×10^{-10} statcoulombs, or 1.602×10^{-19} coulombs.

Element. An atom or group of atoms that is stable and cannot be broken down by ordinary chemical methods.

Entrenched meander. The meandering course of a river that flows in a deep valley.

Entrenched stream. A stream that flows in a channel cut below the general level of the surrounding land and often into bedrock. Cliffs may occur along one or both banks.

Epeirogenic. A type of regional deformation characterized by broad uplifting or downwarping of a large area.

Epicenter. Point on the ground directly over the focus of an earthquake where motion is initiated.

Epoch. A division of geologic time. Geologic periods are often divided into epochs on the basis of unconformities.

Era. The largest division of the geologic time scale. Time since the Precambrian is subdivided into the Paleozoic, Mesozoic, and Cenozoic eras.

Erosion. In broad context, includes the effects of all processes that loosen and remove earth materials: weathering, solution, corrosion, and transportation. Many geologists prefer to restrict use of the term to processes involving removal of material, such as by streams, glaciers, wind, waves, and groundwater.

Erratic. A rock carried by a glacier and laid down on a different type of rock.

Esker. Sinuous ridge composed of till and glaciofluvial deposits and thought to originate in stream channels and sometimes under glaciers.

Eugeosyncline. Many geosynclines can be subdivided into two major parts, *miogeosyncline* and *eugeosyncline.* The eugeosyncline is located generally on the oceanward side of the miogeosyncline, and is usually characterized by greater mobility and the presence of volcanic products and graywacke as well as other sediments.

Exfoliation. Scaling on exposed rocks.

Extension. Stretching or pulling apart.

Extrusion. Igneous rock bodies and lava that crystallized or solidified on the surface of the earth.

Fault. A shear zone along which displacement has occurred.

Fault block. A mass bounded by faults on at least two sides.

Fault block mountain. Mountain formed as a result of the displacement of sections of the earth's crust along steeply inclined faults.

Fault plane. The zone along which movement occurs when faulting takes place.

Fault scarp. An escarpment marking the position of displacements where one side moved up in relation to the other.

Feldspar. One of a group of silicate minerals, including potash feldspars (orthoclase and microcline) and the plagioclase feldspars, which contain Ca and/or Na.

Felsite. A general term applied to igneous rocks of fine-grained texture and light color, indicating a granitic composition.

Ferruginous. Containing iron.

Festoon cross-bedding. A type of cross-bedding formed when channel-filling takes place. Beds have a characteristic curve (concave upward) appearance in cross section.

Firn. Compacted, granular snow.

Fiord. Glaciated valley whose floor is submerged below sea level.

Flagellate. A microscopic plant with one of a variety of shapes, ranging from that of a balloon with a string attached to that of a pot or fancy mask; all have whiplike projections that move.

Flat-pebble conglomerate. A conglomerate or breccia layer formed where a newly deposited sediment layer is broken by wave action, producing platy pebbles.

Flexural fold. A fold formed where flow or slip is restricted by the layer boundaries. The layering exercises an active control on the deformation; the resulting folds represent a true bending of layers.

Flood plain. The flat area along a stream that is occasionally flooded.

Flood tide. The rising, or flow, of water toward the shore as the tide approaches the coast.

Flow casts (*or* load casts). Rolls, lobate ridges, or other raised features that may be produced in a sandstone bed as a result of the flow of sediment in an underlying bed, usually composed of some soft sediment.

Flowstone. Calcium carbonate deposits formed from water in caves, often taking the shape of a cascadelike mass.

Focal depth. The depth below the earth's surface at which an earthquake occurs.

Focus. The point at which an earthquake originates, or where motion is initiated.

Fold. Bend or distortion of rock masses; most easily recognized when bedding is deformed.

Foliation. A secondary rock fabric in which platy minerals or shear surfaces are in parallel alignment. Schistosity and rock cleavage are types of foliation.

Footwall. The block below a dipping fault.

Foraminifers. Single-celled marine animals with small shells composed of calcite.

Foreland. The portion of an orogenic belt located on the flanks of the highest and most severely deformed parts of the mountain system.

Foreland fold. Fold formed in a belt located on the flanks of a mountain system's core.

Fore-set bed. Layer of inclined sediment on the front slope of a delta.

Foreshore. The zone between low tide and the area where the beach either becomes horizontal or slopes landward.

Fossil. The remains of plants or animals that lived in the geologic past, including tracks and trails left by living organisms.

Fossiliferous limestone. Limestone formed from the shells of marine animals or in which fossils make up a significant part of the rock.

Fractional crystallization. Separation of a magma into two or more phases.

Fracture (*or* joint). Break in rock along which displacements have not occurred.

Free-air gravity anomaly. A gravity anomaly that remains after the reading has been corrected for elevation above the reference surface (geoid).

Fringing reef. Reef formed along the edge of an island or continent at the shoreline.

Frost action. The effects caused by freezing water, particularly as an agent of mechanical weathering.

Frost heaving. The process in which the soil or rocks are lifted up as a result of the freezing of ice under them.

Fumarole. A hole or vent through which fumes or vapors issue in an area underlain by volcanic materials.

Gabbro. Rock composed of coarse- to medium-grained plagioclase feldspar, pyroxene, hornblende, and biotite.

Gal. A unit used to express gravitational acceleration. 1 gal = 1 cm/sec^2.

Gamma. A unit of measure of magnetic field strength. 1 gamma = 10^{-5} oersted.

Gauss (*or* oersted). A measure of magnetic field strength. 1 gauss = 1 dyne/unit pole.

Geodesy. The field of geophysics concerned with the shape and dimensions of the earth.

Geoid. The shape of a sea-level surface on earth extended through continents.

Geomorphic agents. The agents, such as running streams, groundwater, wind, waves, and mass wasting, that bring about changes in the shape of the earth's surface.

Geomorphology. The study of the shape of the earth's surface.

Geophone. Instrument designed to pick up vibrations of the ground; used in seismic exploration.

Geosyncline. Long, narrow belt of long-term subsidence and sedimentation.

Geyser. The forceful emission of water from beneath the ground as a result of the sudden change of part of the water to steam.

Geyserite. A chemical sediment composed of SiO_2 formed around the vents of geysers.

Glaciofluvial deposit. Deposit formed from water on, in, or under a glacier, as well as beyond the glacier terminus, by meltwater.

Glassy. The texture of amorphous materials; such materials show no crystals and thus are glasslike.

Globigerina. Single-celled marine animal with a globular snail-like shell.

Globigerina ooze. Calcareous deposits on the sea floor composed of the shells of the protozoans globigerina.

Gneiss. A medium- to coarse-grained metamorphic rock characterized by compositional layering that produces a banded appearance.

Gondwanaland. The name of the southern continental mass thought by some to be the source of Africa, India, South America, and Australia.

Gossan. Deposits of hydrated oxide of iron formed in the upper parts of mineral veins and masses containing pyrite as a result of oxidation and leaching.

Gouge. Soft powdery material resulting from grinding and abrasive action along a fault.

Graben. Elongate crustal block, bounded by normal faults on either side, that has moved down in relation to these sides.

Gradation. A term applied to the combined effects of aggradation and degradation of the earth's crust; sometimes restricted to effects of stream action.

Graded bedding. Sedimentary structure in which there is a systematic vertical change in sediment grain size.

Graded stream. Stream that, over a long period of time, is adjusted so that the slope of the stream channel is sufficient to give the stream the velocity needed to move its load.

Gradient (stream). The rate of change of elevation of the stream with distance along the stream (that is, slope).

Granite. A coarse- to medium-grained equigranular rock composed of potassium feldspars and quartz, with some biotite mica and hornblende usually present in small amounts.

Granite pegmatite. A very coarse-grained rock with granitic composition.

Granitization. The metasomatic processes by which granitic rocks are produced from sedimentary rocks.

Granulation. The crushing of rock under such conditions that no visible openings result.

Gravity separation. The process by which the first crystals that form in a magma, if heavier or lighter than the melt, will sink or rise through the melt and become segregated either at the bottom or as a raft of floating minerals near the top of the magma chamber.

Graywacke. An impure sandstone consisting of quartz and feldspar fragments and small fragments of rocks.

Greenschist facies. Metamorphic facies produced by temperatures and pressure conditions that cause the original sedimentary minerals, including clay and zeolite, to be altered. Rocks assume a greenish appearance as a result of the formation of chlorite and sericite.

Ground moraine. Glacial deposit resulting from melting so that debris on the surface, in the ice, and caught near the bottom of the ice is slowly lowered.

Groundwater. Water moving beneath the ground surface.

Groundwater basin. An area bounded by natural geologic

conditions that cause ground water within it to be more or less separated from adjacent basins.

Guyot. Submerged flat-topped seamount. Most are thought to be volcanoes that have been eroded by wave action.

Half-life. The constant amount of time required for one-half the mass of a radioactive isotope to break down.

Hanging valley. A valley having a floor higher than the valley or shore into which it leads; usually formed by glaciation.

Hanging wall. The surface or block of rock above an inclined fault zone.

Hardness. (1) In water, the amount of calcium carbonate and magnesium carbonate in solution. (2) The resistance a mineral offers to scratching.

Hard pan (*or* caliche). A hard, impervious layer of soil containing insoluble materials and clay.

Head. The upper reaches of a stream or glacier.

Headland. A part of the shore that protrudes into the ocean.

Headward erosion. Progressive extension of the head of a stream as a result of erosion.

Hogback. A sharp ridge formed as a result of the outcrop of a resistant layer.

Holocene. A division of geological time encompassing the last ten to eleven thousand years. Also known as recent time.

Homocline. General name for any block of bedded rocks that has a single, relatively uniform direction of dip (one limb of a large fold may be described as homoclinal).

Hornfels. Fine-grained, nonfoliated, dense, usually dark rock formed near the contacts of igneous intrusions.

Horn. Sharp peak that projects above the surrounding area and was shaped by glacial erosion.

Horst. An elongate block, bounded by normal faults on either side, that has moved up in relation to the sides.

Hot spot. In plate tectonics theory, a center of volcanic activity thought to form as a result of the rise of hot materials from deep in the mantle. Hot spots may remain fixed for long periods of time while the overlying plates move.

Hundred-year flood. A flood that may be expected to recur once every hundred years on the basis of past records of flood stages.

Hydration. Absorption and combination of water with other compounds, as when anhydrite absorbs water to become gypsum.

Hydraulic gradient. The rate of pressure head change per unit of distance at a given place and in a given direction.

Hydraulic head. The pressure at a point in a body of moving water, expressed in terms of the height of the column of water that can be supported by the pressure.

Hydrologic cycle (*or* water cycle). The cycle of phenomena through which water passes.

Hydrolysis. A reaction between water and another compound.

Hydrostatic head. An expression of the hydrostatic pressure in terms of the height of a column of water that can be supported by the pressure.

Hydrostatic level. The level to which water will rise in a sealed well.

Hydrostatic pressure. The pressure exerted at a given point in a body of water at rest by the weight of the overlying body of water.

Hypothermal deposits. The deepest ore deposits associated with metamorphic rocks; usually lenticular in shape and include such minerals as native gold, chalcopyrite, tin, and molybdenite.

Iceberg. Big block of ice broken off a glacier into the sea.

Ice sheet. Large body of ice not confined to a valley; may reach continental dimensions and bury whole mountain ranges.

Igneous. A term meaning "born from fire," applied to both volcanic rocks and rocks that crystallize slowly within the earth's crust.

Index minerals. Minerals used to identify the different grades of metamorphism. From low grade to high grade, these minerals are: chlorite, biotite, garnet, staurolite, kyanite, and sillimanite.

Influent. A condition existing when the water table is depressed below the level of stream and lake bottoms, and water moves from lake or stream toward the water table.

Insolation. The weathering effect of the sun's heat on exposed rocks.

Intensity. A measure of the effects experienced by people at the surface of the ground when earthquakes occur. A standard scale is used.

Internal drainage. A drainage system with no streams leading out of it.

Intrusion. An igneous body emplaced and crystallized beneath the surface of the ground.

Isograd. Line connecting points of equal metamorphic grade; forms a boundary between zones.

Isostasy. The principle that the upper parts of the earth's outer layers are in a state of flotational equilibrium on a lower level.

Isostatic anomaly. A gravity anomaly that remains after the reading has been corrected for density variations thought to exist to the level of isostatic compensation.

Isostatic equilibrium. The condition of the earth's outer layers when they are in flotational equilibrium, with no tendency to move up or down.

Isotopes. Atoms of an element having the same number of protons in the nucleus as generally found in the element, but different numbers of neutrons. Isotopes of an element have the same atomic numbers but different atomic weights.

Joint. A crack or break in rock masses across which the rock has lost cohesion; also called a fracture.

Kame. Glacial deposit laid down in contact with ice. Although used in various ways, the term is often applied to small conical hills formed of stratified drift.

Kame terrace. Terrace built along the side of a glacial valley by load that is dropped by water between the glacier and the adjacent valley wall.

Karst topography. Topography shaped in part by solution and diversion of surface water underground in areas of limestone and dolomite bedrock.

Kettle. An enclosed depression left in drift as the last remnants of ice melt away.

Klippe. An isolated block of material separated from underlying rocks by a fault; associated with low-angle thrust faults.

Lag gravel. Loose gravel-sized particles left in arid and semiarid regions as wind removes finer materials.

Laminar flow. Flow in which the fluid behaves as though it were made up of very thin layers that glide over one another; streamline flow.

Landslide. Perceptible downslope movement of earth, rock, soil, and mixtures thereof.

Lapilli. Small fragments (4 to 32 millimeters in diameter) erupted from volcanoes.

Lateral moraine. A deposit from a valley glacier composed of debris that accumulated on the glacier surface as a result of normal downslope movement of debris from the sides of the valley.

Laterite. A type of clay-rich soil that has a tendency to harden into a bricklike solid.

Laurasia. The name of the northern continental mass thought by some to be the source of North America, Europe, and Asia.

Lava. Magma that reaches the ground surface.

Law of faunal succession. Principle stating that the fossil fauna of rocks of different age are different and provide a key to identification of rocks of that age.

Law of superposition. Principle stating that in a normal stratified sequence each stratum lies on one of greater age.

Limb. The flank or side of a fold.

Liquidus. A curve (in a two-component system) connecting points in a temperature-composition diagram representing the saturation of the liquid phase.

Lithification. The process by which unconsolidated sediments become consolidated rocks.

Lithosphere. The relatively rigid outer rind of the earth. The exact limits are not specified, but it is distinguished from a plastic zone below called the asthenosphere.

Lithostatic pressure. The pressure at a point in the earth caused by the weight of the overlying body of rock.

Load of a stream. The material in transport in a stream. Usually the load is of three parts: material moved along the bottom of the stream channel, material in suspension, and material in solution.

Loess. An unconsolidated, unstratified, and homogeneous sediment composed of small, angular particles derived from a variety of rock types.

Longitudinal dune. A sand dune that is elongate in the direction of the prevailing wind.

Longshore current. A component of motion in the water directed along the shore line.

Low-velocity layer (*or* Gutenberg low-velocity zone or channel). A zone in the upper mantle in which seismic waves move with reduced velocity. The depth of the zone is between 100 and 200 kilometers.

Luster. Appearance taken on by a mineral in reflected light.

Mafic rock. An igneous rock rich in minerals containing iron and magnesium.

Magma (*or* rock melt). Multicomponent silicate melt that usually contains some solid minerals, liquids at high temperature, and gases under pressure.

Magmatic differentiation. Processes by which a magma becomes altered and yields different rock types during crystallization.

Magmatic water. Water produced by magmatic activity; also called juvenile water.

Magnetic anomaly. Abnormally high or low values for the strength, declination, or inclination of the earth's magnetic field.

Magnetic field. The area around a magnet within which the influence of the magnet can be detected.

Magnetic-field strength. The force (in dynes) that would be exerted on a unit pole at a point in a magnetic field.

Magnitude. A measure of the amount of energy released at the focus of an earthquake.

Mantle. The zone in the earth's interior below the crust and above the core.

Marble. The metamorphic equivalent of calcite, limestone, or dolomite.

Mare. The large, flat areas on the moon formed by extensive outpourings of lava; also known as lunar seas.

Marl. Mixture of shells and shell fragments with muds, clay, particles of calcite or dolomite, or sand; an impure limestone found in a semiconsolidated state.

Mass wasting. The downslope movement of surficial materials as a result of the effects of gravity.

M discontinuity (*or* Moho). The major seismic discontinuity used to define the base of the earth's crust. It occurs at a depth of 20 to 40 kilometers beneath continents and 10 kilometers under oceanic crust.

Meandering stream. A stream that flows in a strongly curved course consisting of loops in which the stream almost doubles back on itself.

Mechanical concentration. Concentration of minerals through mechanical processes, such as gravity separation in streams, along coasts, or by wind.

Medial moraine. Ridge of till left in the middle of a valley after a glacier melts.

Mesozoic era. The interval of geologic time following the Paleozoic and preceding the Cenozoic.

Metamorphic aureole. The concentric zones of alteration usually found surrounding igneous intrusions.

Metamorphic facies. Facies comprised of all rocks exhibiting a unique and characteristic correlation between chemical and mineralogical composition in such a way that rocks of a given chemical composition always have the same mineralogical composition. Differences in chemical composition from rock to rock are reflected in systematic differences of their mineralogical composition.

Metamorphic rock. Rock formed through the process of metamorphism.

Metamorphic zone. An area that was exposed to similar metamorphic conditions. Zones are recognized by certain diagnostic minerals.

Metamorphism. The processes acting beneath the ground by which rock is so extensively altered that it obtains a new texture and/or a new mineral composition.

Metasomatism. A process of simultaneous solution and deposition that replaces old minerals.

Meteoric water. All forms of precipitation, including rain, sleet, snow, and hail.

Meteorite. A body of stone or metal that has fallen to the earth from space.

Mica. One of the most distinctive of all mineral groups, exhibiting single perfect cleavage and having a sheet structure. Included are biotite, muscovite, and other less common varieties.

Micaceous (*or* lamellar, foliated). Descriptive of minerals that consist of thin, flat plates.

Microseism. Small-amplitude seismic disturbance caused by wind, waves, and other earth vibrations and recorded on seismograms.

Migmatite. Rock of mixed nature contain metamorphic- and igneous-appearing rock.

Mineral. A naturally occurring element or compound formed by inorganic natural processes and having generally a definite chemical composition and a characteristic atomic structure, which is expressed in an external crystalline form and in other physical properties.

Miogeosyncline. A long, relatively narrow belt in which subsidence and accumulation of sediment (notably shallow-water sand, mud, and limestone) occur. See also *eugeosyncline*.

Moho. See *M discontinuity*.

Monocline. A flexure across which the beds maintain a constant direction of dip.

Moraine. A sedimentary deposit formed along the edges and beneath glaciers from material carried by the glacier.

Morass. See *swamp*.

Mud. A mixture of water and the finer particles of earth and soil.

Mud cracks. Cracks that commonly form in mixtures of clay, sand, and silt when the sediment is dried.

Mudflow. A flow of a mixture of mud and water.

Mylonite. The name given to breccia when it is crushed so completely that it is reduced to a lithified powdery material.

Nappe. A large mass of folded rock that has been moved a long distance from its point of origin.

Natural bridge. Rock bridge formed by the diversion of surface streams underground, remnant of a cave, or a natural tunnel.

Natural levee. A low ridge formed on either side of a stream channel from deposits laid there during flood.

Neap tide. Low tidal range occurring about the time of quadrature of the moon.

Nebula. An intersteller cloud of gas or dust.

Net slip. The straight-line distance after fault movement between two points that were originally adjacent on opposite sides of the fault.

Nonconformity. An ancient erosion surface found in rock bodies where the rock below the erosion surface is crystalline rock, either igneous or metamorphic, and that above the erosion surface is sedimentary.

Normal fault. Steeply inclined fault along which the hanging wall has moved downward in relation to the footwall.

Nuée ardente. A type of volcanic eruption in which incandescent gases and dust are erupted.

Nunatak. Peak or isolated hill that protrudes through a snow field or glacier.

Obduction. In plate tectonics, the process by which part of the sea floor is uplifted and faulted onto other crust, sometimes onto continental crust.

Oblique-slip fault. A fault on which the displacement is oblique across the fault plane.

Obsequent stream. A stream that flows in the direction opposite that of the dip of the underlying rocks.

Oersted. A magnetic force designated in terms of dynes per unit pole strength.

Offshore. From the position of low tide seaward.

Olivine. A mineral series varying in composition from forsterite (Mg_2SiO_4) to fayalite (Fe_2SiO_4).

Oölite. Round, egglike body usually having concentric internal layering.

Ooze. Marine sediment consisting largely of shell fragments of microscopic marine animals.

Ore. Rock or mineral that can be economically mined to produce a metal.

Orogenic belt. A mobile belt in the earth's crust that has been subjected to folding and other deformation.

Orogeny. Deformation of the crust; usually mountain-building processes.

Outwash deposit. Sediment deposited beyond the ice margin by meltwater produced as an ice sheet or valley glacier wastes.

Outwash plain. An area of low relief located beyond the edge of a glacier; the site of deposition of glaciofluvial deposits.

Overturned fold. A fold in which the axial surface is inclined so one limb is over part of the other limb.

Oxbow lake. A lake formed as a result of the cutoff of a meander.

Oxidation. The process by which oxygen combines with elements or compounds.

Pack ice. Floating pieces of ice driven together.

Paleozoic era. A division of geologic time covering that interval after the Precambrian and before the Mesozoic.

Pangaea. Name of the single protocontinent thought by some to be the original source of all continents.

Passive fold. Fold in which flow or slip crosses the layer boundaries, the layers exercising little or no control on the deformation. Layer boundaries serve merely as markers, parts of which are displaced relative to other parts to produce an apparent bending.

Peat. An accumulation of vegetable matter that is in the initial stages of disintegration and decomposition.

Pediment. A gently sloping surface formed by erosion of cliffs or steeper slopes; often a rock surface with a thin veneer of alluvial covering.

Pedology. Study of the soil, its origin, character, and uses.

Pegmatite. Rock composed of very coarse crystals, usually in the form of a dike or lense, formed in the late stage of crystallization of a magma.

Peneplain. A nearly plane surface formed as a result of erosion by stream action.

Perched water table. A condition found where a zone of saturation is located above an unsaturated zone and above the regional water table.

Peridotite. A dark, coarse-grained, equigranular rock containing a large quantity of olivine and/or pyroxene or hornblende, but no quartz or feldspar.

Period. (1) The time between arrival of adjacent peaks in a wave train. (2) One of the major divisions of time used in the geologic time scale.

Permafrost. Permanently frozen ground.

Permanent stream. Stream that flows throughout the year.

Permeability. A measure of the ability of a stratum or rock unit to transmit fluids.

Phaneritic. Textural-description in which crystals in a rock are large enough to be seen with the unaided eye. Phaneritic textures may be coarse-, medium-, or fine-grained.

Phenocryst. Large crystal in a glassy or finer-grained igneous rock.

Photochemical dissociation. The breakdown of water vapor by ultraviolet radiation, releasing free oxygen.

pH scale. Measures the acidity of a solution (the concentration of hydrogen ions). Values range from 0 to 14, with 7 neutral, less than 7 acid, and greater than 7 alkaline.

Phyllite. Fine-grained schistose rock identified by the lustrous, silky sheen that characterizes light reflected from the chlorite and muscovite micas of which it is composed.

Piedmont glaciers. Glaciers that coalesce along a mountain front.

Piezometric level. The static level to which water in an artesian aquifer will rise if it is penetrated by a well.

Piezometric surface. An imaginary surface that everywhere coincides with the static level of the water in the aquifer.

Pillow lava. Lava characterized by ball- or pillow-shaped masses, formed by eruption of lava underwater.

Placer deposit. Accumulation of minerals that are heavy, resistant to chemical weathering, hard, malleable, and durable enough to survive the mechanical concentration processes active in streams or in waves.

Planetisimal. A small, planetlike body.

Plastic flow. Deformation characterized by continuous nonreversible strain.

Playa lake (*or* salina). A shallow lake located in a basin, usually having internal drainage. Such lakes commonly contain water only intermittently and are generally high in salt content as a result of water evaporation.

Pleistocene. A division of geological time also known as the Ice Ages. The Pleistocene started 2 million years ago and continued until the most recent interglacial stage when ice began to melt.

Plucking (*or* quarrying). Process by which pieces of bedrock are lifted out of place by glaciers.

Plunge. The angle (measured in a vertical plane) between a line and the horizon.

Plunging axis. The angle of inclination measured from the horizon when the axis of a fold is not horizontal.

Pluton. A large igneous intrusion.

Polar covalent compound. Compound made of molecules that behave somewhat like small magnets.

Pole. (1) Geographic. The ends of the earth's axis of rotation. These rotational points are the *north* and *south* geographic poles. (2) Magnetic: Any of several points on the surface of the earth where the lines of magnetic force are vertical. Magnetic poles do not coincide with geographic poles, though the earth has *north* and *south magnetic poles.*

Porosity. The volume of pore space in a rock.

Porphyry. An igneous rock composed of large crystals embedded in a fine-grained or glassy groundmass.

Pothole. Oval or round hole formed in the channel of a stream by concentration of abrasion, impact, and eddy current action.

Primary wave. See *dilational wave.*

Progressive wave. Wave in which the form moves laterally.

Pumice. A light-colored, porous, glassy, highly vesicular rock.

P wave. See *dilational wave.*

Pyroclastic. A general term applied to fragmental materials of volcanic origin.

Pyroxene. A group of minerals that are single-chain silicates.

Quadrature. Resultant condition when lines from the sun to the earth and from the moon to the earth are at right angles.

Quarrying. See *plucking.*

Quartzite. The metamorphic equivalent of quartz sandstones.

Quaternary. A division of geological time including the Pleistocene and Recent time.

Radioactive isotope. Isotope that is unstable and breaks down spontaneously by the emission of radiant energy.

Radiolarian. Single-celled marine animal with an internal skeleton composed of radiating siliceous projections; important constituent of sediment in some areas of the ocean.

Range. Variation in level of water surface resulting from tidal influences.

Recessional moraine. A ridgelike mass of drift accumulation from debris dropped by the glacier when the margin of the ice remains in one position for any period of time during retreat of a valley glacier.

Recrystallization. Reorganization of the elements of the original minerals in a rock that may take place when the rock is subjected to high temperatures and pressure, particularly when water is present.

Recumbent fold. A fold with a nearly horizontal axial surface.

Red clay. The second most extensive deep-sea deposit, brown to reddish in color and containing films of manganese and manganese nodules.

Regional metamorphism. The name applied to those metamorphic alterations that affect rocks over large areas and are indicative of widespread environmental changes rather than localized deformation or magmatism.

Rejuvenation. Condition whereby the behavior of a stream and its effects on the topography appear to revert to those of a more youthful stage in the erosion cycle.

Remanent magnetization. Magnetic effects unrelated to the existing external field but due to permanent magnetism of the substance.

Residual concentration. A process by which some valuable mineral constituents accumulate while undesirable constituents are removed by weathering and erosion.

Reverse fault. Steeply inclined fault along which the hanging wall has moved up in relation to the footwall.

Ripple mark. An undulating surface marking on sediment formed as a result of movement of a granular sediment in the medium of deposition.

Roche moutonnée. Rounded hill that has been streamlined by glacial abrasion and scouring.

Rock. A consolidated aggregate of minerals; sometimes large masses of a single mineral.

Rock cleavage. The tendency of rock to break along closely spaced subparallel surfaces.

Rock flour. A powdery substance composed of fine (silt-sized) particles of rock produced by the grinding, abrasive action of glacier ice moving over bedrock.

Rock flowage. Flow in which material remains solid, yet undergoes continuous deformation. Flowage of this type is accomplished by movements along planes of slip within the component crystals of the rock and by intergranular rotation; usually it is accompanied by recrystallization of minerals.

Rock glacier. Mass of rock fragments that assumes a form similar to a valley glacier as a result of slow downslope movement.

Rock structure. The configuration of rock bodies, as they are expressed both at and beneath the ground surface.

Rotational fault. A scissorslike motion involving rotation of the fault blocks relative to one another.

Runoff. The portion of precipitation that flows off the ground surface in streams.

Safe yield. The amount of water that can be withdrawn from a groundwater basin without producing undesired results.

Sag pond. Depression, often filled with water, located along a fault where material in the rock has weakened as a result of deformation and shearing.

Saltation. The bouncing movement of materials in transport in streams and in the air. The material is heavy enough to return to the bottom of the channel or to the ground surface, where it hits other rock and may bounce back into suspension while it rolls in the direction of the current.

Sand drift. Sand deposit formed where a gap between obstructions acts as a funnel.

Sand shadow. Sand accumulation located downwind from and sheltered by an obstruction, such as a boulder, grove of bushes, or cliff, that interrupts the streamlined flow of wind and reduces its velocity.

Sandstone. Any rock composed of fragments between $\frac{1}{16}$ to 2 millimeters in size regardless of composition.

Schistosity. Ease of parting in metamorphic rocks, such as schist and gneiss, resulting from the parallel or subparallel alignment of platy minerals.

Schist. Strongly foliated metamorphic rock of medium to coarse crystalline texture.

Scoria. A vesicular volcanic rock formed by the expansion of gases in lava as it is ejected; compositionally equivalent to basalt.

Scour-and-fill. Small-scale bottom scour or channels that are subsequently filled.

Sea-floor spreading. The theory that the sea floor is growing by means of intrusion and formation of new crust along oceanic ridges.

Sea level. The mean level of the sea after tidal and wind effects are removed.

Seamount. Isolated mountain-shaped rise that usually protrudes more than 500 fathoms (915 meters) above the ocean floor.

Secular change. Long-term variation that is observed as progressive change in declination, intensity, and inclination of the magnetic field; rates of change vary with time at any given point.

Sediment. Materials that settle out of water, wind, or ice, as well as materials precipitated from solution and deposits of organic origin.

Sedimentary facies. Parts of the sedimentary rock unit reflecting the conditions under which it originated. Often these are characterized by distinctive rock types or primary features.

Sedimentary rock. Hard, solid rock formed when sediments become consolidated.

Seif. Longitudinal dune characterized by a long, sharp ridge crest, one side of which is rounded and the other of which falls abruptly as a slip face.

Seismic sea wave. See *tsunami.*

Seismic shear zone (*or* Benioff zone). The zone that dips from deep-sea trenches under island arcs or adjacent continental margins; defined by the concentration of earthquake foci within it.

Seismic wave. The vibration produced by earthquakes or other disturbances that are propagated through the earth's interior as a result of the elastic response of the earth to the disturbance.

Seismograph. An instrument designed to detect and record seismic waves.

Seismology. The study of the origin and propagation of wave motion through the earth's interior.

Shadow zone. The zone between 103 and 143 degrees distance from an earthquake epicenter in which direct waves do not arrive because of refraction of the earth's core.

Shale. Consolidated mud and clay, often containing sand or lime in large amounts and having an earthy odor.

Shear fractures. Breaks or fractures in rock caused by shear failure.

Shear stress. A stress applied in such a way that two adjacent parts of the material slip past one another.

Shear wave (*or* secondary, S-, *or* transverse wave). Seismic wave that involves a shearing of the material; the wave vibrates perpendicular to the direction in which the wave front is moving.

Sheeting. A set of joints or fractures that may develop nearly parallel to the surface of the ground as a response to the release of confining pressure when the weight of the overlying rock is removed.

Shield. Those areas of Precambrian rock exposed at the surface or shallowly buried that have not been folded or complexly deformed since the end of the Precambrian.

Shield volcano. The largest type of volcanic cone, which resembles a shield or low, sloping dome in profile.

Shore. The zone between mean low tide and the landward edge of wave-transported sand.

Shoreline. The line along which the sea and the land meet.

Sialic rock. An igneous rock rich in silicon and aluminum.

Siderite. A meteorite composed primarily of metal, especially nickel and iron.

Silicate. Mineral that contains SiO_4 tetrahedra. The most common rock-forming minerals belong to this group.

Siliceous. Containing silica.

Sill. Concordant, sheetlike intrusive body with large lateral extent relative to its thickness.

Similar fold. Fold within which the shape remains the same from one layer to another, but the thickness varies from place to place in each layer.

Simple eutectic system. A type of behavior of some two-component melts when the components do not form a solid–solution series. These systems are plotted on temperature–composition plots.

Sink hole. Enclosed depression formed by solution, subsidence, or collapse of the surface into a cave.

Slate. The most perfectly foliated metamorphic rock, thought to result from dynamic metamorphism of argillaceous material.

Slaty cleavage. Rock cleavage like that of slate, in which the micaceous minerals show a strong alignment in the plane of the cleavage.

Slickensides. Striations and grooves that appear where rocks on either side of a fault have moved across one another.

Slip face. The downwind face of a sand dune formed as a result of slip and mass movement of sand downslope from the crest of dune.

Slump. Mass downslope movement of material on a curved surface of failure.

Snow field. The area above the snow line where snow remains from year to year.

Snow line. The lower limit of the area of perennial snow.

Soil. The product of rock disintegration and decomposition by weathering, modified by biological agents; capable of supporting plant life.

Solid-solution series. Minerals that can undergo solid-melt reactions within certain limits in such a way that the composition and physical properties vary continuously with varying amounts of the components, as exhibited by plagioclase feldspar.

Solidus. The curve in a temperature–composition diagram (for two-component systems) connecting points above which the solid and liquid are in equilibrium and below which only a solid phase exists.

Solifluction. Soil flowage, particularly that induced by saturation of the soil as a result of partial thawing of the ground.

Solubility. The maximum amount of an ionic compound dissolvable in a given amount of solvent under a given set of conditions of temperature and pressure.

Sorting. Separation of particles according to size, shape, or weight.

Spatter cones. Mounds or towerlike structures built up by successive addition of layers of lava ejected from small openings.

Specific gravity. The ratio of a mineral's weight to that of an equal volume of water.

Spit. A bar and beach built out from the tip of the headland beach forming a projection out from the land mass.

Spring. A place where water issues from beneath the ground surface onto the ground.

Spring tide. Unusually high and low tides that result

when the sun and moon lie in line with the earth and the attraction of the two bodies is cumulative.

Stacks. Columnar masses of rock standing isolated as islands just offshore.

Stalactite. Deposit of calcite that hangs down from the ceiling of a cave.

Stalagmite. A stump-shaped cave deposit, usually of $CaCO_3$, formed on the floor of a cave.

Standing wave. Wave motion in which the form does not move laterally but the crest and troughs reverse.

Stick-slip movement. A type of fault movement characterized by sudden and intermittent movements interspersed by periods when the moving surfaces are stuck together.

Stock. An igneous intrusion that is less than 40 square miles in surface area. Often these are small intrusions of larger plutonic bodies.

Stony meteorite (*or* chondrite). Meteorite composed of nonmetallic materials, especially olivine.

Strain. The response of the rock to an applied stress, especially the distortion or change in the volume of the rock.

Stratification. Rock layering.

Stratum or bed. Layer of sediment or rock of varying color, texture, or composition.

Streak. The color of the powder of a mineral.

Streaming flow. Flow of water moving on medium and low slopes. The water is smooth in some places, and exhibits mild, slow eddying in others.

Stream piracy. The process by which one stream extends in a headward direction, cutting into the valley of a second stream and eventually diverting the flow of the second stream.

Striation. Any linear scratchlike mark. For example, striations are produced where glaciers drag rocks over one another, along fault surfaces, and where cleavages intersect the surface of a mineral.

Strike. The compass direction (bearing) of any horizontal line on a plane.

Strike-slip fault. A fault on which the direction of movement is horizontal along the strike of the fault.

Structural terrace. A flexure that results in a flat, nearly horizontal surface.

Structure. Form and orientation of parts of rock masses and their relation to the whole. Structural geology is concerned with the origin as well as the geometry of rock masses.

Subduction. In plate tectonics, the process by which one plate sinks or moves beneath another and into the earth's mantle where it is assimilated or remelted.

Sublimation. The process by which a material passes from a solid to gaseous state without becoming a liquid.

Subsequent stream. A stream that flows in a channel that is adjusted to underlying rock structure (generally follows weaker rock outcrop belts, faults, and the like).

Subsoil. The B horizon; normally contains fewer organisms than do the overlying layers, but more than the C horizon. Weathering has reduced all rock to fine material.

Superimposed stream. A stream that cuts down through a cover and establishes itself in the underlying bedrock. Superimposed streams frequently have courses that bear little relationship to the variations in the resistance to erosion of the bedrock across which they flow.

Surface wash. The flow of water following a rain or melting snow or ice that brings about downslope movement of surficial materials at the surface of the ground; also called sheet wash.

Surface wave. Elastic wave that is propagated near the surface of a body.

Surge. Rapid (that is, meters/day) movement of valley glaciers.

Suspension. Used in hydrology to refer to material that is moved as solid particles within the water.

Suture. In plate tectonics, the line or zone within a plate along which two plates have come together in the past.

Swallow hole. Opening in a stream channel leading into caverns and solution cavities.

Swamp. An area in which the ground is wet, usually having low relief and containing lakes or remnants of lakes.

Swell. Wind-generated wave that has advanced into regions of weaker winds or calm.

Syenite. A coarse- to medium-grained equigranular igneous rock composed primarily of orthoclase with hornblende or biotite.

Symmetrical fold. A fold in which the two limbs are nearly mirror images of one another.

Syncline. A downfold in which the limbs dip toward the axis.

Talus (*or* scree). Rock fragments that accumulate as a heap or sheet at the base of a steep rock surface.

Tarn. A small lake in the mountains, often in a depression gouged out by a glacier.

Tectonics. The study of the formation and deformation of the earth's crust that result in large-scale structural features.

Temporary base level. A level to which the reduction of topography of an area is restricted for a limited time.

Terminal moraine. Ridge of drift (largely till) formed along the edge of a glacier at the position of the glacier's furthest advance.

Terra rosa. Red and clayey soil found in areas of karst topography.

Tertiary. A period of geologic time, including approximately 67 million years from the end of the Mesozoic Era to the beginning to the Quaternary.

Thrust fault. A fault inclined at a low angle along which reverse-type movement has occurred.

Thrust sheet. A thin mass of rocks carried or moved laterally above a thrust fault.

Tidal bore. A turbulent wave that moves with high velocity up streams as a result of high tides.

Tidal flat. The flat area located along some coasts that is periodically covered with water during high tide.

Tide. The periodic rise and fall of sea level.

Till. Unsorted, usually unstratified or poorly stratified, glacial drift.

Tombolo. Sand bar or beach that connects offshore islands with the mainland or another island.

Top-set bed. Bed composed of flat-lying sediments deposited on top of a delta by streams flowing across the delta, marshes, or low areas on either side of the streams.

Transform fault. A fault that terminates sharply at a place where the movement is transformed into a structure of another type.

Transverse dune. A sand dune oriented with its long axis perpendicular to the prevailing wind.

Transverse stream. A stream that flows across the structural features of the underlying bedrock.

Travel-time curve. The plot of travel time versus distance of seismic waves.

Travertine. Compact calcareous cave deposits.

Truncated spur. Ridge between tributaries entering a major glaciated valley, cut back as a result of glacial erosion in the main valley.

Tsunami. A Japanese term applied to any gravity-wave system formed in the sea as a result of a large-scale, short-duration disturbance of the surface.

Tufa. Porous or cellular spring deposit; applied to spongy, fragile deposit with an earthy texture.

Tuff. Ash or dust compacted to form a rock.

Turbidity current. A mass of water highly charged with material in solution and suspension that flows with turbulent motion down slopes through normal waters.

Turbulent flow. Fluid flow in which the paths of particle motion are irregular, with eddies and swirls.

Ultimate base level. The elevation toward which degradational processes tend to reduce the earth's surface; the elevation the sea would have if all land areas were removed and the material deposited below water level.

Unconfined groundwater. Groundwater not restricted by impervious material in its movement.

Uniformitarianism. The theory that the operation of physical and chemical principles and processes has been unchanged throughout geologic time.

Unit magnetic pole. One pole exerting a force of 1 dyne on another like pole when the two are 1 centimeter apart in a vacuum.

Valley glacier. Glacier confined to a valley, generally on a mountain.

Varves. Banded sequences of sediment in which a particular sequence of beds has been deposited repeatedly. The beds are caused by seasonal variations; each sequence represents one year's deposit.

Vein. A thin, sheetlike intrusion or a filling of a fracture.

Vent. A hole where the pipe or feeder systems for a volcano break through to the ground surface.

Ventifact. Smooth, polished stone resulting from repeated abrasion of sand and silt blowing over a rock for a long time.

Viscosity. Measure of the resistance a fluid offers to shear or flow; the inverse of fluidity.

Water budget. An expression of the total water resources of an area: that introduced into the area and that used or lost by any means.

Water gap. A pass through a mountain ridge through which a stream flows.

Water table. The surface that separates a zone where the pore spaces in the rocks or ground are not completely filled with water (that is, they are unsaturated) from a zone below in which they are saturated.

Wave. The up and down movement on the surface of a body of water.

Wave height. The vertical distance between the top of a wave crest and the bottom of an adjacent trough.

Wave length. The distance between two crests or two troughs.

Weathering. Changes that occur where rocks and minerals come in contact with the atmosphere, surficial waters, and organic life under conditions normally found at the surface of the earth.

Welded tuffs (*or* welded pumice *or* ignimbrite). Volcanic ash deposit in which the ash and glassy particles are welded together.

Wind gap. A gap through a mountain originally formed by water passing through the gap, but now dry.

Window. The place where, as a result of erosion, the rocks beneath a thrust sheet may be seen surrounded by rocks that make up the thrust sheet.

Xenolith. Literally, "foreign" rock; applied to rock fragments found in an igneous rock (for example, inclusions of wall rock).

Yield point. The stress level beyond which a material strains continuously with no additional stress.

Zone of aeration. That part of the soil and rock in which both air and percolating water occur.

CREDITS

Title page: U.S.G.S.

Part I: NASA

Fig. 1.1: U.S.G.S. Fig. 1.4: NASA. Fig. 1.5: Courtesy Kennecott Minerals Company. Fig. 1.6: U.S.G.S.

Fig. 2.1: Smithsonian Institution. Fig. 2.3(*a*): U.S.G.S. (*b*): V. C. Browne. Fig. 2.5(*a*): U.S.G.S. (*b*): U.S.G.S. Fig. 2.13: Smithsonian Institution. Fig. 2.14: Smithsonian Institution. Fig. 2.20(*c*): Smithsonian Institution. (*d*): Smithsonian Institution. (*e*): Courtesy of Bausch & Lomb. Fig. 2.21(*b*): Smithsonian Institution. Fig. 2.25(*a*): Smithsonian Institution. Fig. 2.27(*b*): Photo courtesy of New Mexico Tourism & Travel Div. (*c*): From Wise & deVilliers, 1971.

Fig. 3.1: NASA. Fig. 3.4: Marsh and Dozier, *Landscape: An Introduction to Physical Geography*. © 1981. Reading, Mass.: Addison-Wesley. Figures 5.1, 5.3, and 12.2. Reprinted with permission. Fig. 3.8: Lamont-Doherty Geological Observatory. Fig. 3.9: Redrawn and modified after maps by Heezen and Tharp and charts of the U.S. Navy Hydrographic Office. Fig. 3.11: From Heezen, 1959. Fig. 3.14: From R. von Huene, "Deep sea drilling project Mid-America trench transect off Guatemala," *G.S.A. Bulletin* 91:7 (1980), *431*. Fig. 3.15: Redrawn after B. C. Heezen, G.S.A. Special Paper 65. Fig. 3.16: G.S.A. Fig. 3.17: From B. F. Windley, *The Early History of the Earth*. © 1976 by John Wiley & Sons. Reprinted by permission of John Wiley & Sons, Ltd. Fig. 3.18: Rocky Mountain Association of Geologists. Fig. 3.19: Reprinted with permission of the Geological Association of Canada from GAC Special Paper No. 6: Structure of the Southern Canadian Cordillera, edited by J. O. Wheeler.

Part II: Smithsonian Institution.

Fig. 4.1: Mount Wilson and Las Campanas Observatories, Carnegie Institution of Washington. Fig. 4.3: Yerkes Observatory photograph. Fig. 4.4: Mount Wilson and Las Campanas Observatories, Carnegie Institution of Washington. Fig. 4.5: Mount Wilson and Las Campanas Observatories, Carnegie Institution of Washington. Fig. 4.6: NASA. Fig. 4.8: Palomar Observatory photograph. Fig. 4.9: Palomar Observatory photographs. Fig. 4.11: NASA. Fig. 4.12: Yerkes Observatory photograph. Fig. 4.13: Mount Wilson and Las Campanas Observatories, Carnegie Institution of Washington. Fig. 4.14: NASA. Fig. 4.15: NASA.

Fig. 5.1: Union Pacific Railroad photo. Fig. 5.2: Dr. James L. Carter. Fig. 5.3: NASA. Fig. 5.4: Japan National Tourist Organization. Fig. 5.5: NASA. Fig. 5.6: From the aerial film library of Jack Ammann Photogrammetric Engineers — now Petroleum Information Corp., San Antonio, Tx. Fig. 5.7(*b*): Courtesy of Bausch & Lomb. (*c*): Smithsonian Institution. Fig. 5.8(*a*): Photo courtesy of New Mexico Tourism & Travel Div. (*b*): Smithsonian Institution. Fig. 5.9: Turtox, Chicago, Illinois. Fig. 5.11(*b*): U.S.G.S. Fig. 5.14: U.S.G.S. Fig. 5.15(*b*): Photo by Deep Sea Drilling Project, International Phase of Ocean Drilling. Fig. 5.16: Turtox, Chicago, Illinois. Fig. 5.17: Courtesy: Ward's Natural Science Establishment, Inc. Fig. 5.18(*b*): Reprinted by permission of A.A.P.G. Fig. 5.19: U.S.G.S. Fig. 5.20(*b*): A.G.U. Fig. 5.21: U.S.G.S. Fig. 5.22: From B. C. Heezen et al., "Shaping of the continental rise by deep geostrophic contour currents," *Science* 152(1966), *502–508*. Copyright 1966 by the American Association for the Advancement of Science. Fig. 5.23: Union Pacific Railroad photo. Fig. 5.24: Reproduced by permission of the Director, Institute of Geological Sciences: NERC copyright. Fig. 5.25: Smithsonian Institution. Fig. 5.26(*a*): Reproduced by permission of the Director, Institute of Geological Sciences: NERC copyright. Fig. 5.28: Smithsonian Institution.

Fig. 6.1: Reproduced by permission of the Director, Institute of Geological Sciences: NERC copyright. Fig. 6.2: Reproduced by permission of the Director, Institute of Geological Sciences: NERC copyright. Fig. 6.3: Reproduced by permission of the Director, Institute of Geological Sciences: NERC copyright. Fig. 6.4(*a*): Smithsonian Institution. (*b*): Smithsonian Institution. (*c*): Smithsonian Institution. (*d*): Smithsonian Institution. (*f*): Smithsonian Institution. Fig. 6.7(*a*): U.S.G.S. Fig. 6.8: U.S.G.S. Fig. 6.10: Adapted from N. L. Bowen in *American Journal of Science* 40 (1915), *161–185.* Fig. 6.11: Adapted from N. L. Bowen in *American Journal of Science* 40 (1915), *161–185.* Fig. 6.14: National Publicity Studios, New Zealand. Fig. 6.15: Frederick P. Young Jr. Fig. 6.16(*a*): NASA. Fig. 6.17(*a*): U.S.G.S. (*b*): U.S.G.S. (*c*): G.S.A., Robert R. Compton. Fig. 6.19: Santa Fe Railroad. Fig. 6.20: U.S.G.S. Fig. 6.21: U.S.G.S. Fig. 6.22: W. H. K. Lee, "The thermal history of the Earth." Ph.D. diss., University of California, Los Angeles, 1967.

Fig. 7.1: Reproduced by permission of the Director, Institute of Geological Sciences: NERC copyright. Fig. 7.9: Harker and Tuttle, 1956 (*American Journal of Science,* vol. 254, *239–256*). Fig. 7.10: *G.S.A. Memoir* 73, 1959. Fig. 7.12: Aureole of the Onawa pluton, Maine. (After S. S. Philbrick, *Am. Jour. Sci.* 5th ser., vol. 31, pp. 1–40, 1936.) Fig. 7.14:(*b*): Reproduced by permission of the Director, Institute of Geological Sciences: NERC copyright. Fig. 7.15: Reproduced by permission of the Director, Institute of Geological Sciences: NERC copyright. Fig. 7.16: Winkler, Helmut G. F., *Petrogenesis of Metamorphic Rocks.* New York: Springer-Verlag New York, Inc. Fig. 7.17: From B. F. Windley, *The Early History of the Earth.* © 1976 by John Wiley & Sons. Reprinted by permission of John Wiley & Sons, Ltd.

Part III: Geological Survey of Canada, Ottawa.

Fig. 8.1: Reproduced from U.S. Program for the Geodynamics Project, National Academy Press, Washington, D.C., 1973. Fig. 8.3: Reproduced from U.S. Program for the Geodynamics Project, National Academy Press, Washington, D.C., 1973. Fig. 8.5: Reproduced from U.S. Program for the Geodynamics Project. National Academy Press, Washington, D.C., 1973. Fig. 8.6: Steinbrugge, K. V. and Cloud, W. K. By permission of the Seismological Society of America. Fig. 8.7(*a*): U.S.G.S. Fig. 8.10: U.S.G.S. Fig. 8.11: U.S.G.S. Fig. 8.12: U.S.G.S. Fig. 8.13: U.S.G.S. Fig. 8.14: U.S.G.S. Fig. 8.15: U.S.G.S. Fig. 8.16: U.S.G.S. Fig. 8.17: U.S.G.S. Fig. 8.18: City of Los Angeles, Department of Building and Safety. Fig. 8.19: U.S.G.S.

Fig. 9.13(*b*): J. Lamar Worzel and C. A. Burk. From *A.A.P.G. Memoir* 29. Fig. 9.23: A.G.U. Fig. 9.25: Anderson, Don L., "Latest information from seismic observations." In *The Earth's Mantle,* T. F. Gaskill (ed.). New York: Academic Press, 1967. Fig. 9.29: A.G.U. Fig. 9.31: U.S. Hydrographic Office. Fig. 9.32: Data from D. L. Anderson and M. Smith, "Mathematical and physical inversion of gross Earth geophysical data," *Amer. Geophys. Union Trans.* vol. 49 (1968), *283.*

Fig. 10.1: U.S.G.S. Fig. 10.2: National Publicity Studios, New Zealand. Fig. 10.3: U.S. Air Force. Fig. 10.4: U.S.G.S. Fig. 10.5: A.G.U. Fig. 10.6: A.G.U. Fig. 10.7: Photo by John S. Shelton. Fig. 10.8: After P. B. King, Am. Assoc. Retired Geologists. Reprinted by permission of A.A.P.G. Fig. 10.24(*a*): *Tectonophysics,* vol. 2, no. 1, pp. 1–27. Fig. 10.29(*a*): U.S. Air Force. (*b*): U.S. Air Force. Fig. 10.30: E. L. Fitzgerald. Fig. 10.32(*a*): U.S. Air Force. Fig. 10.34: From *G.S.A. Memoir* 79. Fig. 10.35: From *G.S.A. Memoir* 79. Fig. 10.36: From Nadai, A. *Theory of Flow and Fracture of Solids.* New York: McGraw-Hill, 1950.

Fig. 11.1: NASA. Fig. 11.3: NASA. Fig. 11.4: U.S.G.S. Fig. 11.6: After U.S.G.S. maps. Fig. 11.7: After U.S.G.S. maps. Fig. 11.8: After U.S.G.S. maps. Fig. 11.9: NASA. Fig. 11.10(*a*): After D. Roeder, O. Gilbert, and W. Witherspoon, *Evolution and Macroscopic Structure of Valley and Ridge Thrust Belt.* Tenn. & Va., University of Tennessee Studies 2, 1978. (*b*): After U.S.G.S. maps. Fig. 11.11: After U.S.G.S. maps. Fig. 11.12: From *G.S.A. Bulletin,* vol. 83, no. 3. Fig. 11.13: After U.S.G.S. maps. Fig. 11.14: After R. A. Price and E. W. Mountjoy, Fig. VIII-38 in "Geology and economic minerals of Canada," Geological Survey of Canada Economic Geology Report No. 1; R. J. W. Douglas, editor, 1970. Fig. 11.16: U.S.G.S. Fig. 11.17: After U.S.G.S. maps. Fig. 11.18: U.S.G.S. Fig. 11.19: From *AAPG Memoir* 29. Fig. 11.20: After W. P. Dillon in A.A.P.G. *Memoir* 29. Fig. 11.21: Modified from Klitgord and Behrendt (1979, Fig. 1) in *AAPG Memoir* 29. Fig. 11.22: A.G.U. Fig. 11.26: Cross section of earthquakes located by central Aleutians seismograph network. Apparent bend in seismicity at about 100 km depth is an effect of the plate on seismic waves to the local network. (From Engdahl, E. R., "Seismicity and plate subduction in the central Aleutians," *AGU Monograph; Ewing Symposium, Am. Geophys. Union, 1, 259–273,* 1977. Fig. 11.27: Von Huene after Seely and Dickinson in *AAPG Memoir* 29. Fig. 11.28: A.G.U. Fig. 11.30(*a*): A.G.U. Fig. 11.31: After A.G.U. maps. Fig. 11.32(*a*): A.G.U. Fig. 11.33: Lamont-Doherty Geological Observatory of Columbia University, Palisades, NY 10964.

Fig. 12.3: Flint, R. F., *Glacial and Pleistocene Geology.* New York: John Wiley & Sons, 1957. Reprinted by permission of John Wiley & Sons, Inc. Fig. 12.4: G.S.A. Fig. 12.6: *The Origin of Continents and Oceans,* by A. Wegener, translation of the 1962 4th rev. ed. pub. by Wieweg Verlag. Eng. trans. by John Biram, 1966 Dover Publications, Inc., NY. Fig. 12.7: S. W. Carey. Fig. 12.8: From E. Bullard, J. E. Everett, and A. G. Smith, "The fit of the continents around the Atlantic," *Philosophic Transactions of the Royal Society of London,* A258, 1965. Fig. 12.9: From P. M. Hurley et al., "Test of continental drift by comparison of radiometric ages," *Science* 157 (August 4, 1967), *495–500.* Copyright 1967 by the American Association for the Advancement of Science. Fig. 12.10: From M. W. McElhinny, *Paleomagnetism and Plate Tectonics,* New York: Cambridge University Press, 1973. Fig. 12.11: G.S.A. Fig. 12.12: G.S.A. Fig. 12.13: From W. Lowrie and W. Alvarez, "One hundred million years of geomagnetic polarity history," *Geology* 9:9, 1981. Fig. 12.14: By permission of R. G. Mason. Fig. 12.15: By permission of F. J. Vine and *Journal of Geological Education.* Fig. 12.17: By permission of J. G. Sclater and A.G.U. Fig. 12.20: By permission of B. L.

Isacks and A.G.U. Fig. 12.22: From J. F. Dewey, "A model for the lower Paleozoic evolution of the southern margin of the early Caledonides of Scotland and Ireland," *Scottish Journal of Geology* 7 (1971) *219–240*.

Fig. 13.1: National Publicity Studios, New Zealand. Fig. 13.2: By permission of R. A. Duncan, D. H. Green and G.S.A. Fig. 13.3: Woods Hole Oceanographic Institution. Fig. 13.4: From J. F. Dewey, "A model for the lower Paleozoic evolution of the southern margin of the early Caledonides of Scotland and Ireland," *Scottish Journal of Geology* 7 (1971) *219–240*. Fig. 13.8: By permission of the authors and *Earth and Planetary Science Letters*. Fig. 13.10: M. Kay, "Tectonic evolution of Newfoundland." In K. A. DeJong and R. Scholten, *Gravity and Tectonics*. New York: John Wiley & Sons. Reprinted by permission of John Wiley & Sons. Fig. 13.12: Redrawn and simplified by permission of T. M. Atwater. Fig. 13.13: A.G.U.

Part IV: Oregon Department of Transportation.

Fig. 14.1: National Archives. Fig. 14.2: U.S. Park Service. Fig. 14.3(*b*): U.S.G.S. Fig. 14.4(*a*): Reproduced by permission of the Director, Institute of Geological Sciences: NERC copyright. (*b*): National Publicity Studios, New Zealand. Fig. 14.5(*a*): Royal Canadian Air Force. (*b*): U.S.G.S. Fig. 14.6(*a*): U.S. Department of Agriculture. (*b*): U.S. Air Force. Fig. 14.7(*a*): U.S. Park Service. (*b*): Reproduced by permission of the Director, Institute of Geological Sciences: NERC copyright. Fig. 14.8: Montana Travel Promotion Bureau. Fig. 14.9: U.S. Air Force. Fig. 14.11: U.S.G.S.

Fig. 15.1(*a*): National Archives. (*b*): National Archives. (*c*): U.S.G.S. (*d*): U.S.G.S. Fig. 15.3: U.S.G.S. Fig. 15.4: U.S.G.S. Fig. 15.5(*a*): U.S.G.S. (*b*): National Archives. Fig. 15.6(*a*): Photo by John S. Shelton. Fig. 15.7: U.S.G.S. Fig. 15.8: E. A. Vincent. Fig. 15.10: U.S.G.S. Fig. 15.11: U.S.G.S. Fig. 15.12: U.S.G.S. Fig. 15.13: National Publicity Studios, New Zealand. Fig. 15.14: U.S. Park Service. Fig. 15.15: University of Oregon Press. Fig. 15.16: U.S.G.S. Fig. 15.17: NASA. Fig. 15.18: NASA. Fig. 15.19: From the Royal Society's Report on "The eruption of Krakatoa," 1888. Fig. 15.23: U.S.G.S. Fig. 15.24: U.S.G.S. Fig. 15.25: U.S.G.S. Fig. 15.26: U.S.G.S. and after A.A.P.G. maps. Fig. 15.28: Permission granted by *Geotimes* and P. E. Hammond. Fig. 15.29: U.S.G.S. Fig. 15.30: U.S.G.S. Fig. 15.32: U.S. Government.

Fig. 16.2(*a*): U.S. Air Force. Fig. 16.4: U.S.G.S. Fig. 16.5: U.S.G.S. Fig. 16.6: Royal Canadian Air Force. Fig. 16.7: Royal Canadian Air Force. Fig. 16.8(*a*): U.S.G.S. Fig. 16.9: Photo by John S. Shelton. Fig. 16.10: Map, copyright permission: Hammond Incorporated, Maplewood, N.J. Fig. 16.17(*a*): NASA. Fig. 16.18: U.S.G.S. Fig. 16.19: U.S.G.S.

Fig. 17.1(*a*): U.S. Park Service. Fig. 17.2: McGraw-Hill. Fig. 17.6: Marsh and Dozier, *Landscape: An Introduction to Physical Geography*. © 1981. Reading, Mass.: Addison-Wesley. Reprinted with permission. Fig. 17.7(*b*): U.S.G.S. Fig. 17.8: U.S.G.S. Fig. 17.10: U.S.G.S. Fig. 17.14: U.S. Department of Agriculture.

Fig. 18.1: U.S. Forest Service. Fig. 18.4: National Archives. Fig. 18.5: Courtesy Canadian Pacific Railroad. Fig. 18.6: National Archives. Fig. 18.8: U.S.G.S. Fig. 18.11: Courtesy of Transportation Research Board. Fig. 18.12: Courtesy of Transportation Research Board. Fig. 18.14: U.S.G.S. Fig. 18.15: U.S.G.S. Fig. 18.16: U.S.G.S. Fig. 18.17: U.S.G.S. Fig. 18.18: From G. A. Kiersch, "The Vaiont Reservoir disaster," *Civil Engineering* 4:3, 1964. Fig. 18.19: From G. A. Kiersch, "The Vaiont Reservoir disaster," *Civil Engineering* 4:3, 1964. Fig. 18.20: From G. A. Kiersch, "The Vaiont Reservoir disaster," *Civil Engineering* 4:3, 1964. Fig. 18.21(*a*): National Archives. (*b*): National Archives. Fig. 18.22(*a*): U.S.G.S. Fig. 18.23: U.S.G.S. Fig. 18.24: U.S.G.S.

Fig. 19.2: NASA. Fig. 19.3: U.S.G.S. Fig. 19.4: U.S.G.S. Fig. 19.5: Courtesy of C. P. Berkey; G.S.A. Fig. 19.6: Royal Canadian Air Force. Fig. 19.10: National Archives. Fig. 19.14: Uppsala. Fig. 19.15: National Archives. Fig. 19.16: G.S.A. Fig. 19.20: NASA. Fig. 19.21: NASA. Fig. 19.23: U.S.G.S. Fig. 19.26: Royal Canadian Air Force. Fig. 19.27: U.S. Army.

Fig. 20.1: Reproduced by permission of the Director, Institute of Geological Sciences: NERC copyright. Fig. 20.5: U.S. Air Force. Fig. 20.6: U.S. Air Force. Fig. 20.7: U.S. Air Force. Fig. 20.10: U.S.G.S. Fig. 20.14: Photo by John S. Shelton. Fig. 20.15: U.S.G.S. Fig. 20.16: NASA. Fig. 20.18: U.S. Park Service. Fig. 20.19: U.S.G.S. Fig. 20.20: After U.S.G.S. maps. Fig. 20.21(*a*): U.S.G.S. Fig. 20.25: U.S. Air Force. Fig. 20.26(*a*): NASA. (*b*): From D. E. Frazier, "Recent deltaic deposits of the Mississippi River, their development and chronology," Gulf Coast Geol. Soc. Trans. 17th Annual Meeting, San Antonio, Texas, 1967. Fig. 20.32: After U.S.G.S.

Fig. 21.1: Marsh and Dozier, *Landscape: An Introduction to Physical Geography*. © 1981. Reading, Mass.: Addison-Wesley. Reprinted with permission. Fig. 21.9(*a*): U.S.G.S. Fig. 21.10: U.S.G.S. Fig. 21.13: After U.S.G.S. Fig. 21.14: U.S.G.S. Fig. 21.15: National Archives. Fig. 21.16(*a*): National Archives. Fig. 21.17: U.S.G.S. Fig. 21.19: Photo by Richard Deuerling. Fig. 21.21: Picture of Natural Bridge is complimentary of Natural Bridge of Virginia, Inc. Fig. 21.22: By Wilbur E. Garrett. Copyright © 1979 National Geographic Society. Fig. 21.23: U.S.G.S. Fig. 21.24: Photo by James F. Quinlan.

Fig. 22.1: Virginia Department of Highways and Transportation. Fig. 22.2: U.S.G.S. Fig. 22.3: U.S.G.S. Fig. 22.4: U.S.G.S. Fig. 22.5(*a*): NASA. Fig. 22.6: U.S.G.S. Fig. 22.7: U.S.G.S. Fig. 22.8: U.S.G.S. Fig. 22.13: U.S.G.S. Fig. 22.14(*a*): U.S.G.S. Fig. 22.15(*a*): U.S.G.S. Fig. 22.16(*a*): NASA. (*b*): NASA. (*c*): NASA. Fig. 22.17: N.O.A.A. Fig. 22.18: U.S.G.S. Fig. 22.19(*a*): U.S.G.S. Fig. 22.20: U.S.G.S. Fig. 22.21: U.S.G.S. Fig. 22.22: U.S. Army Corps of Engineers. Fig. 22.23: U.S. Army Corps of Engineers.

Fig. 23.1: U.S. Air Force. Fig. 23.2: From Marsh and Dozier, *Landscape: An Introduction to Physical Geography*. © 1981.

Reading, Mass.: Addison-Wesley. Reprinted with permission. Fig. 23.4: N.O.A.A. Fig. 23.5: U.S.G.S. Fig. 23.6: U.S.G.S. Fig. 23.7: Photo courtesy New Mexico State Tourism & Travel Div. Fig. 23.8: Reproduced by permission of the publisher from *The Physics of Blown Sand and Desert Dunes* by R. A. Bagnold (Chapman & Hall, Ltd.). Fig. 23.9: Reproduced by permission of the publisher from *The Physics of Blown Sand and Desert Dunes* by R. A. Bagnold (Chapman & Hall, Ltd.). Fig. 23.11(*b*): Photo courtesy of the Oregon Department of Transportation. Fig. 23.13: After U.S.G.S. Fig. 23.14: After U.S.G.S. Fig. 23.15: NASA. Fig. 23.16: NASA.

Fig. 24.1: U.S. Navy. Fig. 24.2: U.S. Navy. Fig. 24.3: U.S. Park Service. Fig. 24.4: U.S. Park Service. Fig. 24.5: A.G.U. Fig. 24.6: A.G.U. Fig. 24.7: U.S. Air Force. Fig. 24.8: U.S. Air Force. Fig. 24.9: G.S.A. Fig. 24.10: National Publicity Studios, New Zealand. Fig. 24.12: University of Oregon Books. Fig. 24.13: U.S.G.S. Fig. 24.19: U.S. Air Force. Fig. 24.20: Royal Canadian Air Force. Fig. 24.21: U.S.G.S. Fig. 24.22: Trans World Airlines. Fig. 24.26: U.S.G.S. Fig. 24.28: U.S.G.S. Fig. 24.30: U.S.G.S. Fig. 24.31: U.S. Air Force. Fig. 24.33: U.S.G.S. Fig. 24.36: J. D. Hays et al., "Variations in the earth's orbit: Pacemaker of the Ice Ages," *Science* 194 (1976) *1121–1132*. Copyright 1976 by the American Association for the Advancement of Science. Fig. 24.37: By permission of J. G. Sclater and A.G.U. Fig. 24.38: Federal Council on Science and Technology; A.G.U. Fig. 24.39: From Van der Hammen and others, 1967; published by permission of *Geologie en Mijnbouw*. K. K. Turekian, ed., "Late Cenozoic glacial ages," *Concepts of Physical Geology*, Yale University Press copyright. Reprinted with permission from *Deep-Sea Research and Oceanographic Abstracts* 19, A. McIntyre, W. F. Ruddiman, and R. Janzten, "Southward penetration of the North Atlantic polar front; faunal and floral evidence of large scale surface water mass movements over the last 225,000 years." Copyright 1972, Pergamon Press, Ltd.

Fig. 25.1: NASA. Fig. 25.3: J. D. Milliman and K. O. Emery, "Sea levels during the past 35,000 years," *Science* 162 (1968) *1121–1123*. Copyright 1968 by the American Association for the Advancement of Science. Fig. 25.4: U.S. Navy. Fig. 25.12(*a*): U.S.G.S. Fig. 25.13: U.S.G.S. Fig. 25.14: Reproduced by permission of the Director, Institute of Geological Sciences: NERC copyright. Fig. 25.16(*a*): Reproduced by permission of the Director, Institute of Geological Sciences: NERC copyright. (*d*): Oregon Department of Transportation photo. Fig. 25.18: U.S.G.S. Fig. 25.20: American Society of Civic Engineers. Fig. 25.23: National Publicity Studios, New Zealand. Fig. 25.25: Miles Hayes. Fig. 25.26: NASA. Fig. 25.27(*a*): NASA. (*b*): U.S. Park Service. Fig. 25.29: U.S.G.S. Fig. 25.30: NASA. Fig. 25.31: After U.S.G.S. and A.A.P.G. Fig. 25.32: Scripps. Fig. 25.33: Embassy of France. Fig. 25.34: G.S.A. Fig. 25.35(*a*): Queensland Government Railways. Fig. 25.36: U.S. Navy. Fig. 25.37: U.S. Navy. Fig. 25.38(*a*): G.S.A. Fig. 25.39: U.S.G.S.

Part V: Courtesy Kennecott Minerals Company.

Fig. 26.1: National Archives. Fig. 26.3: U.S.G.S. Fig. 26.5: Arkansas Geological Survey. Fig. 26.6: U.S.G.S. Fig. 26.7: U.S.G.S. Fig. 26.8: U.S. Department of Energy. Fig. 26.9: U.S.G.S. Fig. 26.10: U.S.G.S. Fig. 26.11: West Virginia Geological Survey. Fig. 26.14: From *I.P.E.* 1967. Fig. 26.15: From *I.P.E.* 1967. Fig. 26.16: U.S.G.S. Fig. 26.17: U.S.G.S. Fig. 26.18: U.N.

INDEX